Integrative and cellular aspects of autonomic functions: temperature and osmoregulation

Integrative and cellular aspects of autonomic functions: temperature and osmoregulation

Proceedings of the International Symposium on Integrative and Cellular Aspects of Autonomic Functions held in Bad Nauheim (Germany) on 29-30 July 1993

Sponsored by Deutsche Forschungsgemeinschaft, Max-Planck-Gesellschaft, Hessisches Ministerium für Kunst und Wissenschaft, Giessener Hochschulgesellschaft

Edited by
Klaus Pleschka
Rüdiger Gerstberger

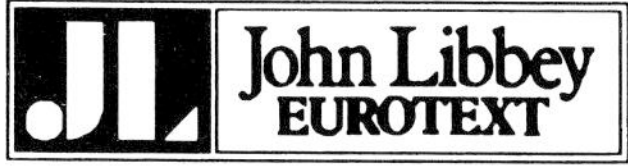

British Library Cataloguing in Publication Data
A catalogue record for this book
is available from the British Library

ISBN 2-7420-0042-9

Éditions John Libbey Eurotext
6, rue Blanche, 92120 Montrouge, France.
Tél. : (1) 47.35.85.52

John Libbey and Company Ltd
13, Smith Yard, Summerley Street, London SW18 4HR,
England. Tel. : (01) 947.27.77

John Libbey CIC
Via L. Spallanzani, 11, 00161 Rome, Italy.
Tel. : (06) 862.289

Foreword

Contained within this single volume are the proceedings of the International Symposium “Integrative and Cellular Aspects of Autonomic Functions: Temperature and Osmoregulation”. This symposium was conducted over a two-day period (July, 1993) at the W.G. Kerckhoff-Institute in Bad Nauheim, Germany, in honor of Professor Eckhart Simon on the occasion of his sixtieth birthday. Professor Simon has been the director of the Max-Planck-Institute for Physiological and Clinical Research in Bad Nauheim since 1974.

Scientists from around the world, representing various institutions and diverse disciplines, participated in the open forum symposium. It is an understatement to allege that scientific researches into the intricate details of thermal and osmolar balances have enjoyed a decade of significant advances. Addressing these problems from the organismic to cellular levels, innovative techniques and new conceptual frameworks have resulted in exciting and new developments.

As organizers for this symposium, it is our hope and expectation that the discussions and illustrations contained herein will serve as catalysts for further research into these important regulatory systems. One must be impressed with the diversity of approaches that are, like an ensemble, all working toward a common goal. To this end, the six sessions organized on central control, effector mechanisms, and endocrine regulatory processes promote the theme of systems integration. These are indeed complex physiological systems which command attention, appreciation, and further inquiry.

We wish to express our sincerest gratitude to all to those institutions, which provided generous financial support in making this conference possible. We are particularly indepted to the W.G. Kerckhoff-Foundation for backing the publication of these proceedings.

Klaus Pleschka
Rüdiger Gerstberger

Contents

Introduction

Integrative and cellular aspects of autonomic functions : temperature and osmoregulation. Eds K. Pleschka, R. Gerstberger. John Libbey Eurotext, Paris © 1994, pp. 3-11.

The SI standard in thermal physiology*

Claus Jessen

Physiologisches Institut der Universität, Aulweg 129, D-35392 Giessen, Germany

In the last thirty years, the SI standard has become a reference in the field of integrative physiology. It is named after the host of our meeting and had its first appearance on the stage of physiology in 1959, when a medical student named Eckhart Simon submitted a thesis on the Windkessel function of the human aorta. The aorta and other large arteries have elastic properties, which convert the highly pulsatile outflow of the left ventricle into a more continuous one in the peripheral arteries. The thesis contained the first quantitative determinations of the elasticity of the aorta, and its key results are now standard textbook knowledge (Fig.1). The way this work was done tells something about the SI-standard. The aorta has some sixty-odd branches, and most of them are terribly tiny. Tie them up properly so that they don't leak, apply pressures and measure volumes, and out comes a pressure-volume diagram. Successful research in physiology requires more than great ideas at the office desk: it needs craftmanship to translate them into reality.

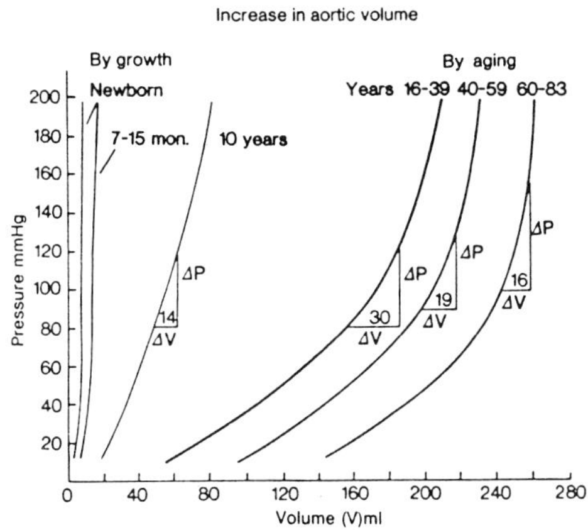

Fig. 1. Pressure-volume curves of the human aorta. Simon, E. (1959): Untersuchungen über das Fassungsvermögen und die Volumendehnbarkeit des gesamten aortalen Windkessels beim Menschen und über Länge und Umfang des Aortenrohrs in Abhängigkeit vom Aorteninnendruck. MD-Thesis, Marburg. From: *Human Physiology*, ed. R.F. Schmidt & G. Thews. New York: Springer (1983)

*Abrigded version of the opening address

This was the prelude to the Kerckhoff-career, which began in 1962 and soon led to the spinal cord. The years before had brought plenty of evidence that the body core of mammals must contain thermosensitive sites other than the hypothalamus. However, it needed a flash of brilliance to locate them. I don't know how the process of intuition, which according to the Merriam-Webster's most physiological definition, is "a highly personal intellectual capacity for passing directly from stimulus to response", has worked in this case. However, I can see two factors, which may have been involved. The first is the inversion of the concept of cephalisation: mechanisms which have attained the highest degree of complexity in the brain, may exist in an elementary form also at the spinal level, and the second, equally important, was to see the technical feasibility of the thermal isolation of the spinal cord. The combination of concept and skill led to the discovery of the spinal thermosensitivity (Fig. 2), which marked the decisive turning point from Benzinger's warm-eye dogma of the hypothalamus as the one and only internal temperature sensor to the present concept of multiple inputs.

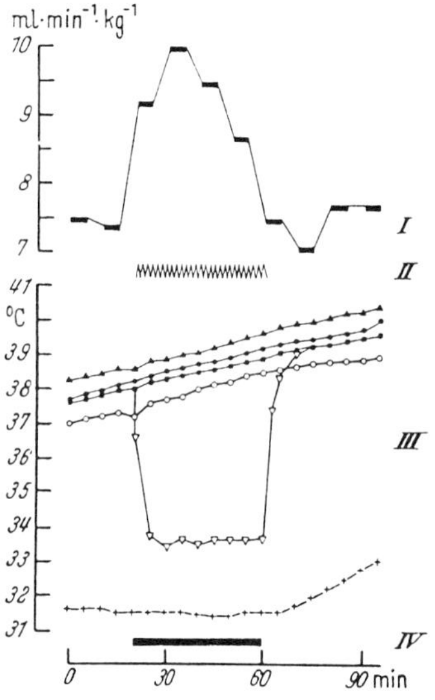

Fig. 2. Shivering and increase in oxygen consumption in response to spinal cord cooling in an anesthetized dog.
Simon, E., Rautenberg, W., Thauer, R., Iriki, M. (1964): Die Auslösung von Kältezittern durch lokale Kühlung im Wirbelkanal. *Pflügers Arch.* 281, 309-331

With hindsight, it must be said that the new landmark on the thermoregulatory map was quite readily accepted, - at least by those who understood. Many subsequent studies from several laboratories have shown that not only shivering but all other thermoregulatory effector mechanisms of mammals and birds are influenced by the temperature of the spinal cord. However, even today 30 years later, the importance of the spinal temperature sensors may still be undervalued. Our current methods are inadequate, and the spinal cord may eventually turn out to be *the* most important source of extra-hypothalamic temperature signals.

It soon became clear that at least some of the responses to spinal thermal stimulation required the afferent transmission of spinal temperature signals to supra-spinal levels of the central nervous system. In 1970, Simon & Iriki described ascending units in the anterior spino-thalamic tract, which showed proportional and dynamic responses to changing spinal cord temperature.

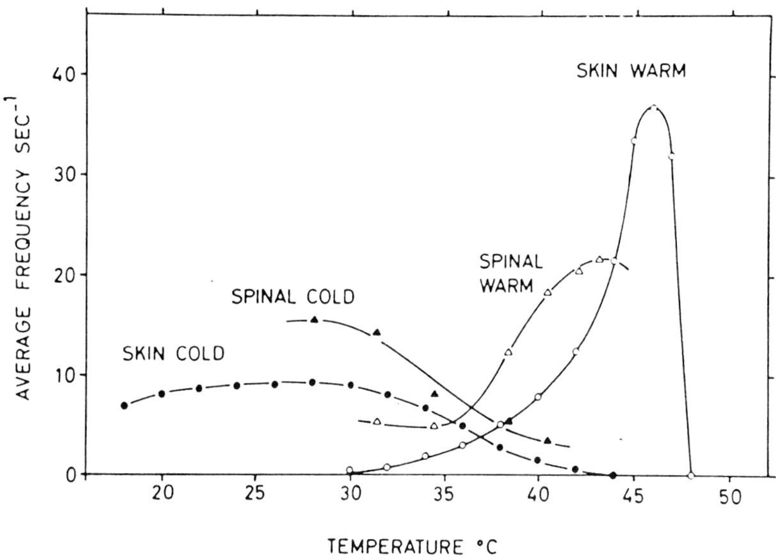

Fig. 3. Average static frequencies of populations of thermosensors in the cat as a function of temperature. Cutaneous receptors from Hensel & Kenshalo, spinal neurons from Simon & Iriki.
Hensel, H. (1974): Thermoreceptors. *Annu. Rev. Physiol.* 36, 233-249.

Fig. 3 contrasts, or better points out the homology between the spinal units and cutaneous thermoreceptors. Like the skin, the spinal cord has two clearly separated sets of sensing elements, one carrying spinal cold signals and the other one warm signals. Later experiments showed that at least those units, which were sensitive to spinal cooling, were also excited by skin cooling. The important implication was that part of the interaction between skin and core temperature signals occurs already at the level of the spinal cord.

There is a study from that period which may not have found the attention it deserved. Its result was simple and clear: spinal cord cooling as well as general cooling were able to generate shivering even in spinalized animals, and the EMG records obtained in spinalized and intact animals were qualitatively indiscernible (Simon, Klussmann, Rautenberg, Kosaka, 1966). This was wrongly misunderstood by some that the authors believed that shivering in intact animals was generated entirely at the spinal level. However, the discussion of the original paper reveals the real intention and is still worth keeping in mind: "Apparently, shivering in homeotherms relies on a property of the central nervous system, which has developed at its segmental level. This observation confirms for shivering, what has been postulated before by Thauer for the temperature regulating system in general: the process of regulation, which in its simplest sense, is the generation of a response counteracting a disturbance, can in principle be accomplished at any level, and in reality most likely involves multiple levels of the central nervous system. Regulation in biology is not the domain of a center even if the hypothalamus is often most important".

In the review on the "Spinal cord as a site of extrahypothalamic thermoregulatory functions" (Simon, 1974) the argument was later refined and broadened. Not just multiple inputs, but also multiple controllers, meshed in a simultaneously parallel and hierarchic organization, with the hypothalamus as the highest level, that was the message of 1974. In the field of temperature regulation, it is still today the most detailed attempt to lend some content to the black box approach to regulation. Interestingly, at about the same time Evelyn Satinoff arrived, via a different line of experimental evidence, at a rather similar concept.

In the mid-seventies, the cooperation with Ted Hammel was resumed, and another key player added her name Oppermann to the standard and our list of references. First in penguins and later in ducks, a most intriguing feature of temperature regulation in birds versus mammals was described and analyzed. Hypothalamic cooling in mammals induces shivering, and that makes sense if the hypothalamus is seen as a temperature sensor. Cooling the hypothalamus in a bird, however, inhibits shivering (Fig. 4) and that did not seem to make any sense at all. Furthermore, cooling caused a vasodilation in the skin and, as shown by I. Schmidt, even panting.

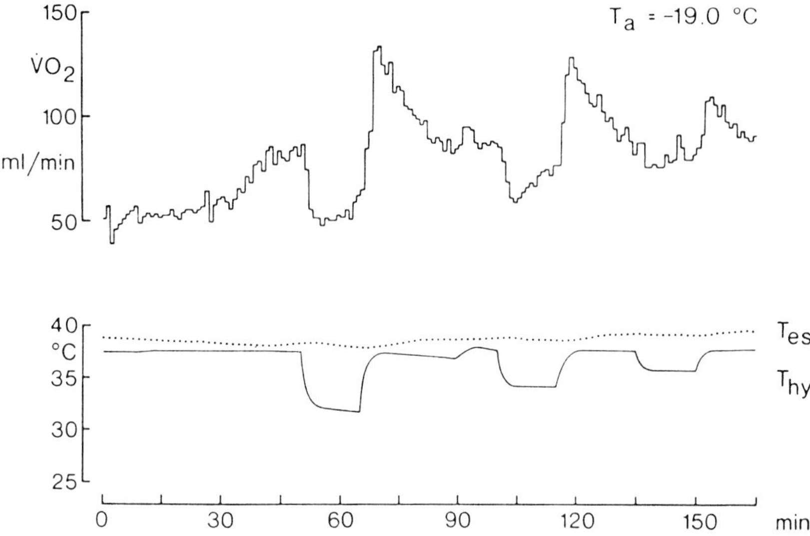

Fig. 4. Oxygen consumption and core temperature at cold ambient conditions as influenced by cooling of the rostral brain stem.
Simon, E., Simon-Oppermann, Ch., Hammel, H.T., Kaul, R., Maggert, J. (1976): Effects of altering rostral brain stem temperature on temperature regulation in the Adelie penguin. *Pflügers Arch.* 362, 7-13

Others had seen isolated aspects of the phenomenon before, but the credit for putting the perspective right goes to Eckart Simon. The clue was the "temperature dependence of signal transmission". The strange results in birds can be explained by two assumptions, which are supported by several lines of evidence. The first is that the avian hypothalamus (Fig. 5, upper panel) generates hardly

any specific temperature signals for autonomic control, and the second is that the pathways linking peripheral thermal inputs with effectors, travers the hypothalamus, and this hypothalamic signal transmission depends non-specifically on brain temperature.
In a reversal of the original Hammel concept, a larger Q_{10} was assigned to the cold pathway, the cooling of which would then inhibit the peripheral drive for shivering.

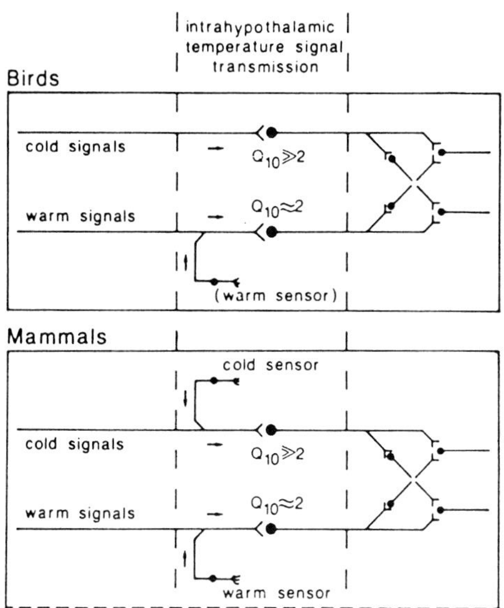

Fig. 5. Temperature dependence of signal transmission and hypothalamic temperature sensors: birds versus mammals.
Simon, E. (1981): Effects of CNS temperature on generation and transmission of temperature signals in homeotherms. A common concept for mammalian and avian temperature regulation. *Pflügers Arch.* 392, 79-88.

If the signal transmission depends on hypothalamic temperature in birds, it should also do so in mammals. Several studies have demonstrated that this is indeed the case, and the lower panel of Fig.5 shows the wiring diagram of a mammal: the pathways of signal transmission have the same unspecific temperature dependencies as in the bird. However, in the mammal they are complemented by specific cold and warm sensors. Not unexpectedly, the interaction between two specific and two unspecific temperature effects becomes rather complex and touches the limits of communicability. However, a serious student of the model is rewarded by a most beautiful set of contour lines. The model is the first and only coherent concept of the temperature regulating circuits in mammals and birds, and gave us an idea how the black box could work, or to put it in the words of its author, the black box is now a little less black.

An element of uneasiness arises from the little sensors (Fig.5). This model and many others require specific thermosensory properties of neurons or neuronal circuits in the brainstem of a mammal, but not so, or at least much less so, in a bird. In fact, numerous authors have shown thermosensitive neurons in the rostral brainstem of mammals, and spent much persuasive power on assigning the specific function to them. However, the neurophysiological approach of those days lost much of its impact when Simon, Hammel and Oksche (1977) found precisely the same, so-called specific types of thermosensitive neurons in the hypothalamus of the duck, in spite of the fact that this species didn't know better than to show unspecific, paradoxical effector responses to hypothalamic temperature. The last sentence of the summary of their paper reads: "Thermosensitivity of a neuron in the rostral brainstem is not a sufficient criterion for presuming a thermoreceptive role for it in the duck, and probably in the mammal either"

There is a modern version of the same argument. More recently, the hypothalamic slice technique had been developed, and persistence of a neuron's thermosensitivity in a calcium-free/high magnesium medium became the new key criterion of primary warm sensitive neurons. In 1987, Nakashima, Pierau, Simon and Hori compared hypothalamic thermoresponsive neurons from duck and rat slices, and the result was that *in vitro* evaluation of hypothalamic neuronal thermoresponsiveness in a bird and in a mammal did not reveal differences at the single unit level, which might explain the diverging contributions of avian and mammalian hypothalamus to deep body temperature perception. Some of us would certainly have welcomed a more positive conclusion at the end of this study. However, it revealed a distinct feature of the SI-standard, which also implies a critical attitude towards fashionable and perhaps prematurely accepted explanations.

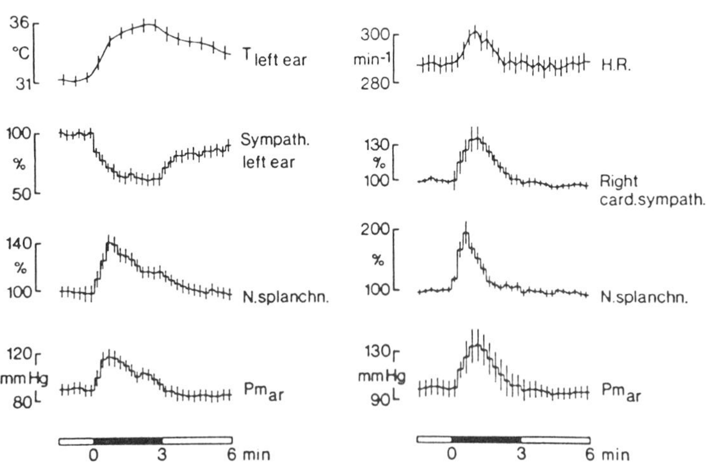

Fig. 6. Antagonistic patterns of regional sympathetic activity.
Simon, E., Riedel, W. (1975): Diversity of regional sympathetic outflow in integrative cardiovascular control: patterns and mechanisms. *Brain Res.* 87, 323-333

Another example stems from another field and another time. In the early seventies it was standard textbook knowledge, mainly based on Cannon's concepts, that, at a given moment, the sympathetic nervous system would be uniformly excited or inhibited in all body regions. However, Thauer knew about the results of Dastre & Morat in 1884, which required a more complex explanation. In a way, the decision against Cannon and for Dastre & Morat spun off the research on the spinal role in temperature regulation.

Fig. 6 shows the responses of rabbits to spinal cord warming. Ear skin temperature rose, and this was, as everybody expected, due to inhibition of its sympathetic supply. However, the activities in the splanchnic and cardiac branches increased considerably. These simultaneous but antagonistic responses marked the end of the mass action as the one and only type of sympathetic control. Fig. 6 stems from an early paper. Later, the problem was analyzed in more detail, and in 1983 the results were summarized by Rowell: "Experiments by Simon and his colleagues revealed that a differentiated pattern of regional sympathetic vasomotor outflow is a specific component of autonomic control. differentiation appears to occur at the level of single unit activity, so different fibers in a single nerve can have different patterns of activity". - What once began as a side-track from temperature regulation, has become a break-through in the understanding of regional circulatory control.

In the late seventies, another line of research developed and soon established itself as the hot topic of the department. It commenced with Hammel and his penguins, and is still going strong with many coworkers and their salt-water ducks (Simon & Gray, 1989). These creatures can survive for several years on 2% saline as their only fluid supply, and they do so by employing their nasal salt glands, which in a great number of birds, represent a second effector organ of fluid homeostasis besides the kidneys.

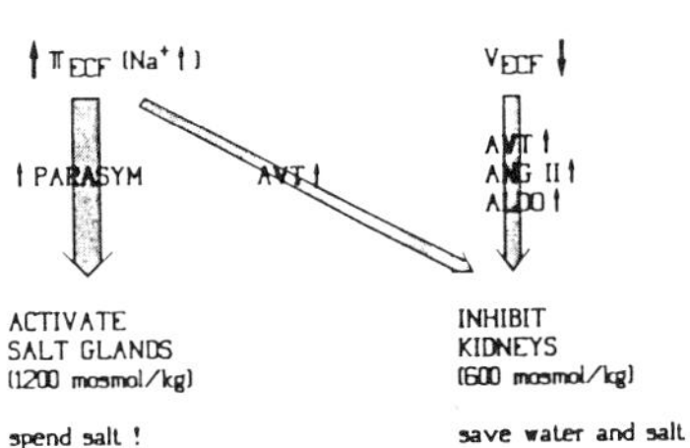

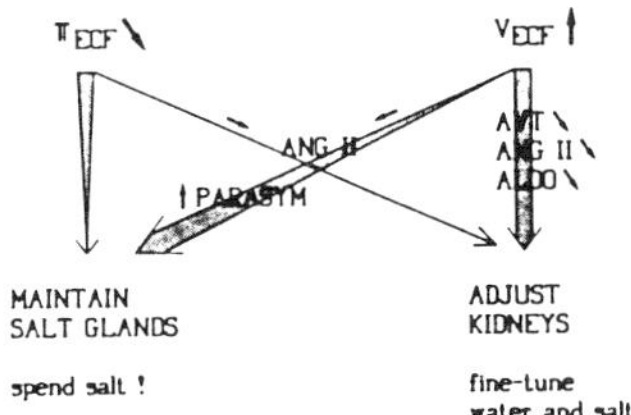

Fig. 7. Block diagram showing responses of salt-water ducks to high salt intake (top) and combined high salt and water intake (bottom)

Fig. 7 shows a simple block diagram as a guide through this section. The primary disturbance is an increase of extracellular osmolality of up to 350 mosmol/kg, which is transduced by hypothalamic osmoreceptors into parasympathetic driving signals for the salt glands. The glands can concentrate up to 1200 mosmol/kg, and deliver salt at a rate of 0.4 mMol/min. There is another aspect: the kidneys can't concentrate very well - not more than 600 mosmol/kg. It's imperative to inhibit them and to prevent them from wasting water. Large increases of AVT - the birds' ADH - , of Angiotensin II and Aldosterone do it so that the renal excretion rates fall to just one tenth of those of the salt glands. These hormonal responses are triggered by osmoreceptors and by volume receptors, which sense a contraction of the ECF volume following dehydration.

This led to an important observation. As we all know, the plasma level of Angiotensin II in mammals is inversely related to the level of plasma sodium. But this is not so in the salt-adapted duck. Here, the higher plasma osmolarity and plasma sodium are, the higher is plasma Angiotensin. The key is that Angiotensin is mainly controlled by ECF volume: the smaller it is, the larger is the level of Angiotensin. This is instrumental in adjusting the birds' reaction to simultaneous loads of salt and water.

In that case, the upward deviation of ECF osmolality becomes smaller and smaller, and so does the osmotic drive to the salt glands. However, salt gland activity should be maintained since it is still the most effective means of mass excretion of salt. This is indeed accomplished. With increasing ECF volume, the Angiotensin II level grows smaller, and so does its inhibiting central effect on salt gland activation. The salt glands remain active, and the parallel release of the inhibition of the kidneys fine-tunes the outputs of water and electrolytes in order to restore euhydration.

It's a very intricate system. Considering the high sodium concentrations, high levels of AVT, Angiotensin II and Aldosterone, these birds are confronted with a combination of all humoral risk factors for hypertension. And yet, the salt ducks have, if anything, a lower blood pressure than their fresh-water sisters, in spite of a seven-fold increase of Angiotensin II.

One can easily imagine the director's joy at this insight in another beautifully organized physiological process, and his pleasure in seeing once more how a species has escaped from environmental constraints and not the least, from the constraints of the models the physiologists have built.

In this last section, I have refrained from mentioning the names of all members and guests of the department who have made important contributions to this work. However, if somebody should feel neglected, I would recommend her or him to wait for the own anniversary and tell what I have experienced myself more than once: like few others, Eckart Simon has the gift to inspire his scholars to work on new projects they had never thought of before, so that they finally love the projects as their own brain children. It is perhaps this single feature which makes the leader of a successful group.

We thank you for what you have done for the field of integrative physiology, and we hope that you will continue to imprint your standard on the team of promising young scientists, which has to carry the Kerckhoff spirit into the next century.

References

Nakashima, T., Pierau, F.K., Simon, E., Hori, T. (1987): Comparison between hypothalamic thermoresponsive neurons from duck and rat slices. *Pflügers Arch.* 409, 236-243.

Rowell, L.B. (1983): Cardiovascular adjustments to thermal stress. In *Handbook of Physiology, Sec. 2, Vol. III/2, Peripheral Circulation and Organ Blood Flow,* ed. J.T. Shepherd, F.M. Abboud, S.R. Geiger, pp. 967-1023. Bethesda Md: American Physiological Society.

Simon, E. (1974): Temperature regulation: the spinal cord as a site of extrahypothalamic thermoregulatory functions. *Rev. Physiol. Biochem. Pharmacol.* 71, 1-76.

Simon, E., Gray, D.A. (1989): Control of salt glands as a second-line defense in body fluid homeostasis of birds. In *Progress in Avian Osmoregulation*, ed. M. Hughes, A. Chadwick, pp. 23-40. Leeds UK: Leeds Philosophical and Literary Society Ltd.

Simon, E., Hammel, H.T., Oksche, A. (1977): Thermosensitivity of single units in the hypothalamus of the conscious Pekin duck. *J. Neurobiol.* 8, 523-535.

Simon, E., Klussmann, F.W., Rautenberg, W., Kosaka, M. (1966): Kältezittern bei narkotisierten spinalen Hunden. *Pflügers Arch.* 291, 187-204.

I. Temperature regulation

A. Control of body temperature and acclimatization

Integrative and cellular aspects of autonomic functions : temperature and osmoregulation. Eds K. Pleschka, R. Gerstberger. John Libbey Eurotext, Paris © 1994, pp. 17-22.

Neuropeptides and body temperature control : integration of *in vivo* and *in vitro* approaches

Ladislav Jansky, Friedrich-Karl Pierau*, Eckhart Simon*, Walter Riedel*, Herbert Schmid*, Christa Simon-Oppermann*, Jutta Schenda*, Holger Sann*, Stanislav Vybiral, Jan Moravec, Martin Konrad, Haddi Muhamad Ehymayed, Jaromir Nachazel and Ludek Cerny

Department of Comparative Physiology, Faculty of Science, Charles University, Vinicna 7, 128 00 Prague 2, Czech Republic
** Max-Planck-Institut für physiologische und klinische Forschung, W.G. Kerckhoff-Institut, Parkstrasse 1, D-61231 Bad Nauheim, Germany*

SUMMARY:

A number of neuropeptides which affect body temperature upon intrahypothalamic application influence firing rate and thermosensitivity of hypothalamic thermoregulatory neurons specifically. However, the controller of the body thermostat represents a neuronal network with complex interactions, which makes it difficult to propose the physiological meaning of these observations. Nevertheless, the effect of these neuropeptides on hypothalamic neurons appears as a shift in the threshold for the activation of thermoregulatory effector mechanisms and/or as a change in thermosensitivity of the hypothalamic controller. Activation of the "warm pathway" in the hypothalamus due to an increase in firing rate and/or due to increase in thermosensitivity of neurons could lower the threshold for the induction of thermoregulatory effectors, whereas inhibition of the "warm pathways" would increase the threshold. Activation of thermoinsensitive interneurons might decrease the thermosensitivity of the hypothalamic controller. These interpretations are not consistent with previous conclusions of Bligh (1972).

INTRODUCTION:

In an attempt to elucidate the mode of action of neuropeptides on the control body temperature the paper summarizes data previously published (Jansky et al., 1986, 1987,1992; Pierau et al., 1989 a,b; Moravec and Pierau, 1992; Schmid et al., 1993; Vybiral et al., 1986, 1987, 1988; Vybiral and Jansky 1989; Ehymayed and Jansky, 1992). It is the aim of the paper to integrate data obtained under "in vivo" and "in vitro" conditions. For the former, body temperature of rabbits was changed by intestinal cooling and warming and the effect of neuropeptides on effector mechanisms of temperature regulation was investigated by microinjections into different hypothalamic areas of the rabbit. In vitro experiments were performed on rat hypothalamic slices by recording the effect of certain neuropeptides on the activity and thermosensitivity of hypothalamic neurons with conventional electrophysiological methods.

RESULTS AND CONCLUSIONS:

The most significant findings of the studies are the following:

a) Regulatory effects of neuropeptides:
   1. Neuropeptides, when injected into the hypothalamus modulate the activity of thermoregulatory centers differently but specifically.
   2. These thermoregulatory effects of neuropeptides are specific to the preoptic area of the

hypothalamus.

3. The effect of neuropeptides is long-lasting and persists for several hours.
4. Some neuropeptides are effective in normothermic animals (bombesin, neurotensin), while other neuropeptides (ACTH, α-MSH, AVP) are effective in febrile animals, only.
5. The effect of neuropepties becomes apparent as a shift of the temperature threshold for the activation of thermoregulatory effector mechanisms (shivering, respiratory evaporative heat loss - REHL, peripheral vasomotor tone - PVMT) either increasing (neurotensin) or decreasing (bombesin) the threshold temperature. Changes in thermosensitivity of the hypothalamic controller may also occur (bombesin).

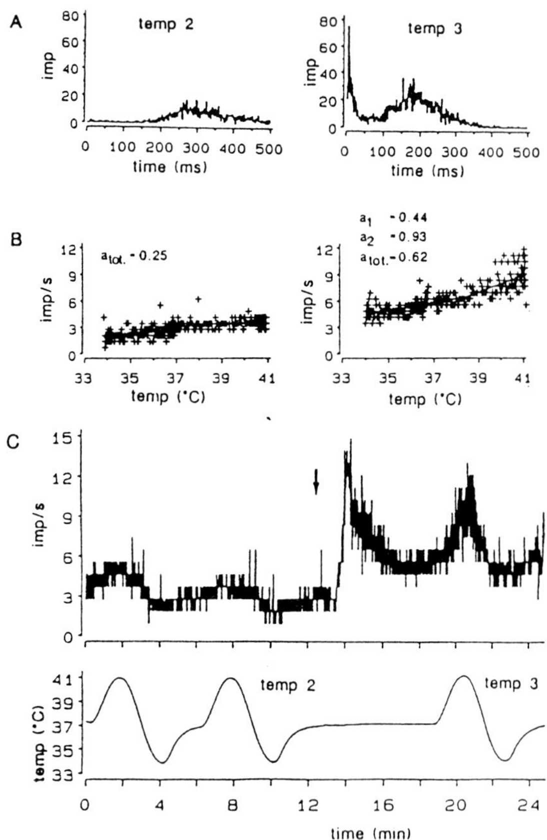

Fig. 1. *Transformation of a temperature insensitive neuron into a warm sensitive neuron after bombesin. A = interval histogram, B = frequency temperature relationship, C = ratemeter recordings and protocol of the slice temperature (Schmid et al., 1993)*
arrow: bolus injection of bombesin (0.1 μg in 0.1 ml); temp 2 and temp 3 : data acquisition during the periodical temperature change indicated in the bottom panel.

6. Antipyretic peptides (ACTH, α-MSH, AVP) act by separating the threshold for shivering from that for panting and vasomotion, thus evoking an effect during the early feverphase which usually occures only in the late phase.
7. It might be suggested that transient increase in the production of neuropeptides during the beginning of fever represents a negative feedback mechanism preventing an uncontrolled increase of body temperature.

b) Cellular effects of neuropeptides:

1. Hypothalamic neurons are mostly warm sensitive or temperature insensitive. Cold sensitive neurons are scarce.
2. Neuropeptides do not only influence the tonic activity (static discharge) of neurons but also influence their thermosensitivity (= temperature coefficient - TC = imp/s/°C) (Fig. 1). The effect of neuropeptides persists for a long time.
3. Neuropeptides can influence the activity of different types of neurons in the hypothalamus independent of their location or thermosensitivity.
4. The effect of neuropeptides on neuronal activity differs in the same type of neurons (Fig. 2) suggesting that the resulting thermoregulatory responses reflect the statistical distribution of neuronal effects.

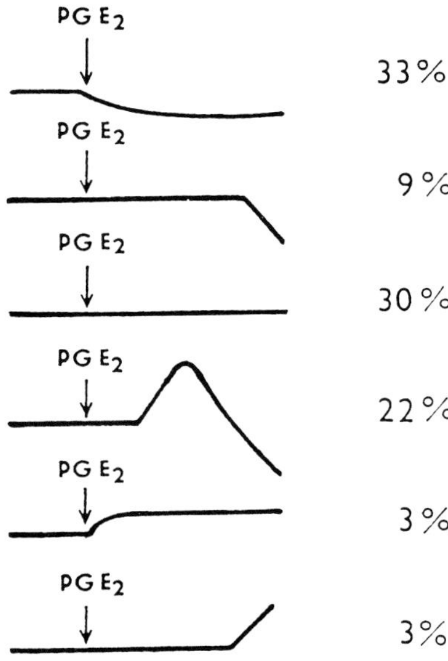

Fig. 2. *Different effects of PGE_2 on spontaneous activity of hypothalamic neurons (frequency/time). The proportion of different types of responses is indicated at the right site.*

5. Hypothermic neuropeptides (bombesin) appear to increase tonic activity and thermosensitivity (TC) of most of the neurons in the anterior hypothalamus. The effect on the TC is more apparent in temperature insensitive neurons than in warm sensitive ones (Fig. 3). Probably, the hypothalamic controller becomes more warm sensitive.
6. The pyretic mediator PGE_2 appears to decrease thermosensitivity (TC) and tonic activity of hypothalamic neurons (Fig. 4), probably resulting in a decrease of the warm sensitivity of the hypothalamic controller.
7. Antipyretic neuropeptides (AVP), when acting on neurons of nonfebrile rats, decrease thermosensitivity (TC) of warm sensitive neurons, while leaving the thermosensitivity (TC) of temperature insensitive neurons unaffected. The tonic activity is increased in the majority of neurons.
8. Substance P, which does not influence thermoregulation under in vivo conditions, increases tonic activity but decreases thermosensitivity (TC) of warm sensitive neurons, while thermoinsensitive neurons are not influenced (Fig. 3).
9. Warm adaptation decreases the number of warm sensitive neurons probably resulting in a decrease of the warm sensitivity of the hypothalamic controller (Konrad and Pierau, 1993).

10. The modulation of temperature coefficients of neurons by endogenous substances suggest that the thermosensitivity of hypothalamic neurons is not an unchangeable inherited property of individual neurons.

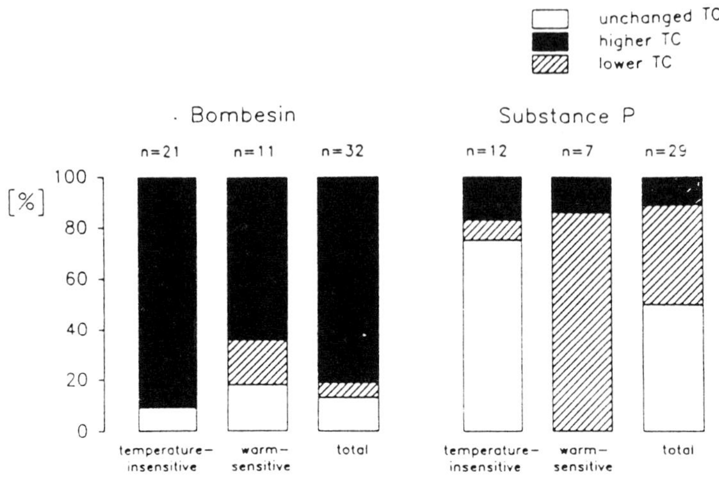

Fig. 3. *Changes in temperature coefficients (TC) of hypothalamic neurons after bombesin and after substance P. Bombesin generally increases TC, while substance P reduces TC in warm sensitive neurons, predominantly (Schmid et al., 1993).*

The question to be solved is how to integrate the "in vivo" and the "in vitro" data, so far obtained. Summarizing the data it could be suggested:

1. Substances which increase firing rate and thermosensitivity of warm sensitive neurons (e.g. bombesin) may lower the thermoregulatory thresholds.
2. Substances which decrease tonic activity and thermosensitivity of warm sensitive neurons may increase the thermoregulatory thresholds (PGE_2).
3. Substances which increase firing rate but lower thermosensitivity (TC) of warm sensitive neurons (substance P, AVP) have no effect on thermoregulation under nomothermic conditions. This might indicate that changes in spontaneous activity and thermosensitivity of neurons are, to a certain extent, complementary and may promote changes or resetting of the body thermostat.
4. Increase of firing rate and temperature coefficient (TC) of thermoinsensitive interneurons may also explain the reduced threshold temperature and alteration of gain of the hypothalamic controller (bombesin). No change of the TC of interneurons could contribute to the lack of substances to change the thermosensitivity of the controller (substance P).

One could use models of the neuronal control of temperature regulation to speculate about the detailed mode of neuropeptide action. For this purpose the Bligh model of thermoregulation (1972) appears to be most suitable, because of its simplicity (Fig. 5). The presumptions of the model can be summarized as follows:

1. Warm and cold signals are transmitted by separate but interlinked neural pathways.
2. The interlinking pathways consist of interneurons, which are relatively temperature insensitive.
3. Activation of the warm or the cold pathway, respectively, can reciprocally inhibit the function of the other one.

Because of the scarcity of cold sensitive neurons in the hypothalamus it can be concluded that neuropeptides act predominantly on warm sensitive neurons in the warm pathway or on thermoinsensitive interneurons in interlinked pathways. Since bombesin generally enhances

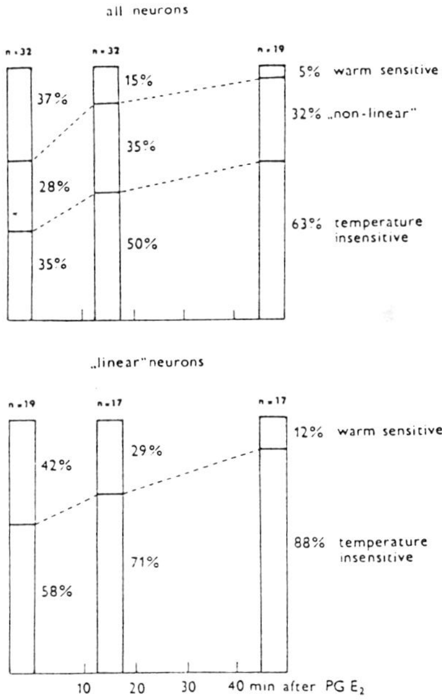

Fig. 4. *The decrease of the temperature sensitivity of warm sensitive neurons by PEG2 results in an increase of the proportion of temperature insensitive neurons (Jansky et al., 1992).*

warm sensitivity of hypothalamic neurons, an activation of the warm pathway and, hence, stimulation of heat loss mechanisms would be expected. Concominantly, an increased inhibition of the cold pathway and, consequently, of heat production mechanisms via interneurons would also occur, resulting in a decrease of the threshold for the induction of all thermoregulatory effector mechanisms. The combination of hypothermia and change of gain of the controller after bombesin may be due to the increased activity of interneurons at sites close to the warm pathway. This view is in contrast with the suggested role of temperature insensitive neurons given by Bligh (1972).

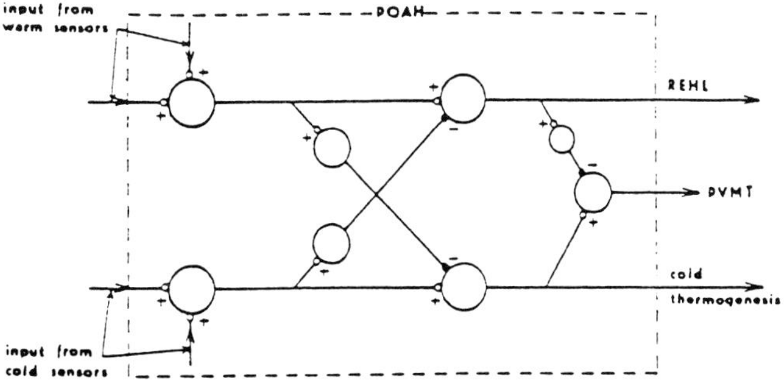

Fig. 5. *POAH: preoptic area anterior hypothalamus; REHL: respiratory, evaporative heat loss; PVMT: peripheral vasomotor tone; + : excitatory, - : inhibitory interconnections.*

In contrast to bombesin, prostaglandin E_2 decreases the neuronal warm sensitivity. This may attenuate the activity of the warm pathway and release the inhibitory action of interneurons on the cold pathway, the result being an activation of the cold pathway. Consequently the threshold for the activation of thermoregulatory effectors would be shifted upwards which in turn would increase heat production and body temperature due to shivering and vasoconstriction, as typically seen during fever.
The reason for the observation that substance P and AVP are without effect on thermoregulation in nonfebrile animals might be, that an activation of the tonic activity of neurons is compensated by their decreased thermosensitivity.

REFERENCES

Bligh, J. (1972): Neuronal model of mammalian temperature regulation. In *Essay on Temperature Regulation*, eds J. Bligh and R.E. Moore, pp. 105-120. Amsterdam, London: North-Holland Publ. Comp.

Ehymayed, H.M., and Jansky, L. (1992): A discrete mode of the antipyretic action of AVP, α-MSH and ACTH. *Physiol. Res.* 41: 57-61.

Jansky, L., Pierau, Fr.-K., and Schenda, J. (1992): The effect of PGE_2 on activity and thermosensitivity of hypothalamic neurones in rat brain slices. *Physiol. Res.* 41: 85-88.

Jansky, L., Riedel, W., Simon, E., Simon-Oppermann, C., and Vybiral, S. (1987): Effect of bombesin on thermoregulation of the rabbit. *Pflügers Arch.* 409: 318-322.

Jansky, L., Vybiral, S., Moravec, J. Nachazel, J., Riedel, W., Simon, E., and Simon-Oppermann, C. (1986): Neuropeptides and temperature regulation. *J. Therm. Biol.* 11: 79-83.

Konrad, M., and Pierau, Fr.-K. (1993): Temperature sensitivity of neurons in slices of the cold and warm adapted rat PO/AH area: effect of bombesin. In *Proc. 3th Discussion on Cellular Neurophysiology*, Chlum, pp. 72-74.

Moravec, J., and Pierau, Fr.-K. (1992): Does the effect of AVP on hypothalamic neurones support its proposed role in endogenous antipyresis? *Physiol. Res.* 41: 89-90.

Pierau, Fr.-K., Schmid, H., and Jansky, L. (1989a): Recruitment of warm sensitive hypothalamic neurones by peptides and changes of extracellular calcium. In *Living in the Cold II*, eds A. Malan and B. Canquilhem, pp. 255-264. Colloque INSERM/John Libbey Eurotext, Ltd.

Pierau, Fr.-K., Schmid, H., Jansky, L., and Schenda, J. (1989b): Long term modulation of hypothalamic neurons by endogenous peptides. In *Thermal Physiology 1989*, ed J.B. Mercer, pp. 111-116. Amsterdam, New York, Oxford: Excerpta Medica.

Schmid, H.A., Jansky, L., and Pierau, Fr.-K. (1993): Temperature sensitivity of neurons in slices of the rat PO/AH area: effect of bombesin and substance P. *Am. J. Physiol.* 264: R449-R455.

Vybiral, S., Cerny, L., and Jansky, L. (1988): Mode od ACTH antipyretic action. *Brain Res. Bull.* 21: 557-562.

Vybiral, S., and Jansky, L. (1989): The role of dopaminergic pathways in thermoregulation of the rabbit. *Neuropharmacology* 28: 15-20.

Vybiral, S., Nachazel, J, and Jansky, L. (1986): Hyperthermic effect of neurotensin in the rabbit. *Pflügers Arch.* 406: 312-314.

Vybiral, S., Szekely, M., Jansky, L., and Cerny, L. (1987): Thermoregulation of the rabbit during the late phase of endotoxin fever. *Pflügers Arch.* 410: 220-222.

Integrative and cellular aspects of autonomic functions : temperature and osmoregulation. Eds K. Pleschka, R. Gerstberger. John Libbey Eurotext, Paris © 1994, pp. 23-31.

Cholecystokinin : its possible role in temperature regulation

Zoltán Szelényi, Miklós Székely and Márta Balaskó

Department of Pathophysiology, University Medical School, H-7643 Pécs, Szigeti út 12, Hungary

INTRODUCTION

Neuropeptides have been found to be mediators of autonomic functions either as substances locally released in various peripheral tissues (endocrine, paracrine or transmitter roles) or as transmitters/modulators produced in, released from and acting in the central nervous system (CNS). One of the most intensively studied of these peptides are those belonging to the cholecystokinin (CCK) family. In the rat brain CCK has been shown to be one of the most, if not the most abundant among all neuropeptides (Crawley, 1985); especially high concentrations of CCK and its receptors were found in the cerebral cortex and hypothalamus, the latter localization making this peptide a good candidate as a central mediator of autonomic functions (Day et al., 1986).

Peripherally administered CCK has been established to possess satiety inducing effect in most species investigated so far (Gibbs et al.,1973). A possible thermoregulatory role of this peptide has also been inferred from studies in which dose-dependent hypothermia was induced after its peripheral injection in rats exposed to subneutral ambient temperatures (Kapás et al., 1987, Morley et al., 1981). Since CCK - as most of the peptides with similar range of molecular mass - does not seem to be able to cross the blood-brain-barrier (Oldendorf, 1981), a specific central site of action for the hypothermia to develop may be questioned. Earlier studies from this laboratory have shown that intracerebroventricular (ICV) injection of CCK-octapeptide (CCK-8) in the rat induced dose-dependent hyperthermia with characteristics similar to those of a typical febrile response (Szelényi and Barthó, 1989). The present experiments have been carried out to characterize thermoregulatory effects of CCK-8 (hypothermia or fever on its peripheral or ICV administration, respectively) and to learn the receptor types of CCK which may be involved in these responses or in those observed after ICV injection of prostaglandin E_1 (PGE_1).

METHODS

Female Wistar rats were implanted under general anaesthesia with chronic lateral cerebroventricular cannula and with subcutaneous polythene cannula for central and peripheral injections, respectively. Conscious, partly restrained rats were put into a metabolic chamber and colonic temperature (Tc), tail skin temperature (Ts) as well as metabolic rate (MR) were continuously measured - the latter with indirect calorimetry (open system) while the animals have been exposed to standard ambient temperatures (Ta) between 18 to 30°C. Sulphated CCK-8, PGE_1, the CCK B- and A-receptor ligands (L-365,260 and L-364,718, respectively) were injected without disturbing the animal (the authors are indebted to Dr. V. Lotti, Merck Sharp and Dohme, for supplying the receptor ligands mentioned). Further details on the methods used will be found shortly (Szelényi et al., 1993).

RESULTS

1. Fever-like response induced by centrally injected CCK-8

As shown by Fig. 1, CCK-8 microinjected intracerebroventricularly (ICV) induces an increase in Tc starting with a short latency which is accompanied by a decrease in heat loss (indicated by a fall in skin temperature) along with a rise in metabolic rate; a co-ordinated thermoregulatory response characteristic of fever. Indeed, this response was similar to the rise of body temperature induced by an ICV injection of PGE_1 and even the dynamics of the two fevers were similar. Fever induced by ICV injected CCK-8 showed the same dependence on the initial body temperature as those elicited by centrally injected PGE or by peripherally injected endotoxin, that is, the lower the initial Tc, the higher the fever induced (see Székely and Szelényi, 1982).

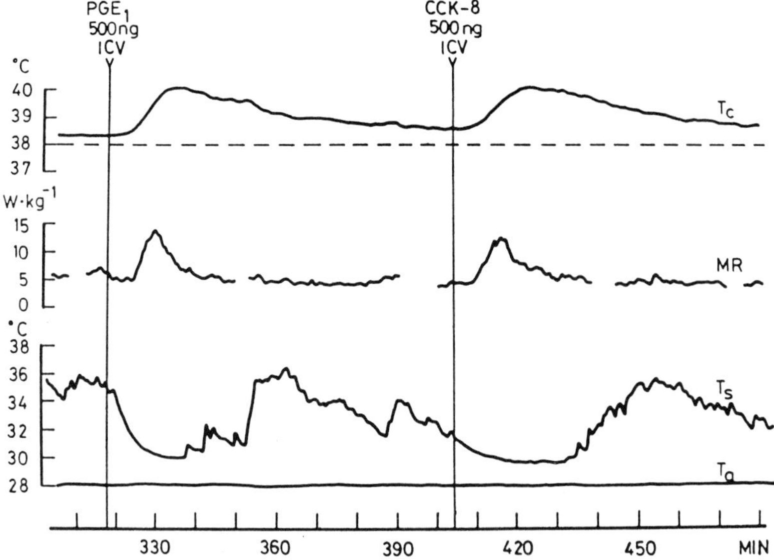

Fig. 1. Effects of PGE_1 and CCK-8, both injected intracerebroventricularly (ICV) on colonic temperature (Tc), metabolic rate (MR) and tail skin temperature (Ts) in a rat (from an original recording, Szelényi et al., 1993).

Also, the behaviour of thermoregulatory effectors accompanying febrile responses of the same magnitude showed some dependence on ambient temperature; in the majority of cases both vasoconstriction and a rise in metabolic rate contributed to fever (A in Fig. 2), while either only the metabolic response (B in Fig. 2) or vasoconstriction (C in Fig.2) could be observed at the start of fever when injections of CCK-8 were applied to rats having been already vasoconstricted or slighty hyperthermic, respectively (Fig. 2). Not only CCK-8 but also a derivative of it, ceruletide, after central application, proved to be effective in inducing a febrile response with characteristics similar to those of CCK-8-induced fever (Szelényi and Barthó, 1989).

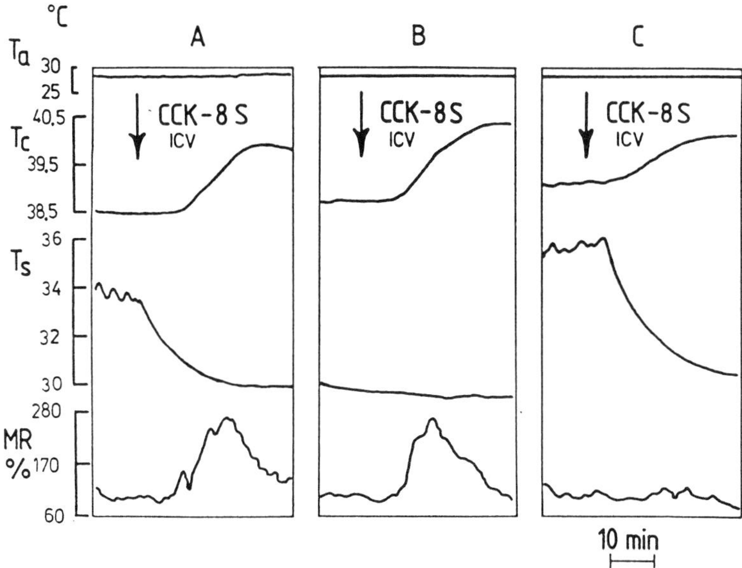

Fig. 2. Three patterns of the response of thermoregulatory effectors following an ICV injection of CCK-8 in rat experiments. For abbreviations see legends to Fig. 1 (Unpublished figure of Szelényi et al., 1993)

2. Hypothermic response to peripherally injected CCK-8

In the same experiments referred to above, peripheral (subcutaneous, SC or intraperitoneal, IP) injections of CCK-8 in rats exposed to subneutral ambient temperatures led to the development of a slight hypothermia (Fig. 3) as observed also by others. To induce hypothermia the dose of CCK-8 to be used was about two orders of magnitude higher than that producing fever in the same animals (Fig. 3). This dose of the peptide might have been enough to induce intensive vasodilatation directly at the site of injection together with the development of a "shock-like state" (Savory, 1987), both effects allowing the development peripheral changes leading to a negative heat balance independent or even against a central thermoregulatory drive in an animal exposed to cold and trying to defend homeothermia. These data and the difficulty of CCK-8-like peptides to cross the blood-brain-barrier make it unlikely that the hypothermia induced by high doses of peripherally injected CCK-8 is a genuine thermoregulatory response mediated or controlled by central mechanisms.

3. Specificity of action of CCK-8 to induce fever or hypothermia

The fever induced by ICV injection of CCK-8 was further characterized by using ligands known to specifically inhibit either the central or the peripheral type of CCK-receptors. In one of these series of experiments after having been tested with ICV injection of PGE_1, rats were injected with CCK-8 before and about one hour after a SC injection of a ligand for CCK type- receptors (central type, as described by Lotti and Chang,1989) inhibitor of type-B CCK receptors (L 356,260) significantly attenuated the fever induced by an ICV injection of CCK-8, while the maximum

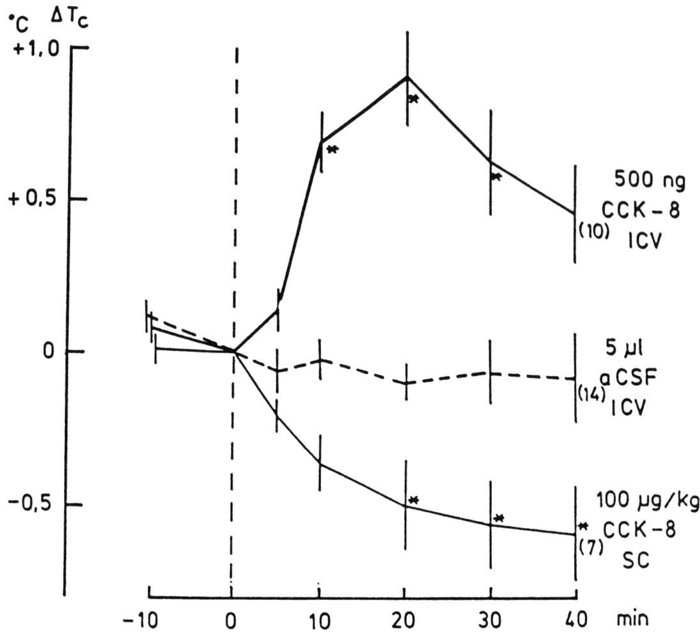

Fig. 3. Changes in Tc after ICV or subcutaneous (SC) injection of CCK-8 or ICV injection of artificial CSF (aCSF) in the same group of rats. Mean ±S.E.M.

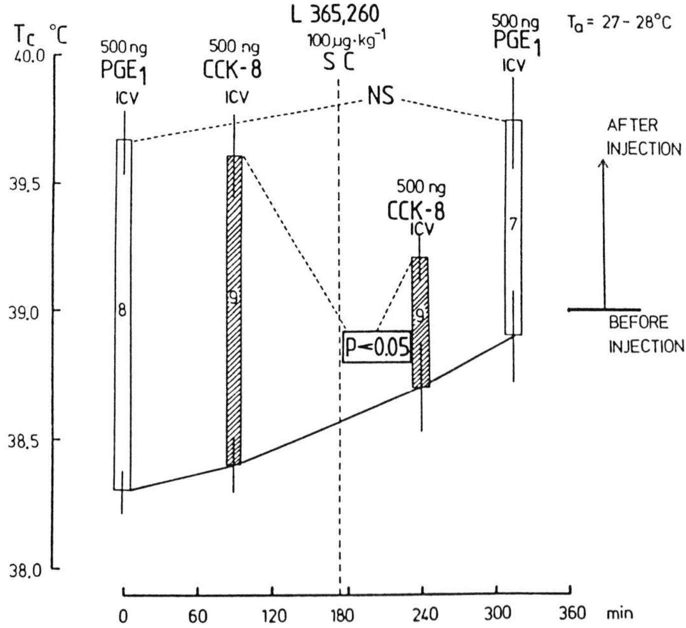

Fig. 4. Effects of an inhibitor of type-B (central) CCK receptors, L-365,260, on PGE_1- and CCK-8-induced fevers in rats. Figures within the columns indicate the number of injections. For abbreviations see legends to Figs 1 and 3.

rise of Tc after ICV injection of PGE_1 remained unchanged. In another similar series of experiments the fever as well as the hypothermia induced by ICV or SC injection of CCK-8 was tested for their site of effect using again the inhibitor for type B CCK receptors. The febrile effect of ICV injection of CCK-8 was significantly attenuated even about two hours after the injection of the ligand, while the hypothermia induced by a SC injection of the peptide remained unchanged. These data, therefore, give further support to the idea that the rise in Tc after CCK-8 (or fever) is indeed the specific and central effect of the peptide, while the hypothermia induced after peripheral adminsitration is not mediated by this set of CCK receptors. In addition, the PGE-fever does not seem to contain a CCK-ergic component (at least not one testable by the inhibitor used in these studies).

To see if the fever or hypothermia induced by central or peripheral injection of CCK-8 could be mediated by the so called peripheral type of CCK receptors, the ligand, L-364,718 (Chang and Lotti, 1986) was tested in a similar way as in the experiments mentioned previously. Fig. 6 summarizes the results obtained showing that this time the hypothermia, but not the fever could be attenuated (or even blocked) by the type A receptor blocker. It should be emphasized that the time-course of injections was very similar to that applied in experiments with type B receptor blocker. These data corroborate the suggestion that in rats the CCK-8-induced hypothermia is most likely a direct peripheral effect of the peptide, while the ICV route seems to be the one affecting central type of CCK receptors at least under the conditions of the present studies.

4. A CCK-ergic mechanism in endotoxin fever (?)

In a study carried out on rats equipped with chronic intravenous (IV) polyethylene cannula (in one

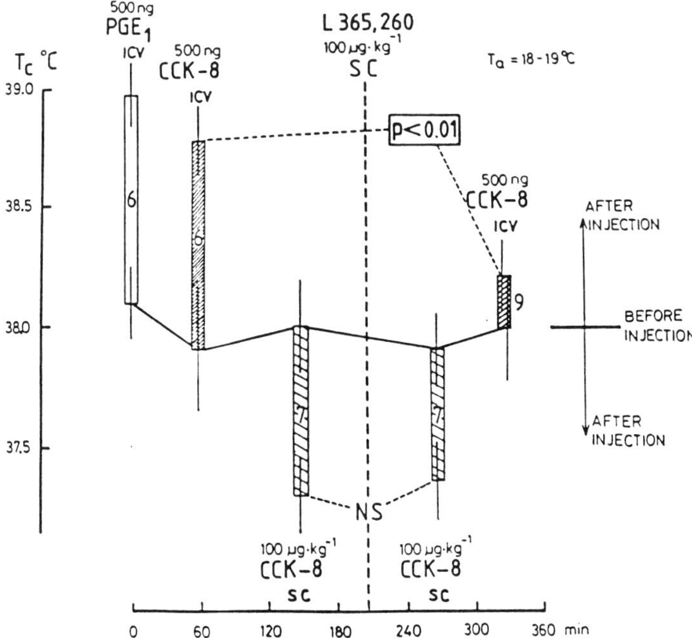

Fig. 5. Effects of an inhibitor of type B (central) CCK receptors on fever or hypothermia induced by an ICV or SC injection of CCK-8. For abbreviations see legends to Figs 1 and 3. Figures within or near the columns indicate the number of injections.

of the external jugular veins) as well as a SC cannula and acclimatized to 4-5°C for at least three weeks, fever induced by an IV injection of Escherichia coli endotoxin (LPS) was observed with or without a preceding SC injection of the inhibitor for type-B CCK receptors. Fig. 7 demostrates that the ligand used did not inhibit LPS fever as a whole, even the overall size of the febrile response remained the same. The characteristic two-phasic nature of LPS-fever was, however, modified to a monophasic one, in that after the inhibition of type-B CCK receptors the first phase of fever was abolished (or at least showed a significant attenuation for about 30 min). The second phase of LPS-fever remained but was delayed (see minutes 140-150). Although the latter experiments have not yet been completed, it is likely that LPS-fever has a CCK-ergic component probably timed at the first phase, the one which has also been proved to possess a strong prostaglandinergic mediation in several species (Skarnes et al., 1981).

DISCUSSION AND CONCLUSIONS

It has been established that autonomic functions could be intimately connected to neuropeptides and neuropeptide systems both in individual peripheral organs, organ systems and in the CNS. The way in which the various families of peptides might influence basic homeostatic mechansism and if they could be regarded as essential components of any of the central regulatory circuits studied so far, are still in debate. Data discussed in the foregoing paragraphs and gained in chronic experiments on rats have been connected to the possible role, if any, of CCK in influencing thermal homeostasis, another regulatory system of basic character in addition to that of feeding/satiety, the latters having been seriously considered as targets of CCK-ergic regulation both in the gastrointestinal tract and in the CNS.

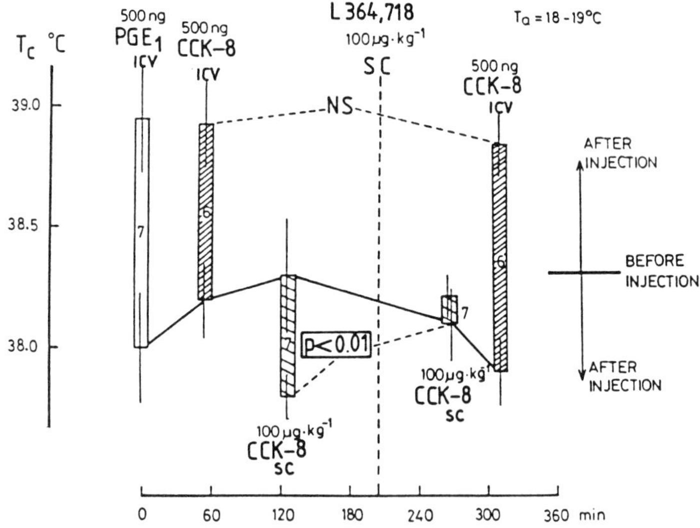

Fig. 6. Effects of an inhibitor of type A (peripheral) CCK receptors (L-364,718) on fever or hypothermia induced by ICV or SC injection of CCK-8. Figures within and near columns indicate the number of injections. For abbreviations see legends to Figs 1 and 3.

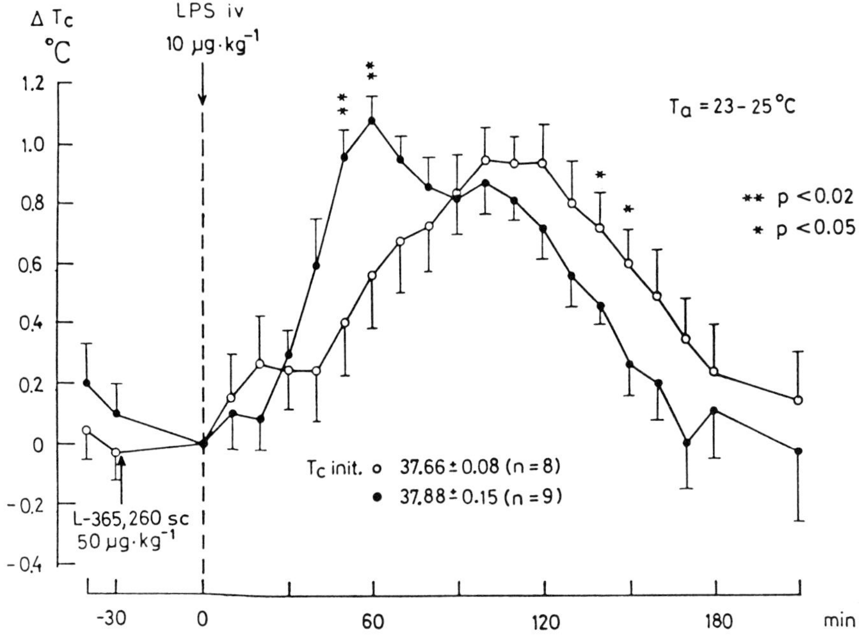

Fig. 7. Fever induced by an intravenous (IV) injection of Escherichia coli endotoxin (LPS) with or without a preceding SC administration of the inhibitor for type B CCK receptors (L-365,260) in cold-acclimated rats. Asterisc indicates statistically significant difference between Tc-values measured the same time after LPS-injection. Tc init.= initial colonic temperature. For other abbreviations see legend to Figs. 1 and 3. (Unpublished from Székely, Balaskó, Szelényi, 1993).

In addition to supporting a hypothermic effect of CCK-8 when injected peripherally data were collected in favour of another effect of CCK-8 in the rat, that is a fever-like response observed regularly after central (ICV) microinjection of the peptide. On the basis of experimental data collected in this laboratory concerning the possible existence and nature of a CCK-ergic thermoregulatory mechanism (or mechanisms) the following conclusions can be drawn:

1. CCK-8 induces a hyperthermic response (the characterisics of which are comparable to those of a fever induced either by PGE or LPS), which
   a. has a short latency after ICV administration of the peptide
   b. is dose-dependent
   c. is accompanied by a rise in MR and/or peripheral vasoconstriction depending on thermal conditions and on the initial body temperature
   d. is the larger, the lower the initial Tc
   e. occurs in a fairly wide range of Ta-s
2. The febrile effect of CCK-8 given ICV seems to be a specific one, since
   a. it can be induced by CCK-8 itself and by a CCK-analogue (ceruletide) but not by some other peptides belonging to a different family (such as neurotensin with its hypotermic potency)

b. it can be attenuated by a type-B receptor blocker, but not by a type-A receptor blocker, although the type-A blocker used can readily cross the blood-brain barrier as shown by Pullen et al. (1987).

3. The CCK-induced hypothermia after its peripheral administration can be explained on the basis of local mechanisms effecting heat balance without the contribution of central temperature control *per se*, since
   a. CCK is a local vasodilatator and (in high concentration) even induces a fall in local tissue metabolism
   b. CCK-8 cannot cross the blood-brain-barrier in appreciable amounts
   c. CCK-induced hypothermia cannot be influenced by inhibitors of CCK receptors of central type (type-B)
   d. CCK-induced hypothermia can be attenuated by an inhibitor of CCK receptors of peripheral type (type-A)
4. Evidence for the possibility of a CCK-ergic mediation of temperature regulation and/or fever is as follows:
   a. highly reproducible fever-like effect of CCK-8 microinjected ICV in rats
   b. CCK-8 stimulates the hypothalamus-pituitary-adrenal axis in the same way as does interleukin-1, one of the cytokines with strong fever-inducing characteristics (Kamilaris et al., 1992)
   c. neither inhibitors of CCK receptors affected PGE-induced fever
   d. (see 2. b)
   e. first phase of LPS-fever (IV injection) is reduced by the type-B receptor blocker (type-A receptor blocker not tested)

ACKNOWLEDGEMENTS

These studies have been supported by OTKA 84, 95 and 472.

REFERENCES

Chang, R.S.L. and Lotti, V.J. (1986): Biochemical and pharmacological characterization of an extremely potent and selective non-peptide cholecystokinin-antagonsist. *Proc. Natl. Acad Sci. USA 83,* 4923-4926.

Crawley, J.N. (1985): Comparative distribution of cholecystokinin and other neuropeptides. Why is this peptide different from all other peptides? *Ann. N.Y. Acad. Sci. 448*, 1-8.

Day, N.C., Hall, M.D., Clark, C.R and Hughes, J (1986): High concentrations of cholecystokinin receptor binding sites in the ventromedial hypothalamic nucleus. *Neuropeptides 8*, 1-18.

Gibbs, J., Young, R. and Smith, G.P. (1973): Cholecystokinin decreases food intake in rats. *J. Comp. Physiol. 84,* 488-493.

Kamilaris, T.C., Johnson, E.O., Calogero, A.E., Kalogeras, K.T., Bernardini, R., Chrousos, G.P. and Gold, P.W. (1992): Cholecystokinin-octapeptide stimulates hypothelamic-pituitary-adrenal function in rats:role of corticotrophin-releasing hormone. *Endocrinology 130,* 1764-1774.

Kapás L., Obál Jr., F., Penke, B., Obál, F. (1987): Cholecystokinin-induced hypothermia in rats: dose-effect and structure-effect relationships, effect of ambient temperature, pharmacological interactions and tolerance. *Neuropharmacology, 26,*131-137.

Lotti, V.J. and Chang, S.L. (1989): A new potent and selective non-peptide gastrin anatagonist and brain cholecystokinin receptor (CCK-B) ligand: L 365260. *Eur. J. Pharmacol. 162,* 273-280.

Morley, J.E., Levine, A.S., Lindblad, S. (1981): Intraventricular cholecystokinin-octapeptide produces hypothermia in rats. *Eur. J. Pharmacol. 74,* 249-251.
Oldendorf, W.F. (1981): Blood-brain barrier permeability to peptides: pitfalls in measurement. *Peptides 2*, 109-111.
Pullen, R.G.L. and Hodgson, O.J. (1987): Penetration of diazepam and the non-peptide CCK antagonist, L 364718 into rat brain. *J. Pharm. Pharmacol. 39,* 863-864.
Savory, C.J. (1987): Analternative explanation for apparent statiety properties of peripherally administered bombesin and CCK in domestic fowls. *Physiol. Behav. 39*, 191-202.
Skarnes, R.C., Brown, S.K., Hull, S.S., McCraken, J.A. (1981): Role of prostaglandin E in the biphasic fever response to endotoxin. *J. Exp. Med. 154,* 1212-1224.
Székely, M. and Szelényi, Z. (1982): The pathophysiology of fever in the neonate. *Handb. Exp. Pharmacol. 60*, 479-528.
Szelényi, Z., Barthó, L. (1989): Can the hypothermia elicited by peripheral injection of CCK-8 in the rat be regarded as a centrally mediated thermoregulatory response ? In *Thermal Physiology 1989,* Ed. J.B. Mercer, pp.285-290.
Szelényi, Z., Barthó, L., Székely, M., Romanovsky, A.A. (1993): Cholecystokinin-octapeptide (CCK-8) injected into a cerebral ventricle induces a fever-like thermoregulatory response mediated by type-B receptors in the rat. *Brain Res (in preparation)*

Integrative and cellular aspects of autonomic functions : temperature and osmoregulation. Eds K. Pleschka, R. Gerstberger. John Libbey Eurotext, Paris © 1994, pp. 33-36.

Prostaglandin I_2 receptor in the brain : its possible involvement in autonomic regulation and nociceptive function

Kiyoshi Matsumura*, Yumiko Watanabe, Hirotaka Onoe and Yasuyoshi Watanabe

Subfemtomole Biorecognition Project, Research Development Corporation of Japan and Department of Neuroscience, Osaka Bioscience Institute, Suita, Osaka 565, Japan

Prostaglandin I_2 (PGI_2) is one of the cyclooxygenase products in the arachidonic acid cascade. The biological significance of PGI_2 has been well established in the peripheral tissues, especially in the circulatory system. It is biosynthesized in the endothelial cells and exerts potent vasodilatory and anti-platelet aggregatory effects. Since other prostaglandins, i.e., prostaglandin E_2, D_2, and $F_{2\alpha}$, act on both peripheral organs and the central nervous system (CNS), this could also be the case for PGI_2. Little is known, however, whether PGI_2 plays any role(s) in the CNS. In the present study, we performed *in vitro* receptor autoradiography for PGI_2 in rat brain using [^{3}H]iloprost, a stable agonist for the PGI_2 receptor. We demonstrate herein that a high concentration of [^{3}H]iloprost binding sites is present in the brain regions involved in the regulation of autonomic reflexes and nociceptive function.

METHODS

Male Wistar rats were anesthetized with thiopental and perfused with ice-cold 20 mM phosphate-buffered saline. The brains were removed and frozen in dry-ice powder. Cryostat sections of the brain (10 µm thickness) were thaw-mounted on gelatin-coated glass slides. The sections were preincubated in 50 mM Tris/HCl buffer (pH 7.4) containing 10 mM $MgCl_2$ for 30 min and then incubated with 10 nM [^{3}H]iloprost for 30 min. After having been washed, the sections were dried and juxtaposed to [^{3}H]sensitive films for 6 months. The films were subsequently developed and analyzed by computer-assisted densitometry. Nonspecific binding of [^{3}H]iloprost binding was evaluated by addition of 10 µM unlabelled iloprost.

RESULTS AND DISCUSSION

General view of iloprost binding sites

The highest density of iloprost binding sites (111 fmol/mg tissue) was found in the medial and commissural parts of the nucleus tractus solitarius (NTS) and in the area postrema (AP; 48 fmol/mg tissue) as seen in Fig. 1. Within the NTS, the dorsal part was especially high in density (Fig. 1B). Superficial layers of the spinal trigeminal nucleus (TRIG; 22 fmol/mg tissue) and the dorsal horn (DH; 15 fmol/mg tissue) of the spinal cord were also rich in the binding sites (Fig. 2). A moderate density (5-10 fmol/mg tissue) was found in the thalamus, hippocampus, cerebral cortex, and dorsal cochlear nucleus.

* To whom correspondence should be addressed.

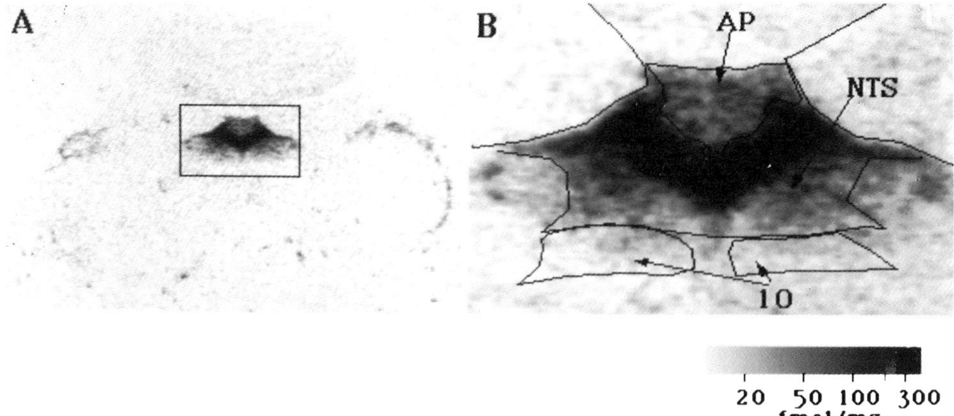

Figure 1 Gray scale coding of specific binding sites for iloprost. The area in the square in A is magnified (x 4) in B. AP, area postrema; NTS, nucleus tractus solitarius; 10, dorsal motor nucleus of the vagus.

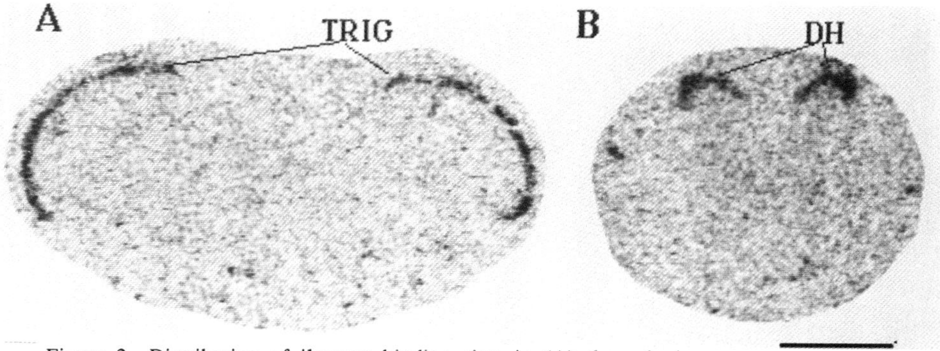

Figure 2 Distribution of iloprost binding sites in (A) the spinal trigeminal nucleus (TRIG) and (B) the dorsal horn (DH) of the spinal cord. These are images of total binding and have not been converted to the value of binding density. Scale bar=1 mm

Specificity of iloprost binding sites

The binding pattern of iloprost was different from that reported for other prostaglandins(i.e., PGE_1, PGE_2, PGD_2, and $PGF_{2\alpha}$), except for that of PGE_1 and of PGE_2 in the NTS, TRIG, and DH (Matsumura *et al.*, 1990, 1992; Watanabe *et al.*, 1988; Watanabe *et al.*, 1989). Thus, there arose a possibility that iloprost crossbound to PGE_2 receptors in the NTS, TRIG, and DH. This possibility is, however, excluded by the following experimental results: First, precise comparison of binding sites for iloprost, PGE_1, and PGE_2 within the NTS in serial sections revealed that iloprost preferentially bound to the dorsal part of the NTS, whereas PGE_1 and PGE_2 bound more broadly and preferentially in the ventral part (Fig. 3). Although no clear distinction of the binding sites for these prostaglandins was found in the TRIG and DH, the following lines of evidence further support the specificity of iloprost binding sites: In the course of development, iloprost binding sites in these three regions are already expressed in the prenatal stage (at late embryonic day 20), whereas PGE_2 binding sites are in the background level at this stage (unpublished observation). Finally, addition of high concentration (10 μM) of unlabelled iloprost to the incubation mixture almost completely abolished [^{3}H]iloprost binding, whereas the same treatment with 10 μM PGE_2 only partially (30 %) reduced

[^{3}H]iloprost binding. These results indicate that iloprost bound to a specific receptor, that is, one distinct from those receptors for other prostaglandins.

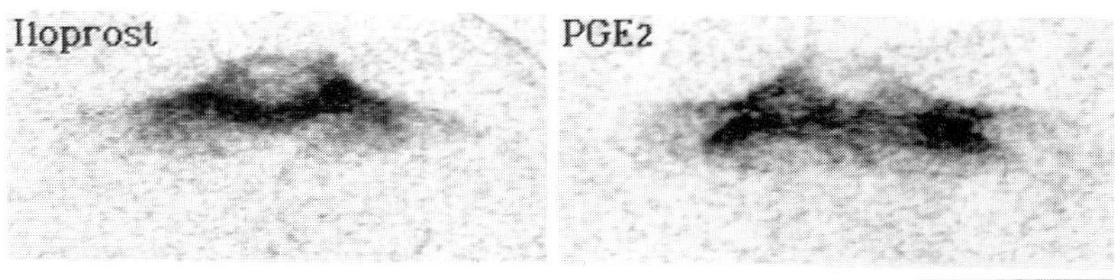

Figure 3 Comparison of iloprost binding sites and PGE2 binding sites in the NTS. These are images of total binding and have not been converted to the value of binding density. Scale bar=1 mm

Origin of iloprost binding sites in the sensory systems

Since the distribution of iloprost binding sites within the NTS nearly overlaps with the terminals of the vagal afferent neurons (5), the iloprost binding sites might be associated with the central terminals of the primary sensory neurons. To examine this possibility, we made a unilateral lesion of the vagus between the ganglion of vagal sensory neurons (the nodose ganglion) and the brain. Three weeks later, the rats were perfused under anesthesia and their brains were prepared for autoradiography. At the rostro-caudal level of the NTS shown in fig. A, the density of iloprost binding to the lesioned side decreased to 46 % of that to the intact side. When iloprost binding to the DH was also examined after lesioning the dorsal root between the dorsal root ganglion and spinal cord, significant reduction of iloprost binding was again observed on the lesioned side.

To confirm the origin of iloprost binding sites, we ligated the vagus either rostral or caudal to the nodose ganglion with a fine silk surgical suture. Twenty-four hours thereafter, the rats were perfused and the vagal nerves were prepared for autoradiography. In both cases, an accumulation of iloprost binding sites was observed on the proximal side (the nodose ganglion side) of the ligation site.

These results indicate that iloprost binding sites (i.e., PGI_2 receptors) in the NTS are produced in the cell body of primary sensory neurons and are moved to their central and peripheral terminals by axonal flow. Considering the anatomical and functional similarities of the nodose ganglion to the trigeminal and dorsal root ganglia together with the present lesion study in the dorsal root, the idea mentioned above could also be applied to the trigeminal and spinal sensory systems.

Functional consideration of iloprost binding sites

It is well established that the NTS receives viscerosensory inputs from the peripheral organs and plays a key role in the regulation of autonomic reflexes. The superficial layers of the TRIG and DH are known to be involved in nociception. The functional significance of the iloprost binding sites in these systems, as well as in other brain regions, remains to be elucidated. There are, however, some studies implicating a neuromodulatory function of PGI_2 in the peripheral terminals of the sensory neurons. PGI_2 has been reported to facilitate neuronal activity of baroreceptor fibers (Chen et al., 1990) and nociceptive fibers (Devor et al., 1992) by acting on their peripheral terminals and is considered to be involved in the regulation of baroreflex and post-injury hyperalgesia. These results suggest that the iloprost binding sites, presumably PGI_2 receptors, are involved in the modulation of neuronal excitability not only in the peripheral but also in the central terminals of the primary sensory systems.

REFERENCES

Chen, H. I., Chapleau, M. W., McDowell, T. S., and Abboud, F. M. (1990): Prostaglandins contribute to activation of baroreceptors in rabbits: possible paracrine influence of endothelium. Circ. Res. 67, 1394-1404.

Devor, M., White, D. M., Goetzl, E. J., and Levine, J. D. (1992): Eicosanoids, but not tachykinins, excite C-fiber endings in rat sciatic nerve-end neuromas. NeuroReport 3, 21-24.

Matsumura, K., Watanabe, Yu., Onoe, H., Watanabe, Y., and Hayaishi, O. (1990): High density of prostaglandin E_2 binding sites in the anterior wall of the 3rd ventricle: a possible site of its hyperthermic action. Brain Res. 533, 147-151.

Matsumura, K., Watanabe, Yu., Imai-Matsumura, K. Connoly, M. Koyama, Y. Onoe, H., and Watanabe, Y. (1992): Mapping of prostaglandin E_2 binding sites in rat brain using quantitative autoradiography. Brain Res. 581, 292-298.

Norgern, R., and Smith, G. P. (1988): Central distribution of subdiaphragmatic vagal branches in the rat. J. Comp. Neurol. 273, 207-223.

Watanabe, Y., Watanabe, Yu., Hamada, K., Bommelaer-Bayt, M.-C., Dray, F., Kaneko, T., Yumoto, N., and Hayaishi, O. (1989): Distinct localization of prostaglandin D_2, E_2, and $F_{2\alpha}$ binding sites in monkey brain. Brain Res. 478, 143-148.

Watanabe, Yu., Watanabe, Y., and Hayaishi, O. (1988): Quantitative autoradiographic localization of prostaglandin E_2 binding sites in monkey diencephalon. J. Neurosci. 8, 2203-2010.

Integrative and cellular aspects of autonomic functions : temperature and osmoregulation. Eds K. Pleschka, R. Gerstberger. John Libbey Eurotext, Paris © 1994, pp. 37-46.

Effects of prostaglandin E_2 on the thermosensitivity of anterior and posterior hypothalamic neurones *in vitro*

V. Kulchitsky*, H. Sann and F.K. Pierau

Max-Planck-Institut für physiologische und klinische Forschung, W.G. Kerckhoff-Institut, Parkstrasse 1, 61231 Bad Nauheim, Germany

SUMMARY

The effects of prostaglandin E_2 (PGE_2), one of the possible mediators of fever, on the electrical activity and the temperature sensitivity of neurones in the nucleus periventricularis (PER) and the nucleus arcuatus (ARC) of the rat hypothalamus were studied using an *in vitro* slice preparation. In agreement with previous studies, the proportion of temperature-sensitive neurones (16% warm-sensitive, 2 % cold-sensitive) was lower in the ARC as compared to the PER (41% warm-sensitive). PGE_2 (10 nM) superfused for 10 min had various effects on the spontaneous activity of the neurones in both nuclei which could outlast the application time for more than 15 minutes. During the application of PGE_2, warm-sensitive neurones (WS) were predominantly inhibited, temperature insensitive neurones (IS) were mainly activated or showed no change in firing rate. The average discharge rates of WS from both nuclei were significantly reduced either during or after PGE_2 superfusion, whereas no change was seen in IS. Similarly, the temperature sensitivity expressed by the mean temperature coefficient (TC) was significantly reduced in WS, and was unchanged in IS. The effects of PGE_2 on firing rate and TC were similar in the two nuclei. No correlation between the PGE_2 induced change in firing rate and in TC could be established for the WS. The data indicate that PGE_2 in a low concentration decreased the firing rate and reduced the temperature sensitivity of WS in the hypothalamic areas close to the third ventricle. These changes in the functional properties of WS might be regarded as the neuronal basis for pyretic effects of PGE_2 *in vivo*.

INTRODUCTION

Endogenous prostaglandins are considered to be possible mediators of the fever reaction in mammals (Stitt, 1986, Kluger, 1991). Evidence for a role of prostaglandin E_2 (PGE_2) in the induction of fever is derived from experiments: (1) showing an increased level of PGE_2 in the hypothalamus or the cerebrospinal fluid during fever, (2) the ability of low doses of PGE_2 injected locally into the rostral hypothalamus or in the third ventricle to induce fever and (3) the antipyretic effect of cyclo-oxygenase inhibitors (see Stitt, 1986).

* Present address: *Institute of Physiology, Academy of Sciences of Belarus, 220072 Minsk, F. Skarina str. 28, Belarus*

The mechanism of body temperature rise during hyperthermia induced by prostaglandin E_2 and during fever is generally ascribed to an elevation of the set point of body temperature (Stitt, 1986). Changes of the activity of neurones in the anterior hypothalamus, which are known to participate in temperature regulation, could be the neuronal basis for such a mechanism. Electrophysiological studies on hypothalamic temperature-sensitive neurones, however, demonstrated inconsistent responses to PGE_2 (Eisenman, 1968, 1982, Schoener and Wang, 1976, Gordon and Heath, 1980, Boulant and Scott, 1986, Watanabe et al., 1987, Ono et al., 1987, Morimoto et al., 1988, Matsuda et al., 1992) . Most of these studies have only regarded the effect of PGE_2 on the spontaneous activity and some have used rather high concentrations. On the other hand, recordings of warm-sensitive neurones *in vivo* (Eisenman, 1968) as well as different models of temperature regulation (Mitchell et al., 1970, Simon, 1981) suggest that the temperature sensitivity (gain) of the regulatory system might be altered during fever. A change of the temperaature sensitivity of hypothalamic neurons by the hypothermic neuropeptide bombesin has been recently demonstrated in hypothalamic slices of rats (Schmid et al., 1993).
Therefore it appeared to be appropriate to study the effects of PGE_2 on both firing rate and temperature sensitivity of hypothalamic neurones. Injections of PGE_2 into the third ventricle and into the preoptic area were similarly effective although differences in the effect of PGE_2 between the anterior and the posterior hypothalamus have been reported (Stitt, 1986). For this reason the effect of PGE_2 on neurones of the nucleus periventricularis (PER) and the nucleus arcuatus (ARC) was compared. Both nuclei are adjacent to the third ventricle but belong to the anterior and posterior hypothalamus, respectively. A low concentration of PGE_2 similar to values measured in the hypothalamus during fever (see Stitt, 1986) was used in this study. The experiments pursue previous experiments demonstrating that bolus injections of PGE_2 reduce the temperature sensitivity of warm-sensitive neurones in the anterior hypothalamus (Jansky et al., 1992).

METHODS

Materials and preparation

Male Wistar rats (170-230 g body weight) were decapitated and the brain quickly removed. Slices of 350-400 μm were prepared as described previously (Schmid et al., 1993). Usually 4-5 slices containing the anterior hypothalamic periventricular area and 2-3 slices containing nucleus arcuatus were obtained. The slices were further trimmed to a size of 3 x 3 mm to contain only structures around the third ventricle. After an equilibration period of at least 2 h in artificial cerebrospinal fluid (aCSF) at 38 °C, the slices were placed onto the silicon base of a recording chamber and fixed with platinum weights. The aCSF containing (mM) NaCl 124.0, KCl 5.0, NaH_2PO_4 1.2, $MgSO_4$ 1.3, $NaHCO_3$ 26.0, $CaCl_2$,1.2, glucose 10.0 was continuously oxygenated with 95% O_2 and 5% CO_2 and perfused at a rate of 1.8 ml/min.
The recording chamber was attached to a Peltier thermoassembly, which enabled constant temperature and reproducible feedback controlled temperature changes. The sinusoidal temperature stimuli used had an amplitude of 6 °C and started from 38 °C to complete a full cycle between 35 and 41 °C within 5 min. A subsequent cycle began not earlier than 5 min after the end of a previous stimulus. The temperature was monitored using a small thermocouple placed on the floor of the recording chamber next to the slice.
Single units were recorded extracellularly using glass-coated platinum-iridium electrodes. After amplification and discrimination, the events and the temperature were recorded on a personal computer using a CED 1401 interface and the CED software spike2.

Experimental protocol

A typical experimental protocol is illustrated in Fig. 1. At the beginning the temperature sensitivity of each unit was determined using 2 - 3 sinusoidal temperature stimuli. The temperature sensitivity was calculated by a special computer program relating the discharge rate of the unit (bin width = 5 s) to the average temperature and fitting either one linear or two piecewise regression lines to the data (Vieth, 1989). The slope of the steepest regression line was used as the temperature coefficient (TC) of the unit. Neurones with a TC ≥ 0.6 imp/s/°C were considered as temperature sensitive, with a positive sign for warm sensitivity and a negative sign for cold sensitivity. All other neurones are by this definition temperature-insensitive (Schmid et al., 1993). After the temperature sensitivity was established, PGE_2 was superfused in a concentration of 10 nM for 10 minutes. About 3 - 4 min after the onset of the superfusion the temperature sensitivity was again evaluated. Starting from 5 min after offset of PGE_2, the temperature sinus was repeated in 10 min intervals. To avoid possible interactions with previous PGE_2 stimuli, only one neurone per slice was tested.

Data analysis and statistics

To define the direct effects of PGE_2 superfusion on the firing rate, the maximal effect (inhibition or excitation, bin width = 10 s) was determined during the first 3 minutes and compared with the mean activity 30 s before PGE_2 application. Only increases or decreases in firing rate ≥ 20% of the spontaneous activity were considered as activation or inhibition, respectively. To evaluate the long-lasting effect of PGE_2, the discharge rates were calculated over 1 min just prior to each temperature stimulus.
The control TC of a unit was defined as the mean of the 2-3 TC determinations before PGE_2 superfusion. In order to establish significant changes in TC, the slopes of the regression lines either during or after PGE_2 application were compared with that of the control data according to Brownlee (1965). Since the TC changes occurred either during or after the superfusion, the maximum or minimum TC for an individual neurone was calculated for the appropriate time period. All data are presented as means ± S.E.M. For statistical comparison either paired or unpaired *t*-test were used.

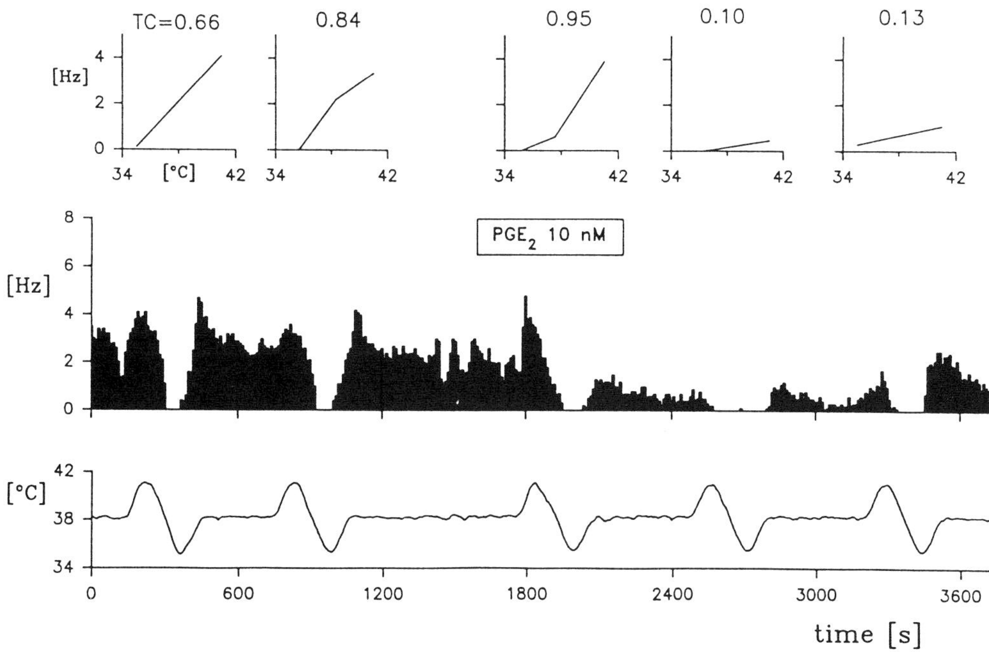

Fig. 1: Original recordings of firing rate and temperature of a warm-sensitive neurone from the nucleus arcuatus (lower part). The relation between firing rate and temperature and the resulting TC during each sinusoidal temperature change are illustrated in the upper part of the graph. In this unit superfusion of PGE_2 10 nM produced a long-lasting decrease in firing rate. The TC was slightly increased during PGE_2 superfusion but substantially decreased during the two following stimuli. Similar responses were also obtained from the nucleus periventricularis.

RESULTS

General characteristics

In 45 experiments the electrical activity of 91 PER neurones and 50 ARC neurones was recorded. The proportion of warm-sensitive neurones (WS) was much lower in the ARC compared to the PER. Thus, 41% of the PER neurones were warm-sensitive and 59% were temperature-insensitive neurones (IS), whereas in the ARC 16% WS, 82% IS and 2% cold-sensitive neurones were found. Of these neurones, 96 were used for superfusion with 10 nM PGE_2 (PER: 31 WS/ 38 IS; ARC: 6 WS/ 21 IS). The average spontaneous activity of the PER - WS at 38 °C was not significantly different from that of the IS, while WS in the ARC expressed an about 3 times higher mean spontaneous discharge compared to the IS (Fig. 2 and 3). On the other hand, comparison of the average TC of WS or IS of the two nuclei showed no significant differences.

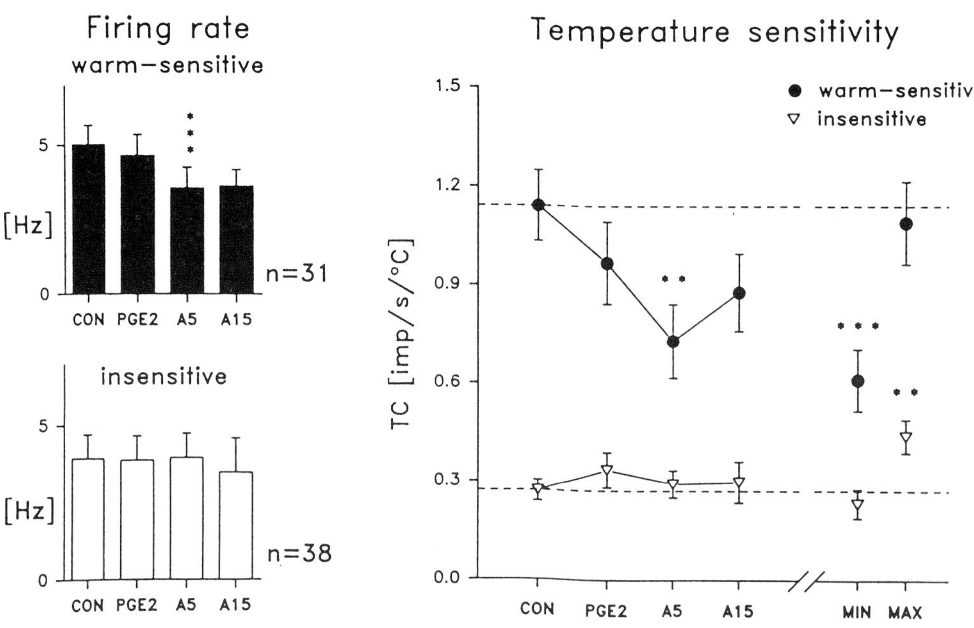

Fig. 2: Effect of superfusion of PGE_2 (10 nM) on the mean firing rate (left side) and the mean temperature sensitivity (right side) of warm-sensitive (upper part) and temperature insensitive (lower part) neurones in the **nucleus periventricularis**. *CON: control values, PGE2: values during PGE_2, A5: values 5 min after PGE_2, A15: values 15 min after PGE_2, MIN, MAX: minimal or maximal TC calculated during the appropriate periods. Mean ± S.E.M.. Significant differences to control values: * = $p < 0.05$. ** = $p < 0.01$.*

Effect of PGE_2 superfusion on spontaneous activity

The direction of changes in discharge rate during the first 3 minutes of PGE_2 superfusion was variable (Table 1). In the PER the dominant response of the WS was an inhibition (48 %), while the discharge of the IS was mainly increased (37%) or unchanged (37%). In the ARC the activity of about half of the WS was unchanged and almost 50 % of the IS showed increased discharge rates. The average maximal frequency after PGE_2 in the WS or IS group, however, was not significantly different from the mean spontaneous activity of the control period (Table

1). Only if the mean discharge during the third minute of PGE_2 superfusion was compared to the control, a significant reduction ($p<0.05$, paired T-Test) of the discharge of the WS was observed in the ARC but not in the other groups (Fig. 2 and 3).
The effect of PGE_2 on the spontaneous activity was often more obvious after the superfusion was stopped. Significant reductions in the mean firing rate of WS were measured 5 (PER, A5) or 15 (ARC, A15) minutes after to offset of PGE_2 superfusion, respectively (Fig. 2 and 3). Five minutes after the end of the PGE_2 superfusion, 42% (13/31) of the WS in the PER exhibited a decreased firing rate, while only 3 WS (10%) were excited and 48% not affected. In the ARC half of the WS (3/6) were inhibited at the end of PGE_2 superfusion, the others were not affected. Some WS (PER: 6/31; ARC:2/6) showed a reduced discharge rate even 15 min after PGE_2. The changes of the mean discharge rate of the WS units were more consistent. They were significantly reduced 5 or 15 min after PGE_2 in PER or ARC, respectively (Fig. 2 and 3).
In regard to IS, the firing rates 5 min after PGE_2 were mainly unchanged (10/21) or increased (9/21) in the ARC and unchanged (17/38) or decreased (11/38) in the PER. Fifteen minutes after PGE_2 the majority of the IS in both areas (PER: 66%; ARC: 86%) exhibited a firing rate similar to the control level of spontaneous activity. In agreement with these data, the mean firing rate of IS from both areas was not significantly changed after the offset of PGE_2 (Fig. 2 and 3).

Table 1:
PGE_2 (10 nM) induced maximal changes in the discharge rate of WS and IS in the PER and ARC during the first 3 minutes of superfusion. The effect of the PGE_2 superfusion (PGE_2) was calculated as the mean of the maximal or minimal discharge (bin width 10 s) observed during the first 3 min of superfusion. Control: mean spontaneous activity averaged over 30 s previous to the stimulus.

	PER		ARC	
	WS	IS	WS	IS
decrease	15 48 %	10 26 %	2 33 %	7 33 %
no change	9 29 %	14 37 %	3 50 %	4 19 %
increase	7 23 %	14 37 %	1 17 %	10 48%
total	31	38	6	21
control [Hz] S.E.M.	4.6 ± 0.7	4.0 ± 0.8	5.4 ± 1	4.8 ± 0.9
PGE_2 [Hz] S.E.M.	4.5 ± 0.7	4.2 ± 0.8	4.8 ± 0.9	3.1 ± 0.5

Effect of PGE_2 superfusion on TC

Similar to the effect of PGE_2 superfusion on the firing rate, the effects on the TC were more pronounced after the offset of the PGE_2 superfusion (Fig. 2 and 3). Since the maximal changes in TC of the individual neurones occurred either during, 5 (A5) or 15 min after (A15) PGE_2 application, the TC change was determined at the time when significant changes were evident. According to this criterion, the TC was significantly decreased in the highest proportion of the WS (13/31) in the PER, whereas it was increased or unchanged in 10 WS, respectively. Included in these numbers are 2 WS that did increase and decreased their TC during the observation period. In the ARC, the TC of 3/6 WS were unchanged, but decrease or increased in 2 or 1 WS, respectively. The TC of the IS in the PER were decreased in 14 of 38 neurones, increased in 12, changed in both directions in 2 and unchanged

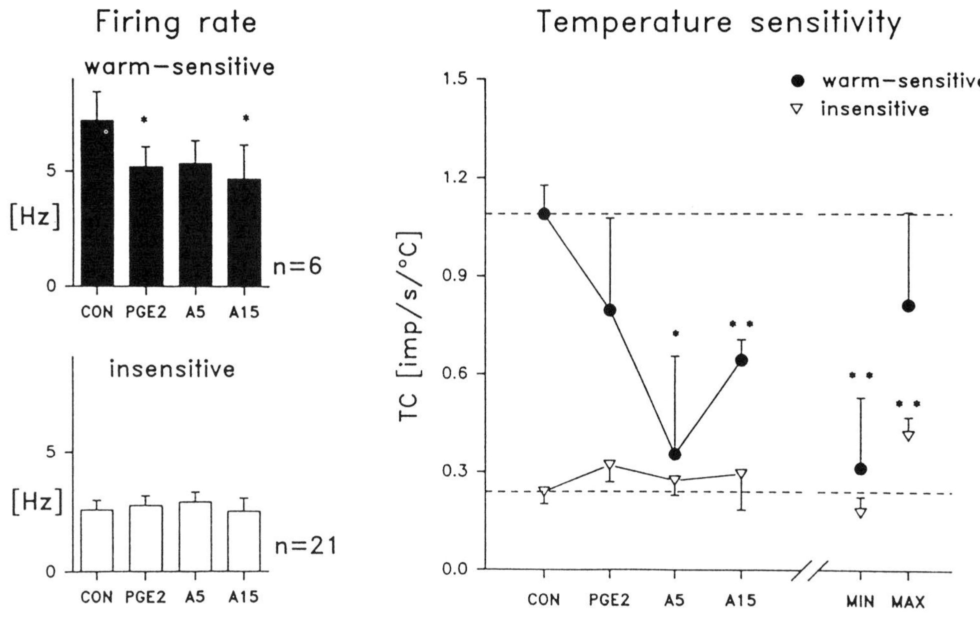

Fig. 3: Effect of superfusion of PGE_2 (10 nM) on mean firing rate and mean temperature sensitivity of warm-sensitive and temperature insensitive neurones recorded in the ***nucleus arcuatus****. Abbreviations see Fig. 2; *** = $p < 0.001$.*

in the rest (10). In the ARC out of 21 IS the TC was increased in 8, decreased in 3, changed in both directions in 2 and unchanged in 8. Although these data appear to be rather variable and might suggest differences between the nuclei, the mean changes in the TC values were similar in both areas. Thus, the average of the minimal TC of the WS in both nuclei was significantly ($p<0.001$) different from the control value, whereas the mean maximal TC was not (Fig. 2 and 3). The mean TC of the WS was reduced from 1.14 ± 0.11 to 0.61 ± 0.09 in the PER and 1.09 ± 0.09 to 0.311 ±0.22 in the ARC, respectively. In contrast, the maximal change in the average TC of the IS but not the mean minimal TC values were significantly different from the control (Fig. 2 and 3). The TC of the IS was elevated from 0.27 ± 0.03 to 0.44 ± 0.05 in the PER and 0.24 ± 0.04 to 0.41 ± 0.06 in the ARC.

As shown in Figures 2 and 3, the averaged TC values of all WS in the PER and ARC during, 5 (A5) or 15 min after (A15) PGE_2 application show a clear tendency to decrease with some indication of recovery (A15). For the IS the tendency towards increased TC due to PGE_2 is not significant. In contrast to the WS, the maximal effect occurred during the application period. In this respect it is noteworthy, that most WS and IS that increase their TC due to PGE_2 show the maximal increase during the superfusion of PGE_2. As shown in Figure 4, WS in which the TC was increased were predominantly found in the lower TC range of 0.6 to 1.2 imp/s/°C, whereas the WS in which the TC was decreased had control TC's between 0.6 and 3.2 imp/s/°C.

The degree of TC change due to PGE_2 superfusion was substantial enough to induce a temporary conversion of 10/31 or 2/6 WS into IS and 6/38 or 4/21 IS into WS in the PER or ARC, respectively (see Fig. 4 filled triangles in the clear fields, neurones that were not converted are shown in the hatched areas).

N. arcuatus

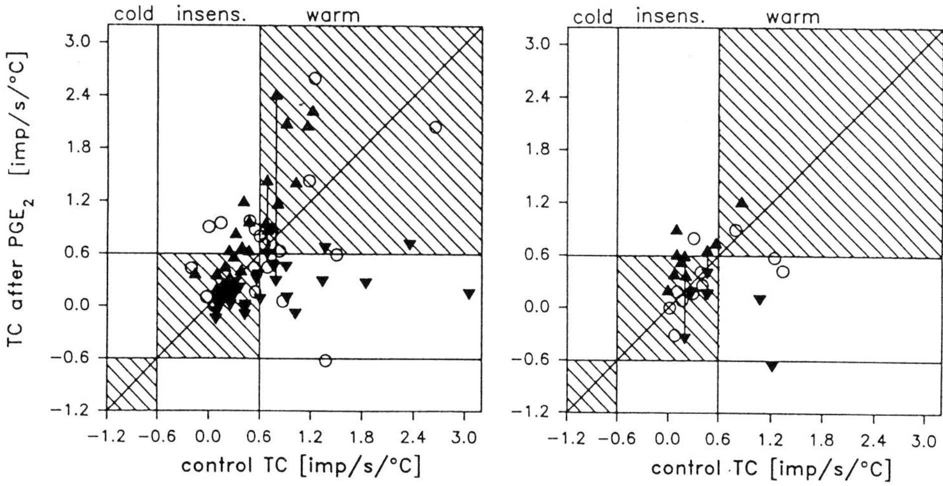

Fig. 4: Effect of PGE_2 on the temperature sensitivity of neurones recorded in the nucleus periventricularis (left) and nucleus arcuatus (right). The TC of individual neurones during the most pronounced PGE_2-induced change are plotted against the TC of the control period. Filled triangles represent significant changes ($p<0.05$), open circles show not significant changes. Neurones showing both increase and decrease in TC are connected by a vertical line. The distance from the line of identity indicates the degree of change. Triangles in the clear areas indicate transformation of a neurone from IS to WS or from WS to IS, respectively.

Relation of the effects of PGE_2 on firing rate and TC

To evaluate whether the changes of TC found in the different groups were related to the changes in firing rate, the relative change in firing rate was plotted against the relative change of the TC for each individual neurone from both nuclei (Fig. 5). Regression lines with significant positive correlation between changes in firing rate and in TC could be obtained for WS ($y=-0.49+0.68x$) and IS ($y=-0.02+0.83x$) during and for the IS also 5 min after PGE_2 superfusion ($y=0.09+1.63x$). Such a correlation could be expected, since neurones with high spontaneous activity were more likely to have a higher TC than those expressing low ongoing activity. However, no correlation could be detected for the WS 5 min ($y=-0.96+0.1x$) and 15 min after PGE_2 ($y=-0.39+0.02x$), although the mean decrease in firing rate and TC of the WS were most pronounced 5 min after the offset of the superfusion.

DISCUSSION

One of the aims of this study was to compare the responses of neurones located close to the third ventricle in the anterior and posterior hypothalamus to PGE_2. The main difference between the two areas investigated was the much lower proportion of WS in the ARC, which is in agreement with previous studies (Schenda et al., 1990). There were, however, no substantial differences in the responses of WS and IS of the anterior or posterior hypothalamus to PGE_2. This is in agreement with autoradiographic studies which show that both areas express similar

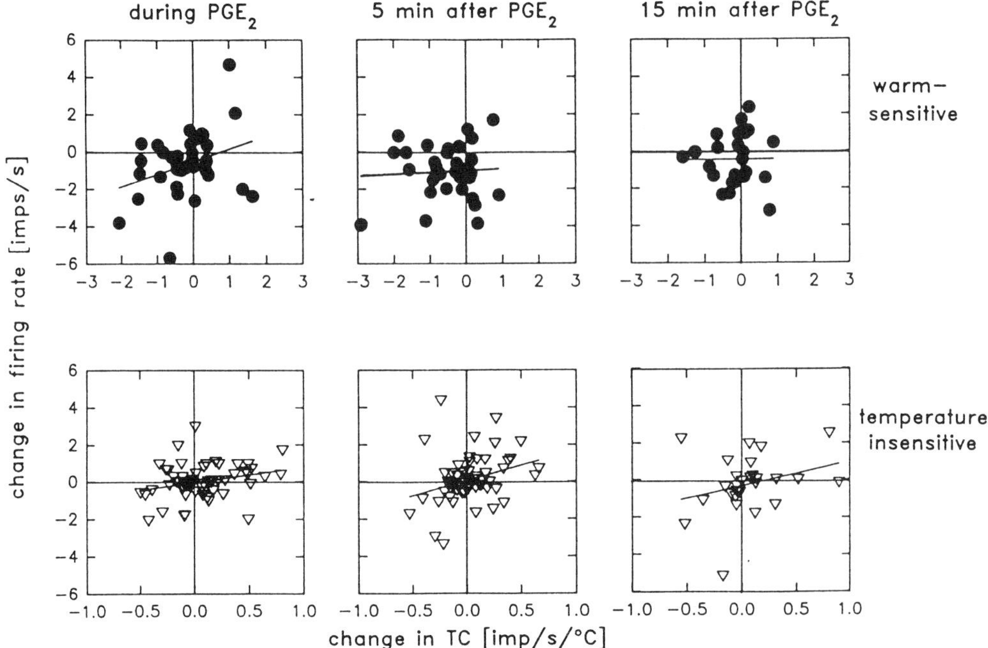

Fig. 5: Relation of the effect of PGE_2 on firing rate and the temperature sensitivity of the individual warm-sensitive and temperature insensitive neurones in both nuclei. The change in firing rate is plotted against the change in TC. Linear regression lines were calculated. Note: although the strongest effect of PGE_2 on both parameters occurred 5 min after the end of the superfusion in WS, there was no significant correlation between these parameters in the individual neurones.

levels of PGE_2 binding-sites (Matsumura et al., 1990). The induction of fever after injection of PGE_2 into the third ventricle might thus involve neurones of the anterior as well as the posterior hypothalamus.

According to models of temperature regulation, a change in the set-point of the regulating system might be the mechanisms of fever induction (see Kluger 1991). The models suggest that warm-sensitive neurones would reduce their firing rate, while cold-sensitive neurones might increase their activity during fever. In agreement with this assumption, the mean activity of WS in the PER and ARC were decreased, although more than half of the cells exhibited no changes in activity or even an increased firing rate. Studies performed *in vitro* and *in vivo* in the preoptic area which only considered the direction but not the intensity of change in neuronal activity had produced variable results. The predominant effect in most studies was excitation (Watanabe et al., 1987, Matsuda et al., 1992, Morimoto et al., 1988, Boulant and Scott, 1986) although inhibition of WS had been reported (Eisenman, 1969 and 1982, Schoener and Wang, 1976, Gordon and Heath, 1980). Warm-sensitive neurones in the ventromedial hypothalamus (Morimoto et al., 1988) or the organum vasculosum laminae terminalis (Matsuda et al., 1992), however, exhibited mainly an inhibition of activity due to PGE_2 and were therefore considered to be more appropriate sites for the PGE_2 induced fever. The present study suggests that warm-sensitive neurones in hypothalamic areas close to the 3th ventricle might be also involved in the thermoregulatory action of PGE_2.

The pronounced reduction of thermo-sensitivity in WS observed in the present experiments is in agreement with data from Eisenman *in vivo* (1969). Such changes in temperature-sensitivity might also lead to an increase in body temperature. Indeed, several authors have proposed models of thermoregulation which include changes in gain due to PGE_2 or fever (Mitchell et al

1970, Simon, 1981). Our data support the notion that PGE_2 increases body temperature via both mechanisms: change in set point and change in gain.

Interestingly, the responses to PGE_2 started in many cases with some delay and also outlasted the stimulus for several minutes, as has been also described by Matsuda et al. (1992). It is worthwhile to note, that in the present study the increase of TC of WS was usually only transient, whereas the decrease in TC was long lasting.

In addition, some units showed biphasic responses, as described previously by Schoener and Wang (1976) for PGE_1 in *in vivo* experiments. This might indicate different mechanisms by which prostaglandins of the E family affect the hypothalamic neuronal network. In agreement with this hypothesis, the firing rate and the TC showed no clear correlation in the individual neurones. Further studies with synaptic blockade techniques are necessary to differentiate the direct effect on the neurone from changes induced by synaptic modulation from other neurones.

In regard to IS, our study has demonstrated that about half of the IS are also affected by PGE_2, which is at variance with previous studies. However, the average values for firing rate and temperature sensitivity were not significantly different from the control, although there was a tendency for an increase of both parameters during the period of superfusion with PGE_2. Thus, it appears unlikely that IS contribute to PGE_2 actions on body temperature. In contrast, the neuropeptide bombesin, which has a strong hypothermic effect, preferentially changed ongoing activity and temperature sensitivity of IS, thus recruiting WS by transformation of IS to WS (Schmid et al., 1993). This indicates that endogenous substances which are able to induce either hypo- or hyperthermia exert their effects by different neuronal mechanisms.

REFERENCES

Boulant, J.A., and Scott, I.M. (1986): Comparison of Prostaglandin E_2 and leukocytic pyrogen on hypothalamic neurons in tissue slices. In *Homeostasis and thermal stress*, eds K.E. Cooper, J. Lomax, E. Schönbaum and W.L. Veale, pp. 78-80. Basel: Karger.

Brownlee, K.A. (1969): Statistical theory and methodology in science and engineering. New York: John Wiley and Sons.

Eisenman, J.S. (1969): Pyrogen-induced changes in the thermosensitivity of septal and preoptic neurons. *Amer. J. Physiol.* 216, 330-334.

Eisenman, J.S. (1982): Electrophysiology of the hypothalamus: thermoregulation and fever. In *Pyretics and antipyretics*, ed A.S. Milton, pp 187-217 Berlin: Springer.

Gordon, C.J., and Heath, J.E. (1980): Effects of prostaglandin E_2 on the activity of thermosensitive and insensitive single units in the preoptic/anterior hypothalamus of unanesthetized rabbits. *Brain Res.* 183: 113-121.

Jansky, L., Pierau, Fr.-K., and Schenda, J. (1992): The effect of PGE_2 on activity and thermosensitivity of hypothalamic neurones in rat brain slices. *Physiol.Res.* 41: 1-3.

Kluger, M.J. (1991): Fever: Role of pyrogens and cryogens. *Physiol. Rev.* 71: 93-127.

Matsuda, T., Hori, T., and Nakashima, T. (1992): Thermal and PGE_2 sensitivity of the organum vasculosum lamina terminalis region and preoptic area in rat brain slices. *J. Physiol.* 454: 197-212.

Matsumura, K., Watanabe, Y., Onoe, H., Watanabe, Y., and Hayaishi, O. (1990): High density of prostaglandin E_2 binding sites in the anterior wall of the 3rd ventricle: a possible site of its hyperthermic action. *Brain Res.* 533: 147-151.

Mitchell, D., Snellen, J.W., and Atkins, A.R. (1970): Thermoregulation during fever: change of set-point or change of gain. *Pflügers Arch.* 312: 293-302.

Morimoto, A., Murakami, N., and Watanabe, T. (1988): Effect of prostaglandin E_2 on thermoresponsive neurones in the preoptic and ventromedial hypothalamic regions of rats. *J.Physiol.* 405, 713-725.

Ono, T., Morimoto, A., Watanabe, T., and Murakami, N. (1987): Effects of endogenous pyrogen and prostaglandin E_2 on hypothalamic neurones in guinea pig brain slices. *J. Appl. Physiol.* 63: 175-180.

Schenda, J., Matsumura, K., and Pierau, Fr.-K. (1990): Diverging effects of bombesin on the temperature sensitivity of neurones in different hypothalamic areas. *Pflüger's Arch. Suppl.* 415: R93.

Schmid, H.A., Jansky, L., and Pierau, Fr.-K. (1993): Temperature sensitivity of neurons in slices of the rat PO/AH area: effect of bombesin and substance P. *Amer. J. Physiol.* 264: R449-R455.
Schoener, E.P., and Wang, S.C. (1976): Effects of locally administered prostaglandin E_1 on anterior hypothalamic neurons. Brain. Res. 117:157-162.
Simon, E. (1981): Effects of CNS temperture on generation and transmisson of temperature signals in homeotherms. A common concept for mammalian and avian thermoregulation. *Pflügers Arch* 392: 79-88.
Stitt, J.T. (1986): Prostglandin E as the neural mediator of the febrile response. *Yale J. Biol. Med.* 59: 137-149.
Vieth, E. (1989): Fitting piecewise linear regression functions to biological responses. *J. Appl. Physiol.* 67: 390-396.
Watanabe, T., Morimoto, A., and Murakami, N. (1987). Effect of PGE_2 on preoptic and anterior hypothalamic neurones using brain slice preparation. *J. Appl. Physiol.* 63, 918-922.

Integrative and cellular aspects of autonomic functions : temperature and osmoregulation. Eds K. Pleschka, R. Gerstberger. John Libbey Eurotext, Paris © 1994, pp. 47-55.

Body temperature of rats selected by heat and inbred

Fujiya Furuyama, Michiko Kumazaki and Hitoo Nishino

Department of Physiology, Nagoya City University Medical School, Kawasumi-cho, Mizuho-ku, Nagoya 467, Japan

ABSTRACT It is well known that body temperature of heat adapted animal is lower than those of un-acclimated animal. Body temperature was studied in special relation to genetic resistance adaptation to ambient heat. Rats were selected for the longest survival time (ST) under heat, sibmated for 30 generations and given the name of FOK rat. Tco in the thermoneutral zone gradually decreased in the developing inbreds. Though FOK rat had no experience of hot environment, Tco of FOK was the lower than the 3 established rat strains which resisted heat shorter period than FOK rat. These findings suggested that hypothermia is a common resultant of many kinds of heat adaptation.

INTRODUCTION

Fox et al. (1963) reported that oral temperature of subjects acclimated to heat for 12-24 days was 0.2°C lower than non-acclimated subjects. Body temperature of heat-acclimated rats were lower than those of non-acclimated rats (Sod-Moriah & Yagil, 1973; Gwosdow & Besch, 1985; Shido & Nagasaka, 1990). Roberts & Chaffee (1976) reported that colonic temperature (Tco) of heat-acclimated deer mice was 1°C lower than the control groups. Esophageal temperature of long-distance runner was 0.6°C lower than those of physically untrained controls (Baum et al. , 1976). Shido et al. (1991) exposed rats daily during a special period of days. Slight hypothermia was observed in thermoneutral zone during the period of previous heat exposure time for 2 days following the intermittent heat exposure.

On the other hand, adaptation is defined to be a change which reduces the physiological strain produced by stressful components of the total environment. The adaptation may occur within the lifetime of an organism (phenotypic adaptation) or be the result of genetic selection in a species or subspecies (genotypic adaptation). Two categories of phenotypic adaptations are recognized: capacity and resistance (Prosser, 1982). Capacity adaptation allows animals to maintain their internal environment within a normal range when exposed to varying external conditions. Resistance adaptation permits animals to withstand extreme environmental variation even if their internal environment has changed. We directed our attention in this study to genetic resistance adaptation to a hot environment.

In the present study, rats were selected for the longest survival time (ST) under heat and sibmated for 30 generations. Tco in the thermoneutral zone was determined for the developing inbreds. Subsequently, Tco was compared among the developed inbred and 3 established strains.

MATERIALS AND METHODS

Animals

In experiment 1, male and female ACI rats were used. In experiment 2, an inbred line was developed from an outbred colony which was synthesized from origins described in our previous report (Furuyama, 1993). In experiment 3, males of FOK, the developed inbred strain, were compared with male ACI, Donryu and Sprague-Dawley (SD) rats. The ACI used was ACI/N which had been sent from the National Institute of Genetics, Japan. The Donryu used was HOS: Donryu purchased from Japan SLC, Inc. The SD used was Crj: CD from Charles River Japan Inc. Rats were always reared at an ambient temperature (Ta) of 24±1°C. The photoperiod was maintained at LD 12:12 (lights on: 7: 00 a. m.) with a light intensity of 300 lux. Rats were kept on small wood shavings in polycarbonate cages and allowed continual access to pellet food and tap water ad libitum.

Development of the Inbred Strain and its Generation Means

Individual rats were selected for breeding on the basis of the longest ST. Animals were sibmated and always maintained in a thermoneutral zone. ST at a Ta of 42.5°C was determined only once for each rat at 13 weeks of age. It was attempted to inbreed up to F30 to increase thermoregulation-related genes in the gene pool.

Heat Exposure Rats were exposed to a Ta of 42.5°C. Tco was continuously recorded using a Cu-Co thermocouple inserted 5-6 cm into the rectum. Rats were able to move freely during heat exposure. However, food and water were unavailable. Other stressful measurements which would shorten the ST of conscious rats were avoided (Frankel, 1959). The ST was calculated on the basis of the heat response curve of Tco as described in our early report (Furuyama et al., 1984)). Thus, no rat was killed by heat during ST determination. When heat exposure was terminated, rats were immediately transferred to a Ta of 24°C and allowed access to a hypotonic saline (0.45-0.90 % NaCl) (Nose et al., 1985) to avoid possible damage from long-lasting voluntary dehydration. These measures insured that a hyperthermia induced by acute but limited heat exposure did not result in considerable heat damage in spite of monitoring over 1 year (Furuyama et al., 1992).

Colonic Temperature Measurements Tco was measured daily at 11: a.m. using or mercury bar thermometer inserted into colon from rectum. In experiment 2 and 3, mercury bar thermometer and Cu-Co thermocouple were used for first 10 days and following 4 days, respectively. The measurements in the 14th day were put in use for data.

Statistics Values are expressed as means and standard deviations (s. d.). Non parametric tests were used for statistical analysis. Differences between the 2 groups were tested using the Mann-Whitney Analysis of variance was carried out using the Kruskal-Wallis test.

Experimental Design

Experiment 1 It was examined how the stress-induced fever decrease in 7 male and 7 female ACI rats as days go by. Tco was measured once a day for 14 days by glass bar mercury thermometer.

Experiment 2 Tco at a Ta of 24±1°C and ST at a Ta of 42.5°C were determined every generation of the developing inbred. These rats had not been exposed to heat until the day.

Experiment 3 Strain differences in Tco at a Ta of 24±1°C and ST at a Ta of 42.5°C were determined in 13 week old conscious rats. Tco and ST in the developed FOK line were compared with those of ACI, Donryu and SD rats.

RESULTS

Thermoregulatory Ability

Generation means of ST of F1 rats were 119.4±24.9 min and 128.0±16.1 min in males and females, respectively (Table 1) and were similar to means of ST values of 18 rat strains that we had previously reported

(Furuyama, 1982; Furuyama et al., 1988). ST extended rapidly until F8-10, and thereafter increased gradually. ST were 330.0±29.1 min and 309.6±38.7 min in males and females of F30, respectively. These values were longer than those of F1 ($p < 0.01$) for both males and females. There was already no heat sensitive individual after F8 in the developing inbred line.

Colonic Temperature during Daily Measurements

As a consequence of daily measurement, Tco in thermoneutral zone decreased gradually (Fig. 1). In male ACI rat Tco was 37.91±0.26°C in the first day. However, it was 37.40±0,14°C and significantly lower than Tco in the 14th day ($p<0.002$). In female ACI rat, Tco was 37.96±0.21°C in the first day. It was 37.69±0.15°C and significantly lower than Tco in the 14th day ($p<0.002$).

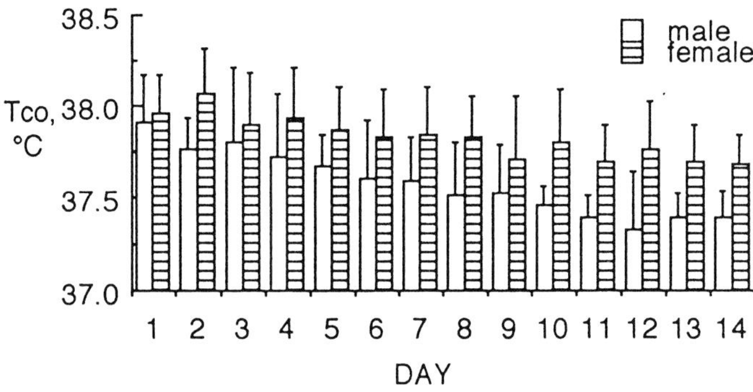

Fig. 1. Daily changes in Tco of male and female ACI rats at 24±1°C. Each column represents mean value; vertical bars represent s. d.

Colonic Temperatures of the Developing Inbred

Tco of the developing inbred gradually decreased during the repeated heat selection and inbreeding (Table 1). Tco in male F3 and F6 were 37.23±0.36°C and 36.84±0.29°C and significantly lower than Tco in male F1, ($p<0.006$ and $p<0.011$, respectively). Tco in female F3 was 37.41±0.66°C. It was slightly lower than Tco in female F1, though its significancy was poor ($p<0.054$). Tco in female F3 was 37.29±0.51°C and significantly lower than Tco in female F1, ($p<0.011$). Tco in male and female F30 were 36.84±0.26°C and 37.16±0.23°C and significantly lower than Tco in male and female F1 ($p<0.001$ and $p<0.003$, respectively).

Generation	male		female	
	mean	s.d.	mean	s.d.
1	37.63	0.16	37.98	0.37
2	37.69	0.40	37.95	0.19
3	37.23	0.36	37.41	0.66
4	37.38	0.36	37.39	0.81
5	37.20	0.24	37.55	0.15
6	36.84	0.29	37.29	0.51
7	36.46	0.34	37.29	0.42
8	36.69	0.35	37.08	0.22
9	36.50	0.22	37.10	0.30
10	36.85	0.21	37.15	0.14
20	36.81	0.43	37.20	0.40
30	36.84	0.26	37.16	0.23

Table 1. Colonic temperature of the developing inbred rat at an ambient temperature of 24±1°C.

Tco decreased in relation to extension in ST in males (Fig. 2) and females (Fig. 3) of the developing inbreds. Tco in males of which mean ST was 156 min was significantly lower than Tco in males of which mean ST was 119 min ($p<0.001$). Tco in males of which mean ST was 223 min was lower than Tco in males of which mean ST was 156 min ($p<0.004$). Tco in females of which mean ST was 163 min was slightly lower than Tco in females of which mean ST was 128 min ($p<0.054$). Tco in females of which mean ST was 250 min was lower than Tco in females of which mean ST was 128 min ($p<0.002$). Tco of heat resistant generation was lower than Tco in generation of relatively short ST.

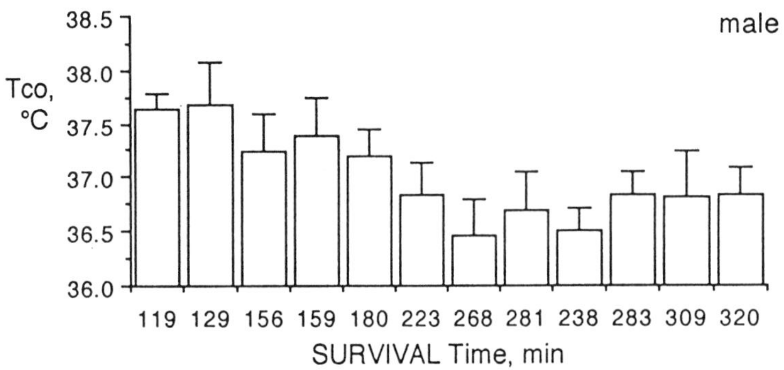

Fig. 2 Tco of male rats of various ST values at 24±1°C. Vertical axis represents Tco. Horizontal axis represents ST. Each column represents mean value; vertical bars represent s. d.

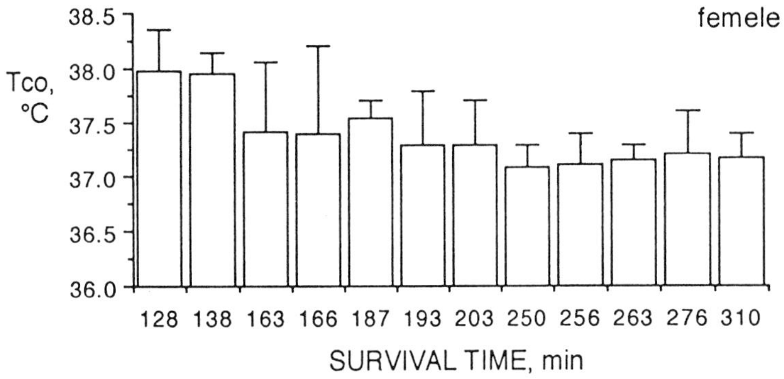

Fig. 3 Tco of female rats of various ST values at 24±1°C. Vertical axis represents Tco. Horizontal axis represents ST. Each column represents mean value; vertical bars represent s. d.

<u>Strain Differences among FOK and 3 Other Rat Strains</u>

Differences in the thermoregulatory ability among ACI, Donryu, SD and FOK rats were studied. ST varied significantly among the 4 rat strains ($p < 0.001$). The ST of the SD rat was 161.0±37.5 min, longer than those of ACI and Donryu ($p < 0.01$) which were 78.5±7.1 min and 80.0±8.8 min, respectively. The ST of FOK was 323.7±22.8 min, longer than the SD rat ($p < 0.01$) which had one of the longest ST among the other 18 strains previously studied (Furuyama, 1982; Furuyama et al., 1988). Tco at a Ta of 24±1°C also varied significantly among the 4 rat strains ($p < 0.001$) (Fig. 4). The Tco value of the SD rat was 37.18±0.37°C, greater than that of ACI and Donryu ($p < 0.01$ and $p<0.05$, respectively). That of FOK was 36.80±0.25°C, significantly lower than the SD rat ($p < 0.009$).

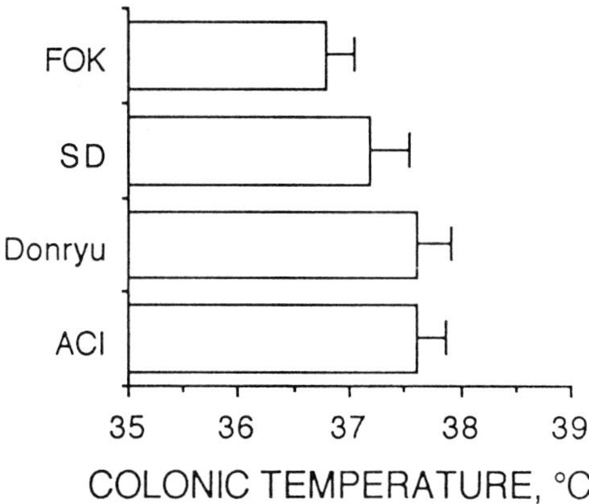

Fig. 4 Strain difference in Tco measured at 24±1°C. Each column represents mean value; vertical bars represent s. d.

DISCUSSIONS

It was reported that hypothermia was observed in thermoneutral zone in phenotypically heat-adapted subjects (Fox et al., 1963) and in long-distance runner (Baum et al., 1976) whose body temperature increases during running. There are many reports on hypothermia in phenotypically heat-adapted animals (Sod-Moriah & Yagil,1973; Roberts & Chaffee,1976; Gwosdow & Besch,1985; Shido & Nagasaka, 1990). It is note worthy that hypothermia was observed in thermoneutral zone during the period of previous heat exposure time for 2 days following intermittent heat exposures (Shido et al., 1991). The findings suggested that the hypothermia did not arise from simple response to hyperthermia but intermediated by possible memory.

On the other hand, this study describes partly the development of an inbred rat termed FOK (Furuyama, 1992) which is chiefly characterized by its genotypic resistance adaptation (Prosser, 1982) to a hot environment. This rat can be used for phenotypic comparison with other strains and for molecular genetic studies on thermoregulation, osmoregulation and resistance adaptation to heat using several numbers of control lines, recombinant inbreds and recombinant congenic lines.

In the present study, the longer the rat resisted heat, the lower its Tco was. Hypothermia in the FOK rat was a consequence of heat selection which had become fixed over 30 generations. It is worth noting that rats were kept throughout their life time at a thermoneutral zone in the present study to avoid disorders in blood pressure regulation resulting continuous heat stress (Fujita & Ymanouchi, 1984). They were in fact exposed to high Ta only once in a lifetime. These findings suggests that hypothermia in FOK is genetically determined and that hypothermia is a common resultant of many kinds of heat adaptation.

ACKNOWLEDGEMENT

This study was partly supported by a Grant-in Aid for General Scientific Research (C) (No. 05670072) from the Ministry of Education, Science and Culture of Japan.

REFERENCES

Baum, E., Bruck, K. & Schwennicke, H. P. (1976): Adaptive modifications in thermoregulatory system of long-distance runners. J. Appl. Physiol. 40, 403-410.

Fox, R. H., Goldsmith, R., Kidd, D. J. & Lewis, H. E. (1963): Blood flow and other thermoregulatory changes with acclimation to heat. J. Physiol. 166, 548-562.

Frankel, H. M. (1959) Effect of restraint on rat exposed to high temperature. J. Appl. Physiol. 14, 997-999.

Fujita, S. & Yamanouchi, C. (1984): Influence of hot environment on body temperature, heart rate and blood pressure on pregnant rats. Exp. Anim. 33, 61-67.

Furuyama, F. (1982): Strain difference in thermoregulation of rats surviving extreme heat. J. Appl. Physiol. 52, 410-415.

Furuyama, F., Ohara, K. & Ota, A. (1984): Estimation of rat thermoregulatory ability based on body temperature response to heat. J. Appl. Physiol.: Respirat. Environ. Exercise Physiol. 57, 1271-1275.

Furuyama, F., Yoshida, T., Kumazaki, M. & Ohara, K. (1988): Thermal salivation and body water economy among Wistar rat strains. Jpn. J. Vet. Sci. 50, 415-423.

Furuyama, F., Kumazaki, M. & Nishino, H. (1992): Overshoot and high equilibrium in body temperature responses of rats to ambient heat: relation to thermoregulatory ability. J. Vet. Med. Sci. 54, 915-921.

Furuyama, F. (1993): Genetic development of an inbred rat strain with increased resistance adaptation to a hot environment. Am. J. Physiol.: (Regulatory Integrative Comp. Physiol.) In Press.
Gwosdow, A. R. & Besch E. L. Effect of thermal history on the rat's response to varying environmental temperature. J. Appl. Physiol. 59, 413-419.
Nose, H., Yahata, T. & Morimoto, T. (1985): Osmotic factors in restitution from thermal dehydration in rats. Am. J. Physiol.: (Regulatory Integrative Comp. Physiol.) 249, R166-R177.
Prosser, C. L. (1982): Theory of adaptation. In Biological adaptation, ed. Hildebrandt, G. & Hensel, H. pp. 2-22. Stuttgart: Georg Thieme Verlag.
Roberts, J. C. & J. Chaffee, R. R. (1976): Metabolic and biochemical aspects of heat acclimation in the deer mouse, Peromyscus maniculatus sonoriensis. Comp. Biochem. Physiol. 53A, 367-373.
Shido, O. & Nagasaka, T. (1990): Heat loss responses in rats acclimated to heat load intermittently. J. Appl. Physiol. 68, 66-70.
Shido, O., Sakurada, S. & Nagasaka, T. (1991): Effect of heat acclimation on diurnal changes in body temperature and locomotor activity in rats. J. Physiol. 433, 59-71.
Sod-Moriah, U. A. & Yagil, R. (1973): Hyperthermia in heat-adapted female rats. Comp. Biochem. Physiol. 46A, 487-490.

Integrative and cellular aspects of autonomic functions : temperature and osmoregulation. Eds K. Pleschka, R. Gerstberger. John Libbey Eurotext, Paris © 1994, pp. 57-66.

Temperature dependence of synaptic transmission in hippocampal slices of hibernating hamsters, warm-acclimated hamsters and rats

P. Igelmund, U. Heinemann and F.W. Klussmann

Zentrum Physiologie und Pathophysiologie der Universität zu Köln, Institut für Neurophysiologie, Robert-Koch-Strasse 39, D-50931 Köln, Germany

SUMMARY

As an indication for cold resistance of neuronal function, we measured the temperature below which synaptic transmission was blocked (*threshold temperature*) in area CA1 of hippocampal slices *in vitro*.

1. In slices from golden hamsters the threshold temperature for synaptic transmission was at about 17°C, which is 2 - 3°C higher than in rat slices. This indicates that hippocampal function at lower temperatures than in nonhibernators is not a prerequisite for hibernation.
2. In slices from hibernating hamsters synaptic transmission at low temperatures was slightly better than in slices from warm-acclimated hamsters. This indicates a hibernation-dependent modification of intrinsic neuronal processes.
3. Modifications of the ionic milieu, like those presumably occuring in the brain with hibernation, raised the threshold temperature. This indicates that hibernation-related variations in the ionic microenvironment result in depression of synaptic transmission at low temperatures. This modulation may be important during entrance into hibernation.

INTRODUCTION

In a cold environment, hibernators periodically reduce their body temperature to values near the freezing point of water for several days or weeks and spontaneously rewarm to the euthermic state. To survive the deep hibernation as well as the entrance and arousal phases, there must be mechanisms which, in contrast to other mammals, provide ordered cell function also at low temperatures and during rapid temperature changes. Such mechanisms, which discriminate hibernators from non-hibernators, may be invariable species properties, they may be circannually modified with hibernation-acclimation, or they may be developed with entrance into hibernation.

Before entrance into hibernation, a preparation period (*hibernation-acclimation*) is indispensible with modifications of, for example, body weight, hormonal state, and lipid composition in cell membranes. In golden hamsters, hibernation-acclimation is exclusively triggered by external factors like short photoperiod and cold. With entrance into hibernation, the body temperature decreases and after several hours adjusts to values of 1-3°C above the ambient temperature. Arousal from hibernation and heating to the euthermic state takes 2-3 hours.

In contrast to ectothermic animals, the adjustment of body temperature of hibernating mammals is actively regulated during hibernation (Heller 1979). So at least some parts of the brain have to work at low temperatures. Thermosensitive neurons in the hypothalamus have been shown to be active at far lower temperatures in the golden hamster than in nonhibernators (Wünnenberg et al. 1976). Specific species differences in the cold tolerance of neuronal tissue between hibernators and non-hibernators have also been found in peripheral nerve conductance (Chatfield et al. 1948) and in hippocampal synaptic transmission (Hooper et al. 1985). Synaptic transmission at low temperatures in the hippocampus of golden hamsters has been reported to be modified with hibernation acclimation and in deep hibernation (Thomas et al. 1986).

According to EEG studies (reviewed by Mihailovic 1972), activity in the different parts of the brain vanishes in an ordered sequence during entrance into hibernation and is resumed in the reverse order during arousal. In deep hibernation, ongoing electrical activity was only seen in the hypothalamus and some parts of the limbic system. The hippocampus seems to be electrically silent in deep hibernation (Chatfield and Lyman 1954). Since the hippocampus is one of the last structures to become silent during cooling and one of the first to resume activity during rewarming, it is thought to play a role in the organization of brain function during entrance into and arousal from hibernation (Mihailovic 1972, Heller 1979). The fact that thyrotropin-releasing hormone (TRH), injected into the hippocampus of hibernating ground squirrels, acts as a potent arousing agent (Stanton et al. 1980) also indicates an important role of the hippocampus.

Apart from potential direct effects of hibernation on neuronal elements such as modifications in density of transmitter receptors, ion channels or ionic pumps, neuronal function may be strongly influenced by changes of the ionic microenvironment occurring during hibernation. In hamsters, distinct rises in the blood plasma concentrations have been found for K^+ (Raths 1962, Tempel and Musacchia 1975), Ca^{2+} (Raths 1962, Ferren et al. 1971), Mg^{2+} (Riedesel 1957, Ferren et al. 1971) and H^+ (Malan 1986, Malan et al. 1988).

In the present study, we investigated the influence of temperature and hibernation on synaptic transmission in hippocampal slices. We compared slices from hibernating golden hamsters with slices from warm-acclimated hamsters and from rats (as non-hibernators). To reveal hibernation related intrinsic modifications of neuronal elements, we measured the temperature below which synaptic transmission was blocked (*threshold temperature*). In a second series, we measured the temperature thresholds in varying extracellular concentrations of K^+, Ca^{2+} and H^+ in order to test the influence of hibernation-related modifications of the ionic milieu on synaptic transmission.

Apart from their importance for our understanding of hibernation, the investigation can be of general interest in the context of the beneficiary action of hypothermia on neural damage induced by seizures and stroke, and in the context of clinical hypothermia during heart or brain stem operations. Parts of this work have been published in abstract form (Igelmund et al. 1990a,b, 1992).

MATERIAL AND METHODS

The experiments were performed on hippocampal slices prepared from hibernating golden hamsters **(HH)**, from warm-acclimated golden hamsters **(HW)** and from Wistar rats **(R)**. Hamsters were purchased from the Zentrale Versuchstieranstalt, Hannover, FRG. Rats were taken from our institute's breed. The animals were provided with food and water *ad libitum*. Warm-acclimated hamsters were housed in individual cages at 22 ± 2°C in a long-day photoperiod (LD 17:7). To induce hibernation, hamsters with a minimum age of 10 weeks were transposed to a short-day photoperiod (LD 7:17), first at 22 ± 2°C for 4 to 8 weeks and then at 7 ± 2°C. After normally 2 - 6 weeks in the cold the animals entered hibernation. Every day the hibernation state of each animal was controlled with the "saw-dust method": Saw-dust was placed on the back of each sleeping animal. The next day, presence or absence of the saw-dust indicated if the animal had moved in the last 24 hours. Hibernating hamsters were chosen for experimentation only when they were in deep hibernation for at least one day and when their respiration was slower than 1 cycle per 3 minutes.

After decapitation (for rats and HW under ether anesthesia), the brain was quickly removed and chilled in ice-cold artificial cerebrospinal fluid (ACSF) for 2 to 5 minutes. The hippocampus was isolated and transverse slices (400 µm) were prepared with a McIlwain tissue chopper and immediately transferred into an interface type recording chamber. The slices were continuously perfused with an aerated (95% O_2, 5% CO_2) ACSF usually containing (in mM): NaCl 124, NaH_2PO_4 1.25, $NaHCO_3$ 26, KCl 5, $CaCl_2$ 2, $MgSO_4$ 2, glucose 10. In the experiments investigating the influence of ionic modifications, experimental solutions differed with respect to K^+ (3 to 7 mM), $CaCl_2$ (2 to 4 mM), and $MgCl_2$ (2 to 8 mM). The temperature of the chamber and of the humid gas above the slices was varied between 37 and 8°C. The pH was 7.45 and 7.28 at 37 and 15°C, respectively. In some experiments, the pH was varied by modification of the CO_2 concentration in the aerating gas to 10% (ΔpH = -0.3) or 2% (ΔpH = +0.4). To allow recovery from the preparation, slices were incubated for at least 1 hour prior to a recording session. In each animal group, a part of the slices was incubated at 37°C and another part at 23°C, but as there seemed to be no influence of incubation temperature, data were pooled.

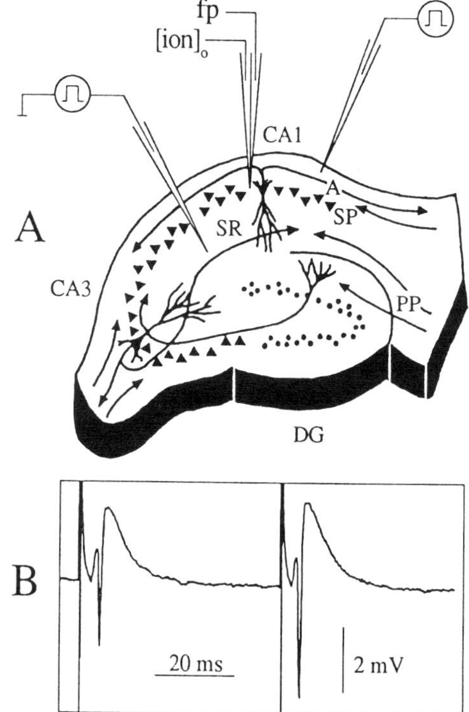

*Fig. 1: **A)** Schematic drawing of a hippocampal slice showing the neuronal connections and the experimental situation. Activity of perforant path (PP) axons synaptically activates granular cells (dots) in the dentate gyrus (DG). The axons of the granular cells are synaptically connected to pyramidal cells (triangles) in area CA3. Axon collaterals of these cells - the Schaffer collaterals - project, together with commissural fibers coming from the contralateral hippocampus, to the dendritic region of CA1 pyramidal cells in the stratum radiatum (SR). The axons of the pyramidal cells leave the hippocampus via the alveus (A). With stimulus electrodes positioned in (SR) the CA1 pyramidal cells were synaptically activated, with stimulus electrodes in the alveus they were antidromically activated. Field potentials and modifications of ion concentrations were recorded with ion selective / reference electrodes in the cell body layer of the pyramidal cells (stratum pyramidale, SP). **B)** Example of a field potential recorded from SP in response to paired pulse stimulation at 37°C. Each stimulus (visible as stimulus artifact) induces a positive going field EPSP superimposed by a negative going population action potential.*

Extracellular field potentials were recorded with microelectrodes positioned into stratum pyramidale of area CA1 (Fig. 1). In a part of the experiments double-barreled ion sensitive / reference microelectrodes were used for simultaneous measurement of the actual concentrations of K^+, Ca^{2+}, or H^+. Pyramidal cells were synaptically activated by stimulation of the Schaffer collateral/commissural fiber bundle with paired pulses (pulse duration 100 µs, intervals between the two pulses of a pair 50 ms, between pairs 30 s) via bipolar glass-insulated platinum wire electrodes inserted into the stratum radiatum. For direct antidromic stimulation of pyramidal cells electrodes were placed in the alveus. The temperature was continuously recorded with thin Cu-CuNi thermocouples (diameter 0.2 mm) placed in the chamber next to the slices. Data were conventionally amplified and recorded on chart writer for on-line control. Simultaneously they were digitized with a CED 1401 interface (Cambridge Electronics) and stored on a personal computer for further analysis.

Values are given as mean ± S.D.. Statistical significance was calculated with the paired t-test for paired differences (first and second population spike in the paired pulse paradigm; changes in response to modification of ionic milieu) and with the one way analysis of variance (SYSTAT) for differences between animal groups.

RESULTS

In transverse hippocampal slices, stimulation of the Schaffer collateral/commissural fiber bundle synaptically activates CA1 pyramidal cells. With extracellular electrodes positioned in the stratum pyramidale, this activity can be recorded as a positive going field EPSP superimposed by a negative going population action potential (*population spike*) (Fig. 1B). Apart from the monosynaptic activation, the pyramidal cells are influenced by interneurons providing feed-forward and feed-back inhibition and excitation (Andersen et al. 1964, Alger and Nicoll 1982). In a time window of some 100 ms after an action potential, elevated calcium concentration in the cell plasma of presynaptic terminals resulting from previous activity and activation of the interneuronal pathways influence the response to a second stimulus. In our experiments, we used paired pulse stimulation with intervals of 50 ms between the two pulses of a pair and 30 s between pairs. With the first stimulus we tested the direct glutamatergic transmission from the Schaffer collaterals/commissural fibers to the pyramidal cells without activity-dependent modulation, whereas the response to the second stimulus included presynaptic facilitation as well as feed-back and feed-forward modulation.

At 37°C, we recorded population spikes with amplitudes of up to 15 mV in rat slices and up to 10 mV in HW and HH slices. With paired pulse stimulation, the slices normally showed facilitation, i.e. the second population spike was bigger in amplitude than the first (Fig. 1B). With cooling, starting from 37°C, the latency and duration of population spikes in response to constant stimulus intensities continuously increased. The population spike amplitude, in contrast, first increased, reaching a maximum between 30 and 25°C, and then decreased (Fig. 2). Finally the population spikes reversibly disappeared. As a measure of the cold resistance of synaptic transmission in the three animal groups, we recorded in the following the temperature below which no population spikes could be evoked, the *threshold temperature*.

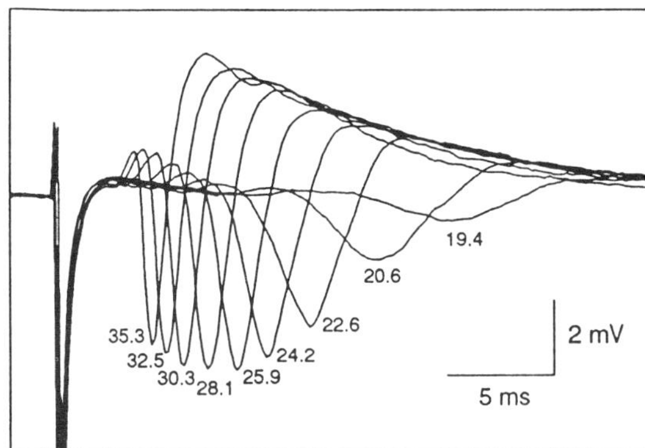

Fig 2: Population spikes at different temperatures. The record shows population spikes induced by stimulation of Schaffer collaterals/commissural fibers in a hippocamppal slice of a warm-acclimated hamster. Records at different temperatures are superimposed (see labels [°C]). Only the first spike in the paired pulse stimulation is shown.

The threshold temperature strongly varied with stimulus intensity. Hence, in order to obtain comparable conditions for the measurement of threshold temperatures, we first had to determine the influence of temperature on stimulus efficiency. Therefore we measured the minimal stimulus intensity which was sufficient to elicit a just visible population spike as a function of temperature. Typical results are shown in Fig. 3. In the temperature range between 37 and ca. 30°C, the stimulus threshold proved to be constant. With cooling below 30°C, the threshold slowly started to rise, and below about 18°C in hamster slices and 15°C in rat slices it finally steeply rose to infinite, with hamster and rat curves running parallel. Based on these results, we measured the threshold temperatures with high stimulus intensities which were about 20fold of the threshold stimulus intensity and 5-10 times the maximal stimulus intensity at 37°C.

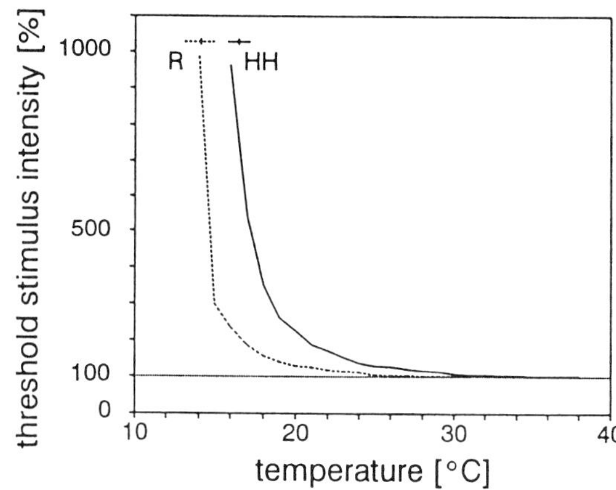

Fig. 3: Stimulus intensity threshold as a function of temperature. The figure shows the values (as % of threshold at 35°C) of a hibernating hamster slice (HH) and a rat slice (R). Above the curves the threshold temperatures ± S.D. for HH slices (n = 10) and rat slices (n = 12) are indicated.

Orthodromically induced population spikes vanished in the temperature range of 13 to 18°C. Antidromically induced population spikes, in contrast, which are independent on synaptic transmission, persisted down to temperatures below 10°C in all 3 animal groups. Also the afferent volley, which reflects the population action potentials of the stimulated Schaffer colaterals/commissural fibers, persisted to some degrees below the threshold temperatures of the pyramidal cell population spikes. This indicates that synaptic transmission is blocked at temperatures at which the cells are still excitable.

The first spike in the paired pulse situation disappeared at 17.1 ± 0.8°C in HW slices, at 16.5 ± 0.7°C in HH slices and at 14.1 ± 0.8°C in rat slices. The difference between hamster slices and rat slices was highly significant ($p<0.001$). The difference between the two hamster groups, in contrast, failed significance ($p>0.1$). For the second spike in the paired pulse paradigm, the threshold temperatures were at 17.5 ± 0.8°C (HW), 16.6 ± 0.6°C (HH) and 15.2 ± 1.0°C (R). In HW slices and rat slices the threshold temperature of the second spike was higher than that of the first (paired t-test, $p < 0.01$). In HH slices, the threshold temperature was equal for both spikes of a pair. The results are summarized in Table 1.

Table 1: Temperatures below which synaptic transmission was blocked (threshold temperatures) in hippocampal slices from warm-acclimated hamsters (HW), hibernating hamsters (HH) and rats (R). The values for the first (ps 1) and the second population spike (ps 2) in response to paired pulse stimulation are separately given. Statistical significance of the difference between adjacent values is marked by +++ ($p < 0.001$), ++ ($p < 0.01$), + ($p < 0.05$) and -- (not significant, $p > 0.1$).

		ps 1		ps 2
HW	(n = 11)	17.1°C ± 0.8°C	++	17.5 ± 0.8°C
		--		+
HH	(n = 10)	16.5°C ± 0.7°C	--	16.6 ± 0.6°C
		+++		++
R	(n = 12)	14.1°C ± 0.8°C	+++	15.2 ± 1.0°C

To investigate the effects of hibernation-related changes of the ionic microenvironment, we modified the ionic composition of the ACSF. $[K^+]$ was varied between 3 and 7 mM, $[Ca^{2+}]$ between 2 and 4mM, $[Mg^{2+}]_o$ between 2 and 8 mM, and the pH was shifted by -0.3 and + 0.4 units.

Modification of the extracellular K^+ concentration ($[K^+]_o$) strongly influenced synaptic transmission at low temperatures. In hamster slices as well as in rat slices, the threshold temperature considerably rose with increasing $[K^+]_o$ (Fig. 4). With elevation of $[K^+]_o$ from 3 to 7 mM, the threshold temperature was elevated in HH slices by 5.1 ± 0.7°C and 5.5 ± 1.0°C for the first and the second spike in the paired pulse paradigm, respectively. In rat slices, the effect was less dramatical (2.8 ± 0.6°C and 4.3 ± 0.9°C for the first and the second spike, respectively). Warm-acclimated hamsters have not been investigated.

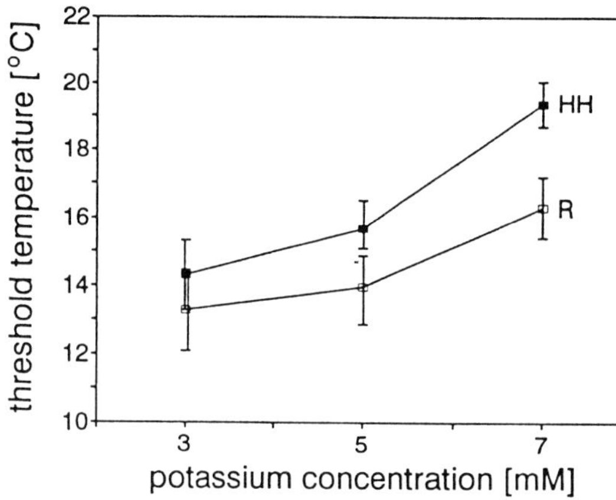

Fig. 4: Temperature below which synaptic transmission is blocked (threshold temperature) as a function of $[K^+]_o$. Mean values ± S.D. from slices of hibernating hamsters (HH, n = 8) and rats (R, n = 6).

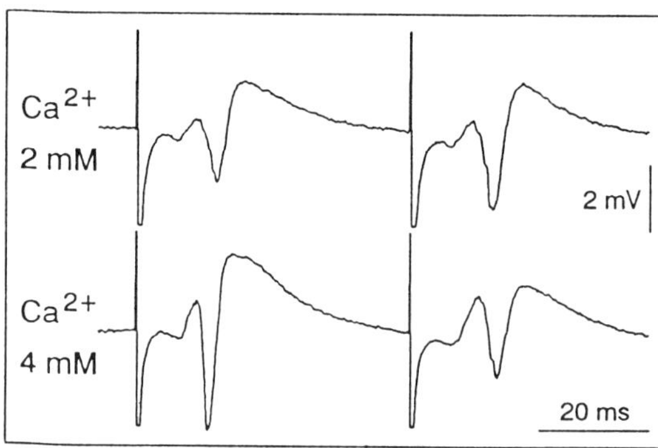

Fig. 5: Paired pulse behaviour in dependence on calcium concentration. Population action potentials from pyramidal cells in response to paired pulse stimulation at 19°C in ACSF with 2 mM (upper trace) and 4 mM Ca^{2+} (lower trace). Elevation of $[Ca^{2+}]_o$ potentiates the first spike and depresses the second.

While in 5 mM K^+ the threshold temperature of the second spike was similar to (in HH) or higher (in rats) than that of the first spike, it was lower in 3 mM K^+ in both animal groups (not shown). Accordingly, the slices showed paired pulse facilitation at just suprathreshold temperatures in 3 mM K^+, whereas they displayed equal spike amplitudes or paired pulse depression in HH and rats, respectively, in 5 mM K^+. Some degrees above the threshold temperature, paired pulse facilitation developed also in HW and rat slices.

Raising $[Ca^{2+}]$ from 2 to 4 mM shifted the threshold temperature downwards by 1 - 3°C. However, in the paired pulse situation this threshold shift mainly concerned the first spike. The threshold temperature for the second spike remained, on average, constant. Accordingly, in the just suprathreshold temperature range the amplitude of the first spike was increased with elevation of $[Ca^{2+}]_o$, whereas the second spike remained nearly constant or was even depressed (Fig. 5).

In some preliminary experiments we raised $[Mg^{2+}]_o$ from 2 to 8 mM in rat slices. In all cases, the threshold temperature was elevated.

The effect of pH was investigated by modification of the CO_2 concentration of the aerating gas (in the ACSF as well as in the gas phase above the slices). With a change from 5% (= pH 7.32 at 22°C) to 2 % and 10% CO_2, the pH in the ACSF was shifted by +0.4 units and -0.3 units, respectively. This resulted in shifts of interstitial pH, as measured with H^+ sensitive microelectrodes, of approximately +0.26 and -0.28 units, respectively, in slices of hibernating hamsters at 22°C. The alkaline shift did neither change the threshold temperature nor the spike amplitude at just suprathreshold temperatures. The shift of 0.3 pH units in acidic direction, in contrast, raised the threshold temperature by about 1°C and decreased the amplitude of half-maximal population spikes at 22°C by approximately 50% in HW and HH slices.

DISCUSSION

In the present study we investigated the cold resistance of synaptic transmission in hippocampal slices of hibernating hamsters, warm-acclimated hamsters and rats. The aim of the study was to reveal 1.) species specific, acclimation-independent differences in cold tolerance between hibernators and nonhibernators, 2.) hibernation-related intrinsic modifications of neuronal function, and 3.) effects of hibernation-related microenvironmental variations on neuronal function.

Differences between hibernators and nonhibernators

During cooling, stimulus-induced activity in hippocampal slices disappeared at temperatures far above the brain temperature of hibernating hamsters. This is consistent with EEG studies (reviewed by Mihailovic 1972) which indicate that the hippocampus is electrically silent in deep hibernation. In our study the temperature below which synaptic transmission was blocked proved to be generally higher in hamster slices than in rat slices. This confirms results of Hooper et al. (1985), which compared the threshold temperatures in hippocampal slices of euthermic chipmunks, golden hamsters and rats. In chipmunk slices, in contrast, the threshold temperatures were considerably lower than in rat and hamster slices. These results indicate that there is a higher intrinsic cold resistance of hippocampal function in some hibernator species, but this is not a general prerequisite for the ability to hibernate.

Hibernation-related intrinsic modifications of neuronal function

The comparison of our results on hippocampal slices from hibernating hamsters (HH) with the data obtained from warm-acclimated hamsters (HW) reveals a difference between the first and the second population spike in the paired pulse paradigm. For the first spike, we did not find a significant difference between HW and HH slices. For the second spike, in contrast, the threshold temperature was significantly lower in HH than in HW slices. This suggests a hibernation-related modification of processes which influence the second stimulus response but not the first. With our stimulation protocol (intervals of 50 ms between the two pulses of a pair and 30 s between pairs), the response of the pyramidal cells to the second pulse is, compared to the first spike, strongly influenced by different activity-dependent processes including presynaptic and postsynaptic mechanisms. In the presynaptic terminals of the Schaffer collaterals/commissural fibers, residual calcium resulting from previous activity enhances transmitter release in the second stimulus response and thus induces facilitation. The fact that paired pulse facilitation at low temperatures is increased in HH slices may point to changes in Ca^{2+} extrusion, internal Ca^{2+} release or Ca^{2+} entry via voltage gated channels. Postsynaptically, modifications in synaptic inhibition may contribute to this effect. Inhibitory GABAergic interneurons which are activated as a consequence of the first stimulus elicit inhibitory potentials in the pyramidal cells (Andersen et al. 1964, Alger and Nicoll 1982) and thus may depress the second response. Our results, therefore, indicate that in the hippocampus of hibernating hamsters the glutamatergic part in synaptic transmission from the Schaffer collaterals/commissural fibers to the CA1 pyramidal cells is not or not importantly improved at low temperatures as compared to warm-acclimated hamsters. There seem, however, to be hibernation-related modifications in facilitating presynaptic processes and/or in the feed-back and feed-forward pathways which lead to a lowered threshold temperature for the second spike in HH slices. This is in agreement with the observation that in HW slices the threshold temperature was significantly higher for the second spike than for the first whereas in HH slices it was similar for both spikes of a pair. Accordingly, HW slices displayed paired pulse depression at just suprathreshold temperatures whereas in HH slices the first and second spike were, on average, of equal size.

Hibernation-related modifications of the threshold temperature of synaptic transmission in hippocampal slices of golden hamsters have previously been reported by Thomas et al. (1986). In that study the threshold temperatures were determined to 15.8 ± 2.7°C (mean ± S.D.) in warm-acclimated golden hamsters, 13.9 ± 1.2°C in cold-acclimated euthermic hamsters and 12.3 ± 1.3°C in hibernating hamsters. The generally lower threshold temperatures and the more prominent effect of cold-acclimation and hibernation as compared to our data may result from the different stimulus protocol. Thomas et al. (1986) stimulated the Schaffer collateral/commissural fiber bundle with single pulses in 1 s intervals. At temperatures in the range of 15°C to 20°C the stimulation with 1 pulse per second has a strong facilitating effect compared to stimulation intervals of 30 s. It is conceivable that the time constants of the mechanisms modified with hibernation or hibernation-acclimation are such that they are most effective at action potential intervals within the range of 1 s.

Influence of hibernation-related modifications of the ionic milieu

Several studies have been undertaken to investigate hibernation-related modifications of homeostasis (for references see Wünnenberg 1990). In these studies electrolyte concentrations have been determined in the blood plasma and in different tissues. To our knowledge, there are no reports on ion concentrations in brain interstitium in the context of hibernation. As a consequence of the blood-brain-barrier the ion concentrations in the cerebrospinal fluid are different to those in the plasma. In addition, the ionic concentrations in the extracellular space are somewhat different from the composition of the cerebrospinal fluid obtained from the ventricular system (Ames III et al. 1964, Lux et al. 1986). Therefore, the absolute ion concentrations in the brain cannot be inferred from the plasma, but it seems reasonable to assume that changes in plasma concentrations are associated with comparable changes in the cerebrospinal fluid. With this assumption we can take the reported plasma modifications as an indication for hibernation-related ionic modifications in brain extracellular space. There is much divergence in the reports on hibernation-related ionic changes (rev. Wünnenberg 1990) which may to a great deal result from intrinsic differences between species. For the following discussion we therefore take into account only studies concerning hamsters (golden hamsters and/or European hamsters).

In the plasma of golden hamsters, $[K^+]$ has been reported to be elevated from 6.5 ± 0.8 mM in the euthermic state to 9.6 ± 3.2 mM in hibernation (Tempel and Musacchia 1975). A comparable elevation has already been indicated by Riedesel and Folk (1958). In European hamsters Raths (1962) found $[K^+]$ to be transiently elevated during entrance into hibernation with no changes, however, during deep hibernation and arousal. In brain interstitium of euthermic mammals, $[K^+]_o$ is in the order of 3 mM (Heinemann and Lux 1975, Heinemann et al. 1977). In our study, elevation of $[K^+]_o$ dramatically raised the temperature below which synaptic transmission was blocked. Rise of $[K^+]$ in brain interstitium during entrance into hibernation will therefore considerably reduce hippocampal activity at low temperatures. If indeed $[K^+]_o$ is returned to normal values before arousal, $[K^+]_o$ may act as a neuronal modulator which depresses neuronal activity only in the beginning phase of a hibernation cycle and not during arousal.

For golden hamsters and European hamsters the calcium concentration in the blood plasma was found to be elevated in hibernation by about 12% (Ferren et al. 1971, Raths 1962). The elevation seems to start with entrance into hibernation and to end with arousal (Raths 1961). The baseline level of $[Ca^{2+}]_o$ in the extracellular space of cats was found in the range of 1.1 and 1.3 mM (Heinemann et al. 1977, Nicholson et al. 1978). In our experiments we raised $[Ca^{2+}]$ in the ACSF from 2 to 4 mM. According to our measurements with Ca^{2+} sensitive microelectrodes, this corresponds to ca. 1.6 and 3 mM of free Ca^{2+} in the extracellular space. As in bicarbonate buffered solutions Ca^{2+} is partially chelated (Eisenman 1967), the effective concentrations of free Ca^{2+} are reduced by ca. 25%. Raising Ca^{2+} in our experiments resulted in improvement of synaptic transmission at low temperatures. This concerned both spikes in the paired pulse stimulation at temperatures several degrees above threshold. In a narrow range just suprathreshold, in contrast, only the first spike was markedly enhanced with increasing $[Ca^{2+}]_o$, whereas the second spike remained constant or was even reduced. These results indicate that elevation of $[Ca^{2+}]_o$ amplifies excitatory as well as inhibitory effects. Elevation of extracellular Ca^{2+} will lead to enhanced presynaptic Ca^{2+} entry and thereby to enhanced transmitter release. At the synapses of the Schaffer collaterals this will increase excitation. At the GABAergic interneuron synapses, it will induce stronger inhibition. While the first spike in the paired pulse paradigm is influenced only by the excitatory action, the second spike is depressed by feed-forward and/or feed-back inhibition. Hence, in the hamster hippocampus during entrance into hibernation, low frequency activity may be amplified whereas high frequency activity rather seems to be reduced by elevation of $[Ca^{2+}]_o$.

The plasma concentration of magnesium has been found to be elevated during hibernation (Riedesel 1957, Ferren et al. 1971). In our preliminary experiments, elevation of $[Mg^{2+}]_o$ resulted in a rise of the temperature threshold and in a decrease in population spike amplitude at suprathreshold temperatures. This is consistent with previous findings on $[Mg^{2+}]$ effects on synaptic transmission in the temperature range above 30°C (Rausche et al. 1990).

With entrance into hibernation, hamsters develop a respiratory acidosis which is thought to have a depressive action on metabolism (Malan 1986, 1988) and which is not compensated in the brain (Malan et al. 1985). With the beginning of arousal, acceleration of respiration induces an alkaline shift which releases the depression (Malan et al. 1988). In our experiments acidosis was a potent depressant of synaptic transmission at low temperatures. It raised the temperature below which transmission was blocked, and at suprathreshold temperatures acidosis reduced the amplitude of population spikes. Hence the variations of the acid-base state *in vivo* will lead to a depression of hippocampal activity during entrance into hibernation which is released at the beginning of arousal. Depressing effects of acidification on synaptic transmission have repeatedly been reported for hippocampal slices of rats at higher temperatures (Balestrino and Somjen 1988, Krnjevic and Walz 1990). Eckerman et al. (1990), who investigated hamster slices at 20 and 25°C, in contrast, measured no change in population spike amplitude during acidification. The lack of effect in that study is probably due to the use of high stimulus intensities which induced maximal spikes (J. M. Horowitz, pers. comm.).

In conclusion, synaptic transmission in the hippocampus at low temperatures is improved in golden hamsters with hibernation or hibernation-acclimation. It is, however, in hibernating hamsters not more resistant to cold than in nonhibernators. Variations in the microenvironment that occur with hibernation are apt to depress hippocampal activity during entrance into hibernation.

Acknowledgements

We are grateful to Erika Walde and Marianne Scheid for technical assistance and to Holger Spangenberger and Dr. Barbara E. Nixdorf-Bergweiler for valuable discussions. This work was supported by grants from the DFG (Ig10/1-1, Ig10/1-2) and the Stiftung Maria Pesch.

REFERENCES

Alger, B.E. & Nicoll, R.A. (1982). Feed-forward dendritic inhibition in rat hippocampal pyramidal cells studied in vitro. *J. Physiol. (London)*. **328**, 105-123.

Ames III, A., Sakanoue, M. & Endo, S. (1964). Na, K, Ca, Mg, and Cl concentrations in choroid plexus fluid and cisternal fluid compared with plasma ultrafiltrate. *J. Neurophysiol.* **27**, 627-681.

Andersen, P., Eccles, J.C. & Loyning, Y. (1964). Pathways of postsynaptic inhibition in the hippocampus. *J. Neurophysiol.* **27**, 608-619.

Balestrino, M. & Somjen, G.G. (1988). Concentration of carbon dioxide, interstitial pH and synaptic transmission in hippocampal formation of the rat. *J. Physiol.* **396**, 247-266.

Chatfield, P.O., Battista, A.F., Lyman, C.P. & Garcia, J.P. (1948). Effects of cooling on nerve conduction in a hibernator (golden hamster) and non-hibernator (albino rat). *Am. J. Physiol.* **155**, 179-185.

Chatfield, P.O. & Lyman, C.P. (1954). Subcortical electrical activity in the golden hamster during arousal from hibernation. *EEG Clin. Neurophysiol.* **6**, 403-408.

Eckerman, P., Scharruhn, H. & Horowitz, J.M. (1990). Effects of temperature and acid-base state on hippocampal population spikes in hamsters. *Am. J. Physiol.* *258*, R1140-R1146.

Eisenman, G. (1967). *Glass electrodes for hydrogen and other cations*. Dekker New York,

Ferren, L.G., South, F.E. & Jacobs, H.K. (1971). Calcium and magnesium levels in tissues and serum of hibernating and cold-acclimated hamsters. *Cryobiol.* **8**, 506-508.

Heinemann, U., Lux, H.D. & Gutnick, M.J. (1977). Extracellular free calcium and potassium during paroxysmal activity in cerebral cortex of the cat. *Exp. Brain Res.* **27**, 237-243.

Heinemann, U. & Lux, H.D. (1975). Undershoots following stimulus induced rises of extracellular potassium concentration in cerebral cortex of rat. *Brain Res.* **93**, 63-76.

Heller, H.C. (1979). Hibernation: Neural aspects. *Ann. Rev. Physiol* **41**, 305-321.

Hooper, D.C., Martin, S.M. & Horowitz, J.M. (1985). Temperature effects on evoked potentials of hippocampal slices from euthermic chipmunks, hamsters and rats. *J. Therm. Biol.* **10**, 35-40.

Igelmund, P., Heinemann, U. & Klußmann, F.W. (1990a). Effects of temperature on synaptic transmission in rat hippocampal slices. Pflüger's Arch. 415, Suppl. 1, R90.

Igelmund, P., Heinemann, U. & Klußmann, F.W. (1990b). Effects of temperature on stimulus induced and spontaneous activity in rat hippocampal slices. In *Brain - Perception - Cognition* (Elsner, N. and Roth, G., eds.), Georg Thieme Verlag, Stuttgart, pp. 410.

Igelmund, P., Heinemann, U. & Klußmann, F.W. (1992). Synaptic transmission in hippocampal slices of hibernating hamsters. *Europ. J. Physiol. Suppl.* **5**, 74.

Krnjevic, K. & Walz, W. (1990). Acidosis and blockade of orthodromic responses caused by anoxia in rat hippocampal slices, at different temperatures. *J. Physiol. (London).* **422**, 127-144.

Lux, H.D., Heinemann, U. & Dietzel, I. (1986). Ionic changes and alterations in the size of the extracellular space during epileptic activity. In *Advances in Neurology Vol. 44. Basic Mechanisms of the Epilepsies: Molecular and Cellular Approaches* (Delgado-Escueta, A.V., Ward, A.A., Woodbury, D.M. and Porter, R.J., eds.), Raven Press, New York, pp. 619-639.

Malan, A., Rodeau, J.L. & Daull, F. (1985). Intracellular pH in hibernation and respiratory acidosis in the European hamster. *J. Comp. Physiol. B* **156**, 251-258.

Malan, A. (1986). pH as a control factor in hibernation. In *Living in the cold* (Heller, H.C., Musacchia, X.J. and Wang, L.C.H., eds.), Elsevier, Amsterdam, London, New York, pp. 61-70.

Malan, A. (1988). pH and hypometabolism in mammalian hibernation. *Can. J Zool.* **66**, 95-98.

Malan, A., Mioskowski, E. & Calgari, C. (1988). Time-course of blood acid-base during arousal from hibernation in the European hamster. *J. Comp. Physiol. B* **158**, 495-500.

Malan, A. & Mioskowski, E. (1988). pH-temperature interactions on protein function and hibernation: GDP binding to brown adipose tissue mitochondria. *J. Comp. Physiol. B* **158**, 487-493.

Mihailovic, L.T. (1972). Cortical and subcortical electrical activity in hibernation and hypothermia. In *Hibernation - Hypothermia. Perspectives and Challenges* (South, F.E., Hannon, J.P., Willis, J.R., Pengelley, E.T. and Alpert, N.R., eds.), Elsevier, Amsterdam, London, New York, pp. 487-534.

Nicholson, C., ten Bruggencate, G., Steinberg, R. & Stöckle, H. (1978). Calcium and potassium changes in extracellular microenvironment of cat cerebellar cortex. *J. Neurophysiol.* **41**, 1026-1039.

Raths, P. (1962). Über das Serum-Natrium, -Kalium und -Kalzium des winterschlafenden und hypothermischen Hamsters (Cricetus cricetus L.). *Z. Biol.* **113**, 173-204.

Rausche, G., Igelmund, P. & Heinemann, U. (1990). Effects of changes in extracellular potassium, magnesium and calcium concentration on synaptic transmission in area CA1 and the dentate gyrus of rat hippocampal slices. *Pflügers Arch.* **415**, 588-593.

Riedesel, M.L. (1957). Serum magnesium levels in mammalian hibernation. *Trans. Kans. Acad. Sci.* **60**, 99-141.

Riedesel, M.L. & Folk, G.E.J. (1958). Serum electrolyte levels in hibernating mammals. *The Am. Nat.* **XCII,866**, 307-312.

Stanton, T.L., Winokur, A. & Beckmann, A.L. (1980). Reversal of natural CNS depression by TRH action in the hippocampus. *Brain Res.* **181**, 470-475.

Tempel, G.E. & Musacchia, X.J. (1975). Renal function in the hibernating, and hypothermic hamster Mesocricetus auratus. *Am. J. Physiol.* **228**, 602-607.

Thomas, M.P., Martin, S.M. & Horowitz, J.M. (1986). Temperature effects on evoked potentials of hippocampal slices from noncold-acclimated, cold-acclimated and hibernating hamsters. *J. Therm. Biol.* **11**, 213-218.

Wünnenberg, W., Merker, G. & Speulda, E. (1976). Thermosensitivity of preoptic neurones in a hibernator (golden hamster) and a non-hibernator (guinea pig). *Pflügers Arch.* **363**, 119-123.

Wünnenberg, W. (1990). *Physiologie des Winterschlafes.* Paul Parey, Hamburg, Berlin

Integrative and cellular aspects of autonomic functions : temperature and osmoregulation. Eds K. Pleschka, R. Gerstberger. John Libbey Eurotext, Paris © 1994, pp. 67-75.

Temperature-dependent burst discharges in magnocellular neurons of the paraventricular and supraoptic hypothalamic nuclei recorded in brain slice preparations of the rat*

H.A. Braun, M.Ch. Hirsch, M. Dewald and K. Voigt

Department of Physiology, Philipps University of Marburg, Deutschhausstrasse 1-2, D-3550 Marburg, Germany

Phasic impulse activity is a characteristic of a larger proportion of arginin-vasopressin (AVP) synthesizing cells in the paraventricular and supraoptic nuclei (PVO, SON) and it has been shown that such grouping of impulses is essential for the amount of hormone released (Brimble and Dyball, 1977; Wakerley et al. 1978, Hatton, 1983). Compared to a random discharge or a regular impulse pattern of single spikes the secretion rate of vasopressin considerably increases when the neurons discharge with the same mean firing rate but in impulse groups (bursts) which are seperated by silent periods of (Dutton and Dyball 1979; Bicknell and Leng, 1981; Shaw et al. 1984; Cazalis et al. 1985).

Indeed, it could be demonstrated that different stimuli which are known to cause neurohypophysal hormone release enhance bursting activity and particularily increase the proportion of bursting cells. Water deprivation, for example, is one of the most potent stimuli to increase peripheral and central vasopressin release (Gray and Simon, 1983; Szczepanska-Sadowska et al., 1983; Roth et al., 1990) and was also found to induce burst discharges in vasopressinergic cells (Arnauld et al., 1975; Brimble and Dyball, 1977; Wakerley et al., 1978). These and many other experiments clearly demonstrated that vasopressin release from magnocellular bursters plays an important role in osmoregulation (for rev. see Hatton, 1990). Vasopressin, however, also contributes to thermoregulation as is well demonstrated by its antipyretic effects (Cooper et al., 1979; Kasting et al., 1979; Roth et al. 1990; Zeisberger, 1990). Much effort has been done in recent years to elucidate the efferent thermoregulatory pathways of vasopressin (Kasting et al., 1979; Naylor et al. 1988; Zeisberger 1990, Roth and Zeisberger, 1992) but less is known about the afferent mechanisms responsible for the enhanced AVP-release during a febrile temperature increase. For example, it might be that circulating cytokines from immune cells posess multiple biological activities, being not only involved in the febrile temperature increase but also modulate neuroendocrine functions (Hooghe-Peters et al., 1991) which counteract the temperature change. Pyretic and antipyretic effects would thereby run in parallel.

Our approach was, first of all, to consider another possibility which relates to the characteristic burst pattern of the vasopressinergic cells. Assuming temperature dependent modifications of the burst pattern, the possibility exists that febrile temperatures, per se, can contribute to the enhanced vasopressin release corresponding to a negative feedback loop.

Temperature dependent burst discharges, indeed, are known from other types of cells as different as, for example, molluscan pacemaker neurons (Junge and Stephens, 1973) and peripheral cold receptors of a variety of species (Iggo and Young, 1975; Braun et al., 1980; Schäfer et al., 1988; Heinz et al., 1990). The temperature

* Supported by Deutsche Forschungsgemeinschaft and the Stiftung P. Kempkes (Marburg).

dependencies of the burst pattern in principle were always the same and it can be assumed that these temperature effects are governed by rather general membrane properties which also might be involved in burst generation in the mammalian brain, including magnocellular bursters of the PVN and SON.
To our knowledge, there is, until now, no systematic analysis of the temperature dependencies of the bursting activity of these vasopressinergic cells. We, therefore, examined the impulse patterns of magnocellular bursters in the paraventricular and supraoptic nuclei in brain slice preparations of the rat to evaluate whether significant modifications of the impulse pattern dependent on local temperature changes can be obtained, and whether these modifications might be in accordance with the assumption that temperature effects, per se, can contribute to an enhanced vasopressin release.

METHODS

The brains of 15 Wistar rats (both sex, 170-250g) were carefully removed within 2 minutes after decapitation and frontal hypothalamic slices (400μm thick) containing the paraventricular and supraoptic nuclei were derived using a vibratom (FTB Vibracut).
The slices were kept separately in a chamber with aCSF (in mM: 124 NaCl, 5 KCl, 1.24 KH_2PO_4, 1.3 $MgSO_4$, 0.9 $CaCl_2$, 26 $NaHCO_3$, 10 glucose) at 35°C, oxygenated by bubbling with 95%O_2-5%CO_2. For recording, the slices were placed on a gold plated thermode, continously superfused by oxygenated aCSF. Temperature stimuli were applied by a home-made thermo-stimulator which provided precise temperature ramps between 0.001°C/s and 2°/s but also sinusoidal stimuli of variable frequency and amplitude. Spikes of spontanously active neurons were recorded with glass coated platin-iridium electrodes which had an impedance of about 5MΩ. The APs were amplified (World Precisions Instruments, DAM80), filtered (notch 50Hz) and displayed on a storage oscilloscope. A spike-level-discriminator sends one TTL-pulse per spike to the computer, which stored the interspike intervals in ms and the temperature in °C/100. These data were analysed and displayed off-line by a computer programm.
Magnocellular neurons were identified, as far as possible, by the location of the electrode tip, their particular spike amplitudes and their characteristic burst pattern (Fig.1).

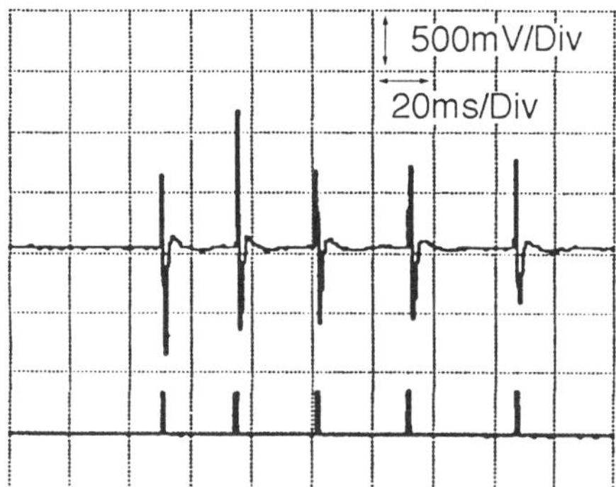

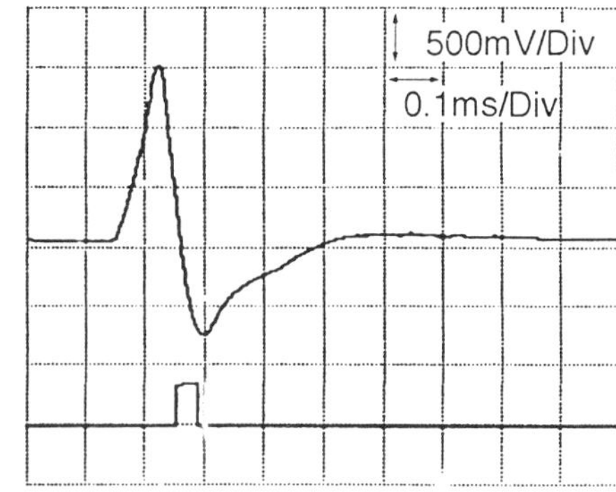

Fig.1
Extracellular impulse recordings from a supraoptic neuron printed from a digital storage oscilloscope during the experiment. The location of the neuron, the enormous spike amplitudes and the typical burst pattern (left) indicate that the impulses are probably generated by a magnocellular neuron.
The upper trace show the original spike-recordings after amplification (1000x) and filtering (50Hz notch), the lower traces show the corresponding TTL-pulses obtained from a spike-level-discriminator. The recordings are from the same neurone, but the spike amplitudes on the left were more or less cut due to the reduced time resoultion which means a reduced sampling rate for digitization.

RESULTS

Bursting activity was observed in about 2/3 (n=22) of those recordings which might be attributed to magnocellular neurons of the SON (80%) and PVO (20%) because of their location in the slice and their enormous spike amplitudes (see methods, Fig.1). But also impulse grouping was sometimes found when the spike amplitudes were much smaller and when the recordings were obtained from other hypothalamic areas (n=4, recorded during another series of experiments dealing with a different question). With respect to the temperature dependencies of the burst discharges, however, no essential differences could be observed. In all recordings of phasic activity, the burst pattern was always altered in principally the same way and exhibited systematic temperature dependencies. The most essential characteristcs of these temperature dependencies are illustrated by the example in Fig.1.
The plots of impulse sequences in Fig.2 were drawn from three different temperatures and illustrate burst discharges which are considerably changed. The continuous plot of successive interspike-intervals (ID) demonstrate that these changes smoothly follow the temperature increase (T). The corresponding changes of the discharge rate are shown in the peri-stimulus-time histogram (F).

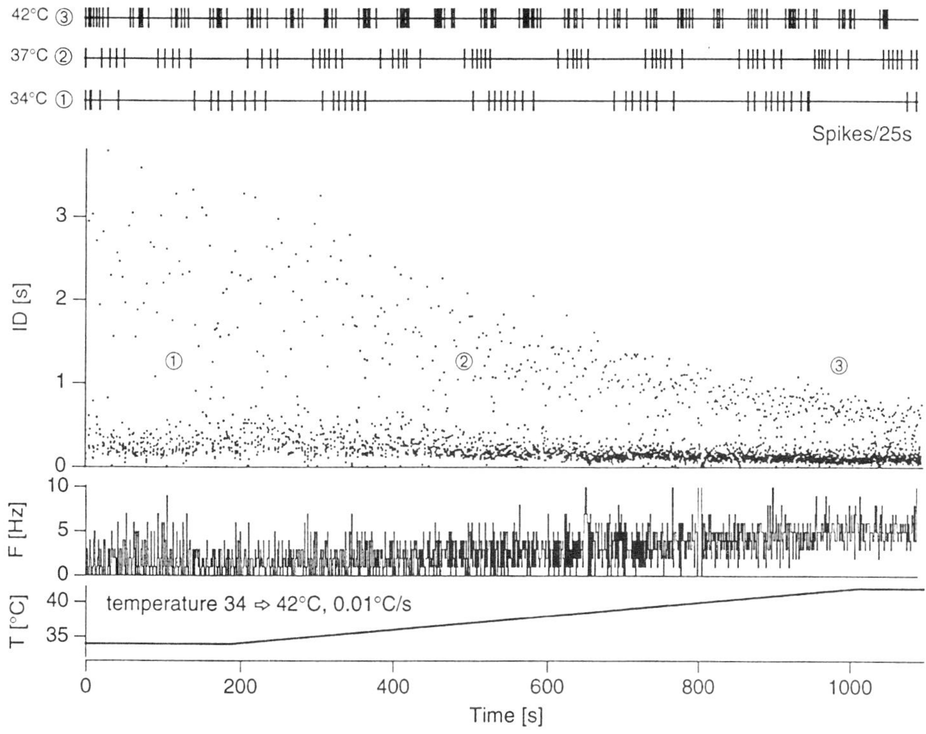

Fig.2
Modification of bursting activity in a single neuron of the N. supraopticus in dependence on a slowly increasing temperature stimulus (T, lowest trace). The frequency changes (F) of the discharge are shown by means of a peri-stimulus-time histogram (bin width: 1s). The modifications of the burst pattern are illustrated by a continuous plot of successive interspike-intervals (ID). The impulse sequences 1 to 3 in the upper traces are drawn from three different temperatures as indicated by the corresponding numbers in the ID-plot.

According to the burst pattern, the interval plot (ID) exhibits two seperate distribution components of longer and shorter intervals. The dots of longer intervals represent the burst pauses (interburst intervals) whereas the line of shorter values corresponds to the intraburst-intervals. Inter- and intraburst intervals are rather broadly distributed and are generally less clearly seperated from each other as, for example, in bursting cold receptors from the skin. Nethertheless, both parameters are clearly observed to dependend on the temperature increase. The distribution of both longer and shorter intervals is continuously narrowed and shifted to shorter values indicating that the burst pauses as well as the intraburst intervals became shorter and more regular at higher temperature. While the burst pauses at 34°C mainly varied between two and four seconds, including the occurence of still much longer intervals, these periods of silence were reduced to only about one second at 43°C. The temperature dependencies of the intraburst intervals were of comparable amount: at highest temperature (42°C), the intraburst intervals in the mean were reduced to less than one third of their values at lowest temperature (34°C).
In parallel to these modifications of the inter- and intraburst intervals, also the cycle period of the burst discharges was reduced when temperature increased. The shortening of the burst cycle is not only associated with a shortened burst pause but also affects the burst duration. However, in spite of the shortened burst duration, the mean number of spikes per burst (SB) remained practically the same, varying in this example rather randomly than systematically between 9, 10.5, and 7.5 impulses per burst at 34, 38, and 42°C, respectively. Similar effects were observed in other recordings, where the number of spikes per burst sometimes slightly increased or slightly decreased or, as in the above example, even turned from increasing to decreasing values or vize versa. The temperature dependencies of the number of spikes per burst thus appeared to be rather arbitrary in contrast to the systematic and most consistent modifications of all the other burst parameters which, however, are not based on the counting of impulses but on the measuring of time-intervals.
The number of of spikes per burst, of course, depends not only on the burst duration but also on the discharge rate within the burst and the shortened burst duration on increasing temperatures appeared to be more or less compensated by the shortening of the intraburst intervals: an increased discharge rate during the bursts produced the same number of spikes in shorter time intervals.
As the number of spikes per burst remained practically unchanged, the mean discharge frequency (F in Fig.2) was essentially determined by the frequency of the bursts. In the example of Fig.2, the discharge rate thereby increased from about 2 imp/s at 34°C to about 5 imp/s at 42°C, which means a frequency change of about 3 imp/s within 8°C, i.e. a temperature coefficient of only 0.38 $imp \cdot s^{-1} \cdot {}^{\circ}C^{-1}$. The conventional criteria for the classification of the cells as warm or cold sensitive, however, demand temperature coefficients higher or lower plus or minus 0.6 $imp \cdot s^{-1} \cdot {}^{\circ}C^{-1}$, repectively. According to this, the cell of Fig.2 would uniquely have to be classified as "temperature insensitive".
Indeed, more than half of the bursters had to be attributed as temperature insensitive although all of them exhibited temperature-dependent modifications of the discharge pattern in the way as examplified in Fig.2. But practically the same relation between temperature sensitive and temperature insensitive cells was obtained when we calculated the temperature coefficients of the burst frequency. Only when we considered the frequency changes of the intraburst discharges, had the majority of these bursters uniquely to be classified as warm sensitive. The temperature dependencies of these intraburst discharges are additionally shown in Fig.3. Despite the enormous variability within a sequence of bursts, these averaged values still elucidated the typical "parabolic" time course of the intraburst intervals as it is also known from many other bursters.

Regardless of such rather arbitrary classifications as referred to above it should be noticed that all of these bursters exhibited positive temperature coefficients irrespective of whether they were classified as temperature sensitive or temperature insensitive with respect to the one or other criteria. No cold sensitive burster was found and even no temperature coefficient of zero or below zero was observed. This qualitative unity is clearly associated with the extaordinarily consistent temperature dependencies of the burst pattern. Increasing temperatures, without any exception, accelerated the burst and enhanced the intraburst discharge without any remarkable effects on the number of spikes per burst which, altogether, mean increasing discharge rate.

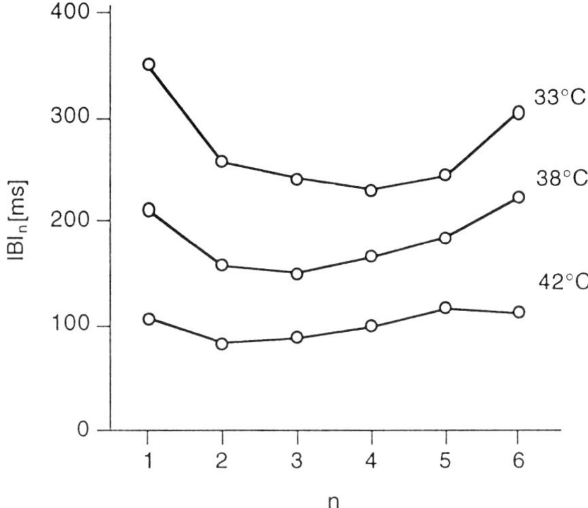

Fig.3
Time course and temperature dependent modifications of intraburst sequences (same cell like in Fig.2). Averaged values of intraburst intervals (ordinate) from about 30 burst each were calculated according to their position in the burst (abscissa).
Only six intervals from the beginning of the burst were considered because more than six intervals per burst only occasionally occured.

The results until now, however, only described the temperature dependencies of burst discharges. But we additionally have to consider that almost half of the cells (n=9) lost their bursting activity under certain stimulus conditions and changed to nonbursting and rather random activity particularly at lower temperatures.

Gradual transitions between bursting and nonbursting activity could repeatedly be induced when sinuoidal temperature stimuli were applied as in the example of Fig.4. At the beginning and at the end of this record, when the temperature was kept constant at 37°C, the cell discharged in bursts. The interval plot (ID) again exhibit separate lines of short intraburst intervals and longer burst pauses. A considerable amount of dots of intermediate intervals indicate a somewhat higher degree of random fluctuations of the pattern which also can be noticed in the bursting spike-sequences. The bursts were rather irregular and occasionally single spikes between successive bursts occured.

When the temperature increased during the sinuoidal stimulus, the effects were practically the same as already known from the example in Fig.1. The burst rhythm was accelerated and the lines of both inter- and intraburst intervals were shifted to shorter values in close relation to the temperature change. A higher concentration of the interval dots in the line of intraburst intervals indicate an enhanced intraburst discharge. When the sinuoidal temperature stimulus turned to decreasing values, the opposite effects were observed which, however, finally led to not only quantitative but also qualitative modification of the firing pattern. The lengthening of the inter- and intraburst intervals was associated with a broadening of the interval distributions and with a reduced number of intraburst intervals which, finally, completely dissappeared. At temperatures slightly below the base-line of 37°C, the burst pattern changed to a rather random firing pattern of mainly single spikes. Only occassionally, doublets still occured. Two of these cells even transiently stopped their activity completely during cooling after a rather narrow temperature range (1 to 2°C) of irregular activity was passed. Transitions from bursting to nonbursting activity were preferably observed at temperatures below 37°C but in two of these "conditional bursters", grouping of impulses could only be induced at temperatures above 38 or 39°C, respectively.

In contrast to the comparably low temperature coefficients of those cells which exhibited bursting activity over the whole stimulus range (generally 32-42°C), a clear majority of those cells which changed from bursting to nonbursting activity (7 of 9) uniquely exceeded the critical value of 0.06 imp·s^{-1}·°C^{-1} required for classification as warm sensitive. This is also the case in the example of Fig.4, where the frequency curve (F) smoothly follow the sinusoidal temperature changes between 34 and 40°C and, thereby, covered a frequency range from about 2 to 10 imp/s which means a temperature coefficient of about 1.3 imp·s^{-1}·°C^{-1}.

In contrast to the most systematic temperature dependencies as described above, the impulse pattern exhibited an enormous variability on comparison of different cells. This interindividual variability is most obviously manifested in the regularity of the pattern and in the cycle period of the bursts. As to the regularity of the pattern, a minority of the cells generated most regular burst which were clearly separated by periods of silence. Another, but also a minority of the cells, seemed to produce the bursts only occasionally superimposed on rather irregular background activity of single spikes. Most of the cells, however, discharged burst of moderate irregularities comparable to those as illustrated in the examples of Fig.2 and Fig.3.

These two examples also represent the majority of the cells considering the cycle period of the bursts. Most of

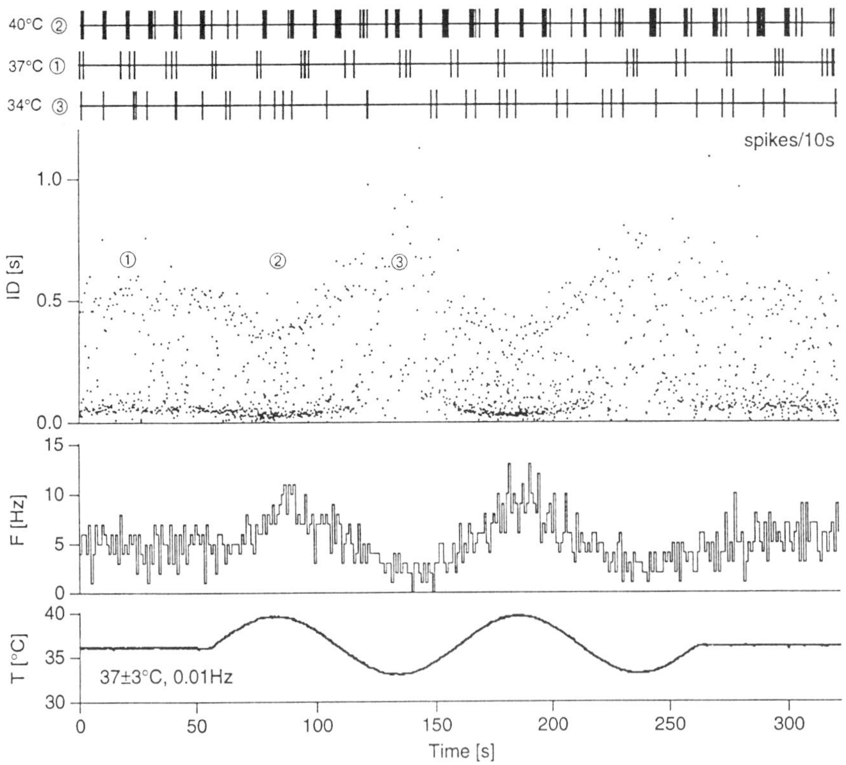

Fig.4
Transitions between burst discharges and random firing in a single neuron of the N. supraopticus extracellularly recorde during a sinusoidal temperature stimulus (T, lowest trace). The frequency changes (F) of the discharge are shown by means of a peri-stimulus-time histogram (bin width: 1s). The modifications of the burst pattern are illustrated by a continuous plot of successive interspike-intervals (ID). The impulse sequences 1 to 3 in the upper traces are drawn from three different temperatures as indicated by the corresponding numbers in the ID-plot.

the cells produced their bursts in periods between one and twenty seconds but burst periods up to one minute or more were also observed.
Despite these and some other differences, however, the temperature dependencies were always the same: increasing temperatures enhanced bursting activity and increased the tendency to bursting discharge in those cells which exhibited only single spike discharges at lower temperatures.

DISCUSSION

The data suggest that bursting activity in the PVO and SON in principle can considerably be enhanced during a febrile temperature increase. These modifications involves two distinct temperature effects: 1) alterations of the burst patterns as manifested, for example, in an acceleration of the burst rhythms and higher intraburst frequencies and 2) transitions to bursting activity in previously randomly discharging cells. Both effects are associated with an increasing mean discharge rate.
As to the possible effects on vasopressin release, it is well documented that the secretion rate of vasopressin depends not only on the mean discharge rate but is considerably enhanced when the cells discharge in impulse groups instead of regular or random sequences of single spikes. (Dutton and Dyball, 1979; Cazalis et al. 1985). The more pronounced effects thus can be expected from the impulse pattern modifications rather than from the increasing mean discharge rate.
Considering the possible mechanisms underlying these interrelations between impulse grouping and enhanced vasopressin release, calcium currents are regarded as an important factor for burst generation and are also assumed to be involved in hormone release (Andrew, 1987; Bourque et al. 1985; Poulain and Carette, 1987). These calcium currents were particularly activated during the intraburst sequences, probably due to a broadening of the spikes during such a high frequent discharge (Bourque et al. 1985, Dyball et al. 1988; Andrew and Dudek, 1985). But the secretion rate during prolonged high frequent discharges rapidly fatigue (Ingram et al., 1982). Altogether, the most important parameters to enhance vasopressin release seem to be relative short impulse sequences of high frequency. And just such a pattern was found, in our study, to be dominant when temperature increased: relative short bursts with high frequent intraburst discharges.

To evaluate the possible effects of these temperature-dependent pattern modifications on vasopressin release, we can also refer to comparable data from other authors who did experiments, for example, with hyper- and hypotonic solutions (e.g. Brimble and Dyball, 1977) or with neurotransmitters and neuromodulators which are known to modulate vasopressin release (e.g. Gähwiler and Dreifuss, 1980; Hatton et al., 1983). The results from such studies are not always consistent and often not easy to compare with each other or with the present data because most publications give mainly qualitative information about the pattern variations. Nevertheless, it is clearly to recognize that the pattern variations which we observed during temperature changes are at least comparable to those of other stimuli which are known to enhance vasopressin release. In consequence, it can be expected that also a temperature increase might be a potent stimulus for vasopressin release.
These results, however, do not necessarily suggest that temperature stimuli have specific effects on vasopressinergic cells. In accordance to Simon and Nolte (1990), we rather consider temperature dependence as a non-specific property of excitable cells. The particular interrelations of ionic mechanisms underlying burst generation possibly facilitates temperature-dependent modifications. Moreover, these specific functional conditions might also account for the extraordinarily consistent temperature effects of bursting cells irrespective of their location and physiological function as, for example, pacemaker cells in vegetative ganglia of molluscs or peripheral cold receptors in vertebrates (see introduction). In the present study, such assumptions are additionally supported by burst patterns which obviously originate from other

hypothalamic cells than from magnocellular neurons of the PVN or SON but, nevertheless, exhibited the same temperature characteristics.

However, although the modifications of the impulse patterns can be attributed to rather general and unspecific temperature effects, they nevertheless might be of distinct physiological significance depending on the functional interconnections between different autonomic control areas in the brain and their effector systems. Thermosensitive neurons of the POAH, for example, are also modulated by a variety of endogenous factors, such as pyrogenic and antipyretic substances (Hori et al., 1988; Shibata and Blatteis, 1991), neurotransmitters (Scott and Boulant 1984; Watanabe et al. 1986) and neuropeptides (Schmid et al. 1993) including vasopressin (Moraveç and Pierau, 1992). Moreover, many of these thermosensitive cells in the POAH respond to osmolality as well (Silva and Boulant, 1984; Nakashima et al., 1985). The present study now suggests that osmosensitive bursters in the paraventriculaar and supraoptic nuclei in turn can be modulated by temperature changes.

On the systemic level, such interrelations between osmoregulation and temperature control are not only manifested in the antipyretic effects of AVP (Zeisberger 1990) but, vice versa, also in temperature dependent changes of the osmosensitivity of AVP release (Simon and Nolte 1990). It therefore can be speculated that temperature-dependent pattern variations as described above might be also involved in the modulation of osmoregulatory effects. In any case, the mutual interaction between different types of stimuli also on the cellular level demands particular attention. For further elucidation, additional studies with special emphasis on such intergrative aspects in neuronal transduction are required.

REFERENCES

Andrew R.D. (1987): Endogenous bursting by rat supraoptic neuroendocrine cell is calcium dependent. J. Physiol. 384: 451-465.

Andrew R.D., Dudek F.E. (1985): Spike broadening in magnocellular neuroendocrine cell of rat hypothalamic slices. Brain Res. 334: 176-179.

Arnauld E., Dufy B., Vincent J.D. (1975): Hypothalamic supraoptic neurones: rates and patterns of action potential firing during water deprivation in the unanaesthitized monkey. Brain Res. 100: 315-325.

Bicknell R.J., Leng G. (1981): Relative efficiency of neural firing patterns for vasopressin release in vitro. Neuroendocrinology 33, 295-299.

Bourque C.W., Randle J.C.R., Renaud R.P. (1985): Calcium-dependent potassium conductance in rat supraoptic nucleus neurosecretory neurons. J. Neurophysiol 54: 1375-1382.

Braun H.A., Bade H., Hensel H. (1980):: Static and dynamic discharge patterns of bursting cold fibers related to hypothetical receptor mechanisms. Pflügers Arch. 386: 1-9.

Brimble M.J., Dyball R.E.J. (1977): Characterization of thr responses of oxytoxin- and vasopressin-secreting neurones in the supraoptic nucleaus to osmotic stimulation. J. Physiol. 271: 253-271.

Cazalis M., Dayanithi G., Nordmann J.J. (1985): The role of patterned burst and intraburst interval on the excitation-coupling mechanisms in the isolated rat neural lobe. J. Physiol. 369: 45-60.

Cooper K.E., Kasting N.W., Lederis K., Veale W.L. (1979): Evidence supporting a role for vasopressin in natural suppression of fever in sheep. J. Physiol. 295: 33-45.

Dyball R.E.J., Grossmann R., Leng G., Shibuki K. (1988): Spike-propagation and conductance failure in the rat neural lobe. J. Physiol. 401: 241-256.

Dutton A., Dyball R.E.J. (1979): Phasic firing enhances vasopressin release from the rat neurohypophysis. J. Physiol. 290: 433-440.

Gähwiler B.H., Dreifuss J.J. (1980): Transition from random to phasic firung induced in neurones cultured from the hypothalamic supraoptic area. Brain Res. 193: 415-425.

Gray D.A., Simon E. (1983): Mammalian and avian antidiuretic hormone: studies related to possible species variations in osmoregulatory systems. J. Comp. Physiol. 151: 241-246.

Hatton G.I. (1983): The hypothalamic slice approaoch to neuroendocrinology. Quart J. Exp. Physiol. 68: 483-489.

Hatton G.I., Ho Y.W., Mason W.T. (1983): Synaptic activation of phasic bursting in rat supraoptic nuceus neurones recorded in hypothalamic slices. J. Physiol. 345: 297-317.

Heinz M., Schäfer K., Braun H.A. (1990):: Analysis of facial cold receptor activity in the rat. Brain Res. 521: 289-295.

Hori T., Shibata M., Nakashima T., Yamasaki M., Asami A., Asami T., Koga H. (1988): Effects of interleukin-1 and arachidonate on the preoptic and anterior hypothalamic neurones. Brain Res. Bull. 20: 75-82.

Iggo A., Young D.W. (1975): Cutanuous thermoreceptors and thermal nociceptors. In: The somatosensory system, ed H.H. Kornhuber, pp. 5-25. Stuttgart: Thieme.

Junge D., Stephens C.L. (1973): Cyclic variations of potassium conductance in aburst generating neuron in Aplysia. J. Physiol. 235: 155-181.

Kasting N.W., Cooper K.E., Veale W.L. (1979): Antipyresis following perfusion of brain sites with vasopressin. Experientia 35: 208-209.

Moravec J., Pierau F.K. (1992): Does the effect of AVP on hypothalamic neurones support its proposed role in endogenous antipyresis? Physiol. Res. 41: 89-90.

Nakashima T., Hori T., Kiyohara T., Shibata M. (1985): Osmosensitivity in preoptic thermosensitive neurons in hypothalamic slices in vitro. Pflügers Arch 405: 112-117.

Naylor A.M., Pittman Q.J., Veale W.L. (1988): Stimulation of vasopressin release in the ventral septum of the rat brain suppresses prostaglandin E1 fever. J. Physiol. 399: 177-189.

Poulain P., Carette B. (1987): Low-threshold calcium spikes in hypothalamic neurons recorded near the paraventricular nucleus in vitro. Brain Res. Bull. 19: 453-460.

Roth J., Schulze K., Simon E., Zeisberger E. (1990): Alteration of endotoxin fever and release of arginine vasopressin by dehydration in the guinea pig. Neuroendocrinology 56: 680-686.

Roth J., Zeisberger E. (1992): Evidence for antipyretic vasopressinergic pathways and their modulation by noradrenergic afferents. Physiol. Res. 41: 49-55.

Schäfer K., Braun H.A., Kürten L. (1988): Analysis of cold and warm receptor activity in vampire bats and mice. Pflügers Arch. 412: 188-194.

Schmid H.A., Jansky L., Pierau F.K. (1993): Temperature sensitivity of neurons in slices of the rat PO/AH area: effect of bombesin and substance P. Am. J. Physiol 264: R449-R455.

Scott I.M., Boulant A. (1984): Dopamine effects on thermosensitive neurons in hypothalmic tissue slices. Brain Res. 306: 157-163.

Shaw F.D., Bicknell R.J., Dyball R.E.J. (1984): Facilitation of vasopressin release from the neurohypophysis by application of electrical stimuli in bursts. Neuroendocrinology 39: 371-376.

Shibata M., Blatteis C.M. (1991): Differential effects of cytokines on thermosensitive neurons in guinea pig preoptic area slices. Am. J. Physiol. 261: R1096-R1103.

Silva N.L., Boulant J.A. (1984): Effecta of osmotic pressure, glucose and temperature on neurons in preoptic tissue slices. Am. J. Physiol. 247: R335-R345.

Simon E., Nolte P. (1990): Temperature dependence of thermal and nonthermal regulation: Hypothalamic thermo- and osmoregulation in the duck. In: Thermoreception and Temperature Regulation, eds. J. Bligh and K. Voigt, pp. 191-199. Berlin, Heidelberg: Springer.

Szczepansky-Sadowska E., Gray D., Simon-Oppermann C. (1983): Vasopressin in blood and third ventricle CSF during dehydration, thirst and hemorrhage. Am. J. Physiol. 245: R541-R584.

Wakerley J.D., Poulain D.A., Brown D. (1978): Comparison of firing pattern in oxytoxin and vasopressin-releasing neurones during progressive dehydration. Brain. Res. 148: 425-440

Watanabe T., Morimoto A., Murakami N. (1986): Effect of amine on temperature responsive neuron in slice preparation of rat brain stem. Am. J. Physiol. 250: R553-R559.

Zeisberger E. (1990): The role of septal peptides in thermoregulation and fever. In: Thermoreception and Temperature Regulation, eds. J. Bligh and K. Voigt, pp. 273-283. Berlin, Heidelberg: Springer.

Integrative and cellular aspects of autonomic functions : temperature and osmoregulation. Eds K. Pleschka, R. Gerstberger. John Libbey Eurotext, Paris © 1994, pp. 77-85.

Local temperature sensitivity of spinal cord neurons recorded *in vitro*

Ulrich Pehl, Herbert A. Schmid and Eckhart Simon

Max-Planck-Institut für physiologische und klinische Forschung, W.G. Kerckhoff-Institut, Parkstrasse 1, 61231 Bad Nauheim, Germany

Summary

The local temperature sensitivity of the spinal cord (SC) of various homeothermic species has been clearly established in *in vivo* experiments. Aim of the present study was to investigate electrophysiologically the inherent temperature responsiveness of neurons recorded from longitudinal and transversal SC slices. 54% of the neurons recorded from transversal slices (n=79) were warm-sensitive and 3% cold-sensitive; 68% of the neurons recorded from longitudinal slices (n=22) were warm-sensitive, the remaining temperature-insensitive.
A phasic overshoot in activity was observed in about half of the warm-sensitive neurons in response to step-like temperature changes. This phasic response, however, could also be elicited in half of the neurons, which were otherwise, according to their static temperature response, characterized as being temperature-insensitive. The temperature sensitivity of at least warm-sensitive neurons could be reconfirmed in media containing reduced [Ca^{2+}] and elevated [Mg^{2+}]. In summary, the properties of temperature-sensitive SC neurons share response characteristics with temperature-sensitive PO/AH neurons and with peripheral temperature receptors. Functionally these neurons can be regarded as the cellular basis for the in vivo unequivocally characterized temperature sensory function of the spinal cord.

Introduction

The first report describing the local temperature sensitivity of the spinal cord (SC) was an in vivo study performed by Simon et al. 1964 in dogs. From this and following studies (Simon, 1974 for ref.; Jessen and Simon-Oppermann, 1976; Lin et al., 1972 in rats) it became obvious, that within the spinal cord there are primary sensory elements which induce appropriate behavioral (Cabanac, 1972; Schmidt, 1978) and autonomic regulatory mechanisms in response to local temperature changes, like shivering and vasoconstriction during cooling and panting or vasodilation during warming. In contrast to the cold induced shivering which might partially be due to the local cold sensitivity of spinal motoneurons (Pierau et al., 1969; Klussmann and Pierau, 1972), the panting of dogs induced by local warming of the lumbosacral SC can only be explained by the existence of ascending neuronal temperature-sensitive elements within the SC. These temperature-sensitive neurons have been characterized electrophysiologically in the cat by Simon and Iriki (1970, 1971a,b) by recording single unit activity in the ascending anterolateral tract of the cervical SC (C_2-C_4) after local temperature stimulation of the lumbal SC.

Supported by DFG Si 230/8-1.

The aim of the present study was to investigate electrophysiologically the inherent temperature responsiveness of neurons recorded from SC slices. In order to determine the influence of intersegmental synaptic input on the properties of temperature-sensitive neurons, the responses of neurons recorded from longitudinal and transversal SC slices were compared. Furthermore, the *in vitro* slice preparation offered the possibility to block all remaining synaptic input by superfusing the slice with artificial cerebrospinal fluid of reduced calcium- $[Ca^{2+}]$ and elevated magnesium-concentration $[Mg^{2+}]$.
In order to gain more information on the phasic and tonic responsiveness of warm- and cold-sensitive SC neurons, different stimuli (ramp, sinus and step-like temperature changes) were applied to the cells in the temperature range between 33 and 41°C.

Material and Methods

Male Wistar rats weighing between 150 and 320 g were anesthetized with 1.5 mg/kg body weight Urethane (Sigma) i.p.. After laminectomy the lower thoracal and upper lumbal spinal cord region was exposed and quickly superfused with ice-cold artificial cerebrospinal fluid (aCSF) of the following composition (in mM): NaCl, 124; KCl, 5; NaH_2PO_4, 1.2; $MgSO_4$, 1.3; $CaCl_2$, 1.2; $NaHCO_3$, 26; glucose, 10; pH: 7.4, equilibrated with 95% O_2 and 5% CO_2, 290 mOsmol/kg).
After cutting the dorsal and ventral roots, a section (10 mm) of the spinal cord comprising the segments L_2 to L_4 was quickly removed and transferred into ice-cold aCSF for about 1 min. Two or three longitudinal or 5-10 transversal slices (500 μm thick) were cut with a custom-made tissue slicer and preincubated at 35°C for at least 1 hour, before the first slice was transferred to the recording chamber and fixed to the bottom with a small platinum weight.
The gold-plated recording chamber was made from solid brass and contained a fluid volume of about 0.7 ml. The temperature was kept constant at 37.0°C by means of a Peltier-element. The chamber was continuously perfused with aCSF at a rate of 1.6 ml/min. ACSF entering the recording chamber was prewarmed to the same temperature as the solution already present in the chamber.
Extracellular recordings were made with glass-coated platinum/iridium (Pt-Ir) electrodes from various layers of the spinal cord. In transversal slices spontaneously active neurons from the superficial dorsal horn (lamina I and II) and around the central canal (lamina X) were investigated for their local temperature sensitivity. Only in those cases where no neurons could be found in these two regions, spontaneously active neurons were investigated in other regions (laminae III-VI) of the lumbal spinal cord as well. Recordings from longitudinal slices were exclusively made from the dorsal horn, but the exact recording site was not determined further.
After a stable recording from a single neuron had been established, the temperature sensitivity of a neuron was determined by changing the temperature between 33 and 41°C either sinusoidally within 7 min or ramp-like with a velocity of 0.02 °C/s. Fast step-like temperature changes (0.5°C/s, steps of 2°C) were also performed in order to detect possible dynamic temperature responses. The slope of the regression line, which resulted from plotting the spontaneous activity (FR) of a neuron against the sinusoidally or ramp-like altered temperature, determined the temperature coefficient (TC) of a cell. A statistical program, which either fits one or two regression lines (Vieth, 1989) to a set of data, calculated the set-point of a temperature response without requiring any further preconditions. Cells were classified as being warm- or cold-sensitive, when either the one (linear regression) or the steeper of the two regression lines (piecewise regression) had a slope (i.e. TC) exceeding +0.6 or -0.6 imp/s/°C, respectively.
In order to block the synaptic input, slices were superfused with aCSF of reduced Ca^{2+} (0.3 mM) and elevated Mg^{2+} (9.0 mM) concentration for at least 10 min before their temperature sensitivity was determined again.
The recorded action potentials were amplified and displayed on a storage oscilloscope (Gould, 1602) and were, after passing a window discriminator (World Precision Instruments, WPI), fed by aid of an interface (CED 1401) into a personal computer for further analysis.
The spontaneous activity of each neuron was evaluated by averaging its activity for 60 seconds prior to the first temperature stimulus.

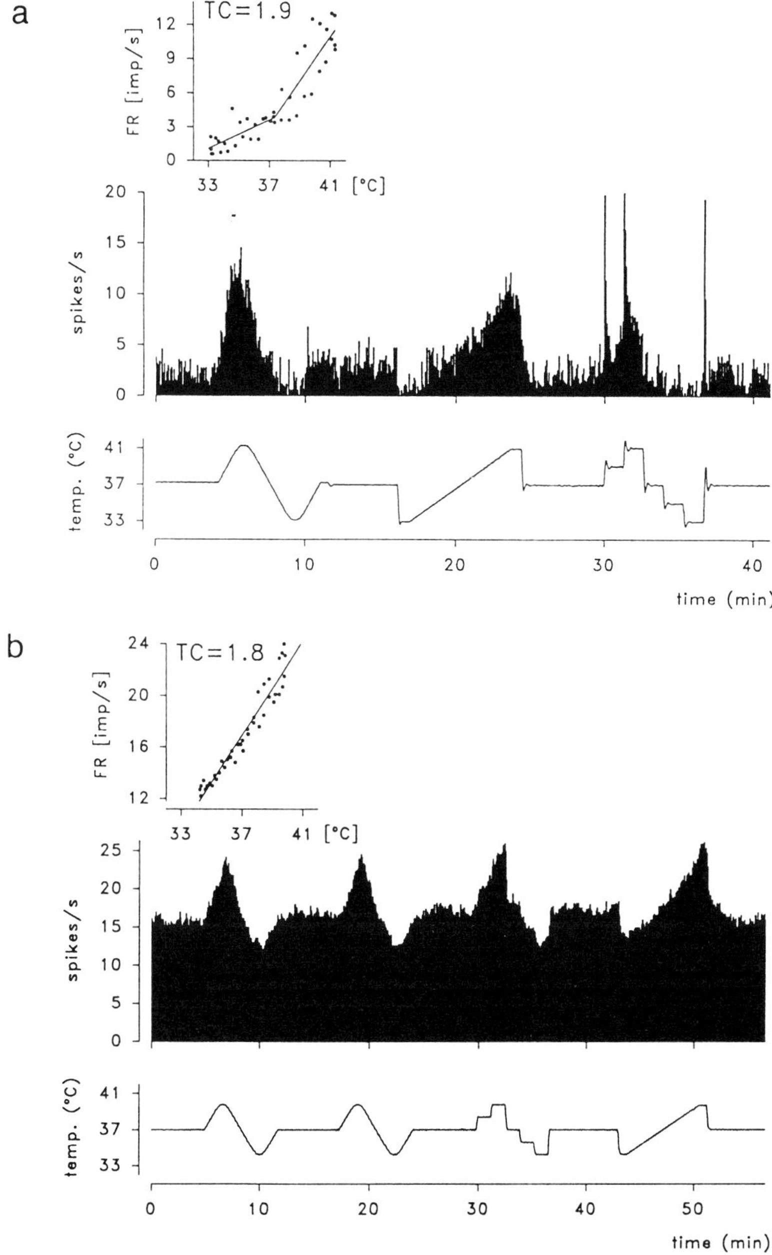

Fig. 1. Continuous rate meter recordings from two warm-sensitive dorsal horn neurons. The neuron in Fig. 1a was recorded from a transversal, the neuron in Fig. 1b from a longitudinal spinal cord slice. Insets: Frequency/temperature relationships for the first temperature stimulus of each neuron. The temperature coefficient (TC) of the neuron in Fig. 1a was 1.9 imp/s/°C for the upper temperature range and for the neuron in Fig. 1b 1.8 imp/s/°C over the entire temperature range. In contrast to the neuron in Fig. 1a, which displayed a tonic and a phasic temperature responsiveness, the neuron in Fig. 1b showed just tonic responses to various temperature stimuli.

If the transient increase or decrease in activity in response to a step-like temperature change exceeded ±30% of the corresponding tonic activity, the neuron was considered as being phasic.
Mean values in the text are given with the standard error of the means (SEM). Significant differences of mean values were determined by using a t-test ($p<0.05$).

Results

In this study successful recordings could be obtained from 101 spontaneously active neurons from the dorsal horn of longitudinal and transversal slices. No obvious difference was observed in the survival time or the amount of spontaneously active neurons found in 500 μm thick longitudinal vs. transversal slices. Out of 79 neurons recorded from transversal SC slices, 54% were warm-, 3% cold-sensitive, 43% were temperature-insensitive. Of the 22 neurons recorded from longitudinal SC slices 68% were warm-sensitive, the remaining were temperature-insensitive. No cold-sensitive neuron could be found so far.
Fig. 1a shows the continuous rate meter recording of a warm-sensitive neuron recorded from the dorsal horn of a transversal SC slice and Fig. 1b demonstrates an example of a neuron recorded from the dorsal horn of a longitudinal SC slice. The two examples were chosen in order to demonstrate the variability of the responses to different temperature stimuli and were not specific for recordings from slices with a different orientation.

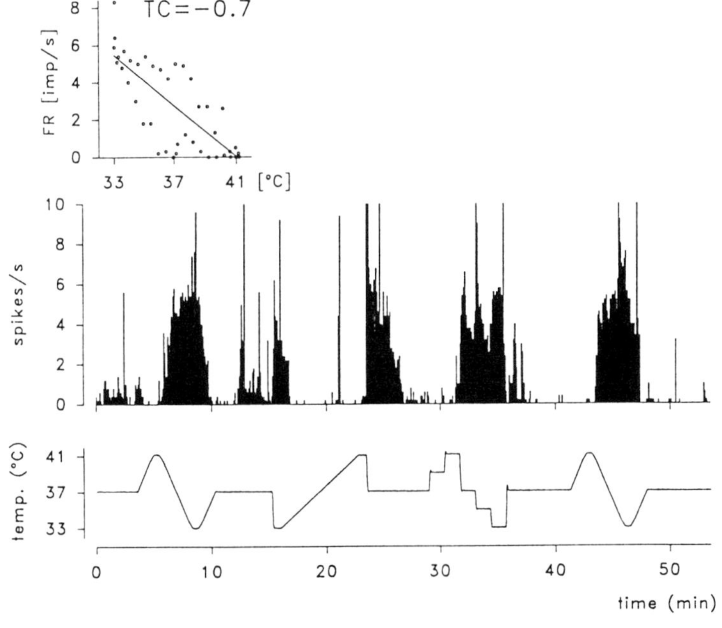

Fig. 2. Cold-sensitive neuron recorded from a transversal SC slice showing tonic and phasic responses to various temperature stimuli. Inset: Frequency/temperature relationship for the first temperature stimulus. The temperature coefficient (TC) of this neuron was -0.7 imp/s/°C and was linear over the entire temperature range tested.

The warm-sensitive neuron in Fig. 1a has a low basal activity in contrast to the cell in Fig. 1b and step-like temperature changes elicit overshooting "phasic" responses only in the former cell. Furthermore, the maximum temperature sensitivity of the neuron in 1a is observed between 37 and 41 °C, while the neuron in Fig. 1b is characterized by a linear temperature response curve (i.e. TC) over the entire temperature range tested.
An example of a cold-sensitive neuron is shown in Fig. 2. This neuron is characterized by a tonic cold-sensitivity (linear TC of -0.7 imp/s/°C), determined from the first sinusoidal temperature change. The cell displays a phasic overshoot in its activity during rapid cooling at the beginning and at the end of ramp-like temperature changes and during the step-like temperature changes, but only in the hypothermic direction.
Both cold-sensitive neurons were found in transversal slices close to the central canal, i.e. in lamina X. The absence of cold-sensitive neurons in recordings from longitudinal SC slices is most likely due to the lower number of recordings. The basal activity of both cold-sensitive neurons was low (<1 imp/s) and showed phasic responses to rapid cooling steps.

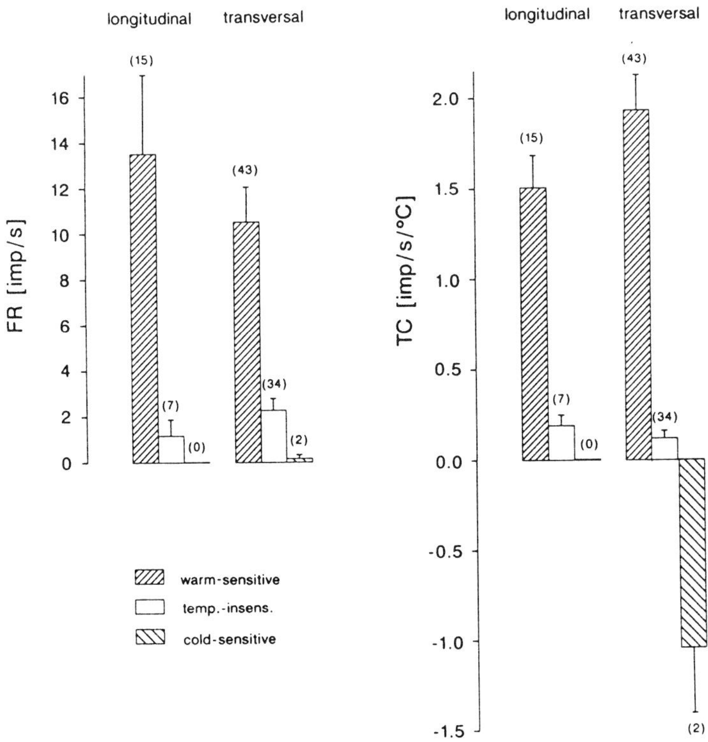

Fig. 3. Average firing-rate (FR) and temperature coefficient (TC) of all warm-sensitive, temperature-insensitive and cold-sensitive neurons recorded from longitudinal and transversal SC slices.

The average spontaneous firing rate (FR) of all neurons from transversal slices at 37°C (6.7 ± 1.0 imp/s) was not significantly different from the mean FR of neurons from longitudinal slices (9.6 ± 2.6 imp/s). The comparison of the mean FR and the mean TC of warm-sensitive neurons revealed neither a difference in the FR (10.5 ± 1.5 imp/s in transversal vs. 13.5 ± 3.4 imp/s in longitudinal slices) nor in the TC (1.9 ± 0.2 imp/s/°C in transversal vs. 1.5 ± 0.2 imp/s/°C in longitudinal slices). Similarly no difference existed between the FR and TC of temperature-insensitive neurons from both preparations (Fig. 3). The basal FR of warm-sensitive neurons, however was significantly higher ($P < 0.001$) in both preparations when compared to the FR of temperature-insensitive neurons (Fig. 3).

The temperature sensitivity of 18 spontaneously active SC neurons (8 warm-sensitive, 10 temperature-insensitive) could be investigated additionally after blocking their synaptic input by superfusing the cells with blocking solution containing reduced $[Ca^{2+}]$ and elevated $[Mg^{2+}]$. 3 of the warm-sensitive neurons ceased 4-6 min after superfusion with blocking solution. The other 5 neurons remained spontaneously active for at least 20 - 52 min. 6 of the 8 warm-sensitive neurons retained their sensitivity in blocking solution. The FR and the TC was increased in 3 and was reduced, but still above 0.6 imp/s/°C in 3 of the remaining neurons.

Fig. 4 shows a warm-sensitive neuron which retains its temperature sensitivity in blocking solution. Although this neuron stopped its spontaneous activity after 9 min in blocking solution the temperature sensitivity of this neuron could clearly be demonstrated. Its TC was even elevated (3.5 imp/s/°C) and the temperature response curve shifted to higher temperatures as compared to the control (0.9 imp/s/°C).

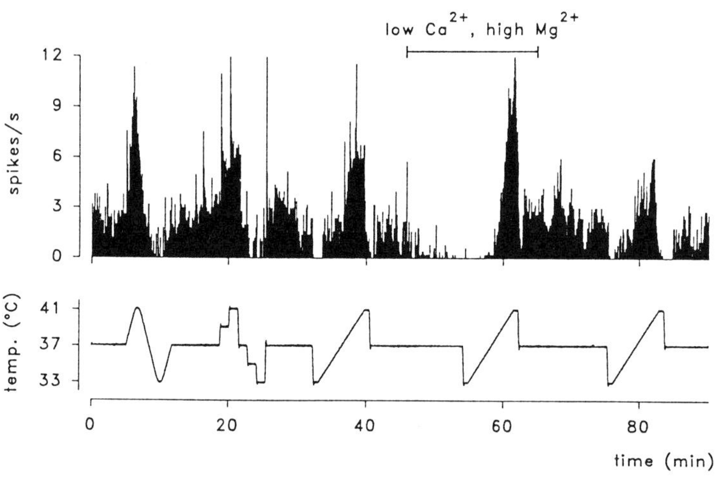

Fig. 4. Warm-sensitive neuron recorded from a transversal SC slice which retains its temperature sensitivity after blocking synaptic transmission by superfusing the slice with solution of reduced Ca^{2+} (0.3 mM) and elevated Mg^{2+} (9.0 mM) concentrations.

With only one exception, all temperature-insensitive neurons stopped their spontaneous activity in blocking solution between 2 and 11 min. This effect was reversible in 7 of the temperature-insensitive neurons. The viability of the recorded neurons could still be validated by occasionally occurring action potentials. Two former temperature-insensitive neurons were transformed to warm-sensitive neurons after superfusion with blocking solution. All other cells remained temperature-insensitive.

66% of all tonically warm-sensitive neurons tested (n=29) responded to step-like temperature elevations with a transient increase in activity and usually with a transient inhibition during cooling (table 1). Only one warm-sensitive neuron showed the reverse responsiveness.
One cold-sensitive neuron showed phasic overshoots in response to rapid cooling (example in Fig. 2), the second showed a "paradoxical" overshoot caused by rapid warming.

Table 1

Number of warm-sensitive, cold-sensitive and temperature-insensitive neurons with phasic overshoots in their activity due to rapid temperature changes

temperature-sensitivity	phasic overshoot induced by rapid warming	phasic overshoot induced by rapid cooling	no effect	total
warm-sens.	18	1	10	29
cold-sens.	1	1	0	2
temp.-insens.	15	2	10	27

Phasic responses, however, were not restricted to temperature-sensitive neurons, but could also be observed in 63% (n=27) of the neurons which were otherwise, according to their TC, characterized as being temperature-insensitive. Two of these neurons showed "reversed" phasic responses, i.e. a phasic increase during cooling and a phasic decrease during warming. Bursts, which have been described predominantly for peripheral cold receptors (Hensel, 1981), could also be observed in some neurons from SC slices, but burst were similarly not specific for temperature-sensitive neurons.

Discussion

This study is the first of its kind using the spontaneous firing rate of neurons from SC slices as a parameter in order to characterize their response to temperature stimuli. The results provide evidence, that the number and temperature sensitivity of warm- and cold-sensitive neurons recorded from SC slices is comparable to the number and temperature sensitivity of PO/AH neurons recorded in slices (Boulant and Dean, 1986; Hori et al.,1980; Nakashima et al., 1987; Schmid and Pierau, 1993). Following the line of the argumentation that the *in vitro* recorded temperature sensitivity of neurons from PO/AH slices reflects their physiological function *in vivo* as central temperature sensors, our result on SC slices underline the functional importance of the spinal cord for temperature regulation.
The characteristics of ascending warm-sensitive fibers in the anterolateral tract *in vivo* (Simon and Iriki, 1971) were comparable to that of warm-sensitive SC neurons with respect to basal firing rate and average temperature sensitivity in the range of 33 - 41°C. Another consistent finding in about 50% of all warm-sensitive neurons from both preparations is the occurrence of a phasic response to rapid temperature changes. The number of cold-sensitive neurons from SC slices is still too small in order to allow a meaningful quantitative comparison with the *in vivo* study.
It could be demonstrated, that at least the warm-sensitivity of neurons persists after blocking the synaptic input, suggesting an inherent and not synaptically induced static temperature

sensitivity. Even the phasic overshoots of some of the warm-sensitive neurons in response to rapid temperature elevations do not seem to be mediated by synaptic input, because the phasic overshoot of the response is preserved in blocking solution.

One major difference between the study done in cats *in vivo* (Simon and Iriki, 1970, 1971 a,b) and the present *in vitro* study in rat SC slices is the proportion of cold-sensitive neurons. While Simon and Iriki (1970) found 7 cold-sensitive units among 25 temperature-sensitive units, we only observed two cold-sensitive neurons in a total of 60 temperature-sensitive neurons in SC slices. This difference, however, is not specific for investigations on temperature-sensitive SC neurons, but is generally observed when comparing numbers of central cold-sensitive neurons from slice preparations with results obtained from *in vivo* recordings, especially when the recordings were done in unanesthetized animals (Boulant and Dean, 1986). This effect may be due to the absence or the reduction of an inhibitory input from warm-sensitive neurons impinging upon spontaneously active temperature-insensitive neurons in anesthetized animals or in slice preparations. This hypothesis, however, needs confirmation in future experiments.

Intersegmental synaptic interaction, as far as it is still intact in longitudinal SC slices, does not seem to be required for the expression of the temperature sensitivity of SC neurons, because no obvious difference could be observed in the number of temperature-sensitive or -insensitive neurons, their spontaneous activity, temperature sensitivity or firing pattern.

The absence of cold-sensitive neurons in longitudinal SC slices might just be due to the lower number of recordings. Following the above mentioned rationale for the higher number of cold-sensitive neurons found *in vivo* as compared to slice studies, the number of cold-sensitive neurons in longitudinal SC slices should be higher and not lower, due to a more preserved synaptic interaction.

Irrespective of all suggestive evidence correlating physiological function with local temperature responsiveness of central neurons, one has to bear in mind that, unlike for peripheral temperature receptors (Hensel, 1981), specific criteria for neurons acting as temperature-sensors in the CNS are generally missing (Barker and Carpenter, 1970; Simon et al., 1986). Comparing the properties of the temperature-sensitive SC neurons with the characteristics of peripheral temperature receptors (Hensel, 1981) reveals, that the tonic and phasic temperature sensitivity, the occurrence of bursts in some neurons and the location of the recorded neurons in the dorsal horn is in line with the notion that these neurons might serve a sensory function for supraspinal CNS regions (Simon, 1972).

In summary, these data underline the functional significance of the local temperature sensitivity of SC neurons. Although specific criteria for temperature-sensitive SC neurons are, like in the PO/AH, still missing, the spinal cord offers a better clue to characterize temperature-sensitive neurons, due to its segmental structure and due to the known morphological separation of neurons belonging to the afferent or efferent signal transduction pathway.

References

Barker, J.L. & Carpenter, D.O. (1970): Thermosensitivity of neurons in the sensorimotor cortex of the cat. *Science* 169, 597-598.

Boulant, J.A. & Dean, J.B. (1986): Temperature receptors in the central nervous system. *Ann. Rev. Physiol.* 48, 639-654.

Cabanac, M. (1972): Thermoregulatory behavior. In *Essays on temperature regulation*, ed. J. Bligh & R. Moore, pp. 19-36. Amsterdam: North Holland Publ. Comp.

Hensel, H. (1981): *Thermoreception and temperature regulation*. London: Academic Press.

Hori, T., Nakashima, T., Kiyohara, T. & Hori, N. (1980): Effect of calcium removal on thermosensitivity of preoptic neurons in hypothalamic slices. *Neurosci. Lett.* 31, 171-175.

Jessen, C. & Simon-Oppermann, C. (1976): Production of temperature signals in the periphally denervated spinal cord of the dog. *Experientia* 32, 484-485.

Klussmann, F.W. and Pierau, F.-K. (1972): Extrahypothalamic deep body thermosensitivity. In *Essays on temperature regulation*, ed. J. Bligh & R. Moore, pp. 87-104. Amsterdam: North Holland Publ. Comp.

Lin, M.T., Yin, T.H. & Chai, C.Y. (1972): Effects of heating and cooling of spinal cord on CV and respiratory responses and food and water intake. *Am. J. Physiol.* 223, 626-631.
Nakashima, T., Pierau, Fr.-K., Simon, E. & Hori, T. (1987): Comparison between hypothalamic thermoresponsive neurons from duck and rat slices. *Pflügers Arch.* 409, 236-243.
Pierau, Fr.-K., Klee, M.R. & Klussmann, F.W. (1976): Effect of temperature on postsynaptic potential of cat spinal motoneurons. *Brain Res.* 114, 21-34.
Schmid, H.A. & Pierau, F.-K. (1993): Temperature sensitivity of neurons in slices of the rat PO/AH hypothalamic area: effect of calcium. *Am. J. Physiol.* 264 (Regulatory Integrative Comp. Physiol. 33), R440-R448.
Schmidt, I. (1978): Behavioral and autonomic thermoregulation in heat stressed pigeons modified by central thermal stimulation. *J. Comp. Physiol.* 127, 75-87.
Simon, E. (1974): Temperature regulation: The spinal cord as a site of extrahypothalamic thermoregulatory functions. *Rev. Physiol. Biochem. Pharmacol.* 71, 1-76.
Simon, E. & Iriki, M. (1970): Ascending neurons of the spinal cord activated by cold. *Experientia* 26, 620-622.
Simon, E. & Iriki, M. (1971a): Ascending neurons highly sensitive to variations of spinal cord temperature. *J. Physiol. Paris* 63, 415-417.
Simon, E. & Iriki, M. (1971b): Sensory transmission of spinal heat and cold sensitivity in ascending spinal neurons. *Pflügers Arch.* 328, 103-120.
Simon, E. (1972): Temperature signals from skin and spinal cord converging on spinothalamic neurons. *Pflügers Arch.* 337, 323-332.
Simon, E., Pierau, Fr.-K. & Taylor, D.C.M. (1986): Central and peripheral thermal control of effectors in homeothermic temperature regulation. *Physiol. Rev.* 66, 235-300.
Simon, E., Rautenberg, W., Thauer, R. & Iriki, M. (1964): Die Auslösung von Kältezittern durch lokale Kühlung im Wirbelkanal. *Pflügers Arch.* 281, 309-331.
Vieth, E. (1989): Fitting piecewise linear regession functions to biological responses. *J. Appl. Physiol.* 67, 390-396.

Integrative and cellular aspects of autonomic functions : temperature and osmoregulation. Eds K. Pleschka, R. Gerstberger. John Libbey Eurotext, Paris © 1994, pp. 87-95.

Heat stress and heat acclimation : the cellular response-modifier of autonomic control

Michael Horowitz

Department of Physiology, Hadassah Schs of Medicine and Dental Medicine, The Hebrew University, POB 1172, Jerusalem 91010, Israel

SUMMARY

The mechanism of heat acclimation comprises an initial phase (STHA) during which [effector output]/[autonomic signal] (EO/AS) ratio decreases compared to pre-acclimation levels and a late phase, long-term heat acclimation (LTHA) during which EO/AS increases, suggesting increased efficiency of the effector response. Our data imply that with acclimation, basal EO/AS ratios may be modified by peripheral effector responsiveness. During STHA effector organ responsiveness to neurotransmitters decreases, possibly due to decreased neurotransmitters binding affinity. This decreased responsiveness superimposed on thermoregulatory demands leads to accelerated autonomic activity to compensate for decreased effector responsiveness. Preliminary studies on salivary cell-line undergoing simulated acclimating conditions imply compensatory cell response to override impaired responsiveness. Thus, in the heat stressed intact body the overall effector output is the sum of opposing effects, resulting from cellular strain and central thermoregulatory demands. With LTHA, improved cellular responsiveness replaces the need for accelerated autonomic response. They are mediated by quantitative and qualitative changes at the effector cellular level.

HEAT ACCLIMATION - A CONCEPTUAL MODEL

All animals can adjust their physiological mechanisms to cope with long term shifts in ambient temperature. This process of adaptation, when acquired under laboratory induced conditions, is called acclimation. Extensive studies carried out in our laboratory on heat dissipation effectors of acclimating rodents provide evidence that heat acclimation is a continuum of processes, varying temporally and differing in their efficiency and optimal performance (Horowitz et al.1983, Horowitz & Meiri, 1985; Horowitz, 1989; Horowitz, 1991; Horowitz & Meiri, 1993). In the intact animal, an apparent acclimated state is displayed after a very short period of exposure to the acclimation regimen (Horowitz et al., 1983). However, it is brought about by a cascade of transiently recruited mechanisms to alleviate the initial strain, disregarding body homeostasis. Later on these mechanisms are replaced by long-lasting features working in concert to produce optimized thermoregulatory responses while maintaining body homeostasis. This is the stage when acclimation has been achieved. Our data allow us to suggest a conceptual model of the acclimation process. For simplicity, we postulate in the model two major distinct phases of acclimation (Fig. 1). In the initial phase of short term heat acclimation (STHA), the [organ output] /[autonomic signal] ratio decreases transiently, so that accelerated efferent activity is required to produce adequate output. This suggests that at this phase of acclimation, the autonomic nervous system plays the major role in alleviating the sustained strain. In contrast, during long term heat acclimation (LTHA) the evoked intrinsic peripheral adaptative responses reduce the need for accelerated excitation. This is manifested by increased effector output despite decreased autonomic stimulation (increased output/signal ratio) (Horowitz and Meiri, 1985; Horowitz and Meiri, 1993).

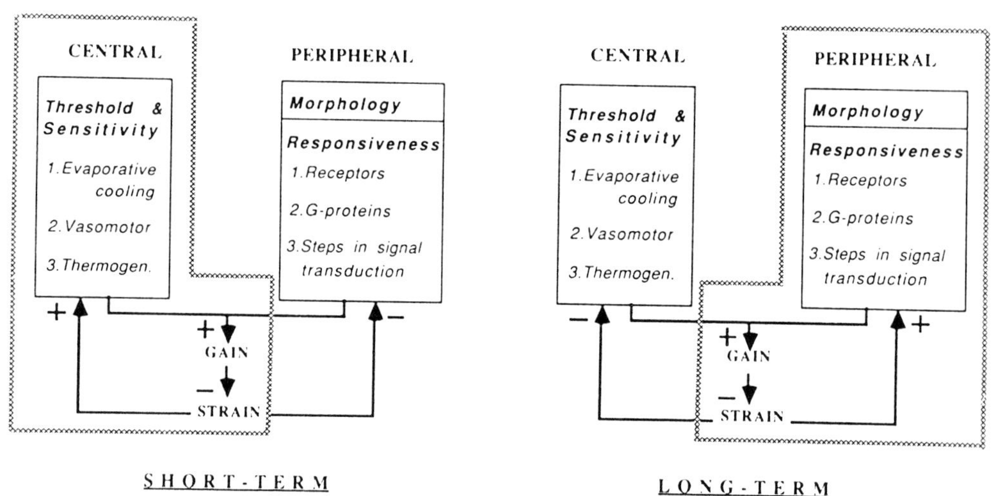

Fig. 1. Heat acclimation - a conceptual model.
Left panel -STHA. During this phase accelerated autonomic discharge overrides impaired organ responsiveness. Right panel - LTHA phase, when acclimation has been achieved. During this phase the share of peripheral cellular adaptive features in homeostatic control under the environmental constraint increases, thus accelerated autonomic discharge diminishes. Changes in the thermoregulatory center (CENTRAL) include decreased temperature thresholds (Tsh) for activation of evaporative cooling and vasomotor responses with increased sensitivity of these effectorial loops. In contrast, heat production (Thermogen.) Tsh increases. For further details on Tsh changes and effector adaptation see explanations in the text and references:Horowitz et al., 1983; Horowitz, 1989;Horowitz and Meiri, 1993).

THE MODIFIER ROLE OF CELLULAR RESPONSIVENESS - EXPERIMENTAL EVIDENCE

The available data suggest that the described autonomic response is the outcome of the combined effect of both ambient stress and the resultant internal strain. The following experimental evidence supports the above concept.

1. Autonomic pathway - peripheral responsiveness interactions: During short term heat acclimation (STHA; 2-5 days) an impairment is observed in effector organ responsiveness to direct stimulation. This decreased responsiveness can be explained, at least in part, by decreased sensitivity of the target organs to transmitters as displayed by the shift to the right of the [neurotransmitter]-[organ output] dose response curves (Fig. 2). In the evaporative cooling system of the rat, in 2 - day acclimated animals, the submaxillary glands require a higher pilocarpine concentration to attain salivation than do non-acclimated glands (Kloog et al., 1985). Likewise, in the atria of 2 - and 5 - day acclimated rats, higher carbamylcholine and noradrenaline (NE) concentrations are required for depression and acceleration , respectively, of the beating rate, than were required pre-acclimation (Horowitz and Meiri, 1993). As demonstrated in Fig. 2, decreased sensitivity to norepinephrine is the least pronounced. This suggests that, at least for the NE transduction pathway, more than one mechanism is responsible for the decreased responsiveness observed. Despite impaired peripheral organ responsiveness, in intact heat stressed rats during STHA, saliva flow rate and volume are remarkably elevated, and in the heart, heat acclimation induced bradycardia occurs. These observations suggest accelerated autonomic response to override impaired peripheral responsiveness. This suggests that during STHA the autonomic motor response evoked by stimulated thermoreceptors is modified by organ responsiveness.

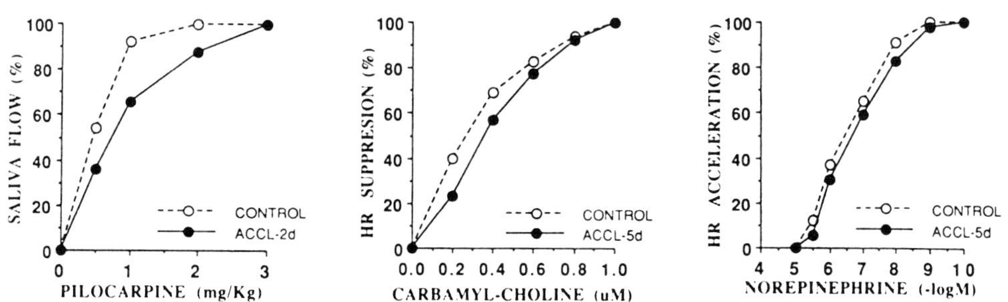

FIG 2. Evidence for decreased organ sensitivity to transmitters during STHA. This is exhibited by the shift to the right of the dose response curves to pilocarpine (activation of submaxillary gland) and carbamyl choline and noradrenaline (chronotropic response of the atria). Figs are redrawn from Kloog et al.,(1985) and Horowitz and Meiri 1993).

In LTHA, responsiveness returns to its preacclimation level. However, increased organ capacity is attained due to increased efficiency of various steps along the effector activation pathway. This increased efficiency decreases the need for accelerated autonomic response.

2. Possible mechanisms underlying altered organ/cellular responsiveness during STHA

The mechanisms underlying altered cellular responsiveness on acclimation are not yet fully elucidated. However, the data accumulated thus far indicate that during STHA changes at the membrane level predominate. In contrast, during LTHA a variety of intracellular, possibly biochemical changes take place. Kloog et al. (1985) were the first to demonstrate that during the initial phase of the heat acclimation regimen upregulation of the cholinergic muscarinic receptors occurred, achieving 100% elevation on day 10 of the acclimation. Such elevation was observed in the atria of the heart as well as in the intestine, suggesting that non thermoregulatory organs also respond to the initial ambient stress (Horowitz and Meiri, unpublished results). Kloog et al. (1985) provided evidence for a change in the ratio of two receptor subtype populations in the two - days acclimated rats. Recently, using the quaternary (hydrophilic) non specific muscarinic ligand N ^{3}H methylscopolamine (^{3}H NMS), we validated these findings and showed that in the rat, the STHA regimen induces submaxillary gland upregulation of one (of two) receptor subtype populations. Concomitantly, a remarkable decrease was observed in the affinity of the muscarinic receptor populations to the ligand (Horowitz, Kaspler, Oron; in preparation). Acute ambient heat stress for several hours neither produced muscarinic receptor upregulation nor altered the basal difference observed between preacclimation and STHA, suggesting that the changes observed during STHA are associated with chronic heat. The presence of decreased receptor affinity may explain, at least in part, decreased glandular responsiveness to the transmitter. It is noteworthy that Fujinami et al. (1991), studying the effects of STHA on adrenergic receptors (AR) in the parotid glands, showed significant transient upregulation of cell surface receptors concomitantly with desensitized activity. Desensitization was attributed to a blunted coupling between the adrenergic receptors and GTP binding protein. Data from our laboratory on the binding affinity of β AR in the heart of STHA rats are in agreement with the above reported data. (Kaspler, Gross, Horowitz, personal communication). Unfortunately, data on the effect of chronic elevated temperature on the G proteins of the muscarinic receptors are not yet available.

3. Autonomic vs cellular compensatory defense strategies during STHA

In the intact rat, STHA is characterized by hyperthermia, accompanied by body dehydration (Horowitz, 1976). During this period, accelerated activity of the salivation cooling mechanism doubles heat tolerance on the fifth day of acclimation (Horowitz et al., 1983). This effectorial system shows post-synaptic desensitization, coinciding with two membranal events: muscarinic receptor upregulation and decreased muscarinic ligand binding affinity. The occurrence of receptor upregulation coincidentally with persistent accelerated excitation during the STHA contradicts the acknowledged concept of the mechanisms underlying receptor upregulation (i.e. denervation). On the contrary, prolonged exposure to the transmitter leads to desensitization which is usually explained by either receptor downregulation or changes in receptor-G-protein interaction.

We suggest that during STHA a variety of processes, having opposing effects, interact. While thermal demands lead to accelerated persistent autonomic stimulation and in turn desensitization, the strained cell compensates for decreased affinity by receptor upregulation. Thus, in the intact body the overall response is determined by the sum of these opposing effects. Similarly, transient upregulation in βAR in the STHA rat parotid gland, observed by Fujinami et al. (1991) is possibly a compensatory response to the impaired coupling between AR and the G protein.

The uniform response of decreased transmitter binding affinity during STHA in all organs studied may imply a non organ-selective heat effect possibly leading to changes in the cell membrane. Indeed, both hyperthermia and dehydration are known as potent biological stressors, leading to physical and possibly chemical alterations in biological membranes (Cossins, 1987), expressed by decreased binding affinity. It is also noteworthy that various stress hormones, known to affect both the cell membrane and receptor density are released during the initial phase of heat acclimation (Gwossdow et al., 1985). Such hormones may play a role in the phenomena observed. However, their role have not yet been elucidated.

4. Cellular response - studies on cell line

To assess our hypothesis of a cellular strategy for coping with desensitization we developed recently a new experimental paradigm of a salivary acinar cell line (HSY) undergoing simulated acclimating conditions. Preliminary experiments on the effect of heat (with hormonal deprivation) on these cells have shown that STHA produces muscarinic receptor upregulation with decreased binding affinity particularly of the 1 receptor subtype (Kaspler, Moran, Horowitz, personal communication). These findings support our hypothesis of a cellular compensatory strategy for coping with membranal impairments.

LONG TERM HEAT ACCLIMATION - EVIDENCE FOR INCREASED EFFICIENCY AND POSSIBLE MECHANISM

LTHA is characterized by increased efficiency of the thermoregulatory organs. Mitchell et al.,(1976) showed in the human perspiratory system that acclimation to humid heat increases the cooling power of perspiratory system by augmentation of the wetted surface area (actual sweat evaporated/evaporation from fully wet skin surface) so that it exceeds 100% with initiation of heat stress and continues being equal to 100% during the sweating session. Non-acclimated humans fail to do so. Increased sweating was associated with temperature induced changes in sweat gland function. Senay et al. (1976), Wyndham et al. (1973) and Rowell et al. (1968), also in humans, reported increased cardiovascular reserve following an acclimation regimen. In rodents, we can estimate that during LTHA the salivation cost (calorie/ml) decreases by more than two fold compared to pre-acclimation. In contrast, salivation cost in STHA rodents increased (Horowitz, 1990) while the potential capacity of the gland to secrete saliva when stimulated by heat increases. Cardiac muscle of LTHA rats also exhibits similar phenomenon. Similar pressure is generated at 20% reduction in cardiac oxygen consumption (Horowitz et al., 1993).

These adaptive changes may attenuate the need for the accelerated autonomic activity observed during STHA. Evidence accumulated thus far indicates that increased effector efficiency is brought about by processes occurring at both organ and cellular levels. Both morphological (e.g. augmented sweat glands and salivary glands size, Okuda et al., 1980, Sato st al., 1990, Horowitz and Soskolne, 1978) and biochemical alterations, leading to quantitative and qualitative changes, may occur.

Quantitative vs qualitative changes in evaporative cooling organs: There are indications that at least in the salivation cooling system, thermal acclimation (LTHA) is associated with a more efficient signal transduction pathway for water secretion (Oron et al., 1991). Using ^{36}Cl as a marker for water secretion and ^{86}Rb as a marker for receptor-transmitter coupling, we showed augmented ^{36}Cl and ^{86}Rb efflux with pre-acclimation muscarinic receptor density (Kloog et al.,1985). This improved cellular performance may work in concert with increased gland size occurring with heat acclimation (Elmer and Ohlin, 1970; Horowitz and Soskolne, 1978) to improve the secretory capacity of the entire organ. Sato et al.(1990) studying long term heat acclimated patas monkey sweat glands provided evidence for increased glandular responsiveness to central nervous system stimulation and more efficient functioning of the gland because of greater sweating per unit length of the secretory tubule. The dose-response curve of sweat rate to methylcholine before and after acclimation, however, did not show increased sensitivity to the neurotransmitter. In contrast, increased sensitivity was observed in heat acclimated humans (Chen and Elizondo, 1974; Sato and Sato, 1983). Data on post synaptic heat-acclimation induced cellular changes in the evaporative cooling system are sparse. Nevertheless, these data contribute to the understanding of the mechanisms underlying the increased [effector output]/[autonomic signal] ratio at the cellular level on LTHA.

Long term heat acclimation changes in the cardiovascular system: Our data suggest that there are similar effects in the adaptation of the cardiovascular system. For example, our data on splanchnic blood flow are discussed. Following LTHA, splanchnic blood flow, the major circulatory reservoir for thermoregulatory skin blood flow, is maintained at an elevated level for a longer period (as opposed to immediate vasoconstriction prior to acclimation) despite skin vasodilatation (Horowitz and Samueloff, 1988, Haddad and Horowitz, in preparation). This significant adaptation allows better temperature regulation of the splanchnic organ and in turn would abolish excessive elevation of metabolic rate in this organ due to the vant Hoff effect. This feature is apparently associated with the increased endurance of the heat stressed acclimated animals (Horowitz and Samueloff, 1988; Shochina et al., 1990). In an attempt to examine our hypothesis on the role played by the target organs as modifiers of the autonomic discharge we studied isolated blood vessels excised from acclimated rats and the splanchnic blood flow of acclimated heat stressed innervated and denervated rats (Haddad, 1991; Horowitz et al., 1993; Haddad and Horowitz personal communication). The data provide evidence that heat acclimated aortic rings can generate greater pressure than vessels excised from non acclimated animals either by depression of β AR activity or by increased efficiency of the α AR signal transduction pathway in young and aged rats respectively. We also found non adrenergic vasodilating component in this vascular bed. These mechanisms, brought into play in the intact, heat stressed animal despite sympathetic innervation, suggest that in this organ too, post synaptic changes in cellular responses work in concert to produce increased organ responsiveness, namely greater vasoconstriction once Tsh for vasoconstriction has achieved.

CONCLUDING REMARKS

Our data, together with data accumulated by other investigators provide evidence for temporal varying central - peripheral interaction with heat acclimation. Cellular responsiveness seem to play a role in this mode of interaction. During STHA impaired cellular responsiveness superimposed on the thermoregulatory demands lead to accelerated autonomic excitation compared to pre-acclimation conditions. During LTHA it is the increased organ efficiency that decreases the need for extra-accelerated autonomic signals. This hypothesis is illustrated in Fig. 3.

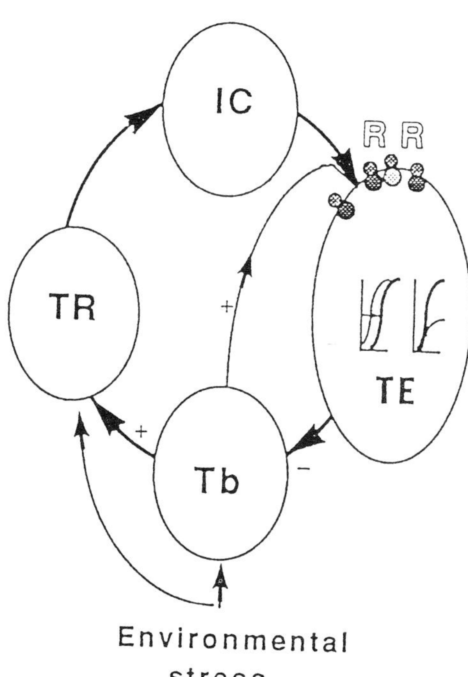

Fig. 3. Central-peripheral interaction during heat acclimation. During STHA an abrupt rise in body temperature lead to impaired membrane function (decreased transmitter binding affinity) and autonomic accelerated excitation. During LTHA increased effector efficiency (due to-long term heat effect on organ function) reduces the need for accelerated autonomic dischatge. IC-Integration center, Tb-body temperature, TE-Thermoregulatory effector, TR-Thermal receptors. R-receptor population.

REFERENCES

Chen, W. Y. & Elizondo R.S. (1974) Peripheral modification of thermoregulatory function during heat acclimatization. J. Appl. Physiol. 37, 367-373.

Cossins, A.R. & Bowler, K. (1987) Temperature biology of animals. Chapman & Hall, London.

Fujinami, H., Komabayshi, T., Izawa, T., Nakamura, T., Suda, K., Tsuboi,M. (1991) Adaptive control of beta-adrenoceptors and adenylate cyclase during short term heat exposure in rat parotid glands. Comp Biochem. Physiol. 98C, 411-416.

Gwosdow, A.R., Besh, E.L., Chen, C.L. (1985) Aclimation of rats following stepwise or direct exposure to heat. J. Appl Physiol. 59, 408-412.

Haddad W. (1992) Acclimation to heat: Blood vessels responsiveness to adrenergic stimulation. M.Sc. Thesis, submitted to the Hebrew University.

Horowitz, M. (1976) Acclimatization of rats to mild heat: Body water distribution and adaptability of submaxillary salivary gland. Pflugers Arch. 366, 173-176.

Horowitz, M. (1989) Heat acclimation: A continuum of processes. In: Thermal Physiology. ed. J. Mercer, pp, 445-450, Elsevier.

Horowitz, M. (1990) Prolonged exposure to heat stress: acquired physiological adaptations - cost and benefits. Archives of complex environmental studies. 2, 11-14

Horowitz, M. (1991) Autonomic and intrinsic mechanisms of heat acclimation as delineated from studies on thermoregulatory effectors. in:Advance in Physiological Science. ed. S.K. Manchanda, pp.352-357, Macmilan India Lmt.

Horowitz, M. & Soskolne, W.A. (1978) The cellular dynamics of the rat's submaxillary salivary gland during heat acclimation. J. Appl. Physiol. 44, 21-24.

Horowitz, M. & Meiri, U. (1985) Altered responsiveness to parasympathetic activation of submaxillary salivary gland in the heat acclimated rat. Comp. Biochem. Physiol. 80A, 57-60.

Horowitz, M. & Samueloff, S. (1988) Cardiac output distribution in thermally dehydrated rodents. Am. J. Physiol. 254, R109-R116.

Horowitz, M. & Meiri, U. (1993) Central and peripheral contributions to control of heart rate during heat acclimation. Pfl. Arch. 422, 386-392.

Horowitz, M., Argov, D. & Mizrahi, R. (1983) Interrelationships between heat acclimation and salivary cooling mechanism in conscious rats. Comp. biochem. Physiol. 74A, 945-949.

Kloog, Y., Horowitz, M., Meiri, U., Galron, R., Avron, A. (1985) Regulation of submaxillary gland muscarinic receptors during heat acclimation. Bioch et Biophys. Acta, 845, 428-435.

Mitchell, D., Senay, L.C., Wyndham. C.H. (1976) Acclimatization in a hot, humid environment: energy exchange, body temperature, and sweating. J. Appl. Physiol. 40, 768-778.

Okuda, N., Kanani, M., Watari, N., Ohara, K. (1980) Morphological changes of the eccrine sweat glands of Japanese monkey after heat acclimation: the mechanisms of peripheral adaptation. In Contributions th Thermal Physiology. eds. Szeleny, Z. and Szekely, M. New York, Pergamon pp.293-295.

Oron, Y., Falach, O., Marmari, I. & Horowitz, M. (1989) Modulation in Rb and Cl efflux in submaxillary gland slices with the course of heat acclimation. Comp. Biochem. Physiol. 94A, 673-676.

Rowell, L.B., Kraning II, K.K., Kennedy, J.W. Evans, T.O. (1967) Central circulatory responses to work in dry heat before and after

acclimatization. J. Appl. Physiol. 22, 509-518.
Sato, F. Owen, M., Matthes, R., Sato, K., Gisolfi, C.V. (1990) Functional and morphological changes in the eccrine sweat gland with & heat acclimation. J. Appl. Physiol. 69, 232-236.
Sato, K., & Sato F. (1983) Individual variations in structure and function of human eccrine sweat gland. Am. J. Physiol. 245, R203-R208.
Shochina, M. & Horowitz, M. (1989) Central venous pressure, arterial pressure and hypovolemia: their role in adjustment during heat stress. J. Therm. Biol. 14, 109-113.
Wyndham, C.H., Rogers, G.G., Senay, L.C., Mitchell, D. (1976) Acclimatization in a hot humid environment: cardiovascular adjustments. J. Appl. Physiol. 40, 779-785.

Integrative and cellular aspects of autonomic functions : temperature and osmoregulation. Eds K. Pleschka, R. Gerstberger. John Libbey Eurotext, Paris © 1994, pp. 97-102.

Interaction of acclimation and fever

Udo Beckmann and Jürgen Werner

Institut für Physiologie, Ruhr-Universität Bochum, MA 4/59, D-44780 Bochum, Germany

Long-term thermo-adaptive modifications enable the organism to cope with environmental stress more efficiently. These consist on the one hand of morphological adaptations which primarily modify the passive system and, on the other hand, of functional adaptations which may be present on the receptor, the controller and/or the effector level, i.e. we are dealing with an adjustment of the whole control loop (see e.g. Brück, 1990, Werner, 1988, 1990).
Fever, too, is rather an adjustment than a disturbance of the system, usually explained by the concept of a set-point shift (see e.g. Simon & Riedel, 1982, Graener & Werner, 1986). Both processes, acclimation and fever, interfere dramatically with thermoregulatory control.

The questions arise as to how far acclimation interferes with the febrile process and whether preceding acclimation may be suited to limiting a higher body temperature increase during fever.

1. METHODS

The interrelations were investigated by studies on
1) thermoneutrally adapted rabbits (8 weeks at T_a=22°C, n=13)
2) cold adapted rabbits (8 weeks at T_a=2°C, n=20)
3) heat adapted rabbits (8 weeks at T_a=30°C, n=20).
The progress of the adaptation process was studied once a week under resting conditions at an ambient temperature equal to that of the adaptation procedure. The fever course after injection of an exogenous pyrogen (LPS, 0.5 µg/kg i.v.) was investigated both before and after adaptation in a climatic chamber at ambient temperatures (T_a) of 22°C, 2°C and 30°C. Half of the cold and heat adapted groups were taken for control experiments with an injection of isotonic NaCl-solution (n=10, each).

We recorded rectal temperature (T_{rect}), mean skin temperature and ear temperature ($T_{s\ ear}$), metabolic heat production (M), respiratory evaporative heat loss (REHL), heart frequency, respiratory frequency (F_{resp}), tidal volume, EMG, and ear blood flow.

2. COLD ACCLIMATION

Cold acclimation evoked a considerable improvement in fur insulation as could be seen from an increase of hair mass of a standard area by about 30 % and a significant decrease of fur conductance. Mean skin temperature rose by about 1.5°C, whereas rectal temperature remained unchanged. During the adaptation process, metabolic heat production at T_a=2°C was decreased by 10 %, i.e. the same core temperature was obtained at a lower energy cost. There were no increases of M or T_{rect} after injection of noradrenaline (0.2 mg/kg i.m. at T_a=22°C), and no brown adipose tissue could be found by random autopsies. So, in the cold adapted rabbits, no development of non-shivering thermogenesis could be detected. F_{resp} was decreased, accompanied by a slight elevation of tidal volume, resulting in a reduction of REHL by about 30 %. $T_{s\ ear}$ was decreased by nearly 2°C during cold adaptation, leading to a reduction of ear convective heat loss of about 40 %.

2.1 Cold acclimation and fever

2.1.1 Fever in cold environment after cold acclimation.

At T_a=2°C, primarily the local fever minimum of the biphasic core temperature course was essentially lowered by cold acclimation. In addition, the second maximum was slightly lower and delayed. In the cold, metabolic heat production is the dominant febrile effector. It was lowered after cold adaptation, but the increase due to fever was unchanged. The strong decrease after the first maximum, even below the level of afebrile values before injection of pyrogens, and the long lasting rise to the second maximum explain the behaviour of core temperature. The maximum of core temperature was hardly lowered, but the metabolic cost in fever was reduced by the same amount as seen in the afebrile control group. The decline of heat production after the first maximum was initiated at a lower core temperature. The same holds for the rise to the second maximum. This suggests an effect on fever by a decrease in the shivering threshold developed during cold adaptation. Only during fever was the effector characteristic of metabolic rate shifted to a lower mean body temperature. In the afebrile control group, primarily a downward shift of the characteristic of the passive system could be observed, due to the improvement of body shell insulation as a morphological effect of cold adaptation.

2.1.2 Fever in a neutral environment after cold adaptation

In thermoneutral environments both maxima of the biphasic core temperature course were decreased after cold adaptation and the local minimum was prolonged. The same characteristic changes could be seen in the time course of metabolic heat production, due to the shift of the effector characteristic.

After cold adaptation the respiratory frequency was lowered by nearly 50/min. For the febrile process in a neutral environment, variations of respiratory evaporative heat loss were of nearly no significance in the cold adapted rabbits. Respiratory frequency was already reduced in the afebrile status to an extent which did not allow substantial reductions during the phases of febrile temperature rise. Actually, a decrease of the threshold for an elevated F_{resp} as a compensation for the advanced body heat accumulation after cold adaptation might have been expected, leading to the panting threshold in fever decline at T_a=22°C being exceeded. However, adaptive effects developed in the cold do not seem to comprise functional mechanisms which may be advantageous in environments different from the adaptation climate.

2.1.3 Fever in the heat after cold acclimation

In fever at high ambient temperature, a large overshoot in the rise of T_{rect} and an elevated load-error after cold adaptation could be observed, sometimes leading to severe hyperthermia. The morphological adjustments during cold adaptation promote heat accumulation, an effect which is obviously maladaptive in the heat. The high core temperature caused a strong circulatory load as indicated by the heart frequency which in comparison to the neutrally adapted group increased by about 50/min. Metabolic heat production remained unchanged with the exception of a slightly amplified motor activity of the animals due to extreme hyperthermia.
During cold acclimation, heat loss mechanisms adapted in a way which, in climatic conditions inverse to the adaptation conditions, were clearly detrimental to the efficiency of control. The effector characteristic of respiratory frequency was shifted to a higher mean body temperature. This caused the panting threshold to be exceeded in the decline phase of fever at a T_{rect} elevated by 0.5°C. The start of panting was not changed in time, showing that the time course of effector control in the fever process was unaffected by adaptation. Furthermore, it was found that during decline of fever, i.e. in a situation requiring maximal vasodilatation, ear blood flow was not markedly elevated, as compared to the afebrile values before injection of pyrogens. The temperature difference between body core and ear at this time was 0.8°C greater than in the thermoneutrally adapted animals, showing that maximum ear blood flow was reduced after cold adaptation. Taking into account the findings of a simultanously reduced minimum ear blood flow during vasoconstriction in a cold environment, this hints at the assumption that, as a morphological adjustment, alterations of the vascular network of the ears took place during cold adaptation, leading to a reduced heat transfer. A functional adaptation such as shift of threshold or change of gain for vasomotor control could not be observed.

2.1.4 Effects on the total fever height

A global survey of the influence of cold acclimation on the total extent of temperature elevation in fever is given by the fever index, an integration of core temperature increase for a given time period after injection of pyrogens. The fever index for 300 minutes (FI_{300}) decreased significantly after cold adaptation for fever in a neutral or cold environment. This holds in comparison both to the same group before adaptation and to the thermoneutrally adapted control group. In a warm environment, however, FI_{300} was higher after cold adaptation. In a cold and neutral environment, where metabolic heat production is the dominant effector for the fever process, the effector activities during fever are adjusted to the properties of the passive system altered by cold adaptation, an effect which is evidently absent in a warm environment.

3. HEAT ACCLIMATION

In a second study, the influence of heat adaptation was investigated. Regulatory parameters measured in one fever-group and one afebrile control-group of heat adapted rabbits were compared to those of rabbits adapted to thermoneutral conditions. In the course of the adaptation process, there were no changes in hair mass, mean skin temperature, or thermal conductance of the fur, which would indicate a change of body shell insulation. Heat adaptation resulted in a significant decrease of rectal temperature and mean body temperature of about 0.4°C. No changes in metabolic rate, ear blood flow, and ear convective heat loss could be observed at T_a=30°C. Respiratory frequency decreased with unchanged tidal volume, resulting in a significant decrease of respiratory evaporative heat loss.

3.1 Heat acclimation and fever

For all environmental conditions applied, the fever index FI_{300} was significantly lower after acclimation than before. In the cold it even had negative values. In neutral and cold environments the differences are also significant in comparison to the non-adapted febrile control group.

3.1.1 Fever in the heat after heat acclimation

In a hot environment, rectal temperature of control as well as of febrile animals was about 0.5°C lower after heat adaptation. After both temperature maxima of the biphasic fever course, there was an accelerated temperature decline, and the total load error was minimized, so that an enhanced efficiency of heat loss mechanisms has to be concluded. Yet the maximum values of F_{resp} were decreased by 80/min, REHL by 0.5 W/kg, and in the effector characteristic of F_{resp} only a slight downward shift of threshold could be detected. Above all, the maximum panting values were lowered. The strong reduction of respiratory frequency during panting brings about an essential conservation of water. The reduction of total respiratory water loss by 50 % (15 ml) should already be an important reduction of the osmoregulatory load. At maximal ear vasodilatation, the temperature difference between body core and ear was unchanged after heat adaptation, but at all ambient temperatures, there was, on average, a higher ear skin temperature and a higher ear blood flow at the same body temperature.

3.1.2 Fever in neutral and cold environment after heat acclimation

In a thermoneutral environment and also in the cold, rectal temperature was significantly lowered in control and in febrile heat adapted rabbits. The time course of the biphasic fever process was changed, and fever duration was shorter. These effects could be attributed to a decrease of metabolic rate by 0.5 W/kg on average at T_a=22°C, (at T_a=2°C by about 1.5 W/kg). The threshold for an elevation of heat production was shifted to a lower body temperature, and the gain of the effector characteristic was smaller after heat adaptation. In the onset phase of fever, the start of shivering was significantly delayed. In fever phases with non-elevated heat production, metabolic heat production was lowered below resting values, causing steeper temperature declines after both fever maxima at thermoneutral temperature.

In the cold, even during fever, rectal temperature fell far below its resting values in the heat adapted rabbits. The minimum ear blood flow during vasoconstriction was elevated, resulting in an elevation of ear skin temperature of 1.3°C and a nearly doubled convective heat loss. These results indicate an enhanced vascularization of the ears as an effect of heat adaptation.

Similarly to the cold adapted rabbits at high ambient temperature, the regulatory processes in fever in this case were not adjusted to compensate for the large occurring load-error caused by adaptive effects.

4. FINAL CONCLUSIONS

As far as cold adaptation is concerned, it can be concluded that in thermoneutral and cold environments, where metabolic heat production is the major fever effector, the regulation in the fever process is well adjusted to the morphological adaptive changes, and the metabolic costs of fever are reduced. Furthermore, the functional controller adjustments of cold adaptation, concerning the effector control of metabolic heat production, can attenuate the fever-induced elevation of core temperature. Whereas, in the heat, cold acclimation favours the development of a fatally high increase of temperature, so in a warm environment, cold adaptation turns out to be maladaptive during the fever process.

After heat adaptation, the total level of temperature elevation in fever is significantly reduced at all ambient temperatures tested. This effect can be attributed to both a shift in the effector characteristic of metabolic heat production and an enhanced efficiency of heat loss mechanisms, which reduce the regulatory load-error during the dynamic fever process at high ambient temperatures. So in conclusion, it can be stated that heat adaptation is best suited to avoiding higher febrile temperatures, however, cold ambient temperatures may encounter hypothermic problems even during fever.

Supported by Deutsche Forschungsgemeinschaft (We 919/1).

5. REFERENCES

Brück, K. (1990): Long-term and short-term adaptive phenomena in temperature regulation. In *Thermoreception and Temperature Regulation*, eds. J. Bligh and K. Voigt, pp. 211-223. Heidelberg: Springer-Verlag.

Graener , R. & Werner, J. (1986): Dynamics of endotoxin fever in the rabbit. *J. Appl. Physiol.* 60, 1504-1510.

Simon, E. & Riedel, W. (1982): Pathophysiologie des Fiebers. *Therapiewoche* 32, 1418-1444.

Werner, J. (1988): Functional mechanisms of temperaure regulation, adaptation, and fever: Complementary system theoretical and experimental evidence. *Pharmac. Ther.* 37, 1-23.

Werner, J. (1990): Models of cold and warm adaptation. In *Thermoreception and Temperature Regulation*, eds. J. Bligh and K. Voigt, pp. 224-228. Heidelberg: Springer-Verlag.

Integrative and cellular aspects of autonomic functions : temperature and osmoregulation. Eds K. Pleschka, R. Gerstberger. John Libbey Eurotext, Paris © 1994, pp. 103-112.

Mechanisms of heat acclimatization due to thermal sweating. Comparison of heat-tolerance between Japanese and Thai subjects

Mitsuo Kosaka[(1)], Takaaki Matsumoto[(1)], Masaki Yamauchi[(1)], Katsuhiko Tsuchiya[(1)], Nobu Ohwatari[(1)], Masakatsu Motomura[(1)], Kinuyo Otomasu[(1)], Guo-Jie Yang[(1)], Jia-Ming Lee[(1)], Udom Boonayathap[(2)], Chucheep Praputpittaya[(2)] and Anchalee Yongsiri[(2)]

[(1)] *Department of Environmental Physiology, Institute of Tropical Medicine, Nagasaki University, 1-12-4 Sakamoto, Nagasaki 852, Japan*
[(2)] *Department of Physiology, Faculty of Medicine, Chiang Mai University, Chiang Mai 50002, Thailand*

Abstract: Heat tolerance and sweat response to heat load of tropical subjects in Chiang Mai and temperate subjects in Nagasaki were compared under identical conditions. Male students in Chiang Mai(n=10) and in Nagasaki(n=10) volunteered for this study. The Thai subjects were a little shorter and slightly leaner than the Japanese. Heat load was applied on the legs by immersion into hot water(43℃) for 30 min in the room at 26.6℃ and 33%rh. Sublingual(oral) temperature was measured with a thermistor probe and local sweat rate was measured by the capacitance hygrometer-sweat capsule method. Change in oral temperature, sweat onset time and local sweat volume were compared between Japanese and Thai. Initial oral temperatures (36.76±0.11℃ in Japanese, 36.71±0.23℃ in Thai) were identical, and no sweat was observed before heat load. Mean sweat onset time (9.3±2.1 min chest in Japanese, 16.6±5.6 min chest in Thai) were significantly longer and local sweat volume (10.19±5.00 mg/cm^2, chest in Japanese, 1.39±0.91 mg/cm^2, chest in Thai) was significantly smaller in Thai subjects than Japanese, however, oral temperature (37.18±0.32℃) of Thai subjects was kept slightly lower than oral temperature (37.42±0.10℃) of Japanese even under a 30 min heat load. Sweat volume on the abdomen was larger than on the chest in 9 of 10 Thai subjects. On the contrary, sweat volume on the chest was larger than that on the abdomen in 7 of 10 Japaese subjects. These results suggest that heat tolerance of tropical subjects is due to a more efficient evaporative ability due to a greater heat loss brought about by their long term exposure to heat. Furthermore, the habituation phenomenon related to the reduction of thermoregulatory effector mechanisms were also considered so as to clarify the mechanisms of thermal acclimatization.

INTRODUCTION

Temperature regulation of the human body is affected in different ways by climatic environments such as air temperature, humidity and air velocity. In "Human Perspiration" , Kuno (1956) described that adaptation to temporary exposure to heat and acclimatization to a tropical climate by permanent residence were distinguishable from each other. Transitory heat acclimation by repeated exposure to heat has been intensively investigated by many researchers. Up to now, it is generally accepted that tropical natives begin to sweat more slowly than temperate natives and the salt concentration in sweat in the former is much lower than in the latter (Kuno, 1956; Yoshimura, 1960; Hori et al., 1976; Ohwatari et al., 1983; Sasaki and Tsuzuki, 1984; Fan, 1987, Matsumoto et al., 1991). We have studied sweat responses to heat load in Japanese and Thai subjects under identical, thermo-neutral conditions, obtaining the preliminary results, which support the previous findings. It was speculated that the slow rise in oral temperature in tropical subjects might be attributed to a lower sweat rate and an increase of dry heat loss through conduction and convection. (Matsumoto et al., 1991). As far as the autonomic nervous control of reduction of thermoregulatory effector gain (Kosaka et al., 1988,1989, Ohwatari et al.,1992) is concerned, most theories consider that central and peripheral temperature signals interact at the level of the preoptic area and the anterior hypothalamus(PO/AH) in the diencephalon, of which the effector mechanism of thermoregulation indicated little or no increase of cerebral blood flow in thermally acclimated animals (habituation phenomenon). In this paper, therefore, precise measurement and analysis of sweat onset time as well as sweat volume due to heat load were carried in order to clarify the mechanism of long term heat acclimatization.

MATERIALS AND METHODS

Twenty healthy male students, 18-21 years old, 10 Japanese from Nagasaki (32° 44'N, 129° 52'E), Japan and 10 Thai from Chiang Mai (18° 47'N, 98° 59'E), Thailand volunteered for this study (Fig.1.A, Table 1). Nagasaki is located in a temperate zone, with hot summers and cold winters, while Chiang Mai is located in a tropical zone, with dry- and wet-seasons. Mean annual ambient temperature is 16.6℃ and 25.9℃ in Nagasaki and Chiang Mai, respectively (Fig.1.B). The experiments for Thai subjects were carried out from January to February in 1991 in Chiang Mai, and those for the Japanese subjects from January to February in 1992 in Nagasaki. All experiments were performed between 02:00-04:00 p.m. to avoid the influence of circadian variation. Experimental procedures were almost the same as the previous paper described by Matsumoto et al., (1991). Namely, each subject wore only shorts and was seated on a chair in the experimental room. The air temperature and relative humidity in the experimental room in Chiang Mai were 26.6±2.0℃ and 33.0±5.1%, respectively. In Nagasaki, the same experimental conditions as in Chiang Mai were simulated in an environmentally-controlled chamber. After staying at rest in the experimental room for at least 30 min, heat load was applied by immersing the lower legs in a hot water bath (43℃) for 30 min. After cessation of the heat load, the subject was allowed to sit in the same condition for a further 30 min.

Fig.1.(A)
A map of Chiang Mai where locates 18°47'N and 98°39'E.(Nagasaki 32°44'N and 129°52'E). Mean annual ambient temperature is 26.6℃ and 16.6℃ in Chiang Mai and Nagasaki, respectively.

Fig.1.(B) Mean monthly ambient temperatures in Nagasaki (32°44'N, 129°52'E, Japan) and in Chiang Mai (18°47'N, 98°59'E, Thailand). The experiments were carried out in the hottest months in both places, from August to September in Nagasaki and from March to April in Chiang Mai.

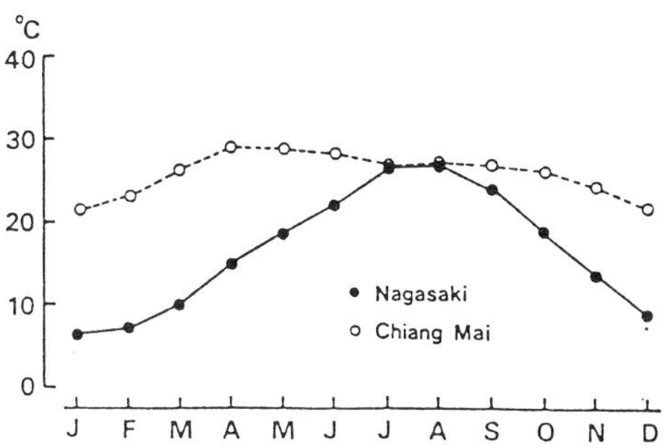

Table 1. Physical characteristics of the subjects

	n	Age. years	Height. cm	Weight. kg
Japanese in Nagasaki	10	20.3±0.6	174.2±4.0	63.7±6.3
Thai in Chiang Mai	10	20.8±0.9	169.7±3.9	58.1±5.9

Values are mean±SD. There was no significant difference between two groups, except for weight. ($P<0.05$)

Oral temperature was measured with a thermistor probe (Model 2100, YSI), placed into the sublingual space before, during and after heat load every 5 min in Chiang Mai.

In Nagasaki, oral and tympanic temperatures as well as skin temperatures on the chest, forearm, thigh and lower leg were measured by thermistor probes (K923, TAKARA Instruments Co.) connected to a personal computer (PC-8801, NEC). Mean skin temperature was calculated according to Ramanathan's formula (Ramanathan, 1964). Furthermore, local sweat rates on the chest and the abdomen were continuously recorded by capacitance hygrometer-sweat capture capsule method (Fan, 1987; Matsumoto et al., 1988, 1989, 1991). Briefly, dry nitrogen gas flowed into the capsule ($10.18cm^2$) attached to the skin with a constant flow rate of 1ℓ/min, and the change of relative humidity of effluent gas was detected by a hygrometer (H211, TAKARA Instruments Co.) connected to a DC-pen-recorder.

Statistical significance was assessed by Student's t-test at 0.05 level and the values were presented as mean±SE or mean ± SD.

RESULTS

The physical characteristics of the subjects were well matched as shown in Table 1. Mean values of age, height and weight in the Japanese subjects were slightly different compared to those in the Thai subjects, however, the differences were not significant except for weight ($p<0.05$). These results are in agreement with Hori et al.(1977), who measured physical characteristics and basal metabolism in 30 young male Thai and 20 young male Japanese, and reported that Thai subjects were a little shorter and more slender (a smaller skinfold thickness) than Japanese. Also, Thai subjects had a slightly lower basal metabolism per unit body surface reported by Matsumoto et al.(1991).

Comparison of changes in oral temperature during and after heat load in Japanese and Thai subjects were demonstrated in Table 2. Mean initial oral temperature at ambient temperature of 26.6℃ and 33% of relative humidity before heat load was 36.76±0.11℃ (mean±SD) in the Japanese and 36.71±0.23℃ in the Thai subjects (NS). After 30 min application of heat load, oral temperature rose and reached 37.42±0.10℃ in the Japanese and 37.18±0.32℃ in the Thai subjects (NS). However oral temperature of Thai subjects were kept slightly lower than oral temperature of Japanese. No significant difference in the oral temperature was found between the two groups throughout the whole experiment. Threshold oral

Table 2. Oral temperature before and after heat load

	before	at 30 min
Japanese (Nagasaki)	36.76±0.1	37.42±0.10 (0.70±0.15)
Thai (Chiang Mai)	36.71±0.23	37.18±0.32 (0.47±0.31) ℃
	N.S.	N.S.

(Mean±SD)

temperatures (including temperature differences from the initial oral temperature) for sweating were summerized in the Table 3. Both at the chest and abdomen, threshold temperatures for sweating in Chiang Mai subjects were higher than those of Japanese subjects which indicate stronger heat tolerance in Thai subjects compared to Japanese subjects.

A typical recording of a Japanese subject in Nagasaki is shown in Fig.2. At 26.6℃ and 33% of relative humidity no sweating was observed. Tympanic temperature followed oral temperature well, although tympanic temperature was slightly higher

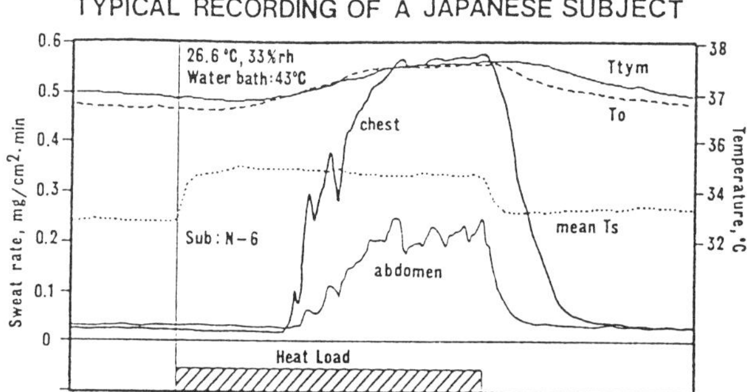

Fig. 2. A typical recording of a Japanese subject in Nagasaki, oral and tympanic temperatures as well as mean skin temperature colmulated by a personal computer and local sweat rates on the chest and abdomen were recorded before during and after heat load by immersion of the lower legs in a hot water bath (43℃). Sub: N-6. No. 6 subject in Nagasaki Ttym: tympanic temperature, To: mean oral temperature Ts: mean skin temperature calculated, of the chest, forearm, thigh and leg, after Ramamathan 1964

Table 3. Threshold oral temperature for sweating

	on the chest	on the abdomen
Japanese (Nagasaki)	36.84±0.09 ($\triangle$T=0.12±0.06)	36.9±0.09 ($\triangle$T=0.17±0.06)
Thai (Chiang Mai)	25.87±0.27 ($\triangle$T=0.27±0.29)*	36.92±0.27 ($\triangle$T=0.21±0.31)*
	N.S. P<0.05*	N.S. N.S.*

(Mean±SD)

than oral temperature before and after the application of heat load on the legs. In the previous paper, local sweat rates on the chest and the abdomen were transiently suppressed by the application of heat load on the legs, and then markedly increased in 9 of 10 Japanese subjects (Matsumoto et al., 1991). However, the transient suppression of sweating by heat load was not induced in this case. Generally, sweat volume on the chest was larger than that on the abdomen in 7 of 10 Japanese subjects. A typical recording of a Thai subject in Chiang Mai is demonstrated in Fig.3. Mean sweat onset time (16.6±5.6 min at the chest) in Thai was significantly longer than the value (9.3±2.1 min) in Japanese as shown in Table 4. Local sweat volume (1.39±0.91 mg/cm^2, at the chest) in Thai was significantly less than the value (10.19±5.00 mg/cm^2 at the chest) in Japanese, as shown in Table 5. and Fig 5.

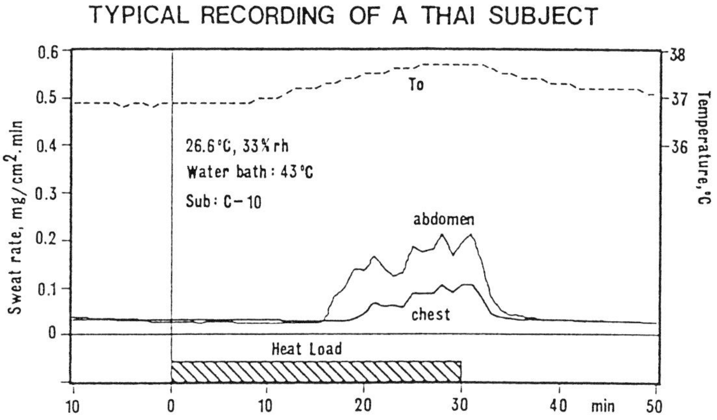

TYPICAL RECORDING OF A THAI SUBJECT

Fig. 3. A typical recording of a Thai subject in Chiang Mai. Oral temperture and local sweat rate on the chest and abdomen were recorded before, during and after heat load by immersion of the lower legs in a hot water bath (43℃).
Sub C-10: No. 10 Subject in Chiang Mai
To: oral temperature

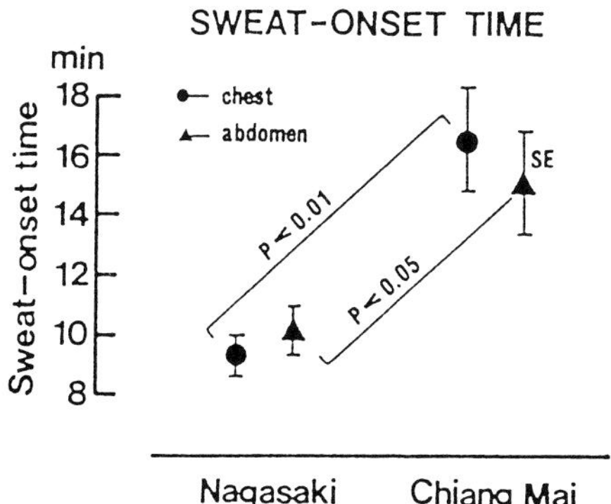

Fig. 4. Comparison of sweat-onset time between Nagasaki and Chiang Mai subjects. (Mean±SE)

Although sweat volume both on the chest and abdomen in Chiang Mai subjects were less than those in Japanese subjects, sweat volume on the abdomen was larger than sweat volume on the chest in 9 of 10 Thai subjects as shown in Fig.6. The central and peripheral mechanism of sweat volume dissociation between the chest and abdomen observed in Chiang Mai subjects(Fig.6) is discussed below. Sweat onset time between chest and abdomen was well correctated for both Japanese and Thai subjects, however, the longer sweat onset time of Thai subjects compared to Japanese subjects and distribution difference of sweat onset time between Japanese and Thai (Fig.7) is also discussed below.

Table 4. Sweat onset time

	chest	abdomen
Japanese (Nagasaki)	9.3±2.1	10.2±2.4
Thai (Chiang Mai)	16.6±5.6	15.1±5.7 (min)
	P<0.01	P<0.05

(Mean±SD)

Table 5. Local sweat volume

	chest	abdomen
Japanese (Nagasaki)	10.19±5.00	6.86±4.16
Thai (Chiang Mai)	1.39±0.91	2.37±1.22 (mg/cm²)
	P<0.01	P<0.01

(Mean±SD)

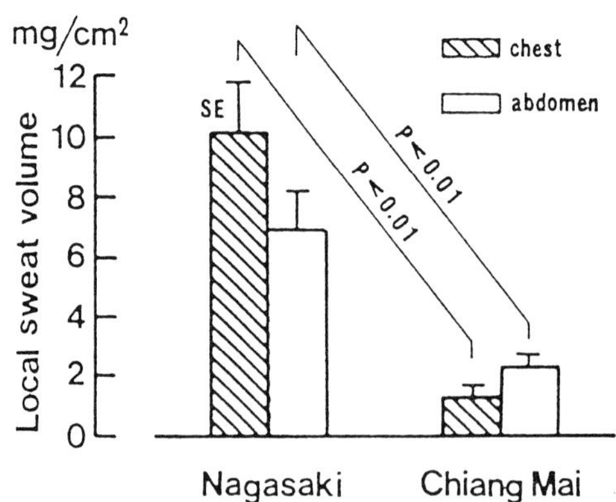

Fig. 5. Comparison of local sweat volume between Nagasaki and Chiang Mai subjects. (Mean±SE)

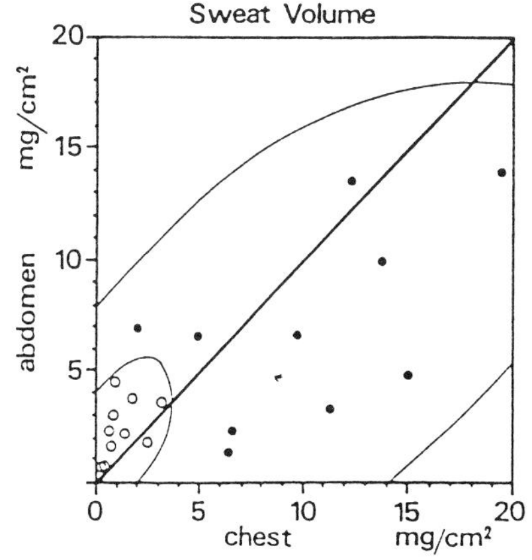

Fig.6. Correlation of sweat volume between chest and abdomen, and dissociation of sweat volume between Japanese and Thai.

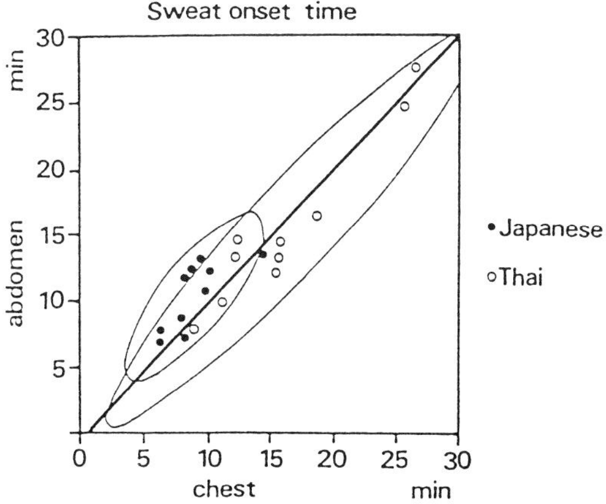

Fig.7. Correlation of sweat onset time between chest and abdomen and dissociation of sweat onset time between Japanese and Thai.

DISCUSSION

The inhabitants in Chiang Mai are expected to be more acclimatized to heat compared to those in Nagasaki, which is located in a temperate zone. In Chiang Mai, mean monthly ambient temperatures are above 20℃ throughout the year, and mean annual ambient temperature is 25.9℃. In order to clarify the mechanisms of long term heat acclimatization to tropical climates, in the present investigation, changes in oral temperature due to 30 min heat load on the legs were compared between the subjects in Chiang Mai (37.18$\pm$0.32℃) and those in Nagasaki (37.42$\pm$0.10℃). However, contrary to expectation, no significant difference was statistically observed at least in the oral temperature between the two groups of subjects.

Hori et al.(1976) and Sasaki and Tsuzuki(1984) studied sweat respones to heat load on the legs in residents of Okinawa (subtropical zone) and reported no significant difference in the rise of core temperature between residents in a subtropical zone and those in a temperate zone. Mean oral temperature measured under basal conditions (at 28℃) in 30 Thai subjects was identical to that in 20 Japanese subjects (Hori et al., 1977). Those resuts were in agreement with our results obtained in the previous study (Matsumoto et al., 1991)

On the other hand, Wyndham et al. (1964) reported conflicting results stating that rectal temperatures before and during heat load in Bantu were lower than those in Causasians; this agrees in part with the results of our study (see Table 2). Their experiments were carried out in winter, while our previous experiments were performed during the hottest month of the year (Matsumoto et al., 1991). Since we have fairly hot summers in Nagasaki, the Japanese subjects in this study are considered to be acclimatized to heat (Matsumoto et al., 1990, 1991). Hori et al., (1976) reported smaller seasonal variations of sweat rate in subtropical residents compared to temperate residents. Therefore, in the present study, experiments were performed from January to February. As we cannot fully explain the discrepancy between our results and Wyndham's, this may be a possible explanation.

In the present study, we selected subjects about 20 year old subjects. Kuno (1956) suggested that natives in the torrid zones acquire the ability to avoid excessive sweating by acclimatization. For settlers of less than 3 years, the sweat reflex is similar to that of newcomers. It has been suggested that more than 6 years of residence in the tropics is necessary to acquire the same capacity as the natives (Morimoto, 1978).

Hori et al., (1977) reported anthropometric data, body temperature and basal metabolic rates in Thai subjects as compared to Japanese people. He concluded that the body shape in Thai people is considered to be more suitable for heat dissipation in hot environments. There is no doubt that the functional, behavioural and anthropometric characteristics of tropical people are better adapted in their capacity to maintain homeostasis in a hot environment.

Regarding the longer sweating onset time of Thai subjects (see Fig.4. Table 4.) as compared to Japanese subjects. The following explanations are possible: a) a shift of the threshold core temperature of sweating onset time, b) a setting of the core temperature to a lower level, and c) a decline in the level of the rising curve of the core temperature during heat load. The factors which determine the set-point of body temperature are lower basal metabolism related to heat production and the physical constitution related to heat dissipation of tropical inhabitants (Hammel et al., 1963).

The generalized outbreak of sweating (Kuno, 1953) was confirmed not only on the chest and abdominal skin but also on the face, neck, back and extremities of our subjects. However, a phenomenon of sweat volume dissociation between the chest and abdomen was observed (see Fig. 2, 3, 6). Although the origin and thermoregulatory mechanism of sweat dissociation between the chest and abdomen is not clear, among others, the following factors have to be considered: a) regional differentiation of sympathetic efferents on the chest and abdominal skin blood flow during thermal stimulation (Iriki, 1983), b) active vasoconstriction of abdominal skin vessel, c) differences in thickness of subcutaneous fatty layers between the chest and abdomen, d) the effect of skin pressure reflex caused by sitting position (Takagi and Sakurai, 1950), e) differences of concentration and distribution of active sweat glands between the chest and abdominal skin, f) air flow of the environmental control chamber, g) the phenomenon of hidromeiosis due to exhaustion of sweat gland activity (Brown and Sargent, 1965; Ogawa, 1987), h) after heat load off, an increase in evaporative heat loss capacity due to the decrease of sweat rate on the abdominal skin. So far as the mechanisms of reduction of thermoregulatory effector gain is concerned, central and peripheral temperature signals interact at the level of the anterior hypothalamus where effector mechanisms of temperature regulation actually exist in thermally acclimated animals.(Kosaka et al.,1988, 1989, Ohwatari et al., 1992)

In conclusion , the various levels at which the acclimatization process takes place may be summarized as follows; (1) central, together with numerous input/output; (2) neuro-glandular junction (with attendant modification); (3) at the end-organ level, particularly the sweat glands and skin.

NOTE

This proceeding is a modified form of our previous paper (Matsumoto et al., 1993: Trop. Med., 35(1), 23-34).

REFERENCES

1)Brown, W.K. & Sargent, F.(1965): Hidromeiosis. Arch. Environ. Health., 11, 442-453.

2)Fan, Y.J. (1987): Determination of heat acclimatization by capacitance hygrometer sweat capture capsule method. Trop. Med., 29(2), 107-121.

3)Hammel, H.T., Jackson, D.C., Stolwijk, J.A.J., Hardy, J.D. & Stromme, S.B. (1963): Temperature regulation by hypothalamic proportional control with adjustable set temperature. J. Appl. Physiol., 18, 1146-1154.

4)Hori, S., Ihzuka, H. & Nakamura, M. (1976): Studies on physiological responses of residents in Okinawa to a hot environment. Jap. J. Physiol., 26, 235-244.

5)Hori, S., Ohnaka, M., Shiraki, K., Tsujita, J., Yoshimura, H., Saito, N. & Panata, M. (1977): Comparison of physical characteristics, body temperature and basal metabolism between Thai and Japanese in a neutral temperature zone. Jap. J. Physiol., 27, 525-538.

6)Iriki, M. (1983): Regional differentiation of sympathetic efferents during thermal stimulation. J. Therm. Biol., 8, 225-228.

7)Kosaka, M., Ohwatari, N., Matsumoto, T. & Yang, G.J. (1988): Changes in hypothalamic blood flow due to thermal acclimation. Trop. Med., 30(2), 163-174.

8)Kosaka, M., Ohwatari, N., Tsuchiya, K. & Matsumoto, T. (1989): Cerebral blood flow change due to thermal acclimation. Thermal physiology 1989, Ed. by J. B. Mercer. Elsevier Science Publishers B.V. pp.463-468.

9)Kuno, Y. (1956): Human Perspiration. Charles C. Thomas, Springfield.

10)Matsumoto, T., Yamauchi, M., Kosaka, M. & Nakamura, K. (1989): Effect of fine alteration in ambient temperature on thermal sweating at rest and during exercise under thermo-neutral conditions. Trop. Med., 31(3), 131-139.

11)Matsumoto, T., Kosaka, M., Yamauchi, M., Nakamura, K., Yang, G.J. Lee, J.M., Tsuchiya, K. & Amador Velazquez, J.J. (1990): Seasonal variation of thermal sweating. Trop. Med., 32(2), 73-80.

12)Matsumoto, T., Kosaka, M., Yamauchi, M., Yang, G.J., Lee, J.M., Tsuchiya, K., Amador Velazquez J.J., Praputpittaya, C., Yongsiri, A. & Boonayatha, U. (1991): Analysis of the mechanisms of heat acclimatization - Comparison of heat-tolerance between Japanese and Thai subject. Trop. Med., 33(4), 127-133.

13)Ohwatari, N., Fujiwara, M., Iwamoto, J., Fan, Y.J., Tsuchiya, K. & Kosaka, M. (1983): Studies on heat adaptation. - Measurement of the sweating reaction of a tropical inhabitant -. Trop. Med., 25(4), 235-241.

14)Ohwatari, N., Kosaka, M. & Matsumoto, T. (1992): Effect of thermal acclimation on change in cerebral blood flow during LPS-pyrogen fever in rabbits. Trop. Med., 34(1), 29-38.

15)Ramanathan, N.L.A. (1964): A new weighing system for the mean surface temperature of the human body. J. Appl. Physiol., 19, 531-533.

16)Sasaki, T. & Tsuzuki, S. (1984): Changes in tolerance to heat and cold after migration from subtropic to temperate zone. pp.479-482. In Hales, J.R.S.(ed.). Thermal Physiology. Raven Press, New York.

17)Takagi, K. & Sakurai, T. (1950): A sweat reflex due to pressure on the body surface. Jap. J. Physiol., 1, 22.

18)Wyndham, C.H., Strydom, N.B., Morrison, J.F., Williams, C.G., Bredell, G.A.G., Von Rahden, M.J.E., Van Rensburg, A.J. & Munro, A. (1964): Heat reactions of Caucasians and Bantu in South Africa. J. Appl. Physiol., 19(4), 598-606.

19)Yoshimura, H. (1960): Acclimatization to heat and cold. pp.61-106. In:Yoshimura, H., Ogata, K. & Itoh, S. (ed.). Essential Problem in Climatic Physiology. Nankodo, Tokyo.

B. Fever and other disorders

Integrative and cellular aspects of autonomic functions : temperature and osmoregulation. Eds K. Pleschka, R. Gerstberger. John Libbey Eurotext, Paris © 1994, pp. 115-124.

Effects of superimposed thermal stress on the febrile response of sheep

Helen P. Laburn, Deborah-Lynne Steinberg, Manoj Raga and Kathleen Goelst

Department of Physiology, University of the Witwatersrand Medical School, 7 York Road, Parktown, Johannesburg, South Africa

INTRODUCTION

Hyperthermia is an accompaniment of infection, exercise and heat stress (Hellon *et al.*, 1991). During the generation of a fever, heat production and heat conservation are stimulated, and body temperature rises. During all other hyperthermias, the rise in body temperature results from an excess of heat generated (as in exercise), or gained (as in exposure to hot environments) over the physiological heat loss mechanisms, and hyperthermia ensues. The hyperthermia of fever is unlike other hyperthermias, in the additional respect that the body temperature rise during fever is regulated, and is consistent with the idea of the resetting of the body's set-point temperature (Mitchell & Laburn, 1985). Physiological and biochemical events which follow bacterial infection, lead to a rise in hypothalamic set-point temperature and thermoregulation proceeds as if the body is cold, until that set-point temperature is reached.

Some of the biochemical events which occur in infection, also occur during exercise. Cannon & Kluger (1983) first showed that an endogenous pyrogen released during infection also was present during and after submaximal exercise in humans. More recently, others have shown that the endogenous pyrogens, interleukins 1 and 6 (Northoff & Berg, 1991) and interferon alpha (Viti *et al.*, 1985) are accompaniments of exercise, and may have a role to play in exercise hyperthermia. We have performed experiments to ascertain whether the hyperthermias of simulated infection and exercise are additive, by superimposing bouts of treadmill exercise on two stages of a developing febrile response in sheep. In the first series of experiments, we studied the effect of exercise during the latent period of onset of the response to an injection of a bacterial pyrogen, a time in which endogenous pyrogen allegedly is initiating the fever (Kluger, 1991). In the second series of experiments we introduced exercise approximately mid-way through the developing febrile episode. Our results suggest that a bout of exercise modifies the normal pattern of the body temperature response during fever only for the duration of the exercise. The overall fever response, including the peak body temperature attained, is not affected by superimposed exercise.

METHODS AND MATERIALS

Surgery
A total of 7 Merino or Dorper-cross adult sheep (mean body weight 47.2 ± 2.3 kg, mean ± SEM) had sterilized radio-telemeters implanted into the abdomen, after sedation and under local anaesthesia. Each animal was premedicated with 0.2mg/kg Rompun (Bayer-Miles, Germany) intramuscularly, and 8 - 10 ml 2% Lignocaine hydrochloride (Centaur Laboratories, Johannesburg) was infiltrated into the paraventral area of the abdomen. The radiotelemters were inserted into the abdominal cavity through a small incision in the abdominal wall. The incision was closed by layered suturing. Yohimbine (Kyron-Lab, Johannesburg) 2 ml, was injected intramuscularly to reverse the effects of the sedative, and 5ml PENI LA (Phenix, S.A.) was give intramuscularly at the time of surgery as antibiotic cover. The surgery was completed in approximately 35 minutes, after which the animals were housed in individual indoor pens. The ambient temperature at which the sheep were housed, as well as that temperature at which the experiments were carried out, varied between 21 and 23°C. A natural light/dark cycle was maintained, water and hay were provided *ad lib*, and commercial sheep concentrate pellets (about 0.5 kg/day) were provided once daily at approximately 15:00. At the completion of the experiments, the animals underwent surgical procedures as above, for removal of the radio-telemeters.

Temperature measurements
Abdominal temperature was measured using temperature-sensitive radio-telemeters (Mini-Mitter, Sunriver, USA). The telemeters were calibrated before insertion into the animals, and the calibration checked after removal from the animals. Details of the calibration procedures and of temperature monitoring from the outputs of the telemeters have been described previously (Laburn ***et al.***, 1992).

Exercise training
About a week following surgery, the sheep commenced a training regimen on a treadmill. The training regimen consisted of standing only on the treadmill for a 30 minute period initially, thereafter each animal was made to exercise at progressively greater intensities over a 2 week period, until a fast walking pace of 2.1 km/hour at a 5° or 10° gradient could be maintained for 30 minutes.

Fever induction
Fever was induced by an injection into the jugular vein of a 1 ml saline solution containing the purified lipopolysaccharide (LPS) of *Salmonella typhosa* (Difco, West Molesey, England) in a dose of 0.4μg/kg. Fevers were produced which had a latency of approximately 20-30 minutes and a duration of several hours. We observed a biphasic pattern of fever response, where the second phase commenced about 100 minutes after injection. Control injections consisted of the same volume of sterile saline. No animal was given LPS injections at intervals of less than one week. The mean body temperature at the time of injection was 39.6 ± 0.07°C (SEM, for n=55 injections).

Experimental protocol
Two series of experiments were carried out, using identical procedures for fever induction. In the first series, the sheep exercised during the latent period of the fever, and in the second series, at a point coinciding with the onset of the second phase of the fever response. In control experiments sterile saline was injected instead of LPS. To assess the effects of the injectates, the sheep stood still on the level treadmill, and injections of saline or LPS were made as appropriate. Exercise always was for a period of 30 minutes with a one minute break at the end of each 10 minute period

to allow for body temperature measurements to be made. Treadmill speed was 2.1 km/hr, and the treadmill was set at an incline of 5° or 10°.

Each experiment commenced between 08:00 and 09:00. The animals were led onto the treadmill, and baseline body temperatures were measured for not less than a 40 minute period. Thereafter, an intravenous injection of saline, or LPS solution was given.

In the first series of experiments, the sheep were made to exercise immediately after LPS or saline injection, and to 30 minutes post-injection. We assessed the effect of two levels of exercise imposed at the start of a fever response; 2.1 km/hour, 10° gradient and 2.1km/hour, 5° gradient. In each case, after exercise, the treadmill was lowered to a level position and the animal stood on the treadmill until temperature monitoring was completed (4 hours post-injection).

In the second series of experiments, the sheep stood on the level treadmill, until 100 minutes post-injection. Thereafter, the animal exercised for a 30 minute period at 2.1 km/hour and 5° gradient. After exercise, the treadmill was lowered into the level position again, and the sheep stood out the remainder of the experimental period.

Statistical procedures and Ethics
We used the paired and unpaired Student's *t*-test to assess differences in responses, with the Bonferroni correction for repeated comparisons (Glantz, 1987). Values of $P<0.05$ were considered significant. All the experimental procedures were approved by the Animal Ethics and Screening Committee of the University of the Witwatersrand, under protocol 92/22/4.

RESULTS

Exercise during the latent period of onset of fever
Figures 1A and 1B show mean responses of sheep to exercise at two intensities, or standing on the treadmill, immediately after the intravenous injection of LPS or sterile saline. The exercise itself induced an immediate rise in body temperature of the sheep which was intensity-dependent. Forty minutes after the start of exercise at a 10° gradient, body temperature had risen by a mean of 0.95 ± 0.12°C, and 40 minutes after the start of the lower level of exercise, mean body temperature of the animals had risen by 0.57 ± 0.12°C.

The injection of LPS too caused a rise in body temperature (Fig. 1), such that within 20 minutes of injection, mean body temperature of the sheep was significantly different to that of the sheep into which saline was injected ($P<0.02$). The fever reached an initial plateau about 100 minutes after injection, but continued to attain a maximum rise of 2.20 ± 0.15°C.

Irrespective of the level of exercise, when the sheep exercised immediately after the LPS injection, body temperature rose more rapidly than when the sheep did not exercise (Fig. 1). The effect is most apparent at the higher of the two levels of exercise. Fig. 1A shows that the mean body temperatures of the sheep during exercise at 10° gradient after saline injection compared to after LPS injection were not significantly different for the first 20 minutes after injection and commencement of exercise. However, after a period corresponding to the latency of the LPS effect, that is at 30 minutes and for the following 40 minutes, the rise in body temperature in LPS-injected, exercising sheep was significantly greater ($P<0.05$) than that of both the LPS-injected sheep standing on the treadmill, and the saline-injected exercising sheep (unpaired *t*-test with Bonferroni correction).

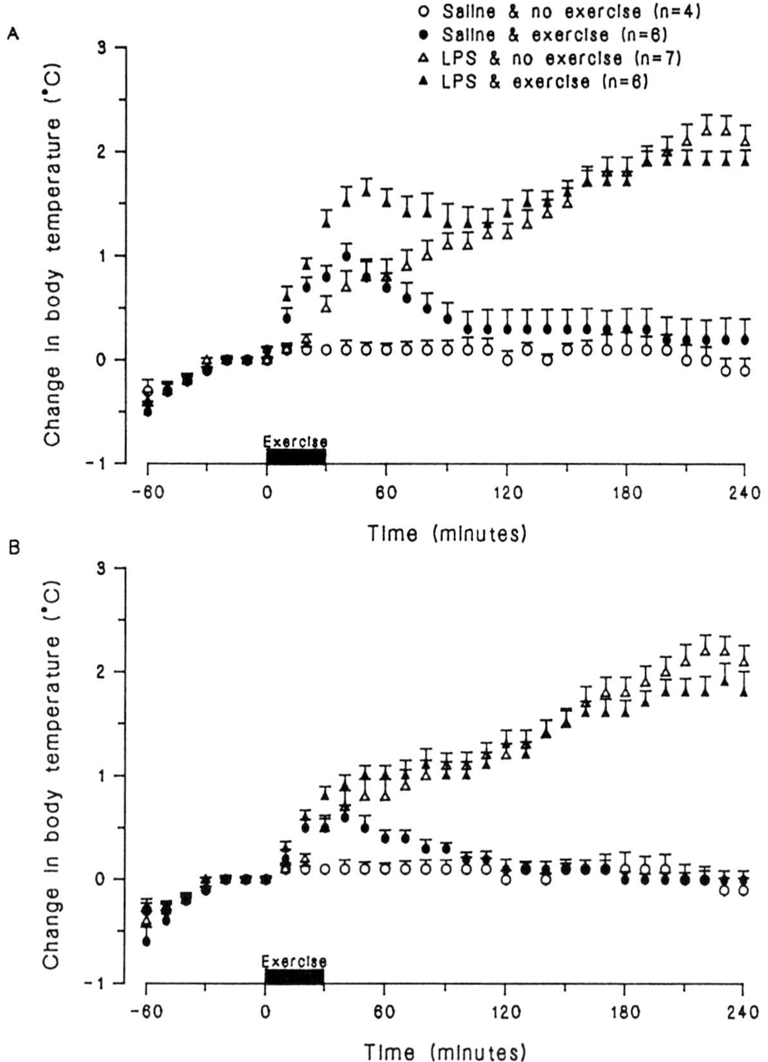

Fig. 1. Changes in body temperature of sheep, after injection of LPS or sterile saline at time zero. Exercise was at a 10 ^{o} gradient (**A**) or 5 ^{o} gradient (**B**) and commenced at time zero and was for 30 minutes. Values shown are means $\pm$ SEMs.

Depending on the level of exercise in this series of experiments, body temperature of the saline-injected animals returned to a level indistinguishable from that of non-exercising sheep 80 minutes after commencement of exercise (Fig. 1A), or sooner, for 5° gradient exercise (Fig. 1B). It is at these approximate times too that the mean responses of the sheep injected with LPS with and without exercise followed almost identical courses (Fig. 1) until the end of our monitoring period. There was no significant difference in the peak febrile temperature attained after LPS injection, in exercising versus non-exercising sheep, for both intensities of exercise.

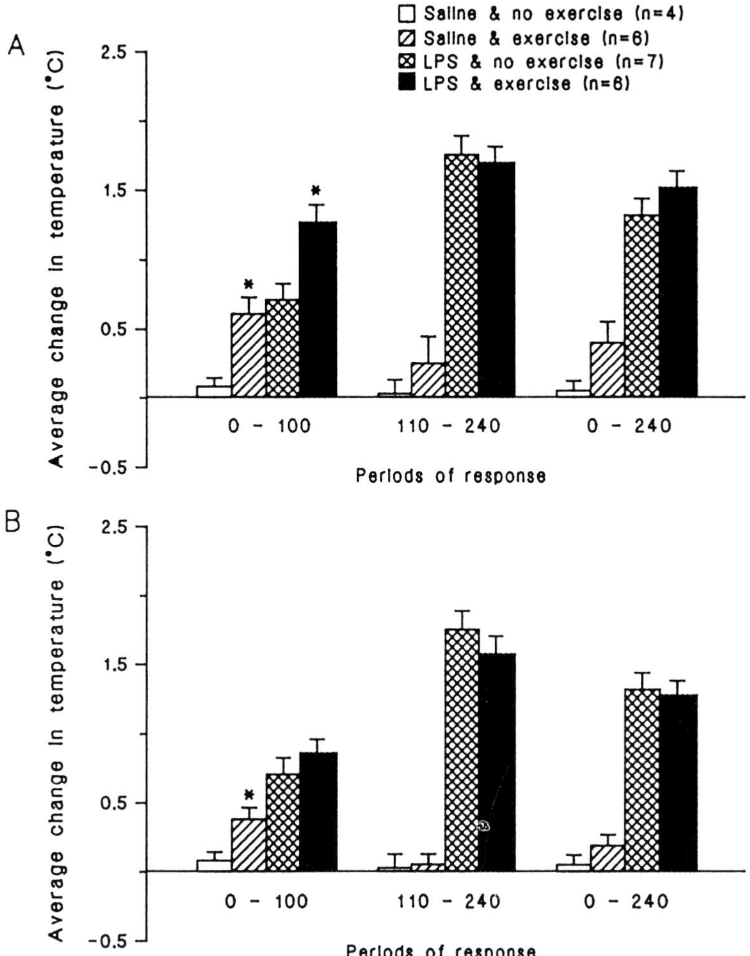

Fig. 2. Average change in body temperature (in °C) calculated for various periods of the temperature responses of sheep after saline or LPS injection. Exercise was during time 0 - 30 minutes of the measurement period. Average changes were calculated for the first 100 minutes (0 - 100), the remaining minutes of the monitoring period (110 - 240), and for the total four hours (0 - 240 minutes). **A** - exercise at 10° gradient; **B** - exercise at 5° gradient. * Denotes value significantly different from the non-exercising control.

Figure 2 shows average body temperature changes (in °C) for various periods post-injection. In the period 0 -100 minutes during which exercise occurred, the average change in temperature after saline injection was significantly increased in exercising sheep ($P<0.05$) both at the 5° and 10° gradients.

In this period too the body temperature change of the LPS-injected sheep exercising at the higher intensity only, was significantly different (P<0.01) to that of the febrile, non-exercising sheep. Neither exercise regimen had a significant effect on the average temperature calculated from 110 minutes after start of exercise or injection, nor indeed over the entire 4 hour period.

Exercise at the commencement of the second phase of fever

Figure 3 shows the mean temperature responses of a group of 5 sheep to either saline or LPS injection at time zero, and with and without a 30 minute bout of exercise commencing 100 minutes post-injection. As in the previous series of experiments, exercise, this time only at 2.1 km/hr and a 5° gradient, induced an immediate rise in body temperature of saline-injected sheep, and mean body temperature rose 0.72 ± 0.10°C (mean ± SEM). After LPS injection, body temperature rose to 2.20 ± 0.25°C above baseline levels after exercise midway during the development of the fever, and to 2.20 ± 0.13°C when there was no exercise.

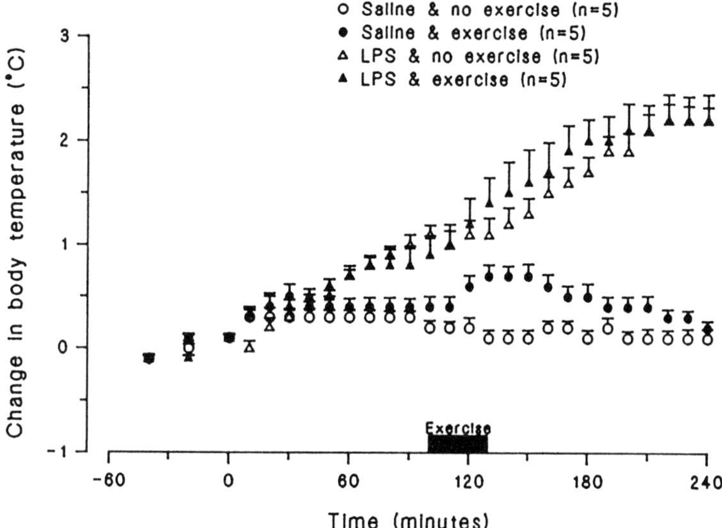

Fig. 3. Changes in body temperature of 5 sheep after injection of LPS or sterile saline at time zero, with exercise, at a 5° gradient, commencing 100 minutes after injection and for 30 minutes thereafter. Values are means ± SEMs.

Figure 4 shows calculations of average body temperature changes for the first (0 - 100 minutes), and second (110 - 240 minutes) phases of the febrile response. For the period inclusive of exercise, we calculated these changes taking as baseline temperature, that body temperature of each sheep prevailing at the start of exercise. Exercise itself in saline-injected animals induced a significant rise in body temperature (P<0.02, paired *t*-test), as shown by the value for the period inclusive of the exercise, that is, 110-240 minutes after injection . The average temperature change over the period (110 - 240 minutes) inclusive of the exercise bout, was significantly different for febrile, exercising sheep compared to the febrile, non-exercising sheep (P<0.02, paired *t*-test). No significant difference could be detected in the average temperature change over the four hour period of the exercise versus non-exercise LPS-injected sheep.

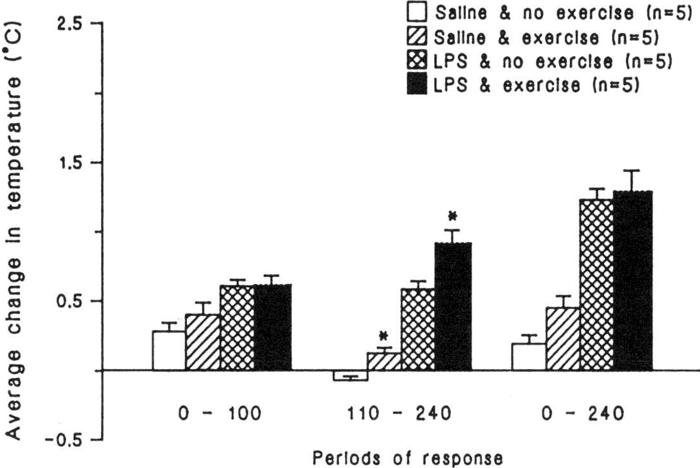

Fig. 4. Average change in body temperature (in °C) calculated for various periods of the temperature response of sheep to injection of LPS or saline at time zero, and exercise, when applicable, at a 5° gradient commencing 100 minutes after injection. Changes are shown for the first phase of the response (0 - 100 minutes), for the phase including the exercise (110 -240) and for the whole measurement period (0 - 240 minutes). Means $\pm$ SEMs are shown for 5 sheep. * Denotes value significantly different from the non-exercising control.

DISCUSSION

We have performed experiments to determine the effect on the pattern and extent of rise of body temperature during fever of a superimposed hyperthermia, that induced by exercise. The exercise regimen we used was not particularly strenuous for the sheep; according to a report by Bell *et al.* (1983), also working on this species, our regimen would have increased oxygen consumption about four-fold. Nevertheless, the bout of exercise induced a significant rise in body temperature of the sheep which was intensity-related, as exercise is in humans (Nadel, 1985).

We were particularly interested in two aspects of the fever response when a non-febrile surge in body temperature was imposed at two stages of fever development. Firstly, would the course of the fever response be affected, and secondly, what would be the effect on the peak febrile temperature?

We found that the course of the febrile response was significantly affected by the exercise-induced hyperthermia, but for the duration of the exercise only. The effect of exercise was most pronounced when exercise was carried out immediately after LPS injection, that is, during the latency to and initial rise in temperature, and when the exercise was more intense. During the latent period for onset of the febrile response, the animals showed rises in body temperature almost identical to the rises which exercise induced when saline only was injected (Fig. 1A). The febrile response in exercising versus non-exercising sheep differs for the duration of the exercise-induced hyperthermia.

Thereafter (80 min post-injection and start of exercise) the fever responses of the two groups of animals are almost identical (Fig 1A).

We found that exercise carried out between the first and second phases of the fever (Fig. 3) also affected the course of the fever response. The exercise itself induced a significant hyperthermia in saline-injected sheep, and significantly increased the average body temperature change in LPS-injected animals for the period 110-240 minutes after exercise commenced (Fig. 4).

Thus our results suggest that when exercise is introduced into the course of a developing febrile episode, the febrile rise in body temperature is enhanced significantly. Tanaka ***et al.*** (1990) showed that the febrile rise in body temperature induced by injection of interleukin 1, about 0.5°C, is superimposed in an additive manner, onto the hyperthermia in exercising rats. In the converse situation, the enhanced fever is not due to a simple additive effect of the two hyperthermias. The calculation of the nett effect of exercise in saline and LPS-treated animals can be made by subtraction of the average temperature changes for procedures which differ only by the presence or absence of exercise. The calculations reveal a simple additive increase in body temperature only when animals are exercised at 10° gradient at time of injection of saline compared to LPS (0.56 vs 0.53 °C). Exercise, at 100 minutes post-injection raised the average temperature change after saline injection by 0.20°C, but by 0.34°C, after LPS injection. Thus when exercise occurred mid-way through the rising phase of the fever, the increase in temperature in febrile animals was about 1.5 times the average rise in temperature caused by exercise in saline-injected animals. It is thought that during the latent period after LPS injection, release of endogenous pyrogens, such as interleukins 1 and 6 occurs (Kluger 1991), and by the actions of these cytokines outside or within the hypothalamus (Morimoto ***et al.***, 1987), activation of heat production and inhibition of heat loss occurs to initiate the fever. Much less is known about the events occurring later in the febrile episode, between the first and second phases of the response, and of the role of cytokines during this period, but endogenous pyrogens are thought to be present to sustain the fever. Endogenous pyrogens (Cannon & Kluger, 1983), including interleukins 1 and 6 are thought to be released during exercise too (Northoff & Berg, 1991; Taylor ***et al.***, 1987). Thus one can postulate that both in the latent period of development of the fever, and midway during the rising phase of fever, a similar high concentration of cytokine mediators of hyperthermia is present in animals which are exercising and have had an injection of LPS. On the basis of the evidence presented, we are not able to explain why the exercise, superimposed on the fever induces different degrees of additional hyperthermia at the two different stages of the febrile response.

Our second interest was in the effect on the ultimate body temperature attained after LPS injection, of superimposed exercise. Figures 2 and 4 show that even at the higher intensity of exercise, there was no significant increase in the average temperature change over four hours of febrile animals which exercised compared to those which were sedentary throughout the febrile period. Moreover, the peak febrile temperatures attained 4 hours after injection, were not significantly different in any of the exercising- versus non-exercising febrile groups of animals.

The rise in body temperature of a fever is thought to be regulated by a set-point temperature which may be that of the hypothalamus, where integration of thermoregulatory effects occurs (Hellon ***et al.***, 1991). It has been shown previously that an elevation of hypothalamic temperature induced by forced tracheostomy breathing in sheep (Laburn ***et al.***, 1988), at the start of a febrile episode, significantly attenuated the rise in rectal temperature in response to LPS injection. In our experiments we raised body temperature by exercise, immediately after LPS injection. A rise in body temperature, by raising arterial temperature, normally would lead to a rise in hypothalamic temperature in sheep, and so we expected that in our experiments too, the response to LPS injection may be inhibited. We found that the overall febrile response was not affected significantly in sheep which exercised during the latent period of the fever response compared to non-exercising sheep. Our results do not conflict with the idea that the hypothalamic temperature plays a central role in the regulation of febrile rises in temperature; rather, we postulate that during the bout of exercise in the 30 minutes after injection of LPS, brain cooling occurred as panting, stimulated by the rise in body temperature, cooled nasal venous and thereafter arterial blood coursing to the brain (Mitchell *et al.*, 1987). As long as normal panting is allowed to take place, hypothalamic temperature of sheep does not change during exercise, despite a rise in rectal temperature (Laburn *et al.*, 1988).

In conclusion, we have shown that exercise bouts during fever disrupt the pattern of fever temporarily, causing acute, additional increases in body temperature in LPS-injected sheep. Exercise, whether imposed during the latent period of onset of the fever, or midway during the development of the fever, has no significant effect on the height of the febrile temperature attained. We suggest that normal thermoregulatory strategies used by the sheep to minimise exercise-induced hyperthermia, are active during exercise in the latent period of the fever. These strategies also prevent rises in hypothalamic temperature as body temperature rises, so avoiding a possible effect on the set-point temperature which apparently governs the fever response. Exercise later in the febrile episode, appears to cause a greater rise in body temperature than can be attributed to the exercise alone.

ACKNOWLEDGEMENTS

We thank Drs S Maeder and A Leonard for assistance with the surgical procedures, and staff of the Central Animal Unit for care of the animals. Steve Cartmell assisted with treadmill training of the sheep. The research was funded by the University of the Witwatersrand, and Foundation for Research Development.

REFERENCES

Bell, A.W., Hales, J.R.S., King, R.B., & Fawcett, A.A. (1983): Influence of heat stress on exercise-induced changes in regional blood flow in sheep. *J. Appl. Physiol.* 55, 1916-1923.

Cannon, J.G., & Kluger, M.J. (1983): Endogenous pyrogen activity in human plasma after exercise. *Science* 220, 617-619.

Glantz, S.A. (1987): *Primer of Biostatistics*. New York: McGraw-Hill.

Hellon, R., Townsend, Y., Laburn, H.P., & Mitchell, D. (1991): Mechanisms of fever. In *Thermoregulation: Pathology, Pharmacology and Therapy*, eds. E. Schönbaum & P. Lomax, pp. 19-54. New York: Pergamon Press, Inc.

Kluger, M.J. (1991): Fever: role of pyrogens and cryogens. *Physiological Reviews* 71, 93-127.

Laburn, H.P., Mitchell, D., & Goelst, K. (1992): Fetal and maternal body temperatures measured by radiotelemetry in near-term sheep during thermal stress. *J. Appl. Physiol.* 72, 894-900.

Laburn, H.P., Mitchell, D., Mitchell, G., & Saffy, K. (1988): Effects of tracheostomy breathing on brain and body temperatures in hyperthermic sheep. *J. Physiol. (Lond.)* 406, 331-344.

Mitchell, D., & Laburn, H.P. (1985): Pathophysiology of temperature regulation. *The Physiologist* 28, 507-517.

Mitchell, D., Laburn, H.P., Nijland, M.J.M., Zurovsky, Y., & Mitchell, G. (1987): Selective brain cooling and survival. *S. Afr. J. Sci.* 83, 598-604.

Morimoto, A., Murakami, Ň., Nakamore, T., & Watanabe, T. (1987): Evidence for separate mechanisms of induction of biphasic fever inside and outside the blood-brain barrier in rabbits. *J. Physiol. (Lond.)* 383, 629-637.

Nadel, E. R. (1985): Recent advances in temperature regulation during exercise in humans. *Federation Proc.* 44, 2286-2292.

Northoff, H., & Berg, A. (1991): Immunologic mediators as parameters of the reaction to strenuous exercise. *Int. J. Sports Med.* 12, S9-S15.

Tanaka, H., Kanosue, K., Yanase, M., & Nakayama, T. (1990): Effects of pyrogen administration on temperature regulation in exercising rats. *Am. J. Physiol.* 258, R842-R847.

Taylor, C., Rogers, G., Goodman, C., Baynes, R.D., Bothwell, T.H., Bezwoda, W.R., Kramer, F. , & Hattingh, J. (1987): Hematologic, iron-related and acute-phase protein responses to sustained strenuous exercise. *J. Appl. Physiol.* 62, 464-469.

Viti, A., Muscettola, M., Paulesu, L., Bocci, V., & Almi, A. (1985): Effect of exercise on plasma interferon levels. *J. Appl. Physiol.* 59, 426-428.

Integrative and cellular aspects of autonomic functions : temperature and osmoregulation. Eds K. Pleschka, R. Gerstberger. John Libbey Eurotext, Paris © 1994, pp. 125-129.

Immune cytokines modulate peripheral cellular immunity through the hypothalamo-sympatethic nervous system

Tetsuro Hori, Toshihiko Katafuchi, Sachiko Take, Yasuo Kaizuka, Tomoyasu Ichijo and Nobuaki Shimizu

Department of Physiology, Kyushu University, Faculty of Medicine, Fukuoka 812, Japan

The brain and the immune system may act together in an integrated way in maintaining the homeostasis of the body through their bidirectional communication. While the brain modulates immunity through the autonomic nervous innervation as well as the neuroendocrine communication, the immune system may signal the brain by means of chemical messengers like cytokines and some other neuroactive substances derived from immunological cells. Cytokines such as interleukin-1 (IL-1) and interferon α (IFN α), which are synthesized from glial cells as well as immunological cells, act directly or indirectly on the brain cells to elicit a wide variety of autonomic, neuroendocrine and behavioral responses such as fever, anorexia, sleep and various hormonal responses like an activation of the hypothalamo-pituitary adrenocortical system (Hori et al., 1991). Immunosuppressive responses have also been described by raising the concentration of IL-1 β and IFN α in the brain. Central injection of both cytokines inhibits the cytotoxicity of natural killer (NK) cells and the mitogenic responses of lymphocytes (Sundar et al., 1990; Take et al., 1992, 1993).

The brain has long been known to induce high levels of IFN α during infection of neurotropic viruses. In addition to the CNS-mediated whole body responses stated above, IFN α produces analgesia and catalepsy (Blalock & Smith, 1981) and suppression of naloxone-induced abstinence in morphine-dependent rats (Dafny et al.,1988). Direct application of IFN α changes the activity of hypothalamic thermosensitive and glucose-responsive neurons in appropriate ways to explain the febrile and anorexic responses to central IFN α, respectively (Hori et al., 1991; Nakashima et al., 1988; Kuriyama et al., 1990). These neuronal responses as well as the central IFN α-induced fever and analgesia have been demonstrated to be attenuated or blocked by an opioid antagonist, naloxone or naltrexone, suggesting the involvement of opioid receptor mechanisms. In accordance with these, IFN α binds to opioid receptors in the brain of rats and mice (Blalock & Smith, 1981; Menzies et al., 1991).

The aim of this paper is to review our recent findings briefly on (1) the central and peripheral mechanisms of IFNα-induced suppression of cytotoxic activity of splenic NK cells and (2) to demonstrate the evidence of the involvement of the hypothalamo-sympathetic nervous system in the central modulation of peripheral immunity.

<u>The reduction of splenic NK activity after central injection of IFNα in the rat.</u>

The effects of intracerebroventricular (ICV) injection of recombinant human IFNα on the cytotoxic activity of splenic NK cells were studied in male Wistar King A rats (Take et al., 1993). The injection was done through a chronically implanted guide cannula into the lateral cerebral ventricle. Thirty min after the end of intracerebroventricular (ICV) injection unless otherwise noted, the animals were deeply anesthetized with ether, sacrificed and splenectomized. The cytotoxic activity of NK cells in the spleen was measured by a 4-hrs standard chromium release assay using ^{51}Cr labeled YAC-1 murine lymphoma cells as targets. The splenic NK activity decreased by about 20-50 % after ICV injection of IFNα (1000-20000 U). However, intraperitoneal (IP) injection of 20000 U IFNα did not affect the splenic NK activity, suggesting that the site of actions of the ICV-injected IFNα (1000-20000 U) is within the brain, but not in the periphery. The measurement of splenic NK activity at different postinjection periods revealed that this immunosuppression was still observed 1 and 2 hrs after ICV injection of 10000 U IFNα but the NK activity almost recovered to control levels at 3 hrs. The reduced NK activity after ICV injection of IFNα is unlikely to be caused by selective egress of NK cells from the spleen, because the NK activity of the peripheral blood also decreased significantly 30 min after ICV injection of 10000 U IFNα. Since no remarkable change in rectal temperature was noted 30 min after ICV injection of 10000 or 20000 U IFNα, the decreased NK activity was not caused by its pyrogenic action.

To determine the site of actions of IFNα in the brain, we injected 200 U IFNα into the different sites of hypothalamus (the preoptic area, ventromedial hypothalamus, lateral hypothalamus, paraventricular nucleus) , neocortex and thalamus in the rat and observed the splenic NK activity 30 min after injection. An injection of 200 U IFNα into the preoptic area decreased the splenic NK activity by about 50 %, which was of the similar degree to that observed after ICV injection of 20000 U IFNα. The injection of 200 U IFNα into the other brain sites tested was without effect. It thus appears that the preoptic area is one of the most sensitive sites, at least, within the hypothalamus to microinjected IFNα to reduce the splenic NK activity.

Naltrexone (NLTX, 50 μg) which was given ICV 30 min before the ICV injection of IFNα (10000 U) completely abolished the ability of IFNα to reduce the splenic NK activity, but the pretreatment with NLTX which was followed by saline injection had no effect on it. This suggests that ICV IFNα suppresses the NK activity through an activation of opioid receptor mechanisms, and agrees well with the previous findings that intracerebral, but not intravenous, injection of morphine or β-endorphin induced a suppression of splenic NK activity in a NLTX preventable way (Mori et al., 1989; Shavit et al., 1986). Shavit et al (1986) showed that the splenic NK activity was inhibited when rats received an inescapable type of electric footshock known to induce opioid-dependent analgesia, but not when they were given an escapable footshock causing opioid-independent analgesia. The same authors also reported that the opioid-dependent shocks increase the growth of the implanted tumor in a naloxone reversible way. In view of the important role of NK cells in immunosurveillance, the present findings, together with the previous ones, might offer one explanation for the reduced immunosurveillance which is associated with a certain type of stress.

<u>The involvement of brain corticotropin releasing factor (CRF) in IFN α -induced decrease in the splenic NK activity.</u>
A neuropeptide CRF has been postulated to be a key mediator of stress-induced responses (Dunn & Berridge, 1990). Central injection of a CRF antagonist, α -helical CRF_{5-41}, attenuates many stress-induced responses. Central injection of CRF mimics some stress responses. Recently it was shown that an ICV injection of CRF reduced the splenic NK activity in the rat (Irwin et al., 1988). In order to determine whether brain CRF is involved in the development of central IFN α -induced suppression of NK activity, we injected α -helical CRF_{9-41} (25 μ g, ICV) 30 min before ICV injection of 10000 U IFN α (Take et al., 1993). Pretreatment with α -helical CRF_{5-41} which by itself had no effect on the NK activity completely abolished the IFN α -induced immunosuppression. Furthermore, we confirmed the reduced NK activity after ICV injection of CRF (2.5 μ g) in agreement with the previous report (Irwin et al., 1988), and this immunosuppression was not affected by NLTX (50 μ g, ICV) which could block the IFN α (10000 U, ICV)-induced inhibition of NK activity. These results suggest that the reduced NK activity as a result of stimulation of opioid receptors by brain IFN α is dependent on the activity of central CRF systems.

<u>Role of the splenic sympathetic nerve in the mediation of central IFN α -induced NK suppression</u>.
The reduced NK activity after ICV injection of IFN α was completely abolished by splenic denervation but not by bilateral adrenalectomy (Take et al, 1993). Furthermore, we found that the reduced NK activity after preoptic injection of IFN α was also dependent on the intact splenic innervation. Since the splenic sympathetic nerve is thought to consist mainly of sympathetic fibers (Bellinger et al., 1989), these findings suggest brain IFN α -induced decrease in NK activity is mediated predominantly through splenic sympathetic innervation. This conclusion is further strengthened by the following observations including ours. First, an intra-third-ventricular injection of 1500-2000 U IFN α produces a long-lasting (for more than 60 min) increase in the splenic sympathetic activity,which was suppressed transiently by an intravenous injection of naloxone (Katafuchi et al., 1993). A microinjection of 200 U IFN α into the preoptic area, where was found most sensitive to microinjected IFN α in reducing the NK activity, also increased the activity of splenic nerve (Katafuchi et al., 1992). Second, an electrical stimulation of splenic nerve efferents produced a reduction of the splenic NK activity, which was blocked by pretreatment with a β -adrenergic receptor antagonist (nadolol), but not by an α -blocker (prazocin) (Katafuchi et al., 1993). Third, neonatal chemical sympathectomy resulted in the increased activity of NK cells (Reder et al., 1989). Fourth, a direct addition of adrenaline into the mixture of the effector/target cells <u>in vitro</u>

reduces the NK activity (Hellstrand et al., 1985).

Moreover, we recently found that an ICV injection of CRF (1 μg) increased the splenic sympathetic activity (Ichijo et al., 1992) and an ICV injection of CRF (2.5 μg) increased the release of noradrenaline in the spleen of rats as determined by a microdialysis.

All these findings, taken together, suggest that IFNα works on the opioid receptor mechanisms most likely in the preoptic area and this results in an activation of the splenic sympathetic nerve through the brain CRF system and thereby suppresses the NK cytotoxicity by β-adrenergic mechanisms.

This work was performed through Special Coordination Funds of the Science and Technology Agency of the Japanese Government (to TH). This work was also supported by in part by Grants-in-Aid for Scientific Research 03557009 and 04454144 (to TH) from the Ministry of Education, Science and Culture, Japan.

REFERENCES

Bellinger, D.L., Felten, S.Y., Lorton, D., and Felten, D.L. (1989): Origin of noradrenergic innervation of the spleen in rats. Brain Behav. Immunity 3, 291-311.

Blalock, J. E., and Smith, E.M. (1981): Human leukocyte interferon (HuIFN): potent endorphin-like opioid activity. Biochem. Biophys. Res. Commun. 101, 472-478.

Dafny, N., and Reyes-Vazquez, C. (1985): Three different types of α-interferons alter naloxone-induced abstinence in morphine-addicted rats. Immunopharmacology 9, 13-17.

Dunn, A.J., and Berridge, C.W. (1990): Physiological and behavioral responses to corticotropin-releasing factor administration: is CRF a mediator of anxiety or stress responses ? Brain Res. Rev. 15: 71-100.

Hellstrand, K., Hermdsson, S., and Strannegard. O. (1985): Evidence for a β-receptor-mediated regulation of human natural killer cells. J. Immunol., 134: 4095-4099.

Hori, T., Nakashima, T., Take, S., Kaizuka, Y., Mori, T., and Katafuchi, T. (1991): Immune cytokines and regulation of body temperature, food intake and cellular immunity. Brain Res. Bull. 27: 309-313.

Ichijo, T., Katafuchi, T., and Hori, T. (1992): Enhancement of splenic sympathetic nerve activity induced by central administration of interleukin-1β in rats. Neurosci. Res Suppl. 17, S157.

Irwin, M., Hauger, R.L., Brown, M., and Britton, K.T. (1988): CRF activates autonomic nervous system and reduces natural killer cytotoxicity. Am. J. Physiol 255: R744-R747.

Katafuchi, T,, Take, S., and Hori, T. (1993): Roles of sympathetic nervous system in the suppression of cytotoxicity of splenic natural killer cells in the rat. J. Physiol. 465: 343-357.

Katafuchi, T.,Take, S., Ichijo, T., and Hori, T. (1992): Hypothalamic control of peripheral immune functions mediated by the sympathetic nervous system. Society of Neuroscience

Abstracts 18, 681.
Kuriyama, K., Hori, T., Mori, T., and Nakashima, T. (1990): Actions of interferon α and interleukin-1 β on the glucose-responsive neurons in the ventromedial hypothalamus. Brain Res. Bull., 24: 803-810.
Menzies, R.A., Rier, S.E., Hall, N.R.S., O'Gardy, M.P., and Oliver, J. (1991): Recombinant interferon-alpha inhibits opioid ligand binding to brain membranes. Society for Neuroscience Abstracts 17:831.
Mori,T., Kaizuka, Y, Hori, T., and Nakashima, T. (1989): Effects of intracerebral injection of β-endorphin on the cytotoxic activity of natural killer cells in the spleen. Jap. J. Physiol. 39(Suppl), S118.
Nakashima, T., Hori, T., Kuriyama, K., and Matsuda, T. (1988): Effects of interferon-α on the activity of preoptic thermosensitive neurons in tissue slices. Brain Res. 454, 361-367.
Shavit, Y., Terman, G.W., Martin, F.C., Lewis, J.W., Liebeskind, J.C., and Gale, R.P. (1985): Stress, opioid peptides, the immune system, and cancer. J. Immunol 135, 834S-837S.
Sundar, S.K., Becker, K.J., Cierpial, M.A., Carpenter, M.D., Rankin, L.A., Fleener. S.L., Ritchie, J.C., Simpson, P.E., and Weiss, J.M. (1989): Intracerebroventricular infusion of interleukin 1 rapidly decreases peripheral cellular immune responses. Proc. Natl. Acad. Sci. 86, 6398-6402.
Take, S., Mori, T., Katafuchi, T., Kaizuka, Y., and Hori, T. (1992): Central interferon α suppresses the cytotoxic activity of natural killer cells in the mouse spleen. Ann. N.Y. Acad. Sci. 650, 46-50.
Take, S,, Mori, T., Katafuchi, T. & Hori, T. (1993): Central interferon α inhibits the natural killer cytotoxicity through sympathetic innervation. Am. J. Physiol. in press.

Integrative and cellular aspects of autonomic functions : temperature and osmoregulation. Eds K. Pleschka, R. Gerstberger. John Libbey Eurotext, Paris © 1994, pp. 131-136.

Fever in rabbits with brain incision between OVLT and PO/AH

Masaaki Hashimoto, Takehiko Ueno* and Masami Iriki

*Department of Physiology and *Department of Neurosurgery, Yamanashi Medical University, Tamaho, Nakakoma, Yamanashi, 409-38 Japan*

Circulating pyrogenic cytokines are considered to access to the brain through anteroventral wall of the third ventricle including organum vasculosum of the lamina terminalis (OVLT). The OVLT region seems to play an important roll for the development of fever induced by those cytokines. Present experiment was designed to investigate the central pathway of the information relating to the development of fever by cytokines. Rabbit preoptic area of the anterior hypothalamus, putative integrator of the pyrogenic information, was disconnected to OVLT region by bilateral transection with retractable wire knife. Febrile responses of the rabbits to intravenous (iv) and intracerebroventricular (icv) interleukin-1β (IL-1β) and icv prostaglandin E_2 (PGE_2) were observed after two weeks of the recovery period. Transection between OVLT and PO/AH extended from the anterior commissure to the ventral end of OVLT dorsoventrally, and to 3 mm laterally from the 3rd ventricular wall. Difference of the febrile response to the pyrogens in the transected group was not remarkable in comparison with sham-operation control group.

INTRODUCTION

Anteroventral wall of the third ventricle including OVLT is considered as an important interface of humo-neuronal transmission of the pyrogenic information conducting from cytokines in the blood to thermoregulatory neurons in the preoptic area of the anterior hypothalamus (PO/AH)(Blatteis et al., 1983; Stitt, 1985). Recently, we reported that circulating interleukin-1 (IL-1), a potent pyrogenic cytokine, acted on endothelial cells of the capillary blood vessels in OVLT region (Hashimoto et al., 1991). It is well known that inhibitor of the arachidonate-metabolism depresses the development of fever induced by endotoxin and pyrogenic cytokines (Hashimoto et al, 1988; Hashimoto, 1991; Watanabe, 1992). Furthermore, OVLT region, in comparison with PO/AH region, has a higher sensitivity to prostaglandin for the development of fever (Stitt, 1991). These evidences agree with the hypothesis that pyrogenic cytokines in the blood access to the brain through OVLT region and activate phospholipase enhancing the arachidonate metabolism somewhere in the AV3V, and the metabolite, that is PGE2 predominantly in rabbits, modifies the activity of thermosensitive neurons in thermoregulatory center, thus result in fever. However, the pathway of the neuronal signal from OVLT to PO/AH region relating to the development of fever by cytokines in the blood was not known. In the present experiment, we investigated the neuronal pathway by means of local brain transection with microknife.

MATERIALS AND METHODS

Animals

Twenty male Japanese white rabbits weighing between 2.5 and 3 kg were used in the experiment. The experiment was performed according to the guideline for the care and uses of animals given by the ethical committee of Yamanashi Medical University. Rabbits were anesthetized with sodium pentobarbital (Nembutal) and placed in a stereotaxic apparatus. An incision along to the midline was made in the scalp, and the skull was exposed and arranged according to the stereotaxic co-ordinates presented by Sawyer et al. (1954). The guide shaft (21-gauge steel tubing) of a retractable wire knife, with the knife blade (27-gauge tungsten wire) withdrawn, was inserted into the brain through a hole (3 mm in diameter) in the skull. The tip of the guide shaft was placed 3.0 mm anterior to bregma, 3 mm lateral to the midline, and 3 mm below the surface of the cortex. The knife blade was then protruded 7 mm across the midline, perpendicular to the medial sagittal plane, and lowered 4 mm ventrally. The blade was retraced and the guide shaft was withdrawn from the brain. This operation was done bilaterally. Six control rabbits underwent similar surgical operation, except neither the knife blade was protruded nor the guide shaft was lowered from the initial set point in the brain 3 mm below the surface of the cortex. Figure 1 shows schematic site of transected area. After this procedure for transection, a polyethylene guide cannula (0.82 mm in outer diameter) was implanted into the lateral cerebral ventricle for the icv administration of drugs. Following surgery, animals were returned

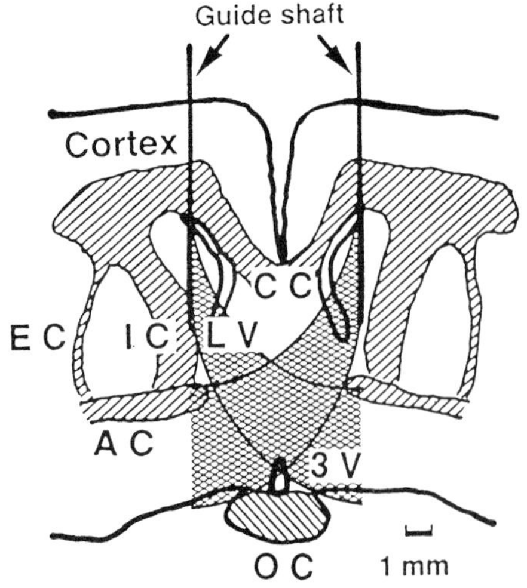

Fig. 1. Schematic site of bilateral transection between OVLT and other PO/AH region with retractable wire knife. Meshed area shows transected plane with the extended knife edge. AC: anterior commissure, CC: corpus callossum, EC: external capsule, IC: internal capsule, LV: lateral ventricle, OC: optic chiasma, 3V: third ventricle

to individual hanging cages till to the beginning of training for adapting the experimental conditions.

<u>Drugs</u>

Frozen stock solution of recombinant rabbit interleukin-1β (rrIL-1β, Hashimoto, 1991) was thawed, and diluted with phosphate buffered saline containing 0.1 %-gelatin (PBS, pH 7.4) into a concentration of 0.3 μg/200 μl and 1 ng/20 μl for iv and icv administrations respectively. Prostaglandin E_2 (PGE_2, Sigma) was dissolved in phosphate buffered saline at a concentration of 2 μg/20 μl for icv administration. A space over the PGE_2 solution of the vial was filled with nitrogen gas to prevent it from oxygenation.

<u>Experimental Protocol</u>

After 10 days of recovering period from surgery, the training for adapting the experimental conditions began. Animals were slightly restrained on the neck in a conventional stock with colonic and pinnal thermocouple in the climatic chamber (24 ± 1 °C, relative humidity of 50 ± 5 %) for several hours per day. A series of experiments were begun 2 weeks after surgery. On a day of the experiment, animals were slightly restrained with the conventional stock in the climatic chamber. Colonic and ear skin temperature were measured with thermocouples, and was continuously monitored and recorded with personal computer. Every animal was challenged to iv (0.3 μg/kg) and icv (1 ng/kg) administration of rrIL-1β and icv administration of PGE_2 (2 μg/kg) taking more than 3 days' intervals for the administration in one route of one drug. Following all recordings, animals were deeply anesthetized, and the brain was fixed with perfusing formalin. Then, the brain was removed from carcass for making thin sections, and the transection area was confirmed with photo microscopy after staining the section with cresylecht violet.

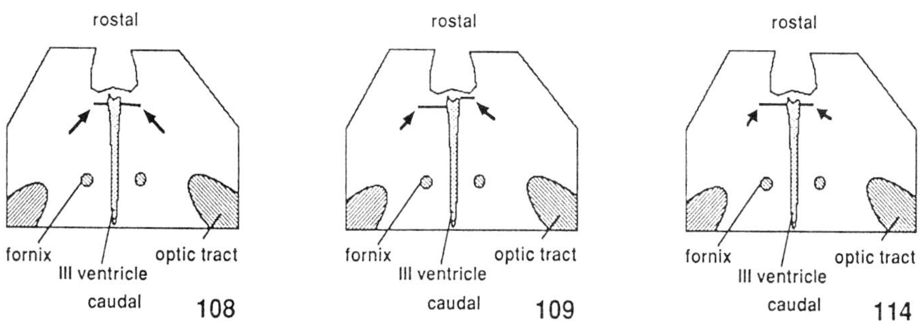

Fig. 2. Schematic site of bilateral transection confirmed on the sagittal section in 3 rabbits (108, 109, 114). The sagittal sections show the level of about 3 mm ventral to anterior commissure. Solid lines pointing by arrows are the site.

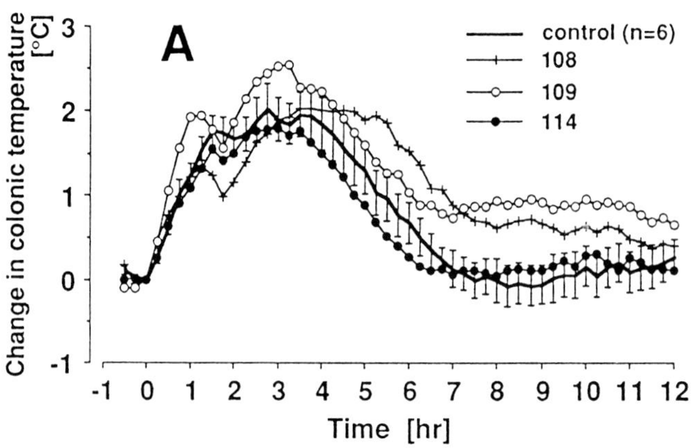

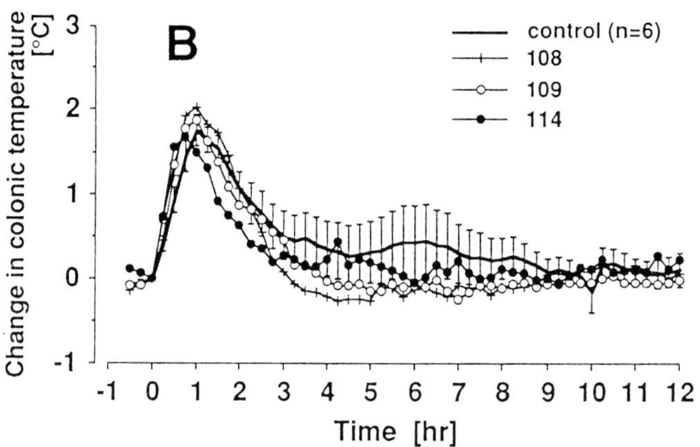

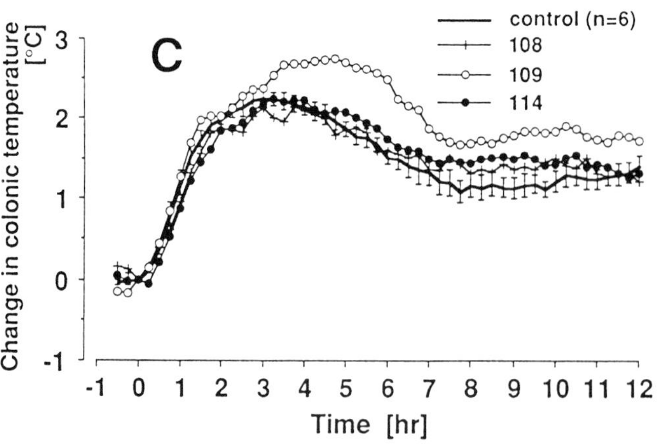

Fig. 3. Means and SEM of the changes in colonic temperature of sham-operated control rabbits and changes in colonic temperature of 3 rabbits (108, 109, 114) with bilateral transection after iv nadministration of rrIL-1β (0.3 μg/kg, A), icv administration of PGE_2 (2 μg/kg, B) and rrIL-1β (1 νg/kg, C) are shown. Drugs were administered at time zero.

RESULTS AND DISCUSSION

Data were acquired, finally, from 7 rabbits out of 14 transected and 6 sham-operation animals. Location of the transection varied. Bilateral transection between OVLT and other PO/AH region was succeeded in 3 rabbits. Figure 2 shows schematic site of the transection observed in the sagittal section at the level of about 3 mm ventral to the anterior commissure. The transection reached about 1.5 to 3 mm laterally from the surface of the 3rd ventricular wall. Changes in colonic temperature induced by the administration of rrIL-1β and PGE_2 in those animals were summarized in Fig. 3. Fever induced by iv and icv administrations of rrIL-1β and iv administration of PGE_2 in control group did not differ from that in animals without surgery with microknife (Hashimoto et al, 1989). Furthermore, any prominent difference in the time course and maximal increase in colonic temperature after administration of the drugs was not observed between control and the 3 rabbits with bilateral transection.

Receptor of PGE_2 is concentrated in anterior wall of the 3rd ventricle around OVLT region (Matsumura et al, 1990), and OVLT has a high sensitivity to arachidonate metabolites with the development of fever (Stitt, 1991). Moreover, existence of PGE_2-sensitive neurons in OVLT region is more than in PO/AH region, and more warm neurons in OVLT region are inhibited by prostaglandin than that in PO/AH region (Matsuda et al, 1992). Therefore, it is hypothesized that pyrogenic information of cytokines in the blood is converted to neuronal signal somewhere nearby the OVLT region, and the signal is conducted to thermoregulatory neurons in the PO/AH region. The present results show that the pathway of the neuronal signal relating to the development of fever may not exist in the transected area in the present experiment. Transection in the present experiment was relatively small, and neuronal fiber pathway from OVLT to PO/AH running through rostro-dorsal area of the anterior commissure is also known. Therefore, there is a possibility that neuronal signal relating to the development of fever is transmitted from OVLT to other PO/AH region. However, it seems that the pathway does not exist in the area within 2 to 3 mm laterally to the wall of the 3rd ventricle.

Electrolytic OVLT lesion significantly diminished febrile response of the rabbit to intravenously administered rrIL-1β (Hashimoto & Iriki, 1992). However, none of the rabbits with entirely destroyed OVLT showed remarkable depression of the febrile response in that experiment. Several lines of evidence imply that OVLT region is involved in a part of the mechanism of fever induced by cytokines. Now, we should consider how much important is OVLT in the mechanism of the development of fever.

REFERENCES

Blatteis, C. M., Bealer, S. L., Hunter, W. S., Llanso-Q, J., Ahokas, R. A., and Mashburn, T. A., Jr. (1983): Suppression of fever after lesions of the anteroventral third ventricle in guinea pigs. *Brain Res. Bull.* 11: 519-526.

Blatteis, C. M., Hunter, W. S., Wright, J. M., Ahokas, R. A., Llanos-Q., J., and Mashburn, T. A., Jr. (1987a): Thermoregulatory responses of guinea pigs with anteroventral third ventricle lesions. *Can. J. Physiol. Pharmacol.* 65: 1261-1266.

Blatteis, C. M., Hales, J. R. S., McKinley, M. J., and Fawcett, A. A.(1987b): Role of the anteroventral third ventricle region in fever in sheep. *Can. J. Physiol. Pharmacol.* 65: 1255-1260.

Hashimoto, M., Bando, T., Iriki, M., and Hashimoto, K. (1988): Effect of indomethacin on febrile response to recombinant human IL-1α in rabbits. *Am. J. Physiol.* 255: R527-R533.

Hashimoto, M., Watanabe, M., and Iriki, M. (1989): Comparison of the pyrogenicity between IL-1 and TNF. In *Thermal Physiology*, ed. J. B. Mercer, pp401-406. Amsterdam: Elsevier.

Hashimoto, M., Ishikawa, Y., Yokota, S., Goto, F., Bando, T., Sakakibara, Y., and Iriki, M. (1991): Action site of circulating interleukin-1 on the rabbit brain. *Brain Res.* 540: 217-223.

Hashimoto, M. (1991): Characterization and mechanism of fever induction by interleukin-1 beta. *Pflügers Arch.* 419: 616-621.

Hashimoto, M., and Iriki, M. (1992): Effects of electrolytic brain lesions on cytokine-induced fever in rabbits. *Jpn. J. Physiol.* 42 (Suppl.): S290.

Matsuda, T., Hori, T., and Nakashima, T. (1992): Thermal and PGE2 sensitivity of the organum vasculosum lamina terminalis region and preoptic area in rat brain slices. *J. Physiol.* 454: 197-212.

Matsumura, K., Watanabe, Y., Onoe, H., Watanabe, Y., and Hayaishi, O. (1990): High density of prostaglandin E2 binding sites in the anterior wall of the 3rd ventricle: a possible site of its hyperthermic action. *Brain Res.* 533: 147-151

Sawyer, C. H., Everett, W., and Green, F. D. (1954): The rabbit diencephalon in stereotaxic coordinates. *J. Comp. Neurol.* 101: 801-824.

Stitt, J. T. (1985): Evidence for the involvement of the organum vasculosum laminae terminalis in the febrile response of rabbits and rats. *J. Physiol.* 368: 501-511.

Stitt, J. T., and Shimada, S. G. (1989): Enhancement of the febrile responses of rats to endogenous pyrogen occurs within the OVLT region. *J. Appl. Physiol.* 67: 1740-1746.

Stitt, J. T. (1991): Differential sensitivity in the sites of fever production by prostaglandin E1 within the hypothalamus of the rat. *J. Physiol.* 432: 99-110.

Watanabe, M. (1992): Characteristics of TNFα-and TNFβ-induced fever in the rabbit. *Jpn. J. Physiol.* 42: 101-116.

Integrative and cellular aspects of autonomic functions : temperature and osmoregulation. Eds K. Pleschka, R. Gerstberger. John Libbey Eurotext, Paris © 1994, pp. 137-141.

Fever and preoptic norepinephrine

Clark M. Blatteis, Osamu Shido and Andrej A. Romanovsky

Department of Physiology and Biophysics, University of Tennessee, Memphis, 894 Union Avenue, Memphis, TN 38163, USA

INTRODUCTION

The literature on the role of hypothalamic norepinephrine (NE) in the febrile response to pyrogens is extensive but inconclusive, and the widely discrepant results remain unreconciled (Cox and Lee, 1982). We found recently (Quan and Blatteis, 1989a, b; Quan et al., 1991, 1992a, b) that NE microdialyzed into the preoptic area of the anterior hypothalamus (POA) of conscious guinea pigs, *i.e.*, into the presumptive site of the thermoregulatory controller, depressed metabolic heat production in a dose-dependent and site-specific manner through an α_2-adrenoceptor-mediated mechanism, thereby causing a fall in core temperature (T_c). If NE indeed played a role in the central modulation of fever, it would appear from these results that, in guinea pigs at least, NE should act as a febrilytic rather than as a febrigenic mediator. The present experiments were performed to investigate this possibility.

METHODS

We conducted three experiments, using male, Hartley-strain guinea pigs weighing 300 - 350g. The animals were housed, fed, and habituated to the experimental procedures under the same conditions as we have described previously; a microdialysis probe was implanted unilaterally into the POA of each guinea pig 2 days before an experiment (Quan et al., 1992b).

In the first experiment, *S. enteritidis* lipopolysaccharide (2 µg/kg, i.v.) was administred either simultaneously with or 3 h after the beginning of the intrapreoptic (iPO) microdialysis of NE bitartrate (10 or 20 µg/µl of acidified pyrogen-free saline [PFSa] at a rate of 2 µl/min for 5-7 h).

In the second experiment, LPS was injected i.v. simultaneously with the onset of the iPO microdialysis of the selective α_2-adrenoceptor antagonists rauwolscine hydrochloride (RAU, 1 or 2 µg/µl) or yohimbine hydrochloride (YOH, 1 µg/µl) in PFSa; the microdialysis was continued for 5 h.

In the third experiment, LPS was administered i.v. 90 min after the beginning of the microdialysis of PFSa into the POA. The effluent microdialysate was collected over 20-min periods just before and at specified intervals after the LPS injection; the samples were analyzed for their NE content by HPLC-EC.

The animals were fully conscious during the experiments, and each guinea pig was used only once. Appropriate control groups were included in all the studies. The ambient temperature was 23.5 ± 1.0° C. T_c was monitored at 2-min intervals throughout all the experiments. The iPO localization of the microdialysis probe was confirmed histologically after the experiments.

RESULTS AND DISCUSSION

We found, in experiment one, that NE microdialyzed continuously into the POA significantly and dose-dependently reduced the biphasic T_c rises characteristically evoked in this species by the i.v. injection of 2 µg of *S. enteritidis* LPS/kg (Blatteis, 1974). This effect was evident whether the NE microdialysis was begun coincidentally with (Fig. 1A) or 3 h before (Fig. 1B) the administration of LPS. The latter finding was particularly noteworthy. When LPS was given in this instance, T_c was depressed by the central action of NE and had stabilized at its lowest level. Under these conditions, if the two drugs were merely counteracting each other's effects, that of LPS should have predominated and fever should have developed. Hence, these data would suggest that the observed attenuation of fever was not simply due to the thermolytic action of NE, but rather that the effect of NE in the POA of these conscious guinea pigs represented a direct, febrilytic action of this transmitter.

The results of the second experiment supported this implication. The specific α_2-adrenergic receptor antagonists RAU and YOH microdialyzed into the POA significantly reduced and suppressed, respectively, the troughs of T_c between the first and second peaks of the LPS-induced

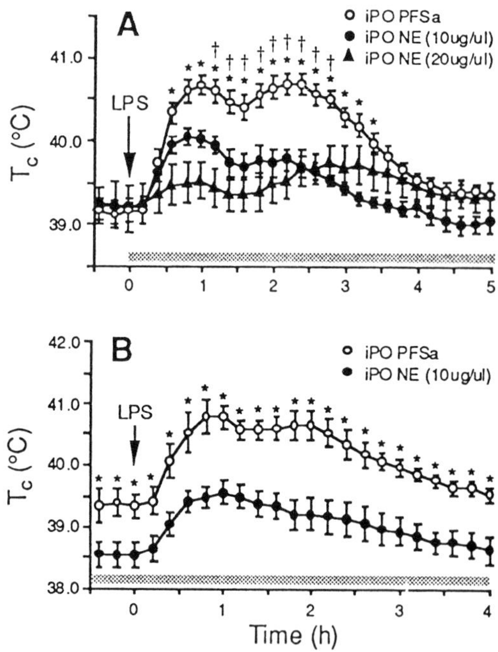

Fig. 1. T_c (means ± SE) before and after the injection of LPS (2 µg/kg, i.v.) into conscious guinea pigs microdialyzed iPO with PFSa (n=5) or NE at 10 (n= 5) or 20 (n=6) µg/ µl. Microdialysis (horizontal bar) started simultaneously (A) or 3 h before (B) the LPS injection (arrow). * and †: significantly different between PFSa and NE groups at 10 and 20 µg/ µl, respectively. (Reprinted with permission from Shido *et al.*, 1993.)

fever. Both also prolonged the elevated T_c after the second peak of fever (Figs. 2A and B). The iPO microdialysis of these α_2-antagonists *per se* had no effect on T_c (not illustrated).

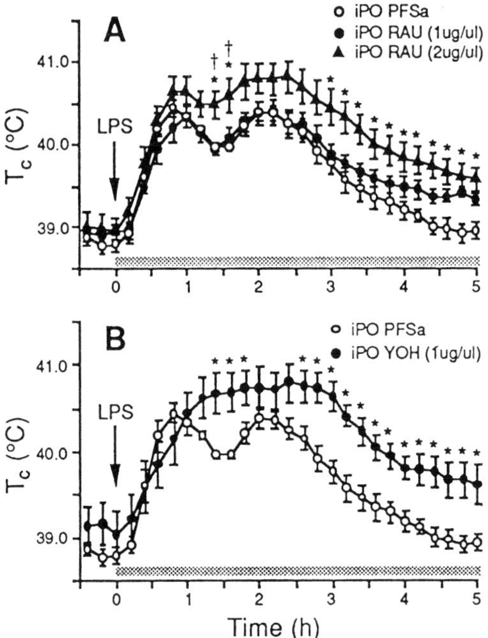

Fig. 2. T_c (means ± SE) before and after the i.v. injection of LPS into conscious guinea pigs microdialyzed with PFSa (n=6), RAU (A) at 1 (n=5) or 2(n=8) µg/ µl, or YOH (B) at 1 µg/ µl (n=5). Microdialysis (horizontal bar) started simultaneously with LPS injection (arrow). * and †: significantly different between PFSa and RAU or YOH groups, respectively. (Reprinted with permission from Shido *et al.*, 1993.)

Finally, in further corroboration, we found in the third experiment that the content of NE in the microdialysate effluent from the POA was elevated coincidentally with the onset of each falling phase of the biphasic febrile response (Figs. 3A and B). These periods corresponded to those when the T_c falls after each peak were abrogated by the iPO microdialysis of the adrenergic receptor antagonists in the previous experiment.

Taken together, these findings suggest that iPO NE may exert a specific febrilytic effect on LPS-induced fever in guinea pigs and, thus, play a physiological role in the initiation of the defervescent phases of the febrile response of this species to LPS. This possibility is compatible with the findings of Morilak et al. (1987) who found that noradrenergic neuronal activity recorded in the locus coeruleus increased only at the peak of muramyl dipeptide-induced fever in freely moving cats. Indeed, there are now various data (Dunn, 1988; Kabiersch et al., 1988) showing that the hypothalamic noradrenergic system becomes activated in response to circulating pyrogens, albeit it is not yet known by what mechanism. One possibility may involve an antagonistic interaction between NE and prostaglandin E_2 (PGE_2) (Quan and Blatteis, 1989), a putative mediator of

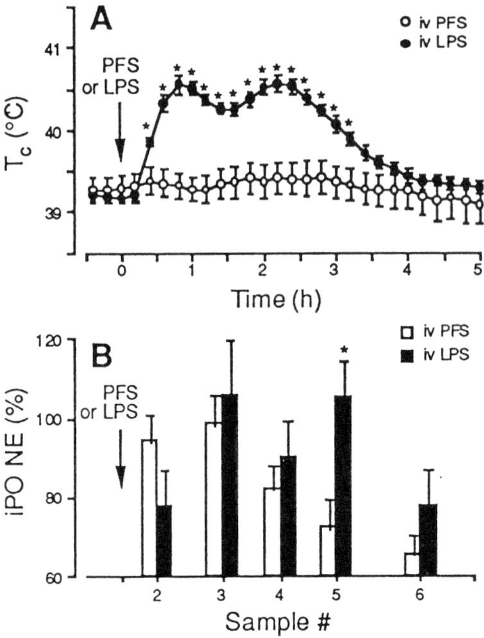

Fig. 3. T_C (A) and iPO NE content (B), the latter expressed as % changes from the pre-injection level (prior to time 0), before and after the i.v. injection of PFS or LPS (arrow) into 9 conscious guinea pigs, respectively. The animals were continuously microdialyzed with PFSa. Sample collection intervals (min after drug): #2 = 20-40, #3 = 70-100, #4 = 130-160, #5 = 170-190, and #6 = 240-260. Values are means ± SE. *: significantly different between groups. (Reprinted with permission from Shido *et al.*, 1993.)

fever in the POA (Coceani, 1991); feedback inhibition by PGE_2 of the presynaptic release of NE in the hypothalamus was documented earlier (Dray and Heaulme, 1984). A possible scenario might be as follows: levels of PGE_2 rising in the POA in response to LPS inhibit the tonic iPO release of NE and drive the febrile rise; subsequently, the decline of PGE_2 disinhibits local NE terminals, allowing iPO NE levels to rise again and triggering febrilysis. However, this is conjectural and awaits further study.

ACKNOWLEDGMENTS

Supported by NIH grant NS22716.

REFERENCES

Blatteis, C.M. (1974): Influence of body weight and temperature on the pyrogenic effect of endotoxin. *Toxicol Appl Pharmacol* 29, 249-258.

Coceani, F. (1991): Prostaglandins and fever: facts and controversies. In *Fever: Basic Mechanisms and Management,* ed. P. Mackowiak, pp. 59-70. New York: Raven.

Cox, B. & Lee, F. (1982): Role of central neurotransmitters in fever. In *Pyretics and Antipyretics,* ed. A.S. Milton, pp. 125-150. Berlin:Springer Verlag.
Dray, F. & Heaulme, M. (1984): Prostaglandins of the E series inhibit release of noradrenaline in rat hypothalamus by a mechanism unrelated to classical α_2 adrenergic presynaptic inhibition. *Neuropharmacology* 23, 457-462.
Dunn, A.J. (1988): Systemic interleukin-1 administration stimulates hypothalamic norepinephrine metabolism paralleling the increased plasma corticosterone. *Life Sci* 43, 429-435
Kabiersch, A., del Rey. A., Honegger, C.G. & Besedovsky, H.O. (1988): Interleukin-1 induces changes in norepinephrine metabolism in the rat brain. *Brain Behav Immun* 2, 267-274.
Morilak, D.A, Fornal, C. & Jacobs, B.L. (1987): Effects of physiological manipulations on locus coeruleus neuronal activity in freely moving cats: I. Thermoregulatory challenge. *Brain Res* 422, 32-39.
Shido, O., Romanovsky, A. A., Ungar, A. L. & Blatteis, C. M. (1993): Role of intrapreoptic norepinephrine in endotoxin-induced frever in guinea pigs. *Am J Physiol* (in press).
Quan, N. & Blatteis, C.M. (1989a): Microdialysis: a system for localized drug delivery into the brain. *Brain Res Bull* 22, 621-625,
Quan, N. & Blatteis, C.M. (1989b): Intrapreoptically microdialyzed and microinjected norepinephrine evoke different thermal responses. *Am J Physiol* 257, R816-R821.
Quan, N., Xin, L. & Blatteis, C.M. (1991): Microdialysis of norepinephrine into preoptic area of guinea pigs: characteristics of hypothermic effect. *Am J Physiol* 261, R378-R385.
Quan, N., Xin, L., Ungar, A. L. & Blatteis , C.M. (1992a): Preoptic norepinephrine-induced hypothermia is mediated by α_2-adrenoceptors. *Am J Physiol* 262, R407-R411.
Quan, N., Xin, L., Ungar, A. L., Hunter, W. S. & Blatteis, C.M. (1992b): Validation of the hypothermic action of preoptic norepinephrine in guinea pigs. *Brain Res Bull* 28, 537-542.

Integrative and cellular aspects of autonomic functions : temperature and osmoregulation. Eds K. Pleschka, R. Gerstberger. John Libbey Eurotext, Paris © 1994, pp. 143-152.

Arginine vasopressin modifies the firing rate and thermosensitivity of neurons in slices of the rat PO/AH area

Jan Moravec and Friedrich-Karl Pierau*

Department of Physiology and Developmental Biology, Charles University Prague, Vinicná 7, 128 00 Prague 2, Czech Republic
** Max-Planck-Institut für physiologische und klinische Forschung, W. G. Kerckhoff-Institut, Parkstrasse 1, D-61231 Bad Nauheim, Germany*

INTRODUCTION

Autonomic thermoregulation in birds and mammals is phylogenetically younger than other homeostatic functions for which the hypothalamus is an important integrative area. It is therefore not surprising that thermosensitive neurons in the preoptic and anterior hypothalamus (PO/AH) appear to be involved not only in the coordination of thermoregulatory mechansims but also of non-thermal homeostatic functions (Hori, 1991). This view is supported by the observation that thermosensitive neurons in the PO/AH are sensitive to osmotic changes and/or to different concentrations of glucose (Silva and Boulant, 1984, Nakashima et al., 1985; Hori, 1991). Experimental data indicate that the antidiuretic hormone arginine vasopressin (AVP) might play an important part for the interrelationship of homeostatic functions, acting either as neurotransmitter or neuromodulator (Riphagen and Pittman, 1986). The concentration of AVP in the cerobrospinal fluid (CSF) was found to be higher than in the blood plasma (Dogsterom et al., 1977).

There is some evidence that AVP may influence the control of body temperature under certain conditions. Changes of ambient temperature as well as local warming or cooling of the PO/AH can affect the release of AVP (Szczepanska-Sadowska, 1974; Forsling et al., 1976; Simon-Oppermann and Jessen, 1977; Zeisberger et al., 1988b). A possible role of AVP as an endogenous antipyretic substance was suggested (Cooper, 1987; Zeisberger, 1990) but the peptide also appears to affect thermoregulation under extreme physiological situations such as thermal adaptation (Zeisberger et al., 1988b), pregnancy (Vaughn et al., 1978; Cooper et al., 1988), dehydration (Szczepanska-Sadowska et al., 1984; Baker and Doris, 1982; Turlejska-Stelmasiak, 1974), hypovolemia (Szczepanska-Sadowska et al,. 1984; Hori, 1991), hypoglycemia (Freinkel et al., 1972; Gale et al., 1981; Kow and Pfaff, 1986), and immobilisation (Zeisberger et al., 1988a). However, the mechanism by which AVP influences the regulation of body temperature remains unclear. In particular the effects of AVP on neurons in the PO/AH have not yet been studied. Therefore, we have investigated the effect of AVP on the activity and thermosensitivity of PO/AH neurons in brain slices from normal rats kept in room temperature (22°C) under "normal" physiological conditions.

METHODS

Preparations and recordings: Coronally oriented slices (400 μm) containing PO/AH tissue from male Wistar rats, weighing 200-250 g, were preincubated in Krebs-Ringer solution for at least 2 h at 35 °C. Subsequently, one slice was placed in a recording chamber (volume 0.5 ml) and continously perfused (1.5 ml/min). Single unit activity in the medial preoptic area (MPA), periventricular nucleus (PE), supraoptic nucleus (SO), and paraventricular nucleus (PAV) were recorded extracellularly with platinum - iridium glass coated microelectrodes.

Temperature stimulation: The temperature of the tissue slice was either kept constant at 38 °C during the search for spontaneously active neurons, or changed periodically by a Peltier thermoassembly within the range of 35-41 °C (rate 0.02 °C/s) to determine temperature sensitivity of the neuron.

Drug application: Concentrated aliquots of AVP (Sigma, Germany) stored at -30 °C were diluted in Krebs-Ringer solution immediately before application . A single bolus injection (150 ng/150 μl) into the inlet tube reached the preparation within 10 s resulting in an instantaneous peptide concentration of about 10^{-7} Mol/l.

Experimental protocol: After isolation of a neuron spontaneous activity was recorded for 10 min at 38 °C, then a temperature cycle (temperature sinus, TS) was performed. After 20 min AVP was applied at 38 °C and neuronal activity was recorded for 15 min before another TS was performed. Spontaneous activity was subsequently recorded till 45 min after AVP application an then another TS was applied.

Data evaluation: The temperature response curve of each neuron was evaluated by a computer program that fitted a piecewise linear as well as a rectilinear regression function to the data over the time period of a complete temperature cycle (Vieth, 1989). Neurons were classified according to the slope of the regression line (temperature coefficient: TC imp/s/°C) in the temperature range where their TC was highest, provided these slopes covered a temperature range ≥ 2 °C. Cells with a positive or negative value of a TC ≥ 0.6 were classified as warm sensitive or cold sensitive, respectively; all other neurons were regarded as temperature insensitive. Changes of the TC after application of AVP were determined using a statistical test for comparison of the two regression lines (Vieth, 1989).
The firing rate (FR) was determined over 1 min intervals preceding the temperature cycle or the bolus injection of AVP. Firing frequency was also calculated over the 15 min following the bolus injection of AVP. The neurons were considered responsive to the drug if the FR changed after the bolus injection by more than 30 per cent over a period of at least 1 minute.

RESULTS

In total 90 single PO/AH neurons were recorded from 28 rats, of which 62 remained stable enough to be used for the entire experiment. The proportion of warm sensitive neurons (WS) was 24 (39 %); 6 neurons (10 %) were cold sensitive (CS), and 32 (51 %) were temperature insensitive (IS). These proportions correspond to those of other studies (Boulant et al., 1989). As indicated in Table 1, all neurons had relatively low firing rates at 38 °C (ranging from 0.02 to 13.0 imp/s), regardless of their thermosensitivity.

	Warm	Cold	Insensitive	Total
Range of firing rates at 38°C	0.4-13.0	0.2-9.8	0.02-8.1	
Firing rate at 38 °C imp/s				
0 -1.5	4	1	17	22
1.5-3.0	6	1	5	12
3.0	14	4	10	28

Table 1. *Single neurons classified according to local thermosensitivity and firing rate at 38 °C.*

In accord with other studies (e.g. Kelso et al., 1982) most of the IS (77%) had low firing rates (≤ 1.5 imp/s), and only 36 percent of the IS had firing rates higher than 3 imp/s.

The main aim of this study was to determine the effect of AVP on PO/AH neurons in vitro. In general, the observed effect was twofold: 1) an immediate change of the firing rate, starting usually within 30 s after application of the drug, and 2) a long term change of the temperature sensitivity of the neuron.

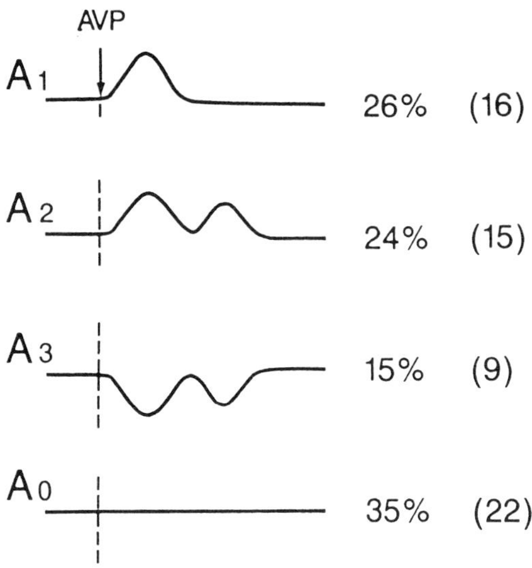

Fig. 1. *Different types of activity changes of PO/AH after bolus injection of AVP (150 ng/150 µl). The Number of reactive neurons is given in brackets.*

Effects of AVP on spontaneous activity.
Of the 62 neurons tested, 31 (50%) increased their firing rate after AVP. The excitatory effect was not uniform and two distinct groups of neurons could be discriminated. The neurons of the first group (A1) responded with only one transient steep increase of FR starting 20-50 s after AVP application (Fig. 3.). The duration of the response varied between 3-7 min. In the second group (A2) the first transient increase of FR was followed by an additional increas of FR within 15 min after peptide administration. Neurons of the third group (A3) were inhibited after AVP injection, exhibiting one or more phases of depression. The depressions were usually not so pronounced and their appearence was not so regular as were the excitatory responses. The repeated activity changes were independent of the type of spontaneous discharge (regular, irregular, bursting). Repeated excitations were more frequent observed in the MPA (47% of the responsive neurons) while in the vasopressinergic nuclei [PE, SO, PAV] 36 % of the neurons demonstrated this pattern. Repeated depressions were almost exclusively observed in the vasopressinergic nuclei and only one of such neurons was observed in the MPA.
An excitatory effect of AVP was observed in most thermoinsensitive neurons (65%) while in the majority of cold neurons (66%) and in almost half of the warm neurons (46%) no change of firing rate was observed (Table 2).
The type of reaction to AVP was related to the firing frequency in WS and IS (Table 3). In both groups the neurons exhibiting high tonic activity were not affected by AVP (A0) while those

	Warm		Cold		Insensitive	
	n	(%)	n	(%)	n	(%)
A1	5	(20)	0		10	(31)
A2	4	(17)	1	(17)	11	(34)
A3	4	(17)	1	(17)	4	(13)
A0	11	(46)	4	(66)	7	(22)
Total	24		6		32	

Table 2. *Classification of hypothalamic neurons in regard to their response pattern to AVP*

with the lowest tonic activity were excited (A1,A2). It may be noted that the differences of the mean firing rates between WS and IS disappeared completely after AVP application in responsive neurons (A1, A2, A3) but remained in the group of unresponsive neurons (A0).

	Warm				Insensitive			
Type of reaction	n	Control		AVP	n	Control		AVP
A1	5	3.6±0.5	*	8.0±1.1	10	2.3±0.3	*	7.9±0.9
A2	4	1.0±0.1	*	2.8±0.2	11	1.3±0.4	*	3.1±0.5
A3	4	4.1±0.3	*	1.3±0.2	4	2.0±0.2	*	1.2±0.1
A0	11	7.5±0.8		7.5±0.6	7	4.3±0.3		3.9±0.3

* $p \leq 0.05$

Table 3. *Mean (±SEM) firing rates before and after application of AVP of warm sensitive and temperature insensitive neurons*

The kind of reaction to AVP appeares to be different in different parts of the POAH. The majority of neurons (86%) localized in the most rostral part of the PO/AH (rostrally to the bregma level -0.3 according to Paxinos and Watson, 1982) were excited by AVP (regardless of their thermosensitivity) (Table 4). In addition, rostrally localized neurons had significantly lower mean firing rates as neurons localized caudally to the bregma level -0.3. Since the mean firing rate of rostral neurons increased significantly after AVP the mean firing rates of rostral and caudal neurons were similar in the presence of AVP.

	Warm				Insensitive			
Localization	n	Control		AVP	n	Control		AVP
rostral	6	3.1±0.3	*	6.0±0.5	6	0.5±0.1	*	4.4±0.4
caudal	18	5.7±0.6		6.3±0.8	26	2.7±0.4	*	4.8±0.6

* $p \leq 0.05$

Table 4. *Mean (±SEM) firing rates before and after application of AVP in warm sensitive and temperature insensitive neurons according to their localization in PO/AH (rostrally or caudally to the bregma level -0.3).*

Effects of AVP on temperature sensitivity of PO/AH neurons.
The TC before and after application of AVP was determined for a total of 56 neurons (23 WS, 5 CS, 28 IS). After AVP application the TC increased significantly in 24 (43%) neurons, decreased in 15 (17%) neurons and remained unaffected in 17 (30%) neurons. Figure 2 summarizes the TC changes observed in different types of neurons. An increased TC after AVP was observed in each of the three types of neurons. This was most prominent in cold sensitive neurons. The changes of the two other types of neurons were depending on their location.

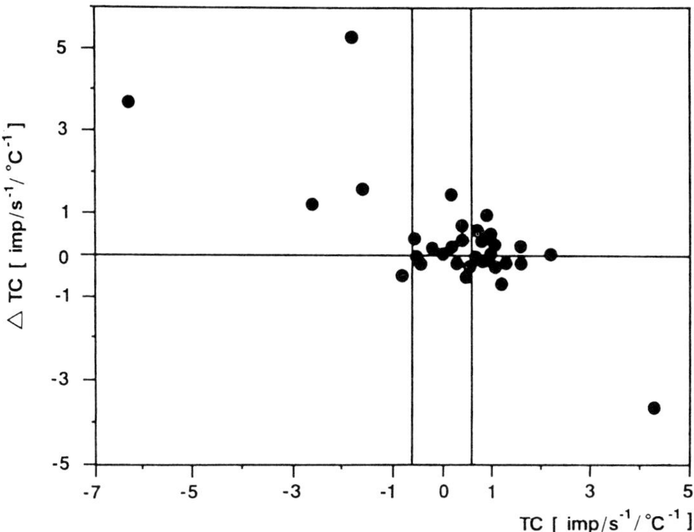

Fig. 2. *Changes of TC of PO/AH neurons after AVP (ordinate) in relation to their TC during the control period (abscissa).*

Effect of AVP on the TC of warm sensitive PO/AH neurons.
The TC of WS ranged from 0.6 to 4.3 imp/s/°C. Of these neurons, 52 % were most thermosensitive above 38 °C, 30 % were most sensitive belowe 38 °C, and only 18 % were equally thermosensitive in both temperature ranges. After AVP application the TC ranged from 0.3 to 3.5 imp/s/°C. Above 38 °C the highest temperature sensitivity was found in 48 % of the neurons, only 17 % of the neurons were most sensitive belowe 38 °C, while the proportion of neurons equally thermosensitive above and below 38 °C increased after AVP by 2 times to 35 %.
A significant increase of the TC (Fig. 4) was only observed in 6 of the 23 warm sensitive neurons (26%). All these neurons were located in the MPA representing 50 % of the warm sensitive neurons in this area. The TC of 3 other MPA neurons decreased, and was unaffected in the rest. The WS of the MPA were not only unique in regard to the AVP effect on their TC but also in respect to the AVP effect on their tonic activity, which was only increased (type A1) or not affected (type A0). On the other hand, in the 6 of the 11 neurons tested in the vasopressinergic nuclei PE, SO, and PAV the TC was decreased after AVP and was unaffected in the rest. The spontaneous activity of 8 of these neurons was decreased (type A3) and unaffected in the rest (type A0) by AVP. This might be an indication for a negative feed-back mechanism for AVP release (Bunn et al., 1986).

Effect of AVP on the TC of cold sensitive PO/AH neurons. The TC of these 5 neurons ranged from -6.3 to -0.8 imp/s/°C. Three of these neurons were most sensitive below and 1 above 38 °C; one neuron was equally thermosensitive in both ranges. After application of AVP the TC ranged from -2.5 to 3.5 indicating a marked upward shift of the TC (4 neurons); this effect was

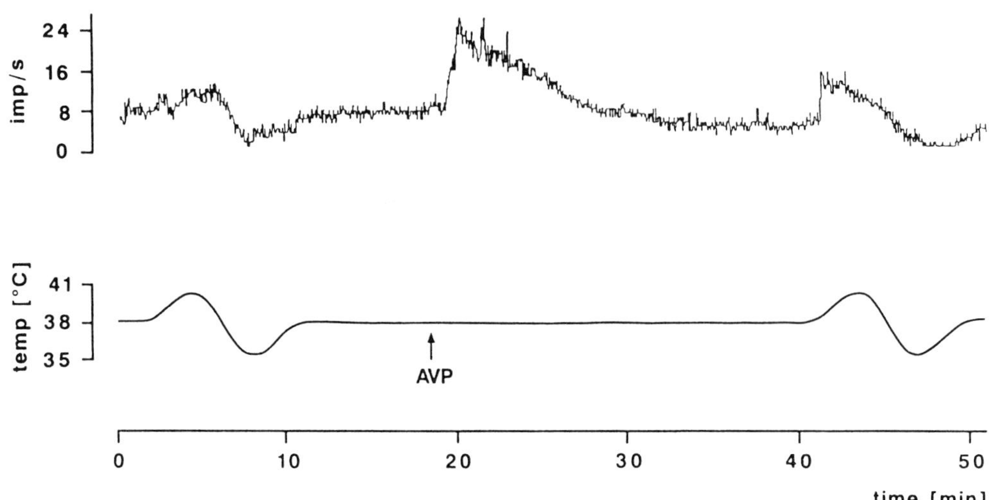

Fig. 3. *Protocol of firing frequency (upper part) of a warm sensitive neuron at 38 °C and during periodical temperature changes before and after bolus application of AVP (150 ng/150 µl). The temperature of the slice is indicated in the lower part.*

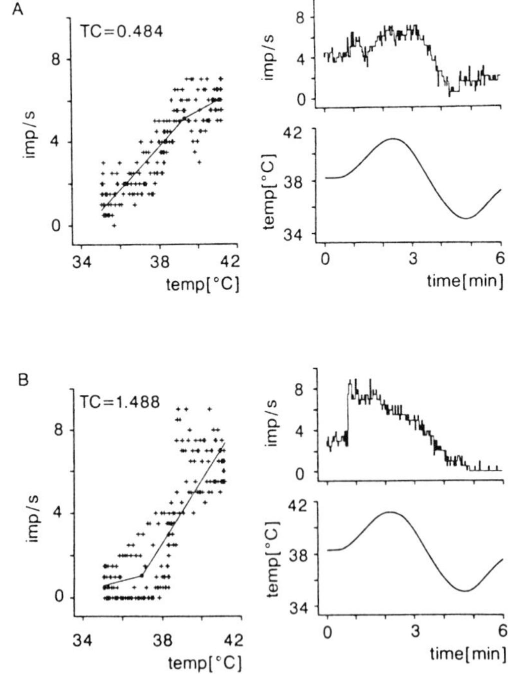

Fig. 4. *Frequency/temperature relationship of a warm sensitive neuron (left site) before (upper part) and after 150 ng/150 µl AVP (lower part). Firing frequency during a temperature cycle before (upper part) and after AVP (lower part) are demonstrated at the right site. Same neuron as in fig. 3.*

large enough to cause one of the neurons to become warm sensitive and another one to become temperature insensitive. The small number of CS does not allow to make regional differences.

Effect of AVP on the TC of temperature insensitive PO/AH neurons. The TC of these 28 neurons ranged from -0.4 to 0.5 imp/s/°C. After AVP application TC values between -0.6 and 1.7 imp/s/°C were observed. A significant increase of the TC was observed in 14 (50%) of the IS whereas the TC decreased in 5 (18%), and was not changed in 9 (32%) neurons. One neuron became cold sensitive while 4 others were transformed into WS. The largest proportion of IS which increased their TC after AVP was observed in the PAV (63%), whereas this reaction was observed in only 36 % of IS in the MPA. It may be noted that no IS were found among the nine neurons tested in the SO (6 WS, 3 CS).

DISCUSSION

The present in vitro study was designed to find possible electrophysiological correlates for thermoregulatory reactions observed after the application of AVP into the brain. Such data might also apply to physiological situations in which the release of AVP is increased. Our data indicate that

(1) the most pronounced effect of AVP on temperature sensitivity was seen in cold sensitive neurons. The TC of these neurons was significantly shifted towards positive values i.e. the cold sensitivity was decreased in 80 % of the CS neurons.

(2) In regard to WS, AVP had different effects in the MPA and the vasopressinergic nuclei. In the former the TC of WS was only increased by AVP (50%); in addition no depression of neuronal activity was observed in these neurons in response to AVP. In the vasopressinergic nuclei, however, AVP mainly increased the TC of IS (63%) whereas the TC in only 36 % of IS was elevated in the MPA. Independent of the area the increased TC was usually accompanied by excitatory reactions (A1, A2) (MPA 100%, other nuclei 67%).

(3) The reaction to AVP were different in neurons located rostrally to bregma level -0.3 to those located caudally of this level. An excitatory reaction to AVP (A1, A2) was typical in rostral neurons (85%) while only 40 % of caudal neurons were excited. The temperature sensitivity was increased in a significantly higher proportion of caudal neurons (47%) than in rostral ones (28%).

(4) AVP excites preferentially neurons with low firing rates, while depression was mainly observed in neurons with higher spontaneous activity.

The preferential activation of neurons with low activity (cf. Table 2, anf Table 3) might indicate that AVP could balance the activity of the neuronal network in situations in which WS may be depressed, e.g. during fever. In fact, such "balancing" effect of AVP might constitute an important feature of the neuromodulatory role of AVP in homeostatic regulatory reactions. This view is supported by other studies on different objects. In in vitro studies Abe et al. (1982) observed in neurons of the SO of guinea-pigs almost exclusively an inhibitory effect of AVP on spike generation which, however, could be reversed to an excitatory one, when AVP was applied to hyperpolarized membranes.

In another study it has been shown that the responses of cereberal vessels to AVP are dependent on the vascular tone. Topical vasopressin produced vasodilation during normotension but vasoconstriction when the vascular tone was decreased by hemorrhagic hypotension (Armstead et al., 1989). The selective excitatory effect of AVP in neurons with low activity might be explained by recent data showing that AVP increases the intracellular calcium concentration by depleting internal stores probably due to the activation of the phosphoinositol system (Lopez-Rivas and Rozengurt, 1984; Hinko et a., 1986). Bombesin, which has also been reported to release calcium ions from intracellular stores (Gallacher et al., 1990) excited more than 60 % of neurons in PO/AH slices in vitro (Schmid et al., 1993). However, in vivo experiments hypothalamic neurons were preferentially inhibited by bombesin (Hori et al., 1986) which might be related to the much higher neuronal activity under these conditions. These results again indicate that "balancing" effects might be a common feature of neuromodulation. In this context, it is of interest that activated cells may induce regenerative calcium waves which may spread from cell to cell, constituting a long-range signal network which may influence distant neuronal circuitries (Berridge, 1993). This appears to be consistent with the repeated excitations or depressions after AVP observed in this study. An alternative explanation would be that neurons might be activated by AVP either directly or indirectly due to physical distortion of the neuronal membrane induced by constriction of nearby microvessels, similar to observations made in hippocampal slices by Albeck and Smock (1988).

In regard to the thermoregulatory effects of AVP its antipyretic action is evident when applied into the ventral septal area (VSA) of febrile animals. However, administration of AVP into the VSA of normothermic animals has no effect on body temperature (Cooper, 1987; Zeisberger 1989; 1990;Fyda et al., 1990). On the other hand, Lin et al. (1983) have observed that administration of AVP into the PO/AH of normothermic rats caused fever and cutaneous vasoconstriction and increased metabolic production. These fever reactions were antagonized by intrahypothalamic pretreatment with yohimbine, propranolol or sodium salicylate.

It is always difficult to relate in vivo data with those obtained in vitro. In the latter the neurons under investigation are exposed to defined concentration of a substance, whereas the actual concentration of substance is difficult to estimate in vivo. In addition, prolonged perfusions into the brain or administration of high doses of a substance might produce desensitization, which is known to develop even at physiological concentrations. Substances may diffuse or may be transported through channels in the interstitial tissue to even distant areas in the brain (Nicholson, 1980). In addition, they may be metabolized or changed by destruction. It has been reported that in the brain the proteolytic product of AVP is even more potent that AVP itself (Burbach et al., 1983). Furthermore, it was a consistent finding in vitro and in vivo studies that WS are depressed and CS are excited by pyrogens while antipyretics reverse this effect (Eisenman, 1982; Boulant et al., 1989; for review). On the other hand, it was found that vasopressinergic terminals of neurons projecting from the bed nucleus of the stria terminalis (BST) inhibit most of the warm responsive neurons in the ventral septal area (Disturnal et al., 1985). Stimulation of BST, however, was reported to suppress prostaglandin E_1 fever (Naylor et al., 1988). Therefore either the WS in the VSA react to pyrogens in the opposite way than do WS in the PO/AH or AVP diffuses from VSA to other places, where it exerts its antipyretic action, as it was supposed by Cooper et al. (1979).

In this study AVP induced a significant decrease of cold sensitivity in 80 % of CS, a significant increase of the TC in 50 % of WS in MPA, and an increase of the tonic activity of WS with low firing rates, which are considered to activate heat loss responses (Boulant et al., 1989). Thus, our results suggest a possible antipyretic action of AVP within the PO/AH. During fever AVP could preferentially increase the firing rate of those WS which were previously depressed by pyrogen. Although increased levels of AVP do not influence body temperature of afebrile animals (Zeisberger, 1989), high levels of AVP in animals exposed to thermal load might balance the depressing effect of cold on WS. An increase of firing frequency could activate heat loss mechansims in pregnant animals which are suppressed by exposure to cold (Vaughn et al., 1978), thus protecting the foetus against hypoxia due to an increased heat production induced in the mothers body by exposure to cold.

REFERENCES

Abe, H., Inoue, M., Matsuo, T., and Ogata, N. (1983): The effect of vasopressin on electrical activity in the guinea-pig supraoptic nucleus in vitro. *J. Physiol. London 337*: 665-685.

Albeck, D., and Smock, T. (1988): A mechanism for vasopressin action in the hippocampus. *Brain Res. 463*: 394-397.

Armstead, W.M., Mirro, R., Busija, D.W., and Leffler, C.W. (1929): Vascular responses to vasopressin are tone dependent in the central circulation of the newborn pig. *Circ. Res. 64*: 136-144.

Baker, M.A., and Doris, P.A. (1982): Effect of dehydration on hypothalamic control of evaporation in the cat. *J. Physiol. London 332*: 457-468.

Boulant, J.A., Curras, M.C., and Dean, J.B. (1989): Neurophysiological aspects of thermoregulation. In *Advances in Comparative and Environmental Physiology Vol. 4*, ed L.C.H. Wang, pp. 117-160. Berlin, Heidelberg: Springer-Verlag.

Bunn, S.J., Hanley, M.R., and Wilkin, G.P. (1986): Autoradiographic localization of peripheral benzodiazepine, dihydroalprenolol and arginine vasopressin binding sites in the pituitaries of control, stalk transected and Brattleboro rats. *Neuroendocrinology 44*: 76-83.

Burbach, J.P.H., Kovács, G.L., DeWied, D., Vannispen, J.W., and Greven, H.M. (1983): A major metabolite of arginine vasopressin in the brain is a highly potent neuropeptide.

Science 221: 1310-1312.
Cooper, K.E. (1987): The neurobiology of fever: Thoughts on recent development. *Ann. Rev. Neurosci. 10*: 297-324.
Cooper, K.E., Blähser, S., Malkinson, T.J., Merker, G., Roth, J. and Zeisberger, E. (1988): Changes in body temperature and vasopressin content of brain neurons, in pregnant and non-pregnant guinea pigs, during fevers produced by Poly I: Poly C. *Pflügers Arch. 412*: 292-296.
Disturnal, J.E., Veale, W.L., Pittman, Q.J. (1985): Electrophysiological analysis of potential arginine vasopressin projections to the ventral septal area of the rat. *Brain Res. 342*: 162-167.
Eisenman, J.S. (1982): Electrophysiology of the anterior hypothalamus: Thermoregulation and Fever. In *Pyretics and Antipyretics*, ed A.S. Milton, pp. 187-217. Berlin, Heidelberg, New York: Springer-Verlag.
Forsling, M.L., Ingram, D.L., and Stanier, M.W. (1976): Effects of various ambient temperatures and of heating and cooling the hypothalamus and cervical spinal cord on antidiuretic hormone secretion and urinary osmolality in pigs. *J. Physiol. London 257*: 673-686.
Freinkel, N., Metzger, B.E., Harris, E., Robinson, S., and Mager, M. (1972): The hypothermia of hypoglycemia. *N. Eng. J. Med. 287*: 841-845.
Fyda, D.M., Mathieson, W.B., Cooper, K.E., and Veale, W.L. (1990): The effectiveness of arginine vasopressin and sodium salycilate as antipyretics in the Brattleboro rat. *Brain Res. 512*: 243-247.
Gale, E.A.M., Bennett, T., Green, J.H., and Macdonald, I.A. (1981): Hypoglycaemia, hypothermia and shivening in man. *Clin. Sci. 61*: 463-469.
Gallacher, D.V., Hanley, M.R., Petersen, O.H., Roberts, M.L., Squire-Pollard, L.G., and Yule, D.I. (1990): Substance P and bombesin elevate cytosolic Ca2+ by different molecular mechanismus in a rat pancreatic acinar cell line. *J. Physiol. London 426*: 193-207.
Hori, T. (1991): An update on thermosensitive neurons in the brain: From cellular biology to thermal and non-thermal homeostatic functions. *Jap. J. Physiol. 41*: 1-22.
Hori, T., Yamasaki, M., Kiyohara, T., and Shibata, M. (1986): Responses of preoptic neurons to poikilothermia-inducing peptides-bombesin and neurotensin. *Pflügers Arch. 407*: 558-560.
Kelso, S.R., Perlmutter, M.N., and Boulant, J.A. (1982): Thermosensitive single-unit activity of in vitro hypothalamic alices. *Am. J. Physiol. 242*: R77-R84.
Kow, L.-M., and Pfaff, D.W. (1986): Vasopressin excites ventromedial hypothalamic glucose-responsive neurons in vitro. *Physiol. Behav. 37*: 153-158.
Lin, M.T., Wang, T.I., and Chan, H.K. (1983): A prostaglandin-adrenergic link occurs in the hypothalamic pathways which mediate the fever induced by vasopressin in the rat. *J. Neural Trans. 56*: 21-31.
Nakashima, T., Hori, T., Kiyohara, T., and Shibata, M. (1985): Osmosensitivity of preoptic thermosensitive neurons in hypothalamic slices in vitro. *Pflügers Arch. 405*: 112-117.
Naylor, A.M., Pittman, Q.J., and Veale, W.L. (1988): Stimulation of vasopressin release in the ventral septum of the rat brain suppresses prostaglandin E_1 fever. *J. Physiol. London 399*: 177-189.
Nicholson, C. (1980): Dynamics of the brain cell microenvironment. *Neurosci. Res. Progr. Bull. 18*: 177-322.
Paxinos, G., and Watson, C. (1982): *The rat brain in stereotaxic coordinates*. Academic Press, Sydney.
Riphagen, C.L., and Pittman, Q.J. (1986): Arginine vasopressin as a central neurotransmitter. *Fed. Proc. 45*: 2318-2322.
Robinson, I.C.A.F. (1983): Neurohypophysical peptides in CSF. *Progr. Brain Res. 60*: 129-145.
Schmid, H.A., Jansky, L., and Pierau, Fr.-K. (1993): Temperature sensitivity of neurons in slices of the rat PO/AH area: effect of bombesin and substance P. *Am. J. Physiol. 264*: R449-R455.
Silva, N.L., and Boulant, J.A. (1984): Effects of osmotic pressure, glucose, and temeprature on neurons in preoptic tissue slices. *Am. J. Physiol. 247*: R335-R345.
Simon, E., and Nolte, P. (1990): Temperature dependence of thermal and nonthermal regulation: Hypothalamic thermo- and osmoregulation in the duck. In *Thermoreception and Temperature Regulation,* eds J.Bligh and K. Voigt, pp. 191-199. Berlin, Heidelberg, New York: Springer Verlag.

Simon-Oppermann, C., and Jessen, C. (1977): Antidiuretic responses to the thermal stimulation of hypothalamus and spinal cord in the conscious goat. *Pflügers Arch. 368*: 33-37.
Szczepanska-Sadowska, E. (1974): Plasma ADH increase and thirst suppression elicited by preoptic heating in the dog. *Am. J. Physiol. 226*: 155-161.
Szczepanska-Sadowska, E., Simon-Oppermann, C., Gray, D.A., and Simon, E. (1984): Plasma and cerebrospinal fluid vasopressin and osmolality in relation to thirst. *Pflügers Arch. 400*: 294-299.
Turlejska-Stelmasiak, E. (1974): The influence of dehydration on heat dissipation mechanisms in the rabbit. *J. Physiol. Paris 68*: 5-15.
Vaughn, L.K., Veale, W.L., and Cooper, K.E. (1987): Impaired thermoregulation in pregnant rabbits at term. *Pflügers Arch. 378*:185-187.
Vieth, E. (1981): Fitting piecewise linear regression functions to biological responses. *J. Appl. Physiol. 67*: 390-396.
Zeisberger, E. (1989): antipyretic action of vasopressin in the ventral septal area of the guinea pig's brain. In *Thermoregulation: Research and Clinical Application,* eds Lomax, Schönbaum, pp. 65-68. Basel: Karger.
Zeisberger, E. (1990): The role of septal peptides in thermoregulation and fever. In *Thermoreception and Temperature Regulation,* eds J. Bligh and K. Voigt, pp. 273-383. Berlin, Heidelberg, New York: Springer Verlag.
Zeisberger, E., Merker, G., and Roth, J. (1988a): Role of septal vasopressin in inhibition of fever during immobilization in guinea pig. *Pflügers Arch. 412*: (Suppl. 1), R89.
Zeisberger, E., Roth, J., and Simon, E. (1988b): Changes in water balance and in release of arginine vasopressin during thermal adaptation in guinea pigs. *Pflügers Arch. 412*: 285-291.

Integrative and cellular aspects of autonomic functions : temperature and osmoregulation. Eds K. Pleschka, R. Gerstberger. John Libbey Eurotext, Paris © 1994, pp. 153-162.

Interleukin-1β production in the circumventricular organs during endotoxin fever in rabbits

Tomoki Nakamori, Akio Morimoto, Kazuhito Yamaguchi*, Tatsuo Watanabe and Naotoshi Murakami

*Department of Physiology, *Institute of Laboratory Animals, Yamaguchi University School of Medicine Ube, Yamaguchi 755, Japan*

Abstract

Interleukin-1ß (IL-1ß) production in the circumventricular organs was investigated in febrile rabbits using *in situ* hybridization and immunohistochemical techniques. The circumventricular organs examined in this study were organum vasculosum lamine terminalis (OVLT), subfornical organ (SFO) and area postrema (AP). Fever was induced by intravenous (i.v.) injection of comparative low dose of bacterial endotoxin [lipopolysaccharide (LPS), (4 µg/kg)], or human recombinant IL-1ß (hIL-1ß) (2 µg/kg). In the OVLT, SFO and AP, significant production of IL-1ß was observed following the LPS injection, but not after hIL-1ß injection.

Introduction

It is widely accepted that bacterial endotoxin, lipopolysaccharide (LPS), causes fever by stimulating monocytes and macrophages to produce pyrogenic cytokine, interleukin-1 (IL-1). IL-1 acts on the central nervous system to synthesize and release prostaglandins which, in turn, induce the febrile response (Dinarello, 1984; Kluger, 1991). Moreover, it is well known that, during LPS fever, cardiovascular, osmotic and hormonal responses occur (Cooper, 1971; Del Rey & Besedovsky, 1992); possible involvement of IL-1 in some of these responses have been suggested (Morimoto et al., 1992; Nakamori et al., 1993; Watanabe et al., 1990). However, to our knowledge, no one has demonstrated elevation of plasma IL-1 level during naturally occurring fever. In addition, it is still controversial whether the peripherally synthesized IL-1, a large protein molecule (17.4 kDa) (Kluger, 1991), enters the blood-brain barrier (BBB). (Banks et al.,1989; Blatteis et al.,1989). Therefore, it is not clear whether peripherally produced IL-1 is actually involved in the development of physiological responses induced by LPS. In contrast, it has been reported that IL-1 is possibly produced by brain parenchymal cells, such as glial cells, when incubated with LPS, *in vitro* (Fontana et al., 1982; Giulian et al., 1986) or following challenges with several stimuli *in vivo*, such as restrain stress (Minami et al., 1991) and administration of extremely high dose of LPS (Fontana et al., 1984).

Furthermore, Coceani et al. (1988) have reported that IL-1ß level in cerebrospinal fluid elevates during fever in cats. Taken together, it is suggested that, during LPS fever, IL-1 synthesis occurs in or near the brain.

Circumventricular organs such as organum vasculosum laminae terminalis (OVLT), subfornical organ (SFO) and area postrema (AP) are known as windows in the brain which are opened to the blood-borne substances. These organs play important roles in physiological responses, such as febrile (Blatteis et al., 1989), cardiovascular (Brody, 1988; Cox et al., 1990; Gross et al., 1990; Konrad et al., 1992; Tramposch et al. 1989), osmotic (Brody, 1988; Iovino et al., 1988; Miselis et al., 1984; Tramposch et al. 1989) and hormonal (Brody, 1988; Ferguson et al., 1990; Komaki et al., 1992) responses. In addition, the OVLT, SFO and the AP are reported to possess wide perivascular spaces around well-developed capillaries with fenestration. Several kinds of reticuloendothelial cells, including macrophages, which might be the source of IL-1, are found in these regions (Faraci, et al., 1989; Gross, 1992; Weindl, 1973; Yamaguchi et al, 1993). Therefore, it is possible that, IL-1 production takes place in circumventriclar organs in response to systemic LPS that enters the fenestrated capillaries of these organs. IL-1 then takes actions near or within the circumaventriclar organs to induce several responses.

In this study, we used *in situ* hybridization and immunohistochemical techniques to determine whether IL-1ß is synthesized in the rabbit's circumventricular organs in response to LPS. We further investigated the IL-1 production in the circumventricular organs following human recombinant IL-1ß (hIL-1ß) administration.

Methods

Male New Zealand white rabbits weighing 2.6 - 3.0 kg were used in this study. They were housed in a room maintained at 21 ± 1 °C with a 12 h light/dark cycle. Food and water were available *ad libitum.*

Rabbits received i.v. injection of either LPS, hIL-1ß or saline. Temperature, *in situ* hybridization and immunohistochemical studies were performed in this study.

Materials

The LPS used in this study was that of *Salmonella typhosa* endotoxin (Difco). LPS or hIL-1ß supplied by Otsuka Pharmaceutical Co. Ltd. (Japan) was dissolved in the sterile saline.

For *in situ* hybridization, proteinase K was purchased from Boehringer (Mannheim, Germany), and polyadenylic acid, heparin sodium salt, yeast total RNA (Type XI) and RNase A (Type III-A) were purchased from Sigma (USA).

For immunohistochemical staining, Vectastain ABC kit (Vector Laboratories Inc. USA) was used.

Rabbit IL-1ß cDNA clone and goat-anti-rabbit IL-1ß polyclonal antibody were a gift from prof. Yoshinaga (Kumamoto Univ. Japan).

Injection doses including appropriate control injections are described in Results.

Measurement of changes in body temperature

On the day of the body temperature experiment, animals were minimally restrained in conventional rabbit stocks, at an ambient temperature of 21 ± 1 °C between 0900 and 1600. Throughout the experiment, the rectal temperature was measured every minute with a copper-constantan thermocouple. The rectal temperature in each animal was allowed to stabilize for at least 90 min before any injections were made. The i.v. injections of LPS, hIL-1ß or saline were made into the marginal ear vein.

Preparation of probes

For the synthesis of antisense and sense RNA probes for *in situ* hybridization study, the rabbit IL-1ß cDNA (about 1.4 kb) was subcloned into pGEM-7Zf(+) vector (Promega, USA). ^{35}S-labeled RNA probes were transcribed with SP6 or T7 RNA polymerase in the presence of uridine 5'-a-[^{35}S] thiotriphosphate (30 TBq/mmol, Amersham, UK) using Riboprobe Gemini system II (Promega, USA). Partial alkaline hydrolysis of the probe was made to shorten the probe size to 100-150 base.

Drug treatment and tissue preparation

For *in situ* hybridization study, rabbits were injected with LPS (4 µg/kg), hIL-1ß (2 µg/kg) or saline (0.1 ml/kg). Each experiment was performed by using 4 rabbits. They were sacrificed at 1 h after i.v. injection of LPS, hIL-1ß or saline by a large dose of pentobarbital. Brains were rapidly removed and frozen in liquid nitrogen. Coronal frozen sections were cut at 12 µm thickness by a microtome cryostat, thaw-mounted onto gelatin-coated slides, fixed in 4% formaldehyde/10 mM phosphate-buffer (PBS) for 30 min at room temperature, rinsed twice in PBS for 5 min, treated with 1 µg/ml proteinase K for 10 min at 37 °C, and rinsed in PBS for 1 min and 3 min again. After being rinsed twice in 2 mg/ml glycine/PBS for 10 min, they were immersed in 0.25% acetic anhydride in 0.1% triethanolamine/0.9% NaCl for 10 min, dehydrated in a series of ethanol (70, 85, 95, 100%) and dipped in chloroform for 5 min. Prehybridization was carried out at 55 °C for 3 h with the prehybridization buffer: 50% formamide, 4xSSC (1xSSC = 0.15 M NaCl/0.015 M sodium citrate, pH7.0) and 20 mM dithiothreitol. Then, ^{35}S-labeled antisense and sense probes were diluted to about 10^6 cpm/section with hybridization buffer for 50 µl/section [Hybridization buffer: 50% formamide, 4xSSC, 5xDenharardt's solution (0.05% each of polyvinylpyrrolidone, bovine serum albumin and Ficoll), 10 mM etylenediaminetetraacetic acid, 33 µg/ml polyadenylic acid, 200 µg/ml heparine, 250 µg/ml yeast total RNA, 500 µg/ml denatured salmon sperm DNA, 20 mM dithiothreitol and 10% dextran sulfate]. The hybridization was carried out at 55 °C for 18 h. Thereafter, the sections were rinsed four times in 2xSSC/10 mM dithiothreitol for 10 min at 55 °C, treated with RNase A (50 µg/ml in 0.5 M NaCl, 10 mM Tris, 1 mM ethylenediaminetetraacetic acid, pH8.0) for 30 min at 37 °C and rinsed twice in 50% formamide/2xSSC/10 mM dithiothreitol for 30 min at 55 °C. After being rinsed in 0.1X SSC/10 mM dithiothreitol for 30 min at 50 °C, the sections were dehydrated with a series of ethanol as above and air-dried. They were exposed to autoradiographic emulsion NTB-3 (Kodak, New York) diluted 1:1 with water for 3 weeks, followed by counterstaining with Hematoxilin-Eosine.

For immunohistochemical study, rabbits were injected with LPS (4 µg/kg) or saline (0.1 ml/kg). Each experiment was performed using 4 rabbits. They were sacrificed at 1 h after i.v. injections of LPS, hIL-1ß or saline by a large dose of pentobarbital. Immediately after pentobarbital injection, they were perfused with 0.1 M cacodylate solution containing 4% paraformaldehyde and 0.1% glutalaldehyde. Subsequently brains were removed and sliced by a microslicer at 40 µm thickness. Thereafter, the slices were pretreated with 5% normal rabbit serum for 30 min to block non-specific binding, then incubated with a 1/800 dilution of goat-anti-rabbit IL-1ß serum at 4 °C overnight. After being rinsed in Tris-HCl containing 0.5 M NaCl, the sections were processed by avidin-biotinylated enzyme complex (Vectastain, ABC kit, Vector Laboratories Inc. U.S.A.). After being rinsed, sections were washed several times in Tris-HCl buffer and reacted with diaminobenzidine containing 0.02% H_2O_2 for 10 min.

Results

Figure 1 shows changes in rectal temperature after i.v. injection of LPS (4 µg/kg; n=5) or hIL-1ß (2 µg/kg; n=5). The i.v. injections of both LPS and hIL-1ß induced biphasic fevers. The first peak of the biphasic fever occurred at 1.25 h with the second peak occurring around 3 h. In contrast, i.v. injection of saline did not induce significant changes (data not shown).

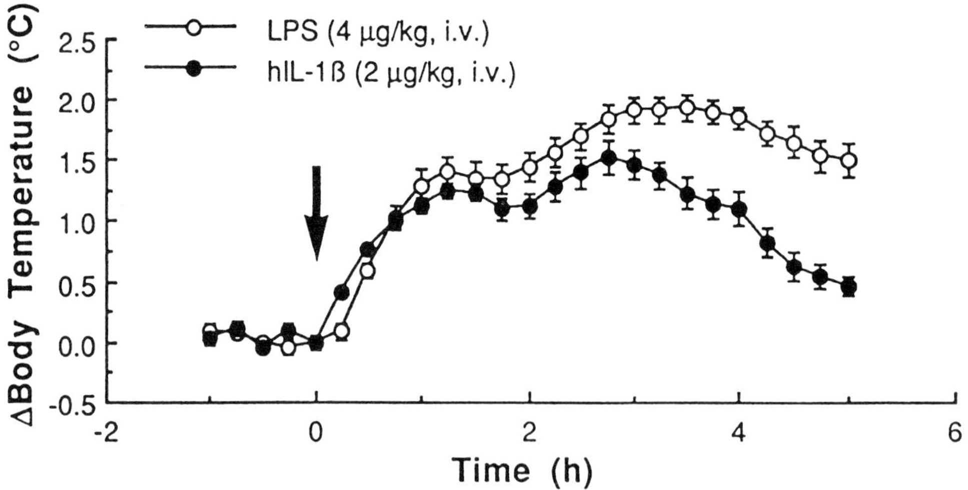

Fig.1. Mean changes (mean ± S.E.) in rectal temperature of 5 rabbits after i.v. injection of LPS (4 µg/kg) or hIL-1ß (2 µg/kg). Arrow indicates the time of injection.

Figure 2 shows the expression of IL-1ß mRNA in the coronal sections of the OVLT (a & b) at 1 h after injection of LPS (4 µg/kg; a) or hIL-1ß (2 µg/kg; b). As shown in Fig. 2a, IL-1ß mRNA was significantly expressed in the OVLT following the i.v. injection of LPS. In contrast, in the OVLT of hIL-1ß injected rabbits, there was no expression of IL-1ß mRNA. Sence RNA, which has the same sequence of nucleotides with IL-1ß mRNA, was used as a control probe, and no hybridization signal to this probe was seen. We further investigated expression of IL-1ß mRNA in the OVLT at 1 h after i.v. injection of saline (0.1 ml/kg). However, there was no hybridization signal in the OVLT (data not shown).

a: LPS (4 µg/kg, i.v.) **b: hIL-1ß (2 µg/kg, i.v.)**

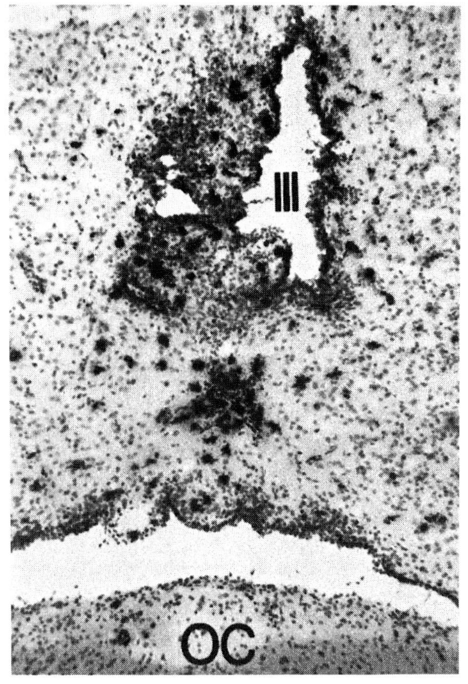

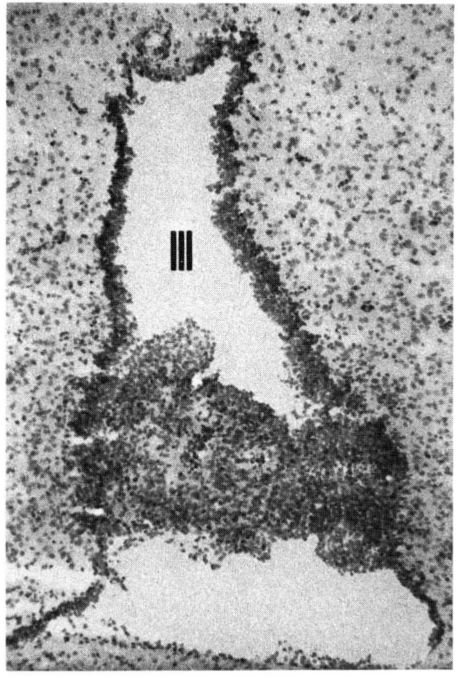

OC: optic chiasma, III: third ventricle
(— 100 µm)

Fig.2. Light-micrographs of IL-1ß mRNA expression in the OVLT (coronal sections) at 1 h after i.v. injection of LPS (4 µg/kg) (a) or hIL-1ß (b).

Figure 3 shows the distribution of IL-1ß-immunoreactive cells in the sections of the OVLT (A), SFO (B) or AP (C) at 1 h after injection of LPS (4 μg/kg). As shown in Fig. 3A, B & C, dendriform IL-1ß-immunoreactive cells were scattered in the OVLT, SFO and AP following the injection of LPS. Throughout the brain, some cells in the pia mater also showed IL-1ß-immunoreactivity, however, the expression was somewhat lower than that seen in circumventricular organs and there was no other regions which showed significant immunoreactivity. In contrast, in the OVLT, SFO or AP of saline-injected rabbits, no immunoreactive cell was observed (data not shown). We further investigated the distribution of IL-1ß-immunoreactive cells in the OVLT, SFO or AP at 1 h after i.v. injection of hIL-1ß (2 μg/kg). However, there was no IL-1ß-immunoreactive cells (data not shown).

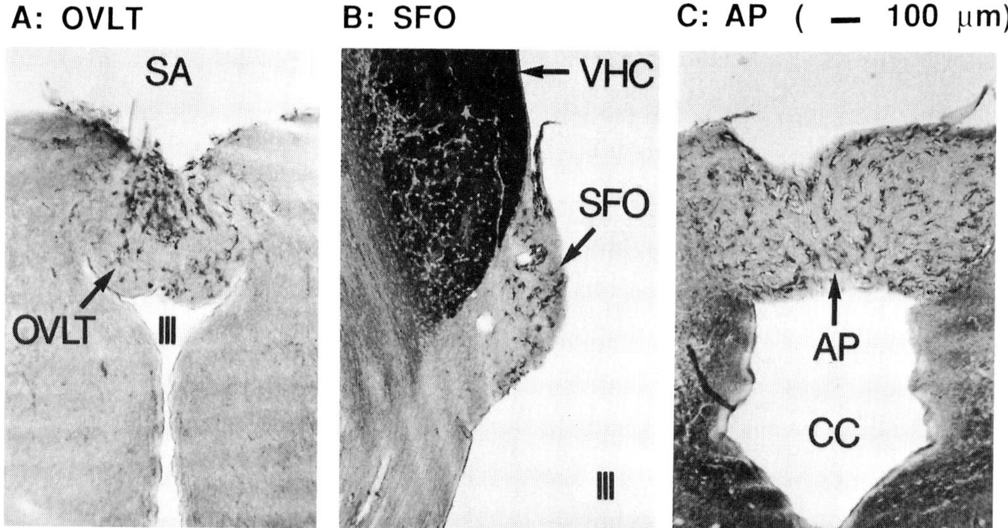

Fig.3. Light-micrographs of IL-1ß-immunoreactive cells in the OVLT (transversal section) (A), SFO (sagital section) (B) or AP (coronal section) (C) at 1 h after i.v. injection of LPS (4 μg/kg).

Discussion

The present results clearly show that, during the febrile response to a physiological dose of LPS, IL-1ß is predominantly synthesized in the cell of circumventricular organs, such as OVLT, SFO and AP. Furthermore, i.v. injection of hIL-1ß did not induce IL-1ß production in the OVLT, SFO and AP. Therefore, it is unlikely that IL-1ß production in these organs during LPS-induced fever is mediated by peripherally produced IL-1ß, at least within 1 h. Since the circumventricular organs are rich in fenestrated capillaries (Faraci et al., 1989; Gross, 1992; Weindl, 1973; Yamaguchi et al, 1993), we cannot rule out the possibility that circulating IL-1 enters the brain across these area. However, despite the efforts of many researchers, a significant increase in plasma IL-1 levels has not been reported during naturally occuring fever. Therefore, it is likely that under physiological conditions, it is the brain level of IL-1 rather than circulating IL-1 that is responsible for the development of physiological responses during endotoxin fever.

Recently, the OVLT is noticed as an interface for induction of fever. We have shown that microinjection of endogenous pyrogen, the active component of which is IL-1ß, into the region near the OVLT causes significant fever and microinjection of sodium salicylate, an inhibitor of prostaglandin synthesis, into the OVLT attenuates the fever induced by i.v. injection of LPS (Morimoto et al., 1988). Moreover, Matsumura et al. (1990) have reported that high distribution of prostaglandin E_2 binding site is observed around the OVLT. In addition, febrile response induced by LPS is significantly reduced after ablation of region including the OVLT (Blatteis et al., 1983). Taken together these facts with the present results, the IL-1ß produced in the OVLT in response to physiological dose of LPS is likely to stimulate prostaglandin release around the OVLT to induce febrile response.

In contrast, several researchers have reported that systemic administration of IL-1 resulted in release of arginine vasopressin (AVP) and corticotropin releasing factor (CRF) (Katsuura et al., 1990; Naito et al., 1991; Nakatsuru et al., 1991; Sapolsky et al., 1987). They suggested that systemic IL-1 passes through the circumventricular organs and acts on hypothalamus to stimulate the secretion of these peptides. CRF (Brown et al., 1982; Fisher, 1989; Morimoto et al., 1993), a factor that stimulates pituitary adrenocortical axis, and AVP (Rohmeiss et al., 1986), a well known hormone involved in osmotic regulation, are currently proposed as a sympathomimetic neuropeptides. Based on the present results, it could be hypothesized that LPS stimulates IL-1 production in the circumventricular organs, which, in turn, induces AVP and CRF release that may play an important roles to cause cardiovascular, osmotic and hormonal regulations during LPS fever.

Acknowledgements.

This study was partly supported by Grant-in-Aid for Scientific Research (A03404018) from the Ministry of Education, Science and Culture of Japan. We thank Prof. M. Yoshinaga (Kumamoto Univ. School of Medicine, Japan) for the kind gift of rabbit IL-1ß cDNA clone and goat anti-rabbit IL-1ß polyclonal antibody, and the Otsuka

Pharmaceutical Co. Ltd. (Japan) for the kind supply of human recombinant IL-1ß.

References

Banks, W.A., Kastin, A.J. & Durham, D.A. (1989): Bidirectional transport of interleukin-1 alpha across the blood-brain barrier. Brain Res. Bull. 23, 433-437.

Blatteis, C.M., Bealer, S.L., Hunter, W.S., Llanos-Q, J., Ahokas, R.A. & Mashburn, Jr.T.A. (1983): Suppression of fever after lesions of the anteroventral third ventricle in guinea pigs. Brain Res. Bull. 11, 519-526.

Blatteis, C.M., Dinarello, C.A., Shibata, M., Llanos-Q, J, Quan, N. & Busija, D.W. (1989): Does circulating interleukin-1 enter the brain? In Thermal physiology, ed. Mercer, J.B., pp. 385-390. USA: Elsevier.

Brody, M.J. (1988): Central nervous system and mechanisms of hypertension. Clin. Physiol. Biochem. 6, 230-239.

Brown, M.R., Fisher, L.A., Spiess, J., Rivier, C., Rivier, J., & Vale, W. (1982): Corticotropin-releasing factor: actions on the sympathetic nervous system and metabolism. Endocrinology 111, 928-931.

Coceani, F., Lees, J. & Dinarello, C.A. (1988): Occurrence of interleukin-1 in cerebrospinal fluid of the conscious cat. Brain Res. 446, 245-250.

Cooper, K.E.(1971): Some physiological and clinical aspects of pyrogens. In *Pyrogens and fever*, ed.G.E.W.Wolstenholme & J. Birch, pp5-21. London: Churchill Livingstone.

Cox, B.F., Hay, M. & Bishop V.S. (1990): Neurons in area postrema mediate vasopressin-induced enhancement of the baroreflex. Am. J. Physiol. 258, H1943-1946.

Del Rey, A. & Besedovsky, H.O. (1992): Metabolic and neuroendocrine effects of pro-inflammatory cytokines. Eur. J. Clin. Invest. 22, 10-15.

Dinarello, C.A. (1984): Interleukin-1. Rev. Infect. Dis. 6, 51-95.

Faraci, F.M., Choi, J., Baumbach, G.L., Mayhan, W.G. & Heistad, D.D. (1989): Microcirculation of the area postrema. Permeability and vascular responses. Circ. Res. 65, 417-425.

Ferguson, A.V., Donevan, S.D., Papas, S. & Smith, P.M. (1990): Circumventricular structures: CNS sensors of circulating peptides and autonomic control centers. Endocrinol. Exp. 24, 19-27.

Fisher, L.A. (1989): Corticotropin-releasing factor: endocrine and autonomic integration of responses to stress. Tr. Pharmaco. Sciences 10: 189-193.

Fontana, A., Kristensen F., Dubs, R., Gemsa, D. & Wber, E. (1982): Production of prostaglandin E and an interleukin-1 like factor by cultured astrocytes and C6 glioma cells. J. Immunol. 129, 2413-2419.

Fontana, A., Weber, E. & Dayer, J. M. (1984): Synthesis of interleukin 1/endogenous pyrogen in the brain of endotoxin-treated mice: a step in fever induction? J. Immunol. 133, 1696-1698.

Giulian, D., Baker, T.J., Shih, L.N. & Lachman, L.B. (1986): Interleukin-1 of the central nervous system is produced by ameboid microglia. J. Exp. med. 164, 594-604.

Gross, P.M.(1992): Circumventricular organ capillaries. Pro. Brain Res. 91, 219-233.

Gross, P.M., Wainman, D.S., Shaver, S.W., Wall, K.M. & Ferguson, A.V. (1990): Metabolic activation of efferent pathways from the area postrema. Am. J. Physiol. 258, R788-797.
Iovino, M., Papa, M., Monteleone, P. & Steardo, L. (1988): Neuroanatomical and biochemical evidence for the involvement of the area postrema in the regulation of vasopressin release in rats. Brain Res. 447, 178-182.
Katsuura,G., Arimura, A., Koves, K. & Gottschall, P.E. (1990): Involvement of organum vasculosum of laminae terminalis and preoptic area in interleukin-1ß-induced ACTH release. Am. J. Physiol. 258, E163-171.
Kluger, M.J. (1991): Fever: Role of pyrogens and cryogens. Physiol. Rev. 71, 93-127.
Konrad, E.M., Thibault, G. & Schiffrin, E.L. (1992): Atrial natriuretic factor binding sites in rat area postrema: autoradiographic study. Am. J. Physiol. 263, R747-755.
Komaki, G., Arimura, A. & Koves, K. (1992): Effect of intravenous injection of IL-1ß on PGE_2 levels in several brain areas as determined by microdialysis. Am. J. Physiol. 262, E246-251.
Matsumura, K., Watanabe, Y., Onoe, H., Watanabe, Y. & Hayaishi, O. (1990): High density of prostaglandin E_2 binding sites in the anterior wall of the 3rd ventricle: a possible site of its hyperthermic action. Brain Res. 533, 147-151.
Minami, M., Kuraishi, Y., Yamaguchi, T., Nakai, S., Hirai,Y. & Satoh, M. (1991): Immobilization stress induces interleukin-1ß mRNA in the rat hypothalamus. Neurosci. Let. 123, 254-256.
Miselis, R.R., Hyde, T.M. & Shapiro, R.E. (1984): Disturbances in water balance controls following lesions to the area postrema and adjacent solitary nucleus. In *The Physiology of Thirst and Sodium Appetite*,ed. G. DeCaro, A.N. Epstein & M. Massi,pp. 279-285. New York: Plenum.
Morimoto, A., Murakami, N., Nakamori, T. & Watanabe, T. (1988): Multiple control of fever production in the central nervous system of rabbits. J. Physiol. Lond. 397, 269-280.
Morimoto, A., Nakamori, T., Morimoto, K., Tan, N. & Murakami, N. (1993): The central role of corticotrophin-releasing factor (CRF-41) in psychological stress in rats. J. Physiol. Lond. 460, 221-229.
Morimoto, K., Morimoto, A., Nakamori, T., Tan, N. Minagawa, T & Murakami, N. (1992): Cardiovascular responses induced in free-moving rats by immune cytokines. J. Physiol. Lond. 448, 307-320.
Naito, Y., Fukuta, J., Shindo, K.,Ebisui, O., Murakami, N., Tominaga, T., Nakai, Y., Mori, K., Kasting, N.W. & Imura, H. (1991): Effects of interleukins on plasma arginine vasopressin and oxytocin levels in conscious, freely moving rats. Biochem. Biophys. Res. Commun. 174, 1189-1195.
Nakamori, T., Morimoto, A. & Murakami, N. (1993): Effect of a central CRF antagonist on cardiovascular and thermoregulatory responses induced by stress or IL-1ß. Am. J. Physiol. (in press).
Nakatsuru, K., Ohgo, S., Oki, Y. & Matsukura, S. (1991): Interleukin-1 (IL-1) stimulates arginine vasopressin (AVP) release from superfused rat hypothalamo-neurohypophyseal complexes independently of cholinergic mechanism. Brain Res. 554, 38-45.

Rohmeiss, P., Becker, H., Dietrich, R., Luft, F. & Unger, T. (1986): Vasopressin: mechanism of central cardiovascular actions in conscious rats. J. Cardiovasc. Pharmacol. 8, 689-696.
Sapolsky, R., Rivier, C., Yamamoto, G., Plotsky, P. & Vale, W. (1987): Interleukin-1 stimulates the secretion of hypothalamic corticotropin-releasing factor. Science 238, 522-524.
Tramposch, A., Lopes, O.U., Chernicky, C.L. & Ferrario, C.M. (1989): Alternative mechanism for attenuated pressor responses in AV3V lesioned dogs. Am. J. Physiol. 257, R431-438.
Watanabe, T., Morimoto, A., Sakata, Y. & Murakami, N. (1990): ACTH response induced by interleukin-1 is mediated by CRF secretion stimulated by hypothalamic PGE. Experientia Basel 46, 481-484.
Weindl, A. (1973): Neuroendocrine aspects of circumventricular organs. In *Frontiers in neuroendocrinology* vol.3, ed. W.F. Ganong & L. Martini, pp.3-32, New York, Oxford.
Yamaguchi, K., Morimoto, A. & Murakami, N. (1993): Organum vasculosum laminae terminalis (OVLT) in rabbit and rat: topographic studies. J. Compara. Neurol. 330, 352-362.

Integrative and cellular aspects of autonomic functions : temperature and osmoregulation. Eds K. Pleschka, R. Gerstberger. John Libbey Eurotext, Paris © 1994, pp. 163-170.

ACTH response induced in rabbits by systemic administration of bacterial endotoxin

Tatsuo Watanabe, Tomoki Nakamori, James M. Lipton* and Naotoshi Murakami

*Department of Physiology, Yamaguchi University School of Medicine, Ube Yamaguchi 755, Japan and *Department of Physiology and Anesthesiology, The University of Texas, Southwestern Medical Center at Dallas, 5323 Harry Hines Boulevard, Dallas, TX 75235-9040, USA*

Introduction

It is widely accepted that one of the potent pyrogenic cytokines, interleukin-1 (IL-1) (Dinarello, 1984), is synthesized and released from monocytes activated by pathological stimuli such as bacterial endotoxin. It has been demonstrated that IL-1 induces an increase in plasma concentration of adrenocorticotrophic hormone (ACTH) (Besedovsky et al., 1986). Therefore, under infectious conditions, endotoxin might activate the pituitary-adrenocortical axis via IL-1 action. In contrast, we have recently shown that the ACTH and febrile responses induced by intravenous (i.v.) injection of IL-1 were suppressed by the pretreatment with prostaglandin synthesis inhibitor, indomethacin (Morimoto et al., 1989; Watanabe et al., 1991b). This suggests the involvement of prostaglandins in the development of IL-1 induced responses. Furthermore, intrahypothalamic injection of prostaglandin E, a final mediator of fever production (Stitt, 1986), resulted in marked increase in plasma level of ACTH (Morimoto et al., 1989), that had been suppressed by the pretreatment with anti-corticotrophin releasing factor (anti-CRF) antibody (Watanabe et al., 1990). These observations suggest that prostaglandin E that produces fever as a final mediator induces the ACTH response by stimulating the release of hypothalamic hormone, CRF.

This study was carried out to investigate the mechanism of induction of ACTH response in rabbits. We have proposed a hypothesis that there exist two separate PG-mediated mechanisms of induction of biphasic fever in rabbits one inside and one outside the blood-brain barrier (Morimoto et al., 1987). The purpose of the present study was, therefore, to examine whether the similar mechanisms are responsible for the induction of the ACTH response in rabbits by systemic administration of bacterial endotoxin. The results showed that the ACTH response induced by i.v. injection of endotoxin was significantly suppressed by systemic pretreatment with indomethacin, while intracerebroventricular (i.c.v.) injection of indomethacin had no effect on this response. Furthermore, immunohistochemical study showed that IL-1 immunoreactive cells were observed in the anterior lobe of the pituitary gland after i.v. injection of bacterial endotoxin. These results suggest that prostaglandins synthesized outside the blood-brain barrier in response to i.v. bacterial endotoxin cause the ACTH response and one of the sites of action of prostaglandins might be the anterior lobe of the pituitary gland.

Methods

In this study, we used male New Zealand white rabbits weighing 3.0-4.0 kg. The rabbits were housed at an ambient temperature of 21±1°C on a 12:12 h light-dark cycle (light on at 7:00 A.M. and off at 7:00 P.M.). Animals had *ad libitum* access to water and standard laboratory rabbit chow. This study consisted of three experiments. In Exp. 1, the rabbits received i.v. injection of endotoxin or saline. In Exp. 2, i.v. injection of bacterial endotoxin followed subcutaneous (s.c.) or i.c.v. administration of indomethacin (INDO). In Exp 1-2, changes in plasma concentration of ACTH were measured after injections. In Exp. 3, i.v. injection of endotoxin or saline was performed to examine the IL-1 production in the anterior lobe of the pituitary gland by immunohistochemistry. For i.c.v. injections, a stainless steel cannula (David Kopf Instruments, No. 201) was surgically placed into a lateral cerebral ventricle. At least 20 days were allowed between surgery and the start of experimentation. During this period, each rabbit was restrained in a conventional rabbit stock for 4 hr a day for at least 3 days to accustom the animals to the restraint, thus minimizing this stress.

The bacterial endotoxin used was lipopolysaccharide (LPS) of *Salmonella typhosa* (Difco), which was dissolved in sterile saline. Sodium indomethacin trihydrate (Merck Sharp & Dohme Research Laboratory Division of Merck & Co., Inc.) was dissolved in sterile saline. For immunohistochemical staining, Vectastain ABC kit (Vector Laboratories Inc. USA) was used. Goat-anti-rabbit IL-1ß polyclonal antibody was a gift from prof. Yoshinaga (Kumamoto Univ. Japan). Doses in each experiment as well as appropriate control injections are described in Results.

ACTH study

On the day of the experiment, each rabbit was subjected to blood samplings. The animal was minimally restrained between 08:00 and 15:00 hr in conventional rabbit stocks, at an ambient temperature of 21±1°C. To collect blood, an intraarterial catheter was inserted into the central artery of the rabbit's ear. To avoid the influences of stresses from the procedures of intraarterial catheterization, a minimum of 90 min was allowed prior to any injections. Injection of LPS or saline was performed at 10:00 hr. INDO was injected s.c. or i.c.v. 15 min before administration of LPS. I.v. injections were made into the marginal ear vein. INDO was injected i.c.v. in a volume of 20 µl followed by a 30 µl saline flush. The blood samples (1.0 ml) were taken 0, 60, 120 and 240 min after injection. The blood, collected into tubes containing EDTA solution (Vacutainer, Becton Dickenson), was centrifuged at 3,000 rpm for 15 min at 4 °C. The plasma was collected into cryovials and then stored at -70 °C until measurement of ACTH. The plasma concentration of ACTH was determined using an ACTH IRMA assay kit (Nichols, CA).

Immunohistochemical study

For immunohistochemical study, rabbits were injected with LPS (1.0 µg/kg, i.v.) or saline (0.05 ml/kg, i.v.). The experiment was performed using 4 rabbits. They were sacrificed at 120 min after i.v. injections of LPS or saline by a large dose of pentobarbitone. Immediately after pentobarbitone injection, they were perfused with 0.1 M cacodylate solution containing 4 % paraformaldehyde and 0.1 % glutalaldehyde. Subsequently pituitary glands were removed and sliced by a microslicer at 40 µm thickness. Thereafter, the slices were pretreated with 5 % normal rabbit serum for 30 min to block non-specific binding, then incubated with a 1/1600 dilution of goat-anti-rabbit IL-1ß serum at 4 °C overnight. After being rinsed in Tris-HCl containing 0.5 M NaCl, the sections were processed by avidin-biotinylated enzyme complex (Vectastain, ABC kit, Vector Laboratories Inc. U.S.A.). After being rinsed, sections were washed several times in Tris-HCl buffer and reacted with diaminobenzidine containing 0.02 % H_2O_2 for 10 min.

Data were analyzed for statistical significance using Sheffe's test (multiple comparison method). Probability values<0.05 were considered to be significant.

Results

Figure 1A shows changes in plasma concentration of ACTH after i.v. injection of LPS (0.1 µg/kg and 1.0 µg/kg) or saline. In Fig. 1B, shown is the effect of s.c. or i.c.v. administration of indomethacin on the ACTH response induced by i.v. injection of LPS (1.0 µg/kg). The i.v. injection of LPS induced significant increases in plasma concentration of ACTH 60, 120 and 240 min after the injection and this response was dose-related. No increase in plasma concentration of ACTH was observed after i.v. injection of saline (Fig. 1A). In contrast, the ACTH response induced by i.v. LPS (1.0 µg/kg) was significantly suppressed by pretreatment with s.c. indomethacin (20 mg/kg), while i.c.v. pretreatment with indomethacin (400 µg) had no effect (Fig. 1B).

Figure 2 shows the distribution of IL-1ß-immunoreactive cells in the transversal sections of the pituitary gland 120 min after i.v. injection of LPS (1.0 µg/kg) or saline. As shown in Fig. 2, no IL-1ß immunoreactive cells were observed in the anterior, intermediate and the posterior lobes of the pituitary gland of rabbits when they were injected with saline (Fig. 2A). However, 120 min after i.v. injection of LPS, IL-1ß immunoreactivity was found throughout the anterior and the posterior lobes of the pituitary gland (Fig. 2B).

Figure 3 shows the IL-1ß-immunoreactive cells in the sections of the pituitary gland 120 min after i.v. injection of LPS (1.0 µg/kg). Two parts of the pituitary gland depicted in Fig. 2B are magnified in Fig. 3. The immunoreactive cells were found in the interstitium of the anterior lobe and, therefore, were not adenocytes. In the posterior lobe, dendriform cells showed immunoreactivity.

Discussion

The present results show that i.v. injection of LPS induced an increase in plasma concentration of ACTH in rabbits. Furthermore, systemic administration of indomethacin suppressed this ACTH response, suggesting the involvement of prostaglandins in development of the ACTH response induced by i.v. LPS. However, the ACTH response was not affected by i.c.v indomethacin. Since the indomethacin at the same doses as used in the present study

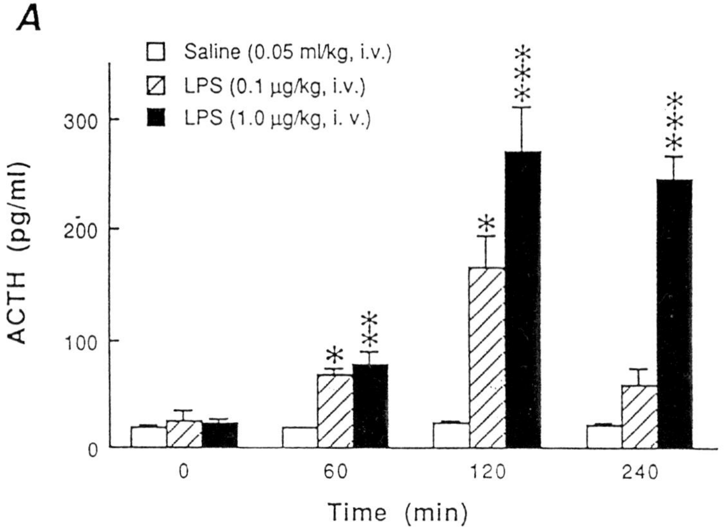

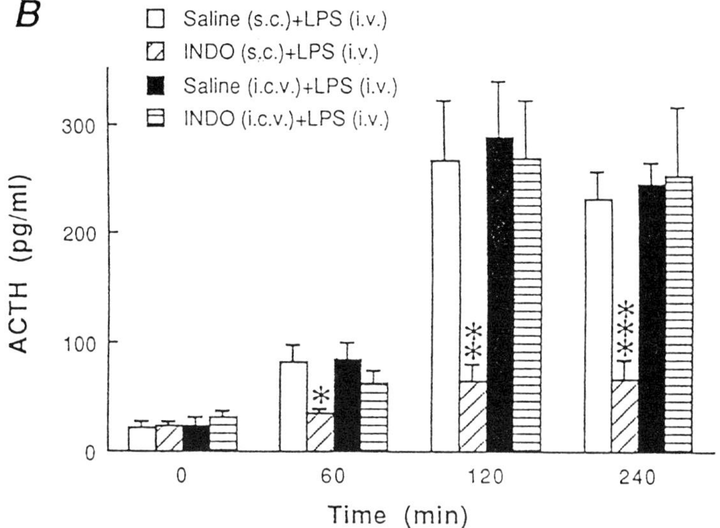

Fig. 1. (A) Mean changes (±S.E.M.) in plasma concentration of ACTH of the same group of rabbits after intravenous (i.v.) injection of LPS (0.1 µg/kg and 1.0 µg/kg) or saline. (B) Mean changes (±S.E.M.) in plasma concentration of ACTH of the same group of rabbits after intravenous (i.v.) injection of LPS (1.0 µg/kg). Rabbits were pretreated with subcutaneous (s.c.) or intracerebroventricular (i.c.v.) administration of indomethacin (INDO) (20 mg/kg, s.c. and 400 µg, i.c.v.) 15 min before injection of LPS.

A: Saline (0.05 ml/kg, i.v.)

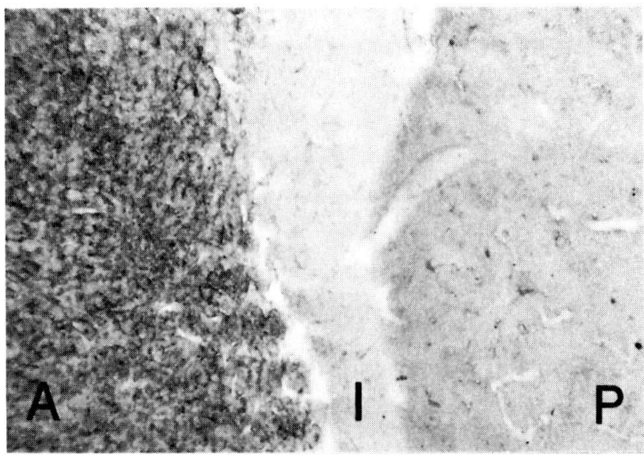

B: LPS (1.0 μg/kg, i.v.)

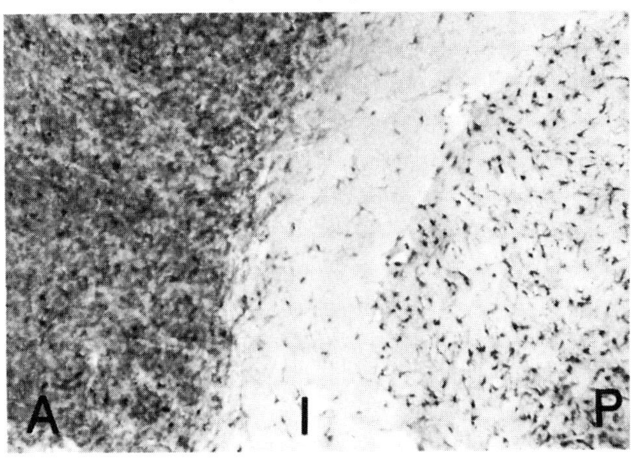

(— 100 μm)

Fig. 2. Light-micrographs of IL-1ß-immunoreactive cells in the pituitary gland at 120 min after intravenous (i.v.) injection of saline (0.05 ml/kg) (A) or LPS (1 μg/kg) (B). A, anterior lobe; I, intermediate lobe; P, posterior lobe.

A: Anterior lobe

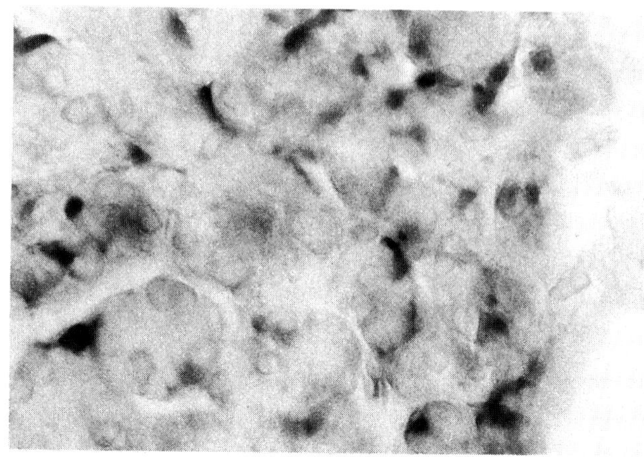

B: Posterior lobe

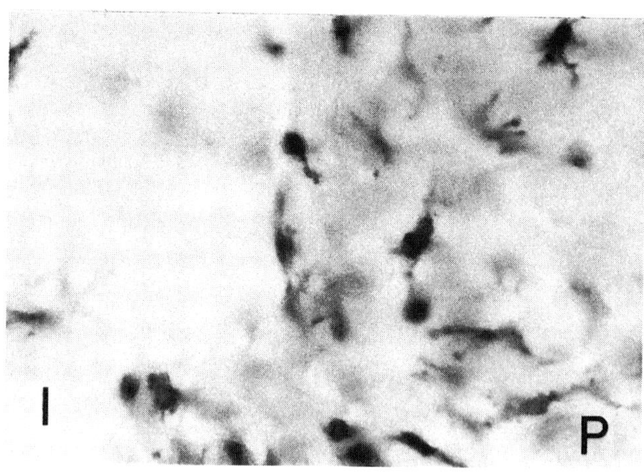

(– 20 µm)

Fig. 3. Light-micrographs of IL-1ß-immunoreactive cells in the pituitary gland at 120 min after intravenous (i.v.) injection of LPS (1 µg/kg). Two parts of the pituitary gland (anterior lobe, A; posterior lobe, B) depicted in Fig. 2B is magnified in Fig. 3 to observe the IL-1ß-immunoreactive cells in detail. I, intermediate lobe; P, posterior lobe.

suppresses fever induced by LPS in rabbits (Morimoto et al., 1987), we believe that this antipyretic drug inhibited prostaglandin synthesis. It is, therefore, inferred that prostaglandins synthesized outside the blood-brain barrier are responsible for the ACTH response induced by i.v. LPS. This finding is in accordance with our previous report that, in rats, interleukin-1 (IL-1) evokes the ACTH increase in plasma through the action of prostaglandins that are synthesized outside the blood-brain barrier (Watanabe et al., 1991b).

Recently, we proposed a hypothesis that there are two separate mechanisms of induction of biphasic fever one inside and one outside the blood-brain barrier (BBB) (Morimoto et al., 1987). The first phase is produced by endogenous pyrogen (EP) (Atkins, 1960)/IL-1, that is synthesized in response to i.v. LPS, acting on structures outside the BBB. The second phase is produced by EP/IL-1 acting on structures inside the BBB. Subsequently, these structures synthesize and release prostaglandins that induce fever. In contrast, the present results suggest the involvement of prostaglandins synthesized outside the BBB in development of the ACTH response. It is, therefore, likely that the mechanisms of induction of febrile and ACTH responses are not the same, although prostaglandins are involved in those responses.

Bernton et al. (1987) reported that IL-1 acts directly on the pituitary gland to stimulate the secretion of ACTH, while other researchers reported that IL-1 induces the ACTH response indirectly via the hypothalamic hormone, CRF (Berkenbosch et al., 1987; Sapolsky et al., 1987; Uehara et al., 1987). We have shown that prostaglandins mediate the induction of the ACTH response to i.v. injection of IL-1 (Morimoto et al., 1989; Watanabe et al., 1991b; Watanabe et al., 1990). Therefore, the sites of action of prostaglandins to induce the ACTH response are still unknown. In the present study, we have investigated whether IL-1ß-immunoreactive cells are observed in the anterior lobe of the pituitary gland after i.v. injection of LPS. The results show that IL-1ß immunoreactivity was found throughout the anterior and the posterior lobes of the pituitary gland in response to i.v. LPS. This result suggests that IL-1ß is produced in the pituitary gland or that circulating IL-1ß binds to its receptor of the pituitary gland. However, latter possibility is unlikely, because circulating IL-1ß has not been detected by many researchers (Kluger, 1991). Therefore, it is inferred that i.v. LPS induces IL-1ß in the pituitary gland and IL-1ß stimulates the secretion of ACTH through the action of prostaglandins in the anterior lobe of the pituitary gland. However, we can not exclude the possibility that the ACTH response induced by i.v. LPS is mediated by the hypothalamic hormone CRF. Recently, we have shown that i.v. or intrahypothalamic injection of prostaglandin E resulted in increases in plasma level of ACTH and this response was suppressed by the pretreatment with anti-CRF antibody (Watanabe et al., 1991a; Watanabe et al., 1990), suggesting that CRF mediates the ACTH response. Therefore, it is likely that prostaglandins stimulate the secretion of ACTH by the direct action on the pituitary gland and the indirect action via the hypothalamic hormone, CRF.

Acknowledgements

This work was partly supported by NIH grant NS10046 from the National Institute of Neurological Divisions and Stroke, and by NATO

Collaborative Research grant 2000467. This work was also partly supported by the Ministry of Education, Science, and Culture Grant-in-Aid for Scientific Research (A)03404018 (Japan). T. Watanabe was a visiting scientist in USA supported by a grant from Uehara Memorial Foundation. The authors are grateful to the Foundation for this financial support.

References

Atkins, E. (1960): Pathogenesis of fever. Physiol. Rev. 40, 580-646.

Berkenbosch, F., Van Oers, J., del Rey, A., Tilders, F., and Besedovsky, H. (1987): Corticotropin-releasing factor-producing neurons in the rat activated by interleukin-1. Science 238, 524-526.

Bernton, E.W., Beach, J.E., Holaday, J.W., Smallridge, R.C., and Fein, H.G. (1987): Release of multiple hormones by a direct action of interleukin-1 on pituitary cells. Science 238, 519-521.

Besedovsky, H., del Rey, A., Sorkin, E., and Dinarello, C.A. (1986): Immunoregulatory feedback between interleukin-1 and glucocorticoid hormones. Science 233, 652-654.

Dinarello, C.A. (1984): Interleukin-1. Rev. Infect. Dis. 6, 51-95.

Kluger, M.J. (1991): Fever:Role of pyrogens and cryogens. Physiol. Rev. 71, 93-127.

Morimoto, A., Murakami, N., Nakamori, T., Sakata, Y., and Watanabe, T. (1989): Possible involvement of prostaglandin E in development of ACTH response in rats induced by human recombinant interleukin 1. J. Physiol. Lond. 411, 245-256.

Morimoto, A., Murakami, N., Nakamori, T., and Watanabe, T. (1987): Evidence for separate mechanisms of induction of biphasic fever inside and outside the blood-brain barrier in rabbits. J. Physiol. Lond. 383, 629-637.

Sapolsky,R., Rivier, C., Yamamoto, G., Plotsky, P., and Vale, W. (1987): Interleukin-1 stimulates the secretion of hypothalamic corticotropin-releasing factor. Science 238, 522-524.

Stitt, J.T. (1986): Prostaglandin E as the neuronal mediator of the febrile response. Yale J. Biol. Med. 59, 137-149.

Uehara, A., Gottschall, P.E., Dahl, R.R., and Arimura, A. (1987): Interleukin-1 stimulates ACTH release by an indirect action which requires endogenous corticotropin releasing factor. Endocrinology 121, 1580-1582.

Watanabe, T., Morimoto, A., Morimoto, K., Nakamori, T., and Murakami, N. (1991a): ACTH release induced in rats by noradrenaline is mediated by prostaglandin E_2. J. Physiol. Lond. 443, 431-439.

Watanabe, T., Morimoto, A., and Murakami, N. (1991b): ACTH response in rats during biphasic fever induced by interleukin-1. Am. J. Physiol. 261, R1104-R1108.

Watanabe, T., Morimoto, A., Sakata, Y., and Murakami, N. (1990): ACTH response induced by interleukin-1 is mediated by CRF secretion stimulated by hypothalamic PGE. Experientia 46, 481-484.

Integrative and cellular aspects of autonomic functions : temperature and osmoregulation. Eds K. Pleschka, R. Gerstberger. John Libbey Eurotext, Paris © 1994, pp. 171-179.

Antipyretic activity of α-melanocyte stimulating hormone and related peptides

Duncan Mitchell and Kathleen Goelst

Brain Function Research Unit, Department of Physiology, University of the Witwatersrand Medical School, Johannesburg, South Africa

SUMMARY

α-Melanocyte stimulating hormone (α-MSH) is a strong contender for the role of endogenous antipyretic and also is a potential substrate for therapeutic antipyretics. The antipyretic activity of the hormone resides in its terminal (11-13) tripeptide Lys-Pro-Val. The tripeptide fragment Lys-Pro-Val is itself a potent antipyretic. We have explored the antipyretic activity of the enantiomer Lys-D-Pro-Val, and an analogue, Lys-D-Pro-Thr, in rabbits given endotoxin 0.1 μg/kg iv. At intravenous doses at which Lys-Pro-Val is antipyretic, neither of the D-proline tripeptides has antipyretic activity, in spite of being more active biologically in other circumstances. At the same dose, neither tripeptide affects normal body temperature. Whether the analogue [D-Pro^{12}] α-MSH also is devoid of antipyretic activity needs to be investigated.

INTRODUCTION

The tridecapeptide α-melanocyte stimulating hormone (α-melanotropin, α-MSH) is a neuropeptide with multiple physiological and behavioural functions (De Wied and Jolles, 1982). The molecule interests thermal physiologists for at least three reasons; it is a strong contender for the role of endogenous antipyretic, it is a useful experimental tool for investigating cytokine action in fever, and it presents a base for the development of potent novel antipyretic agents.

Support for the idea that α-MSH might be an endogenous antipyretic derives from many sources. The synthesis of α-MSH, which occurs largely in the arcuate nucleus, increases in rabbits during fever (Bell and Lipton, 1987), and its concentration in the circulation rises in both rabbits (Martin and Lipton, 1990) and humans (Catania *et al.*, 1991) during experimental fever, and is elevated in febrile patients with HIV infection (Catania *et al.*, 1993). Administration of a specific antiserum to α-MSH in rabbits exacerbates cytokine-induced fever (Shih *et al.*, 1986). Administration of α-MSH itself, whether parenterally or directly into the central nervous system, attenuates fever in all species in which it has been tested (rabbits: Murphy *et al.*, 1983; guinea pigs: Kandasamy and Williams, 1984; rats: Villar *et al.*, 1991; squirrel monkey: Shih and Lipton, 1985) with the anomalous exception of the cat (Rezvani *et al.*, 1986). In experimental fevers, α-MSH exerts antipyretic activity whether the priming pyrogen is a cytokine, a Gram-negative exogenous pyrogen, or a Gram-positive exogenous pyrogen (Goelst and Laburn, 1991). Not only does α-MSH attenuate the thermal response to pyrogen administration, but it also attenuates both peripherally- mediated (Robertson *et al.*, 1988; Goelst *et al.*, 1991) and centrally-mediated (Dao *et al.*, 1988) components of the acute-phase response. In addition, α-MSH inhibits the cytokine-mediated stimulation of the pituitary-adrenal axis and the immunosuppression which occur during fever (Weiss *et al.*, 1991). This suite of properties has not been demonstrated for any other putative endogenous antipyretic.

In the sequence of biochemical events which leads to fever, α-MSH acts after the release of the cytokines but before the release of arachidonic acid. It does not inhibit the release of cytokines from stimulated macrophages (Cannon *et al.*, 1986) and does not attenuate the hyperthermia which follows arachidonic acid (Clark *et al.*, 1985) or prostaglandin (Clark et al., 1985; Davidson *et al.*, 1992) administration, though, paradoxically, prostaglandin-induced, as well as interleukin-induced, hyperthermia is enhanced in α-MSH depleted rats (Martin *et al.*, 1990). The biochemical location of α-MSH action, together with its ability to block both the thermal and acute-phase responses to pyrogens, fit the proposal that α-MSH is a cytokine receptor antagonist (Richards and Lipton, 1984b). Indeed, α-MSH antagonises the activity of at least three of the cytokines potentially involved in fever, namely interleukin-1, interleukin-6 and tumour necrosis factor (Daynes *et al.*, 1987; Martin *et al.*, 1991; Davidson *et al.*, 1992). Because α-MSH is the one known agent thought to be able to block cytokine receptors generally, it is useful in analysing the role, and the site of action, of cytokines in fever (Goelst and Laburn, 1991; Goelst *et al.*, 1991).

Apart from being useful in experimental thermal physiology, α-MSH, and related peptides, may ultimately be useful therapeutic antipyretics. The antipyretic potency of α-MSH in rabbits is 25 thousand times that of paracetamol (acetaminophen), the antipyretic currently used most widely in clinical medicine (Murphy *et al.*, 1987). Unlike paracetamol, and other cyclo-oxygenase inhibitors, therapeutic antipyretics based on peptides related to α-MSH would attenuate more components of the febrile response than just those mediated by prostanoids. At least in rabbits, α-MSH is orally bio-available (Murphy and Lipton, 1982). The parent molecule does contain twelve peptide bonds, however, making it vulnerable to enzymatic degradation. Its pharmacological attributes might be improved, therefore, by substituting some of the amino acid residues with more stable residues, or finding a shorter fragment of α-MSH with antipyretic activity, or both. The antipyretic message in α-MSH resides in its terminal tripeptide Lys-Pro-Val (Deeter *et al.*, 1988), and the substituted tripeptide Lys-D-Pro-Val, and the related substituted tripeptide Lys-D-Pro-Thr, have potent biological activity (Poole *et al.*, 1992). However, a cryptic comment claims that Lys-D-Pro-Thr has no consistent antipyretic activity in rabbits; few experimental details were given and no results were shown (Ferreira *et al.*, 1988). Because of this anomalous report, and because of the potential usefulness of short analogues of α-MSH in experimental thermal physiology and clinical medicine, we have investigated the possible antipyretic activity of Lys-D-Pro-Val, and reinvestigated that of Lys-D-Pro-Thr, in rabbits.

METHODS

We used 24 New Zealand White rabbits, of both sexes, weighing 2.5 to 3.5 kg. The rabbits were housed in individual cages, on 12h light:dark cycles, at an ambient temperature of 19 to 22°C. They had *ad libitum* access to standard rabbit chow and tap water. During measurements, the rabbits were conscious and restrained in conventional rabbit stocks; they were habituated to the stocks before investigations started. Ambient temperature in the test environment was 20 ± 1°C (mean ± SD). We measured the rabbits' temperatures with indwelling copper-constantan thermocouples connected to a data logger (Esterline Angus); the thermocouples were inserted 100 mm into the rectum. The thermocouples were calibrated to an accuracy of 0.1°C by water immersion. We monitored rectal temperatures every 10 minutes for at least 1 hour before, and for 4 hours after, pyrogen injection. All injections were made into an ear lateral vein, at the same time (09:00) each day, to avoid variations resulting from circadian rhythms. Endotoxin injections were separated by at least one week, to prevent the development of tolerance. We induced fever by injecting 0.1 μg/kg of endotoxin (lipopolysaccharide extracted from *Salmonella typhosa*, Difco) in a 0.8-1.2 ml bolus. Peptides were administered 30 minutes later, also in a 0.8-1.2 ml bolus.

The doses of the two peptides were 2.5, 5, 10, 250 and 680 μg/kg; the reasons for the choice of doses will become apparent. Saline was the control agent for both endotoxin and the peptides; each animal acted as its own control.

The two tripeptides were supplied by Cambridge Research Biochemicals, Cheshire, England. The manufacturers characterised the peptides by fast-atom bombardment mass spectrometry, amino acid analysis, and analytical reverse-phase high-performance liquid chromatography. The peptides were kept dry and at -20°C, and were dissolved in saline immediately before use.

Our experimental procedures received the approval of the Animal Ethics Screening Committee of the University of the Witwatersrand, under Clearance Certificate 92/43/3.

RESULTS

Intravenous administration of endotoxin produced the characteristic biphasic thermal response, with a latent period of about 30 min to the rise of the first phase, and with the second phase becoming apparent at about 120 min after endotoxin injection (Fig. 1). Figure 1 shows the effect of four doses of Lys-D-Pro-Val, and of the same four doses of Lys-D-Pro-Thr on the fever. It is apparent that neither tripeptide had any effect on the latency, the amplitude, nor on the pattern of the febrile response to endotoxin.

It is customary to quantify the thermal response to pyrogens in terms of the fever index, the time integral over a fixed period (here 4 hours) of the elevation of temperature above that prevailing at the time of injection of the pyrogen. Figure 2 compares the fever indices for the response to endotoxin injection followed by control saline injection, with those for endotoxin injection followed by all five doses of the two tripeptides. There were no significant effects of the tripeptides on the fever indices at any of the doses we tested.

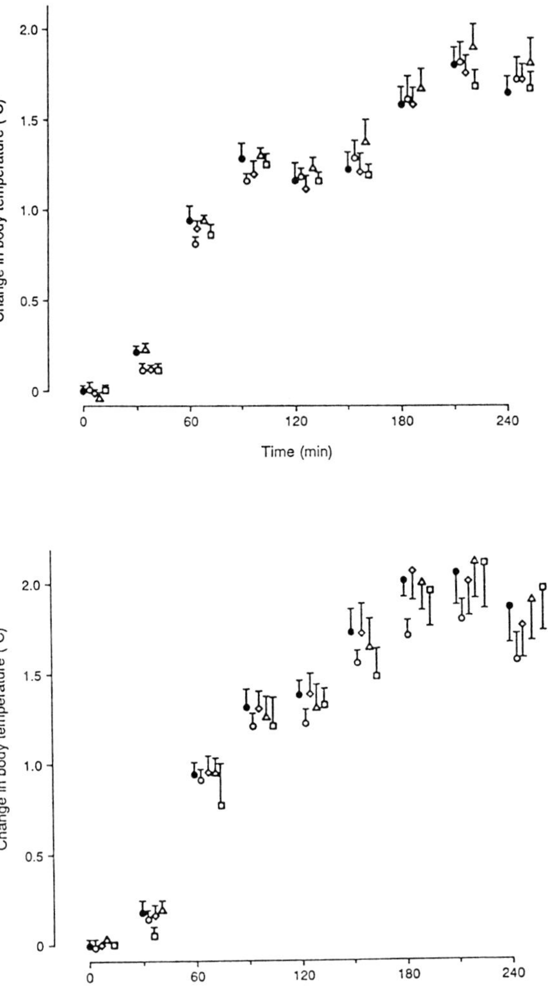

Fig. 1. Mean and SEM change in rectal temperature in 8 rabbits after iv injections of 0.1 µg/kg endotoxin at time 0. Saline (●) or tripeptide at 2.5 µg/kg (o), 5 µg/kg (◇), 10 µg/kg (▵) and 250 µg/kg (□) were given 30 minutes after endotoxin. None of the doses of either Lys-D-Pro-Thr (upper diagram) or Lys-D-Pro-Val (lower diagram) affected the endotoxin fever.

Figure 2 also shows the effects of injecting the tripeptides after a priming injection of saline, rather than endotoxin, a procedure which tests the potential effects of the tripeptides on normal (afebrile) body temperature. Neither tripeptide had any significant effect on normal body temperature at any of the doses we tested.

DISCUSSION

Our results show, first, that, when injected iv at doses between 2.5 μg/kg and 680 μg/kg, neither Lys-D-Pro-Val nor Lys-D-Pro-Thr inhibited the fever induced by iv endotoxin in rabbits, and, secondly, that, given iv at the same doses, neither tripeptide affected the normal body temperature of the rabbits in an ambient temperature of 20°C. Consequently, neither tripeptide was antipyretic, hypothermic, nor hyperthermic in rabbits. Even though a brief earlier report also had claimed that Lys-D-Pro-Thr had no antipyretic activity (Ferreira *et al.*, 1989), we were surprised that the tripeptides were inactive, because other α-MSH-related peptides had exhibited such potent effects on body temperature. It was not the case that our tripeptides were biologically inert. Both have been shown to exhibit potent biological activity in rats (Ferreira *et al.*, 1988; Follenfant *et al.*, 1989; Cunha *et al.*, 1991; Poole *et al.*, 1992).

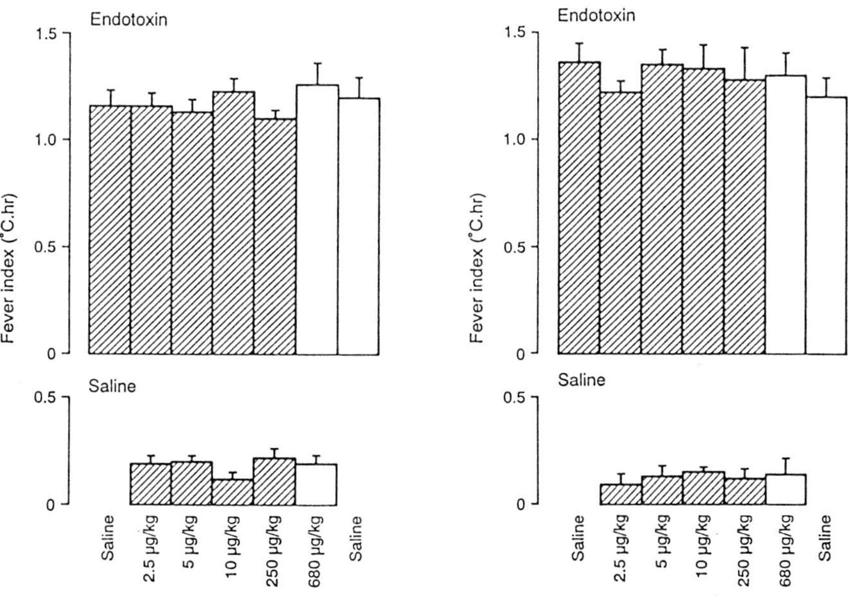

Fig. 2. Mean and SEM four-hour fever indices in 8 rabbits after iv injection of 0.1 μg/kg endotoxin and either saline or tripeptide 30 minutes later (top panel) or saline and tripeptide 30 minutes later (bottom panel). At all doses tested, neither Lys-D-Pro-Thr (left diagram) nor Lys-D-Pro-Val had any significant effect on either the fever index, nor the body temperature of rabbits given a saline injection. The hatched and clear bars refer to two different sets of 8 rabbits, each with their own controls.

The lack of antipyretic activity of the two tripeptides contrasts with the potent antipyretic activity of α-MSH itself. Using identical procedures to those we employed, Goelst and colleagues (1991) have shown that α-MSH attenuates the fever induced by 0.1μg/kg endotoxin in rabbits. Intravenous α-MSH administered 30 minutes after endotoxin brought about an immediate arrest of the developing first phase of the endotoxin fever. The fever was attenuated for 30 minutes, after which the temperatures rose rapidly back to the level evident when endotoxin was administered without an antipyretic. Later, 2 hours after the α-MSH injection, the second phase of the endotoxin fever also was attenuated, and this attenuation lasted until the end of the study period. The antipyretic activity of α-MSH was evident at 2.5 μg/kg, the lower end of the dose range over which we tested both tripeptides. The antipyretic action of α-MSH against endotoxin in rabbits has been confirmed recently, over the same dose range, by Davidson and colleagues (1992).

Having established that neither tripeptide was an effective antipyretic against endotoxin at doses at which the parent molecule α-MSH was antipyretic, we tested the potential antipyretic activity at two much higher doses. The highest dose, 680 μg/kg, was the dose at which Richards and Lipton (1984a) had shown the tripeptide Lys-Pro-Val to be antipyretic against cytokine-induced fever in rabbits. Even at the higher doses, neither of our tripeptides exhibited antipyretic activity against endotoxin. Substituting D-proline for L-proline in Lys-Pro-Val therefore had a much more profound effect than simply stabilising the molecule against enzymatic degradation; it removed its antipyretic activity, while leaving other biological activity intact.

We do not know how the substitution of D-amino acids specifically abolishes antipyretic activity. The explanation does not lie in altered access to the brain, because, even if the D-proline tripeptides could not cross the blood-brain barrier, they would still affect the first phase of endotoxin fever (Goelst *et al.*, 1991). More investigations are necessary before we can advance a possible mechanism. For example, we need to investigate the consequences of substituting D-proline into α-MSH itself, to produce [D-Pro^{12}] α-MSH.

Ours is not the first report of amino acid substitutions in α-MSH producing paradoxical alterations to antipyretic activity. Of all the analogues of α-MSH which have been studied, the analogue most active biologically is [Nle^{4}, D-Phe^{7}] α-MSH (Sawyer *et al.*, 1980; Kobobun *et al.*, 1983; Benelli *et al.*, 1988); it is also the peptide which shows the greatest antipyretic activity (Holdeman and Lipton, 1985). However, [Nle^{4}, D-Phe^{7}] α-MSH appears to be antipyretic only when injected intracerebroventricularly (icv); it was inactive when administered iv to rabbits even at a large dose (Holdeman and Lipton, 1985). The parent α-MSH, though being a less potent antipyretic than [Nl^{4}, D-Phe^{7}] α-MSH when injected centrally, does exhibit antipyretic activity in rabbits whether administered iv or icv (Murphy and Lipton, 1980; Holdeman and Lipton, 1985; Goelst *et al.*, 1991). In Holdeman and Lipton's investigation of [Nle^{4}, D-Phe^{7}] α-MSH, a crude cytokine mixture was used to induce fever. The investigation needs to be repeated with endotoxin, and to include a measure of the acute-phase response (Goelst and Laburn, 1991), before it is possible to speculate on the differential activity of [Nle^{4}, D-Phe^{7}] α-MSH between brain and periphery. Whatever the mechanism, the substitution of two alternative amino acids in positions 4 and 7 of α-MSH also profoundly altered its antipyretic activity.

Not only did the substituted tripeptides we investigated exhibit no antipyretic activity against endotoxin fever in rabbits, but they also had no effect on normal body temperature in the rabbits, at a neutral ambient temperature. The extent to which α-MSH and its analogues affect normal thermal regulation remains controversial. The parent α-MSH causes hypothermia in rabbits when injected iv at a dose of 20 μg (Richards & Lipton, 1984b) a dose much higher than that at which it is antipyretic. Similarly, α-MSH had no effect on the body temperature of normothermic rats at doses of 120 μg/kg intraperitoneally or 4 μg/kg icv, doses at which it is antipyretic in the rats (Villar *et al.*, 1991). The substituted analogue [Nle^4, D-Phe^7] α-MSH caused hypothermia in rabbits in a 21-23°C environment when administered icv at a dose of about 13 ng/kg, but had no effect on normal body temperature when given iv even at a dose of about 5 μg/kg (Holdeman and Lipton, 1985). Anomalously, however, a recent report claims that intrahypothalamic injection of α-MSH and [Nle^4, D-Phe^7] α-MSH induced hyperthermia in rats at doses between 4 and 15 ng (Raible and Knickerbocker, 1993); whether similar intrahypothalamic doses influenced fever in the rats was not investigated.

In conclusion, our attempts to develop a fragment of α-MSH with enhanced antipyretic potency failed. Indeed, all reported attempts to improve pharmacologically on the antipyretic potency of the parent molecule have failed, even though such manipulations have enhanced other biological activities. The amino-acid fragment 1-10 of α-MSH has no antipyretic activity (Lipton *et al.*, 1981), and the fragments 8-13, 10-13, and 11-13 are all less potent antipyretics than α-MSH itself (Deeter *et al.*, 1988). Although the substituted analogue [Nle^4, D-Phe^7] α-MSH is ten times more potent than α-MSH when administered icv, it appears to have little or no antipyretic activity in the periphery (Holdeman and Lipton, 1985), and therefore has little value as a therapeutically useful antipyretic agent. The parent molecule, α-MSH itself, as well as being the strongest contender for the role of endogenous antipyretic, and a useful tool in experimental thermal physiology, remains the most potent of the peptides with both peripheral and central activity, and the agent with the greatest potential as a therapeutic antipyretic.

ACKNOWLEDGEMENTS

We thank Adcock Ingram Pharmaceuticals and the South African Medical Research Council for financial support, Mr D Makoa for assistance with the experiments, and the Central Animal Service of the University of the Witwatersrand for the care of the animals.

REFERENCES

Bell, R.C. & Lipton, J.M. (1987): Pulsatile release of antipyretic neuropeptide α-MSH from septum of the rabbit during fever. *Am. J. Physiol.* 252, R1152-R1157.

Benelli, A., Zanoli, P., Botticelli, A., & Bertolini, A. (1988): [Nle^4, Phe^7] α-MSH improves functional recovery in rats subjected to diencephalic hemisection. *Eur. J. Pharmacol.* 150, 211-219.

Cannon, J.G., Tatro, J.B., Reichlin, S. & Dinarello, C.A. (1986): α-Melanocyte stimulating hormone inhibits immunostimulatory and inflammatory actions of interleukin-1. *J. Immunol.* 137, 2232-2236.

Catania, A., Airaghi, L., Manfredi, M.G., Vivirito, M.C., Milazzo, F., Lipton, J.M. & Zanussi, C. (1993): Proopiomelanocortin-derived peptides and cytokines: relations in patients acquired immunodeficiency syndrome. *Clin. Immunol. Immunopathol.* 66: 73-79.

Catania, A., Suffredini, A.F. & Lipton, J.M. (1991): α-MSH response to endotoxin in normal human subjects. *FASEB J.* 5, A1392.

Clark, W.G., Holdeman, M. & Lipton, J.M. (1985): Analysis of the antipyretic action of α-melanocyte-stimulating hormone in rabbits. *J. Physiol.* (*Lond.*) 359, 459-465.

Cunha, F.Q., Lorenzetti, B.B., Poole, S. & Ferreira, S.H. (1991): Interleukin-8 as a mediator of sympathetic pain. *Br. J. Pharmacol.* 104, 765-767.

Dao, T.K., Bell, R.C., Feng, J., Jameson, D.M. & Lipton, J.M. (1988): C-reactive protein, leukocytes and fever after central IL 1 and α-MSH in aged rabbits. *Am. J. Physiol.* 254, R401-R409.

Davidson, J., Milton, A.S. & Rotondo, D. (1992): α-Melanocyte-stimulating hormone suppresses fever and increases in plasma levels of prostaglandin E_2 in the rabbit. *J. Physiol. (Lond.)* 451, 491-502.

Daynes, R.A., Robertson, B.A., Cho, B., Burnham, D.K. & Newton, R. (1987): α-Melanocyte-stimulating hormone exhibits target cell specificity in its capacity to affect interleukin-1-inducible responses *in vivo* and *in vitro*. *J. Immunol.* 139, 103-109.

Deeter, L.B., Martin, L.W. & Lipton, J.M. (1988): Antipyretic properties of centrally administered α-MSH fragments in the rabbit. *Peptides* 9, 1285-1288.

De Wied, D. & Jolles, J.J. (1982): Neuropeptides derived from pro-opiocortin: behavioral., physiological and neurochemical effect. *Physiol. Rev.* 62, 976-1059.

Ferreira, S.H., Lorenzetti, B.B., Bristow, A.F. & Poole, S. (1988). Interleukin-1β as a potent hyperalgesic agent antagonised by a tripeptide analogue. *Nature* 334, 698-700.

Follenfant, R.L., Nakamura-Craig, M., Henderson, B. & Higgs, G.A. (1989): Inhibition by neuropeptides of interleukin-1β induced prostaglandin-independent hyperalgesia. *Br. J. Pharmacol.* 98, 41-43.

Goelst, K. & Laburn, H. (1991): The effect of α-MSH on fever caused by *Staphylococcus aureus* cell walls in rabbits. *Peptides* 12, 1239-1242.

Goelst, K., Mitchell, D. & Laburn H. (1991): Effects of α-melanocyte stimulating hormone on fever caused by endotoxin in rabbits. *J. Physiol.* (*Lond.*) 441, 469-476.

Holdeman, M. & Lipton, J.M. (1985): Antipyretic activity of a potent α-MSH analog. *Peptides* 6, 273-275.

Kandasamy, S.B. & Williams, B.A. (1984): Hypothermic and antipyretic effects of ACTH and α-melanotopin in guinea pigs. *Neuropharmacology* 23, 49-51.

Kobobun, K., O'Donohue, T.L., Handelmann, G.E., Sawyer, T.K., Hruby, V.J. & Hadley, M.E. (1983): Behavioral effects of 4-Norleucine, 7-D-Phenylalanine-α-melanocyte-stimulating hormone. *Peptides* 4, 721-724.

Lipton, J.M., Glyn, J.R. & Zimmer, J.A. (1981): ACTH and α-melanotropin in central temperature control. *Fed. Proc.* 40, 2760-2764.

Martin, L.W., Catania, A., Hiltz, M.E. & Lipton, J.M. (1991): Neuropeptide α-MSH antagonises IL-6 and TNF-induced fever. *Peptides* 12, 297-299.

Martin, L.W. & Lipton, J.M. (1990): Acute phase response to endotoxin: rise in plasma α-MSH and effects of α-MSH injection. *Am. J. Physiol.* 259, R768-R772.

Martin, S.M., Malkinson, T.J., Veale, W.L. & Pittman, Q.J. (1990): Depletion of brain α-MSH alters prostaglandin and interleukin fever in rats. *Brain Res.* 526, 351-354.

Murphy, M.T. & Lipton, J.M. (1982): Peripheral administration of α-MSH reduces fever in older and younger rabbits. *Peptides* 3, 775-779.

Murphy, M.T., Richards, D.B. & Lipton, J.M. (1983). Antipyretic potency of centrally administered α-melanocyte stimulating hormone. *Science* 221, 192-193.

Poole, S., Bristow, A.F., Lorenzetti, B.B., Gaines Das, R.E., Smith, T.W. & Ferreira, S.H. (1992). Peripheral analgesic activities of peptides related to α-melanocyte stimulating hormone and interleukin-1β. *Br. J. Pharmacol.* 106, 489-492.

Raible, L.H. & Knickerbocker, D. (1993): Alpha-melanocyte- stimulating hormone (MSH) and [Nle^4, D-Phe^7]-α-MSH: effects on core temperature in rats. *Pharmacol. Biochem. Behav.* 44, 533-538.

Rezvani, A.H., Denbow, D.M. & Myers, R.D. (1986): α-Melanocyte-stimulating hormone infused I.C.V. fails to affect body temperature or endotoxin fever in the cat. *Brain Res. Bull.* 16, 99-105.

Richards, D.B. & Lipton, J.M. (1984a): Effect of α-MSH 11-13 (lysine-proline-valine) on fever in the rabbit. *Peptides* 5, 815-817.

Richards, D.B. & Lipton, J.M. (1984b): Antipyretic doses of α-MSH do not alter afebrile body temperature in the cold. *J. Therm. Biol.* 9, 299-301.

Robertson, B., Dostal, K. & Daynes, R.A. (1988): Neuropeptide regulation of inflammatory and immunologic responses. *J. Immunol.* 140, 4300-4307.

Sawyer, T.K., Sanfillippo, P.J., Hruby, V.J., Engel, M.H., Heward, C.B., Burnett, J.B. & Hadley, M.E. (1980). 4-Norleucine, 7-D-phenylalanine-α-melanocyte-stimulating hormone: a highly potent α-melanotropin with ultralong biological activity. *Proc. Natl. Acad. Sci. USA* 77, 5754-5758.

Shih, S.T., Khorram, O., Lipton, J.M. & McCann, S.M. (1986): Central administration of α-MSH antiserum augments fever in the rabbit. *Am. J. Physiol.* 250, R803-R806.

Shih, S.T. & Lipton, J.M. (1985): Intravenous α-MSH reduces fever in the squirrel monkey. *Peptides* 6, 685-687.

Villar, M., Perassi, N. & Celis, M.E. (1991): Central and peripheral actions of α-MSH in the thermoregulation of rats. *Peptides* 12, 1441-1443.

Weiss, J.M., Sundar, S.K., Cierpial, M.A. & Ritchie, J.C. (1991): Effects of interleukin-1 infused into brain are antagonised by α-MSH in a dose-dependent manner. *Eur. J. Pharmacol.* 192, 177-179.

Integrative and cellular aspects of autonomic functions : temperature and osmoregulation. Eds K. Pleschka, R. Gerstberger. John Libbey Eurotext, Paris © 1994, pp. 181-190.

Interactions between the immune system and the hypothalamic neuroendocrine system during fever and endogenous antipyresis

E. Zeisberger, J. Roth and M. J. Kluger*

*Physiologisches Institut, Klinikum der Justus-Liebig-Universität, Giessen, Germany and *Department of Physiology, University of Michigan, Ann Arbor, USA*

SUMMARY

Circulating lymphocytes and macrophages are activated by contact with invading microorganisms or by foreign antigens displayed on cell membranes in tissues. In activated state they produce cytokines stimulating other populations of immune cells to proliferate or to produce antibodies specific to the invaders. Cytokines, like hormones, influence also other organs, among them the brain, where they affect different functions like thermoregulation (fever), water balance (antidiuresis), nutrition (anorexia) and evoke psychological changes, collectively termed "sickness behavior". A higher temperature is beneficial because it supports the immune defence. However, both, an extremely elevated temperature as well as an activated immune system are potentially dangerous, and must be restricted to limited time periods. In this review, we present evidence that hypothalamic suppressive reactions are initiated already during the fever rise. Hypothalamic neuroendocrine cells activating the hypothalamic-pituitary-adrenal axis play an important role in these suppressive reactions. Since this axis is activated also by a number of other stressors, fever and also the immune response may be modified at different physiological states. We speculate on mechanisms of bilateral signal transfer between the immune and central nervous systems and describe the stress responses as complex neuroendocrine reflexes.

INTRODUCTION

Febrile increase in body temperature is currently considered as a beneficial component of the complex immune response (Kluger, 1991). Number of antipyretic mechanisms have also been described that control the febrile rise in body temperature and prevent its increase to lethal values (Zeisberger, 1991). The endogenous antipyretics are proopiomelanocortin-derived hormones, like adrenocorticotrophic hormone (ACTH), α-melanocyte-stimulating hormone (α-MSH) and γ-melanocyte-stimulating hormone (γ-MSH), as well as hormones of the adrenal cortex, like cortisol or corticosterone. They are released due to activation of the hypothalamic-pituitary-adrenal (HPA) axis by hypothalamic hormones like corticotropin-releasing hormone (CRH) and arginine vasopressin (AVP). The responses of the HPA system are well known proto-

types of responses of organisms to internal and external stressors. For details and more complete original literature cf. recent reviews (Harbuz & Lightman, 1992; Lilly & Gann, 1992). Recently, it has been also proposed that a bilateral communication exists between the immune and central nervous systems (Blalock, 1989).
Here, we describe series of experiments in guinea pigs in those we induced an experimental fever by an intramuscular administration of bacterial lipopolysaccharide (LPS) from E. coli (20 µg/kg) and investigated time courses of the release of pyrogenic cytokines (tumor necrosis factor, TNF and interleukin-6, IL-6) and of activation of the HPA-system, considering the activation of CRH- and AVP-producing hypothalamic neurons by means of immunocytochemical methods and the release of cortisol. An attempt is made to regard the stimuli from the immune system as internal stressors and the activation of the HPA-system as a physiologic negative feedback mechanism reestablishing the normal state in the immune system. The interfering influence of concomitant or foregoing activation of other stressors is discussed.

METHODS

Animals. The experiments were performed in several groups of guinea pigs. The animals were equipped either with intraarterial catheters for repeated blood sampling or stereotactically implanted guide cannulae, through which push-pull perfusion or microinfusion cannulae could be introduced into the hypothalamus (for details cf.Roth et al.,1993).
Fever induction and measurement of body temperature. Endotoxin fever was induced by intramuscular injection of bacterial lipopolysaccharide (LPS from E. coli, 0111:B4, 20 mg/kg, Sigma Co., St. Louis, U.S.A.). Abdominal temperature was measured by use of battery-operated biotelemetry transmitters (VM-FH-Discs, Mini-Mitter Co., Sun River, OR, U.S.A.) implanted intraperitoneally. Output (frequency in Hz) was monitored by a mounted antenna placed under each animal's cage (RA 1000 radioreceivers, Mini-Mitter Co., Sunriver, OR, U.S.A.) and multiplexed by means of a BCM 100 consolidation matrix to an IBM personal computer system. A Dataquest IV acquisition system (Data Sciences Inc., St. Paul, MN, U.S.A.) was used for automatic control of data collection and analysis.
Samples. Blood plasma samples or hypothalamic push-pull perfusates were collected 1 h before and 1 h, 2 h, 3 h or 5 h after the injection of LPS or solvent.
Assays. Determination of TNF was based on the cytotoxic effect of TNF on the mouse fibrosarcoma cell line WEHI 164 subclone 13. The percentage of viable cells incubated either with different concentrations of TNF standard or with biological samples were measured by use of the dimethylthiazol-diphenyltetrazolium bromide (MTT) colorimetric assay. Determination of IL-6 was based on the dose dependent growth stimulation of IL-6 on the B 9 hybridoma cell line. After incubation with different concentrations of IL-6 standard or with biological samples, the MTT assay was used to measure the number of living cells (for details and literature c.f. Roth et al., 1993). Cortisol was measured by use of a commercially available radioimmunoassay (Biermann GmbH-Diagnostica, Bad Nauheim, F.R.G.)
Immunohistochemistry. Visualization of AVP- or CRH-containing perikarya or fibers and nerve terminals was performed by immunohistochemical methods which have been described in detail elsewhere (Merker et al., 1989).

RESULTS AND DISCUSSION

The experimental fever induced by systemic injection of LPS may be mediated by stimulation of TNF-production in the blood. The TNF-production depends on the formation of high-affinity complexes of LPS and a 60-kd LPS-binding protein. These complexes interact with mononuclear cells via the CD-14 receptor. TNF induces margination, migration and activation of neutrophils and stimulates differentiation of monocytes precursors. TNF acts also as an endogenous pyrogen and stimulates the acute-phase reaction. Because of these potent and wide-ranging actions it has been proposed that TNF may be the principle causative factor in the initiation of host response to sepsis (Michie & Wilmore, 1990). Similar roles have been ascribed also to other cytokines, such as interleukin-1 (IL-1) and interferon (INF). IL-1 potentiates the proliferation of T-cells, increases the release of granulocytes from bone marrow and promotes myelopoiesis. It is the best known endogenous pyrogen, claimed to mediate the fever caused by exogenous pyrogens, such as LPS (Dinarello et al., 1988). However, during fever, it is difficult to measure increased levels of IL-1 in blood, probably because of its short half-life, which is less then 10 min (Newton et al., 1988), and of its endogenous inhibitors. Also IL-6, produced by both T- and B- lymphocytes and monocytes, has been shown to induce the hepatic acute-phase response and to be a potent endogenous pyrogen (Kluger, 1991). Unlike other cytokines, IL-6 can be produced by several other cell types, including fibroblasts, chondroblasts and endothelial cells. Release of IL-6 by these cells is stimulated by IL-1 and LPS, but also increases in answer to viruses and bacteria.

Thus a whole cascade of cytokines released from immune and non-immune cells precedes and accompanies the febrile response according to activation and suppression of cell populations participating in the immune defence. Another problem concerns the effect of peripherally derived cytokines on the brain structures. To change the thermoregulatory set point, the cytokines should affect either central thermoreceptors or hypothalamic thermointegrative neurons or both. However, all cytokines are large molecules, and under normal conditions unable to cross the blood-brain barrier.

Therefore, we investigated first the time course of peripheral and hypothalamic release of TNF and IL-6 in guinea pigs in different stages of the febrile response. The results of these investigations are summarized in Fig. 1. Whereas IL-6 had a low activity and TNF was not detectable in samples of blood plasma and hypothalamic push-pull perfusates before and after the control injection of saline, the activities of both cytokines were elevated after the injection of LPS. Peak activity of TNF in plasma occurred early in the stage of fever rise. This observation is consistent with the proposed role of TNF as one of the inducers of the immune response. Systemic release of IL-6 strongly correlated with the febrile change of body temperature. The correlation coefficient of nearly 0.9 enables us to classify IL-6 as a true peripheral endogenous pyrogen (Kluger, 1991). Interestingly, increased activities of TNF and IL-6 after LPS injection could not only be observed in plasma but also in hypothalamic push-pull perfusates. In contrast to the values observed in plasma, hypothalamic TNF was highest at the fever peak, IL-6 in the stage of fever rise. This phenomenon indicates that cytokines released in the hypothalamus do not originate from blood and their activities do not correlate exactly with the thermoregulatory changes. Nevertheless, the cytokine activities in hypothalamic perfusates increase soon after fever induction, and questions arise, how hypothalamic cells recognize that they should release cytokines, and what function this release may have.

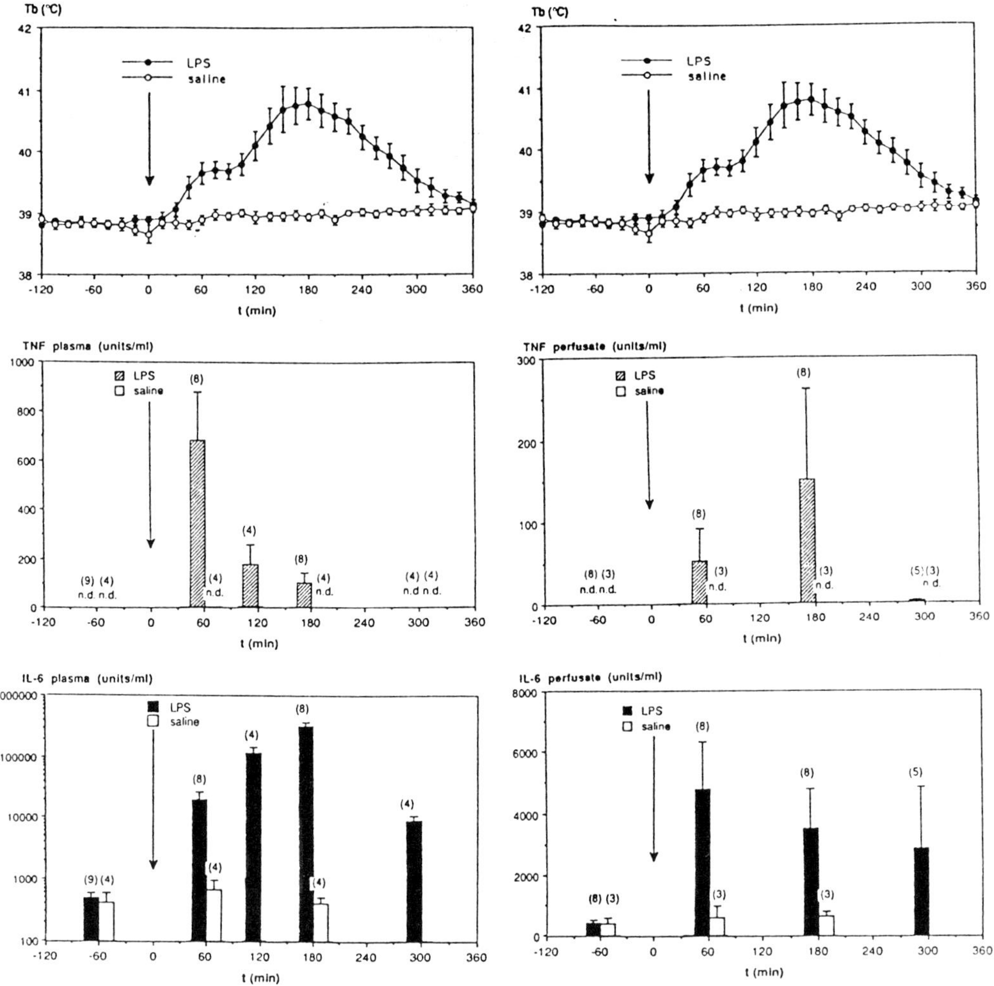

Fig. 1: Changes in abdominal temperature and comparison of activities of TNF and IL-6 in blood plasma (left) and in hypothalamic push-pull perfusates (right) in groups of guinea pigs before and after the injection of LPS (●) or saline (o). Compiled from experiments made in four groups of animals. After data from Roth et al.(1993).

Although the question of the signal transfer has not been solved definitely, the following working hypothesis might be accepted. The signal transfer between the peripheral blood and the brain structures may take place in special neurohemal structures, the circumventricular organs (CVOs) which have fenestrated capillaries allowing larger substances to enter the perivascular space. It is speculated that peripheral cytokines can bind to receptors at astroglial processes protruding into the perivascular space of CVOs and activate these cells to produce prostaglandin E (PGE), as indicated by studies in astroglial cultures (Katsuura et al., 1989). The local release of PGE at the brain side of the barrier may either modulate the activity of specific

ensors in these areas, or activate nearby neurons to influence, via heir axonal projections, the more distant hypothalamic structures. ecently, highly warm-sensitive structures have been described in one f the CVOs, the organum vasculosum laminae terminalis (OVLT), which ere inhibited by cytokines like IL-1 (Matsuda et al., 1992). The importance of the OVLT for the transport of febrile signals has been roposed long ago, and it could be demonstrated that the actions of ytokines at the OVLT are mediated by a PGE-dependent mechanism Morimoto et al., 1988). Our data suggest that cytokines revealed in he hypothalamic push-pull perfusates are produced directly in the rain, and we propose that their release may be initiated by a specific transfer of peripheral immune signals from the CVOs, as also ssumed by others (Katsuura et al., 1990).
entrally released cytokines seem to induce a number of actions, lemonstrated mainly for IL-1, which has been employed in experiments ost frequently. IL-1 receptors have been found in several areas of he brain, including the hypothalamus and the limbic preoptic area. he released cytokines may first influence central thermosensitive and hermointegrative neurons. Second, they influence the HPA-axis by ctivating CRH-release from the hypothalamic paraventricular nucleus PVN). Central application of IL-1 is associated with decreased CRH-content in the median eminence and increased CRH-concentration in ypophyseal portal venous blood. IL-1 also increases release of CRH rom isolated hypothalami of rats. The effect of IL-1 in the PVN may e quite specific because there is evidence that IL-1 activates only arvocellular, CRH-expressing neurons. Also central effects of cytokines appear to be mediated by prostaglandins. CRH-release from hypothalamic cultures can be increased by prostaglandin F, and this action an be blocked by dexamethasone. The suppressive effect of dexamethasone on IL-1-induced fever is mediated by lipocortins which inhibit rostaglandin synthesis. Cyclooxygenase inhibitors also block hypothalamic effects of IL-1 (Morimoto et al., 1989). Similar to IL-1, also NF (Sharp et al., 1989) and IL-6 (Navarra et al., 1991) can activate he HPA-system through CRH-stimulation mediated by prostaglandins.
RH produced in parvocellular neurons of the PVN is transported hrough axons to nerve endings in the external zone of the median eminence. Here it is released into the hypophyseal portal blood and acts n corticotrophs in the anterior pituitary to release ACTH. Several ther peptides interact with CRH and potentiate the stimulatory action f this hormone. The most important of these appears to be AVP which is co-secreted with CRH by a subset of parvocellular neurons and is likewise carried into the anterior pituitary. Pituitary corticotrophs ave membrane receptors for a number of peptides, including CRH, AVP, angiotensin II (A-II), and amines like noradrenaline (NA). CRH stimulates adenylate cyclase in corticotrophs and activates cyclic AMP-dependent protein-kinase leading to ACTH-release. AVP, A-II, and NA potentiate the effects of CRH by increasing levels of intracellular calcium. The effects of CRH and AVP on the corticotroph are additive, and AVP may play and additional role in sustaining ACTH-release during repeated stimuli that desensitize the corticotroph to CRH (Scaccianoce et al., 1991). In addition to the release of ACTH, CRH increases the expression of ACTH-precursor (POMC) messenger-RNA in corticotrophs. The increase in PMOC-transcription induced by CRH can be detected in pituitary cell cultures after 5 min, becomes maximal after 60 min, and can be inhibited by dexamethasone (Roberts et al., 1987).
In order to prove whether these in vitro-derived values correspond to those in vivo, we examined, in the next series of experiments, the time course of activation of the HPA-axis during the febrile response.

For this purpose we investigated the changes in the immunoreactivity of parvocellular PVN-neurons to CRH and AVP, and determined the release of cortisol into the blood plasma in different stages of the febrile response. In the non-febrile state (60 min before LPS-injection) many of the parvocellular neurons in the PVN were stainable by antiserum against CRH, and only a few by antiserum against AVP as shown in Fig. 2.

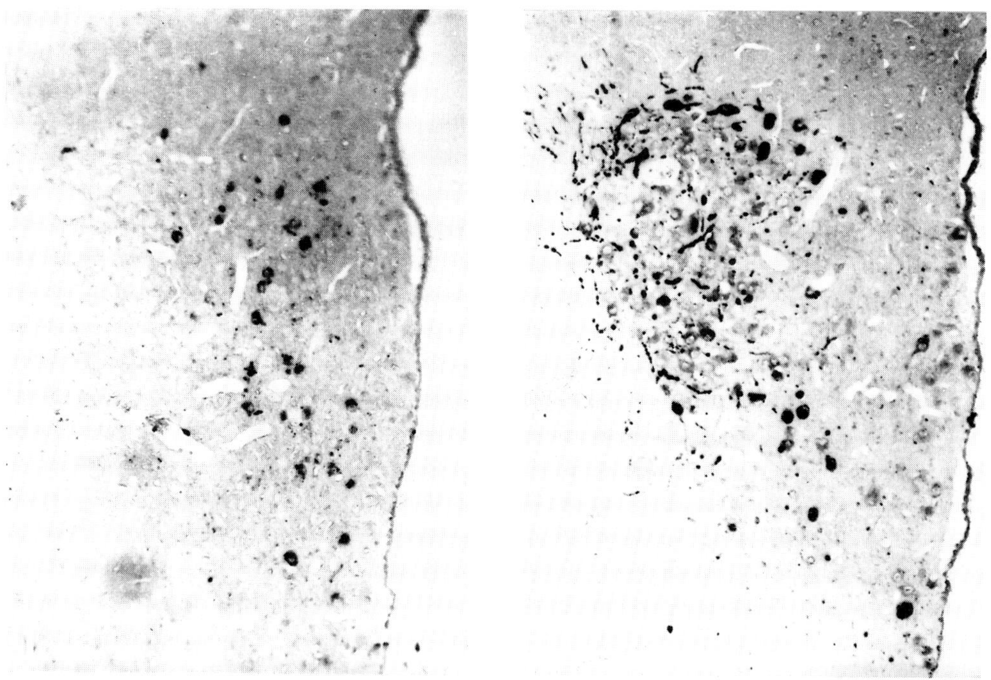

Fig. 2: Immunoreactivity to CRH (left) and AVP (right) on two parallel frontal sections of the brain in a guinea pig before injection of LPS. Note a number of neurons exhibiting the CRH-immunoreactivity in the ventromedial part of the PVN near the 3rd ventricle (left) and the absence of AVP-immunoreactive neurons on the comparable site (right). This is different during the febrile response (cf. text).

Soon after the injection of LPS the distribution of the immunoreactivity changes. Already in the stage of fever rise (1 h after LPS) the immunoreactivity to CRH disappears, and in its place many parvocellular neurons in the ventromedial part of the PVN become stainable by antiserum against AVP as described previously (Zeisberger et al., 1986). In animals killed at the height of the fever, the AVP-reaction was still increased in comparison to controls, but it was slightly reduced in comparison to the stage of fever rise. At the stage of fever decrease the reactivity pattern of AVP neurons again resembled that of the controls although the body temperature had not yet returned to normal. Since similar immunoreactive changes were found also in nerve terminals projecting to the septum and amygdala, we concluded from these experiments that some subsets of parvocellular neurons do not only project to the median eminence, but send their axons to areas of the limbic system participating in endogenous antipyretic reactions.

The early change in the immunoreactivity may be due to suppression of the CRH-production by negative feedback reactions of ACTH or released corticosteroids. Similar changes in the immunoreactivity have been described after adrenalectomy (Kiss, 1988) and documented also by the rise in the relation between AVP and CRH released into the hypophyseal portal blood (Plotsky & Sawchenko, 1987). Since the effects of adrenalectomy represent a long-term feedback mechanism, and we did not know whether the same mechanism can also be activated in a considerably shorter time, we investigated the release of cortisol during fever. The results of these experiments are summarized in Fig. 3.

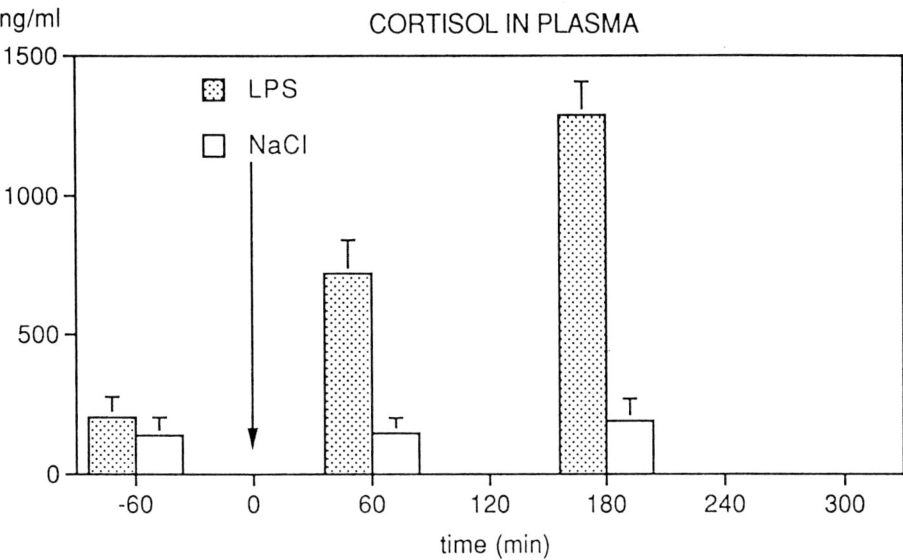

Fig. 3: Cortisol levels in blood plasma in guinea pigs before and after injections of LPS (E.coli, 20 µg/kg) or saline (0.9% NaCl). Columns represent means in 6 animals, the bars indicate SEM.

From the substantial increase in the cortisol concentration in the blood found early after the injection of LPS, we conclude that the changes in the AVP-immunoreactivity of the parvocellular PVN-neurons in the stage of fever rise might be caused by the inhibition of CRH-production and release, via the fast negative-feedback action of the peripherally released corticosteroid. Released corticosteroids are for the most part (90%) rapidly bound to circulating albumin and corticosteroid-binding globulin. The remaining 10% of corticosteroids circulate in unbound, active form. Circulating free corticosteroid can pass through the membrane of target cells and is bound to specific receptors in the cytosol. The steroid-receptor complex then migrates to the nucleus of the cell and interacts with DNA to regulate transcription of effector genes. Through this mechanism corticosteroids have profound effects on proteins and enzymes throughout the body and can influence cellular growth, function, and metabolism. This mechanism, which requires alteration of gene transcription, explains the delay of 1-2 h between the administration of corticosteroids and their maximal effects. However, they can have more rapid actions on cellular membranes, where they inhibit phospholipase-A and decrease eicosanoid production. This may be most important for the effects of corticosteroids on immune function. Corticosteroids have suppressive effects

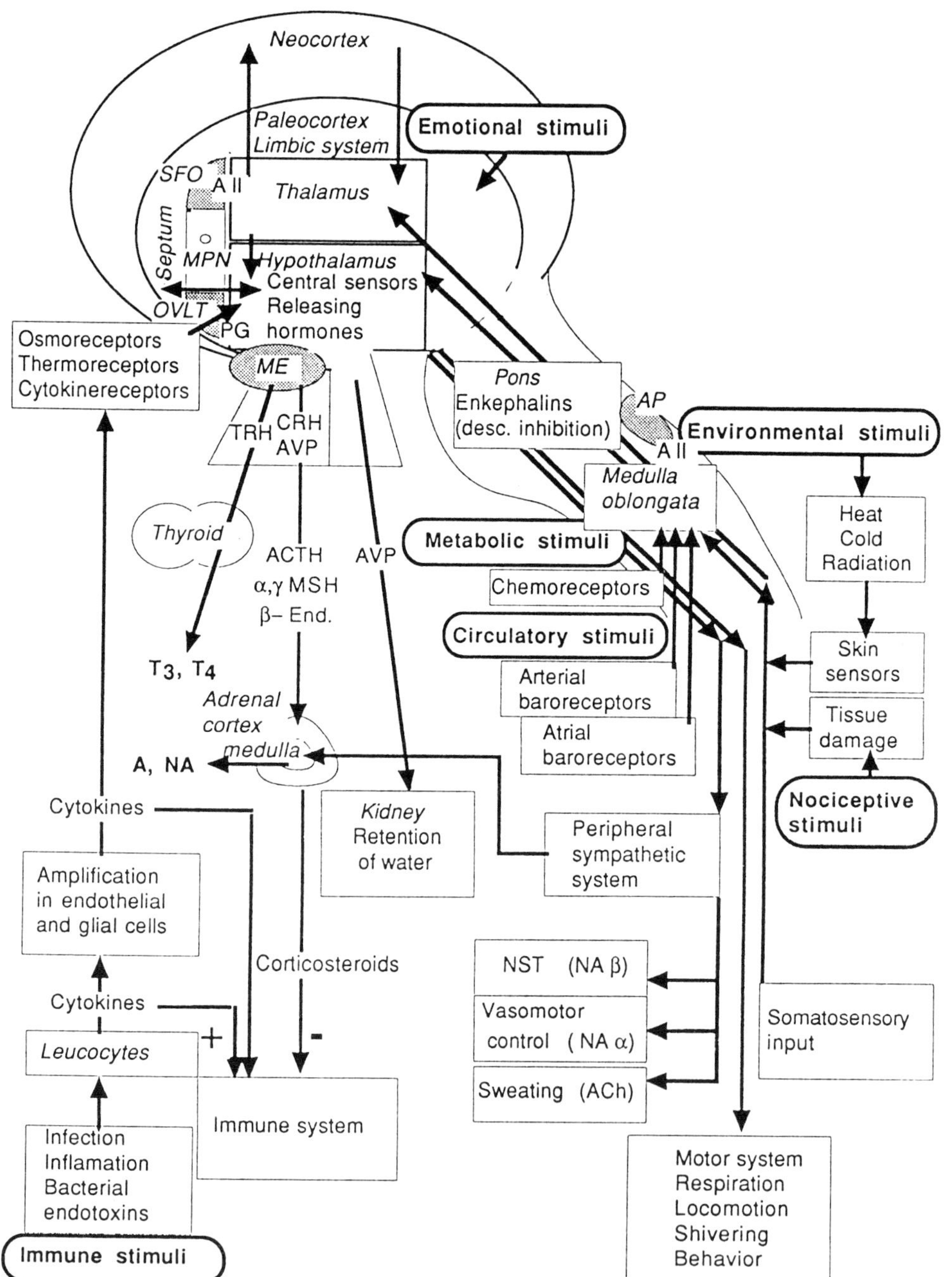

Fig. 4: Schematic diagram showing the convergence of signals evoked by different external and internal stimuli(denoted by bold typing) into the hypothalamus and hypothalamic outputs to peripheral organs mediated by motor nerves, peripheral vegetative nerves, and endocrine systems. The central role of HPA-system in responses to different stressors, including the activation of the immune system, is discussed in the text.

on the immune system. Acting either directly or through the inhibition of cytokine release, corticosteroids can suppress lymphocyte proliferation, antibody formation, and natural killer cell activity. Since the activity of the immune system may be potentiated by a number of positive-, facilitatory-, feedback loops (cf. Fig.4), the inhibition mediated by the HPA system, representing the only negative-feedback loop, is particularly important.

The schematic diagram in Fig.4 is an attempt to summarize different external and internal stimuli (stressors) which may interfere with the homeostasis of the organism. These stimuli act on different external and internal sensors, whose signals are conveyed by afferent pathways to the CNS, where their integration and modulation occur.The hypothalamus contains multisensor neurons sensitive to different stimuli (temperature, osmolality, metabolic substances, hormones, cytokines) which are interconnected to many different neurotransmitter systems carrying information from sensory systems and instructions from superior parts of the CNS. The signals derived from this integration are differentiated in neurons producing releasing hormones and increased in cells of the endocrine system. The resulting efferent endocrine or neural signals alter the peripheral function or systemic metabolism to counteract the disturbance from the original stressor. Thus, external stimuli, or different internal stimuli (nociceptive, metabolic from chemoreceptors, circulatory from baroreceptors, immune and emotional) are all able to activate the HPA-system. Fever is induced by actions of mediators of the immune system on the CNS, and it is soon suppressed by different antipyretic mechanisms which are activated together with the HPA-system. The activation of the HPA-system can be regarded as a negative feedback mechanism in the complex neuroendocrine reflex aiming at reestablishing the normal state in the immune system.

If the HPA-system is activated by non-immune stressors, the febrile response to immune signals can be limited or suppressed. Different conditions tested in our experiments including dehydration, cold adaptation, immobilization stress, electrical stimulation of the PVN or parturition, were all accompanied by an attenuation of the fever response (Roth & Zeisberger, 1992)

ACKNOWLEDGEMENTS

The excellent technical assistance of B. Störr is gratefully acknowledged. The studies were supported by the Deutsche Forschungsgemeinschaft (Ze 183/4-1).

REFERENCES

Blalock, J.E. (1989):A molecular basis for bidirectional communication between the immune and neuroendocrine systems.*Physiol.Rev.* 69, 1-32.

Harbuz, M.S. & Lightman, S.L. (1992): Review: Stress and the hypothalamo-pituitary-adrenal axis: Acute, chronic and immunological activation. *J.Endocrinol.* 134, 327-339.

Katsuura, G., Gottschall, P.E., Dahl, R.R. & Arimura, A. (1989): Interleukin-1 *beta* increases prostaglandin E_2 in rat astrocyte cultures: Modulatory effect of neuropeptides. *Endocrinology* 124, 3125-3127.

Katsuura, G., Arimura, A., Koves, K. & Gottschall, P. (1990): Involvement of organum vasculosum of lamina terminalis and preoptic area in interleukin 1ß-induced ACTH release. *Am.J.Physiol.* 258, E163-E171.

Kiss, J.Z. (1988): Dynamism of chemoarchitecture in the hypothalamic paraventricular nucleus. *Brain Res.Bull.* 20, 699-708.

Kluger, M.J. (1991): Fever: Role of pyrogens and cryogens.*Physiol.Rev.* 71, 93-127.

Lilly, M.P. & Gann, D.S. (1992): The hypothalamic-pituitary-adrenal-immune axis: A critical assessment. *Arch.Surg.* 127, 1463-1474.
Matsuda, T., Hori, T. & Nakashima, T. (1992): Thermal and PGE_2 sensitivity of the organum vasculosum lamina terminalis region and preoptic area in rat brain slices. *J.Physiol.(Lond.)* 454, 197-212.
Merker, G., Roth, J. & Zeisberger, E. (1989): Thermoadaptive influence on reactivity pattern of vasopressinergic neurons in the guinea pig. *Experientia* 45, 722-726.
Michie, H.R. & Wilmore, D.W. (1990): Sepsis, signals, and surgical sequelae: a hypothesis. *Arch.Surg.* 125, 531-536.
Morimoto, A., Murakami, N. & Watanabe, T. (1988): Effect of prostaglandin E_2 on thermoresponsive neurones in the preoptic and ventromedial hypothalamic regions of rats. *J.Physiol.(Lond.)* 405, 713-725.
Morimoto, A., Murakami, N., Nakamori, T., Sakata, Y. & Watanabe, T. (1989): Possible involvement of prostaglandin E in development of ACTH response in rats induced by human recombinant interleukin-1. *J.Physiol.(Lond.)* 411, 245-256.
Navarra, P., Tsagarakis, S., Faria, M.S., Rees, L.H., Besser, G.M. & Grossman, A.B. (1991): Interleukins-1 and -6 stimulate the release of corticotropin-releasing hormone-41 from rat hypothalamus *in vitro* via the eicosanoid cyclooxygenase pathway. *Endocrinology* 128, 37-44.
Newton, R.C., Uhl, J., Covington, M. & Back, O. (1988): The distribution and clearance of radiolabeled human interleukin-1 beta in mice. *Lymphokine Res.* 7, 207-216.
Plotsky, P.M. & Sawchenko, P.E. (1987): Hypophysial-portal plasma levels, median eminence content, and immunohistochemical staining of corticotropin-releasing factor, arginin vasopressin, and oxytocin after pharmacological adrenalectomy. *Endocrinology* 120, 1361-1369.
Roberts, J.L., Lundblad, J.R., Eberwine, J.H., Fremeau, R.T., Salton, S.R.J. & Blum, M. (1987): Hormonal regulation of POMC gene expression in pituitary. *Ann.NY Acad.Sci.* 512, 275-285.
Roth, J., Conn, C.A., Kluger, M.J. & Zeisberger, E. (1993): Kinetics of systemic and hypothalamic interleukin 6 and tumor necrosis factor during endotoxin fever in the guinea pig. *Am.J.Physiol.* (in press)
Roth, J. & Zeisberger, E. (1992): Evidence for antipyretic vasopressinergic pathways and their modulation by noradrenergic afferents. *Physiol.Res.* 41, 49-55.
Scaccianoce, S., Muscolo, L.A.A., Cigliana, G., Navarra, D., Nicolai, R. & Angelucci, L. (1991): Evidence for a specific role of vasopressin in sustaining pituitary-adrenocortical stress response in the rat. *Endocrinology* 128, 3138-3143.
Sharp, B.M., Matta, S.G., Peterson, P.K., Newton, R., Chao, C. & McAllen, K. (1989): Tumor necrosis factor-α is a potent ACTH secretagogue: comparison to interleukin-1ß. *Endocrinology* 124, 3131-3133.
Zeisberger, E., Merker, G., Blähser, S. & Krannig, M. (1986): Role of vasopressin in fever regulation. In *Pharmacology of thermoregulation homeostasis and thermal stress: Experimental & therapeutic advances*, ed. K.E. Cooper, P. Lomax, E. Schönbaum & W.L. Veale, pp.62-65. Basel: Karger.
Zeisberger, E. (1991): Peptides and amines as putative factors in endogenous antipyresis. In *New Trends In Autonomic Nervous System Research*, ed. M. Yoshikawa, M. Uono, H. Tanabe & S. Ishikawa, pp. 86-89. Amsterdam: Elsevier.

Integrative and cellular aspects of autonomic functions : temperature and osmoregulation. Eds K. Pleschka, R. Gerstberger. John Libbey Eurotext, Paris © 1994, pp. 191-199.

Disorders of thermoregulation in the elderly[1]

Peter Lomax

Department of Pharmacology, School of Medicine and the Brain Research Institute, University of California, Los Angeles, California 90024, USA

On reviewing the medical literature it appears generally to be accepted that the older the individual the greater their likelihood of developing disorders of thermoregulation. In standard medical texts "*advancing age*" is considered to be a predisposing factor in the pathogenesis of both accidental hypothermia and classical heat stroke:

HYPOTHERMIA: "*Pathogenesis. Advanced age ...*"

HEATSTROKE: "*Classical heatstroke occurs especially in ... the elderly ...*" (Knochel, 1985)

This view stems largely from reviews, published 15 or more years ago, of the epidemiology of accidental hypothermia in Britain and heat stroke in certain cities in the United States (for references see Cooper & Ferguson, 1983; Kenny & Hodgson, 1987). What was continually stressed by the authors of these reports, and what is frequently overlooked in subsequent discussions, is that socioeconomic concomitants of ageing, rather than pathophysiological changes, largely could account for the higher incidence of disorders of temperature regulation in the older population.

Research into ageing, and associated disorders, has advanced greatly over the past decade or so and it would seem appropriate to reexamine the relationship between age and susceptibility to thermoregulatory disturbances in the light of newer findings and in the changing demographic profile.

Undoubtedly, the world population is currently ageing. To what extent this trend will continue, and whether there is a biologically determined upper limit to life expectancy, are subjects of contemporary experimentation and debate which recently have caused received wisdom to be questioned. But, for the near future at least, it is important to assess the consequences and impact of an ageing population both on the individual and on society as a whole.

[1]Dedicated to Professor Eckhart Simon, Director and Head of the Department at the Max-Plank-Institut für Physiologische und Klinische Forschung, Bad Nauheim in friendship and respect on the occasion of his 60th birthday.

In directing attention to thermoregulatory disorders the following considerations are relevant:

- To what extent is the world population ageing?
- What is the likely upper limit to longevity?
- Do currently available statistics indicate that disorders of temperature regulation are more common in the aged?
  - Accidental hypothermia
  - Heat stroke
- What factors could account for any age related differences?
  - Concurrent disease and drug treatment
  - Degenerative changes in the thermoregulatory system
  - Social conditions
- Does the thermoregulatory system age? If so, are the adverse effects of age preponderantly found in the old?
  - Changes in heat production
  - Changes in heat dissipation
  - Changes in thermal afferent input
  - Changes in the central controlling mechanisms

Before examining these several questions one needs to define the "elderly". Essentially such a definition will be arbitrary but the World Health Organization has suggested the ranges:

Middle age	45 to 59 years
Elderly	60 to 74 years
Old	75 to 90 years
Very old	> 90 years

Thus, for the purposes of the present discussion, it would seem appropriate to classify as elderly those individuals over 60 years of age.

To what extent is the present world population ageing?

Since the mid-19th century there has been a near doubling of the expectation of life at birth from 40 to near 80 y. Most of the decline in mortality in the earlier part of the century reflects the marked reduction in neonatal, infant and maternal deaths. Over the past 25 years gains in life expectancy were achieved largely in the elderly population. Thus, during the present century the rise in life expectancy has resulted in an ageing population (Fig. 1). In order for life expectancy at birth to increase to an average biological limit of 85 y it has been estimated that mortality rates from all causes would need to decline over 60% in those >50 y (Olshansky *et al.*, 1990). Cures for the major degenerative diseases would reduce overall mortality by about 75% which renders it unlikely that life expectancy at birth could exceed 85 y. This is the basis for the 'predicted maximum' in Fig. 1. However, not all evidences concur with this view. Studies with rapidly breeding species, such as flies, suggest that death rates do not automatically increase with age as most researchers have believed (in accordance with the "Gompertz law") (see Barinaga, 1992). On the other hand, life expectancy from birth has barely increased in Denmark during the past 20 y and may have actually decreased in females (News, 1993).

Clearly this debate will continue; for the near term, however, it would seem that we can expect life expectancy to advance and for the proportion of aged individuals in the general population to increase. Age related disorders will also continue to be more prevalent so that it is of some consequence to identify those conditions for which the aged are at risk.

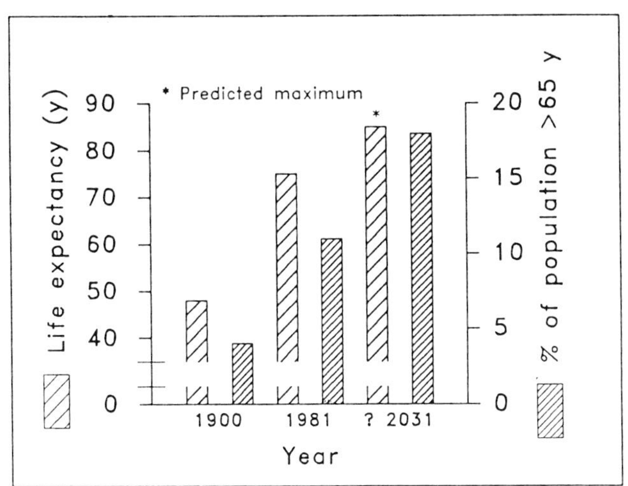

Fig. 1. The ageing population

Are disorders of temperature regulation more common in the elderly?

Accidental hypothermia

The most extensive review of accidental hypothermia was that carried out by Paton (1991). Table 1 summarizes all of the cases Paton could find in the literature in which the ages of the victims were mentioned.

Table 1. Cases of accidental hypothermia in which the ages were reported.

Age (y)	Number of cases	Deaths reported
1-10	24	
11-20	2	
21-30	7	
31-40	1	
41-50	3	
51-60	3	
61-70	3	
71-80	1	
81-90	1	
"young mountain hikers"	100	
53-80	8	7
56-86	23	20
22-86	11	
17-80	27	
"children"	19	9

These reports do not support the suggestion that the aged are more likely to develop accidental hypothermia; if anything the records indicate that it is the young, rather than the elderly, who are especially at risk. This is undoubtedly due to the fact the younger people are more prone to expose themselves to conditions in which hypothermia may occur.

Heat Stroke

Epidemiological data for heat stroke are less frequent. During severe heat waves in the US midwestern states in 1934 and 1966 many cases of classical heat stroke occurred and there was a relatively greater number of deaths in the aged compared to the younger population (Collins, 1934; Schumann, 1972). At the time this preponderance of victims aged >60 y was ascribed to the fact that most of the patients suffered from metabolic, respiratory and/or cardiovascular diseases which were thought to compromise normal thermoregulatory function. Later authors have tended to perpetuate this view, e.g., Cooper & Veale (1983). More recent detailed data have emerged from victims of stress induced heat stroke during the annual pilgrimages to Makkah in Saudi Arabia. Al-Khawashki *et al.* (1983) presented clinical data from 172 cases during the summer of 1982. In 125 the ages were known; Fig. 2 illustrates the age distribution of these individuals.

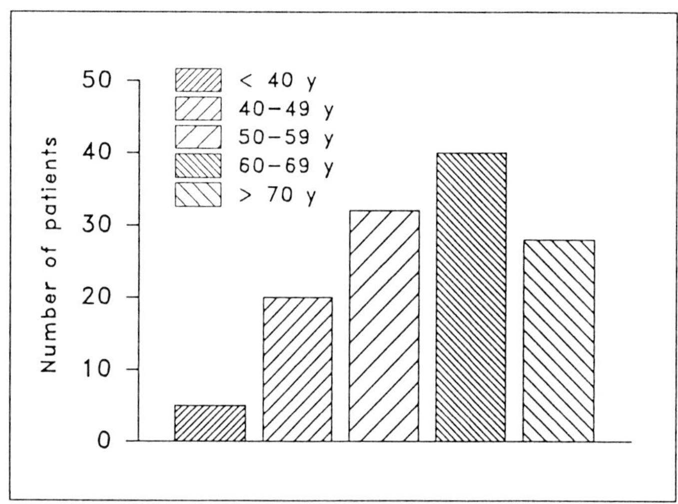

Fig. 2. Age distribution of 125 heat stroke victims Mina and Arafat 1982

In 1982 there were approximately 2,000,000 individuals journeying on the pilgrimage to Makkah and over 1,000 cases of stress induced heat illness. The age distribution of the pilgrims has not been reported but it would seem likely the there were fewer young compared to older age groups. As seen in Fig. 2, even in a population with a bias to the aged, stress induced heat stroke does not seem to be markedly more frequent in those over 60 y.

Thus, while the aged appear to be at a greater risk of developing classical heat stroke during periods of high environmental temperature and humidity there seems not to be an increased susceptibility to stress induced heat stroke in those over 60 y.

Concurrent diseases and thermoregulation

Cardiovascular and metabolic diseases which impair the circulation will usually reduce the activity of thermoregulatory effector systems, most significantly heat loss by radiation from the skin and from sweating. Diseases which reduce muscular activity will lower metabolic heat production. It seems to be accepted that such conditions contribute to heat illness in those over 60 y.

Many cardiac disorders can lead to congestive heart failure. The most commonly encountered in general medical practice are listed in Table 2.

Table 2. Pathogenesis of congestive cardiac failure

Disorder	Age of onset
Congenital heart disease	Birth onwards
Valvular heart disease	Generally < 40 y
Pulmonary hypertension	
Primary	Rare, 10-40 y
Secondary	Depends on underlying condition All age groups
Arterial hypertension	All ages. Malignant hypertension 40-60 y
Coronary artery disease	Increasing incidence with age Less common in females < 50 y
Cardiac arrhythmias	All ages
Primary cardiomyopathy	All ages
Disorders of the pericardium	All ages

The metabolic and endocrine disorders which have been considered as risk factors for thermoregulatory disorders are listed in Table 3.

Table 3. Endocrine diseases

Disorder	Age of onset
Hypopituitarism	
1°	Younger women
2°	30-50 y
Hyperpituitarism	Childhood to middle age
Diabetes insipidus	All ages (usually traumatic)
Hypothyroidism (myxedema)	
1° (95% of cases)	All ages (> in women < 40 y)
Hyperthyroidism (Grave's disease)	30-40 y
1° adrenal insufficiency	Childhood onwards
Cushing's syndrome	~ 40 y at onset
Diabetes mellitus	Childhood onwards

The only one of these diseases which is relatively frequent, and which demonstrates a marked preponderance of victims over 60 y of age, is coronary artery insufficiency; here the determinant for onset could be the duration of associated risk factors (smoking, diet, other diseases, etc.).

Psychiatric disorders constitute another major group of inflictions of mankind. Such patients are frequently treated with a variety of centrally acting drugs, many with peripheral side effects, which could impair temperature regulation under adverse environmental conditions. The role of phenothiazines in the pathogenesis of heat stroke has been stressed in several reports (see Clark & Lipton, 1991). These authors identified 56 victims in which the age was recorded who presented with heat stroke and who were being treated with neuroleptics. Of these, 10 were over 60 y and 44 between 20 and 59 y.

Degenerative changes in the thermoregulatory system

Primary abnormalities of the thermoregulatory system, which lead to relative poikilothermia, are uncommon. They include agenesis of the corpus callosum, cerebrovascular accidents (stroke, basilar occlusion, subarachnoid hemorrhage) cerebral tumors, trauma, encephalitis, granulomata, Wernicke's encephalopathy and childhood progressive idiopathic degeneration. Again none of these conditions is predominantly associated with old age.

Thus, in considering the major clinical disorders which may impair temperature regulation, particularly during exposure to extremes of temperature and humidity, few are significantly more frequent in the aged.

Social conditions

To a certain degree the assessment of the role of poverty and undernutrition in the incidence of thermoregulatory disorders, particularly accidental hypothermia, in the aged has been clouded by political considerations in that governments have not wished to recognize that such adverse factors exist within their societies. There can be no doubt that poor housing, lack of adequate heating and poor caloric diets will predispose to thermal instability and the aged are generally more likely to suffer such deprivations.

Does the thermoregulatory system age?

The concept that physiological regulatory systems become impaired with advancing age seems to be accepted uncritically by both the general and scientific communities. If this assumption is correct and, if so, whether the decrement is of such as magnitude to result in clinical disorders is not so easy to establish.

Changes in heat production

Resting metabolic rates are generally lower in randomly selected older compared to younger subjects. However, the elderly tend to be less active physically than the young. When physical activity is taken into account the differences between age groups tend to disappear (Poehlman *et al.*, 1990) (Fig. 3).

Sweat production

There has been a number of separate studies of the decrement in sweating with age during exposure to elevated environmental temperatures (20-50°C) (see Kenny & Hodgson, 1987). At low relative humidities (4-24%) older individuals exhibit a fairly uniform deficit of 20% compared to the young. However, at a relative humidity > 50% this age difference disappears.

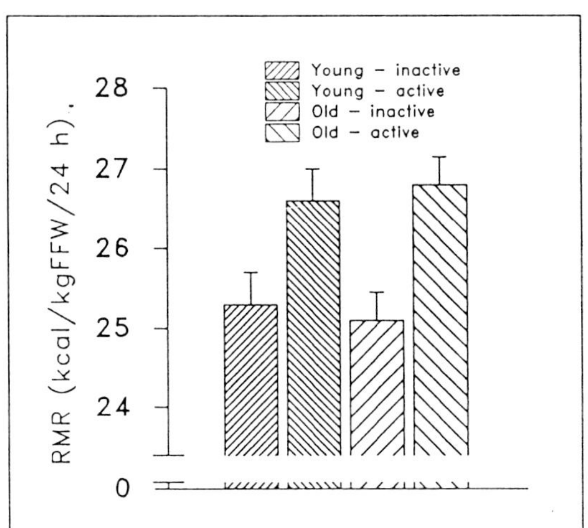

Fig. 3. Comparison of resting metabolic rates (RMR) in young and old subjects adjusted for physical activity. Vertical bars represent 1 SEM

Temperature regulation during exercise

In a study of young and older women exercising at 48°C in a dry environment the increase in rectal temperature was greater in the older age group (Fig. 4).

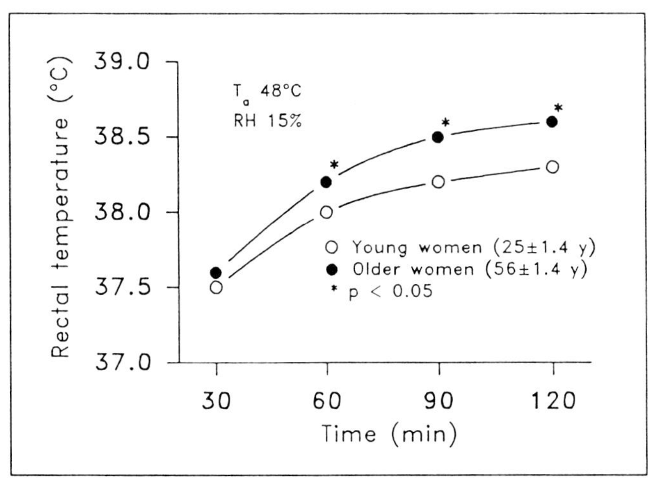

Fig. 4. Effect of exercise at 35% $\dot{V}O_{2\ max}$ on rectal temperature in young and older women. Data from Kenny & Hodgson, 1987.

Since this study was conducted at low humidity (15%) the significantly higher core temperatures in the older subjects could be a reflection of relatively impaired sweating in this group (*vide supra*). Whether the same differences are found under more humid conditions is not known.

Changes in the CNS

The development of techniques in molecular biology over recent years has led to the recognition of heat shock proteins (HSP). These molecules act as a protective, or adaptive, response to cellular stress. Exposure of rats to high environmental temperatures causes the induction of HSP70 mRNAs which correlates with colonic temperature. This suggests that synthesis of HSP70 is an important component of the response to hyperthermic stress. In older rats (comparing 5 and 24 months) the graded increase in HSP70 mRNA expression was less at elevated ambient temperatures and was correlated with the magnitude of the rise in core temperature (Blake *et al.*, 1991). A reduced fever response in old animals is well documented, so that the reduced expression of HSP70 in older individuals could be the result of the attenuated rise in core temperature. Whether the relative lack of stress proteins would render the individual more susceptible to adverse environmental conditions requires further consideration.

Conclusions

From the present evidence the emerging impression is that age is a significant variable in the pathogenesis and epidemiology of accidental hypothermia and heat stroke. However, different groups appear to be at risk under particular circumstances and the blanket assumption that the aged are predominantly likely to develop thermoregulatory disorders is not well founded. In old age molecular, biochemical and histopathological changes occur which could have an adverse effect on temperature regulation; this is an area which should be the subject of further research. However, social and economic conditions are clearly a major factor in determining the well-being of humans residing in regions with extremes of climate and these factors alone could account for many of the instances in which the aged are found to be disproportionately affected.

REFERENCES

Knochel, J.P. (1985): Disorders due to heat and cold. In *Textbook of Medicine*, ed. J.B. Wyngaarden & L.H. Smith, pp. 2304-2306. Philadelphia: Saunders.

Cooper, K.E. & Ferguson, A.V. (1983): Thermoregulation and hypothermia in the elderly. In *The Nature and Treatment of Hypothermia*, ed. R.S. Pozos & L.E. Wittmers, pp. 35-45. Minneapolis: University of Minnesota Press.

Kenny, W.L. & Hodgson, J.L. (1987): Heat tolerance, thermoregulation and ageing. *Sports Medicine* 4, 446-456.

Olshansky, S.J., Carnes, B.A. & Cassel C. (1990): In search of Methuselah: estimating the upper limits of human longevity. *Science* 250, 634-640.

Barinaga M. (1992): Mortality: overturning received wisdom. *Science* 258, 398-399.

News (1993): Danes' life expectancy stagnates. *BMJ* 306, 1226-1227.

Paton, B.C. (1991): Accidental hypothermia. In *Thermoregulation: Pathology, Pharmacology and Therapy*, ed. E. Schönbaum & P. Lomax, pp. 397-454. New York: Pergamon Press.

Collins, S.D. (1934): *Public Health Reports (USA)* 49, 1015.

Schumann, S.H. (1972): Patterns of urban heat wave deaths and implications for prevention: data from New York and St. Louis during July 1966. *Environ. Res.* 5, 59-75.

Cooper, K.E. & Veale, W.L. (1983): The elderly and their risk of heat illness. In *Heat Stroke and Temperature Regulation*, ed. M. Khogali & J.R.S. Hales, pp. 189-196. Sydney: Academic Press.

Al-Khawashki, M.I., Mustafa, M.K.Y., Khogali, M. & El-Sayed, H. (1983): Clinical presentation of 172 heat stroke cases seen at Mina and Arafat - September, 1982. In *Heat Stroke and Temperature Regulation*, ed. M. Khogali & J.R.S. Hales, pp. 99-108. Sydney: Academic Press.

Clark, W.G. & Lipton, J.M. (1991): Drug-related heatstroke. In *Thermoregulation: Pathology, Pharmacology and Therapy*, ed. E. Schönbaum & P. Lomax, pp. 125-177. New York: Pergamon Press.

Poehlman, E.T., McAuliffe, T.L., Van Houten, D.R. & Danforth, E. (1990): Influence of age and endurance training on metabolic rate and hormones in healthy men. *Am. J. Physiol. 258, 256-262.*

Blake, M.J., Fargnoli, J., Gershon, D. & Holbrook, N.J. (1991): Concomitant decline in heat-induced hyperthermia and HSP70 mRNA expression in aged rats. *Am. J. Physiol.* 260, R663-R667.

Integrative and cellular aspects of autonomic functions : temperature and osmoregulation. Eds K. Pleschka, R. Gerstberger. John Libbey Eurotext, Paris © 1994, pp. 201-208.

Induction of hypertension in rats exposed chronically to cold

Melvin J. Fregly

Department of Physiology, University of Florida, College of Medicine, Gainesville, FL 32610, USA

Many studies have now shown that chronic exposure of rats to mild cold (5°C) is accompanied by an elevation of systolic, diastolic and mean blood pressures and cardiac hypertrophy (Chevillard and Bertin, 1965; Fregly, 1954; Fregly and Papanek, 1989; Fregly et al, 1989, 1991; Gilson, 1950; Heroux and Dugal, 1961). These signs characterize the development of the syndrome of hypertension during chronic exposure to cold. Resting heart rate of cold-treated rats was also increased significantly above that of warm-acclimated controls. These studies are significant because they provide, for the first time, a "naturally occurring" hypertension which is induced without either surgical intervention, administration of excessive doses of either hormones or drugs, or genetic manipulation.

Additional studies in which systolic blood pressure was measured indirectly from the tail of the rat have shown that a 2 to 3 week exposure to cold is required to elevate blood pressure significantly above pre-cold exposure level (Fregly and Papanek, 1989). Studies have also shown that blood pressure is not immediately reversible when the animals are removed from cold to air at 26°C. Thus, rats exposed to cold for 7 weeks still had blood pressures that were elevated above the level of controls 4 weeks after return to a thermoneutral temperature (Shechtman et al, 1990b). Thus, the possibility exists that an elevation of blood pressure induced by a longer period of exposure to cold may not be reversible after return to a thermoneutral temperature. Other studies showed an age and ambient temperature dependence; i.e. younger rats developed cold-induced hypertension more readily than older rats, and hypertension developed more readily at ambient temperatures between 5 and 9°C (Shechtman et al, 1990a).

Studies have also been carried out to determine the minimal time of daily exposure to cold (5°C) that would result in a significant elevation of blood pressure in rats (van Bergen et al, 1992). Groups of rats were exposed to cold for 4, 8 and 24 hours daily. After 42 days of either continuous or intermittent exposure to cold, there was a sigmoid relationship between the hours per day exposed to cold and systolic blood pressure. Whether the blood pressures of the intermittently exposed groups would reach that of the continuously exposed group with additional exposures to cold is unknown, but worthy of additional study. Thus, graded elevations of blood pressure occur with increasing duration of daily exposure to cold.

With respect to mechanism by which blood pressure may become elevated, studies of changes in the sensitivity of the baroreflex response (delta heart period/delta mean arterial pressure) during exposure to cold revealed a significant reduction (Papanek et al, 1991). Plasma norepinephrine and epinephrine concentrations increased within 24 hours of exposure to cold. Thus, a reduction in the negative feedback control of the sympathetic nervous system (SNS) allows an increased secretion of norepinephrine to occur during exposure to cold. It is difficult to determine the exact contribution of the SNS to the elevated arterial pressure seen with chronic exposure to cold. However, in addition to a direct vasoconstrictor effect, a hyperactive SNS could increase blood pressure through a variety of indirect mechanisms. An increased activity of the SNS can increase cardiac output and can thereby contribute to an increase in systemic blood pressure. In addition, increased activity of the SNS can increase plasma renin activity (PRA) and the concentration of angiotensin II (AII), a very potent vasoconstrictor agent, in plasma. The renin-angiotensin system also influences the relationship between renal sodium/water balance and arterial pressure.

To date, 16 separate experiments involving 120 rats have been carried out in the cold. A data-base has been established containing cardiovascular and other data from the cold-exposed and warm-acclimated groups used as controls in these experiments. The results from these combined data are presented here.

METHODS

All rats used in these studies were males of the Sprague Dawley strain weighing initially from 200-250 g. They were kept in individual cages in either a room maintained at 25 ± 2 or $5 \pm 2^{o}C$. Each room was illuminated from 7 AM to 7 PM. Finely powdered food (Purina Laboratory Chow) and tap water were provided ad libitum. Food and water intakes, as well as urine outputs, were measured at various periods of time during some, but not all, experiments. Urine was analyzed for its concentrations of norepinephrine and epinephrine by HPLC with electrochemical detection and for its concentrations of sodium and potassium by flame photometry. Systolic blood pressure was measured in each rat in all experiments prior to and at weekly intervals during exposure to cold. However, of the 16 experiments performed, not all of them were carried out for the same length of time. To assess whether responsiveness to angiotensin II (AII) was affected by chronic exposure to cold, the dipsogenic response to administration of AII (150 ug/kg, s.c.) was measured for one hour after treatment. At sacrifice, blood was collected, serum separated, and analyzed for PRA by radioimmunoassay. At death, heart, kidneys, adrenal glands and interscapular brown adipose tissue were removed, trimmed of extraneous tissue and weighed on a torsion balance.

The results described above represent combined data from 16 separate experiments used to form a data-base which might reasonably be expected to profile at least some of the changes occurring during chronic exposure of rats to cold air. The duration of the longest experiment was 12 weeks, with others ranging from 4 to 11 weeks, although all 16 experiments lasted a minimum of 4 weeks. A repeated measures one-way analysis of variance was used to assess the significance of the data.

RESULTS

Systolic blood pressure, measured indirectly from the tail, increased within the first week of exposure to cold and was significantly elevated (8%) above pre-cold exposure level by the second week (Figure 1). Within 5 weeks of exposure to cold, systolic blood pressure reached a maximum at 20% above pre-cold exposure level where it was maintained for 12 weeks (Table 1). Food and water intakes increased approximately 55 and 10%, respectively, above control value within 1 week of exposure to cold and reached

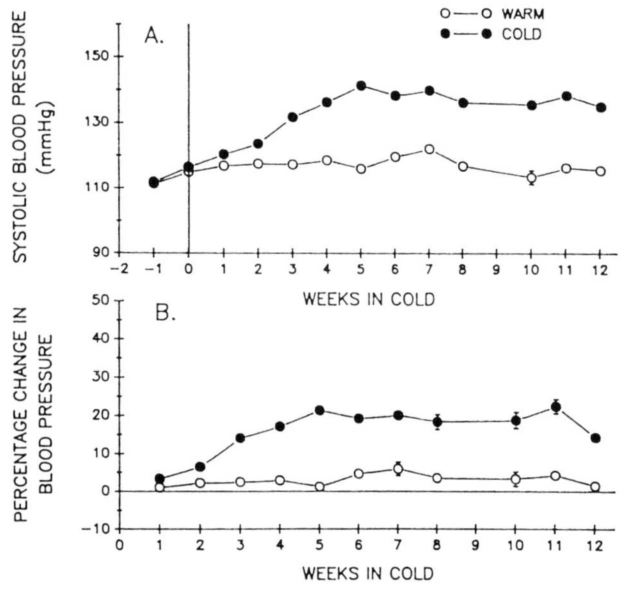

Figure 1

levels of approximately 83 and 50%, respectively, above control value by the sixth week. These levels were maintained for the remainder of the 12 weeks.

Urine output increased 60% within the first week of exposure to cold; reached a maximum (100% above control) within 6 weeks where it was maintained for the remainder of the 12 weeks (Table 1). Urine was analyzed for its concentrations of norepinephrine and epinephrine. Urinary norepinephrine and epinephrine <u>outputs</u> increased approximately 550 and 110%, respectively, above control levels during the first week of exposure to cold. They decreased thereafter to reach levels of 200 and 40% above control by the sixth week of exposure to cold where they remained for the duration of cold exposure (Table 1). PRA increased approximately 90% above control value during the second week of exposure to cold and decreased in a curvilinear fashion to approach the level of the control group by the twelfth week of exposure to cold (Table 1).

Responsiveness to AII, as assessed by percent change in water intake one hour after its administration, increased approximately 90% within the first week of exposure to cold and declined in a curvilinear fashion to approach control level by the nineth week of exposure to cold (Table 1). The percent changes in weights of heart and kidneys both increased approximately 19 and 22%, respectively, above control value by the first week of exposure to cold and increased linearly thereafter with increasing duration of exposure (Table 1). In contrast, the percent change in the weight of the adrenal gland increased about 38% during the first week of exposure and decreased linearly with time thereafter (Table 1).

Table 1

Effect Of Chronic Exposure To Cold On Changes In Certain Cardiovascular Functions In The Rat

Function	Percent Change From Control At:			
	1	3	5	12 Weeks in Cold
Systolic BP	5	15	20	20
Food Intake	55	62	74	83
Water Intake	10	19	41	50
Urine Output	60	82	98	100
Norepi. Output	550	300	230	200
Epine. Output	110	50	42	40
Plasma Renin Act.	40	79	60	5
Dipsogenic Resp. to AII	90	35	25	8
Heart Weight	19	20	21	23
Renal Weight	22	24	25	31
Adrenal Weight	38	36	31	25
Brown Fat Weight	120	130	140	150

DISCUSSION

The results of the combined data presented here show that the systolic blood pressures of cold-treated rats, measured indirectly from the tail, increase above the level of the control group within 1 week of exposure and are significantly ($P<0.05$) different by the second week. The rise in pressure is modest (130-140 mm Hg) and remains at a level about 20% above that of the control group for the duration of exposure to cold.

Of the potential factors that might influence the elevation of blood pressure during exposure to cold, the increase in PRA during the first 5 weeks is noteworthy. This increase occurs at a time when the responsiveness to AII, as assessed by AII-induced dipsogenesis, has also increased. It is of interest that the increase in PRA and the increase in dipsogenic responsiveness to AII parallel each other (Table 1). Thus, PRA and AII concentration in blood are increased when responsiveness to the latter is also increased. This suggests that the renin-angiotensin system may be responsible for induction of hypertension in cold-exposed rats. The increase in PRA may be driven by the increased secretion of norepinephrine and epinephrine which, in turn, is induced by a reduction in the negative feedback control exerted by the baroreflex mechanism (Papanek et al, 1991). An increase in plasma concentration and urine output of these neurohormones has been observed within 24 hours of exposure to cold (Carlberg and Fregly, 1986; Papanek et al, 1991). Whether an increase in secretion of aldosterone also contributes to the development of cold-induced hypertension has not been thoroughly investigated. However, others have reported a 3 fold increase in plasma concentration of aldosterone in rats of the Long Evans strain after 6 weeks of exposure to air at 15°C, and at a time when PRA is unchanged from control (Delost et al, 1989).

The hypothesis stated above that the renin-angiotensin-aldosterone (RAA) system is mainly responsible for the elevation of blood pressure during chronic exposure of rats to cold is supported by recent studies in which this system was blocked at several levels (Baron et al, 1991; Fregly et al, 1993; Shechtman et al, 1991). Administration of captopril, an angiotensin I converting enzyme inhibitor; losartan potassium, an angiotensin II, AT-1, receptor blocker; clonidine, an inhibitor of the release of renin from the kidneys, and spironolactone, an aldosterone receptor blocker, either inhibited or attenuated the elevation of blood pressure induced by exposure to cold. Further, reduction in the content of sodium in the diet also attenuated the elevation of blood pressure induced by exposure to cold (van Bergen et al, 1992). This observation is consistent with an involvement of the RAA system in the development of cold-induced hypertension.

An increase in the weight of the whole heart occurred within 1 week of exposure to cold (Table 1) and remained elevated at a level about 20% above control throughout the period of cold exposure. The elevation of blood pressure accompanied by an increase in the weight of the heart suggests that the cold-exposed rats were hypertensive. Other studies have shown hypertrophy of the left ventricle (Shechtman et al, 1991; van Bergen et al, 1992). A question remains as to whether the increase in the weight of the heart is a direct consequence of the increase in blood pressure or whether elevated circulating levels of catecholamines may play a role (Rossi and Carillo, 1985). In favor of the latter is the observation that cardiac weight of cold-treated rats remained elevated above the level of controls even though their blood pressures had been prevented from rising by the inhibitors of the RAA system mentioned above. Further, cardiac hypertrophy has been demonstrated in rats by chronic injections of the neurotransmitters, norepinephrine (Harri et al, 1982), epinephrine (Fell et al, 1981) and the synthetic beta-adrenoceptor agonist, isoproterenol (Oestman-Smith, 1979). Additional studies are required to account more fully for this observation.

The increase in renal weight of cold-treated rats occurred within the first week of exposure to cold and remained for the duration of the study (Table 1). This may be related to an increase in intake of proteins and electrolytes resulting from the increased food intake induced by exposure to cold (Table 1). The resulting increased work of the kidneys to excrete metabolites and electrolytes may contribute to induction of the hypertrophy.

There was a progressive reduction in adrenal weight during the period of exposure to cold (Table 1). This may be related to a decrease in secretion of glucocorticoid hormones, as well as epinephrine, during the period of exposure to cold (Table 1)(Straw and Fregly, 1967).

An often reported result is confirmed by these combined data. Thus, it has been reported that cold-exposed rats ingest less water for a given food intake than do rats at thermoneutral temperatures (Fregly, 1989, 1991). A similar finding has also been reported for pigeons (Henderson et al, 1992). While this would merely suggest physiologically that the cold-exposed rats are concentrating their urine to a greater degree than their warm-acclimated controls, studies from this laboratory have shown that this is not the case (Fregly and Tyler, 1972). The urines of cold-treated rats are less concentrated than those of controls and output is much greater. The requirement for a greater urine volume in which to excrete metabolic products requires a greater water intake (Table 1). Nonetheless, the ratio of urine output/water intake is still greater for cold-treated rats than for warm-acclimated controls (Fregly, 1991). Further study is required to provide a better understanding of this and its potential consequences.

The increase in the weight of brown adipose tissue was evident by the first week of exposure to cold (Table 1). The increase in weight was maintained throughout the period of exposure to cold. These results are consistent with those reported earlier and suggest the importance of this tissue for non-shivering thermogenesis in rats exposed to cold (Himms-Hagen, 1990).

The results presented here show the changes in certain cardiovascular and hormonal functions during a 12 week exposure of rats to cold. The important observation of the development of hypertension under these conditions underscores the complexity of the changes that occur during the process of acclimation to cold in the rat. It is difficult to understand how the development of hypertension during exposure to cold can be of benefit to the acclimated rat. It is more likely to be the result of a nonlethal compromise arrived at by the organism in order to have the benefit of other mechanisms and processes that maintain body temperature. The development of hypertension during acclimation to cold teaches that acclimation is not the result of a single change, but rather of multiple changes that, in addition to producing their specified effects, also interact to produce unexpected ones.

ACKNOWLEDGMENT

Supported by grant HL39154-07 from the National Heart, Lung and Blood Institute, National Institutes of Health.

REFERENCES

Baron, A.,Riesselmann, A., and Fregly, M.J. (1991): Effect of chronic treatment with clonidine and spirolonactone on cold-induced elevation of blood pressure. Pharmacology 43:173-186.

Carlberg, K.A., and Fregly, M.J. (1986): Catecholamine excretion and beta-adrenergic responsiveness in estrogen-treated rats. Pharmacology 32:147-156.

Chevillard, L., et Bertin, R. (1965): Pression artérielle et réactivité aux catécholamines chez le rat né ou adapté à 5°, 20°, 30°C. C.R. Soc. Biol. (Paris) 159:341-345.

Delost, P., Laury, M-C., Tournaire, C., Zizini, L., Bertin, R., and Portet, R. (1989): Evidence for a cold-induced aldosterone stimulation in the rat. Steroids 54:55-69.
Fell, R.D., Terblanche, S.E., Winder, W.W., and Holloszy, J.O. (1981): Adaptive responses of rats to prolonged treatment with epinephrine. Am. J. Physiol. 241:C55-C58.
Fregly, M.J. (1954): Effects of extremes of temperature on hypertensive rats. Am. J. Physiol. 176:275-281.
Fregly, M.J. (1989): Effect of exposure to cold on fluid and electrolyte exchange. in Hormonal Regulation of Fluid And Electrolytes, eds. J.R. Claybaugh and C.E. Wade, pp. 87-116, New York: Plenum.
Fregly, M.J. (1991): Water and electrolyte exchange during exposure to cold. in Thermoregulation: Pathology, Pharmacology and Therapy, eds. E. Schonbaum and P. Lomax, pp. 455-487, New York: Pergamon.
Fregly, M.J., Kikta, D.C., Threatte, R.M., Torres, J.L., and Barney, C.C. (1989a): Development of hypertension in rats during chronic exposure to cold. J. Appl. Physiol. 66:741-749.
Fregly, M.J., and Papanek, P.E. (1989): Cold-induced hypertension in rats. in Thermal Physiology, 1989, ed. J.B. Mercer, pp. 551-557, Amsterdam: Elsevier.
Fregly, M.J., Rossi, F., van Bergen, P., Brummerman, M., and Cade, J.R. (1993): Effect of chronic treatment with losartan potassium (DuP 753) on the elevation of blood pressure during chronic exposure of rats to cold. Pharmacology 46:198-205.
Fregly, M.J., Shechtman, O., van Bergen, P., Reeber, C., and Papanek, P.E. (1991): Changes in blood pressure and dipsogenic responsiveness to angiotensin II during chronic exposure of rats to cold. Pharmacol. Biochem. Behav. 38:837-842.
Fregly, M.J., and Tyler, P.E. (1972): Renal response of cold exposed rats to pitressin and dehydration. Am. J. Physiol. 222:1065-1070.
Gilson, S.B. (1950): Studies on the adaptation to cold air in the rat. Am. J. Physiol. 161:87-91.
Harri, M., Kunsela, P., and Oksanen-Rossi, R. (1982): Modification of training-induced responses by repeated norepinephrine injections in rats. J. Appl. Physiol. 53:665-670.
Henderson, D., Fort, M.M., Rashotte, M.E., and Henderson, R.P. (1992): Ingestive behavior and body temperature of pigeons during long-term cold exposure. Physiol. Behav. 52:455-469.
Héroux, O., and Dugal, L.P. (1961): Effet de l'acide ascorbique sur l'hypertension expérimentale. Can. J. Med. Sci. 29:164-175.
Himms-Hagen, J. (1990): Brown adipose tissue thermogenesis: role in thermoregulation, energy regulation and obesity. in Thermoregulation: Physiology and Biochemistry, eds. E. Schonbaum and P. Lomax, pp. 327-414, New York: Pergamon.
Oestman-Smith, I. (1979): Adaptive changes in the sympathetic nervous system and some effector organs of the rat following long term exercise or cold acclimation and the role of cardiac sympathetic nerves in the genesis of compensatory cardiac hypertrophy. Acta Physiol. Scand. Suppl. 477:1-116.
Papanek, P.E., Wood, C.E., and Fregly, M.J. (1991): Role of the sympathetic nervous system in cold-induced hypertension. J. Appl. Physiol. 71:300-306.
Rossi, M.A., and Carillo, S.V. (1985): Does norepinephrine play a central causative role in the process of cardiac hypertrophy? Am. Heart J. 109:622-624.
Shechtman, O., Fregly, M.J., and Papanek, P.E. (1990a): Factors affecting cold-induced hypertension in rats. Proc. Soc. Exp. Biol. Med. 195:364-368.

Shechtman, O., Papanek, P.E., and Fregly, M.J. (1990b): Reversibility of cold-induced hypertension after removal of rats from cold. Can. J. Physiol. Pharmacol. 68:830-835.

Shechtman, O., Fregly, M.J., van Bergen, P., and Papanek, P.E. (1991): Prevention of cold-induced increase in blood pressure of rats by captopril. Hypertension 17:763-770.

Straw, J.A., and Fregly, M.J. (1967): Evaluation of thyroid and adrenal-pituitary function during cold acclimation. J. Appl. Physiol. 23:825-830.

van Bergen, P., Fregly, M.J., and Papanek, P.E. (1992a): Effect of a reduction in sodium intake on cold-induced elevation of blood pressure in the rat. Proc. Soc. Exp. Biol. Med. 200:472-479.

van Bergen, P., Fregly, M.J., Rossi, F., and Shechtman, O. (1992b): Effect of intermittent exposure to cold on the development of hypertension in rats. Am. J. Hypertension 5:548-555.

Integrative and cellular aspects of autonomic functions : temperature and osmoregulation. Eds K. Pleschka, R. Gerstberger. John Libbey Eurotext, Paris © 1994, pp. 209-216.

Changes in body temperature and plasma catecholamine concentrations during restraint stress in spontaneously hypertensive rats

K. Tsuchiya[(1)], I. Matsumoto[(2)], K. Shichijo[(3)], T. Matsumoto[(1)], M. Kosaka[(1)], T. Aikawa[(2)], M. Itoh[(3)] and I. Sekine[(3)]

[(1)] Department of Environmental Physiology, Institute of Tropical Medicine, [(2)] Department of Physiology, School of Medicine and [(3)] Department of Pathology, Atomic Bomb Disease Institute, School of Medicine, Nagasaki University, Nagasaki 852, Japan

It is well known that in the rat a rise in core temperature is induced by psychological stress, for example, an exposure to open field on the desk (Briese & Quijada, 1970). There are many kinds of stress stimulations for the rat, such as cage switching, noise, air jet on the face, water immersion, restraint with the loosely fitting cage or tightly fitting cylinder, and immobilization by taping on the plate (Amal & Sanyal,1981). Effects of such stress on cardiovascular activity and body temperature have been reported. Depending on the type, the duration of stress and the sensitivity of the rat, different response patterns can be observed. Recently, the rise in core temperature was intensively studied in relation to fever and its central mechanism (Kluger et al., 1987; Long et al., 1990; Briese & Cabanac, 1991; Briese, 1992, Nakamori et al., 1993). It has been reported that brown adipose tissue (BAT) thermogenesis, which is controlled by the sympathetic nervous system, contributes to stress-induced rise in core temperature in loosely restrained rats (Shibata & Nagasaka, 1982), and that the stress-induced rise in core temperature is augmented by cold-acclimation (Tsuchiya & Kosaka, 1989).

Spontaneously hypertensive rats (SHR) developed from the normotensive Wistar-Kyoto strain (WKY) by Okamoto & Aoki (1963) possess genetic traits of hypertension as well as cardiovascular hyperreactivity and behavioral hyperactivity (Hendley et al., 1983; Knandahl & Hendley, 1990). Changes in blood pressure and heart rate due to psychological stress with air jet on the face were compared between SHR and WKY (Hallback & Folkow, 1974; Ely et al., 1985) and the role of the sympathetic nervous activity on the stress responses was discussed (Lundin et al., 1984). Larger and longer lasting cardiovascular responses to psychological stress in SHR were reported compared to WKY.

Tsuchiya et al. (1989) compared the rises in core temperature between SHR and WKY, and reported that the rise in core temperature during the restraint stress with a loosely fitting cage was higher and more sustained in SHR compared to WKY. Circulating plasma noradrenaline (NA) level is a useful estimation of average sympathetic outflow (Goldstein et al., 1983). In the present study, therefore, changes in plasma catecholamine level and in core temperature due to restraint stress with a loosely fitting cage were determined in SHR, WKY and Wistar rats, in order to clarify the mechanism of hyperreactivity to stress in SHR and the involvement of catecholamines with these responses induced by the stress.

Materials and Methods

Adult male SHR, WKY (Charles River, Japan) and Wistar rats, more than 4 months old, were used in this study. Two or three rats were housed in a plastic cage with wood shavings under constant room temperature and in a 12hr:12hr light-dark cycle (8:00-20:00). Food and water were given ad libitum. Experiments were performed during day time (10:00-17:00).

More than 4 days before the experiment (Popper et al.,1977), a silicone tubing (OD 1.0mm, ID 0.5mm) was chronically cannulated into the jugular vein under pentobarbital anesthesia (50 mg/kg, i.p.). The tip of the cannula was placed in the right atrium. The free end of the cannula was tunneled under the skin and exteriorized behind the neck and filled with heparinized isotonic saline. After the surgical procedure, the rat was injected with sulbenicillin (Takeda) intramuscularly and moved to an individual cage.

On the experimental day, in order to accustom the rat to the laboratory, the home cage containing a rat was transferred to the laboratory more than 4-5 hours before the experiment. The rats of each strain were divided into two groups. One was the restraint group and the other was the control group without restraint. The rats in the former group were suddenly restrained in a loosely fitting wiremesh cage (Amal & Sanyal, 1981) under conscious condition. Control rats were allowed to be in their own home cage. Rectal (Tre), tail skin (Ttail) and ambient air (Ta) temperatures were continuously measured by means of the thermistor probes. For measurement of Tre, a thermistor probe was inserted into the rectum more than 5 cm beyond the anus and taped on the tail. For measurement of Ttail another probe was attached on the tail skin with one layer of adhesive tape. As the index of the tail skin vasomotion, the difference (dTtail) between Ttail and Ta (dTtail = Ttail - Ta) was used. Systolic blood pressure was measured by the tail cuff method (PE-300, Narco). These parameters were simultaneously recorded on a UV oscillograph (Type 3L, SANEI).

Blood samples (0.6 ml) were drawn through the chronically implanted jugular cannula within 2 min before the restraint, every 20 min for 1 hour and every 30 min for another 2 hours during the restraint. After centrifugation (3000 rpm for 5 min), plasma was stored at -80℃ until assayed, and the remaining blood cells were resuspended in 0.3 ml heparinized isotonic saline and returned to the rat through the cannula.

Plasma concentrations of catecholamines, noradrenaline (NA) and adrenaline (A), were measured by coulometric electrochemimcal determination method (Hallman et al., 1978) using high-performance liquid chromatographic analytical system (ESA, INC. U.S.A., Model 5100A, Coulochem Detector). In brief, 0.20 - 0.30 ml of plasma was placed in a disposable column for column chromatography (Sepacol, Seikagaku Kogyo Inc., Tokyo) with 70 mg alumina, one ml of Tris buffer solution (pH 8.6) and one ml of phosphate buffer solution (pH 7.0), and a known amount of 3,4-dihydroxybenzylamine, which served as an internal standard. Each sample was shaken for 20 min to allow amines to be adsorbed onto the alumina and the supernatant was aspirated to near dryness. After 160 μl of 0.2 M perchloric acid

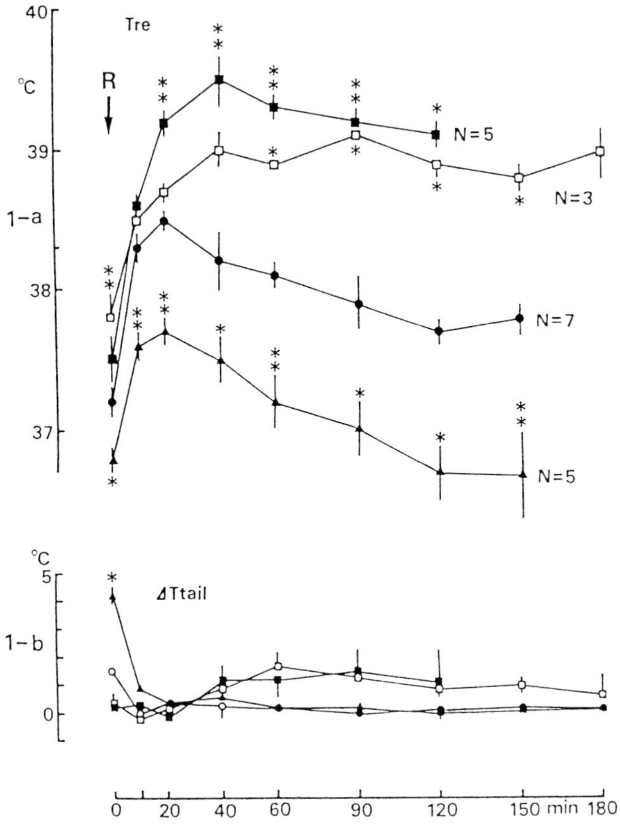

Fig.1 Change in rectal temperature (Tre) (1-a) and the difference (dTtail) between tail skin (Ttail) and ambient air (Ta) temperatures, (dTtail = Ttail - Ta) (1-b), due to restraint stress in conscious spontaneously hypertensive rats (SHR), Wistar-Kyoto rats (WKY) and Wistar rats. An arrow indicates the beginning of restraint stress. Closed rectangle with bars (M.±S.E.): SHR, open rectangle: SHR with blood samplings, closed circle: WKY, and closed triangle: Wistar rats. Asterisks show the statistical significance in comparison with the value in WKY (*:$p<0.05$, **:$p<0.01$).

was added to the alumina phase, the acidic eluate containing the desired catecholamines was centrifuged in the receiver tube, and 80 μl of the final acidic eluate was injected onto a C18 reverse-phase column. The mobile phase contained 0.02 M trichloroacetic acid, 0.075 M sodium dihydrogenphosphate, 0.05% EDTA and 5% menthanol - 10% acetonitrile - 85% H_2O (pH 3.1). Minimal detectable concentration of desired catecholamines were ranged from 3 to 5 pg/vial.

Values were presented as means±S.E.. Statistical significance was tested by Mann-Whitney U test.

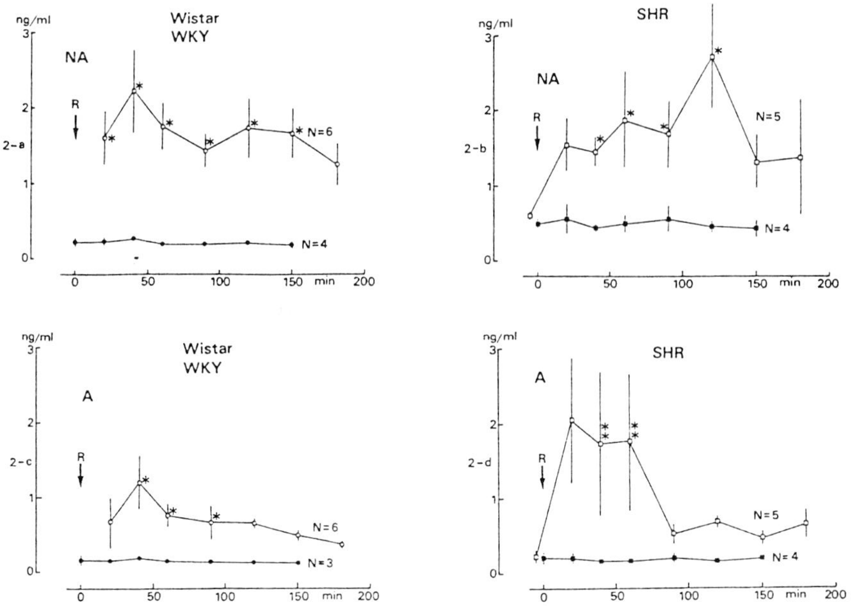

Fig.2 Changes in plasma noradrenaline (NA) concentration of WKY-Wistar group (2-a) and SHR group (2-b), and changes in plasma adrenaline (A) concentration in WKY-Wistar group (2-c) and SHR group (2-d) during restraint stress in the conscious animals. Open circles with bars are mean concentration (M.±S.E.) of plasma NA and A during restraint stress, and solid circles with bars are without restraint stress (control). Arrows indicate the beginning of the restraint. Asterisks indicate the statistical significance in comparison with unrestrained control (*:p<0.05, **:p<0.01).

Results

Systolic blood pressure measured with tail cuff method was 210±7 mmHg in SHR (N=5) and 127±4 mmHg in WKY (N=6). The former was significantly higher (p<0.01) than the latter.

Changes in Tre and dTtail due to restraint stress in SHR, WKY and Wistar rats were shown in Fig. 1. Immediately after the beginning of restraint stress, mean Tre was 37.5±0.2 ℃ in SHR (N=5), 37.2±0.1 ℃ in WKY (N=7) and 36.8±0.1 ℃ in Wistar rats (N=5). In these three strains, by restraint stress, core temperature increased intensely (Fig. 1-a). In WKY and Wistar rats, Tre reached to the peak at the 20th-30th min after the beginning of restraint and then turned to decrease. In SHR, however, Tre remained to rise for 40 min and sustained for 2-3 hours with a slight reduction. The declining slope after the peak in WKY and Wistar rats were steeper than SHR. The mean maximum rise in Tre in SHR (1.8±0.1 ℃) was significantly greater than that in WKY (1.1±0.1 ℃) and Wistar rats (0.7±0.2 ℃), p<0.01.

As shown in Fig. 1-b, dTtail in all groups immediately reduced in a sustained fashion throughout restraint stress.

In Fig. 2, changes in plasma NA and A levels due to restraint stress in SHR (N=5) and WKY-Wistar group (N=6) were shown. After restraint, plasma NA level rapidly increased in both groups. The peak of NA level was observed at the 40th min and followed by gradual decline in WKY-Wistar group (Fig. 2-a). In contrast, the rise in NA was sustained for 120 min in SHR (Fig. 2-b). The peak value of NA level was 2.73±0.70 ng/ml in SHR and 2.25±0.57 ng/ml in WKY-Wistar group. Plasma A level also increased after restraint stress in both groups, whereas, after the peak value obtained around the 30th min, plasma A level decreased to near initial value around the 100th min, as shown in Figs. 2-c and 2-d. Peak level of A was 2.03±0.85 ng/ml in SHR and 1.20±0.37 ng/ml in WKY-Wistar group.

Under free-moving condition in their home cages, no significant change in plasma NA and A level was observed in either SHR or WKY-Wistar group (Fig. 2). Mean value of NA level was 0.49±0.02 ng/ml in SHR and 0.22±0.01 ng/ml in WKY-Wistar, and mean A concentration was 0.18±0.01 ng/ml in SHR and 0.16±0.01 ng/ml in WKY-Wistar. Plasma NA level in SHR was significantly greater ($p<0.01$) than that in WKY-Wistar group, whereas there was no significant difference in plasma A levels between SHR and WKY-Wistar.

Discussion

SHR was developed as an animal model of essential hypertension (Okamoto & Aoki, 1963). However, they possess the characteristics not only of hypertension but also of cardiovascular hyperreactivity to stress and behavioral hyperactivity (Hendley et al., 1983; Knandahl & Hendley, 1990). SHR show the more intense responses to environmental stress compared to WKY: greater cardiovascular responses to air jet (Lundin et al., 1984; Ely et al., 1985) and to light, noise and vibration (Hallback & Folkow, 1974) and greater increase in plasma NA and A by cage switching or foot shock (McCarty & Korpin, 1978; Chiueh & McCarty, 1981), greater excretion of urinary NA and A during exposure to cold stress (Yamori et al., 1985). However, there was no strain difference in levels of NA or A in the resting condition (McCarty & Kopin, 1978; Roizen et al., 1979; Chiueh & McCarty, 1981; Picotti et al., 1982), except that between young SHR and WKY (Pak, 1981).

It had been reported and widely accepted that resting core temperature of SHR was high in comparison with normotensive WKY. Physiological mechanisms and significance of high core temperature in SHR were discussed in many studies concerning with the change in heat production and heat dissipation of SHR (Collins et al., 1987). On the other hand, Hojos and Engberg (1986) suggested that the high core temperature observed in SHR might be contaminated by the effect of the noradrenaline-mediated hyperreactivity to environmental stress e.g. handling of the animal during the temperature measurement procedure. Furthermore, radiotelemetry method of body temperature measurement revealed that there was no significant difference in the resting body temperature between SHR and WKY even after the exposure to the ambient temperatures ranging from 5℃ to 35℃ for 1 hour as well as 24 hours recording under thermoneutral condition (Berkey et al., 1990; Morley et al., 1990).

In this study, initial Tre in SHR which was measured immediately after the placement of the rat into a restraint cage was slightly higher than in WKY and Wistar rats. This slight higher initial core temperature of SHR in this study might be due to the effect of chronic jugular cannulation, because in our previous study, there was no significant difference in the initial core temperature between SHR and WKY which were measured as same as the present study (Tsuchiya & Kosaka, 1992).

After restraint with a loosely fitting cage, Tre sharply rose in all of three strains. The magnitude of the rise in Tre was greater in SHR than in WKY and Wistar rats. In SHR, elevated body temperature was sustained throughout the stress stimulation, whereas, in WKY and Wistar rats the rise in Tre was transient and Tre turned to decline after 20 min of restraint as shown in Fig. 1. Therefore, this sustained rise in core temperature induced by stress might be due to hyperreactivity of SHR to environmental stress.

As mentioned above, the more intense and more prolonged cardiovascular response in SHR compared to WKY, and the contribution of long lasting activation of sympathetic nerves to this response (Lundin & Thoren, 1982; Lundin et al., 1984) were reported. These data on cardiovascular responses to stress also support our present results on body temperature.

In normotensive rats, rise in catecholamines level (Kventansky et al., 1978; DeTurk & Vogel, 1980) and activation of sympathetic nerves (Lundin & Thoren, 1982; Lundin et al., 1984) were reported during the stress, such as handling or immobilization. And it was discussed that mixed venous NA reflects the average contributions from the various vascular beds and therefore estimates overall sympathetic tone (Goldstein et al., 1983).

Shibata and Nagasaka (1982) showed the involvement of BAT in the stress-induced elevation of core temperature, and suggested that BAT thermogenesis during the stress was attributed to an increased tone of the sympathetic nerves to the BAT. On the other hand, Hayashi et al. (1988) reported that there were no significant difference in heat production induced by the injection of β-adrenergic agonist, isoprenaline, between SHR and WKY. Therefore, strain difference of stress-induced rise in core temperature might be attributed not to functional difference of BAT, which is major site of catecholamine-mediated thermogenesis in rats, but to difference in efferent sympathetic activity.

During stress, vasoconstriction in the peripheral region especially in the tail was reported (Tsuchiya & Kosaka, 1989, 1992; Tsuchiya et al., 1989). Rat tail is an important organ for temperature regulation, because of wide glabrous skin area and of well controlled mechanism of blood flow by the sympathetic vasoconstrictor nerves (O'Leary et al., 1985). In this study, in all groups, sustained decrease in tail skin temperature was observed during restraint stress. This fact suggests that during restraint the sympathetic nerves are activated and suppresses heat dissipation through the tail skin and contributes to stress-induced rise in core temperature.

Under free-moving condition in this study, there was no significant difference in resting plasma A level between SHR and WKY, however plasma NA level in SHR was slightly high compared to that in WKY. During restraint, plasma NA and A levels increased in SHR as well as in WKY, however, the rise in A level was transient, and did not correspond with the rise in core temperature. In SHR, the greater and more sustained rise in plasma NA level was observed in comparison with WKY. Furthermore, the patterns of the rise in plasma NA level were similar to those in core temperature in both strains of SHR and WKY as shown in Figs. 1 and 2.

In conclusion, these results suggest that the sustained core temperature rise in SHR during stress is attributed to prolonged excitement of sympathetic nervous system which cause the responses of heat production and heat conservation by peripheral vasoconstriction.

References

Amar,A.,Sanyal A.K.(1981):Immobilization stress in rats: Effect on rectal temperature and possible role of brain monoamines in hypothermia. Psychopharmacology 73,157-160.

Berkey,D.L.,Meeuwsen,K.W. & Barney,C.C.(1990):Measurements of core temperature in spontaneously hypertensive rats by radiotelemetry. Am.J.Physiol.258,R743-R749.

Briese,E.(1992):Cold increases and warmth diminishes stress-induced rise of colonic temperature in rats. Physiol.Behav.51,881-883.

Briese,E. & Cabanac,M.(1991):Stress hyperthermia:physiological arguments that it is a fever. Physiol. Behav.49,1153-1157.

Briese,E. & De Quijada,M.G.(1970):Colonic temperature of rats during handling. Acta Physiol.Latinoam.20,97-102.

Chiueh,C.C. & McCarty,R.(1981):Sympatho-adrenal hyperreactivity to footshock stress but not to cold exposure in spontaneously hypertensive rats. Physiol. Behav.26,85-89.

Collins,M.G.,Hunter,W.S. & Bratteis,C.M.(1987):Factors producing elevated core temperature in spontaneously hypertensive rats. J.Appl.Physiol.63,740-745.

DeTurck,K.H. & Vogel,W.H.(1980): Factors influencing plasma catecholamine levels in rats during immobilization. Pharmacol. Biochem. Behav.13,129-131.

Ely,D.L.,Friberg,P.,Nilsson,H. & Folkow,B.(1985):Blood pressure and heart rate responses to mental stress in spontaneously hypertensive (SHR) and normotensive (WKY) rats on various sodium diets. Acta. Physiol. Scand.123,159-169.

Goldstein,D.S.,McCarty,R.,Polinsky,R.J. & Kopin,I.J.(1983):Relationship between plasma norepinephrine and sympathetic neural activity. Hypertension,5,552-559.

Hajos,M. & Engberg,G.(1986):Emotional hyperthermia in spontaneously hypertensive rats.Psychopharmacology,90,170-172.

Hallback,M. & Folkow,B.(1974):Cardiovascular responses to acute mental "stress " in spontaneously hypertensive rats. Acta Physiol.Scand.90,684-698.

Hallman,H.,Farnebo, L.-O,Hamberger,B. & Jonsson,G.(1978):A sensitive method for the determination of plasma catecholamine using liquid chromatography with electrochemichal detection. Life Sci.23,1049-1052.

Hayashi,S.,Wickler,S.J.,Gray,S. & Horwitz,B.A.(1988):Nonshivering thermogenesis and brown fat in spontaneously hypertensive rats. Proc.Soc.Exp.Biol.Med.188,435-439.

Hendley,E.D.,Atwater,D.G.,Myers,M.M. & Whitehorn,D.(1983): Dissociation of genetic hyperactivity and hypertension in SHR. Hypertension 5,211-217.

Kluger,M.J.,O'Reilly,B.,Shope,T.R. & Vander,A.J.(1987):Further evidence that stress hyperthermia is a fever. Physiol Behav.39,763-766.

Knardahl,S. & Hendley,E.D.(1990): Association between cardiovascular reactivity to stress and hypertension or behavior. Am.J.Physiol.259,H248-H257.

Kvetnansky,R.,Sun,C.L.,Lake,C.R.,Thoa,N.,Torda,T. & Kopin,I.J.(1978):Effect of handling and forced immobilization on rat plasma levels of epinephrine, norepinephrine, and dopamine-β-hydroxylase. Endocrinology 103,1868-1874.

Long,N.C.,Vander,A.J.,Kluger,M.J.(1990):Stress-induced rise of body temperature in rats the same in warm and cool environments. Physiol.Behav.47,773-775.

Lundin S.,Ricksten,S.-E. & Thoren,P.(1984):Interaction between "mental stress" and baroreceptor reflexes concerning effects on heart rate, mean arterial pressure and renal sympathetic activity in conscious spontaneously hypertensive rats. Acta Physiol.Scand.120.273-281.

Lundin,S. & Thoren,P.(1982):Renal function and sympathetic activity during mental stress in normotensive and spontaneously hypertensive rats. Acta Physiol. Scand.115,115-124.

McCarty,R. & Kopin,I.J.(1978): Alternations in plasma catecholamines and behavior during acute stress in spontaneously hypertensive and Wistar-Kyoto normotensive rats. Life Sci.22,997-1006.

Morley,R.M.,Conn,C.A.,Kluger,M.J. & Vander,A.J.(1990): Temperature regulation in biotelemetered spontaneously hypertensive rats. Am.J.Physiol.258,R1064-R1069.

Nakamori,T.,Morimoto,A.,Morimoto,K.,Tan,N. & Murakami,N.(1993): Effects of α- and β-adrenergic antagonists on rise in body temperature induced by psychological stress in rats. Am.J.Physiol.264,R156-R161.

Okamoto,K. & Aoki,K.(1963): Development of a strain of spontaneously hypertensive rats. Jpn.Circ.J.27,282-293.

O'Leary,D.S.,Johnson J.M. & Taylor,W.F.(1985):Mode of neural control mediating rat tail vasodilation during heating. J.Appl.Physiol.59,1533-1538.

Pak,C.H.(1981): Plasma adrenaline and noradrenaline concentrations of the spontaneously hypertensive rat. Jpn.Heart J. 22, 987-995.

Picotti,G.B.,Carruba,M.O.,Ravazzani,C.,Boniolotti,G.P. & Da Prada,M.(1982):Plasma catecholamine concentrations in normotensive rats of different strains and in spontaneously hypertensive rats under basal conditions and during cold exposure. Life Sci.31,2137-2143.

Popper,C.W.,Chiueh,C.C. & Kopin,I.J.(1977):Plasma catecholamine concentrations in unanesthetized rats during sleep, wakefulness, immobilization and after decapitation. J.Pharmacol.Exp.Ther.202,144-202.

Roizen,M.F.,Weise,V.,Grobecker,H. & Kopin,I.J.(1979):Plasma catecholamines and dopamine-β-hydroxylase. Life Sci.17,283-288.

Shibata,H. & Nagasaka,T.(1982):Contribution of nonshivering thermogenesis to stress-induced hyperthermia in rats. Jpn.J.Physiol.32,991-995.

Tsuchiya,K. & Kosaka,M(1989): Effect of thermal acclimation on the stress-induced elevation of core temperature in rats. Trop.Med.31,175-181.

Tsuchiya,K. & Kosaka,M.(1992): Effect on thermal acclimation on blood pressure and stress-induced elevation of core temperature in spontaneously hypertensive rats and Wistar rats. Trop.Med.34,77-89.

Tsuchiya,K.,Kosaka,M. & Ozaki,M.(1989): Effect of loose restraint on body temperature in spontaneously hypertensive rats (SHR) and stroke-prone SHR (SHRSP).Thermal Physiology 1989,ed J.B. Mercer,pp563-568,Elsevier Science Publishers B.V.

Yamori,Y.,Ikeda,K.,Kulakowski,E.C.,McCarty,R. & Lovenberg,W.(1985):Enhanced sympathetic-adrenal medullary response to cold exposure in spontaneously hypertensive rats. J.Hypertesion 3,63-66.

ntegrative and cellular aspects of autonomic functions : temperature and osmoregulation. Eds K. Pleschka, R. Gerstberger. John Libbey Eurotext, Paris © 1994, pp. 217-222.

Animal study on pathogenesis of heat stroke

M.-T. Lin, Z. Szreder and C.-L. Shen*

*Department of Physiology, and *Department of Anatomy, National Cheng Kung University Medical College, Tainan City, Taiwan, Republic of China*

In anesthetized rabbits, the occurrence of hyperthermia (> 42°C), intracranial hypertension (>45 mmHg) and hypotension (<60 mmHg) were taken as the onset of heat stroke. Functional or anatomical abnormalities of organ systems such as liver damage, renal degeneration, muscle damage, circulatory dysfunction, cerebral ischemia as well as degeneration of neurons in different brain regions. On the other hand, during the fever plateau, the febrile rabbits did not show the above findings, although they had a similar level of hyperthermia. The data indicate that the tissue anoxia, rather than thermal injury, is the main cause for the onset of heat stroke syndrome.

The pathophysiology of heat stroke has been extensively studied by investigators of many different disciplines (Clowes & O'Donnell, 1974; Knochel, 1974; Stefanini, 1975). Most experiments have been conducted several hours or days after the onset of heat stroke. Relative little information is available about the pathophysiological changes that occur at the onset of heat stroke. In this study, anestehtized rabbits were exposed to a high ambient temperature (T_a=40°C) to induce heat stroke and to observe alterations in several physiological parameters, blood contents and changes in cranial and extracranial structures at the onset of heat stroke. By identifying the functional or anatomical alterations occurring at the onset of heat stroke, a clearer picture of the pathogenesis of heat stroke emerges.

The animals were individually caged and kept at a T_a of 22±1°C with natural light-dark cycles, and were maintained on laboratory rabbit chow with tap water available _ad libitum._ Cerebroventricular cannulae were implanted under general anesthesia (pentobarbital sodium, 30 mg/kg, i.v.). The stereotaxic coordinate used were from the atlas of Sawyer et al. (1954). The cannulae were placed in the third cerebral ventricle. A period of 2 weeks was permitted to allow the animals to recover before the experimentation. Three groups of animals were studied: 1) heat-stroke rabbits:heat stroke was induced by exposing the animals, under pentobarbital sodium anesthesia, to a T_a of 40°C; the occurrence of hyperthermia (>42°C), intracranial hypertension (>45 mmHg) and hypotension (<60 mmHg) was taken as the onset of heat stroke; 2) febrile rabbits: the rabbits were given by an i.v. dose of polyriboinosinic acid:polyribocytidylic acid (poly I:C; 50 µg/kg; Pharmacia Molecular Biologicals, Uppsala, Sweden) to induce fever (>42°C) at a T_a of 24°C; 3) control rabbits: the rabbits, under pentobarbital sodium anesthesia, were given by an i.v. dose of normal saline (ml/kg) at a T_a of 24°C. Immediately following the onset of heat stroke or during the fever plateau, the mean arterial pressure (MAP), the intracranial pressure (ICP), the cerebral perfusion pressure (CPP = MAP - ICP), relative cerebral blood flow (rCBF) and biochemiacl blood contents were measured or calculated. Then, the animals were killed with an overdose of pentobarbital sodium for histological verification of organ systems including brain, heart, kidney, leg muscle, liver and lung. The systemic circulation was perfused with physiological saline, followed by a 10% fromalin. The fixed brains or organs were then cut in 4-µm sections and stained with hematoxylin and eosin. Sections were studied to evaluate the pathology that occurred at the onset of heat stroke. A close cranial window was made on left parietal skull for measurement of rCBF by a PeriFlux PF-2 Doppler flowmeter (Perimed, Sweden). Doppler blood flowmeter was positioned epidurally for measuring the relative changes in CBF.

Table 1. The mean±SEM values of physiological parameters of 8 control rabbits,8 heat-stroke rabbits and 8 febrile rabbits before their sacrifice for histology

Animal group	T_{co}, °C	MAP,mmHg	ICP,mmHg	CPP,mmHg	rCBF,%
Control	39.5±0.2	92.5±4.4	15.1±1.4	77.4±3.3	100
Heat stroke	42.6±0.3*	62.8±2.1*	45.2±1.6*	17.6±1.4*	15±5*
Febrile	42.5±0.4*	97.4±5.4	17.3±1.9	80.1±3.1	97±10

* $P < 0.05$, significantly different from corresponding control values (control group), Student's _t_-test.

Table 2. The mean±SEM of blood contents of 8 contral rabbits, 8 hear-stroke rabbits and 8 febrile rabbits before their sacrifice for histology

Animal group	Na^+ mM/L	K^+ mM/L	Cl^{-1} mM/L	CO_2 mM/L	Glucose mg/dl	BUN mg/dl	SGOPT U/L	SGPT U/L
Control	143±6	4±1	104±5	25±3	110±6	14±3	19±4	24±5
Heat stroke	148±7	7±1*	105±4	8±1*	270±15*	26±3*	170±22*	115±4*
Febrile	145±5	3±2	103±6	23±4	108±5	16±4	22±5	27±6

* $P < 0.05$, significantly different from corresponding control values (control group), Student's _t_-test.

Both Table 1 and Table 2 contain summaries of the mean and standard error values for each of the physiological and biochemical parameters collected from three groups of animals before their sacrifice for histological verification. Heat stroke was induced by exposing 8 animals to heat (40°C) with a latency of 62±5 min. Pyrogenic fever was induced by administering 50 µg/kg of poly I:C into the ear marginal vein of 8 rabbits kept at room temperature (24°C). Their colonic temperature (T_{co}) reached its fever plateau (42.5°C) about 150 min after the injection. It can be seen from Table 1 that the heat-stroke animals had lower values for MAP, CPP or rCBF, but a higher value for ICP than the control animals or the febrile animals during the fever plateau. The T_{co} values are 39.6, 42.6 and 42.5°C, respectively for the control rabbits, the heat-stroke rabbits and the febrile rabbits. Biochemical determination revealed that, at the onset of heat stroke, the blood levels of K^+, glucose, BUN, SGOT and SGPT increased significantly, whereas the blood CO_2 contents decreased significantly from the control levels (Table 2). In addition, at the onset of heat stroke, the major anatomical findings were as follows: 1) degeneration of neurons (as characterized by simple chromatolysis or disappearance of Nissl bodies of nerve cells) were noted in different brain regions including the hypothalamus(Fig. 1), the cere-

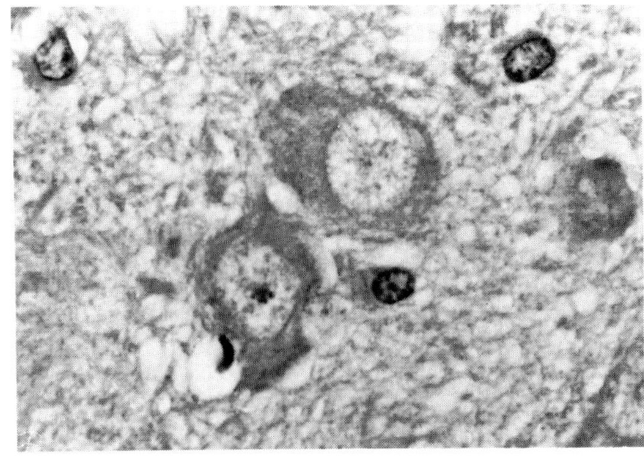

Fig. 1. A photomicrograph of the hypothalamus of a heat-stroke rabbit showing degeneration of neurons. H & E, x 1000.

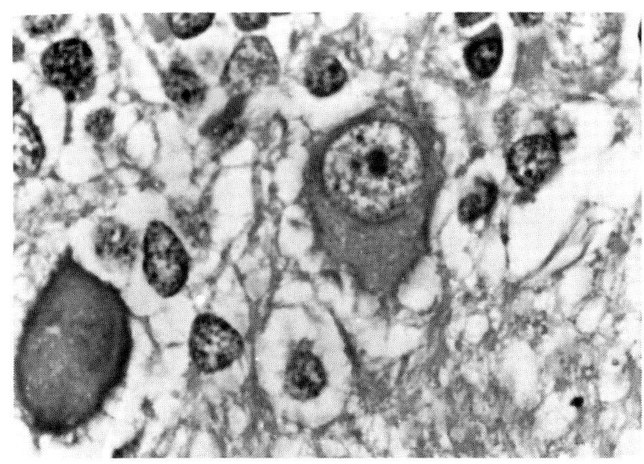

Fig. 2. A photomicrograph of the cerebellum of a heat-stroke rabbit showing deneration of neurons. H & E, x 1000.

bellum (Fig. 2), the hippocampus, the pons and the cortex; 2) hepatocyte swelling with numerous RBC and eosinophils in the compressed sinusoid as well as lymphocyte infiltration in the interstitial tissue (Fig. 3); 3)hemorrhage was common in the interstitial tissue, the glomerulus and the collecting tubes of the kidney (Fig. 4); and 4) the leg muscle displayed coagulation necrosis as well as lymphocyte infiltration (Fig. 5). It should be stressed that no pathological findings were found in either the cranial or the extracranial structures of the febrile rabbits even though tehir colonic temperature had already reached a plateau level of 42.5°C.

According to the findings of many investigators (Austin & Berry, 1956; Ferris et al., 1938; Gottschalk & Thomas, 1966; Vertel & Knochel, 1967), functional abnormalities of organ systems such as acute renal failure, liver damage, muscle damage, circulatory dysfunction associated with disseminated intravascular coagulopathy as well as central nervous system dysfunction were observed in patients with heat stroke. Both our previous (Lin & Lin, 1992; Shih et al., 1984) and present results, using an animal heat-stroke model, demonstrated that hyperthermia, cerebral edema, cerebroventricular congestion, intracranial hypertension, cerebral ischemia, cerebral neuronal degeneration, hypotension, alterations of blood contents and pathological changes of the liver, the muscle and the kidney were

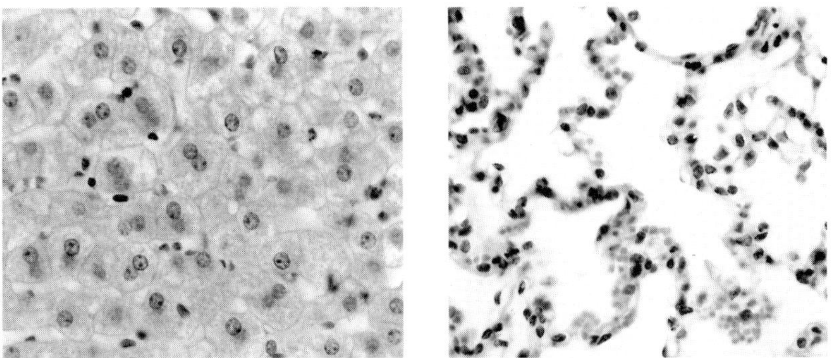

Fig. 3. Photomicrographs of the liver of a heat-stroke rabbit showing hepatocyte swelling with numerous RBC and eosinophils in the compressed sinusoid (left) as well as diffused intersttial hemorrhage with mild lymphocyte infiltration (right). H & E, x 128.

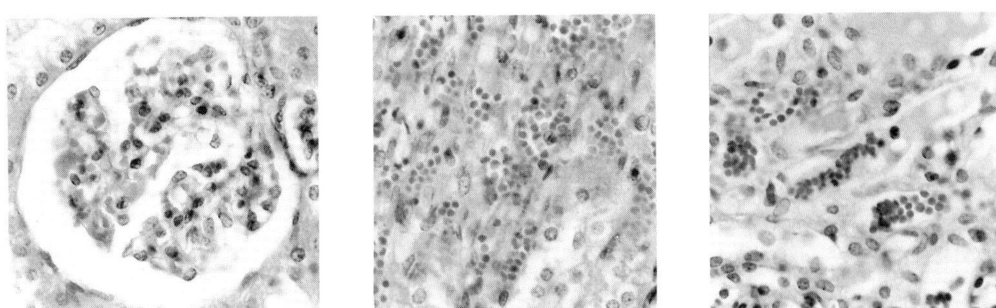

Fig. 4. Photomicrographs of the kidney of a heat-stroke rabbit showing glomerular hemorrhage (left), diffuse interstitial hemorrhage (middle) and hyaline casts in the collecting tubes (right). H & E, x 128.

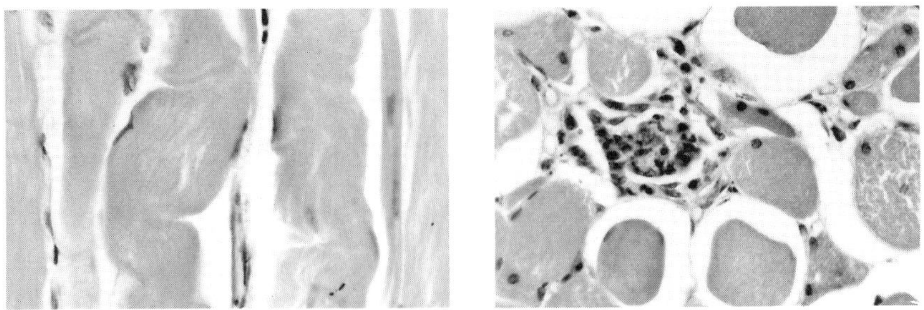

Fig. 5. Photomicrogarphs of the leg muscle of a heat-stroke rabbit showing coagulation necrosis (left) and mild leucocytes infiltration (right). H & E, x 128.

observed in rabbits at the onset of heat stroke. The similarity in the pathophysiology of heat stroke between the patients and the rabbits demonstrates that our animal model is a good choice for future studies on pathogenesis of heat stroke.

In the present results, in addition to liver admage, the serum levels of SGOT, SGPT and BUN were elevated in rabbits with heat stroke. The hepatic damage may be related to (1) tissue hypoxia (due to circulatory collapse), (2) disseminated intravascular coagulopathy, and (3) directly thermal injury (Kew et al., 1970). Consequently, liver damage that occurred at the onset of heat stroke resulted in hyperglycemia found in this animal model. The decreased CO_2 contents in the blood at the onset of heat stroke can be attributed to lactic acidosis (due to hypotensive episode and impaired liver function) and hyperthermia (Emmett & Navies, 1977; Sprung et al., 1980). Furthermore, the hyperkalemia induced by heat stroke may be due to (1) metabolic acidosis causing potassium shift from intrcellular to extracellular fluid, and (2) liver and muscle damage causing potassium release. It has also been suggested that hyperthermia would damage the endothelial wall of the small vessels and induce clumping of the platelets at the damaged sites (Weber & Blakely, 1969). Subsequntly, obstruction of the small vessels eventually leads to increased capillary permeability and to petechial hemorrhage (Sohal et al., 1968). On the other hand, it has been proposed that tissue hypoxia or anoxia is responsible for the cellular damage in heat stroke (Hartman 1937). Our results also showed that heat-stroke rabbits, but not febrile rabbits, displayed both hypotension and cerebral ischemia. No pathological findings were observed in the febrile rabbits even though their colonic temperature had already reached a plateau level of 42.5°C which is equivalent to that of heat stroke. Thus, it appears that tissue hypoxia or anoxia, rather than tissue injury due to hyperthermia, is the main cause for the onset of heat stroke syndrome.

The work reported here was supported by grants from the National Science Council (Taipei, Republic of China).

REFERENCES

Austin, M.G., and Berry, J.W. (1956): Observations on one hundred cases of heat stroke. JAMA 161: 1525-1529.

Cloves, G.H.A., and O'Donnell, T.F. (1974): Heat stroke. N. Engl. J. Med. 291: 564-567.

Emmett, M., and Navies, R.G. (1977): Clinical use of the anion gap. Medicine 56: 38-54.

Ferris, E.B., Blankenhorn, M.A., and Robinson, H.W. (1938): Heat stroke: Clinical and chemical observations in 44 cases. J. Clin. Invest. 17: 249-262.

Gottschalk, P.G., and Thomas, J.E. (1966): Heat stroke. Mayo Clin. Proc. 41: 470-482.

Kew, M., Bersohn, I., and Seftel, H. (1970): Liver damage in heat stroke. Am. J. Physiol. 49: 192-202.

Hartman, F.W. (1937): Lesions of brain following fever therapy. Etiology and pathogenesis. JAMA 109: 2116-2121.

Knochel, J.P. (1974): Environmental heat illness. An eclectic review. Arch. Intern. Med. 133:841-864.

Lin, M.T., and Lin, S.Z. (1992): Cerebral ischemia is the main cause for the onset of heat stroke in rabbits. Experientia 48: 225-227.

Sawyer, C.H., Everett, J.W., and Green, J.D. (1954): The rabbit diencephalon in stereotaxic coordinates. J. comp. Neurol. 101:801-824.

Shih, C.J., Lin, M.T., and Tsai, S.H. (1984): Experimental study on the pathogenesis of heat stroke. J. Neurosurg. 60: 1246-1252.

Sohal, R.S., Sun, S.C., and Colcolough, H.L. (1968): Heat stroke: An electron microscopic study of endothelial cell damage and disseminated intravascular coagulation. Arch. Intern. Med. 122: 43-47.

Sprung, C.L., Portocarrero, C.J., and Fernaine, A.V. (1980): The metabolic and respiratory alterations of heat stroke. Arch. Intern. Med. 140: 665-669.

Stefanini, M. (1975): Heat stroke. In Injuries of the Brain and Skull, Part I. Handbook of Clinical Neurology, Vol. 23, ed. Vinken, P.J., and Bruyn, G.W., pp. 669-681. North-Holland, Amsterdam.

Vertel, R.M., and Knochel, J.P. (1967): Acute reneal failure due to heat injury: An analysis of ten cases associated with a high incidence of myoglobinuria. Am. J. Med. 43: 435-451.

Weber, M.B., and Blakely, J.A. (1969): The hemorrhage disthesis of heat-stroke: A consumption coagulopathy successfully treated with heparin. Lancet 1: 1190-1192.

C. Central control of body temperature

Integrative and cellular aspects of autonomic functions : temperature and osmoregulation. Eds K. Pleschka, R. Gerstberger. John Libbey Eurotext, Paris © 1994, pp. 225-230.

Microdialysis measurement of monoamine release from the preoptic anterior hypothalamus of rats in response to heat or cold exposure

M.-T. Lin, W.-H. Su and Z. Szreder

Department of Physiology, National Cheng Kung University Medical College, Tainan City, Taiwan, Republic of China

The effect of environmental temperature on the extracellular concentration of noradrenaline (NA), adrenaline (A), dopamine (DA), 3,4-dihydrophenylacetic acid (DOPAC), homovanillic acid (HVA), 5-hydroxytryptamine (5-HT) and 5-hydroxyindole-3-acetic acid (5-HIAA) in the preoptic anterior hypothalamus (POAH) measured by in vivo microdialysis coupled to high-performance liquid chromatography with electrochemical detection were conducted in freely moving rats. The mean concentration of 5-HIAA and DOPAC in the POAH decreased during heat exposure (35°C), but increased during cold exposure (4°C). The endogenous release of either NA, A, DA, 5-HT or HVA was not evoked during heat or cold exposure in these rats. The above results on the rat provide additional support for the role of both 5-HT and DA in the POAH in the central control of thermoregulation.

The role of the POAH in the control of thermoregulatory responses is well established (Bligh, 1973; Bligh, 1979). In addition, a neurochemical model involving NA, DA, 5-HT and acetylcholine has been put forward (Bligh, 1979; Feldberg & Myers, 1963; Lin, 1984). In the rat, during the first hour of heat exposure (32°C), the turnover of 5-HT increased in the POAH (Simmonds, 1970). On the other hand, in the rat, the turnover of hypothalamic NA increased during heat (32°C) or cold (9°C) exposure (Simmonds, 1969). It should be mentioned that , in these previous studies, changes in the turnover of 5-HT or NA were estimated mainly by measuring the rate of disappearance of [3H]5-HT or [3H]NA from brain tissues. There are at least two shortages in the interpretation of results from experiments involving radioactive isotopes. First, [3H]5-HT or [3H]NA does not mix homogenously with endogenous 5-HT or NA within the neurons and that those 5-HT-containing or NA-containing neurons in close contact with c.s.f. are preferentially labelled by [3H]5-HT or [3H]NA. Secondly, there may be some accumulation in other neurons. In the present study, a different method was used. Experiments were carried out to assess the effect of environmental temperature on the extracellular concentration of 5-HT, 5-HIAA, DA, DOPAC, HVA, NA and A in the POAH measured by in vivo microdialysis coupled to a HPLA-ECD in freely moving rats.

Male albino rats weighing 350 - 400 g were housed singlely under a 12/12 h light/dark cycle (light on at 6:00 h) with free access to food and water. For surgery, rats were anesthetized with pentobarbital sodium (6 mg/100 g, i.p.) and placed in a Kopf stereotaxic frame in the flat skull position. A 10-mm guide tube (21-gauge stainless-steel tubing) was implanted in the upper corder of anterior commisure (A 8.7 mm, L ±0.8 mm and V 6.7 mm below dura) (Paxinos & Watson, 1982). Following surgery, the guide cannula was plugged with a stylet and rats were returned to their home cages for a minimum recovery period of one week. The evening before an experiment, rats were briefly immobilized with a volatile anesthetic (Metofane, Pitman-Moore Inc., Washington Crossing, NJ) and a dialysis probe was slowly lowered through the guide cannula and cemented in place. The dialysis probe was a CMA-12 microdialysis probe (2 mm in length) from Carnegie Medicine (CMA, Sweden). The length of the dialysis probe was adjusted to place the probe tip at a point 0.5 mm above the base of the preoptic anterior hypothalamus (A 8.7 mm, L ±0.8 mm and V 8.7 mm below dura; see Fig. 1). The CMA-120 system (CMA, Sweden) was used for microdialysis experiments in an awake, freely moving animal. It consisted of a liquid swivel with a rotating stand for the collecting vial, a counterbalancing arm and a cage designed as a hemisphere to prevent the animal from bumping the implant on the head into the walls of the cage. Ringer solution was used to perfuse microdialysis probes at 0.001 ml/min by a high precision pump (CMA-100). Sequential 15 min perfusates were collected by a microfraction collector (CMA-140) into polyethylene microsample tubes containing 0.005 ml of 0.1 M perchloric acid with 10^{-7} M ascorbic acid and 0.005 ml of the perfusates were analyzed for NA, A, DA, DOPAC, HVA, 5-HT and 5-HIAA on a microbore HPLC-ECD system. Performance of each microdialysis probe was calibrated by dialysis of a known amount of the standard mixture and recoveries of all analytes was then determined. Brain concentrations of NA, A, DA, DOPAC, HVA, 5-HT and 5-HIAA were calculated by determing each peak height ratio relative to the internal standard 3-MT and also corrected by each probe performance. Standard stock solutions of NA, A, DA, DOPAC, HVA, 5-HT and 5-HIAA and 3-MT were prepared at concentration of 0.02 ng/ml in 0.1 M perchloric acid and stored at -70°C in the dark and thawed in an ice bath prior ro preparation of a standard mixture. The internal standard 3-MT and standard mixture were prepared fresh every day from a portion of these stock solutions after appropriate dilution with Ringer solution containing 10^{-7} M ascorbic acid in 0.1 M perchloric acid. The mobile phase was prepared by adding 60 ml acetonitrile, 0.42 g SOS (2.2 mM), 200 g monosodium dihydrogen orthophosphate (14.7 mM), 8.82 g sodium citrate (30 mM), 10 mg EDTA (0.027 mM), and 1 ml diethylamine in doubled distilled water. The solution was adjusted to a pH 3.5 by concentrated orthophosphoric acid and its final volume was adjusted to 1 l. The mixture was filtered with a 0.22 micron nylon filter under reduced pressure and degassed by purging helium gas for 20 min. The flow-rate was 0.05-0.06 ml/min maintaing column pressure at approximately 7.6 Mpa.

In vitro tests were run to determine 5-HT or other monoamines recovery in the dialysis solution at the flow rate used. At room temperature, dialysis probes were immersed in dialysis solution containing 5-HT or other monoamines. At a flow rate of 1 µl/min the relative recovery (e.g., 5-HT concentration in the dialysis solution divided by 5-HT concentration in the beaker) was 12±1.5% (n = 9). Basal monoamine levels were considered stable after observation of 4 consecutive samples with no upward or downward trend. These basal values were averaged and all samples were expressed as a percent of the mean baseline. Duncan's multiple range test (multiple time point experiments) were used for post hoc determination of significant difference ($P < 0.05$).All probe placements were in the region of POAH as verified by standard histological procedures. Animals were perfused with physiological saline, followed by 10% formalin. The brains were removed, sliced on a freezing microtome and stained with formyl thionine. A representative diagram of a dialysis probe in the POAH is shown in Fig. 1.

Figure 2 shows the mean concentrations of 5-HIAA, DOPAC and HVA measured before, during and after heat exposure (35°C). During heat exposure mean concentrations of both 5-HIAA and DOPAC in the POAH decreased significantly. The magnitude of these decreases typically varied between 3- and 4-fold compared to basal concentrations. In the one hour period following reduction of the ambient temperature

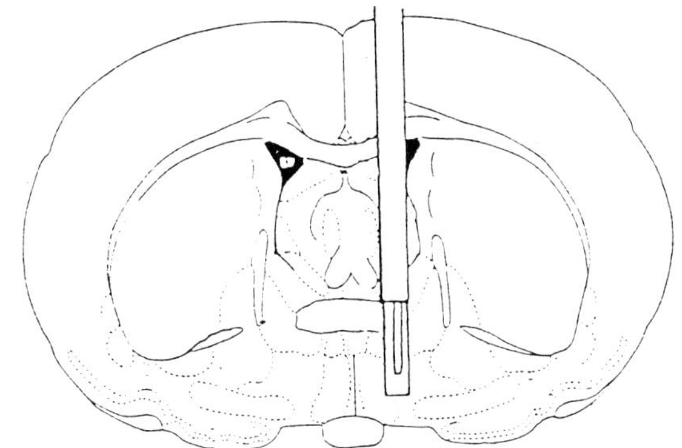

Fig. 1. Coronal section of the rat brain (8.7 mm anterior of interaural zero) showing a schematic diagram of the dialysis probe in the target area of the hypothalamus. The guide cannula is not represented in this figure but would extend to a point just above the anterior commisure (adapted from Paxinos & Watson, 1982).

(T_a) to 22°C concentrations of both 5-HIAA and DOPAC rose significantly to preheat exposure levels. Figure 3 shows the mean concentration of 5-HIAA, DOPAC and HVA measured before, during and after cold (4°C) exposure. During cold exposure mean concentrations of both 5-HIAA and DOPAC in the POAH increased significantly. The magnitude of these increases typically varied between 2- and 3-fold compared to basal concentrations. In the two hour period following elevation of T_a to 24°C concentration of 5-HIAA or DOPAC fell gradually to precold exposure levels. However, the HVA levels in the POAH, where measurable, remained unchanged during heat and cold exposure. No changes in the concentration of 5-HIAA or DOPAC were found in response to heat or cold exposure when dialysis probes were placed in the cerebral cortex. Levels of 5-HT and DA in the POAH could not normally detected and levels of NA and A varied being just detectable in some animals and not in others. Table 1 shows the effect of T_a on the colonic temperature (T_{co}) of rats 1, 2 and 3 h after exposure to the temperature. Only those rats exposed to 35°C showed a significant change ($P < 0.05$) in T_{co} compared to controls. The slight hyperthermia (0.6°C) observed at 35°C was established within the first 1 h of exposure to the temperature and remained approximately constant for at least a further 2 h.

It is well established that, during heat exposure, the rat displays decreased metabolic heat production and increased heat loss (as reflected by cutaneous vasodilation and respiratory evaporative heat loss). By the contrary, during cold exposure, the rat displays increased metabolic heat production and decreased heat loss (as reflected by cutaneous vasoconstriction) (Gordon, 1990). In the present results, mean concentrations of both 5-HIAA and DOPAC in the POAH decreased during heat exposure, but increased during cold exposure in rats. The concentrations of 5-HIAA and DOPAC in the brain, respectively, provide an accurate reflex of activity in serotoninergic and dopaminergic neurons of the brain (Cooper et al., 1978). The above results on the rat provide additional support for the role of serotoninergic and dopaminergic systems in the POAH in the central control of thermoregulation.

Previous studies, using the rate of disappearance of [3H]5-HT or [3H]NA from discrete areas of rat brain as indices of the rates of turnover of endogenous 5-HT or NA in these areas, have demonstrated different pattern of release of 5-HT in response to heat or cold exposure (Simmonds, 1969,1970). They found that the turnover of 5-HT in the POAH or the cortex + hippocampus, but not in the striatum or midbrain, increased with increasing T_a from 9°C to 32°C. In addition, either heat or cold exposure elevated the turnover of NA in the POAH of rat brain. In the sheep, using microdialysis technique, Kendrick et al. (1989) showed that concentration of DA and NA significantly increased, and those of 5-HIAA significantly decreased in the medial preoptic area during heat exposure. Another studies, using push-pull cannulae, have demonstrated release of DA and NA during exposure to heat (Myers & Beleslin, 1971; Myers & Chinn, 1973;

Table 1. Effect of T_a on T_{co} of rats 1, 2 and 3 h after exposure to each T_a

T_a, °C	Time after exposure to each T_a		
	1 h	2 h	3 h
	°C	°C	°C
22	37.7±0.2	37.6±0.2	37.7±0.3
4	37.8±0.2	37.9±0.3	37.9±0.2
35	38.3±0.2*	38.2±0.2*	38.3±0.2*

The values are means±SEM of 9 rats for each group. The T_{co} of untreated control at room temperature (24°C) was 37.7±0.2°C (n = 9).

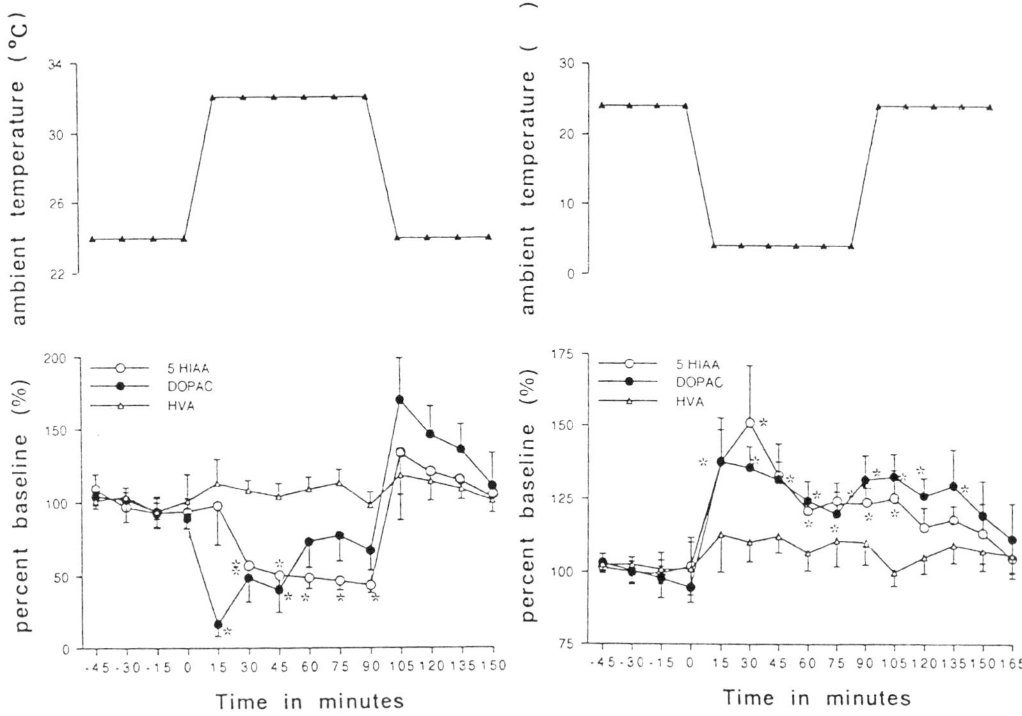

Fig. 2. Effect of heat exposure on extracellular 5-HIAA, DOPAC and HVA in the POAH (n = 9). The data are plotted as means±SEM of percent baseline during 15-min sampling periods. * $P < 0.05$ for difference; Duncan's multiple range test.

Fig. 3. Effect of cold exposure on extracellular 5-HIAA, DOPAC and HVA in the POAH (n = 9). The data are plotted as means±SEM of percent baseline during 15-min sampling periods. * $P < 0.05$ for difference; Duncan's multiple range test.

Ruwe & Myers, 1978; Yaksh & Myers, 1972) , and 5-HT during exposure to cold (Myers & Beleslin, 1971) in monkeys and/or cats. The decrease in 5-HIAA of POAH in both the rat (present results) and the sheep (Kendrick et al., 1989) during heat exposure, as well as the increase in the hypothalamic 5-HT of monkey brain during cold exposure (Myers & Beleslin, 1971) is coincident with the implication that serotonin pathways in the central control of heat gain but have also reported decreases in 5-HT release in the hypothalamus during heat loss (Myers & Beleslin, 1971). These observations tend to indicate that there may be a parallel depression in heat gain mechanisms during conditions when heat loss mechanisms are maximally activated. However, the contention is not supported by the findings of Simmonds (1970). He observed that the turnover of hypothalamic 5-HT of rat brain increased with increasing T_a from 9°C to 32°C. Clearly more experiments are needed to investigate the possibility. On the other hand, the increase in NA/DA release in the hypothalamus of the sheep (Kendrick et al., 1989), the monkey (Myers & Beleslin, 1971) or the rat (Simmonds, 1969) during heat exposure implicates that noradrenergic and/or dopaminergic projections to the POAH are activated during heat loss. Again, the hypothesis is questioned by the findings that the turnover of NA (Simmonds, 1969), DOPAC (present results) in the POAH of rat brain increased also during cold exposure.

The work reported here was supported by grants from the National Science Council of Republic of China (Taipei, Taiwan, ROC). The authors wish to thank Drs. J. S. Kuo and F. C. Cheng of Department of Medical Research, Veterans General Hospital, Taichung, Taiwan, ROC) for helpful consultation during the experimentation.

REFERENCES

Bligh, B. (1973): Temperature regulation in mammals and vertebrates. Amsterdam: North-Holland.

Bligh, B. (1979): The central neurology of mammalian thermoregulation. Neuroscience. 4: 1213-1236.

Cooper, J.R., Bloom, F.E., and Roth, R.H. (1978): The Biochemical Basis of Neuropharmacology. 3rd edn. Oxford Press: New York.

Feldberg, W., and Myers, R.D. (1963): A new concept of temperature regulation by amines in the hypothalamus. Nature 200: 1325.

Gordon, C.J. (1990): Thermal biology of the laboratory rat. Physiol. Behav. 47: 963-991.

Kendrick, K.M., De La Riva, C., Hinton, M., and Baldwin, B.A. (1989): Microdialysis measurement of monoamine and amino acid release from the medial preoptic region of the sheep in response to heat exposure. Brain Res. Bull. 22: 541-544.

Lin, M.T. (1984): Hypothalamic mechanisms of thermoregulation in the rat: Neurochemical aspects. In Thermal Physiology, ed J.R.S. Hales, pp. 113-118. Raven Press: New York.

Myers, R.D., and Beleslin, D.D. (1971): Changes in serotonin release in hypothalamus during cooling or warming of the monkey. Am. J. Physiol. 220: 1746-1754.

Myers, R.D., and Chinn, C. (1973): Evoked release of hypothalamic norepinephrine during thermoregulation in the cat. Am. J. Physiol. 24:230-236.

Paxinos, G., and Watson, C. (1982): The Rat Brain in Stereotaxic Coordinates. 2nd edn. Academic Press: Sydney.

Ruwe, W.D., and Myers, R.D. (1978): Dopamine in the hypothalamus of the cat: Pharmacological characterization and push-pull perfusion analysis of sites mediating hypothermia. Pharamcol. Biochem. Behav. 9: 65-80.

Simmonds, M.A. (1969): Effect of environmental temperature on the turnover of noradrenaline in the hypothalamus and other areas of rat brain. J. Physiol. 203: 199-210.

Simmonds, M.A. (1970): Effect of environmental temperature on the turnover of 5-hydroxytryptamine in various areas of rat brain. J. Physiol. 211: 93-108.
Yaksh,T.L., and Myers, R.D. (1972): Hypothalamic "coding" in the unanesthetized monkey of noradrenergic sites mediating feeding and tehrmoregulation. Physiol. Behav. 8: 251-257.

Integrative and cellular aspects of autonomic functions : temperature and osmoregulation. Eds K. Pleschka, R. Gerstberger. John Libbey Eurotext, Paris © 1994, pp. 231-241.

Modulation of thermoregulatory afferent and efferent signals by brain stem

Haruhiko Sato

Department of Physical Education, Faculty of Physical Education, Chukyo Women's University, 55 Nadakayama, Yokone-cho, Obu-City, Aichi 474, Japan

INTRODUCTION

The temperature control system centering around the preoptic and anterior hypothalamic areas has been well established by many studies previously performed (see Ref. Simon et al., 1986). Several studies have been recently reported indicating the involvement of the lower brain stem nuclei, the nucleus raphe magnus (**NRM**) and locus coeruleus-the subjacent nucleus, subcoeruleus (LC/SC), in thermoafferent system (see Ref. Brück & Hinckel, 1982). Since anatomical findings that the **NRM** ans **LC/SC** project their respective serotonergic and noradrenergic descending collaterals to the ventral and dorsal horns in the spinal cord (e.g. Dahlström & Fuxe, 1965; Nygren & Olson, 1977), many electrophysiological investigations have revealed that these nuclei exert the centrifugal modulation of motor, cardiovascular, respiratory and nociceptive functions (see Ref. Basbaum, 1984). Although some their roles in thermoregulatory system have been pointed out, the concrete evidences on their functions or physiological significances remain uncertain.

Recently, a seires of studies have been aimed to investigate furthermore the contribution of the **NRM** and **SC** to the hypothalmic central thermoregulation. It will be shown that these nuclei centrifugally modulate thermoregulatory afferent and efferent signals, i.e., warm and cold signals from the skin and gamma motoneuron activity to drive cold shivering response, respectively.

RESULTS

1. Modulation of temperature signal transmission

The experiments were performed on adult, male rats of Wistar strain weighing between 300 and 450 g. Animals were initially anesthetized with urethane 1.2-1.5 g/kg (i.p.), supplemented as required. Surgical procedures were essentially similar to those previously described elsewhere (Sato et al., 1990). Laminectomy was performed to expose the lower segments of the spinal cord and a mineral oil pool was formed with skin flaps. During experiments, the temperature of the mineral oil pool was maintained at 37.5°C by a heating lamp and the rectal temperature was maintained at 37.5°C by a heating pad placed underneath the animal. The activity of dorsal horn cells reactive to changes in temperature of the scrotal skin was extracellularly recorded by a pontamin sky blue glass microelectrode. Their spike activity was displayed as a mean firing rate every 2 s on a recorder. Thermal stimulation was performed by a Peltier devise through which D.C. current was finely servo-controlled. Thermal stimulation to the scrotal skin was restricted within the innocuous range between 10 and 40 °C in which the cutaneous thermoreceptors was considered to be maximally activated to elicite the thermoafferent impulses as regulatory input signals. Monopolar, cathodal constant-current pulses (100-300 µA, 300 µs, 50-200 Hz) were given usually for 5-20 s by two stainless steel electrodes which were stereotaxically inserted into the **NRM** (AP 1.5-2.0, L 0, V 2.2-3.0) and **SC** (AP 0.2-0.8, L 1.0-1.5, V 0) according to the stereotaxic coordinates by Paxinos and Watson (1982). An anodal indifferent needle electrode was inserted subcutaneously. Recording and stimulating sites were electrolytically lesioned at the end of experiments and histologically verfied by the ways similar to those described elsewhere (Sato & Hashitani, 1989). The dorsal horn units responsive to warming or cooling of the scrotal skin were classified into two groups: warm-responsive units showing an increase in firing rate by skin warming and cold-responsive units showing an increase in firing rate by skin cooling. The effect of electrical stimulation in both the **NRM** and **SC** was characterized by an incomplete or complete depression of temperature responses of these two types of dorsal horn unit. **Fig. 1** demonstrates typical examples of warm-responsive dorsal horn units of which the temperature-responses were markedly affected, remaining their basal dis-

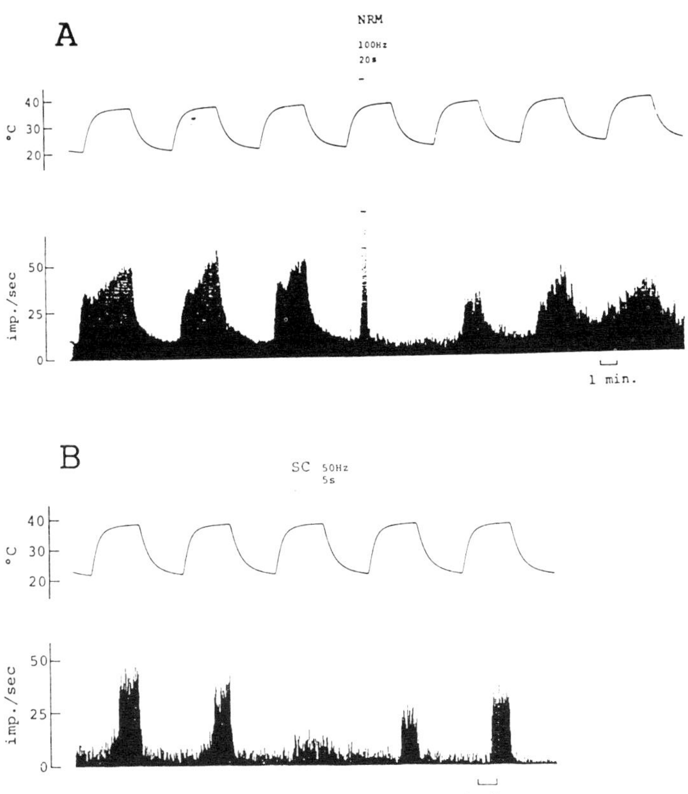

Fig. 1A and B. The effects of electrical stimulation in the **NRM** (A) and SC (B) on the temperature responses of two warm-responsive dorsal horn cells evoked by repeating of warming and subsequent cooling of the scrotal skin in the innocuous range (22-37°C). In A and B, the upper trace, temperature of scrotal skin and the lower trace, mean firing rate counted every 2 s. Stimulation parameters were repetitive pulses (300 µs square pulses at 100 µA current intensity) delivered at 100 Hz (in A) and 50 Hz (in B) for the durations (20 s in A, 5 s in B) indicated by the bars above the traces of skin temperature. Note a complete depression of warm-response followed by marked artifacts of **NRM** stimulation (in A) and an incomplete depression of warm-response by SC stimulation (in B) with a gradual recovery of temperature response after stimulation in both A and B.

charge activity almost unaffected. 95% of 44 warm-responsive units tested were depressed by **NRM** stimulation. By **SC** stimulation, 47% of 32 warm-responsive units tested were depressed, but 50% not affected. Thus, depression of warm-responsive units was evoked predominantly by **NRM** stimulation. On the other hand, cold-responsive dorsal horn units also were depressed by electrical stimulation in both the **NRM** and **SC**. 79% of 14 cold-responsive units tested were depressed by **SC** stimulation, and 63% were depressed by **NRM** stimulation. Thus, this type of unit was depressed predominantly from the **SC**. Furthermore, whether such a **NRM**-evoked depression of warm-responsive dorsal horn cell is mediated by the serotonergic pathways or not was examined in four rats by intravenously injecting a small dose of serotonin-antagonist, Methysergide. Reproducible **NRM**-evoked depressions of warm-response were observed to be clearly prevented by the injection of the drug. Although it was failed to test the antagoning action of alpha-blocker on SC-evoked depression of cold-responsive dorsal horn cell, it seems highly possible that the pathways mediating such depression are noradrenergic.

2. Modulation of gamma motoneuron activity

Gamma efferent activity

Single gamma efferent fibers were dissected in the ventral root of lower spinal segments (L4-5) and their spontaneous discharge was recorded by platinum electrodes. This series of experiments were performed in urethane-anesthetized rats whose bilateral hypothalamic regions were electrolytically lesioned to eliminate the influences from the higher center of temperature regulation to gamma motoneurons (Sato, et al., 1990). The techniques of electrolytic lesion and histological veirification for their sites have been described elsewhere (Sato & Hashitani, 1989). All experiments were performed under a constant room temperature of 25-28°, since gamma motor activity was greatly influenced by changes in skin temperature (Sato & Hasegawa, 1977). The criteria for the identification of gamma fiber were the conduction velocity (less than 40 m/s) and discharge activations reflexly elicited. Conduction velocity was measured by the method as described elsewhere (Sato et al., 1990). Electrical stimulation in the **NRM** and **SC** was performed by a similar type of monopolar electrode with the stereotaxic coordinates and stimulating parameters similar to those described in the previous section. Most 38 (86%) of 44 gamma efferents examined were affected by electrical stimulation in the **NRM**. The effect of **NRM** stimulation was characterized by a strong depression followed by a complete cessation of firing lasting either for a short period or for more than 30 min. Of 38 gamma efferents affected by **NRM** stimulation, 30 (68%) were inhibited, only 8 facilitated. In contrast, SC stimulation produced a brief facilitation of gamma efferent activity in many cases. Of 27 gamma efferents affected by SC stimulation, 25 (78%) were facilitated, 2 inhibited and 5 unaffected. Whether the pathway mediating **NRM**-evoked inhibition of gamma activity is serotonergic or not was examined. Reproducible **NRM**-evoked gamma inhibitions were prevented by the administrationin of serotonin-antagonist, p-chlorophenylalanin (10 mg/kg, i.p.) in all the experiments performed in four rats. On the other hand, whether the pathway mediating SC-evoked facilitation of gamma activity is noradrenergic or not was examined by intravenously administrating of alpha-blocker, Phenoxybenzamin. In all of the experiments performed in five rats, SC-evoked facilitation of gamma activity was antagonized by the drug.

Stretch response of muscle spindle ending

To determine which type of gamma motoneuron, dynamic or static, was activated or inhibited by the **NRM** and SC, the effects of electrical stimulation in these nuclei on the afferent responses of muscle spindle endings to stretch of the muscle were investigated. This series of experiments also were performed similarly in urethane-anesthetized rats whose bilateral hypothalamic areas were electrolytically lesioned. Single Group Ia fibers from the muscle spindle primary ending in the gastrocnemius et soleus (**GAS**) or tibialis anterior (**TA**) were isolated by dissecting the dorsal root filaments of lumbosacral segments. Group Ia fibers were identified principally by the conduction velocity (more than 72 m/s) and prominant dynamic response to stretching of the muscle, which were essentially similar to those in the experiments described elsewhere (Sato, 1983, 1984). A standard ramp-and hold-stretch of a constant length (2-4 mm) with a lengthening speed (3-20 mm/s) and a duration of the total stretching phase of 10-20 s was repeatedly applied to the muscle at intervals of 30-60 s. In each response of single muscle spindle endings to stretch of the muscle, peak frequency (**P**) at the end of stretching, static frequency (**S**) 0.5 s later and the dynamic index (**DI**), the difference in discharge frequency between **P** and **S**, were ele-

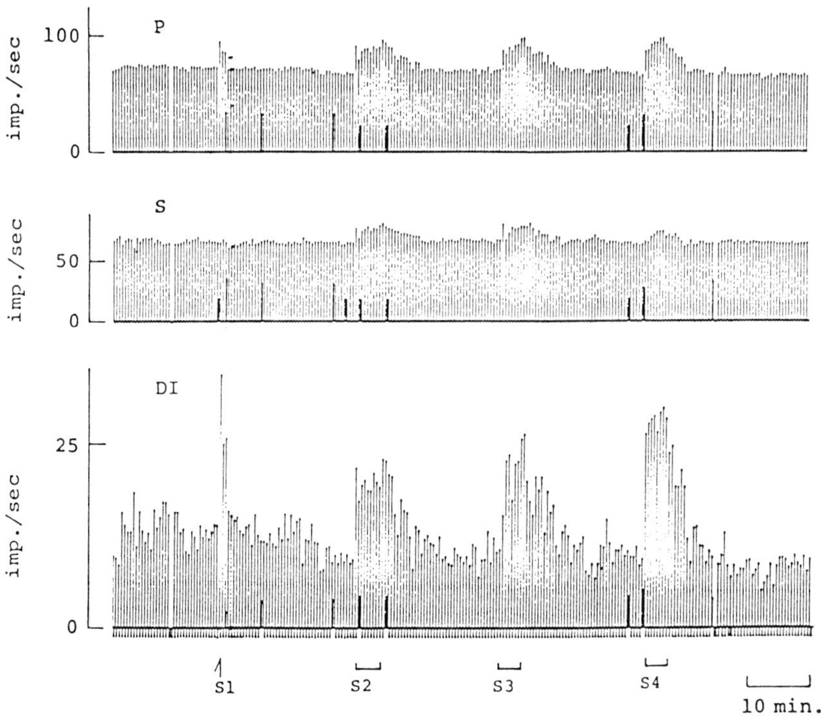

Fig. 2. The effects of electrical stimulation in the NRM on the stretch responses of a muscle spindle primary ending in the gastrocnemius to stretch of the muscle. In each response of the spindle ending to a standard ramp- and hold-stretching of a constant length with a constant lengthening speed and duration which was repeatedly applied to the muscle at intervals of 30 sec, peak frequency (P) at the completion of stretching, static frequency (S) 0.5 sec later and the Dynamic Index, (DI), the difference between P and S, were electronically detected and displayed on a recorder as heights of bars for P, (in the top trace), S (in the middle trace) and DI (in the bottom trace) in the figure. Stimulation parameters were repetitive pulses (200 µs square pulses, at 200 µA current intensity) delivered at 200 Hz (in S1), 50 Hz (in S2), 100 Hz (in S3) and 200 Hz (in S4) for the durations of 10 sec (in S1) and 3 minutes (in S2-4) indicated by the bars in the bottom of the figure. Note frequency-depended responses of P and DI evoked by NRM stimulation.

ctronically detected by a frequency meter equipped with peak value-holding circuits and displayed on a recorder. Which type of gamma motoneuron was activated was evaluated essentially basing on the following criteria: an increase of the DI despite an increase of S was attributed to dynamic gamma activation while an increase of S with the resultant decrease of the DI to static gamma activation (Matthews, 1972). **Fig.** 2 shows a typical example of stretch response of a primary ending in the GAS muscle which was facilitated by NRM stimulation. NRM stimulation was found to facilitate stretch response of the primary endings in the GAS (100%) and TA (80%), and the facilitatory effect was judged to be attributed to dynamic gamma motor activation. In contrast, SC stimulation was found to facilitate stretch response of the primary endings in the GAS (70%) and TA (80%), and the facilitory effect was judged to be attributed to static gamma motor activation.

3. Modulation of cold shivering response

A series of experiments to analyze quantitatively shivering response (SR) were performed in rats whose the hypothalamic areas were left intact. Animals were initially anesthetized with a high dose of numbutal (40-60 mg/kg, i.p.). During experiment, the depth of anesthesia was finely controlled at a constant level by the intermittent administration of a small amount of numbutal solution (4 mg/ml) through an intraperitoneally indwelled catheter. By varying the rate of the intermittent administration (100-500 ms), the anesthetic condition just to initiate shivering were maintained for several hours. The animal was fixed in a stereotaxic apparatus and surrounded with acryl boards as to form a small climatic chamber. The air into the chamber was supplied from a heat exchange element perfused by the water maintained at various temperatures. The motor unit activities were recorded electromyographically by means of coaxial needle electrodes inserted into the GAS and TA muscles. Motor unit activations repeatedly induced by lowering the temperature inside the chamber, usually by 10-20°C below 37°C for a few minutes every 10 minutes (whole body cooling), were fairly reproducible as "shivering responses" (SR) in terms of total number of the spikes elicited. In all experiments, the effects of stimulation in the NRM and SC on the activities of motor units in both the GAS and TA were examined. A marked facilitation (130-300 %) was found to be evoked by NRM stimulation in the majority (75 %) of SR in the GAS examined, while by **SC** stimulation in a larger number of **SR** in the **TA**. No clear tendency of facilitation or inhibition of **SR** was found in **TA** by **NRM** stimulation and in **GAS** by **SC** stimulation. Considering the results on the afferent response of muscle spindle endings obtained in the previous section, it was deduced that such **NRM**-evoked **GAS-SR** facilitation is attributed to dynamic gamma activation and SC-evoked **TA-SR** facilitation to static gamma activation.

DISCUSSION

The present result that the activity of the spinal dorsal horn cells responsive to innocuous thermal stimulation was markedly depressed by electrical stimulation in the **NRM** as well as in the **SC** suggestes that the spinal transmission of warm and cold signals is inhibited by the descending collaterals of these nuclei. The centrifugal modulation by the serotonergic and noradrenergic descending neural systems of the **NRM** and LC/SC, respectively, have been described on various physiological functions. Particularly, in the previous studies on nociceptive transmission, the responses of dorsal horn cells reactive exclusively to noxious heat (above 45-50°C) to the skin were investigated (e.g. Ren et al., 1990). Changes in temperature of the skin in the innocuous range as described in the present study could contribute as thermoregulatory warm and cold input signals to temperature control center. "Switching response" of thalamic neurons which was firstly reported by Hellon and Misra (1973) and extensively investigated (e.g. see Ref. Schingnitz & Werner, 1986) appears a high degree of convergence of thermoafferent signals due to the centrifugal modulation. It was suggested that switching response might be generated in the **NRM** (Hellon & Taylor, 1982) and trigger a signal for warm defence (Werner et al, 1986). However, the present study demonstrated a concret neural mechanism for the centrifugal modulation of thermoafferent processing by the **NRM** and **SC**. Furthermore, the evidences supporting the specific projections of warm signals to the **NRM** (Jahn, 1976; Dickenson, 1977) and cold signals to the **SC** (Hinckel & Schröder-Rosenstock, 1981) might conform to the independent inhibitions of warm signals by **NRM** and of cold signals by **SC** as described in the present study. The roles and physiological significances of the neural mechanism including these nuclei remain for further study. The descending influences of the serotonergic pathways from the **NRM** and the noradrenergic pathways from the **LC/SC** on the spinal alpha motoneurons also have been studied electrophysiologically (e.g. Roberts et al., 1988) and pharmacologically (e.g. White & Neuman, 1983). The activity of gamma motoneurons which is essential to the development of cold shivering is descendingly inhibited by the NRM (Sato et al., 1990) whereas facilitated by the SC as described in the present study. In addition, the present study revealed that the NRM activates dynamic gamma motoneurons and the SC does static gamma motoneurons, and finally suggested that facilitation of GAS-SR is ascribed to the former and that of TA-SR to the latter. These results may suggest the different descending control of cold shivering in the extensor and flexor via two gamma motor systems from the brain stem level. Whether the generation of cold shivering is ascribed to the activation of dynamic (Schäfer & Schäfer, 1973; Sato, 1983) or static gamma motoneuron (Henatsch, 1966; Sato, 1984) might depend upon the thermoreceptive sites activated and the muscle studied, extensor or flexor. The neural mechanism for the generation of cold shivering also remains for further study.

SUMMARY AND CONCLUSION

The summarized results suggest: 1. The raphe- and subcoeruleospinal pathways inhibit the spinal transmission of warm and cold signals contributing to the central thermoregulation, respectively. 2. Gamma motoneurons were descendingly inhibited by the serotonergic pathways of the NRM while facilitated by the noradrenergic pathways of the SC. 3. NRM-evoked facilitation of stretch response of the primary endings in both the GAS and TA muscles was ascribed to dynamic gamma activation while SC-evoked facilitation of stretch response of the primary endings in both the GAS and TA muscles to static gamma activation. 4. NRM-evoked facilitation of shivering response in the GAS was ascribed to dynamic gamma activation and SC-evoked facilitation of shivering response in the TA to static gamma activation.

A neural mechanism was postulated for the centrifugal modulations of transmission of thermoregulatory afferent and efferent signals by the NRM and SC: 1) modulation of thermoregulatory input signals in a manner as the NRM inhibits warm signals while the SC does cold signals and 2) modulation of thermoregulatory output signals in a manner as the NRM facilitates extensor shivering via dynamic gamma system while the SC does flexor shivering via static gamma system.

ACKNOWLEDGEMENTS

Methysergide was graciously supplied by Sandoz Pharmaceuticals. This work was supported by Grants-in-Aid for Scientific Research (No. 63570083, 01570089, 03670082, 05670075) from the Ministry of Education, Japan.

REFERENCES

Basbaum, A. I. and Fields, H. L. (1984): Endogenenous pain control systems: brainstem spinal pathways and endorphin circuity. Annu. Rev. Neurosci. **7**, 309-338.

Brück, K. and Hinckel, P. (1982): Thermoafferent systems and their adaptive modifications. Pharmac. Ther. **1**, 357-381.

Dahlström, A. and Fuxe, K. (1965): Evidence for the existence of monoamine neurons in the central nervous system. II. Experimentally induced changes in intraneuronal amine levels of bulbo-spinal neuron systems. Acta Physiol. Scand. **64** (Suppl. 247), 1-36.

Dickenson, A. H. (1977): Specific responses of rat raphe neurones to skin temperature. J. Physiol. Lond. **273**, 277-293.

Hellon, R. F. and Misra, N. K. (1973): Neurons in the ventro-basal complex of the rat thalamus responding to scrotal skin temperature changes. J. Physiol. Lond. **232**, 389-399.

Hellon, R. F. and Taylor, D. C. M. (1982): An analysis of a thermal afferent pathway in the rat. J. Physiol. Lond. **326**, 319-328.

Henatsch, H. D. (1966): Spinal motor systems and dynamic/static properties of muscle spindles in experimental tremor states of cat. In: Granit R. ed. Muscular Afferents and Motor Control. Almqvist and Wiksells, Stockholm. pp 165-176.

Hinckel, P. and Schröder-Rosenstock, K. (1981): Responses of pontine units to skin-temperature changes in the guinea-pig. J. Physiol. Lond. **314**, 189-194.

Jahns, R. (1976): Different projections of cutaneous thermal inputs to single units of the midbrain raphe nuclei. Brain Res. **101**, 355-361.

Matthews, P. B. C. (1972): Mammalian Muscle Receptors and their Central Actions. Arnold: London.

Nygren, L.-G. and Olson, L. (1977): A new major projection from locus coeruleus: the main source of noradrenergic nerve terminals in the ventral and dorsal columns of the spinal cord. Brain Res. **132**, 85-93.

Paxinos: G. and Watson, C. (1982): The Rat Brain in Stereotaxic Coordinates. Academic Press, New York.

Ren, K., Randich, A. and Gebhart, G. F. (1990): Electrical stimulation of cervical vagal afferents. I. Central relays for modulation of spinal nociceptive transmission. J. Neurophysiol. **64**, 1098-1114.

Robert, M. H., Davies, M., Girdlestone, D. and Foster, G. A. (1988): Effects of 5-hydroxytriptamine agonists and antagonists on the responses of rat spinal motoneurones to raphe obscurus stimulation. Br. J. Pharm. **95**, 437-448.

Sato, H. (1983): Effects of skin cooling and warming on stretch responses of the muscle spindle afferent fibers from the cat' tibialis anterior. Exp. Neurol. **81**, 446-458.

Sato, H. (1984): Effects of changes in preoptic temperature on stretch response of muscle spindle endings in the cat's soleus muscle. Pflügers Arch. **402**, 144-149.

Sato, H. and Hasegawa, Y. (1977): Reflex changes in discharge activities of gamma efferents to varying skin temperatures in cats. Pflügers Arch. **372**, 195-201.

Sato, H. and Hashitani, T. (1989): Facilitation or suppression of fusimotor activity induced by changes in temperature of the midbrain reticular formation of rats. J. therm. Biol. **14**, 115-122.

Sato, H., Hashitani, T., Isobe, Y., Furuyama, F. and Nishino, H. (1990): Descending influences from nucleus raphe magnus on fusimotor activity in rats. J. therm. Biol. **15**, 259-265.

Schäfer, S. S. and Schäfer, S. (1973): The behavior of the proprioceptors of the muscle and the innervation of the fusimotor system during cold shivering. Exp. Brain Res. **17**, 364-380.

Schingnitz, G. and Werner, J. (1986): Significance of scrotal afferents within the general thermoafferent system. J. therm. Biol. **11**, 181-189.

Simon, E., Pierau, F.-K. and Taylor, D. C. M. (1986): Central and peripheral thermal control of effectors in homeothermic temperature regulation. Physiol. Rev. **66**, 235-300.

Taylor, D. C. M. (1982): The effects of nucleus raphe magnus lesions on an ascending thermal pathway in the rat. J. Physiol. Lond. **326**, 309-318.

Werner, G., Schinginitz, G. and Mathei, J. (1986): Analysis of switching neurons with in the thermoafferent system. Exp. Brain Res. **64**, 70-76.

White, S. R. and Neuman, R. S. (1983): Pharmacological antagonism of facilitatory but not inhibitory effects of serotonin and noradrenaline on excitabity of spinal motoneurons. Neuropharmac. **22**, 489-49.

Integrative and cellular aspects of autonomic functions : temperature and osmoregulation. Eds K. Pleschka, R. Gerstberger. John Libbey Eurotext, Paris © 1994, pp. 243-252.

Afferents to nucleus raphe magnus demonstrated by iontophoretic application of unconjugated cholera toxin B in rats and guinea-pigs

Dirk Matthias Hermann, Peter Hinckel, Pierre-Hervé Luppi* and Michel Jouvet*

*Physiologisches Institut der Justus-Liebig-Universität, Aulweg 129, D-35392 Giessen, Germany. *Département de Médecine Expérimentale, CNRS URA 1195, INSERM U52, 8, avenue Rockefeller, 69373 Lyon Cedex 08, France*

SUMMARY

The medullary raphe nuclei nucleus raphe magnus (NRM) and nucleus raphe pallidus (NRP) are known to be implicated into the afferent system of thermoregulation in the brain-stem and are also important for sleep-wake-regulation. We examined the afferents of NRM and rostral NRP in the rat and the guinea-pig with a highly sensitive iontophoretic tracing technique, using unconjugated cholera toxin subunit b (CTb) as a retrograde tracer, in order to better understand adaptive phenomenons of thermoregulation and phenomenons of sleep-wake-regulation. Restricted injections of CTb have been performed into various parts of NRM and rostral NRP. Important afferent structures that have never before been described have been demonstrated. Four major substructures with substantial differences of the afferent projection pattern could be delineated, namely dorsal caudal NRM, lateral caudal NRM, rostral NRP and rostral NRM.

INTRODUCTION

Nucleus raphe magnus (NRM) and nucleus raphe pallidus (NRP) are implicated in the thermoafferent system of the brain-stem (Dickenson, 1977, Brück and Hinckel, 1980, 1982). Hinckel et al. (1982, 1986a, 1986b) have demonstrated thermoadaptive effects to cold temperatures in the NRM. These effects are partly generated by forebrain structures, partly by structures of the lower brain-stem. Only few regions in the forebrain are known to be thermo-responsive to peripheral changes of temperature. Basic structures of this thermoregulatory system are preoptic area and lateral hypothalamus. However, in these regions no thermo-adaptive effects have been observed yet. Recently (1989), Grahn and Heller proposed the hypothesis that activity of thermoresponsive neurons in NRM is mainly dependent on cortical EEG states of arousal. This idea is not consistent with results of the experiments by Hinckel et al., which clearly show that thermoregulatory and thermoadaptive functions in NRM are still partly maintained after complete transsection of the descending afferents of NRM. Up to now few has been known about the cortical afferents of NRM. Thermoregulatory effects have been shown by Hinckel et al. in medial parts of NRM, but not in lateral parts of NRM. It seemed important to know if there is a neuroanatomical correlate to this observation.

The serotonergic system of the brain-stem plays an important role in awakening. In contrast to the other raphe nuclei, for NRM and neighbouring magnocellular reticular formation an additional significant role in the induction of paradoxical sleep has been suggested (Petitjean, 1981, 1985). However, the originating structure inducing paradoxical sleep is yet unknown. It seemed important to investigate NRM by means of a modern neuroanatomical technique.

Few studies have been presented about the forebrain afferents of the NRM. Veening et al. (1990) offered a study about hypothalamic and other diencephalic afferents to rostral NRM in the rat using True Blue applicated by pressure injection as a retrograde tracer. Carlton et al. (1983) published a Horse-Radish-Peroxidase (HRP)-study in the rat, using a transcannula technique for application of the tracer. Peschanski and Besson (1984) presented an investigation about diencephalic afferent connections from the meso-diencephalic junction to the raphe system including NRM in the rat using iontophoretically applicated Wheat-Germ-Agglutinin (WGA)-HRP.

Few literature has been published about the afferents of NRP in the past. Luppi et al. (1987) presented an investigation about hypothalamic afferents of NRP in the cat using unconjugated CTb applicated by pressure injections, Hosoya et al. (1985, 1987, 1989) about hypothalamic, especially dorsal hypothalamic afferents in the rat using WGA-HRP applicated by pressure injections. Up to now, no study has been published that completely described afferents throughout the brain, whereas no publication at all has been presented about afferents to the rostral part of NRP.

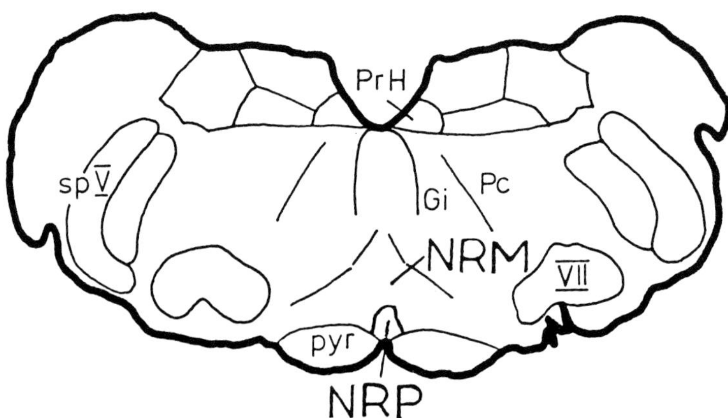

Fig. 1: Localisation of caudal NRM and rostral NRP in the upper medulla oblongata of the rat. Note that NRP is wrapped both dosally and laterally by NRM.

In our study, the afferents of NRM and rostral NRP have been investigated by the highly sensitive iontophoretic cholera toxin b (CTb) technique in rats and guinea-pigs in combination with electrophysiological recording control. Rostral NRP, which has a diameter of less than 300 µm in lateral extension, is wrapped both dorsally and laterally by caudal NRM in the rat (see Fig. 1). Up to now it has not been possible to distinguish caudal NRM and rostral NRP neuroanatomically because of this complex topology.

Iontophoretically applicated unconjugated CTb offers some advantages as a neuroanatomical tracer compared with traditional tracers (Luppi et al., 1990). Unconjugated CTb is taken up by terminal fibres with synaptical contact inside the investigated area, but not by fibres of passage without synaptic contact. In contrast to traditional tracers nearly no significant artefactual labeling of fibres of passage is obtained. Iontophoretic application of the tracer largely guarantees injections without histological damage. Traditional pressure injections cause necrotic damage of the brain-tissue. Necrotic fibres always take up tracer substance, thus also fibres of passage are labeled. Iontophoretically applicated unconjugated CTb permits to perform very small injection-sites. Restricted injection-sites with a diameter up to 250 µm can be obtained. At the same condition, it offers a substantially higher sensitivity compared with all traditional retrograde tracers. Finally, unconjugated CTb is useful as retrograde and anterograde neuronal tracer. Afferent perikarya and efferent terminal dendrites can be made visible.

MATERIALS AND METHODS

Iontophoretic injections of a 1 % solution of unconjugated cholera toxin subunit b (CTb, LIST Biological Laboratories) in 0.1 M phosphate buffer (PB) have been performed using a glass micropipette with a diameter of the tip of 4-5 µm under pentobarbital anaesthesia after purification of the CTb according to the technique of Luppi. Various parts of NRM and NRP of the rat and the guinea-pig have been injected by an alternating positive current of 0.5-1 µA over 10-15 min. Injection-sites of a diameter of 250-500 µm have been achieved. The stereotaxical localisation of the site of injection has been verified using a simple electro-physiological recording system.

After this surgery animals were allowed to survive for 6 to 19 days. They have been sacrified under deep pentobarbital anesthesia by transcardial perfusion of the animal using Hartmann's solution, followed by 500 ml of fixative cooled at 4 °C containing 4 % paraformaldehyde, 0.1 % glutaraldehyde and 0.2 % picric acid in 0.1 M PB. The brain was removed, postfixed overnight at 4 °C in 0.1 M PB containing 2 % paraformaldehyde and 0.2 % picric acid, and rinsed for 48-72 hours in 0.1 M PB containing 30 % sucrose. Coronal 20 µm sections were cut on a freezing microtome and rinsed twice in phosphate buffered saline containing 0.1 % Triton X-100 (PBS, PBS-T) and 0.1 % sodium acid where they were stored until immunohistochemical staining.

Immunohistochemical reactions of free floating sections were carried out by sequential incubation in goat antiserum to CTb (LIST Biological Laboratories) at a dilution of 1:40000 in PBS-T containing 0.1 % sodium-acide for 60 hours at 4 °C, in biotinylated swine antigoat immunoglobulin (DAKOPATTS), diluted 1:2000 in PBS-T, overnight at 4 °C and streptavidin conjugated with horse radish peroxidase, diluted 1:40000 in PBS-

T, for 90 min. at room temperature. Following each step sections were rinsed twice for 30 min. each in PBS-T. The sections were immersed in 0.02 % 3,3'-diaminobenzidine .4HCl (DAB, Sigma), 0.3 % nickel ammonium sulfate and 0.003 % hydrogen peroxide in 0.05 M Tris-HCl buffer pH 7.6, processed under microscopical control, and rinsed twice in PBS-T with 0.1 % sodium acide to stop the staining reaction. The sections were mounted on gelatine-coated glass-slides, dried, dehydrated using graded alcohols and coverslipped.

Abbreviations

AA, anterior amygdaloid nucleus; APTV, ventral part of the anterior pretectal area; BST, bed nucleus of the stria terminalis; f, fornix; FF, fields of Forel; fr, fasciculus retroflexus; Gi, gigantocellular reticular formation; III, nucleus of the oculomotorius nerve; IP, interpeduncular nucleus; LHA, lateral hypothalamic area; LHb, lateral habenular nucleus; LM, lateral nucleus of the mammillary corpora; LPO, lateral preoptic area; MCPO, magnocellular preoptic nucleus; MHb, medial habenular nucleus; ML, lateral part of the medial nucleus of the mammillary corpora; mlf, medial longitudinal fasciculus; MM, medial part of the medial nucleus of the mammillary corpora; mp, mammillary peduncle; mt, mammillothalamic tract; NRM, nucleus raphe magnus; NRP, nucleus raphe pallidus; Pc, parvocellular reticular formation; PHA, posterior hypothalamic area; Pir, piriform cortex; PrH, nucleus prepositus hypoglossi; PV, paraventricular thalamic nucleus; pyr, pyramidal tract; R, rubral nucleus; SI, substantia innominata; SNC, compact part of the substantia nigra; SNR, reticular part of the substantia nigra; spV, spinal tract of the trigeminal nerve; STh, subthalamic nucleus; SuM, supramammillary nucleus; VII, nucleus of the facialis nerve; ZI, zona incerta;

RESULTS

For the rat, injections have been performed in several parts of NRM and rostral NRP, at the caudal level into the dorsocaudal complex of NRM dorsolaterally to NRP (n= 4 cases), into the caudal B3-serotonergic system dorsally to the pyramidal tract about 350 µm laterally to the midline (lateral caudal NRM, n= 4 cases), and into the rostral NRP (n= 1 case, does not touch NRM); three injections have been performed into rostral NRM. Some further cases touched more than one of these regions. Injections of control have been performed outside of NRM and NRP. For the guinea-pig, moreover rostral and rostrolateral NRM have been injected (n=6 cases). NRM and rostral NRP are crossed by dense fibre bundles which originate from motor cortex, prefrontal cortex and extrapyramidal structures. To insure that no artefactual projections were demonstrated, for the rat an injection of control has been performed inside the medial part of the pyramidal tract, which is known to contain dense bundles of the above mentioned fibres. No retrogradely labeled cells at all were seen allover the cortex and inside extrapyramidal nuclei. This shows that iontophoretically applicated unconjugated CTb is not taken up by fibres of passage.

Efferents of NRM and Rostral NRP to the Forebrain

Injections of CTb for the rat which were placed in various parts of the preoptic area and the lateral hypothalamic area showed that NRM afferents to these structures derive only from rostral parts of the nucleus up to the level of the genu of the facialis nerve and not at all from more caudal parts. Confirmatively injections of CTb into NRM revealed NRM-efferents to hypothalamic levels after injections into rostral NRM, but not after injections into caudal NRM. Efferents from rostral, but not from caudal NRM and NRP were furthermore seen inside the paraventricular thalamic nuclei. Efferent projections of NRM to lateral preoptic area, lateral hypothalamic area and paraventricular thalamic nuclei derive only from rostral NRM and not from caudal NRM and NRP.

Forebrain Afferents to NRM and Rostral NRP

For the rostral NRM of the rat and the guinea-pig, intense afferents from medial prefrontal, lateral prefrontal, infralimbic, insular and perirhinal cortex have been observed (Fig. 2). For caudal NRM only very moderate input from these areas was found, a remarkable number of labeled cells was observed only from the medial prefrontal cortex for dorsal caudal NRM. The prefrontal cortical pattern for rostral NRP was quite different from that for rostral NRM. Compared with rostral NRM, labeling was

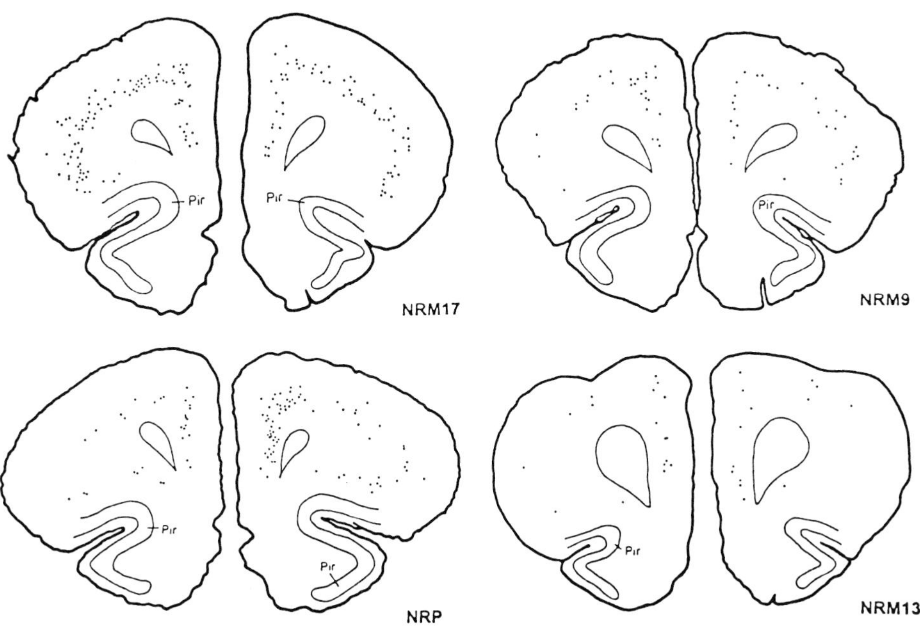

Fig. 2: Prefrontal cortical afferents for rostral NRM (NRM17), dorsal caudal NRM (NRM9), NRP and lateral caudal NRM (NRM13) in the rat. Note the specific projections of medial prefrontal, lateral prefrontal and insular cortex for the four regions.

significantly stronger in the medial prefrontal cortex and the infralimbic cortex for NRP, whereas comparatively very few labeled cells were seen in the lateral prefrontal cortex for NRP. Both rostral NRM and rostral NRP received strong afferent projections from insular and perirhinal cortex. As a control, an injection of the anterograde tracer Phaseolus-Vulgaris-Leucoagglutinin (PHAL) has been performed iontophoretically into the infralimbic cortex for the rat. Efferent fibre bundles could be seen consistently inside rostral NRM and NRP, but nearly not at all inside dorsal caudal and lateral caudal NRM.

For the rat, afferents from lateral preoptic area and bed nucleus to caudal NRM were intense from levels caudally to the anterior commissure; afferents to rostral NRM were less intense (Fig. 3). In contrast to caudal NRM, rostral NRM received additional remarkable afferents from rostral levels of lateral preoptic area and bed nucleus, from substantia innominata and magnocellular preoptic nucleus; for caudal NRM only single cells were seen. Afferents were seen in the paraventricular hypothalamic nucleus for rostral and caudal NRM. Within the lateral hypothalamic area, afferents to caudal NRM were almost restricted to its dorsal part; rostral NRM received additional afferents from the perifornical and the very lateral part. For the hypothalamic structures, results for rostral NRM of the guinea-pig and the rat were almost identical. In contrast to the medial parts of NRM, lateral caudal NRM received significantly less afferents from the preoptic area and the lateral hypothalamic area. For NRP, at preoptic levels a densely clustered string of afferent cells was found bilaterally around the third ventricle in the most rostral preoptic area, a dense afferent projection was also seen in the ventral part of the preoptic area caudally up to the level of the anterior commissure. In contrast to rostral NRM, significantly less labeled cells were found dorsally inside bed nucleus and substantia innominata. For rostral NRP a strong afferent was found dorsally to the dorsomedial hypothalamic nucleus. This projection was significantly stronger than for NRM. For the lateral hypothalamic area, afferent cells were almost organised in the perifornical and lateral part.

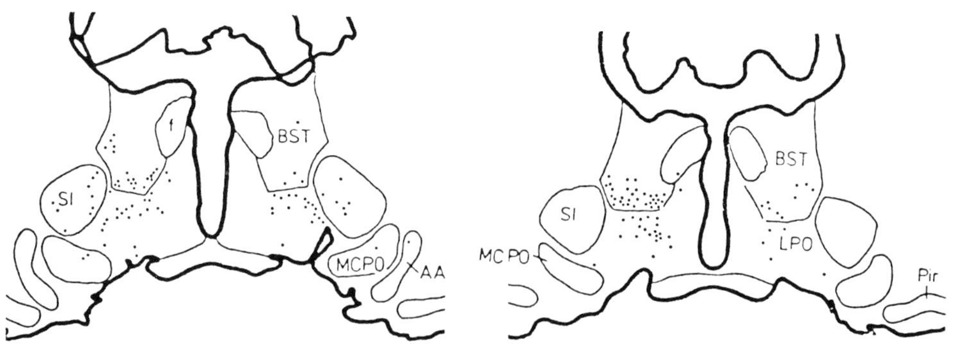

Fig. 3: Afferents at the level of the caudal preoptic area for rostral NRM (left, NRM17) and dorsal caudal NRM (right, NRM9) of the rat.

For rostral NRM, substantial afferents were observed from midbrain periaqueductal gray and mammillary bodies; for dorsal caudal NRM, significantly less labeled cells were seen (Fig. 4). For dorsal caudal NRM, remarkable afferents were observed from fields of Forel and anterior pretectal area, for rostral NRM few afferent cells were seen. For the rat, the central amygdala showed some afferent cells; for the guinea-pig, no cells were seen in this structure. A remarkable projection, which has never before been described, has been shown for the rat within the lateral habenular nuclei in our study. Nearly throughout the periaqueductal gray (also at brain-stem levels), very few afferent cells were found for NRP. A bilateral projection, which was densely clustered by afferent cells and seems to be specific for NRP because it could not be seen for NRM, was found bilaterally in the ventromedial periaqueductal gray dorsally to the nucleus of the trochlearis nerve. A further bilateral projection, which is specific for NRP because it could not be described for NRM, was found ventrolaterally to the nucleus of the oculomotorius nerve in the reticular formation.

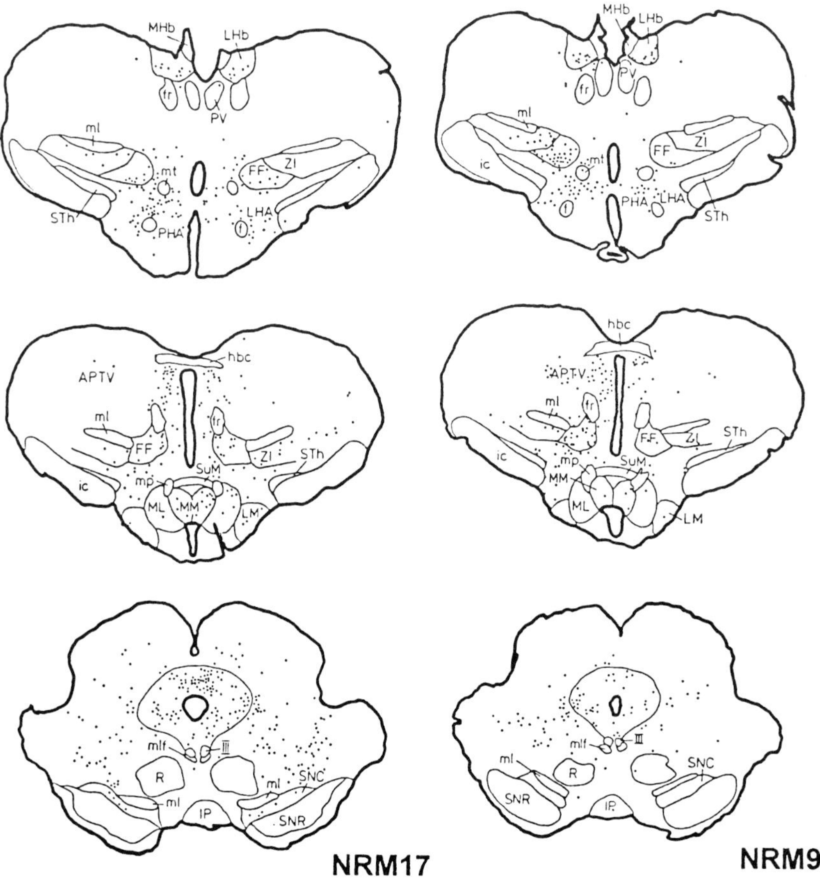

Fig. 4: Afferents at levels of the mesodiencephalic junction and the rostral midbrain for rostral NRM (NRM17) and dorsal caudal NRM (NRM9) of the rat

DISCUSSION

The cases offer restricted injection-sites which do not touch the neighbourhood structures. The sites show no labeling of crossing fibre systems, neither in rostrocaudal, nor in lateral direction. In total, the two cases show a similar number of labeled cells, the two injections seem comparative in size. In comparison to rostral NRM, dorsal caudal NRM offers a more restricted pattern of afferents, with however highly dense afferent projections from the caudal part of the lateral preoptia area and the part of lateral hypothalamic area dorsally to the fornix. In contrast to dorsal caudal NRM, rostral NRM shows a rather dispersed projection pattern from hypothalamic levels and strong afferents from certain cortical structures.

An afferent cortical projection to NRM has already been described by Carlton et al. (1983). Carlton et al. however described a projection from moreover all over the parietal cortex. This observation could not be confirmed. Afferent projections were only found in the insular and perirhinal cortex. The projections from infralimbic, prelimbic, medial precentral, insular and perirhinal cortex have never before been described. Cortical afferents for NRP have never been described.

An excellent study about hypothalamic afferents to NRM has been published by Veening et al. in 1990, who presented a case with an injection-site inside rostral NRM. Veening et al. found a dense afferent projection throughout whole rostral and caudal zona incerta. In none of our cases using this highly sensitive technique this projection could be confirmed in this way. We observed a projection with ipsilateral dominance in the medial part of the caudal zona incerta as well as in the fields of Forel in all of our caudal cases. Our observations within preoptic area and hypothalamus for rostral NRM are similar to the results presented by Veening et al. for NRM, consistently indicating that Veening et al. presented a case of rostral NRM. Our hypothalamic results for NRP are consistent with Luppi et al. (1987) and Hosoya et al. (1985, 1987, 1989).

CONCLUSIONS

Using a highly sensitive technique, it has been shown that NRM consists of at least three distinct nuclei which have different afferent and efferent projection patterns. For the first time, caudal NRM and rostral NRP could be distinguished by means of tracing experiments. In the past, no thermoregulatory effects could be shown in the lateral parts of NRM. We were able to show that lateral NRM receives significantly less afferent input from both cortical and hypothalamic structures.

In the brain, adaptive functions are usually generated by the hippocampus. Ino et al. (1989) described direct projections of non-pyramidal neurons in the hippocampus and the dentate gyrus to dorsal and central superior raphe nucleus in the cat after pressure-injections of WGA-HRP. We were interested, if there was a projection from the hippocampus or the dentate gyrus to NRM and NRP in the rat and the guinea-pig. No labeled cells could be identified in the two regions. However, multiple anatomical structures, which are known to be important autonomic relay-stations of hippocampal efferents could be described as important afferent strutures of NRM, as the prefrontal cortex, the insular cortex, the perirhinal cortex, the bed nucleus of the stria terminalis and the preoptic area, the perifornical part of the lateral hypothalamic area as well as the central nuclei of the amygdala. Electrophysiological experiments will be necessary to determine which of these structures is responsible for adaptive effects of temperature regulation.

Among the four investigated structures of the medullary raphe system, rostral NRM is unique in the fact that it sends off intense efferents to hypothalamus and preoptic area. Compared with dorsal caudal NRM and lateral caudal NRM, rostral NRM receives significantly more intense afferents from prefrontal cortex, insular cortex, perirhinal cortex, bed nucleus of the stria terminalis, preoptic area, substantia innominata, perifornical part of the lateral hypothalamic area and periaqueductal gray. These facts suggest closed loops between these specific efferents and afferents for rostral NRM which might be involved in adaptive effects.

In contrast to the other three raphe structures, rostral NRM sends off intense efferents to the paraventricular nuclei of the thalamus. The paraventricular nuclei of the thalamus are known to be implicated in sleep-wake-regulation. An important role of rostral NRM in the generation of paradoxical sleep has to be considered.

LITERATURE

Brück, K., and P. Hinckel (1980): Thermoregulatory noradrenergic and serotonergic pathways to hypothalamic units. J. Physiol. 304:193-202

Brück, K., and P. Hinckel (1982): Thermoafferent systems and their adaptive modifications. Pharmac. Ther. 17:357-381

Carlton, S. M., G. R. Leichnetz, E. G. Young, and D. J. Mayer (1983): Supramedullary afferents of the nucleus raphe magnus in the rat: A study using the transcannula HRP gel and autoradiographic techniques. J. Comp. Neurol. 214:43-58

Dickenson, A. H. (1977): Specific responses of rat raphe neurones to skin temperature. J. Physiol. (Lond.) 273:277-293

Grahn, D. A., H. C. Heller (1989): Activity of most rostral ventromedial medulla neurons reflect EEG/EMG pattern changes. Am. J. Physiol. R1496-1505

Hinckel, P., and K. Schröder-Rosenstock (1982): Central thermal adaption of lower brain stem units in the guinea-pig. Pflügers Arch. 395:344-346

Hinckel, P., W. T. Perschel, K. Pfizenmaier, and K. Brück (1986a): Neurophysiological correlates of short and long term thermal acclimation. J. Auton. Nerv. Syst. Suppl. 561-565

Hinckel, P., and W. T. Perschel (1986b): Influence of cold and warm acclimation on neuronal responses in the lower brain stem. Can. J. Physiol. Pharmacol. 65:1281-1289

Hosoya, Y. (1985): Hypothalamic projections to the ventral medulla oblongata in the rat, with special reference to the nucleus raphe pallidus: a study using autoradiographic and HRP techniques. Brain Res. 344:338-350

Hosoya, Y., R. Ito, and K. Kohno (1987): The topographical organization of neurons in the dorsal hypothalamic area that project to the spinal cord or to the nucleus raphe pallidus in the rat. Exp. Brain Res. 66:500-506

Hosoya, Y., Y. Sugiura, F.-Z. Zhang, R. Ito, and K. Kohno (1989): Direct projections from the dorsal hypothalamic area to the nucleus raphe pallidus: a study using anterograde transport with Phaseolus vulgaris leucoagglutinin in the rat. Exp. Brain Res. 75:40-46

Ino, T., K. Itoh, H. Kamiya, T. Kaneko, R. Shigemoto, I. Akiguchi, N. Mizuno (1989): Direct projections from Ammon's horn to thr rostral raphe regions in the brainstem of the cat. Brain Res. 479:157-161

Luppi, P.-H., K. Sakai, D. Salvert, P. Fort, and M. Jouvet (1987): Peptidergic hypothalamic afferents to the cat nucleus raphe pallidus as revealed by a double immunostaining technique using unconjugated cholera toxin as a retrograde tracer. Brain Res. 402:339-345

Luppi, P.-H., P. Fort, M. Jouvet (1990): Iontophoretic application of unconjugated cholera toxin B subunit (CTb) combined with immunohistochemistry of neurochemical substances: a method for transmitter identification of retrogradely labeled neurons. Brain Res. 534:209-224

Peschanski, M., and J.-M. Besson (1984): Diencephalic connections of the raphe nuclei of the rat brainstem: An anatomical study with reference to the somatosensory system. J. Comp. Neurol. 224:509-534

Petitjean, F. (1981): Insomnie et hypersomnie chez le chat: Etude des mechanismes monoaminergiques. Thèse de Médecine, Université Claude Bernard, Lyon, France

Petitjean, F., C. Buda, M. Janin, M. Sallanon, and M. Jouvet (1985): Insomnie par administration de parachlorophenylalanine: Reversibilité par injection périphérique ou centrale de 5-hydroxytryptophane et de serotonine. Sleep 8:56-67

Veening, J. G., P. C. Stroeken, P. Posthuma, and H. W. M. Steinbusch (1990): A comparison of the descending hypothalamic and some other diencephalic projections to the nucleus raphe magnus and the mesencephalic periaqueductal gray in the rat. Neurosci. Res. Comm. 7(2):123-132

Integrative and cellular aspects of autonomic functions : temperature and osmoregulation. Eds K. Pleshka, R. Gerstberger. John Libbey Eurotext, Paris © 1994, pp. 253-257.

Hypothalamic network for thermoregulation : old but still unanswered question

Kazuyuki Kanosue, Motoko Yanase-Fujiwara, Takayoshi Hosono* and Yi-Hong Zhang

*Department of Physiology, *Department of Obstetrics and Gynecology, Osaka University Medical School, Yamadaoka 2-2,, Suita, 565 Osaka, Japan*

The body temperature of homeothermic animal is controlled by multiple autonomic and behavioral responses and thermoreceptors are distributed throughout the body. This multiple input-output system is regulated by the central nervous system, on the top of which the hypothalamus including the preoptic area is located. From the 30s to the early 60s, the neuronal network for thermoregulation was investigated by stimulation and ablation of the brain, and the importance of the hypothalamus became apparent. In spite of the flourishing outcome of single unit studies in 60s and afterwards, there is almost no advance in our knowledge of thermoregulatory "network", which makes it difficult to link electrophysiological and pharmachological data directly with thermoregulatory responses observed in whole animals. We don't know yet, for example, what groups of neuron are working for control of each effector response, and where they send efferent signals. This paper will describe a part of our recent study for connecting this "missing link" in the study of thermoregulation.

1.Warm- vs. cold-sensitive neurons. It is generally considered that preoptic warm-sensitive neurons, which increase firing rates with increase in local temperature, send efferent signals for heat loss responses, while preoptic cold-sensitive neurons, which increase firing rates with the decrease in local temperature, send efferent signals for heat production (Boulant, 1980). In the preoptic area, however, warm-sensitive neurons are encountered more frequently during microelectrode exploration both _in vivo_ and _in vitro_ studies (Nakayama, 1985; Hori, 1991). Therefore, heat production might also be controlled by warm-sensitive neurons. To test this possibility, the effect on shivering activity of microinjection of excitatory amino acid into the preoptic area was studied in ketamine anesthetized rats. A canulla for drug microinjection was implanted on one side of the preoptic area and a thermode for local brain warming was implanted on the other side. Electromyogram was recorded with needle electrodes in the thigh muscles. The rat was exposed to a cold environment (15 °C) to produce shivering. Microinjection of sodium L-glutamate (0.2 mM, 0.5 µl) into the preoptic area suppressed shivering as effectively as local preoptic warming. When shivering was weak or absent in a slightly warmer environment (20-22°C), procaine hydrochloride (5 per cent, 0.5 µl) was injected into the same site of which glutamate injection had been effective for suppressing shivering. Procaine injection increased the EMG activity. The glutamate injection would likely suppress shivering by activating cell bodies near the injection site, because procaine injection into the same site had opposite effects. Therefore, the fact that both glutamate injection and local warming suppressed shivering suggests that warm-sensitive neurons in the preoptic area predominantly work for control of shivering (Fig. 1).

2. Intrahypothalamic network for shivering. Warming one side of the preoptic area suppressed shivering on both sides of the body (Kanosue et al, 1991). To examine how the control of shivering is shared between the brain's right and left sides, ketamine anesthetized rats had been implanted thermodes on both sides of the preoptic area and received unilateral transection of the hypothalamus just caudal to the anterior hypothalamus, which severed fibers running rosto-caudal direction. Unilateral preoptic warming on the intact side produced bilateral suppression of cold-induced shivering without predominance of either side of the body but warming on the transected side had no effect on shivering. Therefore, no information seems to be exchanged between the left and right preoptic area for control of shivering and that efferent signals from the preoptic area cross the midline somewhere below the hypothalamus to equally innervate both sides of the body (Fig. 1). When a partial transection covered the lateral part of the hypothalamus (the medial forebrain bundle), the transection had the same effect as did unilateral transections of the whole hypothalamus. Efferent signals for shivering, thus, descend through the medial forebrain bundle (Fig. 1). When a transection covered the dorsomedial region of the posterior hypothalamus bilaterally leaving the medial forebrain bundle intact, rats did not shiver even when rectal temperature fell to 32 °C. Thus, the dorsomedial posterior hypothalamus seems essential for producing shivering in rats, as had been shown in cats (Hemingway, 1963).
To test if the effect of thermal stimulation of the preoptic area is produced by way of the posterior hypothalamus, the posterior hypothalamus was completely isolated from the rostral hypothalamus by microknife cuts and, additionally, the medial forebrain bundle on the left side was transected. Warming the left preoptic area had no effect on shivering, because the efferent fibers on that side had been transected. In contrast, warming the right side suppressed shivering. The preoptic area, thus, may send inhibitory signals directly to the lower network for shivering control, bypassing the posterior hypothalamus, which may generates tonic excitatory signals for shivering (Fig. 1).

3. Intrahypothalamic network for vasomotor activity. Skin vasodilation occurs bilaterally during unilateral thermal or electrical stimulation of the preoptic area (Kanosue et al., 1991). In rats which had received similar unilateral transection of the hypothalamus as in the shivering experiment, unilateral preoptic warming or electrical stimulation not only on the intact side but also on the transected side produced bilateral vasodilation on the hind paw. Regardless of which side of the preoptic area was warmed, the threshold stimulus temperature eliciting vasodilation was always lower for the paw on the intact side. Therefore, information

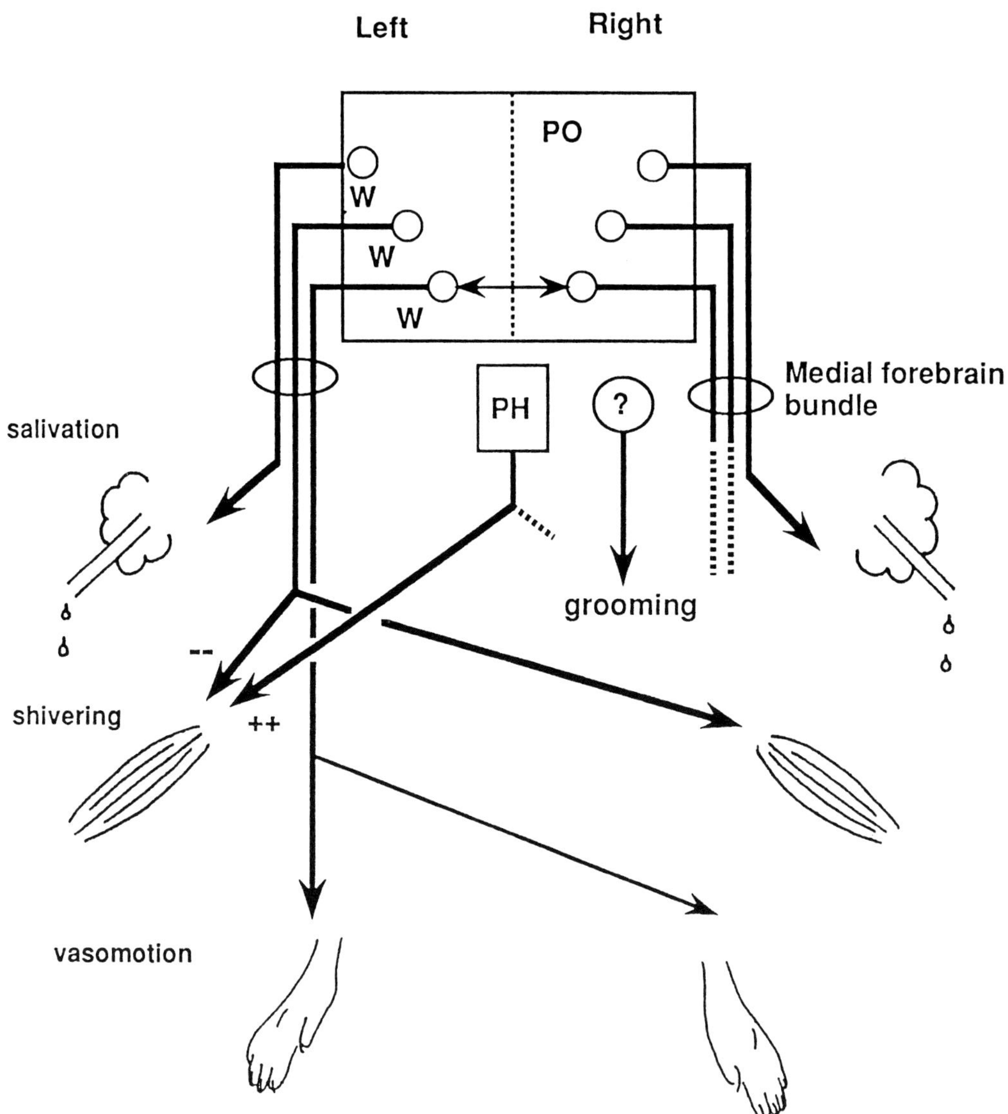

Fig. 1. Schematic diagram of thermoregulatory network. Salivary glands are innervated only by ipsilateral preoptic area (PO), skin blood vessels are predominantly innervated by ipsilateral PO, and shivering is equally innervated by both sides of the PO. Different sets of warm sensitive neruon (W) send efferent signals for these three autonomic responses. Connection between the left and right PO exists only for vasomotor control. Posterior hypothalamus (PH) is essential for maintaining shivering activity (++) and it works independently with the PO, which sends inhibitory signals (--). Thermosensitive site for grooming behavior locates outside the PO and it also works independently with the PO network for salivary secretion.

controlling thermoregulatory vasomotion--unlike that controlling shivering--crosses the midline within the preoptic area. Vasomotor efferent signals innervate both sides of the skin blood vessels and this innervation is stronger on the ipsilateral side (Fig. 1). Unilateral transection of a part of the hypothalamus that included the medial forebrain bundle had the same effect as did unilateral transection of the whole hypothalamus. Efferent fibers for vasomotor control would also decend through the medial forebrain bundle (Fig. 1).

4. Independence of effector responses. The hypothalamus is not a single center for thermoregulation, but locates at the top of a reflexive thermoregulatory hierarchy. In the hypothalamus different neuronal modules seem to exist for different effector responses (Roberts & Mooney, 1974; Satinoff, 1978). Then, are there neuronal links between these modules, which "coordinate" activities of the modules for appropriate regulation of body temperature?

Connections between the left and right preoptic area have some effect on vasomotor control, while shivering is not affected by the connection between the left and right preoptic area (Fig. 1). This means that the preoptic network controlling shivering and that controlling thermoregulatory vasomotion function independently even though they are located close to or intermingled each other. Likewise, we have recently shown that salivary secretion occurred only from the ipsilateral side during unilateral elecrical stimulation of the preoptic area (Kanosue et al., 1990), suggesting that there is no connection between the right and left networks controlling salivary secretion. Even though both salivary and vasodilatatory systems work for heat dissipation, they also seem to be controlled independent of each other (Fig. 1).

As noteed above,the preoptic area and posterior hypothalamus respectively send inhibitory and excitatory signals for controlling shivering. Neuronal connection between these two structures, if any, does not seem essential.

Rats in a hot environment spread saliva on their body surface (grooming) for evaporative heat loss. To make this response effective, salivary secretion and grooming behavior should be synchronized. However, thermosensitive sites for these two structures are not identical: grooming is induced by warming the posterior hypothalamus (Tanaka et al., 1986), while salivary secretion is elicited by warming the preoptic area (Kanosue et al., 1990). To examine the relationship between these two responses, salivary secretion from the submandibular gland of freely moving rat was recorded, together with simultaneous observation of grooming behavior. When grooming appeared frequently and copious salivary secretion continued in a hot (40 °C) environment, there was no correlation between the duration of grooming and the rate of salivary flow (Yanase et al., 1991). Thus, thermally induced salivary secretion and grooming behavior appear to be controlled by independent mechanisms (Fig. 1).

All these results suggest that even though differnt effector responses are functionally coordinated for appropriate thermoregulation, neuronal links connecting modules controlling each response do not exist in the hypothalamus. The hypothalamus would be merely an assembly of networks for different effectors, and these networks would work independently (Roberts et al., 1974; Satinoff, 1978; Yanase et al., 1991).

REFERENCES

Boulant, J. A. (1980). Hypothalamic control of thermoregulation. Neurophysiological basis. In Handbook of the hypothalamus Vol.3 Part A, eds P. J. Morgane and J. Panksepp. Dekker, New york, pp.1-82.

Hemingway, A. (1963). Shivering. Physiological Review 43, 397-422.

Hori, T. (1991). An update on thermosensitive neurons in the brain: from cellular biology to thermal and non-thermal homeostatic functions. Japanese Journal of Physiology **41**, 1-22.

Kanosue, K., Nakayama, T., Tanaka, H., Yanase, M., & Yasuda, H. (1990). Modes of action of local hypothalamic and skin thermal stimulations on salivary secretion in rats. Journal of Physiology 424, 459-471.

Kanosue, K., Niwa, K., Andrew, P. D., Yasuda, H., Yanase, M., Tanaka, H., & Matsumura, K. (1991). Lateral distribution of hypothalamic signals controlling thermoregulatory vasomotor activity and shivering in rats. American Journal of Physiology 260, R485-R493.

Nakayama, T. (1985). Thermosensitive neurons in the brain. Japanese Journal of Physiology **35**, 375-389.

Roberts, W. W., & Mooney, R. D. (1974). Brain areas controlling thermoregulatory grooming, prone extension, locomotion, and tail vasodilation in rats. Journal of Comparative and Physiological Psychology 86, 470-480.

Satinoff, E. (1978). Neural organization and evolution of thermal regulation in mammals. Science 201, 16-22.

Tanaka, H., Kanosue, K., Nakayama, T., & Shen, Z. (1986). Grooming, body extension, and tail vasomotor responses induced by hypothalamic warming at different ambient temperatures in rats. Physiology and Behavior 38, 145-151.

Yanase, M., Kanosue, K., Yasuda, H., & Tanaka, H. (1991). Salivary secretion and grooming behaviour during heat exposure in freely moving rats. Journal of Physiology 432, 585-592.

Integrative and cellular aspects of autonomic functions : temperature and osmoregulation. Eds K. Pleschka, R. Gerstberger. John Libbey Eurotext, Paris © 1994, pp. 259-268.

Involvement of limbic-neuroendocrine interactions in control of hibernation

F. Nürnberger, T.F. Lee*, J.F. Staiger and L.C.H. Wang*

*Department of Anatomy and Cytobiology, Justus-Liebig University of Giessen, Aulweg 123, D-35392 Giessen, Germany and *Department of Zoology, University of Alberta, Edmonton, Alberta, T6G 2E9, Canada*

In hibernating mammals, dramatic changes of all physiological processes can be observed between the euthermic, non-hibernating condition and the torpid hibernation state. During the euthermic state, the physiology of hibernators does not differ from that of other homeothermic mammals unable to hibernate, however, during the torpor-like hibernation state, most of these processes are restrained very conspicuously. Apart from the inactivity of somatic systems, the metabolic rate is drastically diminished, the body temperature drops below 10°C, the heart rate is below 10 systoles per minute, and the respiratory rate is suppressed to less than 1 inspiration cycle per minute (Eisentraut, 1956; Lyman et al., 1982). This decrease in activity of vegetative processes allows hibernators to save up to 90% of the energy, which is normally needed by euthermic animals during the harsh ecological conditions of the palearctic winter (Wang, 1989).

Although hibernating animals are somatically inactive, mammalian hibernation is a precisely controlled physiological condition. By autonomous mechanisms, the body temperature is physiologically regulated above the ambient temperature (Mislin & Vischer, 1942), and, in contrast to hibernators among the lower vertebrates and invertebrates, hibernating mammals arouse from the torpid state only on the basis of internal heat production (Eisentraut, 1934; Tähti & Soivio 1977) Hibernation, thus, can be regarded as a model providing a natural experiment for the activation vs. inactivation of specific autonomous processes.

CENTRAL NERVOUS CONTROL OF HIBERNATION

The autonomous processes accountable for the physiological features related with hibernation are controlled by several nuclei within the diencaphalon and the brainstem. A very important role in this control process most likely plays the neuroendocrine system of the hypothalamus (Azzali, 1954; Suomalainen & Nyholm, 1956; Polenov & Yurisova, 1975). Especially in the hypothalamo-hypophysial systems (e.g., the osmoregulatory, somatotropic, thyreotropic, and adrenocorticotropic system), several hibernation-related

Supported by the Deutsche Forschungsgemeinschaft (Nu 36/2-3)

physiological alterations, which may be causative for the phenomena allowing distinct mammals to hibernate, are well documented (Yurisova & Polenov, 1979; Stanton et al., 1982; Nürnberger, 1983; Nürnberger et al., 1985; 1986; Nürnberger & Heinrichs, 1988; Nürnberger, in press).

In contrast to the rather extensively described hypothalamic output systems to both the pituitary and the autonomous centers of the brainstem, our knowledge on the supervisory mechanisms of hypothalamic input systems is still limited (cf. Swanson, 1987). Accordingly, the elucidation of these mechanisms resembles an attractive field of research; the natural experiment provided by mammalian hibernation may help to solve several of the problems, how the neuroendocrine system is controlled by other brain regions of higher intergative level.

The limbic system represents one important neuronal apparatus, which may control the neuroendocrine system within the hypothalamus. Among the numerous limbic centers, the septum and the amygdala are the main entities hosting the source neurons of hypothalamopetally oriented limbic fiber tracts and, on the other hand, receiving nerve terminals originating from different hypothalamic nuclei (Andy & Stephan, 1976; Swanson & Cowan, 1979). In the present communocation, we mainly focus on the septo-hypothalamic axis.

CONNECTIVITY BETWEEN LATERAL SEPTAL NUCLEUS AND HYPOTHALAMUS

Because of several discrepancies in the connectivity pattern between hypothalamus and septum reported so far (Sofroniew & Weindl, 1978; DeVries & Buijs, 1983; Caffe & van Leeuwen, 1987), the efferent and afferent connections of the septum, particularly of the lateral septum, have been reinvestigated in an hibernator, the Richardson's ground squirrel (Spermophilus richardsonii). By the use of the retrogradely transported tracer horseradish peroxidase and the anterogradely transported tracer Paseolus vulgaris-leucoagglutinin in combination with immunostainings for hypothalamic neuropeptides (vasopressin, met-enkephalin, corticoliberin, somatostatin), perikarya within the paraventricular and periventricular hypothalamic nuclei could be identified as origin for vasopressin-, somatostatin-, corticoliberin-, and met-enkephalin-immunoreactive axon endings within the lateral septum (cf. also Sakanaka et al., 1982; Ishikawa et al., 1986; Sakanaka et al., 1988; Staiger & Nürnberger, 1989, 1991a, 1991b). Most interestingly, these peptidergic fiber systems formed terminal fields, which subdevided the lateral septal nucleus in particular portions. In the ventral portion, vasopressin-immunoreactive axon terminals were found, whereas somatostatin was immunostained in fibers located further dorsally. The met-enkephalin-immunoreactive terminals were observed in the intermediate lateral septum and that immunoreactive for corticoliberin were identified in the dorsolateral septum.

Cell bodies located in the lateral septum, in return, gave rise to fiber tracts terminating predominantly in the anterior and lateral hypothalamic areae. This feedback connection only rarely reached the peptidergic neurons of defined hypothalamic nuclei directly, interneurons in the hypothalamic areae served as relay stations for

the integration of the input from the lateral septal nucleus (Staiger & Nürnberger, 1991b).
In the foreward direction, the lateral septum is connected with the nucl. diagonalis (Broca), which, in continuation, sends efferents to the hippocampus and subiculum. On this level, an extensive neuronal communication with other centers of the inner circle of the limbic system can be suggested. Hippocampofugal projections course, among other targets, back to the lateral septum closing a triangular septo-hippocampal nerve circuit (Staiger et al. 1993).

The knowledge of these anatomical details is of crucial importance for the understanding of the septal functions, viz., the control of a variety of physiological and behavioral processes related to cognitive functions or autonomic regulation, e.g., water and food intake, thermoregulation including fever, osmoregulation etc (Brady & Nauta, 1953; Siegel & Skog, 1970; Fried, 1972; Gordon & Johnson, 1981; Desmontes-Mainard et al., 1986; King & Nance, 1986; Thomas, 1988). The latter autonomic functions controlled by the lateral septum, especially the thermoregulatory effects, rised the question, whether this limbic entity may also be involved in particular control mechanisms of hibernation.

HIBERNATION-RELATED CHANGES IN THE REACTIVITY OF PEPTIDERGIC INPUTS OF THE LATERAL SEPTUM

By application of immunocytochemical techniques, which have been combined with radioimmunological assays, the reactivity pattern of different peptide and amine systems during the various states of hibernation, hypothermia, and euthermia has been investigated. Within the septal terminal regions, we observed an increased reactivity of the vasopressin-immunoreactive fiber terminals in ground squirrels (Spermophilus richardsonii and S. columbianus, Fig. 1a-b.), hedgehogs (Erinaceus europaeus), and dormice (Glis glis, cf. Nürnberger et al., 1982; Nürnberger, 1983). In contrast, in a short-term hibernator, the golden hamster, the septal pool of immunoreactive Vasopressin decreased during hibernation (Nürnberger et al., 1982). The latter finding is in accord with observations of Hermes et al. (1989) in the European hamster (Cricetus cricetus). The staining reactivity of experimentally hypothermic animals did not significantly differ from that of euthermic controls.

Only minor changes in the immunoreactivity of septal peptide systems in relation to hibernation and euthermia could be observed in oxytocin-, somatostatin-, and corticoliberin-containing terminals (Nürnberger et al., 1985; 1986; Nürnberger & Heinrichs, 1988). Similarly, the immunoreactivity pattern to dopamine-beta-hydroxylase representing the guiding enzyme for the synthesis of norepinephrine did not vary significantly, however, a conspicuous increase in immunoreactivity was observed in nerve terminals stained for serotonin.

The most dramatic changes in the reactivity, however, were found in the met-enkephalin-immunoreactive fiber terminals within the intermediate subnucleus of the lateral septum. During hibernation, the numerous fiber profiles were found to be strongly immunoreactive, whereas the euthermic controls displayed only a small number of less intensively immunostained axons (Fig. 1c-d.;

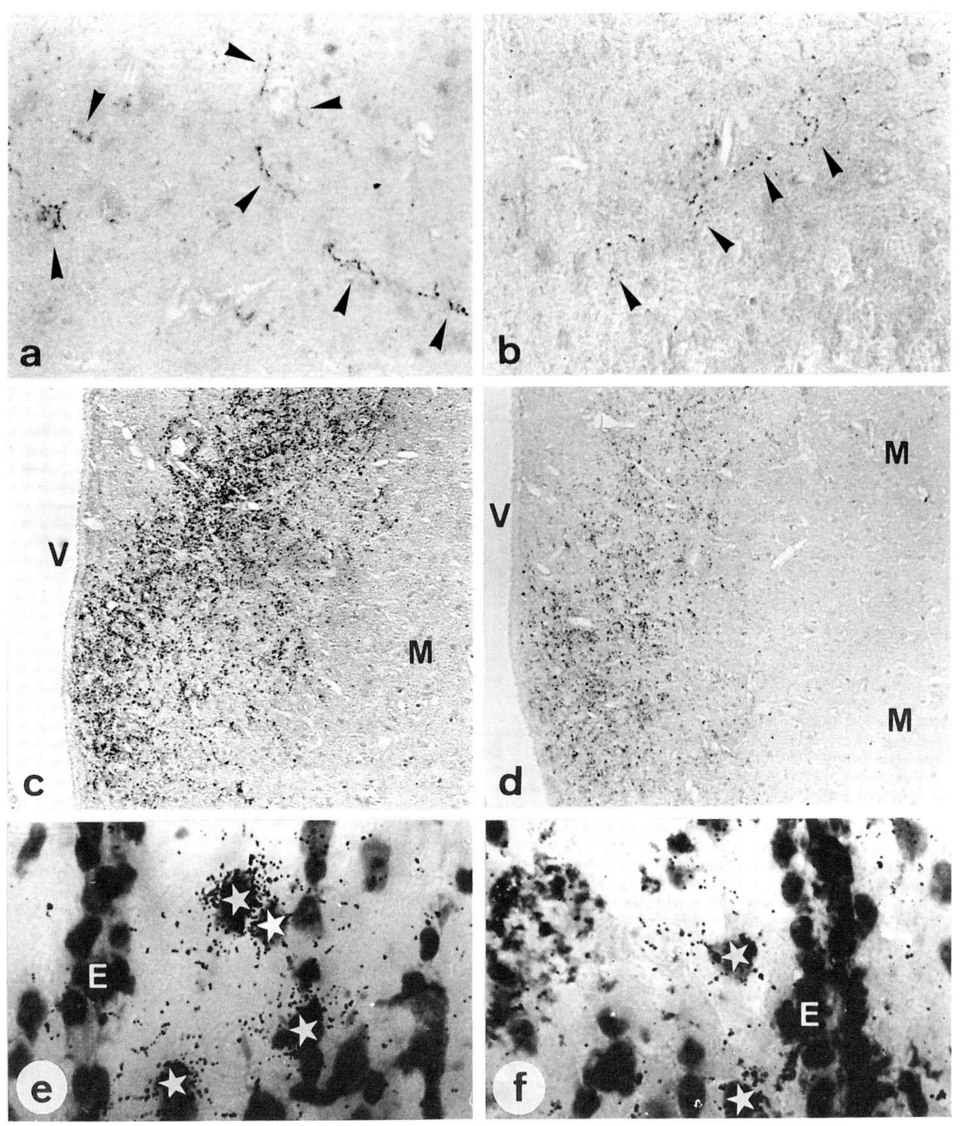

Fig. 1. Vasopressin-immunoreactive nerve fibers (arrowheads) in the ventral portion of the nucl. septi lateralis of hibernating (**a**) and euthermic Spermophilus richardsonii (**b**); x350. Methionine-enkephalin immunoreactive fiber terminals in the intermediate portion of the nucl. septi lateralis of hibernating (**c**) and euthermic Spermophilus columbianus (**d**). V lateral ventricle; M medial septum; x140. Autoradiographic demonstration of hybridized met-enkephalin oligoprobe on neurons (asterisks) in the periventricular nucleus of hibernating (**e**) und euthermic S. columbianus (**f**). E ependyma of the third ventricle; x600

Nürnberger et al., 1991). This prominent change could also be verified by radioimmunoassays performed in (unfortunately) fairly large tissue samples of different parts of the brain by Kramerova et al. (1983) in Spermophilus suslicus.

Since both immunocytochemistry and radioimmunoassays suffer from the problem that no clearcut conclusion can be drawn on the synthesis, storage or release of the peptide investigated, we studied the gene expression and, accordingly, the potential synthesis of met-enkephalin in the brain of hibernating and euthermic Columbian ground squirrels (Spermophilus columbianus) by means of in-situ hybridization. Special emphasis was put on the enkephalinergic perikarya in the periventricular and paraventricular hypothalamic nuclei, which were identified to be the source neurons of the enkephalinergic projections to the intermediate lateral septal nucleus. By analyzing the silver grains in the autoradiograms, a significantly higher level of gene expression for enkephalin could be observed within the perikarya of the periventricular hypothalamus of hibernating vs. euthermic individuals (Fig. 1e-f.), whereas the perikarya in the medial septum and the striatum did not show an increase in hybridization product. This result speaks in favor of an increased rate of synthesis of met-enkephalin in neurons of the hypothalamo-septal pathway, which most likely implies an increase of release of enkephalin from synapses in the lateral septum. The enkephalinergic elements of the extrapyramidal system in the putamen and nucl. caudatus, however, and in other limbic structures do not show hibernation related alterations.

The gene expression for somatostatin, which was investigated by non-radioactive in-situ hybridization, did not significantly change between euthermia and hibernation.

STEREOTAXIC APPLICATION OF VASOPRESSIN AND ENKEPHALIN IN THE LATERAL SEPTAL NUCLEI

In order to eludidate the functional role of the increased immunoreactivity of the septal vasopressin and met-enkephalin system during hibernation, stereotaxicylly centered infusion and injection experiments applying vasopressin or met-enkephalin exactly into the lateral septal portion hosting the fiber terminals of these peptide systems have been performed in ground squirrels: Vasopressin was infused bilaterally into the lateral septal nuclei of Spermophilus richardsonii by the use of osmotic mini pumps (Alzet, Savo, Kissleg, FRG; 500ng/h in 0.5µl artificial CSF). After the onset of the infusion, the colon temperature of euthermic individuals dropped by .1°C for several days before reaching the average value again (Fig. 2a). It was not investigated, whether this recovery of the colon temperature was due to adaptation processes or destruction of vasoprssin within the reservoir of the pump.

Met-enkephalinamide was injected bilaterally via chronically implanted guide cannulas into both nuclei septi laterales of Spermophilus columbianus. If the cannula was placed precisely in the intermediate portion of the lateral septal nucleus, a very drastic hypothermic effect could be observed: The injection of 5µg met-enkephalinamide in each injection site caused a decrease of the brain temperature of more than 3°C. In general, the strong

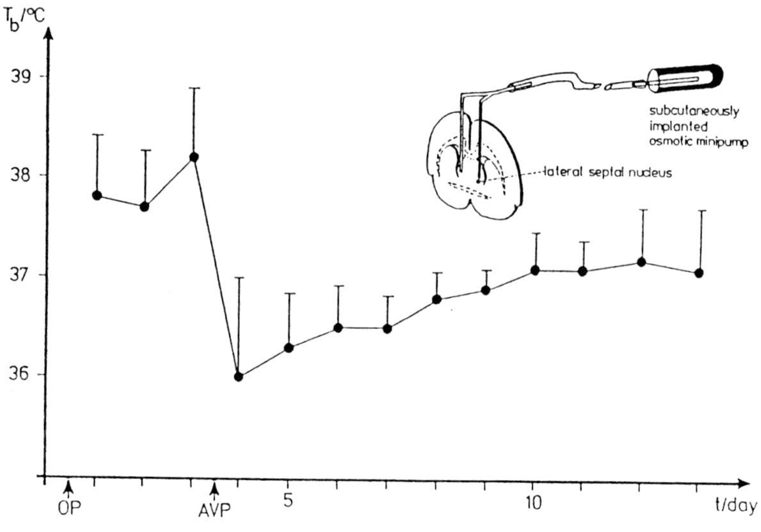

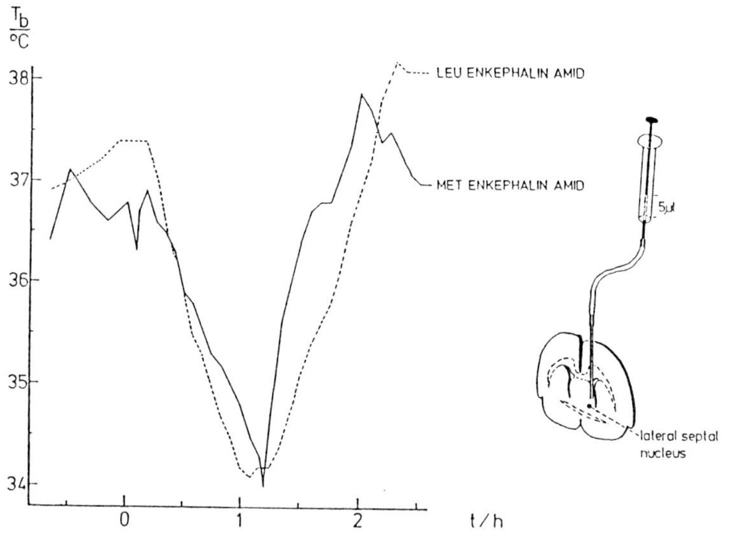

Fig. 2. Effect of infusion of vasopressin (500ng/h in 0.5μl artificial CSF) into the lateral septum of euthermic Spermophilus richardsonii (**a**). T_b colon temperature; OP time of operation; AVP arginine-vasopressin, symbolizes the onset of infusion.
Effect of bilateral injection of met-enkephalinamide into the lateral septal nuclei of euthermic Spermophilus columbianus (**b**). 0 represents the time of injection; T_b brain temperature.

hypothermic reaction was obtained after the application of slightly higher doses (about 20-40μg) probably caused by centering the cannulas not exactly in the terminal field of the met-enkephalin-immunoreactive fibers (Lee et al., 1989). To reach a similar effect in rats, an injection of more than 200μg is needed. The hypothermic reaction was accompanied with a complete lack in motility; the animals rested and dropped their muscle tone for about 2h reaching the minimum after about 1h.

In ground squirrels artificially aroused from hibernation and injected right after reaching the euthermic state, the response to intraseptal injections of met-enkephalinamide was significantly less pronounced or even caused an increase of the brain temperature (Lee et al., 1989). A similar state dependent effect of peptides was described by Belousov & Belousova (1992). These authors found thyreoliberin to be a stimulatory peptide increasing the frequency of the spontaneous activity in slice preparations of the medial septum-diagonal band complex. This increase, however, was significantly more pronounced in slice preparations of hibernating ground squirrels.

CONCLUSIONS

The data described in this communication indicate that the neuroendocrine system of the hypothalamus is linked to the lateral septal nucleus of the limbic system (cf. Herman et al., 1989) and that both abundantly connected systems have a pronounced effect on the control of physiological processes causative for hibernation. Both encepalic centers are reciprocally connected with fiber tracts containing various transmitter and peptide systems (cf. Staiger & Nürnberger, 1989; 1991a; 1991b). Among these transmitter and peptide systems, especially the met-enkephalin and the vasopressin system exert major influences on thermoregulatory processes. In euthermic hibernators, these peptides mainly cause hypothermic reactionsnecessary for the onset of hibernation..

The hypothermic effects integrated by the lateral septal nucleus may be mediated either via the septal efferents to the thermointegrative nuclei of the preoptic hypothalamic region (Wünnenberg et al., 1976) or via limbic pathways reaching motoric systems and the brainstem (Raisman, 1966; Swanson & Cowan, 1979; Marcinkiewicz et al., 1989). The latter axis is of major importance, since hibernation is initiated by passing different stages of sleep including the highly synchronized slow wave sleep (Lyman, 1982).

Mammalian hibernation is - according to our present knowledge - based on the alteration of several transmitter and peptide systems. The balanced availability of all these factors seems to be necessary for entering this natural state of precisely regulated torpor, however, there may be also some physiological specializations within the central nervous system uniquely expressed in mammalian hibernators (cf. Pakhotin et al., 1993).

REFERENCES

Andy, O.J. & Stephan, H. (1976): Septum development in primates. In The septal nuclei, ed.J.de France, pp. 3-36. New Jork: Plenum Press.

Azzali, G. (1954): Ricerche sulla neurosecrezione negli animali ibernanti. Publ. Staz. zool. Napoli 24 Suppl, 32-33.

Belusov, A.B. & Belusova, J.V. (1992): State-dependent and state-independent effects of thyrotropin-releasing hormone on medial septum neuronal activity in the brain slice of waking and hibernating ground squirrels. Neuroscience 50, 857-866.

Brady, J.V. & Nauta, W.J.H. (1953): Subcortical mechanisms in emotional behavior: affective changes following septal forebrain lesion in the albino rat. J. Comp. Physiol. Psychol. 46, 393-346.

Caffe, A.R., van Leeuwen, F.W. & Luiten, P.G.M. (1987): Vasopressin cells in the medial amygdalaof the rat project to the lateral septum and ventral hippocampus. J. Comp. Neurol. 261, 237-252.

Desmontes-Mainard, J., Chauveau, J., Rodriguez, F., Vincent, J.D. & Poulain, D.A. (1986): Septal release of vasopressin in response to osmotic, hypovolaemic and electrical stimulation in rats. Brain Res. 381, 314-321.

DeVries, G.J. & Buijs, R.M. (1983): The origin of the vasopressinergic and oxytocinergic innervation of the rat brain with special reference to the lateral septum. Brain Res. 273, 307-317.

Eisentraut, M. (1934): Der Winterschlaf der Fledermäuse mit besonderer Berücksichtigung der Wärmeregulation. Z. Morphol. Ökol. 29, 231-267.

Eisentraut, M. (1956): Der Winterschlaf mit seinen ökologischen und physiologischen Begleiterscheinungen. Jena: Fischer.

Gordon, F.J. & Johnson, A.K. (1981): Electrical stimulation of the septal area in the rat: prolonged suppression of water intake and correlation with self-stimulation. Brain Res. 206, 421-430.

Herman, J.P., Schaefer, M.K., Young, E.A., Thompson, R., Douglas, J., Akil, H. & Watson, S.J. (1989): Evidence for hippocampal regulation of neuroendocrine neurons of the hypothalamo-pituitary-adrenocortical axis. J. Neurosci. 9, 3072-3082.

Hermes, M.L.H.J., Buijs, R.M., Masson-Pévet, M., van der Woude, T.P., Pévet, P., Brenkle, R. & Kirsch, R (1989): Central vasopressin infusion prevents hibernation in the European hamster (Cricetus cricetus). Proc. Natl. Acad. Sci. USA 86, 6408-6411.

Ishikawa, K., Taniguchi, Y, Kurosumi, K. & Suzuki M (1986): Origin of septal thyrotropin-releasing hormone in the rat. Neuroendocrinology 44, 54-58.

Kramerova, L.I., Kolaeva, S.H., Yukhananov, R.Y. & Rozhanets V.V. (1983): Content of DSIP, enkephalins, and ACTH in some tissues of active and hibernating ground squirrels (Citellus suslikus). Comp. Biochem. Physiol. 74 C, 31-33.

King, T.R. & Nance, P.M. (1986): Neuroestrogenic control of feeding behavior and body weight in rats with kainic acid lesions of the lateral septal area. Physiol. Behav. 37, 475-481.

Lee, T.F., Nürnberger, F., Jourdan, M.L. & Wang, L.C.H. (1989): Possible involvement of septum in seasonal changes in thermoregulatory responses to met-enkephalinamide in ground squirrels. In Thermoregulation: research and clinical application, eds. P. Lomax & E. Schönbaum, pp. 200-203. Basel: Karger.

Lyman, C.P. (1982): Entering hibernation. In Hibernation and torpor in mammals and birds, eds. C.P. Lyman, J.S. Willis, A. Malan & L.C.H. Wang, pp. 37-53. New York: Academic Press.
Lyman, C.P., Willis, J.S., Malan, A. & Wang, L.C.H. (1982): Hibernation in mammals and birds. New York: Academic Press.
Marcinkiewicz, M., Morcos, R. & Chretien, M. (1989): CNS connections with the median rahe nucleus: retrograde tracing with WGA-apo-HRP-gold complex in the rat. J. Comp. Neurol. 289, 11-35.
Mislin, H. & Vischer, L. (1942): Zur Biologie der Chiroptera. II Die Temperaturregulation der überwinternden Nyctalus noctula Schreb. Verh. Schweiz. naturkdl. Ges., Sitten, 131-133.
Nürnberger, F. (1983): Der Hypothalamus des Igels (Erinaceus europaeus L.) unter besonderer Berücksichtigung des Winterschlafes. Marburg: Thesis, Philipps University.
Nürnberger, F. (in press): The neuroendocrine system in hibernating mammals. Cell Tissue Res.
Nürnberger, F. & Heinrichs, M. (1988): GRH-immunoreactive Systeme im Hypothalamus von Winterschläfern. Anat. Anz. 168, 80.
Nürnberger, F., Blähser, S. & Merker, G. (1982): Reactivity pattern of vasopressin in long-term hibernators. Pflügers Arch. Suppl 392, R 31.
Nürnberger, F., Rorstad, O.P. & Lederis, K. (1985): The hypothalamo-neurohypophysial system and hibernation in the ground squirrel, Spermophilus richardsonii. In Neurosecretion and the biology of neuropeptides, eds. H. Kobayashi, H.A. Bern & A. Urano, pp. 518-520. Tokyo: Japan Sci. Soc. Press/Berlin: Springer.
Nürnberger, F., Lederis, K. & Rorstad, O.P. (1986): Effects of hibernation on somatostatin-like immunoreactivity in the brain of the ground squirrel (Spermophilus richardsonii) and European hedgehog (Erinaceus europaeus). Cell Tissue Res. 243, 263-271.
Nürnberger, F., Lee, T.F., Jourdan, M.L. & Wang, L.C.H. (1991): Seasonal changes in methionine-enkephalin immunoreactivity in the brain of a hibernator, Spermophilus columbianus. Brain Res. 547, 115-121.
Pakhotin, P.I., Pakhotina, I.D. & Belousov, A.B. (1993): The study of brain slices from hibernating mammals in vitro and some approaches to the analysis of hibernation problems in vivo. Prog. Neurobiol. 40, 123-161.
Polenov, A.L. & Yurisova, M.N. (1975): The hypothalamo-hypophysial system in the ground squirrels, Citellus erythrogenys Brandt and Citellus undulatus Pallas. Z. mikrosk.-anat. Forsch. 14, 773-777.
Raisman, G. (1966): The connexions of the septum. Brain 89, 317-348.
Sakanaka, M., Senba, E., Shiosaka, S., Takatsuk, K., Inagaki, S., Takagi, H., Kawai, Y., Hara, Y. & Tohyama, M. (1982): Evidence for the existence of an enkephalin-containing pathway from the area just ventrolateral to the anterior hypothalamic nucleus to the lateral septal area of the rat. Brain Res. 230, 240-244.
Sakanaka, M., Magari, S., Shibasaki, T. & Lederis, K. (1988): Corticotropin releasing factor-containing afferents to the lateral septum of the rat brain. J. Comp. Neurol. 270, 405-415.
Siegel, A. & Skog, D. (1970): Effects of electrical stimulation of the septum upon attack behavior elicited in the hypothalamus of the cat. Brain Res. 23, 371-380.
Sofroniew, M.V. & Weindl, A. (1978): Projections from the parvicellular vasopressin- and neurophysin-containing neurons of the suprachiasmatic nucleus.Amer. J. Anat. 153, 331-340.
Staiger, J.F. & Nürnberger, F. (1989): Pattern of afferents to the lateral septum in the guinea pig. Cell tissue Res. 257, 471-490.

Staiger, J.F. & Nürnberger, F. (1991a): The efferent connections of the lateral septal nucleus in the guinea pig: intrinsic connectivity of the septum and projections to other telencephalic areas. Cell Tissue Res. 264, 415-426.
Staiger, J.F. & Nürnberger, F. (1991b): The efferent connectivity of the lateral septal nucleus in the guinea pig: projections to the diencephalon and the brainstem. Cell Tissue Res. 264, 391-413.
Staiger, J.F., Giesselmann, C. & Nürnberger, F. (1993): Is the vasopressinergic tract from the paraventricular hypothalamic nucleus to the lateral septal nucleus a missing link in septal memory functions? Ann. Anat. 175 Suppl. (Verh. Anat. Ges. 88), 218-219.
Stanton, T.L., Winokur, A. & Beckman, A.L. (1982): Seasonal variation in thyreotropin-releasing hormone (TRH) content of different brain regions and the pineal in the mammalian hibernator Citellus citellus. Regulat. peptides 3, 135-144.
Suomalainen, P. & Nyholm, P. (1956): Neurosecretion in the hibernating hedgehog. Bertil Hanström Zool. Papers, 269-277.
Swanson, L.W. (1987): The hypothalamus. In Handbook of chemical neuroanatomy, vol 5, eds. A. Björklund, T. Hökfelt & L.W. Swanson, pp. 125-277. Amsterdam: Elsevier.
Swanson, L.W. & Cowan, W.M. (1979): The connections of the septal region. J. Comp. Neurol. 186, 621-656.
Thäti, H. & Soivio, A. (1977): Respiratory and circulatory differences between induced and spontaneous arousals in hibernating hedgehogs 8Erinaceus europaeus L.). Ann. Zool. Fenn. 14, 197-202.
Thomas, E. (1988): Forebrain mechanisms in the relief of fear: the role of the lateral septum. Psychobiol. 16, 36-44.
Wang, L.C.H. (1989): Ecological, physiological and biochemical aspects of torpor in mammals and birds. in Animal adaptation to cold, ed. L.C.H. Wang, pp. 361-401.
Wünnenberg, W., Merker, G. & Speulda, E. (1976): Thermosensitivity of preoptic neurons in a hibernator (golden hamster) and a non-hibernator (guinea pig). Pflügers Arch. 363, 119-123.
Yurisova, M.N. & Polenov, A.L. (1979): The hypothalamo-hypophysial system in the ground squirrel, Citellus erythrogenys Brandt. Cell Tissue Res. 198, 539-556.

Integrative and cellular aspects of autonomic functions : temperature and osmoregulation. Eds K. Pleschka, R. Gerstberger. John Libbey Eurotext, Paris © 1994, pp. 269-274.

Circadian changes in neuronal thermosensitivity in the rat suprachiasmatic nucleus

Philippe S. Derambure, Jay B. Dean and Jack A. Boulant

Department of Physiology, College of Medicine, Ohio State University, Columbus, Ohio 43210, USA

The suprachiasmatic nucleus (SCN) of the mammalian hypothalamus maintains a circadian pacemaker responsible for the diurnal rhythms displayed in several regulatory systems, including the daily changes in body temperature (Kittrell, 1991). Electrophysiological recordings from hypothalamic tissue slices indicate that suprachiasmatic neurons show a circadian rhythm in their firing rates, even when these tissue slices are maintained *in vitro* for up to three days (reviewed in Gillette, 1991). In tissue slices from the rat, a nocturnal animal, *in vitro* SCN neurons display their highest firing rates during the day and their lowest firing rates during the night.

The suprachiasmatic nucleus is located near the preoptic region and anterior hypothalamus (PO/AH), an area known to be important in thermoregulation (reviewed in Boulant, 1980). Electrophysiological studies show that, like most other hypothalamic regions, about 30-40% of the PO/AH neurons are considered to be thermosensitive (Dean & Boulant, 1989). In addition to the effect of the circadian pacemaker on temperature regulation, some studies suggest that temperature itself can affect circadian rhythms such as sleep (McGinty & Szymusiak, 1990; Parmeggiani, 1990). Much of the present paper reviews the results of a recently submitted study by Derambure & Boulant in which circadian changes in both firing rate and thermosensitivity were recorded in SCN neurons in hypothalamic tissue slices. Also in the present paper, the proportions of thermosensitive SCN neurons are compared with an earlier study (Dean & Boulant, 1989) in which thermosensitive properties were recorded in neurons throughout the diencephalon.

Research supported by the National Institutes of Health (NS14644).
Present address of P.S. Derambure: Faculte de Medecine, Laboratoire de Physiologie, Lille University, 1 Place de Verdun, 59045 Lille, France.
Present address of J.B. Dean, Department of Physiology and Biophysics, Wright State University, Dayton, Ohio 45435, U.S.A.

METHODS

Young male, Spague-Dawley rats were conditioned to a 12-hour:12-hour light-dark cycle. In this review, all indications of circadian time (CT) refer to this subjective day-night cycle for the colony of host animals. CT=0 hours is the beginning of the subjective day, when lights were turned on in the host colony; and CT=12 hours is the beginning of the subjective night, when lights were turned off in the host colony.

Rats were decapitated, the hypothalamus was removed, and 300 µm or 400 µm thick slices were sectioned. Suprachiasmatic neurons were recorded from frontal tissue slices (Derambure & Boulant), and the other diencephalic neurons were recorded from horizontal tissue slices that did not contain the SCN (Dean & Boulant, 1989). Slices were placed in a perfusion-interface incubation chamber constantly perfused with a nutrient medium (pH 7.4; 300 mosmol) which was gassed with 95% O_2-5% CO_2 and which contained (in mM): 5 KCl, 124 NaCl, 2.4 $CaCl_2$, 1.3 $MgSO_4$, 1.24 KH_2PO_4, 26 $NaHCO_3$, and 10 glucose. The perfusion medium was maintained near 36°C or 37°C by a Peltier thermoelectric assembly, and tissue temperature was monitored by thermocouples placed under or on top of the slice. Tissue temperature was manipulated by changing the temperature of the perfusion medium or by using water-perfused thermodes placed directly under the recorded region of the slice. Extracellular neuronal activity was recorded using glass microelectrodes filled with 3 M NaCl. Single unit spikes, having signal-to-noise ratios greater than three, were counted by a ratemeter. Firing rate (impulses/second, $imp \cdot s^{-1}$) and tissue temperature were recorded from different neurons at different circadian times. One-hour running averages of neuronal firing rates and thermosensitivities were used to determine the changes in these parameters as a function of circadian time.

Neuronal thermosensitivity was determined by the slope (or thermal coefficient) of firing rate plotted as a function of tissue temperature, i.e., $imp \cdot s^{-1} \cdot °C^{-1}$. This thermal coefficient was determined over the temperature range (minimum range: 2°C) in which a neuron expressed its greatest thermosensitivity. Using similar criteria to previous studies (Boulant & Dean, 1986; Curras, Kelso & Boulant, 1991), warm sensitive neurons had positive thermal coefficients of at least +0.8 $imp \cdot s^{-1} \cdot °C^{-1}$, and cold sensitive neurons had negative thermal coefficients of at least -0.6 $imp \cdot s^{-1} \cdot °C^{-1}$. All other neurons were defined as temperature insensitive. It is important to note that the basis of these thermosensitivity criteria rests in an early *in vivo* study (Boulant & Hardy, 1974), which showed that these same criteria could effectively identify preoptic neurons that received afferent input from peripheral thermoreceptors; i.e., changes in skin or spinal temperature affected 59% of warm sensitive preoptic neurons, 73% of cold sensitive preoptic neurons, but less than 3% of the temperature insensitive preoptic neurons. These criteria can, therefore, distinguish preoptic neurons that not only sense their own temperature but also integrate hypothalamic thermal information with afferent peripheral thermal signals. For this reason, we believe that it is important not to use criteria less positive than +0.8 $imp \cdot s^{-1} \cdot °C^{-1}$ or less negative than -0.6 $imp \cdot s^{-1} \cdot °C^{-1}$, since such neurons do not act as integrators of hypothalamic and peripheral thermal information.

RESULTS

In the submitted study by Derambure & Boulant, circadian changes in firing rate and thermosensitivity were recorded in SCN neurons in frontal hypothalamic tissue slices. The SCN was divided into dorsomedial and ventrolateral regions, and these regions were further divided into rostral or caudal sections. When plotted as a function of circadian time (CT), no significant changes in averaged neuronal firing rates were evident in the ventrolateral SCN; but, dorsomedial SCN neurons displayed a circadian variation, similar to that described by Gillette (1991). The most prominent firing rate variation was observed in the rostral sections of the dorsomedial SCN. As shown by the averaged firing rates in Fig. 1A, dorsomedial SCN neurons displayed their peak firing rates near CT=5 hours, and this activity decreased to a minimal level before CT=12 hours, which was the beginning of subjective night for the host animals.

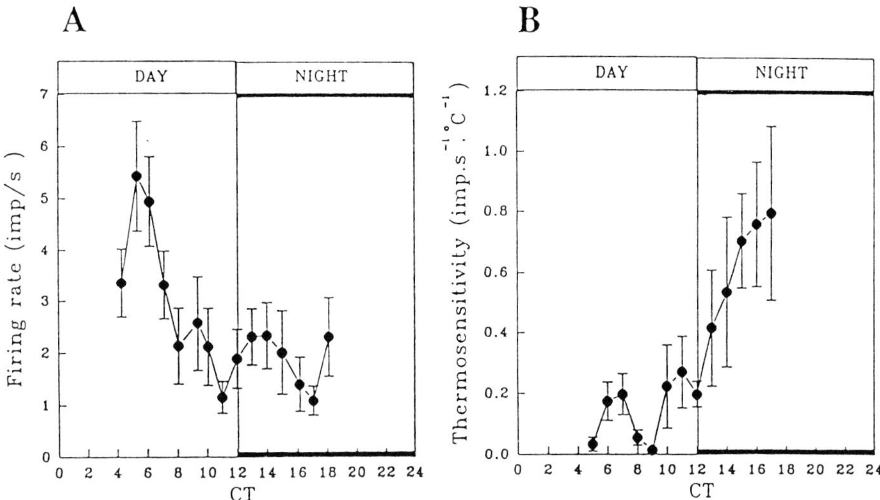

Fig. 1. One-hour running averages (± SEM) of neuronal activity as a function of circadian time (CT). **A.** firing rates of 83 rostral dorsomedial SCN neurons. **B.** thermosensitivities or thermal coefficients of 63 rostral ventromedial SCN neurons. [This figure is taken from a short paper by Derambure & Boulant, Sleep Research 22:617, 1993.]

The submitted study by Derambure & Boulant found circadian variations in the neuronal thermosensitivity in all regions of the suprachiasmatic nucleus. Average neuronal thermosensitivities were low during the subjective day, but they rose at the beginning of the subjective night, reaching peak levels near CT=16 hours. The most prominent circadian variation in neuronal thermosensitivity occurred in rostral sections of the ventrolateral SCN. As shown in Fig. 1B, thermosensitivity in this region rose quickly after the beginning of subjective night, and average thermosensitivity of all neurons recorded at CT=17 hours approached +0.8 $imp \cdot s^{-1} \cdot {}^{\circ}C^{-1}$, which is the minimum criterion for warm sensitive neurons.

Another indication of the variation in SCN neuronal thermosensitivity is illustrated in Table 1, which shows the proportions of warm sensitive, cold sensitve, and temperature insensitive neurons recorded at different circadian times. In Table 1, SCN neurons recorded in frontal tissue slices (Derambure & Boulant, submitted) are compared with neurons recorded in horizontal slices throughout the diencephalon (Dean & Boulant, 1989). Both of these studies used the same criteria to classify neurons according to thermosensitivity.

Table 1. Effect of circadian time (CT, in hours) on the proportions of warm sensitive, cold sensitive and temperature insensitive neurons in the SCN (from a study by Derambure & Boulant, submitted) and in the diencephalon (from a study by Dean & Boulant, 1989).

SCN:	**CT=4-8**	**CT=8-12**	**CT=12-20**
Warm Sensitive	3%	4%	24%
Cold Sensitive	1%	4%	0%
Temperature Insensitive	96%	92%	76%
Total Number (n)	(63)	(73)	(72)

DIENCEPHALON:	**CT=4-8**	**CT=8-12**	**CT=12-20**
Warm Sensitive	51%	45%	32%
Cold Sensitive	6%	2%	7%
Temperature Insensitive	43%	53%	61%
Total Number (n)	(55)	(100)	(84)

As shown in Table 1, during the subjective day (CT=4-8 hours and CT=8-12 hours), the proportion of warm sensitive neurons recorded in the SCN remained at 3-4%. This proportion was extremely low when compared to the 30-40% reported in many studies of the preoptic region (Boulant & Dean, 1986) and when compared to the 45-51% found throughout the entire diencephalon (Table 1). At night (CT=12-20 hours), however, there was a dramatic increase in the proportion of SCN warm sensitive neurons. From CT=12-20 hours, the proportion of SCN warm sensitive neurons rose to 24% (Table 1). The submitted Derambure & Boulant study also found that, from CT=16-20 hours, warm sensitive neurons accounted for more than 30% of neurons in the entire SCN and more than 40% of neurons in the ventrolateral SCN. At night, therefore, the proportion of SCN warm sensitive neurons is comparable to the proportions observed in the preoptic area and other diencephalic regions; however, during the day the proportion of SCN warm sensitive neurons is significantly lower than in other regions.

An increase in neuronal warm-sensitivity at night has not been demonstrated in other diencephalic areas. Table 1 shows the proportions in neurons recorded in tissue slices cut horizontally throughout the entire diencephalon. Because these are horizontal slices, the recorded diencephalic neurons had no intact synaptic connections with suprachiasmatic neurons. When compared at the same times as the SCN, the diencephalon showed no increases in neuronal warm-sensitivity at night. It should also be noted that, like previous preoptic studies (reviewed in Boulant & Dean, 1986), the proportions of cold sensitive neurons remained low in both the SCN and in the diencephalon. Table 1 further suggests that the proportion of cold sensitive neurons is not affected by circadian time.

DISCUSSION

The present paper indicates that suprachiasmatic neurons show circadian changes in their activity which is different from other diencephalic neurons, at least diencephalic neurons that are not synaptically connected with the SCN. In tissue slices from the rat, suprachiasmatic neurons show their peak firing rates during the day, but peak neuronal thermosensitivities occur at night when the noctural animal would normally be awake and active. Because of the synaptic connections from the suprachiasmatic nucleus to the preoptic area (Watts, 1991), these findings suggest an opportunity for interactions between temperature and circadian rhythm.

Previous studies indicate that the suprachiasmatic nucleus is at least partly responsible for the diurnal changes in body temperature (Kittrell, 1991). If synaptic connections exist between warm sensitive neurons in the SCN and preoptic region, increases in SCN neuronal thermosensitivity may enhance the ability of the preoptic region to regulate body temperature. This may even provide an explanation for previous studies which claim that the central control of body tempertature is more effective at certain circadian times, such as during wakefulness compared to different phases of sleep (Parmeggiani and Sabattini, 1972; Tanaka, et al., 1990).

It has also been suggested that changes in body temperature can alter circadian patterns, such as the sleep-wake cycle (Horne and Reid, 1985; McGinty and Szymusiak, 1990). Increases in body temperature may produce slow or delayed neural changes that eventually result in increased sleep. Temperature may, therefore, influence circadian patterns by directly influencing the activity of SCN neurons. If temperature changes can alter circadian patterns, the present study suggests that the effectiveness of temperature would be greatly dependent on the time of day. In the rat, SCN temperature changes should be relatively ineffective during the day since neuronal thermosensitivity is quite low, at least between CT=4-12 hours. On the other hand, SCN temperature changes should be very effective during the night, when neuronal thermosensitivity and the proportion of warm sensitive neurons are high.

REFERENCES

Boulant, J.A. (1980): Hypothalamic control of thermoregulation: neurophysiological basis. In: *Handbook of the Hypothalamus: Behavioral Studies of the Hypothalamus*, vol. 3A, ed. P.J. Morgane and J. Panksepp. pp. 1-82. New York: Dekker.

Boulant, J.A. and Dean, J.B. (1986): Temperature receptors in the central nervous system. *Annu. Rev. Physiol.* 48, 639-654.

Boulant, J.A. and Hardy, J.D. (1974): The efferct of spinal and skin temperatures on the firing rate and thermosensitivity of preoptic neurones. *J. Physiol., (London)* 240, 639-660.

Curras, M.C., Kelso, S.R. and Boulant, J.A. (1991): Intracellular analysis of inherent and synaptic activity in hypothalamic thermosensitive neurons. *J. Physiol., (London)* 440, 257-271.

Dean, J.B. and Boulant, J.A. (1989): In vitro localization of thermosensitive neurons in the rat diencephalon. *Am J. Physiol.* 257 (*Regulatory Integrative Comp. Physiol.* 26), R57-R64.

Derambure, P.S. and Boulant, J.A. (1993): Temperature sensitivity of suprachiasmatic neurons: circadian rhythms in rat hypothalamic tissue slices. *Sleep Research* 22, 617.

Derambure, P.S. and Boulant, J.A. (submitted): Circadian thermosensitive characteristics of suprachiasmatic neurons in vitro. *Am J. Physiol.* (*Regulatory Integrative Comp. Physiol.*)

Gillette, M.U. (1991): SCN electrophysiology in vitro: rhythmic activity and endogenous clock properties. In: *Suprachiasmatic Nucleus The Mind's Clock*, ed. D.C. Klein, R.Y. Moore and S.M. Reppert, pp. 125-143. New York: Oxford.

Horne, J.A. and Reid, A.J. (1985): Night-time sleep EEG changes following body heating in a warm bath. *Electroencephalogr. Clin. Neurophysiol.* 60, 154-157.

Kittrell, E.M.W. (1991): The suprachiasmatic nucleus and temperature rhythms. In: *Suprachiasmatic Nucleus The Mind's Clock*, ed. D.C. Klein, R.Y. Moore and S.M. Reppert, pp. 233-245. New York: Oxford.

McGinty, D.J. and Szymusiak, R. (1990): Hypothalamic thermoregulatory control of slow wave sleep. In: *The Diencephalon and Sleep*, ed. M. Mancia and G. Marini, pp. 97-109. New York:Raven Press, Ltd.

Parmeggiani, P.L. (1990): Homeostatic function of the hypothalamus and control of the wake-sleep cycle. In: *The Diencephalon and Sleep*, ed. M. Mancia and G. Marini, pp. 133-145. New York:Raven Press, Ltd.

Parmeggiani P.L. and Sabattini L. (1972): Electromyographic aspects of postural, respiratory and thermoregulatory mechanisms in sleeping cats. *Electroencephalogr. Clin. Neurophysiol.* 33,1-13.

Tanaka, H., Yanase, M., Kanosue, K. and Nakayama, T. (1990): Circadian variation of thermoregulatory responses during exercise in rats. *Am. J. Physiol.* 258 (*Regulatory Integrative Comp. Physiol.* 27), R836-R841.

Watts, A.G. (1991): The efferent projections of the suprachiasmatic nucleus: anatomical insights into the control of circadian rhythms. In: *Suprachiasmatic Nucleus The Mind's Clock*, ed. D.C. Klein, R.Y Moore and S.M. Reppert, pp. 77-106. New York: Oxford.

D. Effector mechanisms of temperature regulation

Integrative and cellular aspects of autonomic functions : temperature and osmoregulation. Eds K. Pleschka, R. Gerstberger. John Libbey Eurotext, Paris © 1994, pp. 277-284.

Measurement of the intracellular calcium in the brown adipocyte

Masanori Nagai, Kiyoshi Tsuchiya and Masami Iriki

Department of Physiology, Medical University of Yamanashi, Tamaho, Nakakoma, 409-38 Yamanashi, Japan

Brown adipose tissue (BAT) contributes to cold adaptation of rodents and awaking of the hibernators by providing excess heat. In man, however, distribution of BAT is obscure, and thus BAT has been considered to play a minor role except neonates whose mechanism for shivering thermogenesis is not fully developed. However, non-shivering thermogenesis by BAT has recently been confirmed in adults (Himms-Hagen, 1990; Lean, 1989). Further, thermogenic function of BAT contributes to diet-induced thermogenesis (Rothwell and Stock, 1983). These facts indicate that BAT plays a significant role in regulating energy metabolism and body weight.

Brown adipocytes, composing BAT, produce heat by activating lipolysis of their intracellular deposits of fat when stimulated. Noradrenaline released from the sympathetic neurones innervating BAT activates lipolysis of the brown adipocytes most effectively. Transport of free fatty acids and intake of glucose by the brown adipocytes are also activated besides lipolysis by adrenergic stimulation. Intracellular calcium possibly plays a role in these processes, but quantitative and temporal analysis of intracellular free calcium has not been carried out. Thus, the role of calcium in the brown adipocytes has not yet been fully understood. In the preceding studies, fluorescence spectrophotometry was applied to the suspension of cells in order to measure the concentration of the cytosolic free calcium. By this method, however, cell density along the light path in the cuvette of the spectrophotometer is variable. Therefore, long-term and stable recording of the calcium concentration is impossible. Further, repetitive stimulation of the cells by exchanging perfusates with different constituents is not possible, either. Cultured brown adipocytes are so far inadequate for the measurement of calcium concentration, because the activity of thermogenin, uncoupling protein responsible for thermogenesis, is lost in the cell culture.

In the present experiments, we have measured calcium concentration in the freshly-dispersed, single brown adipocytes by using a calcium indicator, fura-2, and a video camera-attached image processor. By the present technique, we have successfully measured a vast and long-lasting increase in the cytosolic free calcium in the single cells by adrenergic stimulation.

MATERIALS AND METHODS

Isolation of the cell

Brown adipocytes were obtained from the interscapular, axillary and dorso-cervical brown adipose tissues of the male Wistar rats of 8 - 12 weeks old. Before the brown adipose tissue were dissected, animals were anaesthetized with dietyl-ether, and bled by transection of the carotid arteries. Dissected brown adipose tissues were rinsed and minced, and then dispersed with Tyrode solution containing 3% collagenase (type III, Sigma, USA) and 0.4% bovin serum albumin (essentially free fatty acid- free, Sigma). Tyrode solution used contained 137 mM NaCl, 2.7 mM KCl, 1.8 mM $CaCl_2$, 1.0 mM $MgCl_2$, 5.56 mM Glucose and 5 mM Tris, and pH of the solution was adjusted to 7.38 by 0.1 N HCl. One dispersing procedure lasted 5 min, and the procedure was usually repeated 2 - 3 times until the isolation of the cell was ascertained by microscopic observation. The suspension of the cells was intermittently bubbled with a gas mixture of 95% O_2 and 5% CO_2.Cell suspension obtained was then incubated with 0.16 mM fura-2-AM (Dojin Lab., Japan) for 30 min at 37 °C.

Selection of the cell

Perfusion chamber of acrylic material was designed to contain 200 ul of cell suspension. A cover glass was attached to the bottom of the chamber by dental impression material (Coltex, Coltene AG, Switzerland). Cover glass was coated with 0.1% polylysine solution (Muto Chemicals, Japan) beforehand, and single cells in the suspension were attached and immobilized on the surface of the cover glass in 20 - 30 min after they were introduced into the chamber. By this procedure, stable processing of the cell image and repetitive stimulation of the cell by exchanging medium became possible. Cells for the measurement were then chosen under fluorescence microscope (IM35, Carl Zeiss, Germany). For the present experiments, we chose cells with 10 - 12 µm diameter. These cells contained 2 - 4 intracellular droplets of fat, and were classified as prematured type of brown adipocytes (Goglia *et al.*, 1992). We chose these cells by following two reasons. 1) These cells were easily attached to the surface of the cover glass coated with polylysine. 2) Fluorescence images of these cells were clear so that calcium indicator, fura-2, was thought to distribute homegeneously in the cytosol.

Measurement of fluorescence

We used excitation wavelength (λex) of 340 and 380 nm, and emission wavelength (λex) of 500 nm. Excitation wavelength was exchanged by computer-assisted filter unit (Carl Zeiss), and excitation by both wavelengths completed in 60 msec. Fluorescence image at each excitation wavelength was then fed into image processor (ARGUS100, Hamamatu Photonics, Japan) through a video camera (SIT camera, Hamamatu Photonics). The ratio of the fluorescence intensity (I_{340}/I_{380}) was calculated at each pixel, and the ratio image of the cell was obtained. One image frame contained 512 x 512 pixels.

Calculation of calcium concentration

Following equation by Grynkiewicz *et al.* (1985) was adopted in order to calculate the concentration of the intracellular free calcium from the ratio of the fluorescence intensity.

$$[Ca^{2+}] = (R - R_{min})/(R_{max} - R) \times Kd \times \beta$$

where R is a ratio of fluorescence intensity (I_{340}/I_{380}) measured, Rmin is R when calcium concentration is minimum, R_{max} is R when calcium concentration is maximum, β is the ratio of I_{380} when calcium concentration is minimum and I_{380} when calcium concentration is maximum (I_{380} with $[Ca^{2+}]_{min}/I_{380}$ with $[Ca^{2+}]_{max}$), and Kd is a dissociation constant between fura-2 and calcium.

We chose 224 nM for Kd at 37 °C. We determined R_{min} in the cells perfused with Tyrode solution with no calcium plus 5 mM EGTA (Sigma), and did R_{max} in the same cells perfused in turn with Tyrode solution with 5 mM $CaCl_2$ plus 5 mM ionomycin (Calbiochem, USA). R_{min}, R_{max} and β determined were 0.48 ± 0.05 (mean ± SE, n=6), 1.93 ± 0.33 and 0.27 ± 0.04, respectively. By this method, we established a ratio-concentration curve reliable between 1 to 10^3 nM of the cytosolic free calcium.

RESULTS

Three-dimensional distribution of the intracellular calcium is shown in Fig. 1A. The conteau of calcium concentration of two single brown adipocytes are obviously shown. Calcium did not distribute homogeneously in the cytosol, but its concentration varied depending on the site within the cell. Concentration of calcium also fluctuated spontaneously. Figure 1B shows this fluctuation of calcium concentration for 10 min at four different sites within the cell shown on the left side of Fig. 1A. The frequencies and amplitudes of fluctuation varied depending the site within the cell. We calculated the ratio of fluorescence intensity at all pixels within the area of the cell, and thus determined the average concentration of the cytosolic free calcium ($[Ca^{2+}]_{in}$). The average $[Ca^{2+}]_{in}$ was 12.4 ± 2.0 nM (mean ± SE, n=13) at rest.

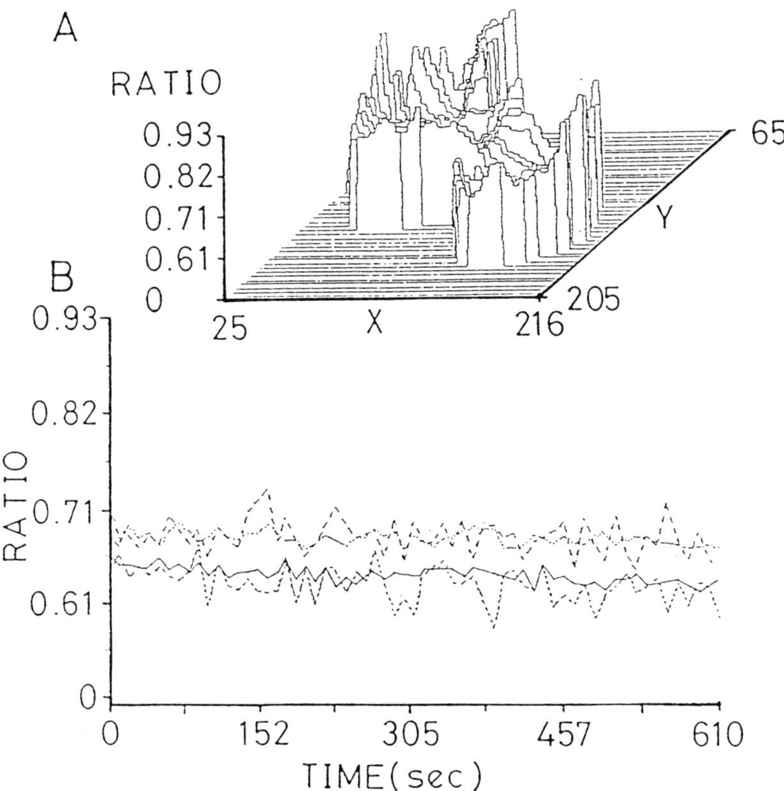

Fig. 1. Three-demensional distribution of intracellular calcium in the single brown adipocytes (A) and fluctuation of intracellular calcium concentration at different sites in a cell during 10 min (B).

Noradrenaline (NA) significantly increased $[Ca^{2+}]_{in}$ of the single brown adipocytes. Figure 2 shows a time course of changes in $[Ca^{2+}]_{in}$ of a cell stimulated by 10^{-5} M NA. $[Ca^{2+}]_{in}$ steeply increased with a latency of about 10 min, and higher level of $[Ca^{2+}]_{in}$ was maintained over 30 min. Three-dimensional representations of the intracellular calcium concentration at times indicated by A, B and C in Fig. 2 were shown in Fig. 3. The amplitude of concentration increase differed from site to site within the cell. The responsiveness to NA greatly varied among cells, however, the cells with resting $[Ca^{2+}]_{in}$ around 10 nM responded more sensitive to NA in the preliminary experiments. Therefore, we examined dose-response relationship especially in these cells. The maximum $[Ca^{2+}]_{in}$ reached in 30 min after NA administration was shown in Fig. 4. $[Ca^{2+}]_{in}$ increased in a dose-dependent manner between 10^{-7} and 10^{-5} M of NA. Stimulatory effect on $[Ca^{2+}]_{in}$ by 10^{-5} M NA was most potent. Average $[Ca^{2+}]_{in}$ increased to 135 nM by 10^{-5} M NA, but there was an individual site in a certain cell where calcium concentration increased to 700 nM.

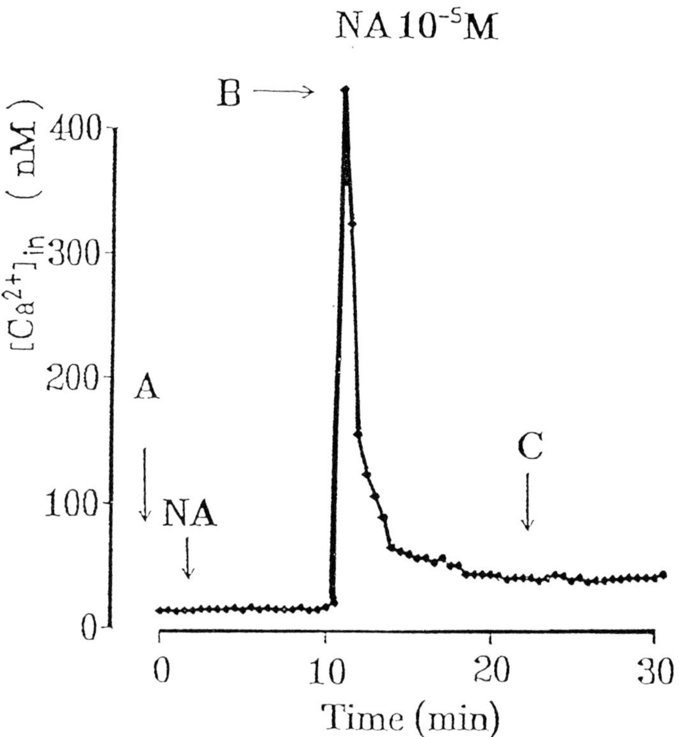

Fig. 2. Changes in the intracellular calcium concentration in a single brown adipocyte by noradrenaline of 10^{-5} M.

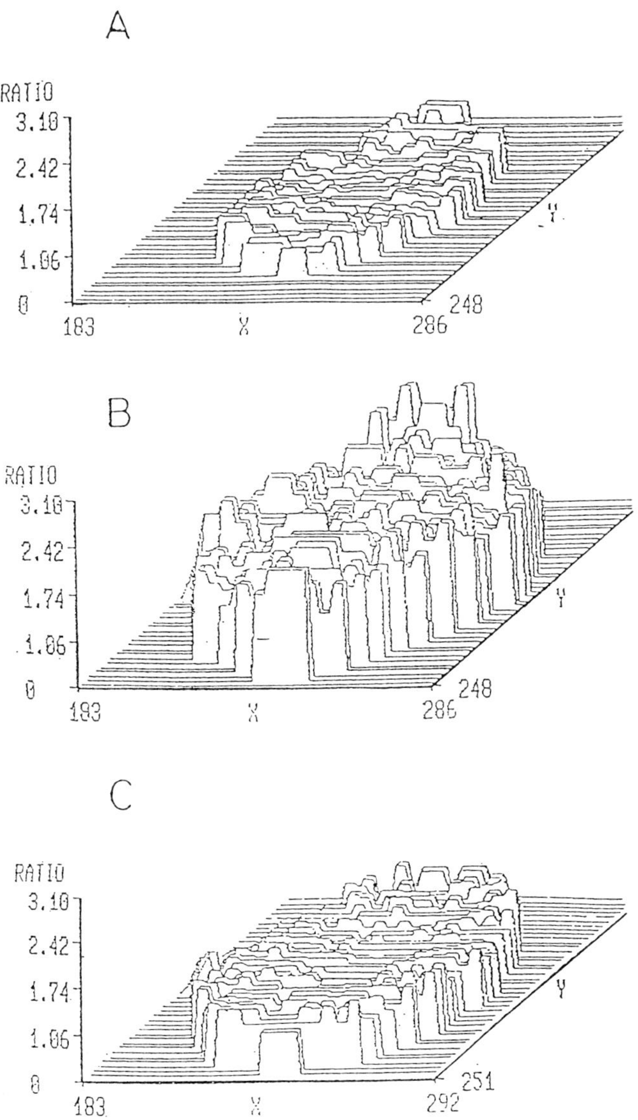

Fig. 3. The effect of noradrenaline (10^{-5} M) on three-demensional distribution of the intracellular calcium in a single brown adipocyte. The same cell shown in Fig. 2. Contour of the calcium concentration is shown at times A.B and C indicated in Fig. 2.

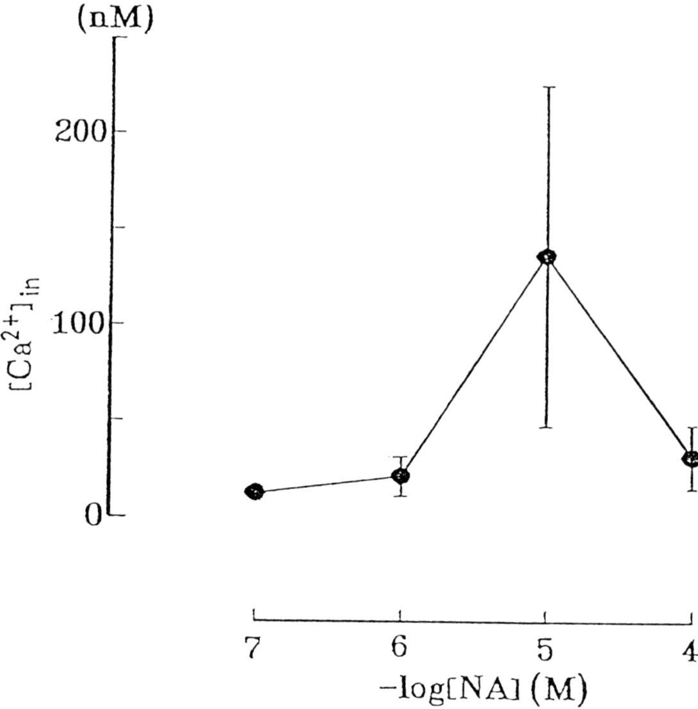

Fig. 4. Dose-response curve for noradrenaline. Mean with standard error are shown. The number of the data is 9 - 13.

DISCUSSION

We have measured the concentration of the cytosolic free calcium in the single brown adipocytes in this study. Wilcke and Nedergaard (1989) reported that NA increased intracellular calcium, using suspension of the brown adipocytes loaded with fura-2 and fluorescence spectrophotometer. However, they didn't try to calculate the actual concentration of calcium from calcium signals of the calcium-fura-2-complex because of the existence of autofluorescence in their preparation. We didn't recognized any significant autofluorescence, and were able to establish a ratio-concentration curve reliable between 1 to 10^3 nM of cytosolic free calcium. In the preceding experiment where spectrophotometer was used, cell density along the light path was variable. Therefore, calcium concentration was not monitored longer than 5 min. By introducing single brown adipocytes, we have successfully measured calcium signals over 30 min, and found that concentration of the cytosolic calcium occasionally reached its maximum more than 10 min after stimulation and high $[Ca^{2+}]_{in}$ was maintained over long period (Fig. 2). This long-lasting increase in $[Ca^{2+}]_{in}$ has never been reported previously.

Brown adipose tissue contains brown adipocytes with different degrees of differentiation (Cinti, 1992; Goglia *et al.*, 1992). The suspension of brown adipocyte as previously employed contains all these types of the cell, and the response of the suspension comes out as a summation of the responses of adipocytes with different degrees of differentiation. We chose brown adipocytes with diameter of 10 - 12 μm and 2 - 4 intracellular deposits of fat for the present experiments by the reasons described in the "METHODS" section. These cells are classified as "preadipocyte" according to Goglia *et al.* (1992). The average $[Ca^{2+}]_{in}$ of these cells was 12.4 ± 2.0 nM at rest. On the other hand, Wilcke and Nedergaard (1989), using cell suspension, have deduced that resting $[Ca^{2+}]_{in}$ is about 200 nM. We have never observed this high concentration in the resting cell even at a site within the cell where calcium concentration is locally higher (Fig. 1).

Dose-response curve for NA in the cell suspension (Wilcke and Nedergaard, 1989) shows that 10^{-8} M NA increases intracellular concentration of calcium. However, in our experiments NA lower than 10^{-7} M had no effect on $[Ca^{2+}]_{in}$. It is of interest to determine whether this right-hand shift of dose-response curve is proper to the premature brown adipocytes.

Respiratory rate of the brown adipose tissue, an indicator for heat production, starts increasing immediately after NA administration and reaches the maximum 1 min after stimulation. Our results have shown that $[Ca^{2+}]_{in}$ increased with a longer latency, 4 - 10 min, after adrenergic stimulation and the maximum increase is usually attained 10 - 30 min after stimulation. Therefore, increase in $[Ca^{2+}]_{in}$ which we have measured does not directly correlate with thermogenic function of the brwon adipocytes. However, some other functions of the brown adipocytes become apparent with a longer latency after adrenergic stimulation. For example, concentration of cAMP, an indicator for lipolysis, reaches its maximum 5 min after adrenergic stimulation, and excretion of free fatty acid, as a result of lipolysis, lasts over 40 min after stimulation (Pettersson and Vallin, 1979). Also, concentration of GDP, an inhibitory modulator of thermogenesis, increased with 10 - 15 min latency after adrenergic stimulation (Nicholls, 1983). Intake of glucose (Marette and Bukowiecki, 1991; Omatsu-Kanbe and Kitasato, 1992) and activation of 5'deionidase, an enzyme which transforms thyroxine into triiodothyronin, of the brown adipocytes are also induced by adrenergic stimulation with different time course (Silva and Larson, 1983). The role of intracellular calcium in these functions will be invistigated in further studies, and our technique have made it possible to record temporal and spatial changes in calcium concentration for a long time in the single brown adipocytes.

REFERENCES

Cinti, S. (1992): Morphological and functional aspects of brown adipose tissue. *Peiatr. Adolesc. Med.* 2, 125-132.

Goglia, F., Geloen, A., Lanni, A., Minaire, Y., and Bukowiecki, L.J. (1992): Morphometric-stereologic analysisi of brown adipocyte differentiation in adult mice. *Am. J. Physiol.* 262, C1018-C1023.

Grynkiewicz, G., Poenie, M., and Tsien, R.Y. (1985): A new generation of Ca^{2+} indicators with greatly improved fluorescence properties. *J. Biol. Chem.* 260, 3440-3450.

Himms-Hagen, J. (1990): Brown adipose tissue thermogenesis: interdisciplinary studies. *FASEB J.* 4, 2890-2898, 1990.

Lean, M.E. (1989): Brown adipose tissue in humans. *Proc. Nat. Soc.* 48, 243-256.

Marette, A., and Bukowiecki, L.J. (1991): Noradrenaline stimulates glucose transport in rat brown adipocytes by activating

thermogenesis. Evidence that fatty acid acitvation of mitochondrial respiration enhances glucose transport. *Biochem. J.* 277, 119-124.
Nicholls, D.G. (1983): The thermogenic mechanism of brown adipose tissue. *Biosci. Rep.* 3, 431-441.
Omatsu-Kanbe, M., and Kitasato, H. (1992): Insulin and noradrenaline independently stimulate the translocation of glucose transporters from intracellular store to the plasma mambrane in mouse brown adipocytes. *FEBS LETT.* 314, 246-250.
Pettersson, B., and Vallin, I. (1976): Norepinephrine-induced shift in levels of adenosine 3':5'-monophosphate and ATP parallel to increased respiratory rate and lipolysis in isolated hamster brown-fat cells. *Eur. J. Biochem.* 62, 383-390.
Rothwell, N.J., and Stock, M.J. (1983): Diet-induced thermogenesis. In *Mammalian thermogenesis*, ed. L. Girardier & M.J. Stock, pp.208-233. New York: Chapman & Hall.
Silva, J.E., and Larsen, P.R. (1983): Adrenergic activation of triiodothyronine production in brown adipose tissue. *Nature* 305, 712-713.
Wilcke, M., and Nedergaard, J. (1989): $\alpha 1$-and $\beta 1$-adrenergic regulation of intracellular Ca^{2+} levels in brown adipocytes. *Biochem. Biophys. Res. Commun.* 163, 292-300.

Integrative and cellular aspects of autonomic functions : temperature and osmoregulation. Eds K. Pleschka, R. Gerstberger. John Libbey Eurotext, Paris © 1994, pp. 285-294.

Do changes of sympathetically stimulated thermogenesis underlie the developmental changes in the cold defense of rat pups ?

Heiko Döring, Gerhard Körtner, Karola Meyer and Ingrid Schmidt

Max-Planck Institut, W.G. Kerckhoff-Institut, Parkstrasse 1, D-61231 Bad Nauheim, Germany

SUMMARY

To determine possible causes for the sluggish activation of cold defense in 1-day-old rat pups, we compared pharmacologically induced increases in oxygen consumption and core temperature with the corresponding responses of 11-day-old pups. Isolated pups resting under thermoneutral conditions were either given subcutaneous injections of norepinephrine (100-1600 µg/kg) or of tyramine (10 mg/kg). Although the increases in the oxygen consumption of 1-day-old pups induced either by exogenously supplied or endogenously released norepinephrine were slightly slower than those of 11-day-old rats, they were considerably faster than the responses induced by sudden cold exposure. The sluggishness of the younger pups' cold defense thus seems attributable neither to a limited availability of endogenous norepinephrine nor to immaturity of their brown adipose tissue, the dominant site of thermoregulatory thermogenesis in rat pups.

INTRODUCTION

The autonomic thermoregulatory abilities of newborn rat pups are severely limited: physically by their poor insulation and unfavorable surface-to-volume ratio, and physiologically by their immature nervous system. The ability to maintain a stable core temperature (Tc) when isolated at room temperature emerges only slowly during the first two weeks of life (Conklin and Heggeness, 1971; Schmidt et al., 1987; Spiers and Adair, 1986), but thermoregulatory behavior greatly reduces the metabolic demand in the normal rearing situation (Alberts, 1978; Schmidt et al., 1986 and 1987). A critical determinant of Tc stability under these conditions of rapidly changing availability of behavioral warming by the mother or littermates is the time needed to activate metabolic cold defense. But despite numerous reports on the steady-state thermoregulatory responses of the immature rat, we still know little about the dynamics of cold defense activation in these pups.

Thompson and Moore (1968) were the first to demonstrate that the rate with which rat pups increase their oxygen consumption ($\dot{V}O_2$) after a step decrease of ambient temperature (Ta) differs markedly between 1-week-old and 2-week-old rat pups. The rate of activating thermoregulatory thermogenesis in juvenile rats received little attention thereafter until its methodological importance for studies of brown adipose tissue (BAT) development was pointed out recently (Körtner et al., 1993). Brown adipose tissue is the main effector organ for metabolic cold defense in suckling-age rat pups (Nedergaard et al., 1986), and measurements of its acute activity as well as propranolol blockade have indicated that the sympathetic activation of BAT is essential for the rapid activation of cold defense in pups 5 days old and older (Körtner et al., 1993). But when 1-day-old pups were subjected to physiologically comparable cold loads (that is, to Ta that eventually also caused a steady-state doubling of metabolic rate) $\dot{V}O_2$ increased only very sluggishly.

The study of developmental changes in cold defense is complicated by circadian changes: during the minimum phase of the juvenile circadian Tc cycle, 5- and 11-day-old pups subjected to a step decrease in Ta show much slower and weaker increases in metabolic rate than do pups exposed to the same stimulus during the maximum phase of the circadian Tc cycle (Nuesslein et al., 1993). At constant Ta, the $\dot{V}O_2$ in the circadian maximum of 11-day-old pups increases with increasing loads, but the $\dot{V}O_2$ in the circadian minimum might fall to or below the thermoneutral level at moderate cold loads (Nuesslein and Schmidt, 1993). And consequently, acute BAT activity differs markedly between the circadian minimum (at the beginning of the light phase) and the circadian maximum (at the beginning of the dark phase) (Redlin et al., 1992). Although the amplitude of the juvenile circadian Tc cycle varies throughout the first 3 weeks of life, the Tc in the circadian maximum remains fairly stable (Nuesslein and Schmidt, 1990). This implies that developmental changes in cold defense and BAT activity can be most clearly evaluated during the maximum phase of the juvenile circadian Tc cycle.

The aim of this study was to determine possible causes for the sluggish activation of metabolic cold defense in 1-day-old pups. We therefore compared the pharmacologically-induced increases in the $\dot{V}O_2$ of 1- and 11-day-old rats, stimulated either by exogenously supplied or by endogenously released norepinephrine, with their responses to ambient cold loads. To prevent distortion of measurements due to circadian variations all measurements were carried out in the stable maximum phase of the juvenile circadian Tc cycle - that is at the beginning of the dark phase. Therefore all the pups from any one litter were tested simultaneously, though this permitted only alternate recording of $\dot{V}O_2$, whereas Tc of all pups could be recorded continuously.

METHODS

Animals. As in our previous studies, we used 1- and 11-day-old Zucker rat pups because the developmental and circadian changes of juvenile thermoregulation have been most thoroughly documented in this strain of hooded rats (Körtner et al., 1993; Mumm et al., 1989; Nuesslein and Schmidt, 1990; Redlin et al., 1992, Schmidt et al., 1986; 1987). The genotype of pups was identified retrospectively after weaning, and only data recorded from lean (Fa/-) pups are reported here. Pups were mother-reared in a 12:12 h light:dark cycle at a Ta of 22°C and a relative humidity of 55-60%.

Experimental set-up. Experiments were carried out in climatized rooms at a relative humidity of 60-70%. The set-up was similar to the one used by us for artificial rearing of pups (Mumm et al., 1989; Nuesslein and Schmidt, 1990). In short, 7 pups from one litter were individually placed in 8x8x9-cm chambers floating in a water bath that was kept 1°C warmer than the air temperature to prevent condensation. Metal tubes loosely fitting through a centered hole in the lid and protruding to the bottom of the chambers aspirated air from each chamber.

Air flows adjusted to 100 ml/min for the 1-day-old pups and to 200 ml/min for the 11-day-old pups resulted in oxygen extraction values between 0.1% and 1%.

Temperature and oxygen consumption measurements. To ensure an accurate record of deep body temperature, we measured colonic Tc continuously with thin copper-constantan-thermocouple probes inserted 1.8 cm in 1-day-old pups and 2.5 cm in 11-day-old pups. $\dot{V}O_2$ was alternately recorded from the 7 animal chambers and one reference chamber by using the multichannel open-flow system described in detail in another article in this volume (Nuesslein et al., 1993). Briefly, the oxygen content of the air from each of the chambers was alternately measured for 2.5-min periods by using an AMETEK S-3A analyzer, while the airflow was measured by a Tylan massflow meter.

Experimental protocol. Data from 3 experimental series are reported here, all carried out at the end of the daily light phase, but with the lights kept on during the experiments (about 4 h). In two experimental series 1- and 11-day-old pups were equipped with silastic catheters (Dow Corning 602-105) that were introduced 15 mm under the loose skin of the neck while the pup was briefly anesthetized with CO_2. The catheter was connected to a PP10 tube fixed with instant glue and tape to the back skin and led to join the thermocouple passing through the hole in the floating animal chamber. Pups were rewarmed to 37°C under a lamp before the animal chambers were placed in the water bath. To maintain the pups above their lower critical temperature, the water bath was set to 35.5°C for the 1-day-old and to 32.5°C for the 11-day-old pups.

One to two hours after they had reached a stable Tc above 37°C, each pup received an injection, with 10-min intervals between injecting one pup and the next. Norepinephrine (NE) or tyramine (TY) were given through the catheters (in volumes of 5-10 µl/g body mass and rinsed in with 20 µl saline). Control pups received only saline. Norepinephrine was given in doses ranging between 100 and 1600 µg/kg. Tyramine was used only at one dose level (10 mg/kg), because pilot experiments with 1- and 11-day-old pups had shown that there were no consistent dose-dependent differences in response velocity or intensity when TY was injected at doses ranging from 2.5 to 30 mg/kg. Oxygen consumption in the 7 pups was alternately measured according to a fixed time schedule that permitted to obtain for each individual about 2-5 measurements during the preinjection period under thermoneutral conditions, and 4-7 measurements at various times between 5 and 60 min after the injection (at time 0).

For comparison, we present the $\dot{V}O_2$ and Tc data from a third experimental series, carried out during a preceeding study in which pups had been exposed to step decreases of ambient temperature to determine the acute cold-induced changes of BAT activity (Körtner et al., 1993). In this series of experiments, after a thermoneutral exposure as described above, Ta was lowered (at time 0) to values resulting roughly in a steady state doubling of oxygen consumption (Ta=30°C for 1-day-old and Ta=27°C for 9- to 11-day-old pups). For further details see Körtner et al. (1993) or Nuesslein et al. (1993) in this volume.

Determination of the time lag in the oxygen consumption measuring system. After a prewarming period zinc-air-batteries (Ralston Energy Systems) consume oxygen at a constant rate proportional to the current drawn from the batteries. To experimentally evaluate the rate of change recorded by the O_2 analyzer when the oxygen consumption in the animal chamber made a step increase from 0 to some constant value (see Fig. 3), prewarmed batteries were introduced at time 0 into an empty animal chamber from which air was being aspirated at either 100 or 200 ml/min, and oxygen extraction was recorded as it was normally. These measurements permit us to evaluate the degree the $\dot{V}O_2$ values measured for the pups were distorted by the time lag of the measuring system - and, if necessary, to calculate correction factors (z) for determination of instantaneous $\dot{V}O_2$ (Bartholomew et. al., 1981).

Evaluation. The $\dot{V}O_2$ of pups was calculated from the massflow rate and the oxygen content of the dried air by using Withers' (1977) equation for a flow through mask and by assuming an RQ of 0.8 (Markewicz et al., 1993). Oxygen consumption per unit body mass was calculated from the body mass measured immediately before each experiment. During the last 30 min before the treatment, mass specific oxygen consumption under thermoneutral conditions was first determined separately for each pup. For each of the 3 experimental groups at each age the average of the values measured at thermoneutrality was used as basis for calculating the percentage by which $\dot{V}O_2$ increased. Oxygen consumption measurements are presented as means over periods of 5-20 min duration as necessary to appropriately dissolve the more or less rapid changes in the different parts of the experiment.

Temporal mean Tc was evaluated in each pup for the last 30 min before the injection, and in 5-min periods thereafter. In text and figures mean values ± SE are presented. The statistical significance of differences between groups was evaluated by Mann-Whitney U-testing, and significance levels of $2P \leq 0.05$ were used throughout this study.

RESULTS

Responses to norepinephrine

When NE (100, 200, 400 or 800 µg/kg) was injected into 1-day-old pups their $\dot{V}O_2$ increased rapidly. The initial rate of increase was highest at the highest dose, which resulted in $\dot{V}O_2$ increasing roughly 100 % above the thermoneutral level (25±0.3 $ml \cdot min^{-1} \cdot kg^{-1}$, N=44) within 20 min after the injection. At all doses the NE-stimulated increases in VO_2 reached their highest levels between 30 and 50 min after the injection, but these maximum levels did not increase with increasing dosages. The highest levels were 58±2.6 $ml \cdot min^{-1} \cdot kg^{-1}$ (N=6) after 200 µg/mg and 57±3.5 $ml \cdot min^{-1} \cdot kg^{-1}$ (N=6) after 400 µg/kg, whereas the maximum response to the 800-µg/kg dose was only 49±1.7 $ml \cdot min^{-1} \cdot kg^{-1}$ (N=8). This value is significantly lower and does not differ from the highest $\dot{V}O_2$ reached after the 100-µg/kg dose. Figure 1 (top) demonstrates the time course of the $\dot{V}O_2$ and Tc increases both for the dose resulting in the highest maximum $\dot{V}O_2$ and for the dose resulting in the highest rate of increase. The increase of Tc was correspondingly faster at the higher dose, and the highest Tc reached after the 800-µg/kg dose (39.1±0.2, N=12) was slightly lower than the highest Tc reached after the 200-µg/kg dose (39.5±0.2, N=7).

When 800 ug/kg NE injections were given to 11-day-old rats, their $\dot{V}O_2$ increased within 20 min to higher maximum values, about 140 % above the thermoneutral level (32±0.6 $ml \cdot min^{-1} \cdot kg^{-1}$, N=20) which was also significantly higher than that of 1-day-old rats (Fig. 1, bottom). After the 200-µg/kg dose, $\dot{V}O_2$ increased by only about 80 % before it started to decrease within 20 min. Correspondingly, Tc increased only to 39.0±0.2°C (N=11) at the lower dose, whereas it reached 40.6±0.1°C (N=11) after the higher dose. Injection of 1600 µg/kg into three 11-day-old pups resulted in slightly larger responses than the 800-µg/kg dose, with Tc leveling off at 41°C. But as these pups appeared severely stressed after the experiment no other pups were given this high dose. A 100-µg/kg dose was also tested in only a few 11-day-old pups, since it resulted in very weak and inconsistent responses at this age.

Responses to tyramine

When tyramine was injected into 1-day-old rats (N=23), the rates with which $\dot{V}O_2$ and Tc increased and the maximum levels reached were similar to those recorded after the injections of 800 µg/kg NE, but the duration of the response was shorter (Fig. 2, top). Similarily, when TY was injected into 11-day-old pups (N=21), $\dot{V}O_2$ and Tc increased at about the same rate as after the injection of 800 µg/kg NE. (Fig. 2, bottom). The maximum $\dot{V}O_2$ reached was with >160 % above the thermoneutral level slightly higher than that reached after the NE injection. But because of the shorter duration of the response, Tc reached only 40.4±0.1°C (N=21) before decreasing rapidly.

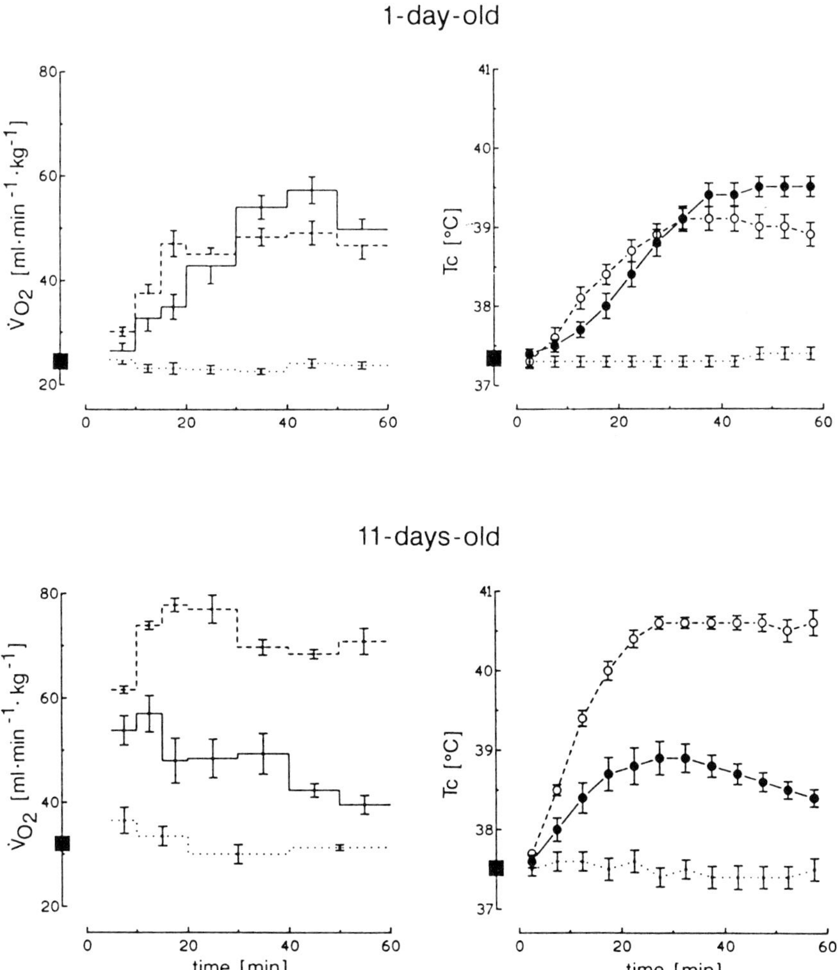

Fig. 1. Oxygen consumption ($\dot{V}O_2$) and core temperature (Tc) of 1-day-old and 11-day-old pups after injection of norepinephrine (NE) at time 0. VO_2 recording was started 5 min after the injection. Solid lines, filled circles: 200 µg/kg. Dashed lines, open circles: 800 µg/kg. Dotted lines: saline. Squares on y-axis indicate the thermoneutral values before NE-injection.

Responses to a step decrease in ambient temperature

In response to the onset of ambient cold loads, $\dot{V}O_2$ of 1-day-old pups increased much slower than after pharmacological sympathetic stimulation (Fig. 3, top). A steady state doubling of $\dot{V}O_2$ was reached only about an hour after the step decrease of Ta. Pups 9-11 days of age, in contrast, increased their $\dot{V}O_2$ immediately in response to the step decrease in Ta (Fig. 3, bottom), doubling their $\dot{V}O_2$ within less than 30 min.

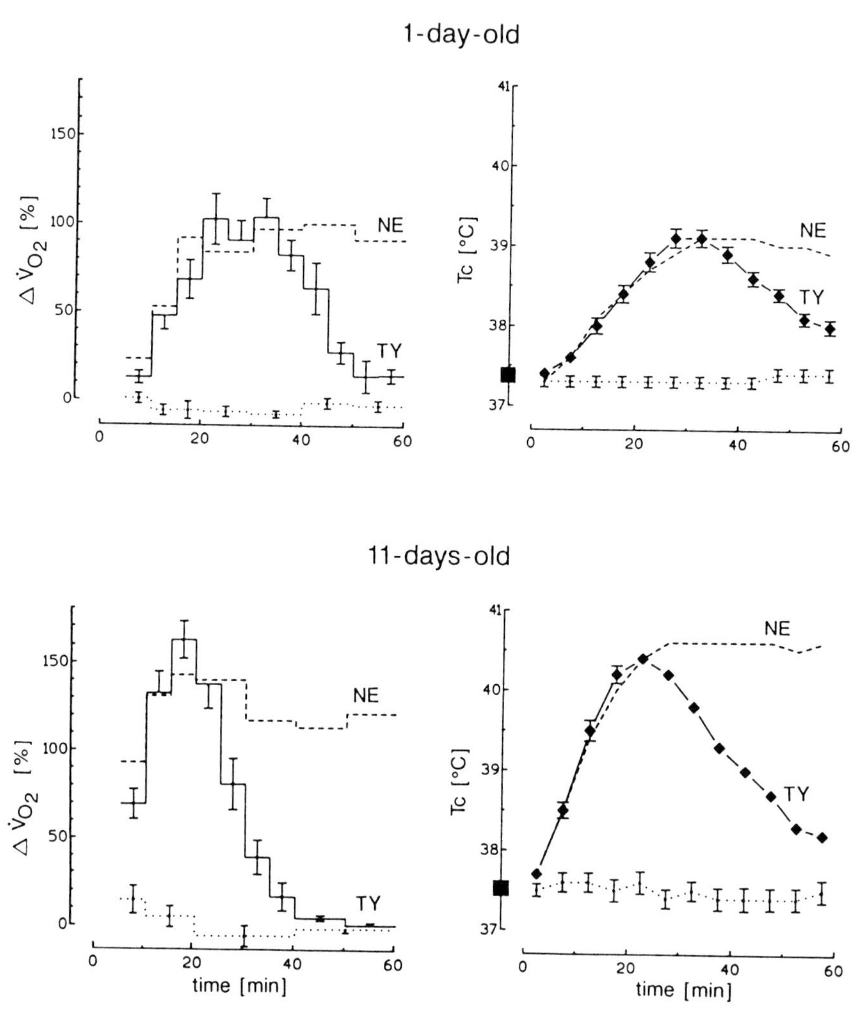

Fig. 2. Increase in the oxygen consumption ($\triangle\dot{V}O_2$ as a percentage of the thermoneutral value) and core temperature (Tc) of 1-day-old and 11-day-old pups after injection of tyramine (TY, 10 mg/kg, solid lines) at time 0. $\dot{V}O_2$ recording was started 5 min after the injection. The responses to norepinephrine (NE, 800 µg/kg, interrupted lines) are indicated for comparison. Dotted lines show responses to saline. Squares on y-axis indicate Tc at thermoneutrality.

DISCUSSION

The main finding of this study is that 1-day-old rats can roughly double their metabolic rate within 20 min of pharmacological sympathetic stimulation (with high doses of NE), whereas they need about an hour to double their metabolic rate after an appropriate step decrease of Ta. This means that the sluggish activation of the newborn pups' cold defense cannot be attributed to the immaturity of the thermogenic apparatus in their BAT. In 9- to 11-day-old pups tested in the maximum phase of the juvenile circadian Tc cycle, $\dot{V}O_2$ increases nearly as promptly in response to the onset of an ambient cold load as to pharmacological sympathetic stimulation. Correspondingly, Tc of 1-day-old pups decreases by about 3°C whereas that of 11-day-old pups decreases by less than 1°C after a step decrease of Ta triggering roughly a doubling of $\dot{V}O_2$ in both age groups (Körtner et al., 1993).

Because the used air flow rates were rather low, we have to consider the degree to which the recorded $\dot{V}O_2$ values were distorted by the dead space in the measuring system. Figure 3, shows that the delay in the response of the measuring system is small in comparison with the physiological delay of the 1-day-old rats'

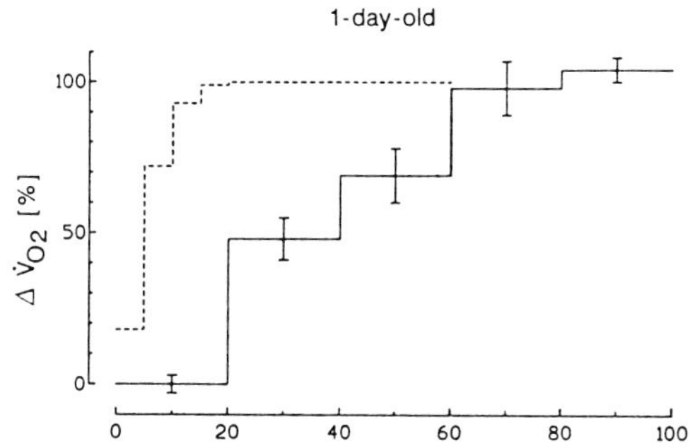

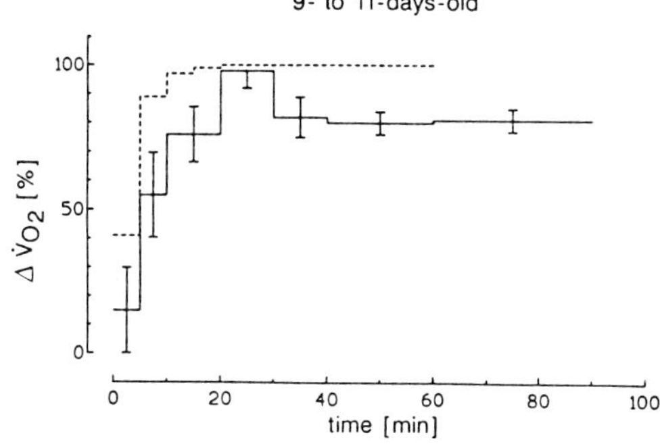

Fig. 3. The solid lines show the increase in the oxygen consumption ($\Delta\dot{V}O_2$ as a percentage of thermoneutral value) of 1-day-old and 9- to 11-day-old pups at the onset of cold exposure. Starting at time 0 ambient temperature had been decreased within 5 min to 30 and 27°C, respectively. For comparison the dotted lines show the $\dot{V}O_2$ measured by the oxygen analyzer when air was aspirated like in the experiments with 100 ml/min (top) or 200 ml/min (bottom) but when oxygen was consumed at a constant rate by zinc-air batteries put into the animal chamber at time 0.

response to a decrease of Ta. For 11-day-old rats, on the other hand, the delay in the $\dot{V}O_2$ increase after the onset of the cold load is only slightly larger than the physical delay of the measuring system. The physiological difference between the responses of 1- and 11-day-old rats to the step decrease of Ta was thus much larger than the distortion caused by the use of the different flow rates. Moreover, the time lag in the system could not distort comparisons of the rate of change between treatment groups, since all pups of the same age were measured with the same flow rates. And because, with the exception of the TY response of 11-day-old pups, the rate of change became very small before $\dot{V}O_2$ started to decrease again, the absolute values given above for the maximum responses are also accurate.

To compare the initial rate of change of $\dot{V}O_2$ between the two age groups, however, we need to correct for the different rates at which the air samples were aspirated. Because this kind of correction requires closely spaced data points, we cannot calculate instantaneous $\dot{V}O_2$ (Bartholomew et al., 1981) based on the mean values presented in Fig. 1-3. Corrections factors "z" were therefore determined from data sampled at 1-min intervals while oxygen was consumed at a constant rate by zinc-air batteries. These factors (z=0.22 for air aspirated at 100 ml/min and z=0.30 for air aspirated at 200 ml/min) were then used to calculate instantaneous $\dot{V}O_2$ from the 4th-order polynominals fitted to the $\dot{V}O_2$ data gathered after tyramine injection (Fig. 4). Obviously, the slightly faster increase of $\dot{V}O_2$ to its maximum in the older pups suggested by the raw data was not an artefact of the higher air flow rate applied, but reflects a physiological difference.

By eliciting the release of endogenous NE, tyramine mimics the effects of sympathetic stimulation (Kiang-Ulrich and Horvath, 1982), and the thermogenic response of both 1-day-old and 11-day-old rats given TY injections closely resembles their responses to high doses of NE, except for being more transitory. The slightly slower response of 1-day-old rats to pharmacological sympathetic

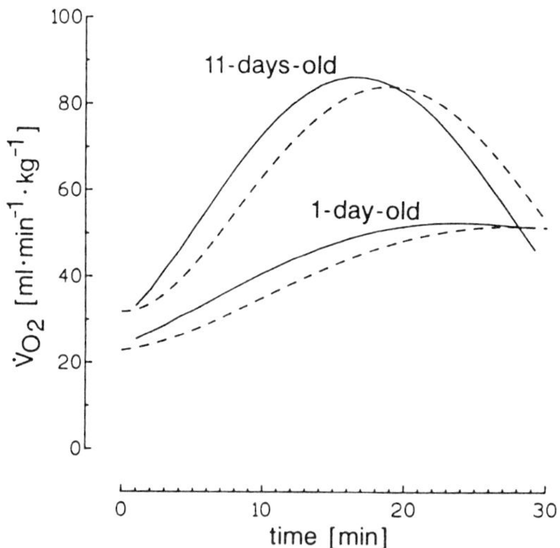

Fig. 4. The solid lines show instantaneous oxygen consumption ($\dot{V}O_2$) calculated from 4th-order polynomials (dashed lines) fitted to the $\dot{V}O_2$ data of 1-day-old rats (r=0.67) and 11-day-old rats (r=0.87) after tyramine injection. (See text for procedure used to calculate instantaneous $\dot{V}O_2$).

stimulation may indicate some postsynaptic differences in the thermogenic apparatus. The difference in the response to TY is, however, much smaller than the developmental change in the response to the sudden onset of cold load. This means that although the NE-content and turn-over in BAT of 1-day-old rats have been reported to be markedly lower that those in the BAT of two-week-old rats (Bertin et al., 1990), the amount of endogenous NE available is also not the limiting factor causing the sluggish response to ambient cold. In addition to their slower thermogenic response, however, the 1-day-old pups have a markedly lower thermoneutral level of metabolic rate and a lower maximum $\dot{V}O_2$ after pharmacological stimulation - even when expressed as a percentage of the thermoneutral value (see Fig. 2). This difference might reflect the lower uncoupling protein content in BAT of 1-day-old pups, supporting the assumption that total thermogenic capacity of BAT increases between 1 and 11 days of age (Porras et al., 1990).

Our results confirmed the observations of Hsieh et al. (1971), who found that the sensitivity to NE is much higher in 1-day-old pups than in older pups whereas their maximum $\dot{V}O_2$ is smaller. Whereas the responses of our 11-day-old rats given 200 µg/kg NE were only weak and transitory, this dose elicited the largest $\dot{V}O_2$ increase in 1-day-old rats (to 2.3 times the thermoneutral level). But in contrast to the toxic effect Hsieh et al. reported to be produced by 400 µg/kg NE injected into an unspecified strain of rats - reared at 28°C and tested in the morning - our 1-day-old Zucker rats - reared at 22°C and tested in the evening - did not show a significantly depressed thermogenic response at NE dosages below 800 µg/kg. Other studies carried out with NE-treated 1-day-old rats are also not suited to determine whether this difference is due to strain, rearing temperature, or circadian phase (Moore and Underwood, 1963; Thompson and Moore, 1968).

Most studies of the juvenile circadian Tc cycle in rats have been carried out in lean (Fa/-) Zucker rat pups (Nuesslein et al., 1990; 1993; Redlin et al., 1992; Schmidt et al., 1986; 1987), but the amplitude of the daily Tc rhythm is similar in other pigmented strains (Long Evans, Brown Norway) and smaller in Wistar albino rats (Nuesslein, 1992). Neglecting the effects of circadian variation will therefore probably have a smaller effect on studies using albino rats than it will on studies using hooded strains. Developmental studies of cold defense abilities are nonetheless difficult to interpret when carried out during the first 6 h of the light phase (that is in the minimum phase of the circadian Tc cycle) because the extent to which the cold defense is suppressed during this time changes with Ta as well as with age (Nuesslein and Schmidt, 1990; 1993).

Possible peripheral and central causes for the developmental changes in response velocity have been discussed previously (Körtner et al., 1993). The results of pharmacological stimulation reported here have now clearly demonstrated that even though there are some age-related differences in sympathetically stimulated thermogenesis, the sluggish response of 1-day-old pups to sudden cold exposure, and the correspondingly large decrease in Tc, are not due to differences at the effector level.

Acknowledgements. This study was supported in part by the DFG (Schm 680). We thank Randy Kaul for his advice as NESP (Native English Speaking Person).

REFERENCES

Alberts, J.R. (1978): Huddling by rat pups: Group behavioral mechanisms of temperature regulation and energy conservation. *J. Comp. Physiol. Psychol. 92*: 231-245.

Bartholomew, G., Vleck, D., and Vleck, C.M. (1981): Instantaneous measurements of oxygen consumption during pre-flight warm-up and post-flight cooling in sphingid and saturniid moths. *J. Exp. Biol. 90*: 17-32.

Bertin, R., Mouroux, I., De Marco, F., Portet, R. (1990): Norepinephrine turnover in brown adipose tissue of young rats: effects of rearing temperature. *Am. J. Physiol. 259*: R90-R96.

Conklin, P., Heggenes, F.W. (1971): Maturation of temperature homeostasis in the rat. *Am. J. Physiol. 220*: 333-336.

Hsieh, A.C.L., Emery, N., and Carlson, L.D. (1971): Calorigenic effect of norepinephrine in newborn rats. *Am. J. Physiol. 221 (6)*:1568-1571.

Kiang-Ulrich, M., and Horvath, S.M. (1982): Age- and strain-related differences in metabolic response to Tyramine in rats. *Proc. Soc. Exp. Biol. Med. 171*: 242-246.

Körtner, G., Schildhauer, K., Petrova, O., and Schmidt, I. (1993): Rapid changes in metabolic cold defense and GDP binding to the brown adipose tissue mitochondria of rat pups. *Am. J. Physiol. 264*: R1017-R1023.

Markewicz, B., Kuhmichel, G., and Schmidt, I. (1993): Contributions of decreased thermogenesis to fueling the onset of obesity in Zucker rats. *Am. J. Physiol. 265, (Endocrinol. Metab.* 28): E000-E000 (in press).

Moore, R.E., and Underwood, M.C. (1963): The thermogenic effects of noradrenaline in new-born and infant kittens and other small animals. A possible hormonal mechanism in the control of heat production. *J. Physiol. London 168*: 290-317.

Mumm, B., Kaul, R., Heldmaier, G., Schmidt, I. (1989): Endogenous 24-hour cycle of core temperature and oxygen consumption in week-old Zucker rat pups. *J. Comp. Physiol. B 159*: 569-575.

Nedergaard, J, Connoly, E., Cannon, B. (1986): Brown adipose tissue in the mammalian neonate. In *Brown adipose tissue*, ed. P. Trayhurn and D.G.Nicolls. Arnold, London: 152-213.

Nuesslein, B., and Schmidt, I. (1990): Development of circadian cycle of core temperature in juvenile rats. *Am. J. Physiol. 259*: R270-R276.

Nuesslein, B. (1992): Der juvenile circadiane Kerntemperatur-Rhythmus von Ratten: Physiologische Grundlagen und synchronisierende Faktoren. Theses, Universität Marburg.

Nuesslein, B., and Schmidt, I. (1993): Is core temperature regulated in the minimum of the rat's juvenile circadian cycle? IUPS Thermal Physiology Commission, Symposium on Temperature Regulation, Aberdeen, Scotland.

Nuesslein, B., Petrova, O., Schildhauer, K., and Schmidt, I. (1993): Morning depression of cold defense in juvenile rats. In *Integr. Cell. Asp. Auton. Funct.*, eds K. Pleschka, R. Gerstberger, and K. Fr. Pierau, London: John Libbey.

Porras, A., Penas, M., Fernandez, M., Benito, M. (1990): Development of the uncoupling protein in the rat brown- adipose tissue during the perinatal period. *Europ. J. Biochem. 187*, 671-675.

Redlin, U., Nuesslein, B., and Schmidt, I. (1992): Circadian changes of brown adipose tissue thermogenesis in juvenile rats. *Am. J. Physiol. 262*: R504-R508.

Schmidt, I., Barone, A., Carlisle, H.J. (1986): Diurnal cycle of core temperature in huddling, week-old rat pups. *Physiol. Behav. 37*: 105-109.

Schmidt, I., Kaul, R., and Heldmaier, G. (1987): Thermoregulation and diurnal rhythms in 1-week-old rat pups. *Can. J. Physiol. Pharmacol. 65*: 1355-1364.

Spiers, D.E., Adair, E.R. (1986): Ontogeny of homeothermy in the immature rat: metabolic and thermal responses. *J. Appl. Physiol. 60 (4)*: 1190-1197.

Thompson, G.E., Moore, R.E. (1968): A study of newborn rats exposed to the cold. *Can. J. Physiol. Pharmacol. 46*: 865-871.

Withers, P.C. (1977): Measurement of $\dot{V}O_2$, $\dot{V}CO_2$, and evaporative water loss with a flow-through mask. *J. Appl. Physiol. 42*: 120-123.

Integrative and cellular aspects of autonomic functions : temperature and osmoregulation. Eds K. Pleschka, R. Gerstberger. John Libbey Eurotext, Paris © 1994, pp. 295-304.

Morning depression of cold defense in juvenile rats

Barbara Nuesslein, Olga Petrova, Kirsten Schildhauer and Ingrid Schmidt

Max-Planck-Institut, W.G. Kerckhoff-Institut, Parkstrasse 1, D-61231 Bad Nauheim, Germany

SUMMARY

The aim of this study was to compare the cold defense responses of 1-, 5-, and 11-day-old rat pups exposed to a step decrease in ambient temperature (Ta) at the beginning of the light phase with the previously investigated responses of pups similarily exposed at the beginning of the dark phase. The core temperature (Tc) and oxygen consumption ($\dot{V}O_2$) of isolated littermates was recorded while the pups were maintained above their lower critical temperature for 1-2 h and after Ta was decreased within 5 min to values that result in a doubling of $\dot{V}O_2$ when pups are tested in subjective evening. The $\dot{V}O_2$ of 11- and especially of 5-day-old pups tested in the subjective morning increased very sluggishly, and reached only 65% above the thermoneutral level after 90 min, whereas their $\dot{V}O_2$ nearly doubles within 30 min when they were tested in the evening. The $\dot{V}O_2$ of 1-day-old pups tested in the morning increased with a time course similar to that of pups tested in the evening, reaching 100% above the thermoneutral level after about 1 h. A markedly different picture of the developmental changes will thus be obtained depending on the circadian phase in which the activation of cold defense is investigated in suckling-age rat pups.

INTRODUCTION

During the first postnatal week rat pups develop an endogenous circadian core temperature (Tc) rhythm characterized by a sharp decrease in Tc around the time of lights-on (the minimum phase) and a high, stable Tc around the time of lights-off (the maximum phase) (Nuesslein and Schmidt, 1990). When 10-day-old rat pups are artificially reared at moderate cold loads, during the minimum phase their oxygen consumption ($\dot{V}O_2$) might drop to 50% of the value measured at thermoneutrality. During the maximum phase, on the other hand, their $\dot{V}O_2$ at the same cold load is twice as high than at thermoneutrality (Redlin et al., 1992). These and those few other developmental studies on the thermoregulatory abilities of suckling-age rat pups in which circadian phase was considered, were restricted to the study of steady-state responses (Spiers, 1988; Planche and Joliff, 1987), whereas no information was available about possible circadian changes in the speed with which the cold defense was activated. After a recent study during the maximum phase of the juvenile circadian Tc rhythm had demonstrated that brown adipose tissue (BAT) activity increases - in parallel with metabolic rate - much faster in 5- and 11-day-old pups than in 1-day-old rat pups (Körtner et al., 1993), the aim of the present study was to compare those responses with the responses to identical cold loads experienced during the minimum phase.

METHODS

Animals. We used lean (Fa/-) Zucker rat pups for these experiments, because the juvenile circadian Tc rhythm has been most thoroughly characterized in this hooded strain (Mumm et al., 1989; Nuesslein und Schmidt, 1990; Redlin et al., 1992). Four litters (7 to 11 pups per litter) were tested at the ages 1, 5, and 11 days (day of birth = day 0), when they had average body weights of 6.4±0.1 (N=32), 10.3±0.2 (N=30), and 19.3±0.3 g (N=38). The litters were reared by their mothers, who had free access to food and water, in colonies maintained at 22°C and 60% air humidity under a 12:12 hour light:dark cycle. To avoid disturbing the animals during their resting phase, the colonies were routinely tended during the last hour of the light phase or under dim red light during the dark phase.

Experimental set-up. The experiments were performed in a set-up otherwise used for artificial rearing (Mumm et al., 1989; Kaul et al., 1990; Nuesslein and Schmidt, 1990). In short, in a climatic chamber kept at a relative humidity of 60-70%, each pup of a litter was isolated on absorbent padding in a 8•8•9-cm plastic chamber. These chambers floated in a water bath kept 1°C warmer than the air temperature in the climatic chamber to prevent condensation. The air temperature inside the animal chambers closely followed the changes of water temperature.

Temperature measurements. Colonic Tc was continuously measured with 0.06-mm copper-constantan thermocouples (California Finewire) sheathed in PP 10 that was sealed with a soft silastic tip. To ensure an accurate record of deep body temperature, thermocouples were inserted 1.8, 2.2 and 2.8 cm into 1-, 5- and 11-day-old pups. Water temperature (Ta) and each animals Tc were continuously recorded on a multichannel point printer.

Oxygen consumption measurements. The oxygen consumption of 7 pups in each experiment was measured successively by a multichannel open-flow system (Fig. 1). Metal tubes loosely fitting through a centered hole in the lid continuously aspirated air from near the bottom of each animal chamber. Airflows adjusted to 100, 150, and 200 ml/min for 1-, 5-, and 11-day-old pups maintained oxygen extraction between 0.1 and 1%. As shown by another study in this volume (Döring et al., 1993), the different flow rates used for each age group produced artifactual differences in the measured response rate that were so much smaller than the physiological differences between the groups that they could be neglected in this study. Every 2.5 min, a timing circuit connected to eight 3-way magnetic valves successively connected the dried airflows from one of the 7 animal chambers or a reference channel to a TYLAN massflow meter and an AMETEK S-3A electrochemical oxygen analyzer. Needle valves in the air stream from each chamber and in each bypass pumping system were preset to keep the pressure drop across each chamber and therefore the airflow through each chamber constant throughout each experiment.

Experimental protocol. With the minor modifications pointed out below, we used the same experimental procedure as in the preceeding study (Körtner et al., 1993), except that the present experiments were started at the end of the dark phase instead at the end of the light phase. Seven to eleven pups from one litter were brought into an approximately thermoneutral environment, equipped with thermocouples, and placed into the animal chambers. Their $\dot{V}O_2$ was successively recorded while they were then maintained above their lower critical temperature for 1-2 h: the ambient temperatures (Ta) used here were 35.5°C for the 1-day-old pups, 34.5°C for the 5-day-old pups and 33.5°C for the 11-day-old

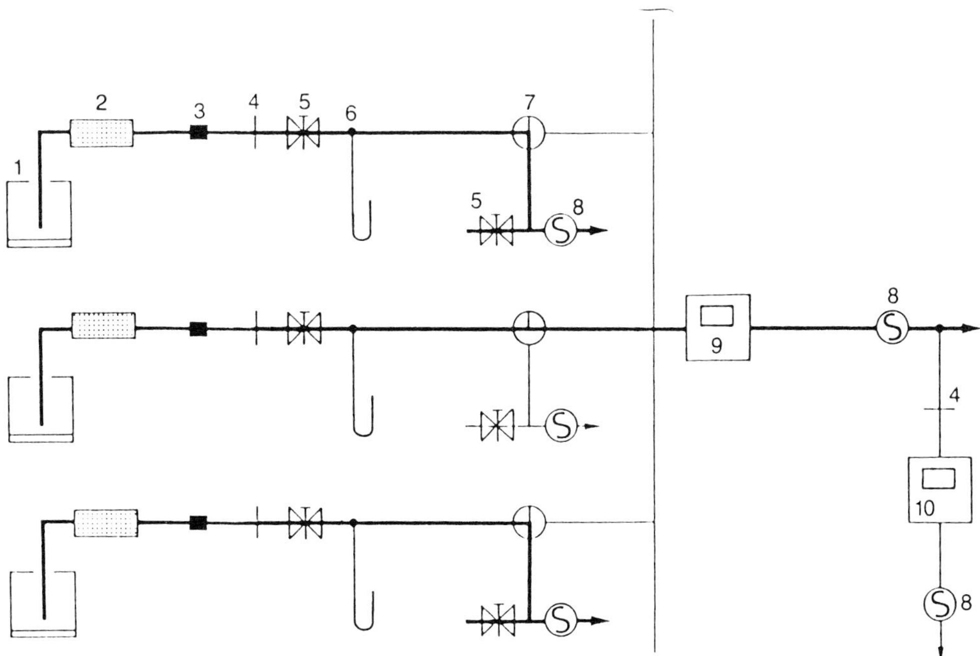

Fig. 1. Oxygen analyzing system (three of eight channels). 1, animal container; 2, anhydrous $CaSO_4$ (Drierite); 3, activated charcoal (to absorb organic gases); 4, membrane filter; 5, needle valve; 6, water manometer for setting and monitorng bypass airflow equal to the measured airflow; 7, magnetic valve; 8, pump; 9, massflow meter; 10, electrochemical oxygen analyzer.

pups. Then Ta was lowered within 5 min to the same values which had resulted roughly in a doubling of oxygen consumption in each age group during the maximum phase of the Tc rhythm (Körtner et al., 1993). These cold-load temperatures were 30°C for 1-day-old pups, 29°C for 5-day-old pups, and 26°C for 11-day-old pups. The pups in these experiments were thus exposed to a step decrease in Ta about 2 h after the start of the light phase in the home colony that is, at the time at which the circadian minimum of Tc occurs in long-term cold exposed pups (Mumm et al., 1989; Nuesslein and Schmidt, 1990, Nuesslein, 1992).

Evaluation. The temporal mean Tc of each pup was evaluated over 30-min periods during the thermoneutral exposure and over 5-min periods during the cold exposure, and average values are presented for each of the 4 litters. We assumed an RQ of 0.8 (Markewicz et al., 1993), and calculated the $\dot{V}O_2$ of each pup from the massflow rate and the oxygen content of the airstream aspirated from its chamber (Withers, 1977). Body mass measured before the start of each experiment was used to calculate mass-specific oxygen consumption per unit body mass. Because the $\dot{V}O_2$ was recorded successively, these data are presented here as the means of all values measured in 30-min-periods during exposure to thermoneutral conditions ($20 \leq N \leq 40$) and in 10-min-periods during cold exposure ($6 \leq N \leq 16$). Mean values are presented ±SE. The statistical significance ($2P \leq 0.05$) of differences between groups was evaluated by using the Mann-Whitney U-test.

We similarly evaluated the cold defense responses of the 1- to 11-day-old pups previously studied at the beginning of the dark phase (Körtner et al., 1993). Because the 5- and 11-day-old pups in that study (3 litters, each with 9 to 10 pups) had been killed at various times after the onset of the cold load for analysis of BAT, few animals were studied for a full hour after the onset of the cold exposure. We therefore included the data from the habituation experiments carried out with the same litters on day 3 and 4 and on days 9 and 10. With the Ta chosen slightly (0.5-1°C) higher to match the greater heat loss of the smaller, younger pups, the Tc and mass-specific $\dot{V}O_2$ values measured during the habituation experiments differed insignificantly from those measured on day 5 and day 11. The data for 1-day-old pups studied at the beginning of the dark phase (N=27 from 4 litters) were taken from an experimental series in which no pups had been killed during the 90-min cold exposure. For each age group, the data shown in the figures are mean values (±SE) of all data gathered within the specified time intervals: for Tc, $15 \leq N \leq 90$; and for $\dot{V}O_2$, $21 \leq N \leq 90$ under thermoneutral conditions and $5 \leq N \leq 28$ at cold-load conditions.

RESULTS

For all groups, the average core temperature during the last 30 min before the cold exposure was close to 37.5°C (Fig. 2). Like the Tc of 1-day-old pups cold exposed at the beginning of the dark phase, the Tc of 1-day-old pups tested at the beginning of the light phase dropped below 35°C within the first half hour and then started to slowly increase again. But whereas the Tc of 3- to 5-day-old pups cold exposed at the beginning of the dark phase dropped only to about 36°C before stabilizing slightly above this

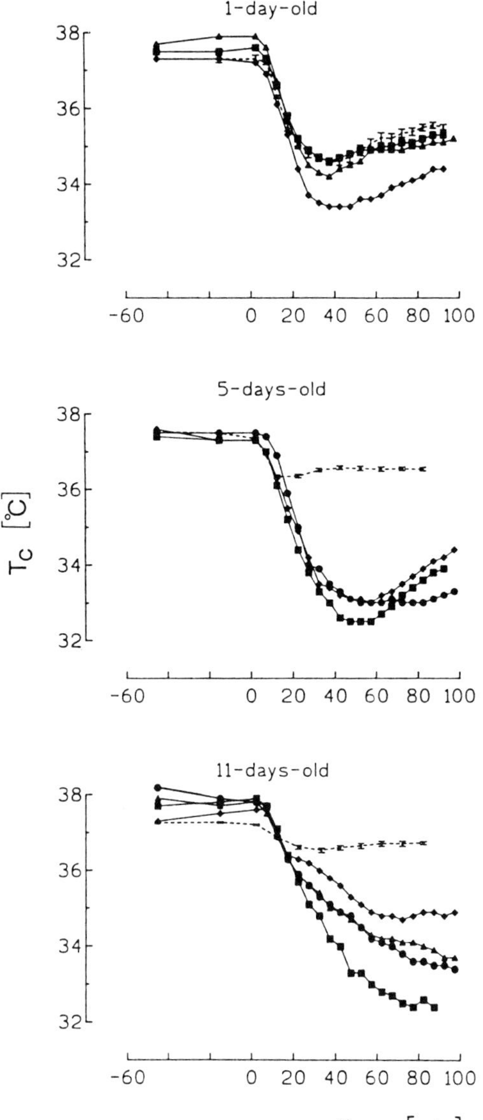

Fig. 2. Core temperatures (Tc) at thermoneutrality and after the onset (time 0) of cold exposure during the beginning of the light phase (solid lines) and dark phase (dotted lines). For the pups tested during the beginning of the dark phase mean values (±SE) derived from Körtner et al. (1993) are shown. For each of the four litters tested during the beginning of the light phase data are shown separately (SE, which for clarity is not shown, was usually 0.1-0.3°C and never exceeded 0.6°C). One of the litters was not tested at 1 day of age, and another not at 5 days of age.

level, the Tc of 5-day-old pups tested at the beginning of the light phase continued to fall for about 1 h to 33°C or less, before increasing again. And the Tc of 9- to 11-day-old pups cold exposed at the beginning of the dark phase stabilized only slightly below 37°C, whereas that of 11-day-old pups tested at the beginning of the light phase decreased continuously to values around 34°C after 100 min of cold exposure.

At the end of the thermoneutral period, the $\dot{V}O_2$ ($ml{\bullet}min^{-1}{\bullet}kg^{-1}$) for each age group was slightly lower at the beginning of the light phase then it was at the beginning of the dark phase: it averaged 22.2±0.3 (N=32) in the 1-day-old pups, 27.2±0.5 (N=27) in the 5-day-old pups and 29.7±0.6 (N=36) in the 11-day-old pups (Fig. 3). During the cold exposure, the $\dot{V}O_2$ of the 1-day-old pups increased similarily during the minimum and maximum phases of the juvenile circadian Tc cycle. It increased significantly within the third 10-min period after the drop of ambient temperature and had reached twice the thermoneutral level after one hour. But in contrast to the near doubling of the $\dot{V}O_2$ of 3- to 11-day-old pups exposed to cold at the beginning of the dark phase, the $\dot{V}O_2$ of 5- and 11-day-old pups exposed to cold at the beginning of the the light phase never reached values even 70% above the thermoneutral level. Whereas the $\dot{V}O_2$ of 3- to 5-day-old pups increased significantly within the second and that of the 9- to 11-day-old pups within the first 10-min period after the onset of the cold exposure at the beginning of the dark phase, the 5-day-old pups responded much more sluggishly than did the 11-day-old pups at the beginning of the light phase. At this time their $\dot{V}O_2$ did not increase significantly until 1 hour after the onset of cold exposure, whereas that of the 11-day-old pups increased significantly during the second 10-min period.

DISCUSSION

The main finding of this study is that the speed with which 5-day-old rat pups activate cold defense changes dramatically with circadian phase. Whereas 3- to 5-day-old pups respond nearly as promptly to the onset of physiologically comparable cold loads as two-week-old pups at the beginning of the dark phase, the responses of 5-day-old pups tested at the beginning of light phase rather resemble those of 1-day-old pups. The influence of circadian phase on the rate of increase of $\dot{V}O_2$ is less dramatic in 11-day-old pups, but they also show a striking circadian difference in their ability to stabilize Tc during cold exposure. Only the 1-day-old pups did not show any circadian changes in their cold defense responses. Circadian differences might thus explain why the pioneering study of Thompson and Moore (1968) found the speed of cold defense activation to increase markedly between 6 and 12 days of age, whereas Körtner et al. (1993) found the major change in the time course of cold defense to occur between 1 and 3 days of age. Because BAT is the crucial effector organ for the juvenile cold defense, circadian differences might also contribute to the discrepancies between different reports on the developmental changes of BAT parameters (Körtner et al., 1993; Mouroux et al., 1990; Porras et al., 1990; Sundin and Cannon, 1980).

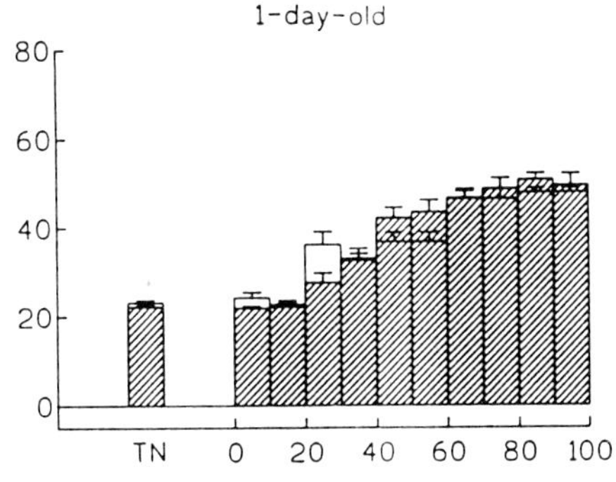

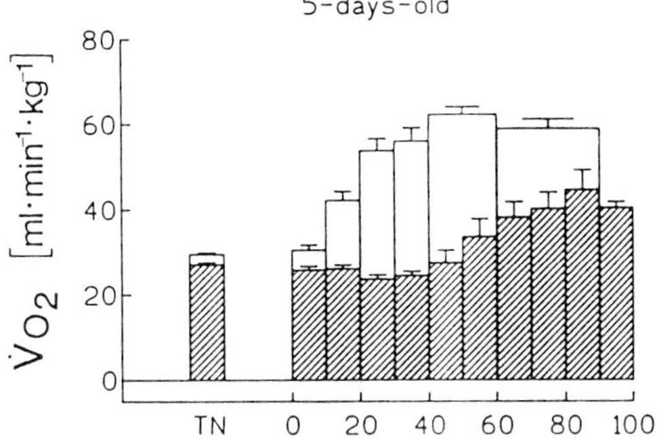

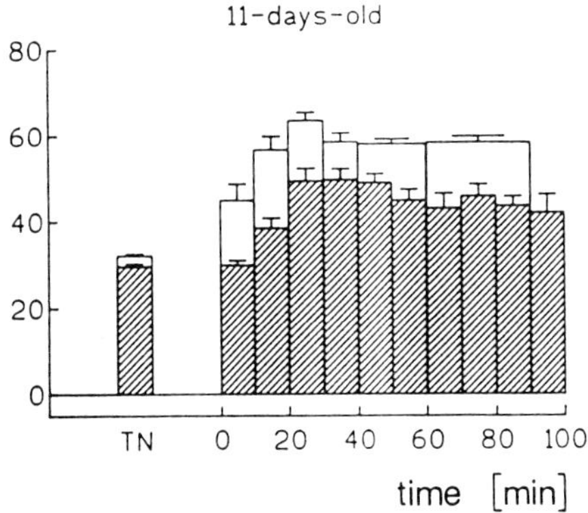

Fig. 3. Mean values (±SE) of oxygen consumption ($\dot{V}O_2$) at thermoneutrality (TN) and after the onset of cold exposure (at time 0). Hatched bars: values measured during the beginning of the light phase. Clear bars: values measured during the beginning of the dark phase (derived from Körtner et al., 1993).

The amplitude of the juvenile circadian Tc cycle changes with age and with Ta (Nuesslein, 1992; Nuesslein and Schmidt, 1990; 1993). If pups of different ages are chronically exposed to physiologically comparable cold loads - i. e. to loads stimulating the same percentual increase of $\dot{V}O_2$ above the thermoneutral level (Körtner et al., 1993) - the greatest amplitude of the rhythm occurs in 10- to 12-day-old pups (Nuesslein and Schmidt, 1990). If isolated artificially reared pups of this age are chronically exposed to ambient conditions that in the circadian maximum result in the $\dot{V}O_2$ increasing 100% above the thermoneutral level, in the circadian minimum their $\dot{V}O_2$ decreases below the thermoneutral level and their Tc drops to about 32°C (Redlin et al., 1992). In 5-day-old pups under corresponding conditions, the Tc in the circadian minimum falls only to about 34°C and $\dot{V}O_2$ remains above the thermoneutral level (Mumm et al., 1989). The present results, however, show that the age dependency of the circadian changes in the acute activation of cold defense might differ from the age dependency of circadian changes in the steady-state. The morning depression of the cold defense response to a rapid decrease in Ta was considerably larger in 5-day-old pups than in 11-day-old pups, though their evening response to the same cold load had been close to a 100% increase of $\dot{V}O_2$ for both age groups. This indicates that already earlier than expected from the investigations of the static thermoregulatory responses, circadian phase exerts a very strong influence on the responses to quickly changing ambient conditions - i. e. conditions that are typical of the microenvironment of pups huddling in the nest.

After the second day postpartum mother rats begin to leave their litters unattended for increasingly long periods, mainly during the dark phases (Croskerry et al., 1976). In the mother's absence, litters control their Tc by a combination of autonomic and behavioral thermoregulation. Under these conditions the huddling pups regularly exchange the warmer inner and the colder outer positions in the nest (Alberts, 1978; Schmidt et al., 1986; 1987). The rapid and intense activation of autonomic cold defense occuring in pups three days old and older at the beginning of the dark phase is therefore an important factor ensuring that huddling pups are able to maintain a high and stable Tc at that part of the day when the mother is least likely to return to the nest soon (Schmidt et al., 1986; 1987). For 1-day-old pups and for older pups at the beginning of light phase, on the other hand, a sluggish activation of autonomic cold defense at times when the mother is likely to soon be available for rewarming can result in a significant saving of energy (Nuesslein, 1992). Whether this response pattern is further strengthened by corresponding circadian changes in the thermoregulatory behavior of the huddling pups remains to be investigated.

Although it is also unclear whether or not the circadian changes of maternal nursing activity result in a decreased milk supply for pups at the end of the dark phase (for discussion see Schmidt et al., 1986), experiments in artificially reared pups have clearly shown that the decreased activation of cold defense around the time of lights-on is independent of circadian changes in food availability (Nuesslein and Schmidt, 1990; Nuesslein, 1992). Infusion of norepinephrine in high doses has demonstrated that the ability of 10-day-old pups to generate heat is not impaired at the time of their circadian minimum (Nuesslein and Schmidt, 1993).

Neither can the circadian-phase-independent sluggish cold defense response of 1-day-old pups be attributed to a deficit in norepinephrine-stimulated thermogenesis (Döring et al., 1993). Central changes thus seem to underlie the economically slow activation of cold defense responses observed throughout the entire day in 1-day-old pups and at the beginning of the light phase in older pups.

Acknowledgements. This study was supported in part by the DFG (Schm 680/1-3). We are grateful to Roswitha Bender, Irene Küchenmeister and Tanja Wolf for their skillful technical help in this as well as in many preceeding studies. Furthermore we like to thank Randy Kaul for his advice as NESP (Native English Speaking Person).

REFERENCES

Alberts, J.R. (1978): Huddling by rat pups: Group behavioral mechanisms of temperature regulation and energy conservation. *J. Comp. Physiol. Psychol. 92*: 231-245.

Croskerry, P. G., Smith, G. K., Leon, L. N., and Mitchell E. A. (1973): An inexpensive system for continuously recording maternal behavior in the laboratory rat. *Physiol. Behav. 16*: 223-225.

Döring, H., Körtner, G., Meyer, K., and Schmidt, I. (1993): Do changes of sympathetically stimulated thermogenesis underlie the developmental changes in the cold defense of rat pups? In *Integr. Cell. Asp. Auton. Funct.*, eds K. Pleschka, R. Gerstberger, and K. Fr. Pierau, London: John Libbey.

Kaul, R., Heldmaier, G., and Schmidt, I. (1990): Defective thermoregulatory thermogenesis does not cause onset of obesity in Zucker rats. *Am. J. Physiol. 259*: E11-E18.

Körtner, G., Schildhauer, K., Petrova, O., and Schmidt, I. (1993): Rapid changes in metabolic cold defense and GDP binding to the brown adipose tissue mitochondria of rat pups. *Am. J. Physiol.*, 264: R1017-R1023.

Markewicz, B., Kuhmichel, G., and Schmidt, I. (1993): Contributions of decreased thermogenesis to fueling the onset of obesity in Zucker rats. *Am. J. Physiol. 265 (Endocrinol. Metab.* 28): E000-E000 (in press).

Mouroux, I., Bertin, R., and Portet, R. (1990): Thermogenic capacity of the brown adipose tissue of developing rats; effects of rearing temperature. *J. Dev. Physiol. 14*: 337-342.

Mumm, B., Kaul, R., Heldmaier, G., Schmidt, I. (1989): Endogenous 24-hour cycle of core temperature and oxygen consumption in week-old Zucker rat pups. *J. Comp. Physiol. B 159*: 569-575.

Nuesslein, B. and Schmidt, I. (1990): Development of circadian cycle of core temperature in juvenile rats. *Am. J. Physiol. 259*: R270-R276.

Nuesslein, B. (1992): Der juvenile circadiane Kerntemperatur-Rhythmus von Ratten: Physiologische Grundlagen und synchronisierende Faktoren. Theses, Universität Marburg.

Nuesslein, B., and Schmidt, I. (1993): Is core temperature regulated in the minimum of the rat's juvenile circadian cycle? IUPS Thermal Physiology Commission, Symposium on Temperature Regulation, Aberdeen, Scotland.

Planche, E., and Joliff, M. (1987): Evolution des dépenses énergétiques chez le rat Zucker au cours de la première semaine de la vie. Effet de l'heure des mesures. *Reprod. Nutr. Dev. 27*: 673-679.

Porras, A., Penas, M., Fernandez, M., Benito, M. (1990): Development of the uncoupling protein in the rat brown-adipose tissue during the perinatal period. *Europ. J. Biochem. 187*, 671-675.

Redlin, U., Nuesslein, B., and Schmidt, I. (1992): Circadian changes of brown adipose tissue thermogenesis in juvenile rats. *Am. J. Physiol. 262*: R504-R508.

Schmidt, I., Barone, A., Carlisle, H.J. (1986): Diurnal cycle of core temperature in huddling, week-old rat pups. *Physiol. Behav. 37*: 105-109.

Schmidt, I., Kaul, R., and Heldmaier, G. (1987): Thermoregulation and diurnal rhythms in 1-week-old rat pups. *Can. J. Physiol. Pharmacol. 65*: 1355-1364.

Spiers, D. E. (1988): Nocturnal shifts in thermal and metabolic responses of the immature rat. *J. Appl. Physiol. 64(5)*: 2119-2124.

Sundin, U., and Cannon, B. (1980): GDP-binding to the brown fat mitochondria of developing and cold-adapted rats. *Comp. Biochem. Physiol. 65B*: 463-471.

Thompson, G.E., Moore, R.E. (1968): A study of newborn rats exposed to the cold. *Can. J. Physiol. Pharmacol. 46*: 865-871.

Withers, P.C. (1977): Measurement of $\dot{V}O_2$, $\dot{V}CO_2$, and evaporative water loss with a flow-through mask. *J. Appl. Physiol. 42*: 120-123.

Integrative and cellular aspects of autonomic functions : temperature and osmoregulation. Eds K. Pleschka, R. Gerstberger. John Libbey Eurotext, Paris © 1994, pp. 305-311.

The frequencies of grouped discharges during cold tremor in shrews : an electromyographic study

Dieter Kleinebeckel, Alfred Nagel[1], Peter Vogel[2] and Friedrich-Wilhelm Klussmann

Institut für Neurophysiologie der Universität zu Köln, Robert Koch Strasse 39, 5000 Köln 41, Germany
[1] Zoologisches Institut der Universität Frankfurt, Siesmayerstrasse 70, 6000 Frankfurt am Main 11, Germany
[2] Institut de zoologie et d'écologie animale, Bâtiment de Biologie, 1015 Lausanne, Switzerland

The frequencies of grouped discharges during cold shivering was determined in several genera of shrews with body masses between 54.2 and 4.7 g, and with different levels of basal metabolic rate. In some animals body temperatures were recorded simultaneously. In shrews with body masses of around 40 g the mean shivering frequency was 39 Hz, a value similar to that of laboratory mice with the same body mass. In shrews with body mass of below 20 g, however, frequencies of grouped discharges of around 50 Hz were recorded, with the highest value of 58 Hz in one individual shrew. The results suggest that the frequency of grouped discharges is primarily correlated with body mass rather than with the level of basal metabolic rate.

INTRODUCTION

Cold tremor, or shivering, is an "involuntary tremor of skeletal muscles as a thermoeffector activity for increasing metabolic heat production" (Simon, 1987). In order to evaluate the intensity of cold tremor during studies of thermoregulatory physiology, the electromyogram frequently is used. Generally, in mammals three different patterns of activity were found during cold tremor: continous tonic muscle activity, burst-like activity and activity in form of grouped discharges. This type of muscle activity, which is characterized by rhythmical synchronization of muscle action potentials, represents the typical cold tremor. Grouped discharges have been described for many laboratory species like dogs, cats, rabbits, guinea pigs, rats, mice (Spaan and Klußmann, 1970) and also for men (Elble and Randall, 1976).

During cold tremor, Spaan and Klußmann (1970) found the frequency of grouped discharges to be higher the smaller the animal. These results have been confirmed by Günther et al. (1983) and, independently, by Russian authors (Kuzmina et al., 1987). The smallest animal species studied so far was the laboratory mouse. The mean frequency of grouped discharges in 24 mice (mean body mass 38 g) was 38 Hz (Spaan and Klußmann, 1970; Günther et al. 1983). According to the above mentioned correlation between body mass and the frequency of grouped discharges, frequencies between 50 and 60 Hz were predicted for the smallest mammals, the shrews, with a body mass of about 5 g (Kleinebeckel and Klußmann, 1990, p.248).

In order to confirm this prediction, electromyographic recordings were taken from 6 different species (4 genera) of shrews during cold tremor: 2 species of shrews with a body mass similar to laboratory mice, and 4 species with a body mass smaller than the laboratory mouse (Table 1). Preliminary results of this work have appeared in abstract form (Kleinebeckel and Vogel, 1991).

MATERIAL AND METHODS

Animals

The following species of shrews with different body mass were investigated (Table 1):

(1) *Suncus murinus* (musk shrew), $\bar{x}$ = 35.4 g (34.0 and 36.8 g)	n = 2
(2) *Crocidura olivieri* (African giant shrew), $\bar{x}$ = 43.2 g (32.2 and 54.2 g)	n = 2
(3) *Crocidura russula* (greater white-toothed shrew), $\bar{x}$ = 10.6 g (6.8, 7.6, 9.8, 10.5, 11.0, 14.1 14.3 g)	n = 7
(4) *Neomys fodiens* (European water shrew), 14.9 g	n = 1
(5) *Sorex coronatus* (Jersey Shrew), 12.0 g	n = 1
(6) *Sorex minutus* (pygmy Shrew), 4.7 g	n = 1

The experiments on species (2-6) were carried out at the University of Lausanne (Switzerland), the experiments on species 1 and 3 at the University of Frankfurt (Germany). In 3 individuals of species 3 experiments were also performed at the University of Cologne (Germany). Species 4 and 6 were wild-trapped (near Lausanne), the other species were captive-born. The origin of the *Suncus murinus* was Japan. The conditions for keeping and breeding shrews in captivitiy have been published by Nagel (1985), Sparti and Genoud (1989) and Sparti (1990).

Recording procedures

In each animal a very small bipolar metal electrode (Sigrist and Kleinebeckel, 1982) was implanted under ethrane/nitrous oxide anesthesia into the medial gastrocnemius muscle. The wires of the electrode were passed through the skin. The myoelectric signals were preamplified (Tektronix TM 503) and filtered (upper cut-off frequency 10 KHz, lower cut-off frequency 1 Hz). The electromyograms were either recorded on magnetic tape (Bell and Howell VR 3200) and later A/D-converted on a floppy-disk-computer (Vuko FDC-2), or directly A/D-converted on floppy-disks.
After awakening from anesthesia the animals were placed into an experimental cage (20 cm deep, 12 cm diameter), which was cooled at the bottom with crashed ice. This induced shivering in the animals, which lasted for several minutes and was then more and more replaced by voluntary locomotor activity. The electromyographic signals were controlled optically (Vuko Digitalscope VKS 220-16) and acoustically. Directly after the recording period the shrews were anesthized again, the electrode removed, and the wounds closed.
In *S. murinus* (n=2) and *C. russula* (n=2) body temperature was recorded additionally to the electromyographic activity. In these animals a small radio-transmitter acting as a temperature sensor (minimitter) was implanted in the abdominal cavity several weeks before the experiment. The temperature signals of the transmitters were recorded simultaneously with the electromyographic activity.

Assessement of shivering frequency

The digitized electromyographic signals were analyzed by means of a personal computer. Grouped discharges in the electromyographic recordings were marked by a cursor, thus determining the beginning and end of a series of grouped discharges and the beginning of any individual grouped discharge in the recording. Thus the freqency of grouped discharges in each series could be determined. Only those series of grouped discharges were analyzed in which at least 3 consecutive grouped discharges occurred. The maximal number of grouped discharges in one series was 22. In all 14 animals which were invesitigated a total of of 2645 grouped discharges were analyzed, in the mean 189 per animal with a maximum of 947 and a minimum of 27 grouped discharges.

RESULTS

Fig. 1 represents typical examples of series of grouped discharges in shrews. This form of cold induced electromyographic activity, i.e. typical cold shivering, amounted to about 10 per cent of the total muscle activity during cold. Ninety per cent of muscle activity appeared as "continuous tonic" activity. Burst-like activity was not found in shrews.

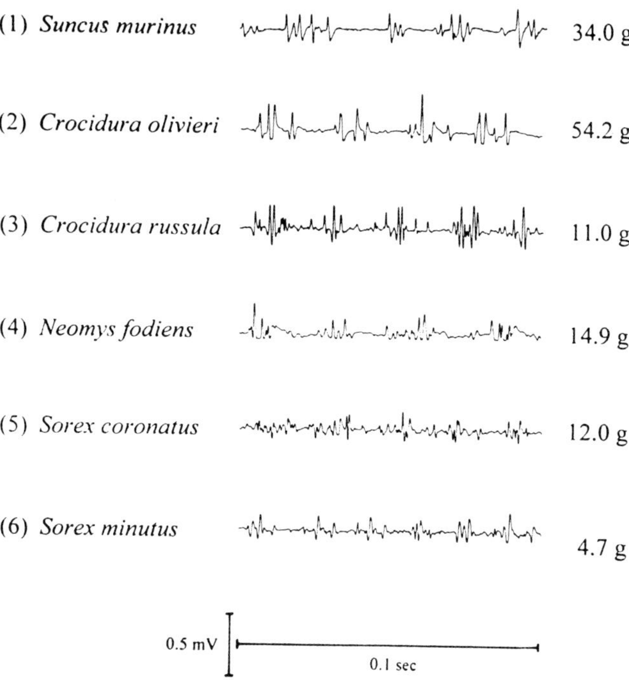

Fig.1. Original EMG-registrations of grouped discharges in 6 species of shrews. In each registration four to six grouped discharges can be recognized.

The analysis of the frequencies of grouped discharges (Hz) led to the following results (Table 2):

(1) *Suncus murinus*		
43.6 ± 8.8 and 43.0 ± 8.4 Hz	$\bar{x}$ = 43.3 Hz	39 Hz
(2) *Crocidura olivieri*		
30.6 ± 7.5 and 37.5 ± 7.6 Hz	$\bar{x}$ = 34.0 Hz	
(3) *Crocidura russula*		
58.4 ± 8.5, 52.6 ± 13.8, 57.0 ± 11.7, 55.6 ± 11.0, 50.1 ± 13.6, 56.7 ± 15.8, 57.2 ± 12.4 Hz	$\bar{x}$ = 55.4 Hz	50 Hz
(4) *Neomys fodiens*		
45.1 ± 10.0 Hz	45.1 Hz	
(5) *Sorex coronatus*		
46.7 ± 12.9 Hz	46.7 Hz	50 Hz
(6) *Sorex minutus*		
53.2 ± 12.2 Hz	53.2 Hz	

The mean frequency of grouped discharges in species 1 and 2 with a mean body mass similar to that of laboratory mice was 39 Hz. In species 3 - 6, however, with significantly smaller body masses, a higher shivering frequency was measured: The mean frequency in these species was about 50 Hz.

In Fig. 2 the correlation between shivering frequency and body mass is shown as published by Kleinebeckel and Klußmann (1990), supplemented by the results of this study. As predicted by these authors the shivering frequencies of shrews do fit very well the regression line for laboratory animals with greater body masses.

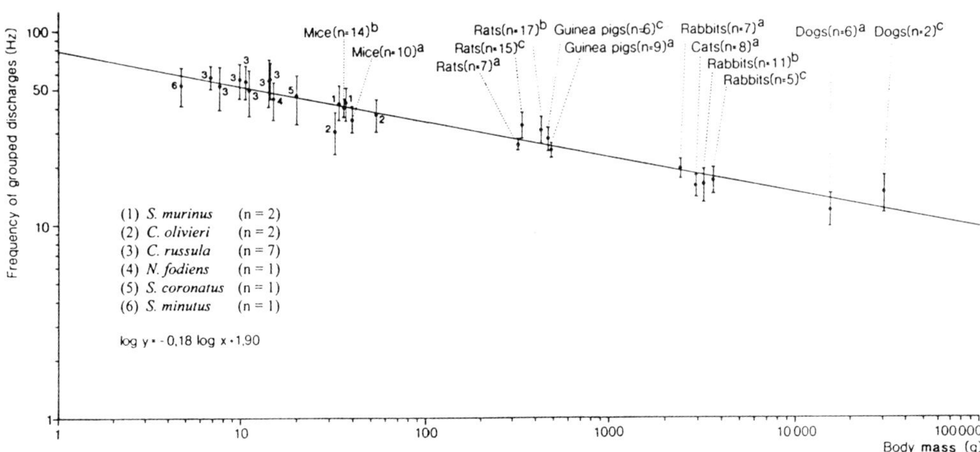

Fig.2. Correlation between frequency of grouped discharges and body mass in various adult laboratory animals and in shrews of different body size. Double logarithmic scale. The values of laboratory animals are summarized from several publications (a: Spaan and Klußmann, 1970; b: Günther et al., 1983; c: Filz, 1987; mean values with standard deviation). The values of 14 individual shrews are from this study. The correlation coefficient (r) of the regression line is 0.97.

Originally this regression was based on measurements in laboratory animals only (without shrews) with the following regression equation (Kleinebeckel and Klußmann, 1990):

$\log y = -0.18 \log x + 1.89$ $(r = 0.94)$ [y: frequency of grouped discharges in Hz; x: body mass in g].

Inclusion of all measurements of shivering frequencies and body mass in 14 shrews (as shown in Fig. 2) does not change the slope of the regression line, however the correlation coefficient is increased from 0.94 to 0.97.

$\log y = -0.18 \log x + 1.90$ $(r = 0.97)$.

In all shrews with a body mass of less than 20 g frequencies of shivering were higher than those in laboratory mice, reaching values higher than 50 Hz.

In *S. murinus* (n = 2) mean body temperature during the experiments was 33.3 ± 1.9 °C while in *C. russula* (n = 2) body temperature was 34.7 ± 2.7 °C.

DISCUSSION

Among mammals the smallest belong to the shrews (Soricidae). According to Repenning (1967) the classification of the Soricidae is as follows:

Family	Subfamily	Genera
Soricidae	Crocidurinae, white-toothed shrews	*Crocidura, Suncus*
	Soricinae, red-toothed shrews	*Sorex, Neomys, Blarina*

The body temperature of the Crocidurinae is moderately high (on average about 35.5 °C; Frey, 1979, Nagel, 1985) and labile. All white-toothed shrews that have been subjected to food shortage in captivity have shown the ability for entering torpor (Nagel, 1977; Genoud, 1988). In contrast, most Soricinae maintain higher normal body temperatures (on average 38.5 °C) and they are not able for torpor (Nagel, 1985; Genoud, 1988). In the present experiments body temperature was measured only in the Crocidurinae *S. murinus* (n = 2) and *C. russula* (n = 2). In comparison to the data of Frey (1979) and Nagel (1985) body temperatures were somewhat lower in our experiments. This may have been caused by the short anaesthesia necessary for electrode implantation prior to the measurement of shivering frequency. Within the other Crocidurinae (*C. russula,* n= 5; *C. oliveri*, n = 2), in which body temperature was not measured, torpor can be excluded, since all animals showed intense locomotor activity after maximal 5 minutes of shivering. The Soricinae (n = 3) too showed no signs of hypothermia. But even if the body temperature would have been lower than normal in some of these animals this would have led to lower shivering frequencies, because according to Spaan and Klußmann (1970) the frequency of grouped discharges during shivering decreases with decreasing body temperature. However body temperature needs to decrease by 3 - 4°C till a measurable decrease in shivering frequency can be seen (own observations).

Besides the differences in body temperature and the ability for torpor, Crocidurinae and Soricinae are different with respect to the Basal Metabolic Rate (BMR). In Crocidurinae BMR values are similar to, or only slightly higher than, one might expect from Kleiber's (1961) allometric relationship between BMR and body mass, varying between 110 per cent and 135 per cent (Nagel, 1985; Sparti, 1990; Geiger and Nagel, 1992). In contrast, BMR in Soricinae is higher than expected. It is very high in *S. minutus, S. araneus* and *S. coronatus* (339, 338 and 299 per cent of the expected value; Nagel, 1985; Sparti, 1989) and more moderate in *N. fodiens* (163 - 169 per cent of the expected value; Nagel, 1985; Sparti, 1990). Correlations of the energy metabolism in shrews with other physiological, ecological, and ethological pecularities are discussed by Vogel (1980), Nagel (1985) and Genoud (1988).

Since BMR could not be measured in the present experiments, in Fig 3. the "levels of BMR" of shrews as published by Sparti (1989, 1990) were compared with the average frequencies of grouped discharges and the average body weight of the genera of shrews from our experiments. The animals were divided into two groups according to their body mass. One group with a body mass of about 10 grams with individual body masses of 5 to 15 g (*C. russula, N. fodiens, S. coronatus, S. minutus*) and a second group with a body mass of about 40 g with individual body masses between 32 and 54 g (*S. murinus* and *C. oliveri*).

In the shrews with a body mass of about 10 g mean frequency of grouped discharges was found to be 50 Hz. For those species with very high BMR (*Sorex minutus, Sorex coronatus*) as well as for those with more moderate (*Neomys fodiens*) and only slightly higher BMR (*Crocidura russula*) this finding seem to indicate that the level of BMR does not determine the frequency of grouped discharges during shivering. However, the findings support the view of Kleinebeckel and Klußmann (1990) that the smaller the animal, the more often one unit of muscle mass has to contract to produce enough heat in order to compensate for the greater and more unfavourable body surface - body mass relationship during exposure to cold. On the other hand smaller muscle fibres in small animals must be active more often than greater ones in greater animals to produce the same quantity of heat. Comparison of the two groups of shrews with body masses of 10 g and 40 g respectively indicate that the shivering frequency is determined mainly by the body size.

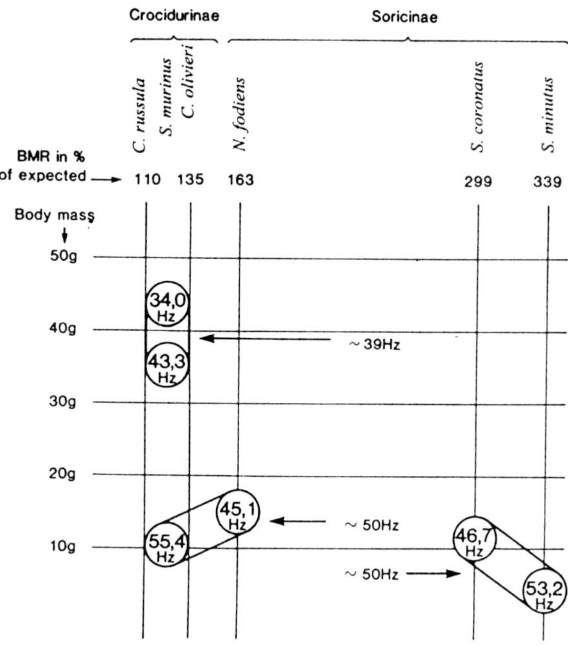

Fig.3. Relationship between the basal metabolic rate (according to Nagel, 1985; Sparti, 1990; Geiger and Nagel, 1992), mean body mass, and mean frequency of grouped discharges in 6 species of shrews (values from this study).

REFERENCES

Elble, R.J. and Randall, J.E. (1976): Motor-unit activity responsible for 8 to 12 Hz component of human physiological finger tremor. *J. Neurophysiology* 39, 370-383.

Filz, H.P. (1987): Vergleichende elektromyographische Analyse des Kältetremors bei Tierarten verschiedener Größe während des postnatalen Wachstums. *Inaugural-Dissertation*. University of Cologne, Germany.

Frey, H. (1979): La température corporelle de *Suncus etruscus* (Soricidae, Insectivora) au cours de l'activité, du repos normothermique et de la torpeur. *Rev. Suisse Zool.* 86, 23 - 36.

Geiger, S. and Nagel, A (1992): Oxygen consumption, respiratory and cardiac activity of the Asian shrew *Suncus murinus* (Soricidae, Insectivora). *Zeitschrift für Säugetierkunde*, *Supplement to Vol.* 57, 25 -26.

Genoud, M. (1988): Energetic strategies of shrews: ecological constraints and evolutionary implications. *Mammal Review* 18, 173-193.

Günther, H., Brunner, R. and Klußmann, F.W. (1983): Spectral analysis of tremorine and cold tremor electromyograms in mammal species of different size. *Pflügers Arch.* 399, 180-185.

Kleiber, M. (1961): *The fire of life*. New York: John Wiley.

Kleinebeckel, D. and Klußmann, F.W. (1990): Shivering. In *Thermoregulation: Physiology and Biochemistry,* eds. E.Schönbaum and P.Lomax, pp. 235-253. New York: Pergamon Press.

Kleinebeckel, D. and Vogel, P. (1991): Frequency of shivering in shrews. *Pflügers Arch.*, Supplement to Vol. 418, R55.

Kuzmina, G.I., Mejgal, A.Ju., and Sorokina, L.V. (1987): Thermoregulatory motor unit activity of the deltoid muscle in man. *Fiziologija celoveka* 13, 432-435.

Nagel, A. (1977): Torpor in the European white-toothed shrews. *Experientia* 33, 1455 - 1456.

Nagel, A. (1985): Sauerstoffverbrauch, Temperaturregulation und Herzfrequenz bei europäischen Spitzmäusen (Soricidae). *Zeitschrift für Säugetierkunde* 50, 249-266.

Repenning, C.A. (1967): Subfamilies and genera of the Soricidae. *Geol. Sur. Prof. Pap.* 565, 1-74.

Sigrist, G. and Kleinebeckel, D. (1982): An implantable myography electrode for recording muscle activity in freely moving small animals - A new technological approach. *IEEE Trans. Biomed. Eng.* 29, 730-736.

Simon, E. (1987): Glossary of terms for thermal physiology. *Pflügers Arch.* 410, 567-587.

Spaan, G. and Klußmann, F.W. (1970): Die Frequenz des Kältezitterns bei Tierarten verschiedener Größe. *Pflügers Arch.* 320, 318-333.

Sparti, A. (1989): Etude comparee de la thermoregulation et du budget de l'eau chez les Soricines et les Crocidurines (Insectivora, Mammalia). *Thèse de Doctorat*, l'Université de Lausanne.

Sparti, A. and Genoud, M. (1989): Basal rate of metabolism and temperature regulation in *Sorex coronatus* and *S. minutus* (Soricidae, Mammalia). *Comp. Biochem. Physiol.* 92A, 359-363.

Sparti, A. (1990): Comparative temperature regulation of African and European shrews. *Comp. Biochem. Physiol.* 97A, 391-397.

Vogel, P. (1980): Metabolic levels and biological strategies in shrews. In *Comparative Physiology: Primitive Mammals*, eds. K.Schmidt-Nielsen, L.Bolis, and C.R. Taylor, pp. 170-180. Cambridge: Cambridge University Press.

ACKNOWLEDGEMENTS

We thank Susanne Schulze and Kirsten Wehner for the preperation of small myography electrodes, and Renate Clemens for drawing the figures. We further thank Dr. Sen-ichi Oda, Nagoya University, for the musk shrews.

Integrative and cellular aspects of autonomic functions : temperature and osmoregulation. Eds K. Pleschka, R. Gerstberger. John Libbey Eurotext, Paris © 1994, pp. 313-322.

Cranial pneumatization and its significance for temperature regulation in juvenile greenfinches (*Carduelis chloris* L.)*

Günther Warncke

Institute of Neurophysiology, University of Cologne, Robert-Koch-Strasse 39, D-50931 Köln, Lindenthal, Germany

INTRODUCTION

The pneumatization of the skullcap of birds leads to a sandwich structure (BÜHLER, 1973; STORK 1976). Bone layers that are connected by short trabeculae are interspaced with air-filled cavities, resulting in a stabilization of the neurocranium according to the principle of lightweight construction. In completely pneumatized adult birds, compared to juveniles, there is no sign of additional stress by cranial kinesis or external forces. It has been suggested, therefore, that the air-filled cavities around the brain serve primarily as insulation against cold (STORK, 1972).
This is indirectly supported by the incomplete pneumatization of the skullcap in African and Australian birds (SERVENTY et al., 1967) and in the swift (Apus apus) that is known to evade cool weather. The incomplete pneumatization in Galapagos finches too may have climatic causes. Studies by WARNCKE & STORK (1977) and WARNCKE (1985a, 1985b) have confirmed that the pneumatic spaces of bird skulls serve indeed as an insulating layer protecting the brain against excessive heat loss. Environmental temperatures control the speed of pneumatization in juvenile birds according to these results.
It is uncertain as to how the process of pneumatization is affected by additional factors like plumage insulation or body habit. Especially plumage insulation may be very important. Juvenile and adult greenfinches, for example, demonstrated by their pilomotor responses to decreasing environmental temperatures that air layer and feathers together serve as good protection against cold (WARNCKE, 1982).
This study deals with the significance of cranial pneumatization and plumage for the heat budget of juvenile greenfinches. Skin and core temperatures, total heat production, shivering, and vasomotor and pilomotor responses of the birds at different environmental temperatures are compared to clarify if and to what extent cranial pneumatization and plumage form insulating layers and have an effect on the heat budget.
In addition, it was to be examined as to whether body temperatures can be kept constant in decreasing environmental temperatures and which mechanisms are thereby used for thermoregulation. To estimate

* Supported by the Deutsche Forschungsgemeinschaft

the insulating effect of cranial pneumatization from that of plumage, almost completely pneumatized birds (only a small area of the frontal was unpneumatized) with juvenile plumage were evaluated in the cold test.

MATERIAL AND METHODS

Data of 70 juvenile greenfinches (mean body mass [b.m.] 25.1 g) were collected (Table 1). Animals received woodland bird food and water ad libitum. They were divided into three groups and kept at approximately natural temperature conditions in outdoor aviaries (Tnat), in warm cages (25 ± 1°C), and in cold cages at 5°C ± 1°C. The birds were temperature acclimated for four weeks at LD 12:12 hours.

Body temperatures: All body temperatures (Tb) were measured with flexible thermocouples (NiCr-Ni; diameter 0.01 mm). Temperatures in the area of the hypothalamus (Thyp) and colonic (Tco) and axillary (Tax) temperatures were taken as core temperatures (Tc). In addition, skin temperatures from the head (Th), back (Td), breast (Tbr), wing (Tw), and leg (tarsometatarsus [Tlg]) as well as the subcranial temperature (Tsc) were recorded at regular intervals. In acp juveniles, the interlaminal temperature (Til) was also measured.

Table 1: Characters of greenfinches used in the experiments. Abbreviations: cup=completely unpneumatized; acp=almost completely pneumatized; jp=juvenile plumage; oc=outdoor cage; wc=warm cage; cc=cold cage

n	degree of pneumatization	condition of plumage	cage conditions
10	cup	jp	oc
10	cup	jp	wc
10	cup	jp	cc
10	cup	jp	oc
30	acp	jp	wc

Electromyogram: Muscle shivering was led electromyographically from the pectoral muscle by differential measurements (KLEINEBECKEL & KLUSSMANN, 1990). Electromyographic activity (EMG) was recorded when the bird shivered while sitting still.

Heat production: Oxygen consumption (VO_2) as a standard for heat production was digitally recorded using an open flow method (ΔO_2 = 0.01%; flow 30 l/h). VO_2 was recalculated to body mass (b.m.) and STPD conditions.

Pilomotor responses: The 'Mean Feather Index' (FI) introduced by MC FARLAND and BAHER (1968) for the Barbary dove (streptopelia risoria) was applied to rate the pilomotor responses of the birds. It was supplemented by the author with one additional plumage state: 0 =plumage decumbent; 1=normal plumage state; 2=plumage half erect; 3= plumage fully ruffled. This classification results in a Mean FI with a minimum of 0 and a maximum of 18.

Surgical techniques and narcosis: The measurement of Thyp and Tsc required operations to insert the thermocouples. Surgery was performed under deep equithesin narcosis (2 µl/g b.m.; BECH & MIDTGÅRD, 1981; WARNCKE, 1993). The head was stereotactically fixed and the thermocouples for the measurement of Thyp and Tsc were inserted 4-5 mm into the brain through a trepanation hole in the frontal. One tip was placed into the third ventricle, caudal of the optic chiasma. The birds were able to move without impairment. They stayed in the chamber for six hours with water ad libitum. All animals were weighed before and after every experiment.

Examination of cranial pneumatization in living birds: By wetting a small feather tract on the top of the head, a good rating of pneumatization is possible through the skin. Unpneumatized regions with the pink brain shimmering through and whitish pneumatized areas reveal the degree of pneumatization. The ontogeny of pneumatization is subdivided into six typical stages (0-V). For the present study, we shall confine ourselves to the three stages given below for the greenfinch (WARNCKE, 1985b):

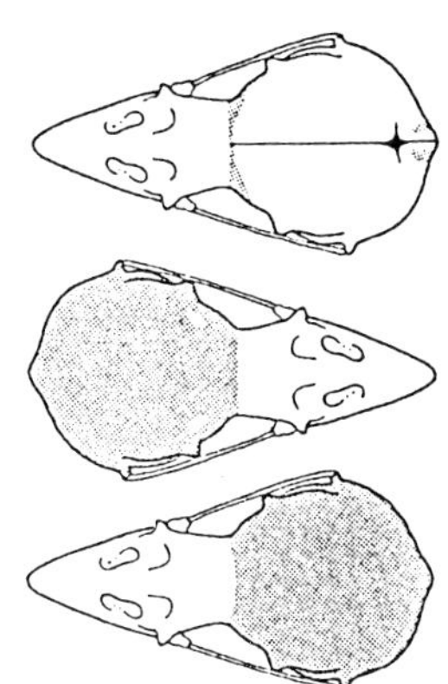

Stage 0: Completely unpneumatized, single-layered neurocranium; no or first signs of the beginning of pneumatization in the basioccipital and exoccipital areas

Stage IV: Almost completely pneumatized vault of the cranium; only two lateral 'windows' next to the orbital rims

Stage V: The neurocranium is completely pneumatized.

The late stage (IV) allows a separate rating of the insulating effects of pneumatization and body plumage. In order to obtain greenfinches with persisting frontals and juvenile plumage, the birds were kept beyond the summer months at long-day conditions (LD 18:6 hours) and at constant Ta = 25 ± 1°C.

R E S U L T S

Body temperatures of unpneumatized greenfinches dependant on environmental temperature.

After warm acclimation: The averages were 41.6°C for Thyp, 41.0°C for Tsc, and 40.4°C for Th measured above. Tc of axilla and colon were 41.6°C and 42.3°C, the skin temperatures of back and breast were 41.5°C and 40.8°C. A Tw of 41.1°C was determined. The lowest temperature was that of the unfeathered tarsometatarsus at 30.4°C.
Under natural temperature conditions: Tco, Thyp, and Td had the highest values of this series at 40.7°C, 40.4°C, and 40.3°C. Tbr, Tax, and Tw were 37.7°C, 39.9°C, and 39.1°C. The results for Tsc and Th were 39.9°C and 38.2°C. The tarsometatarsal measurement was 31.1°C.
After cold acclimation: All birds showed a distinct decrease of both Tc and skin temperatures. Thyp dropped to 38.5°C, Tsc to 36.5°C, and Th to 35.4°C. Tco was 40.3°C, Tax and Td were 38.6° and 38.8°C. While Tbr remained at 38.0°C and Tw at 36.3°C, Tlg showed an extreme decrease to 8.8°C. The differences between the groups were significant in the U test and highly significant for the tarsometatarsal data.

Thermoregulatory responses of unpneumatized greenfinches dependant on environmental temperature

Total heat production after warm acclimation: Warm acclimated greenfinches had an average VO_2 of 5.7 ml O_2/g.h. During the experiments, each lasting six hours, there were only small differences of oxygen consumption. The first 70 minutes of the experiments were an exception, since VO_2 increased by 0.6 ml O_2/g.h.
Under natural temperature conditions: Birds at Tnat had an average oxygen consumption of 6.43 ml O_2/g.h, between the values of acclimated animals. They showed smallest variations during the experiment.
After cold acclimation: All birds showed a distinctly higher O_2 consumption than warm acclimated animals. Their average consumption of 8.7 ml O_2/g.h exceeded that of warm acclimated individuals by 3.1 ml O_2/g.h. An increase of heat production did not start until after 90-100 minutes after the start of experiment. This began quickly and continued for more than three hours until an increase of 2.7 ml O_2/g.h had been reached and the values stabilized. The differences of the oxygen data between groups were significant.

Cold tremor

After warm acclimation: As a rule, the EMG of warm acclimated birds was rather low at an average of 370 µV. At the beginning of the test, the values were still high at 970 µV. They gradually decreased to 150-200 µV and stabilized between 100 and 150 µV after 180 minutes. The intense EMG at the beginning did not represent cold tremor, but was caused by an initial unrest in the birds.
Under natural temperature conditions: The birds of these series had a mean value of 970 µV that was above that of warm acclimated animals, but distinctly below that of cold acclimated birds. As in warm acclimated birds, there was a gradual decrease from 1200 µV at the beginning to 800 µV; the probable cause of this decrease is given above (MILITZER, 1986).
After cold acclimation: Parallel to the VO_2 data in this group, there was a considerable increase of EMG to an average of 2870 µV. EMG values already reached 2100 µV at the beginning of the experiments. They increased slowly to 3400 µV during four hours and had a maximum of 3600 µV.

Pilomotor responses

Besides the previously described effector mechanisms, the birds may also show pilomotor reactions. The 'Mean Feather Index' (FI) extended to four ruffle states by WARNCKE (1985a) was used for the rating of pilomotor responses.

Mean FI after warm acclimation: These birds ruffled their feathers only slightly so that they had on the average a minimum FI of 1.7. It was somewhat increased during the initial phase, but varied only by 0.5 until the end of the experiment.
Under natural temperature conditions: Similar to the VO_2 data, the birds had intermediate FI values of 4.8, falling between those of the temperature acclimated groups. Starting with a FI of 8 at the beginning of the experiment, the animals reduced the ruffling of their feathers to FI = 5 during the next 70 minutes. Sporadically, they still showed pilomotor variations; 90 minutes before the conclusion of the test, there were hardly any more changes.
After cold acclimation: With a FI of 17.8, this group ruffled their

feathers more intensely than all the others individuals examined. The FI increased from 16 at the beginning of the experiment to 18 within 30 minutes and remained constant.

Effect of temperature acclimation on body mass

Body mass (b.m.) remained almost constant during the warm acclimation of the birds. During the cold acclimation, however, there was a small increase of b.m. starting four to five days after exposure to cold and lasting five to seven weeks. During this time, b.m. increased by 2.18 g on the average. Animals under natural day conditions, on the other hand, showed an increase of only 1.4 g. In some post-mortem examinations, it was evident that pectoral muscle mass had increased.

Body temperatures and thermoregulatory responses of juvenile unpneumatized and almost completely pneumatized greenfinches at decreasing environmental temperatures

In addition to temperature acclimation of juvenile greenfinches, cold experiments were also carried out at slowly decreasing environmental temperatures. The aim was to investigate if and by which mechanisms the birds are able to maintain their heat budget. cup juveniles that had been kept at Tnat were used for this experiment, expecting that they would show the strongest thermoregulatory responses. Moreover, the animals of this series were later to be compared to animals with an acp neurocranium. Starting at 25°C, Ta was gradually (1°C/6 min) lowered to 5°C. The result was a drop in all body temperatures, especially those of the skin (Table 2).

Body temperatures: Thyp dropped by 1.4°C, Tsc by 1.3°C, and Th even by 3.9°C. Tco and Tax as core temperatures showed lower decreases by 0.4°C and 0.6°C. Tbr and Td decreased by 1.9°C and 0.7°C. While Tw deviated only by 1.1°C from the initial temperature, the tarsometatarsal temperature dropped by 23.6°C.
Total heat production: The VO_2 of the birds increased immediately after the beginning of the temperature drop. The O_2 consumption increased from 5.8 ml O_2/g.h until a temperature of 18°C was reached. It declined briefly and then continued to increase as from 17°C until the initial temperature had more than doubled, reaching a maximum at 12.8 ml O_2/g.h.
Cold tremor: Parallel to VO_2, the EMG showed a continuous increase with declining Ta. It had a mean value of 3500 µV, with maxima as high as 4100 µV.
Pilomotor responses: The time course of this parameter was similar to those of O_2 consumption and EMG. At 5°C the birds reached a maximal value of FI = 18 .

Head temperatures of almost completely pneumatized greenfinches with juvenile plumage

To allow an evaluation of the insulating effect of pneumatization separately from that of body plumage, the study was complemented by experiments on acp greenfinches with juvenile plumage.
In contrast to the unpneumatized juvenile birds, it was possible in this group, due to the almost complete pneumatization of the neurocranium, to insert two further thermocouples into the head of the animal. The first of these additional thermocouples was placed bet-

ween the laminae, the second directly beneath the still unpneumatized spot of the frontal.

Table 2: Body temperature differences (ΔTb) of unpneumatized (cup) and almost completely pneumatized (acp) greenfinches at environmental temperatures decreasing from 25°C to 5°C. Abbreviations: Thyp=hypothalamus temperature; Tsc=subcranial temperature; Til=interlaminal temperature; Tup=temperature in the unpneumatized frontal region; Tco=colonic temperature; Tax=axillary temperature; Tbr=skin temperature of the breast; Td=skin temperature of the back; Tw=wing temperature; Th=skin temperature of the head; Tlg=skin temperature of the leg

ΔTb	cup	acp	significance level
ΔThyp	1.4°C	0.6°C	$p < 0.01$
ΔTsc	1.3	0.8	$p < 0.01$
ΔTil	-	1.3	-
ΔTup	-	1.0	-
ΔTco	0.4	0.4	$p < 0.10$
ΔTax	0.6	0.6	$p < 0.10$
ΔTbr	1.9	2.2	$p < 0.05$
ΔTd	0.7	0.9	$p < 0.05$
ΔTw	1.1	1.4	$p < 0.05$
ΔTh	3.9	3.6	$p < 0.05$
ΔTlg	23.6	24.0	$p < 0.10$

The head temperatures, as well as the core and skin temperatures of these series, are compared with the data of completely unpneumatized birds in Table 2.
Thyp and Tsc were on the average 0.8°C and 0.5°C higher than in unpneumatized individuals, but Th was 0.3°C lower. Tsc of the unpneumatized region was increased by 0.3°C. There were no differences in Tco and Tax, while Tbr and Td differed by 0.3°C and 0.2°C. The difference in Tf was 0.4°C.
The thermoregulatory responses of acp animals differed only by 0.4 ml O_2/g.h and 300 µV in the EMG from the data of unpneumatized birds (Table 3); FI did not differ. The birds seemed less exhausted at the

Table 3: Differences of oxygen consumption (ΔVO_2), electromyogram (ΔEMG), and FI (ΔFI) of unpneumatized (cup) and almost completely pneumatized (acp) juvenile greenfinches at environmental temperatures decreasing from 25°C to 5°C

thermoregulatory response	cup	acp	significance level
ΔVO_2 (ml O_2/g.h)	7.0	6.6	$p < 0.05$
ΔEMG (µV)	3500	3200	$p < 0.05$
ΔFI	18	18	-

end of the experiment, they hopped around and immediately took in water and food by themselves.

D I S C U S S I O N

The significance of body insulation in juvenile greenfinches

Juvenile birds were distinctly different from each other in all Tb. Cold acclimated individuals showed the lowest and those kept at 25 °C the highest Tb. Animals kept under natural conditions that were exposed to the changing day and night temperatures during spring and summer had intermediate values. In cold acclimated animals, for example, Tco and Tax were 2.0°C and 3.9°C lower, respectively, than in warm acclimated birds. Tbr was remarkably low at 38.0°C, although a considerably higher temperature could be expected (STEEN & ENGER, 1957). The differences are especially striking in Th of cold acclimated birds. Thyp with a value of 38.5°C was as much as 2.9°C lower than in warm acclimated animals. Tsc and Th were 4.4°C and 5.0°C lower giving additional evidence that neither the insulation by the single-layered skullcap nor by the skin of the head with its feathers was sufficient to protect the brain effectively against heat loss (WARNCKE & STORK, 1977; WARNCKE, 1985b). This was also true for the remaining body regions, especially the wrist with its good blood supply and its sparse feathering in all examined juvenile animals. It had a temperature that was 4.8°C lower than in warm acclimated animals.
In contrast to studies on adult birds (DAWSON & CAREY, 1976; BARNETT, 1970; SAARELA et al., 1984) dealing with the dry weight of feathers, MCNABB & MCNABB (1977) looked for the first time at the heat conductance of skin preparations from juvenile quails with downs and feathers. They found an increase of the heat capacity by 22 % after the shedding of down at the first juvenile moulting.
In order to reduce the loss of body heat as far as possible, a maximal vasomotor reaction is put to use particularly in the unfeathered leg region (BERNSTEIN, 1974; BAUDINETTE, 1976). In cold acclimated birds as compared to those kept at natural conditions or warm acclimated animals, this resulted in tarsometatarsal temperatures that were 22.3°C lower. We can assume, however, that this mechanism is totally insufficient; it can only serve as a supplement to other mechanisms (RAUTENBERG, 1980a,b).
These are namely insulation by intact body plumage, seasonal storage of body fat, and complete pneumatization, all of which are specific for adult birds.
Both VO_2 and EMG of juvenile greenfinches indicate that each of these groups uses a different mode of thermogenesis. Warm acclimated animals, experiencing no excessive heat loss at 25°C, had a rather low VO_2 of 5.70 ml O_2/g.h. It differed little from the 6.43 ml O_2/g.h measured in animals kept under natural conditions. Cold acclimated juvenile greenfinches had the highest consumption of all with 8.78 ml O_2/g.h. The EMG intensities were comparable to the oxygen data.
Individuals kept at 25°C did not show any EMG activity, but only electrocardiographic potentials. At natural temperatures, isolated intermittent shivering up to 1000 µV could be observed. Cold acclimated birds again reached maximal values. They increased their muscular activity up to 2870 µV. The slightest changes of their Tc simultaneously reduced or increased their shivering intensity.
It is remarkable that cold acclimated juveniles show a steep increase of EMG and VO_2 during the experiment. This was caused by the observed increase of locomotor activity resulting in an additional heat production by the locomotor system. These birds did not only show a substantially increased EMG, but first of all seemed much more restless than the other two experimental groups. WEST

(1965) was able to show that at low Ta both VO_2 and EMG of the birds increased linearly. He is the same opinion as HOHTOLA & STEVENS (1986) and CAREY et al. (1989) that the locomotor activity and the EMG are the two most important mechanisms of controlled heat production in birds. VO_2 would not increase simultaneously with the shivering if there was a different way of heat production (HOHTOLA, 1982; MARSH & DAWSON, 1988; CAREY et al., 1989).

An insulation of the body by depositing subcutaneous fat stores, as described for the American goldfinch (Carduelis tristis) by DAWSON & CAREY (1986), was not achieved in juvenile greenfinches. Although fat deposits were evident in various body regions of cold acclimated individuals and of those kept under natural conditions, it only occurred in very thin layers and small amounts, in contrast to adult birds. What is more, the feather tracts exactly coincide with regions containing little or no fat. The still apparent difference of body mass, especially in cold acclimated animals, may be due to a mass increase of the pectoral muscle and of the gastrocnemius. AULIE (1976) and DAVISON & LICKISS (1979) arrived at similar results in their studies on juvenile chickens (Gallus domesticus) of different age groups.
If we look at the feather tracts of juvenile birds, we find that they are in general sparsely feathered. Dorsally, there is only a narrow streak of feathers running from the nape to the uropygial gland that is slightly widened towards the middle. The head is almost completely, but not densely feathered in the auricular and throat regions. Wings and sides show only sparse feathering, while the thighs are densely covered. On the ventral side, two narrow lateral feather zones stretch towards the throat. Apart from that, there is only a small feather tract at the wing joint.
Barely 50 % of the total body surface of the bird are covered with feathers. Besides this, the juvenile plumage of greenfinches contains 25-30 % less contour feathers than the adult plumage (WARNCKE, unpublished observations). Juvenile greenfinches almost completely lack the down feathers that are situated in adults between the follicles of contour and wing feathers. This results in large empty spaces between the feathers; together with the totally unfeathered body regions, these exclude an effective insulation. Consequently, the intense piloerection that is displayed by cold acclimated juvenile birds has not almost any effect.
Juvenile greenfinches exposed to low Ta dispose neither by the unpneumatized, single-layered neurocranium nor by the postjuvenile plumage of a sufficient body insulation. They can counter heat losses only by vasomotor reactions and by an increase of thermogenesis appearing in the EMG and the VO_2.

The significance of the almost completely pneumatized neurocranium in juvenile greenfinches

In order to allow a separate rating of the insulating effect of pneumatization and body plumage, juvenile cup and acp greenfinches with juvenile plumage had been examined at decreasing Ta. The acp animals had two small persisting 'windows' in their frontals on both sides (diameter 2-2.5 mm).
Both Tb and the thermogenesis were clearly different in the two groups. Especially the head temperatures of the acp birds exemplified during the temperature decrease from 25°C to 5°C that the sandwich structure of the almost completely pneumatized neurocranium had a considerable influence on the low head temperatures. Thyp and

Tsc decreased by 1.4°C and 1.3°C during the cooling phase in unpneumatized birds, but only by 0.6°C and 0.8°C in almost completely pneumatized animals. This equals a reduction of heat loss from the hypothalamus by more than 55 %. The reduction in the peripheral subcranium still amounts to nearly 38 %. Together with Til and Tup below the persisting unpneumatized region, this indicates a substantial temperature gradient between the peripheral shell and the hypothalamus.

The differences of skin temperatures in acp animals were slightly, but significantly larger. It remains open if this is only due to the lower total heat production or to regulation by the hypothalamus or the thermosensory structures of the spinal cord. In any case it could clearly be demonstrated in this study that regardless of the plumage the cranial pneumatization of greenfinches as an independent mechanism has a crucial influence on the heat balance of the brain. It has therefore an important function for temperature regulation and heat budget in greenfinches.

REFERENCES

AULIE, A.: The effect of intermittent cold exposure on the thermoregulatory capacity of Bantam Chicks (Gallus domesticus). Comp. Biochem. Physiol. A, 53, 346-350 (1976)

BAUDINETTE, R.V., LOVERRIDGE, J.P., WILSON, K.J., MILLS, C.D., SCHMIDT-NIELSEN, K.: Heat loss from feet of herring gulls at rest and during flight. Am. J. Physiol., 230, 4, 920-924 (1976)

BECH, C., MIDTGARD, U.: Brain temperature and the rete mirabile ophthalmicum in the Zebra Finch (Poephila guttata). J. Comp. Physiol., 145, 89-93 (1981)

BERNSTEIN, M.H.: Vascular responses and foot temperature in pigeons. Am. J. Physiol., 226, 6, 1350-1355 (1974)

BÜHLER, P.: Sandwichstrukturen - Leichtbau im Vogelschädel. Der Deutsche Baumeister, 2, 100-103 (1973)

CAREY, C., JOHNSTON, R.M., BEKOFF, A.: Thermal thresholds for recruitment of muscles during shivering in winter acclimatized house finches. in: Mercer, J.B. (ed.): Thermal physiology. Elsevier Science Publishers B.V., 685-690 (1989)

DAVISON, T.F., LICKESS, P.A.: The effect of cold stress on the fasted, water-deprived, neonate chicken (Gallus domesticus). J. Thermal Physiol., 4, 113-130 (1979)

DAWSON, W.R., CAREY, C.: Seasonal acclimatization to temperature in Cardueline Finches. Part I: Insulative and metabolic adjustments. J. comp. Physiol., 112, 317-333 (1976)

HOHTOLA, E.: Thermal and electromyographic correlates of shivering in the pigeon. Comp. Biochem. Physiol., 73 A, 2, 159-166 (1982)

HOHTOLA, E., STEVENS,E.D.: The relationship of muscle electrical activity, tremor and heat production to shivering thermogenesis in Japanese Quail. J. exp. Biol., 125, 119-135 (1986)

KLEINEBECKEL, D., KLUßMANN, F.W.: Shivering. in: Schönbaum, E., Lomax, P. (eds.): Thermoregulation: Physiol. and Biochem., Pergamon Press, Inc., 235-253 (1990)

MARSH, R.L., DAWSON, W.R.: Energy substrates inand metabolic acclimatization in small birds. in: Bech, C., Reinertsen R.E.(eds.): Physiology of Cold Adaption in Birds, Plenum Press. (NATO ASI series, Series A, Life Sciences, vol. 173), 105-114 (1989)

McNABB, F.K.M., McNABB, R.A.: Skin and plumage changes during the development of thermoregulatory ability om Japanese Quail Chicks. Comp. Biochem., 58 A, 163-166 (1977)

MILITZER, K.: Wege zur Beurteilung tiergerechter Haltung bei Labor-, Zoo- und Haustieren. Paul Parey, Berlin and Hamburg (1986)

RAUTENBERG, W.: Temperature regulation in cold environment. Symposium on temperature regulation on birds, XVIIth Congressus Internationalis Ornithologici (IOC), 321-326 (1980a)

RAUTENBERG, W.: The importance of pilomotor response in temperature re-gulation. Proceedings of 28th Int. Congress of Physiol. Sci., Pecs, Akademia Kiado, Budapest (1980b)

SAARELA, S., RINTAMÄKI, R., SAARELA, M.: Seasonal variation in the dynamics of ptiloerection and shivering correlated changes in the metabolic rate and body temperature in the pigeon. J. Comp. Physiol., 154 B, 47-53 (1984)

SERVENTY, D.L., NICHOLLS, C.A., FARNER, D.S.: Pneumatization of the cranium of the Zebra Finch, Taeniopygia castanotis. Ibis, 109, 570-578 (1967)

STEEN, J., ENGER, P.S.: Muscular heat production in pigeons during exposure to cold. Am. J. Physiol., 191, 1, 157-158 (1957)

STORK, H.-J.: Zur Entwicklung pneumatischer Räume im Neurocranium der Vögel (Aves). Z. Morph. Ökol. Tiere, 73, 81-94 (1972)

WARNCKE, G., STORK, H.-J.: Biostatische und thermoregulatorische Funktion der Sandwich-Strukturen in der Schädeldecke der Vögel. Zool. Anz., 199, 251-257 (1977)

WARNCKE, G.: Einfluß von Umgebungstemperatur und Kältereizen auf Pneumatisationsprozesse, Plusterverhalten und Gewichtszunahme beim Grünfinken (Carduelis chloris L.). Verh. Dtsch. Zool. Ges., 290 (1982)

WARNCKE, G.: Die Bedeutung der Schädelpneumatisierung und der Gefiederisolation für den Wärmehaushalt von juvenilen und adulten Grünfinken (Carduelis chloris L.). Verh. Dtsch. Zool. Ges., 78, 327, (1985a)

WARNCKE, G., STORK, H.-J.: Altersbestimmung der Vögel nach dem Grad der Schädelpneumatisation - Möglichkeiten und Grenzen. In: BUB, H. (Ed.): Kennzeichen und Mauser europäischer Singvögel. Neue Brehm-Bücherei (NBB) Wittenberg Lutherstadt, Vol. 570, 146-172 (1985b)

WARNCKE, G.: Methods of pain elimination in birds. Journal of the Association of Avian Veterinarians (JAAV); in press (1993)

WEST, G.C.: Shivering and heat production in wild birds. Physiol. Zool., 38, 2, 111-120 (1965)

Integrative and cellular aspects of autonomic functions : temperature and osmoregulation. Eds K. Pleschka, R. Gerstberger. John Libbey Eurotext, Paris © 1994, pp. 323-328.

Multiple nightly torpor bouts in hummingbirds

C. Bech[(1)], A.S. Abe[(2)], J.F. Steffensen[(3)], M. Berger[(4)] and J.E.P.W. Bicudo[(5)]

[(1)] *Department of Zoology, University of Trondheim, N-7055 Dragvoll, Norway*
[(2)] *Departmento de Zoologia, Universidade Estadual Paulista, 13.500 Rio Claro, Brazil*
[(3)] *Marine Biological Laboratory, University of Copenhagen, Strandpromenaden 5, DK-3000 Helsingør, Denmark*
[(4)] *Westfalisches Museum für Naturkunde, W-4400 Münster, Germany*

SUMMARY

Body temperature was measured in three species of Brazilian hummingbirds during outdoor conditions, ensuring exposure to natural variations in both ambient temperature and photoperiod. All 3 species used torpor. However, in contrast to what has been observed during laboratory studies in hummingbirds, our experiments showed that torpor could be entered several times during a single night. In two of the hummingbird species studied two separate nightly torpor periods could be seen, while in the third, and smallest, species studied even three distinct torpor periods could be observed. The reason, why such a pattern of torpor has not been detected in previous studies, could stem from the absence of external stimuli during laboratory experimentations. During our experiments, on the other hand, the birds were exposed to natural environmental conditiòns. We suggest that some external stimuli (any sound of the tropical night) being the most likely candidate as an initiator of the 'untimely' nightly arousals. The results also indicate that the hummingbirds will have an energetic benefit from even very short torpor periods.

INTRODUCTION

Torpor, defined as a period of inactivity with a regulated low body temperature and depressed metabolic rate during part of the nocturnal phase, has been extensively reported from hummingbirds (*e.g.* Lasiewski and Lasiewski 1967, Lasiewski *et al.* 1967, Hainsworth and Wolf 1970, 1978, Carpenter 1974, Hainsworth et al. 1977, 1981, Beuchat et al. 1979, Krüger et al. 1982, Hiebert 1990, 1992). Although torpor may appear in well fed individuals (Carpenter and Hixon 1988), most studies have concluded that nocturnal torpor in hummingbirds occurs as a response to an energy depletion. The actual initiation of torpor depends on the amounts of energy reserves at roosting time (Hainsworth et al. 1977), the prevailing ambient temperatures, the length of the nocturnal phase and the time elapsed into the dark phase (Hiebert 1992). Thus, the available evidence seems to indicate that hummingbirds integrate these informations and show an appropriate respons if a threshold of energy reserves is met. There is evidence that such a threshold value decreases throughout the night (theory of 'time-dependent threshold'; Hiebert 1992). Entry into torpor can consequently be initiated throughout the entire dark phase, while arousal normally occurs during the last few hours before sunrise (Hiebert 1990). The torpid period, although being of variable length, has thus always been described as one single period during the nocturnal phase. However, during a study on the occurrence of torpor in three species of Brazilian hummingbirds exposed to semi-natural conditions, we found evidence that torpor bouts could occur two or even three times each night. In the present paper we report these findings.

MATERIAL AND METHODS

The study was carried out at the Museu de Biologia at Santa Teresa in the state of Espirito Santo, Brazil (20°S, 41°W, about 700 m above sea level), where we studied three species of resident hummingbirds, viz. the Versicolored Emerald *Amazilia versicolor*, the Black Jacobin *Melanotrochilus fuscus*, and the Swallow-tailed hummingbird *Eupetomena macroura*. The mean body masses of the three species were 4.2, 8.1, and 9.0 grams, respectively. We were unable to determine the sexes in any of the three species, which all are common breeding birds in the study area (Ruschi 1982). The study was conducted during the months of November and December, thus during the normal breeding period of most hummingbirds in the region.

The hummingbirds in the study area are accustomed to feed at artificial feeders. Hence, individuals of the three species were easily caught at the feeders during the afternoon and kept individually caged and deprived of food for variable time until sunset (19:00 hrs). The birds were weighed (Mettler, accuracy 0.01g) and then placed in smaller overnight cages, approximately 2.8 L cardboard-boxes, each provided with a perch. Usually the birds roosted quietly on the perch during the night, although in some cases they had apparently spent the night sitting on the floor of the cage. The boxes were placed outdoors during the night. The walls and top of each box were bored with holes, to ensure that the hummingbirds were exposed to natural variations in both ambient temperature and photoperiod.

The body temperature of the hummingbirds were measured by using a copper-constantan thermocouple (California fine wire, type 00) placed subcutaneously, laterally on the pectoral muscle. The thermocouple was fixed in place with small pieces of adhesive tape. During measurement the subcutaneously placed tip of the thermocouple was covered by the wing. Control experiments made on all three species showed that such measurements of pectoral temperature did not differ by more than 0.2-0.3°C from simultaneous measurements of cloacal temperature. A thermocouple was placed inside one of the cages to record the actual ambient temperature to which the birds were exposed. Nightly ambient temperatures recorded during the period of study were in the range of 20°C to 22°C.

Body temperature was measured at regular intervals throughout the night. The thermocouple were connected to a Data Translation (DT 2805) A/D converter, via a DT-757 Terminal board, and processed by a computer using a Labtech Notebook data acquisition programme. Each night, up to six individuals of the three species were studied simultaneously. After arousal, at about 6:00-7:00 hrs, the birds were removed from the boxeses and, after removing the thermocouples and re-weighing, released again.

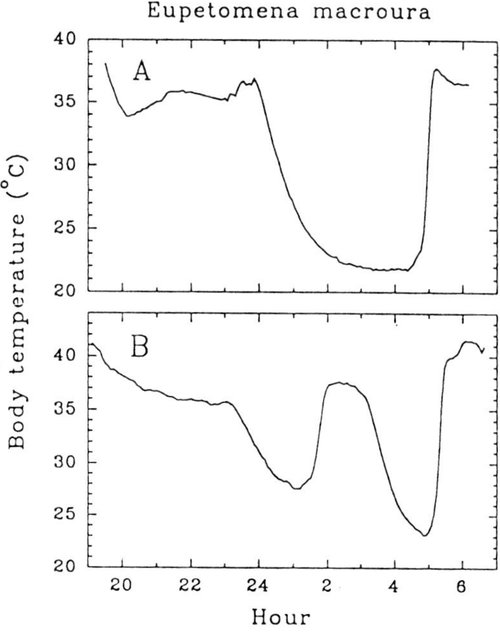

Fig. 1. Body temperature of two *Eupetomena macroura*; one showing the normal pattern with only a single torpor period during the latter half of the night (A), the other having two distinct periods of torpor (B).

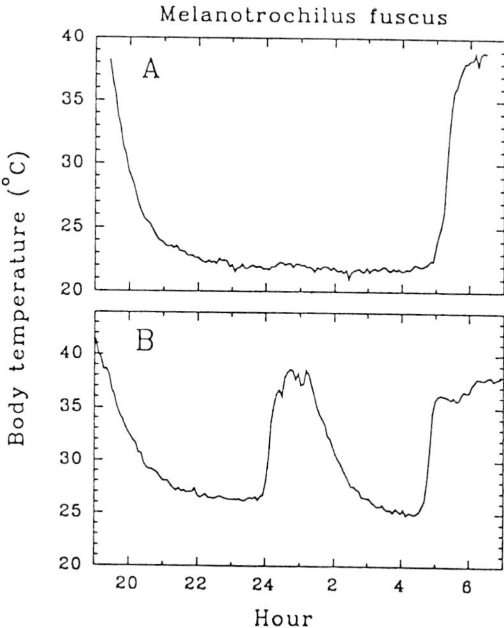

Fig. 2. Body temperature of two *Melanotrochilus fuscus* in which there was either one long-lasting torpor period (A) or two distinct torpid periods (B) during the night.

RESULTS AND DISCUSSION

All three species of hummingbirds used in the present study were able to enter torpor more than once per night, with some of the torpor periods lasting for only a few hours. In *E. macroura* 1 out of 12 torpid individuals had two distinct torpor periods (Fig. 1); in *M. fuscus* 1 out of 9 had two such periods (Fig. 2), while in *A. versicolor* 4 out of 15 torpid individuals showed multiple torpor bouts (3 of these even had three distinct bouts, Fig. 3). Very short torpor periods resembling those found in the present study have been described in hummingbirds before; *e.g.* Hainsworth et al. (1977) reported torpor periods of only 2.5 hours duration in Rivoli's hummingbird, *Eugenes fulgens* and of 3.5 hours duration in the Black-chinned hummingbird, *Archilochus alexandri*, while Hiebert (1990) showed that torpor bouts of 2.5-3.0 hours duration could occur late in the night in the Rufous hummingbirds, *Selasphorus rufus*. It is thus conceivably that most hummingbirds have

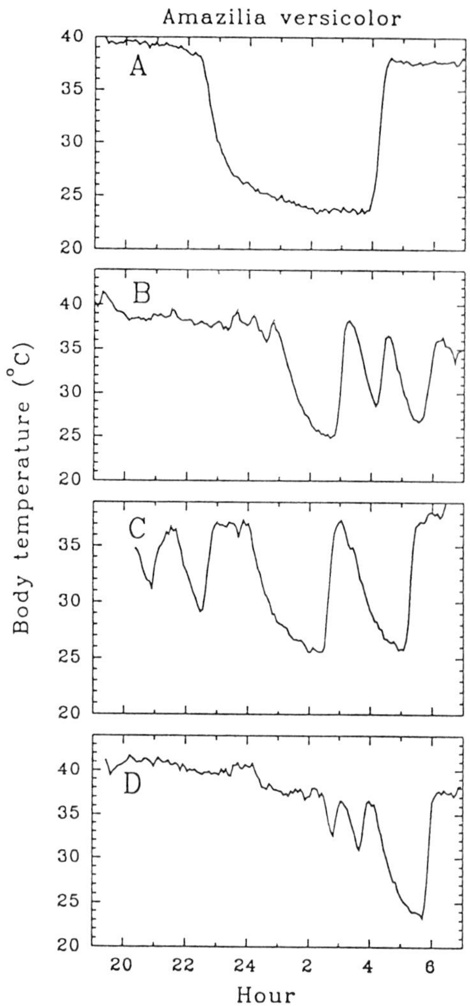

Fig. 3. Body temperature of four *Amazilia versicolor* in which there was either one torpor period during the night (A) or multiple periods of torpor (B, C and D).

the ability to enter into torpor for such short periods of time. However, the present description of multiple periods of nightly torpor in hummingbirds would seem to be the first report of such cases. With multiple torpor bouts during a single night, some of the torpor bouts consequently were of very short duration; merely the time for an entrance immediately followed by an arousal (Fig. 3).

The increased energy requirement for arousal has been said to limit the energy-saving potential of torpor (Lasiewski and Lasiewski 1967, Calder and King 1974, Bartholomew 1981). The results of the present study show that some individuals may not utilize the full time required to enter intō torpor, but may actually arouse from torpor even before their body temperature has reached its lowest level (Fig. 3), *i.e.* body temperature was only lowered to a value between the normothermic and normal torpid values. Such a pattern, involving 'interrupted' periods of torpor, were most often observed in *A. versicolor*. Our results would seem to indicate that even such very short periods of torpor may be of energetic significance for the birds, and that the cost of arousal would not counteract the reduction in metabolic rate resulting from the, even short-term, fall in body temperature. This is also in agreement with a recent study on the Rufous hummingbirds, kept in captivity, where Hiebert (1990) concluded that these birds save energy by entering into very short periods of torpor. Ruf & Heldmaier (1992) likewise found that the daily energy requirement of Djungarian hamsters *Phodopus sungorus* could be reduced by using daily torpor of less than one hours duration.

We can offer only speculations as to the cause of the described pattern of torpor, which seems to be in conflict with the assumption that there is a minimum body weight ('set-point'), below which the hummingbird is obliged to enter into torpor (Hainsworth et al. 1977). However, when the birds in the present study experienced an 'interrupted' torpor, they would inevitably enter torpor again after having reached normothermic body temperature, suggesting that some threshold-value of the body mass still is operating. The multiple bouts is most likely to be related to the fact, that the hummingbirds in the present study were exposed to natural external stimuli, which not only involves temperature and photoperiod, but also all natural sounds of the tropical night. Thus, in contrast to most other studies on torpor in birds, in which laboratory constrained individuals have been exposed to constant ambient conditions throughout the night phase, our findings may probably more resemble the natural pattern in the use of torpor. More studies on hummingbirds during natural conditions will be highly needed to clarify how widespread the described pattern of torpor really is.

ACKNOWLEDGEMENTS

This study was generously supported by the Erna and Victor Hasselblads Stiftelse (JFS), Ib Henriksens Fond (to JFS), the Danish Natural Science Research Council (to JFS), and the Deutsche Forschungsgemeinschaft (grant Be 536 to MB). We also wish to thank the Brazilian National Research Council (CNPq) for its support (grants to ASA, EB & JFS) and for permission to carry out this study in Brazil (permit no. WX-16/87). Thanks are also due to the staff at the Museu de Biologia Prof. Mello Leitao in Santa Teresa for their invaluable support during our stay.

REFERENCES

Bartholomew, G.A. (1981): A matter of size: An examination of endothermy in insects and terrestrial vertebrates. In *Insect Thermoregulation*, ed. B. Heinrich, pp. 45-78. New York: John Wiley & Sons.

Beuchat, C.A., Chaplin, S.B. & Morton, M.L. (1979): Ambient temperature and the daily energetics of two species of hummingbirds, *Calypte anna* and *Selasphorus rufus*. *Physiol. Zool.* 52, 280-295.

Calder, W.A. & King, J.R. (1974): Thermal and caloric relations in birds. In *Avian Biology vol. IV*, ed. D.S. Farner & J.R. King, pp. 259-413. New York: Academic Press.

Carpenter, F.L. (1974): Torpor in an Andean hummingbird: its ecological significance. *Science* 183, 545-547

Carpenter, F.L. & Hixon, M.A. (1988): A new function for torpor in a hummingbird. *Condor* 90, 373-378.

Hainsworth, F.R. & Wolf, L.L. (1970): Regulation of oxygen consumption and body temperature during torpor in a hummingbird, *Eulampis jugularis*. *Science* 168, 368-369.

Hainsworth, F.R. & Wolf, L.L. (1978): The economics of temperature regulation and torpor in non-mammalian organisms. In *Strategies in the Cold: Natural Torpidity and Thermogenesis*, ed. L.C.H. Wang & J.W. Hudson, pp. 147-186. New York: Academic Press.

Hainsworth F.R., Collins, B.G. & Wolf, L.L. (1977): The function of torpor in hummingbirds. *Physiol. Zool.* 50, 215-222.

Hainsworth, F.R., Tardiff, M.F. & Wolf, L.L. (1981): Proportional control for daily energy regulation in hummingbirds. *Physiol. Zool.* 54, 452-462.

Hiebert, S.M (1990): Energy costs and temporal organization of torpor in the rufous hummingbird (*Selasphorus rufer*). *Physiol. Zool.* 63, 1082-1097.

Hiebert, S.M. (1992): Time-dependent threshold for torpor initiation in the rufous hummingbird (*Selasphorus rufus*). *J. Comp. Physiol.* 162, 249-255.

Krüger, K., Prinzinger, R. & Schuchmann, K.-L. (1982): Torpor and metabolism in hummingbirds. *Comp. Biochem. Physiol.* 73A, 679-689.

Lasiewski, R.C. & Lasiewski, R.J. (1967): Physiological responses of the Blue-throated and Rivoli's hummingbirds. *Auk* 84, 34-48.

Lasiewski, R.C., Weathers, W.W. & Bernstein, M.H. (1967): Physiological responses of the Giant hummingbird, *Patagona gigas*. *Comp. Biochem. Physiol.* 23, 797-813.

Ruf, T. & Heldmaier, G. (1992): The impact of daily torpor on energy reqyirements on the Djungarian hamster, *Phodopus sungorus*. Physiol. Zool. 65, 994-1010.

Ruschi, A. (1982): *Hummingbirds of state of Espirito Santo*. Sao Paulo: Editora Rios Ltda.

Integrative and cellular aspects of autonomic functions : temperature and osmoregulation. Eds K. Pleschka, R. Gerstberger. John Libbey Eurotext, Paris © 1994, pp. 329-338.

Blood and brain temperatures of an unrestrained goat during an 11-month period

Claus Jessen

Physiologisches Institut der Universität, Aulweg 129, D-35392 Giessen, Germany

ABSTRACT

The temperatures of the arterial blood (TBLOOD) and the brain (TBRAIN) were continuously recorded in an animal living outdoors in a pen. Air temperature (TAIR) varied between -11°C and +44°C. The annual mean and standard deviation of TBLOOD was 38.34 ±0.42°C. The correlation between TAIR and TBLOOD was weak (r=0.19). A closer correlation existed between the daily ranges of TAIR and TBLOOD (r=0.62). The relationship between TBLOOD and TBRAIN showed a seemingly random distribution between 37.9 and 38.9°C TBLOOD. In this range, the brain could be more than 0.3°C warmer or more than 0.2°C cooler than the blood. Thus, a mechanism affecting TBRAIN independent of TBLOOD was operative in the normothermic zone of body temperature. Its effect was that in all sub-sets of data the variation of TBRAIN was smaller than that of TBLOOD. Both temperatures showed a clear circadian rhythm, the amplitude of which was greatly reduced in December and January. The circadian amplitude of TBRAIN was smaller than that of TBLOOD.

INTRODUCTION

In 1965, Bligh *et al.* reported on the 12-month course of a deep muscle temperature, which was recorded intermittently by radiotelemetry in an unrestrained sheep. Deep muscle temperature was taken as a representative of general body core temperature. Present methods permit continuous recordings of brain and blood temperatures in freely moving animals. This study concentrates on the relationships between a) air temperature (TAIR) and arterial blood temperature (TBLOOD), b) arterial blood temperature and brain temperature (TBRAIN), and c) the circadian rhythms of blood and brain temperatures. The measurements were done in a two-year old castrated male goat with a body weight of approximately 45 kg, which lived outdoors in a pen for more than a year. The recordings comprise 6456 hours and include 248 complete days. The collection of data commenced in August 1992 and ended in June 1993.

METHODS

The pen had a size of 4 x 8 m and provided some shelter from sun, wind and rain. Hay, grain and water were available *ad lib.* The animal was attended once a day in the morning. TAIR was taken once an hour at a shielded site 2 m above the ground.- The animal had guide tubes ending in the hypothalamus and 70 mm deep in a carotid artery. Thermistors connected to a portable data logger (Mini-Mitter, Sunriver) measured TBRAIN and TBLOOD and were corrected against a quartz-thermometer so that their readings at a given temperature agreed within ± one 0.07°C sampling step. The logger sampled at 2-min intervals and was unloaded every third week.

RESULTS

The original records showed periods of relative stability and episodes, in which large variations of both temperatures occurred. The reasons for the fast changes were not always evident (Fig. 1).

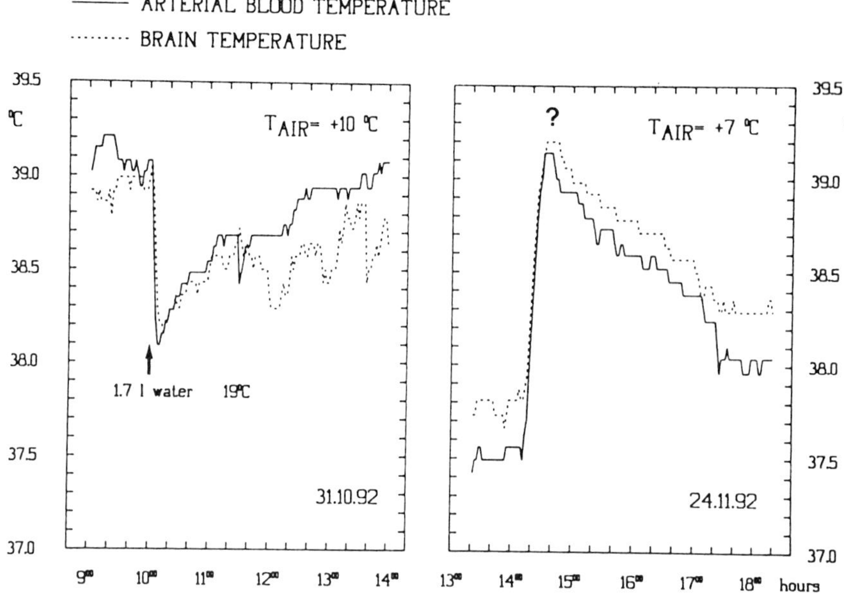

Fig. 1. Short-term variations: After deliberately drinking cold water, TBLOOD fell 1°C within 6 min and required 4 hrs to reach the pre-drinking level (left).- For unknown reasons, TBLOOD rose 1.6°C within 20 min and showed a slow decline afterwards (right).

Fig. 2 shows one-hour means of all temperatures during 24 hours in the spring. From midnight on, TBLOOD decreased to 37.6°C at 11.00 hours, although TAIR had reached its minimum at 7.00 hours. This was followed by a 2°C increase of TBLOOD within 3 hours to 39.6°C in the early afternoon. For the first hours of the night and in the range around 38°C, TBRAIN was lower than TBLOOD. It occurred again in the afternoon, when TBRAIN remained 0.6°C below TBLOOD.

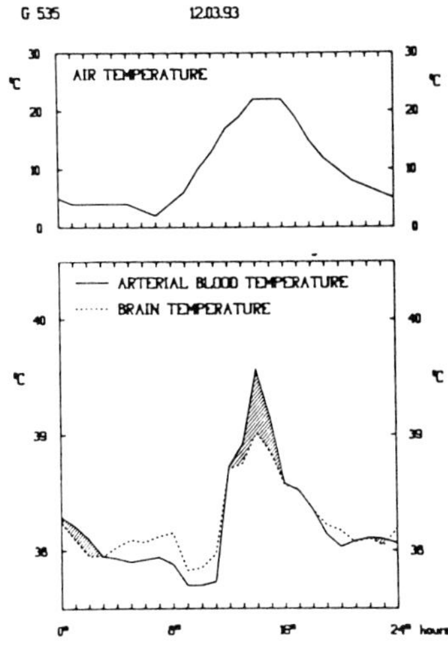

Fig. 2. One-hour means of TBLOOD and TBRAIN during a cool night and a warm day in the spring. Cross-hatched areas: TBRAIN was lower than TBLOOD. This occurred at high TBLOOD during the warm hours of the day, but to some extent also in the early hours, when TBLOOD and TBRAIN were around 38°C.

Relationship between TAIR and TBLOOD

Fig. 3 shows that the influence of TAIR on TBLOOD (24-hour means) was relatively weak. Between +5 and 25°C TAIR , TBLOOD varied between 38 and 39°C and appeared to be mainly determined by non-environmental factors. A closer correlation existed between the day spans (max - min) of TAIR and TBLOOD.

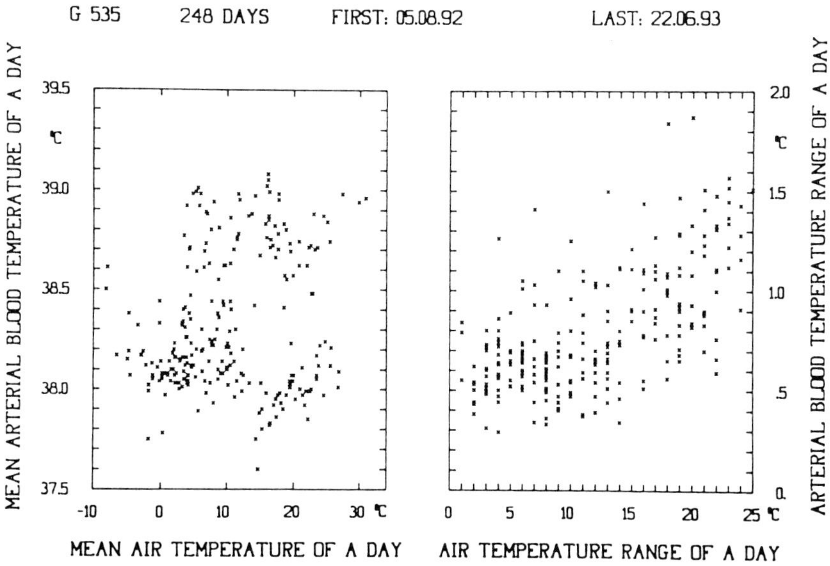

Fig. 3. Loose correlation between the day means of TAIR and TBLOOD (r=0.19), and closer correlation between the day spans (r=0.62).

Relationship between TBLOOD and TBRAIN

All data of a given period were first sorted in 0.1°C-classes of TBLOOD. Means, standard deviations (SD), minima and maxima of TBRAIN were calculated for each class of TBLOOD. Additionally, the frequencies, at which the different classes of TBLOOD occurred, were determined.

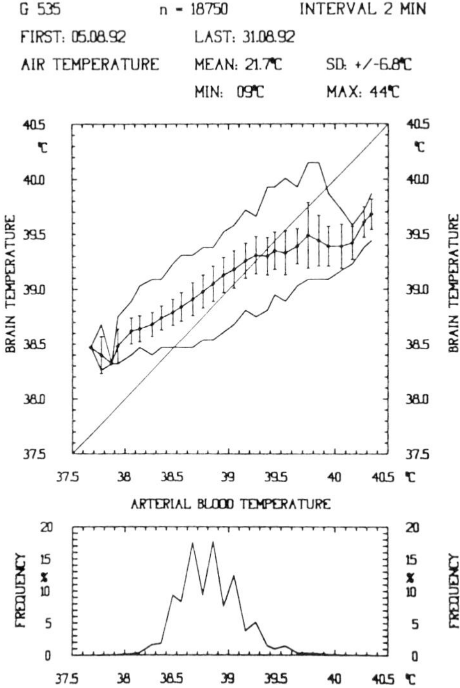

Fig. 4. Upper panel: TBRAIN as a function of TBLOOD calculated on the base of raw data (2 min-sampling interval). Lower panel: frequency at which the 0.1°C-classes of TBLOOD occurred. (*Note:* the saw-toothed pattern follows from the single 0.07°C sampling steps of the data logger, which fill classes of 0.1°C width alternatingly with small [1x] and large [2x] samples)

Fig. 4 analyzes the 2 min-raw data obtained in August 92, which included the hottest day of the year. The range below 39.3°C TBLOOD, which comprised 90% of the data, is characterized by a quasi-linear relationship between the means of both temperatures: the lower TBLOOD, the warmer was the brain relative to the arterial blood. Above 39.3°C, the means of TBRAIN were lower than TBLOOD. In most classes, the SD of TBRAIN was in the order of ±0.2°C. The minima and maxima, however, showed large spans. At 39.6°C TBLOOD, TBRAIN could be as low as 38.9°C or as high as 40.0°C, and also at the lower end of the central range, a 0.8°C span of TBRAIN was observed at 38.5°C TBLOOD. The largest deviations from the line of identity occurred near the ends of the range. At 37.8°C TBLOOD, the brain was once found to be 0.9°C warmer, and at 39.9°C TBLOOD, there was a single 2 min-period when the brain was 0.85°C cooler than the blood. Taken together, the data show that, at a high time resolution, the relationship between TBLOOD and TBRAIN was highly variable: at any given level of TBLOOD in the central range, TBRAIN could be 0.3-0.4°C higher or lower than TBLOOD.

Fig. 5 analyzes the one-hour means of the same period. The processing of data was the same as for Fig. 5. The range of TBLOOD is now reduced to 2°C, and the distribution of the frequencies approaches symmetry around 38.7°C. The spans of TBRAIN at a given class of TBLOOD are now also smaller. However, even on the basis of one-hour means TBRAIN could vary, at a given level of TBLOOD, within a range of 0.55°C. This occurred in the 38.7°C-class of TBLOOD. The largest degree of selective brain cooling was -.67°C (TBRAIN - TBLOOD) at a blood temperature of 39.9°C. At 38.0°C TBLOOD, the brain was 0.5°C warmer than the blood.

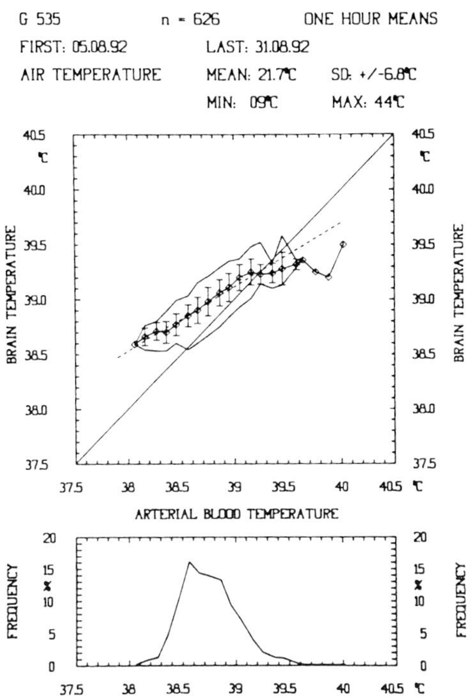

Fig. 5. Upper panel: TBRAIN as a function of TBLOOD calculated on the base of one-hour means. The broken line shows the regression of TBRAIN on TBLOOD. Lower panel: frequency at which the 0.1°C-classes of TBLOOD occurred.

The broken line in Fig.5 shows the regression of TBRAIN on TBLOOD. The good match between the class-means for 95% of the data and the line, and the correlation coefficient of 0.80 indicate that the linear regression provides an adequate description for this set of data. This is in contrast to other studies which have shown that in the laboratory, the relationship between TBLOOD and TBRAIN consists of two linear segments with clearly different slopes and intersection at the line of identity (Kuhnen & Jessen, 1991).

Altogether, 6456 hrs were analyzed, which equals 84% of the theoretical 11-month maximum. The overall mean of TAIR was 10.5 ±9.4°C (SD), and the extremes were -11°C and +44°C. TBLOOD was 38.34 ±0.42°C, and the range extended from 37.1°C to 40.0°C. TBRAIN was 38.47 ±0.39°C, and the range was from 37.4°C to 39.9°C. The largest degree of selective brain cooling (TBRAIN - TBLOOD) was -0.67°C, and the maximum of 'brain warming' was 0.56°C.

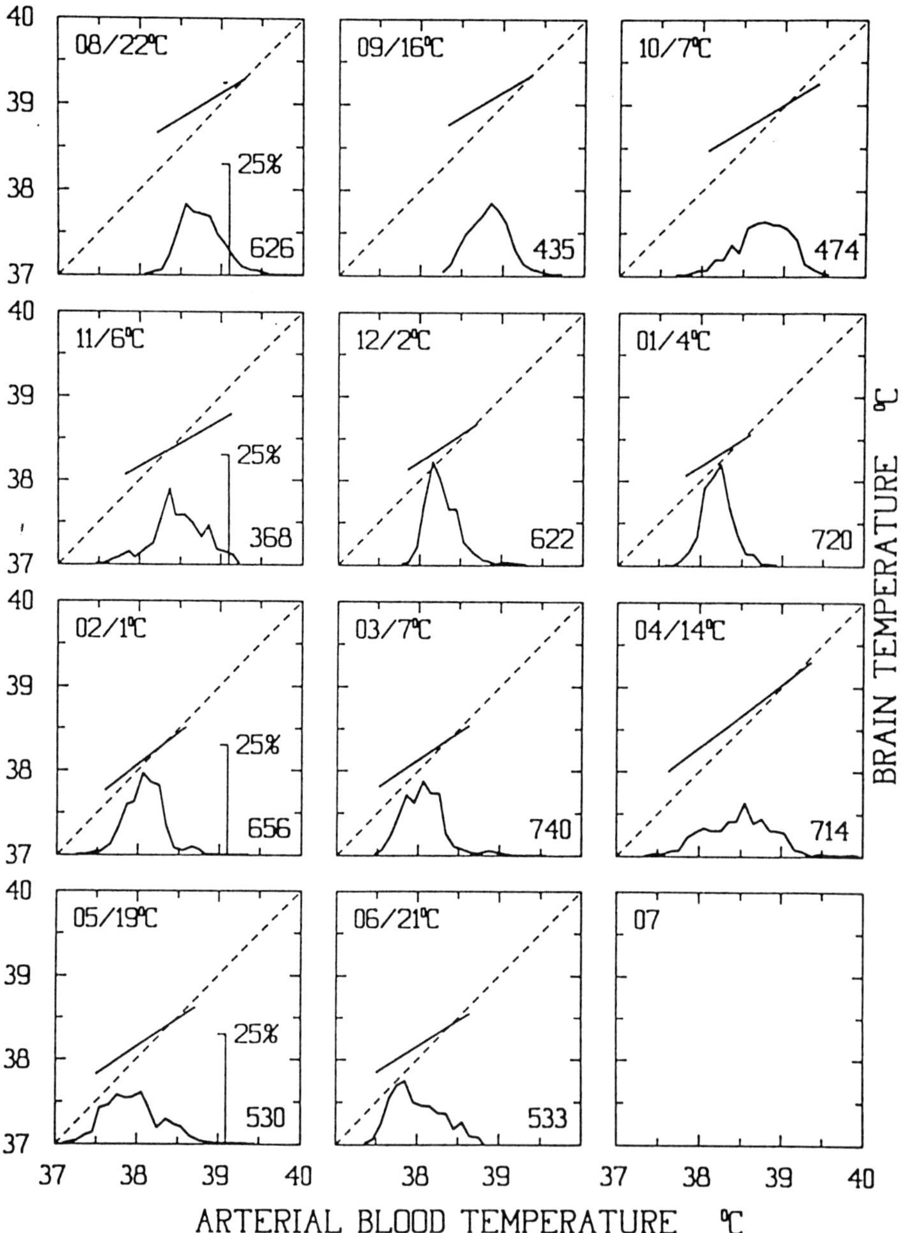

Fig. 6. A single panel shows the frequency of the 0.1°C-classes of TBLOOD and a regression of TBRAIN on TBLOOD for a month. Top left: number of the month and mean TAIR. Bottom right: number of hrs.

Fig. 6 is based on one-hour means and presents all data obtained in the 11-month period. For each month, the frequency is plotted at which the 0.1°C-classes of TBLOOD occurred (inset scale). The monthly means showed a clear seasonal trend. From August to October, the means of TBLOOD remained between 38.74 and 38.84°C. From November to February, the means of TBLOOD declined steadily to 38.09°C and remained near this level until the end of June. Thus, the seasonal trends of TBLOOD and TAIR were not closely related to each other. The standard deviations of the monthly means of TBLOOD varied between the minimum of ±0.19°C in January and the maximum of ±0.43°C in April.

The monthly means of TBRAIN were usually 0.1-0.2°C higher than those of TBLOOD. The noticable exception was November, when mean TBRAIN was 0.05°C lower than mean TBLOOD. In this month, 212 out of 368 hours fulfilled the criterion of selective brain cooling (TBRAIN < TBLOOD), and the mean TBLOOD of this sub-set was 38.46 ±0.26°C. The standard deviations of the monthly means of TBRAIN varied between the minimum of ±0.17°C (September and December) and the maximum of ±0.34°C in April. On average, the standard deviations of TBRAIN were three quarters as large as those of those of TBLOOD.

The regressions of TBRAIN on TBLOOD had correlation coefficients between 0.68 (January) and 0.94 (June). The mean was 0.86 ±0.08 (SD). The regression coefficients varied between 0.56 (November) and 0.75 (April). The mean was 0.64 ±0.07 (SD). Thus, in all months TBRAIN varied less than TBLOOD. On average, the variation of TBRAIN amounted to less than two thirds of the variation of TBLOOD.

In a larger pool of data, the point of intersection between the regression line and the line of identity can be thought to represent the equivalent of the mean threshold of selective brain cooling (Kuhnen & Jessen, 1991). In all months except November, the majority of the data points (80-97%) lay on the left side of the point of intersection. Consequently, the slopes of the regression lines were predominantly determined by data below the mean threshold of selective brain cooling.

The threshold reached its highest level in September (39.3°C) and declined during the winter months. The minimum (38.3°C) was observed in February. Thus, the seasonal trend of the threshold of selective brain cooling grossly paralleled the seasonal trend of TBLOOD.

Circadian rhythms of TBLOOD and TBRAIN

Separately for each complete day, the means of TBLOOD and TBRAIN were calculated. For each hour of the day, the difference between the actual temperature and the daily mean was computed. For a given month, the differences of all corresponding hours were averaged. In Fig. 7, the mean differences and the standard errors of two representative months are plotted versus the hours. In August, both temperatures displayed a clear rhythm with a minimum near 5.00 hours and a maximum at 19.00 hours. The amplitude was approximately 0.5°C. The sinusoidal wave form is distorted between 6.00 and 10.00 hours, which coincides with the time when animal and pen were attended.

In December, the amplitude of the circadian rhythm was greatly reduced. The morning minimum occurred at 7.00 hours, while the maximum was reached past midnight. Again the wave form is distorted between 8.00 and 12.00 hours.

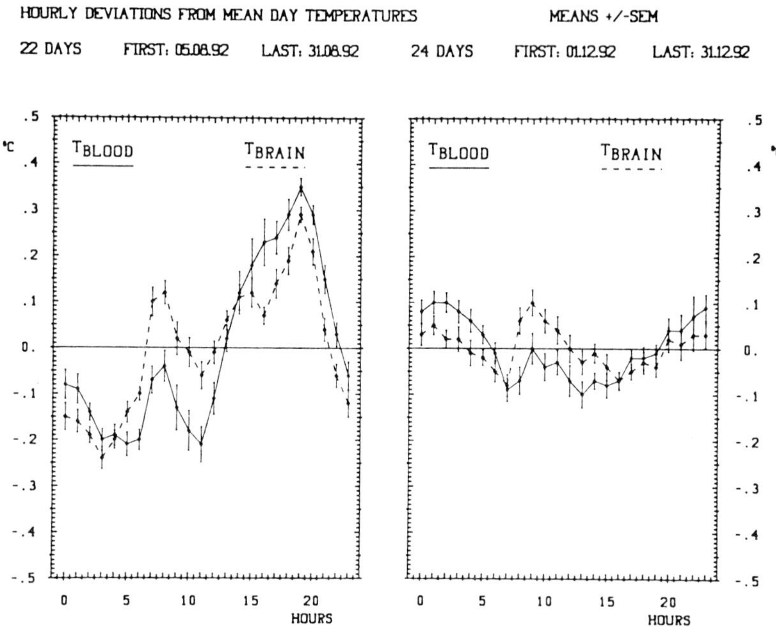

Fig. 7. Circadian variations of TBLOOD and TBRAIN in August and December. For each hour, the means ±SEM of the differences between the actual temperature and the daily mean is shown.

In order to quantify the circadian variations, the hourly deviations from the day means were added regardless of their signs. The result (°C) is roughly equivalent to the sum of the areas (°C x hrs) below and above the curves on both sides of the zero line. From August to November, the sums were between 3.2 and 3.8°C and, contrary to previous findings (Bligh *et al.*,1965), decreased to 1.3°C in December and January. The following months showed a gradual return to larger values with a maximum of 5.3°C in June. In all months, the circadian variation of TBRAIN was smaller and amounted to just 70 ±11% of that of TBLOOD.

Fever

In September the animal fell ill (Fig. 8). The temperatures began to rise on day 3 and peaked near 41.5°C on day 5, when the animal was transferred indoors. In constrast to previous observations on artificial fever in sheep (Laburn *et al.*,1988), TBRAIN exceeded TBLOOD throughout the fever. Even when TBLOOD was 41.4°C, TBRAIN was approximately 0.3°C higher and remained at 41.7°C for several hours. - The animal recovered completely and in the following 9 months, no other episodes of fever were observed.

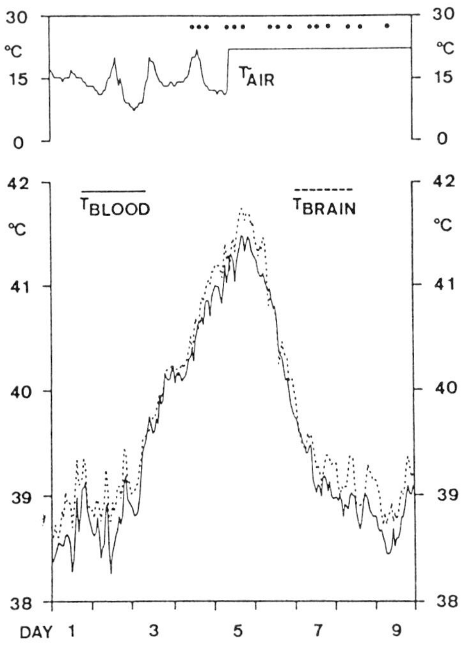

Fig. 8. Four days of fever. The animal was transferred indoors on day 5 and received antibiotics from day 4 to day 9 (filled circles in the top line).

DISCUSSION

Under well-controlled conditions of the laboratory, the body core temperature of resting medium-sized animals attains a mimimum in the thermoneutral zone and increases with higher or lower TAIR (e.g. dog: Hallwachs, 1960). It was not the case in this single unrestrained animal: over the 40°C-range of TAIR, which was covered in the year, only a loose positive correlation between TAIR and TBLOOD was found. The animal panted mildy on warm days, but was never observed to shiver on cold days. Thus, adjustments of behaviour and external insulation were obviously the principal means to cope with the cold season. However, at any daily mean TAIR between +5 and +28°C, the daily mean of TBLOOD could be as low as 38.0°C or as high as 39.0°C. Within these limits, the level of TBLOOD was apparently determined by unidentified factors.-The short-term capability of the temperature regulating system can be seen in the relationship between the daily ranges of TAIR and TBLOOD. A regression of the minimal TBLOOD-ranges on the TAIR-ranges shows that a day range of TAIR comprising 25°C could be handled with a 0.57°C range of TBLOOD (which included the endogenous circadian component). However, on most days the ranges of TBLOOD were much wider. This and the frequent episodes of large short-term variations (Fig.1) are indicative of a generally rather loose type of control: over a 24 hr period, internal body temperature *could* be regulated within narrow limits, but more often the system permitted wider deviations.

Goats possess a carotid rete, by which TBRAIN can be maintained at a relatively lower level, when general body temperature increases. The process is termed selective brain cooling (SBC) and had in laboratory experiments a precise threshold near 38.8°C. Below the threshold, TBLOOD and TBRAIN were tightly coupled and the slope was close to one, while above the threshold, TBRAIN increased much less than BLOOD (Kuhnen & Jessen, 1991).

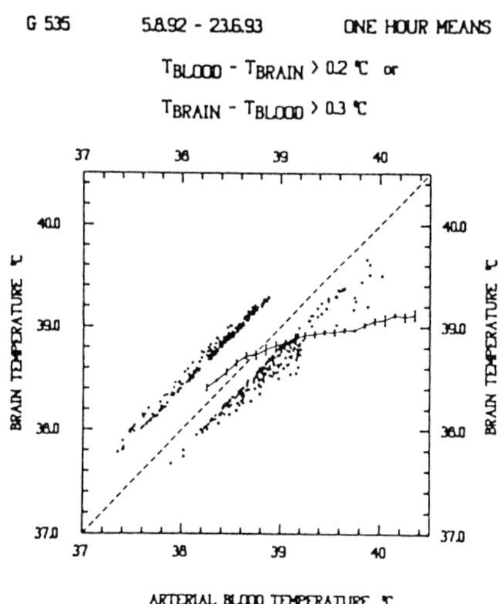

Fig. 9. Sub-set of data with major differences between TBLOOD and TBRAIN. In the range of TBLOOD from 37.9 to 38.9°C, the brain could be more than 0.3°C warmer or more than 0.2°C cooler than the blood. Solid line and bars: means ±SEM of TBRAIN = f(TBLOOD) in laboratory experiments.

This was not the case in the unrestrained animal (Fig. 9). Over a wide range of TBLOOD, data are distributed above and below the line of identity, implying that the threshold of SBC was variable between 37.7 and 39.3°C. Within this range, TBRAIN was randomly distributed around the line of identity. It must be concluded that in the normothermic zone, the relationship between TBLOOD and TBRAIN cannot be explained entirely by the level of TBLOOD. Non-thermal factors appear to play an important role. However, the consequence of the operation of SBC in the normothermic zone was that the variation of TBRAIN over the full range of TBLOOD was clearly less than that of TBLOOD: brain temperature was more stable than general body core temperature (Fig. 6).

REFERENCES

Bligh, J., Ingram, D.L., Keynes,R.D., and Robinson, G.S. (1965): The deep body temperature of an unrestrained Welsh mountain sheep recorded by a radiotelemetric technique during a 12-month period. *J. Physiol. (Lond.)* 176, 136-144.

Hallwachs, O. (1960): Sauerstoffverbrauch und Temperaturverhalten des unnarkotisierten Hundes bei Lufttemperaturen von -10 bis +35°C. *Pflügers Arch.* 271, 748-760.

Kuhnen, G. and Jessen, C. (1991): Threshold and slope of selective brain cooling. *Pflügers Arch.* 418, 176-183.

Laburn, H.P., Mitchell, D., Mitchell, G., and Saffy, K. (1988): Effects of tracheostomy breathing on brain and body temperatures in hyperthermic sheep. *J. Physiol. (Lond.)* 406, 331-344.

Integrative and cellular aspects of autonomic functions : temperature and osmoregulation. Eds K. Pleschka, R. Gerstberger. John Libbey Eurotext, Paris © 1994, pp. 339-346.

Effects of selective brain cooling on thermoregulatory responses during exposure to hot dry or hot humid air

Gernot Kuhnen

Physiologisches Institut der Universität, Aulweg 129, D-35392 Giessen, Germany

SUMMARY

Thermoregulatory responses of five goats were determined in 54 experiments at different levels of selective brain cooling (SBC). SBC resulted in a dissociation of brain and trunk temperatures. The results showed that at lower levels of SBC and relatively higher brain temperatures the heat loss mechanisms were activated stronger compared to higher levels of SBC in spite of the same trunk temperatures. The effect of SBC was to smooth the onset and to reduce the slope of heat loss mechanisms.

INTRODUCTION

The increase of arterial blood temperature (T_{blood}) during heat load should result in a corresponding rise of brain temperature (T_{brain}), but in many mammalian species T_{brain} increases at a much lower rate than T_{blood}. The process which causes a dissociation of T_{brain} and T_{blood} is termed selective brain cooling (SBC). SBC starts at normothermic temperatures and is most evident at hyperthermic temperatures where T_{brain} can be about 2 °C lower than T_{blood}. The heat exchange resulting in SBC takes place in the cavernous sinus by cooling the arterial blood which is supplied to the brain. The arterial blood on its way to the brain passes the carotid rete, a network of small arteries. The carotid rete is situated in the cavernous sinus which receives venous blood cooled by draining the nasal mucosa and the skin of the face. Cavernous sinus and carotid rete form an internal heat exchanger (for review of SBC see Baker, 1982; Mitchell et al., 1987).

The dissociation of brain and trunk temperatures (T_{trunk} represented by T_{blood}) by SBC must affect the general thermoregulation of the animal which is controlled by temperature signals of both brain and trunk. Uncoupling of T_{brain} from T_{trunk} during hyperthermia must reduce the internal signals driving heat loss mechanisms. This effect of SBC seems to be a disadvantage in keeping body temperature constant. The aim of study 1 was to look at this problem by comparing thermoregulatory responses during two thermal conditions

inducing different levels of SBC. Low and high relative humidity of the ambient air were used to induce different levels of SBC.
The aim of study 2 was to determine the thermoregulatory responses to cold and heat load with and without the influence of SBC. SBC could be eliminated by supplying the brain with arterial blood bypassing the carotid rete or by blocking the venous supply of the cavernous sinus. In our experiments another way was used which does not eliminate the heat exchange in the cavernous sinus but was concealing the effect of cooling the brain. The expression of SBC was masked by increasing T_{brain} (by means of heat exchangers controlling carotid blood temperature) at the same rate as T_{trunk}. T_{brain} was kept about 0.2 °C higher than T_{trunk} corresponding to the difference of T_{brain} and T_{trunk} at hypothermic temperatures. Comparing the thermoregulatory responses with and without the influence of SBC should demonstrate the effects of SBC on thermoregulation.

MATERIAL AND METHODS

The experiments were performed on conscious female goats. In sterile conditions and under general anaesthesia (halothane, nitrous oxide) the animals were implanted with hypothalamic guide tubes for monitoring hypothalamic temperatures (T_{hypo}) and devices to alter head and trunk temperatures by means of heat exchangers acting on the arterial blood. Methods have been published before in detail (Jessen et al., 1990; Kuhnen & Jessen, 1991). The animals were in perfect health throughout the period of experiments.
The experimental set up was described recently (Kuhnen & Jessen, 1992). In brief, T_{blood} was measured in the aorta and the carotid arteries representing T_{trunk} and the temperature of the blood supplying the brain (T_{caro}), respectively. T_{brain} was represented by T_{hypo}. Respiratory minute volume ($\dot{V}_E$) was measured by means of a respiration mask and was converted to BTPS. Respiratory frequency (RF) was determined by counting chest movements. Respiratory evaporative heat loss (REHL) and metabolic rate (MR) were determined in an open-circuit system.
SBC was calculated from the difference between T_{caro} and T_{brain}. The slope of SBC (°C/°C) is defined as the increase of SBC per °C increase of T_{caro}.

Study 1

Thirty-six experiments were performed in three goats at an ambient temperature of 35 °C, either at low humidity (LH, 22 %) or high humidity (HH, 84 %). In experiments of type A the arterial blood supplied to trunk and head was increased in parallel from 38.5 to 40.5 °C at a rate of nominally 0.02 °C/min. In experiments of type B T_{trunk} was increased from 39 to 41 °C, while T_{caro} was clamped at 38.8 °C to keep T_{brain} constant. All experiments were performed at LH and HH.

Study 2

In two goats 18 experiments were performed at an ambient temperature of 20 °C and a relative humidity of 55 %. In control experiments T_{trunk} and T_{caro} were increased (0.03 °C/min) identically from about 37 to ca. 40.8 °C, while in other experiments T_{trunk} and T_{brain} were increased at the same rate. T_{brain} was adjusted to T_{trunk} by increasing T_{caro} at a higher rate than T_{trunk}, therefore masking SBC.

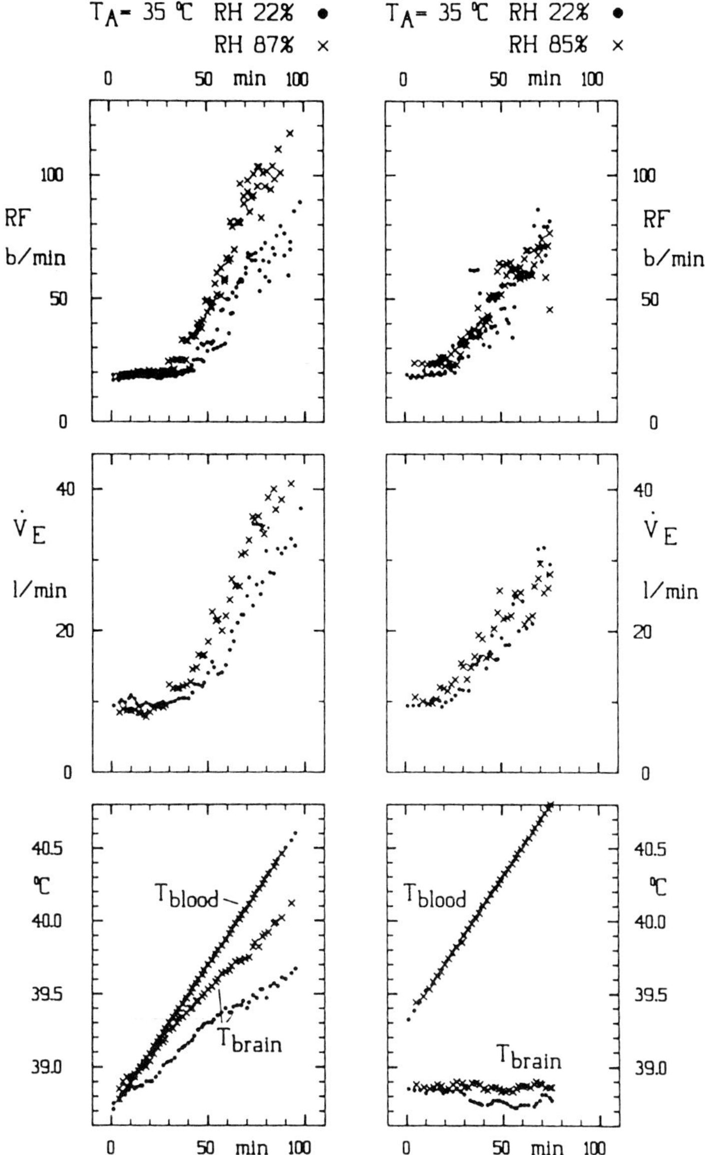

FIG.1. Time course of experiments of type A and B (left and right panels, respectively). In type A blood temperatures (T_{blood}) of the trunk and carotid arteries were increased (0.02 °C/min) by means of extracorporeal heat exchangers. In type B trunk temperature (T_{blood}) was increased (0.02 °C/min) and carotid blood temperature was clamped at 38.8 °C to keep brain temperature (T_{brain}) constant. Experiments were performed at low (•) or high (x) humidity. Smoothed data from 6 experiments in one goat. Ta, ambient temperature; RH, relative humidity; RF, respiratory frequency; $\dot{V}_E$, respiratory minute volume.

RESULTS

Study 1

Fig. 1 shows the data of one animal in which T_{blood} ($T_{trunk}=T_{caro}$) was increased in a ramp-like fashion at HH and LH (type A, left side of Fig.1). T_{blood} increased at nearly the same time course during LH and HH. Owing to the different cooling powers of the inspired air in LH and HH, SBC (difference between T_{blood} and T_{brain}) was either large and T_{brain} at a given T_{blood} was low (LH), or SBC was small and T_{brain} relatively higher (left bottom panel). The higher T_{brain} resulted in a stronger activation of heat loss mechanisms in spite of the same level of T_{trunk} (=T_{blood}), as $\dot{V}_E$ and RF show (left middle and top panels, respectively).
On average in the three goats the increase of REHL per °C T_{trunk} increase (slope) decreased significantly ($p<0.001$) with HH from 1.21 ± 0.01 (LH) to 0.57 ± 0.06 W/kg °C (HH) (weighted mean ± SEM, N=3, n=18) due to the low cooling power at HH. The slope of SBC (SBC per °C increase of T_{caro}) was nearly halved by HH (0.62 ± 0.03 vs. 0.34 ± 0.02 °C/°C, $p<0.001$). $\dot{V}_E$ and RF increased with HH 34.82 ± 2.66 vs. 45.50 ± 2.48 l/min °C ($p<0.05$) and 184.80 ± 13.03 vs. 230.40 ± 11.03 breaths/min °C ($p<0.05$), respectively.
The right side of Fig.1 shows the data of experiments of type B. The values of $\dot{V}_E$ and RF show no systematic differences between HH and LH due to the nearly identical T_{brain}. On average there were no significant differences in $\dot{V}_E$ and RF between LH and HH in spite of the significant decrease of REHL at HH (1.11 ± 0.05 vs. 0.46 ± 0.09 W/kg °C, $p<0.01$).

Study 2

Fig. 2 shows the results of 5 control experiments and 5 experiments in which the expression of SBC was prevented in goat G546. The control experiments show that at T_{blood} below 38.3 °C T_{brain} was about 0.2 °C higher than T_{blood}. At higher T_{blood} T_{brain} increased at a lower rate than T_{blood} resulting in an 1.2 °C lower T_{brain} at T_{blood} of 40.8 °C. In experiments with masked SBC T_{brain} was kept a level on average 0.18 ± 0.01 °C higher than T_{blood} (=T_{trunk}) throughout the experiments. The relatively higher T_{brain} in these experiments resulted in an earlier and stronger activation of heat loss mechanisms ($T_{ear\ skin}$, REHL) in response to T_{blood}. However, only the differences in slopes of REHL reached a significant level ($p<0.01$).
The values are shown in detail in Table 1. Fig. 3 shows the results in another goat confirming the results mentioned above, although in this goat T_{brain} could not be controlled as good as in G546.

DISCUSSION

A panting animal responds to a rise of internal temperature by an increase of $\dot{V}_E$, which is accompanied by a rise of REHL. However, the resultant heat loss depends on the enthalpy of the inspired air: the higher the temperature and humidity of the inspired air, the lower is the respiratory heat loss for a given $\dot{V}_E$ (Albers, 1977). SBC may work as a tool to adjust the activity the respiratory heat loss mechanisms to the cooling power of the inspired air.

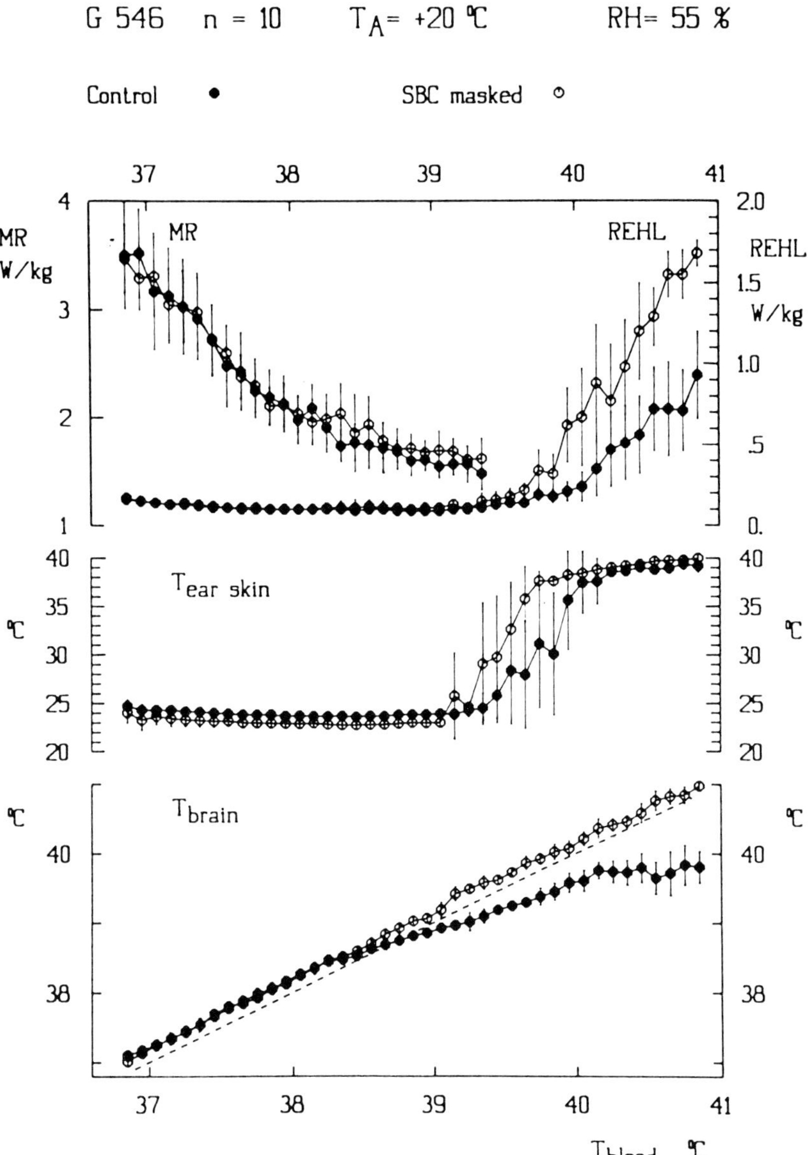

FIG.2. Increase of trunk temperature (T_{blood}) (0.03 °C/min) with normal dissociation of blood and brain temperatures by selective brain cooling (●) and in experiments in which the dissociation was inhibited (o). The dashed line in the bottom panel represents the line of identity between T_{blood} and T_{brain}. The thermoregulatory responses to T_{blood} and T_{brain} are shown in the middle ($T_{ear\ skin}$, ear skin temperature indicating skin blood flow) and top panels (MR, metabolic rate; REHL, respiratory heat loss). Mean values and SD are shown for 5 experiments of each type.

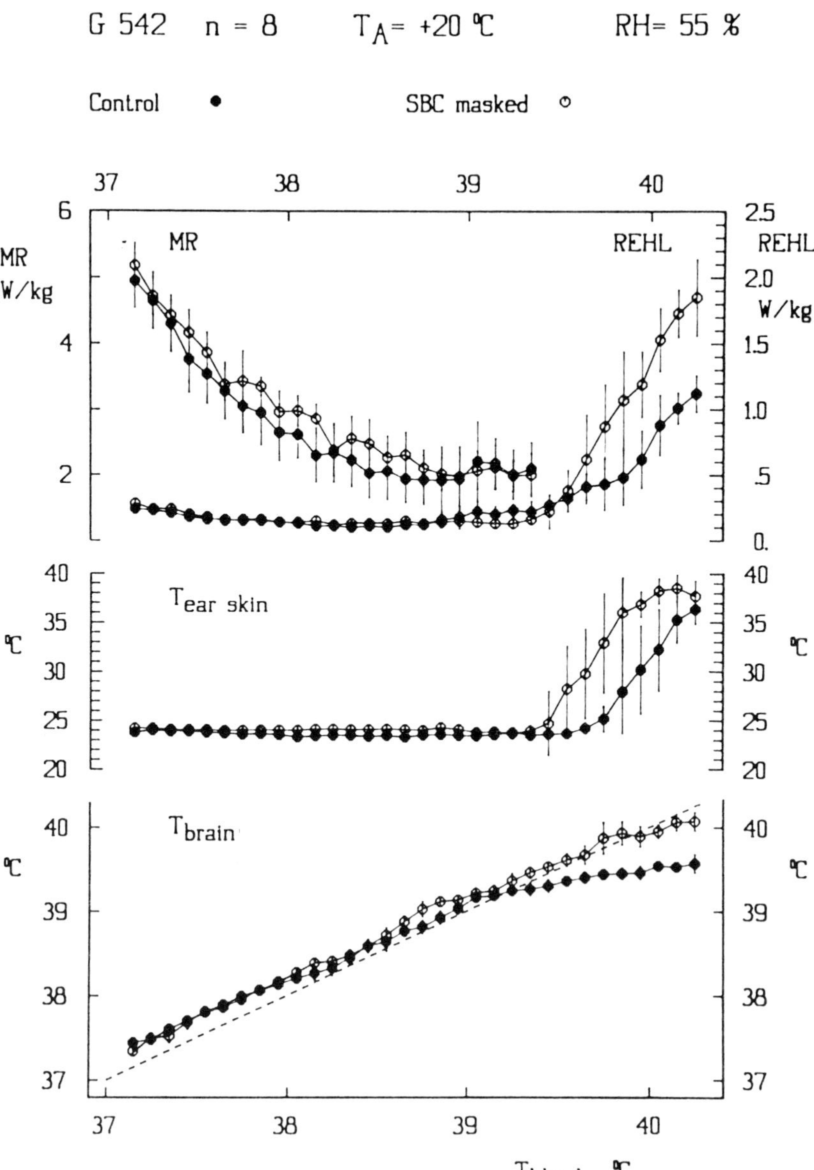

FIG. 3. Increase of trunk temperature (T_{blood}) (0.03 °C/min) with normal dissociation of blood and brain temperatures by selective brain cooling (●) and in experiments in which the dissociation was inhibited (o). The dashed line in the bottom panel represents the line of identity between T_{blood} and T_{brain}. The thermoregulatory responses to T_{blood} and T_{brain} are shown in the middle ($T_{ear\ skin}$, ear skin temperature indicating skin blood flow) and top panels (MR, metabolic rate; REHL, respiratory heat loss). Mean values and SD are shown for 4 experiments of each type.

Table 1

	G 542 control n=4	SBC masked n=4	G 546 control n=5	SBC masked n=5
REHL				
slope	1.18 ± 0.27	1.83 ± 0.30 *	1.01 ± 0.11	1.66 ± 0.37 **
threshold	39.62 ± 0.21	39.43 ± 0.15 ns	40.24 ± 0.33	39.89 ± 0.23 ns
MR				
slope	-2.66 ± 0.51	-2.58 ± 0.54 ns	-1.21 ± 0.27	-1.23 ± 0.37 ns
threshold	38.32 ± 0.31	38.49 ± 0.29 ns	38.17 ± 0.34	38.09 ± 0.24 ns
$T_{ear\ skin}$ threshold	39.72 ± 0.10	39.51 ± 0.22 ns	39.64 ± 0.29	39.33 ± 0.17 ns
T_{brain}-T_{caro}		0.16 ± 0.02		0.18 ± 0.01

Mean values ± SD of thresholds (°C trunk temperature) and slopes (W/kg °C; REHL, respiratory evaporative heat loss; MR, metabolic rate) of two goats are shown. T-tests between control and SBC masked: ns (not significant), * ($p<0.05$), ** ($p<0.01$).

The experiments of study 1 have shown that a higher humidity of the inspired air reduced REHL and SBC, resulting in a relatively higher T_{brain}. The higher T_{brain} caused a stronger activation of $\dot{V}_E$ and RF. The stronger activation of the respiratory heat loss mechanism increases REHL and can partly compensate the lower cooling power of the inspired air.
In type B experiments $\dot{V}_E$ and RF were nearly identical at LH and at HH implying that the activity of the heat loss mechanisms was not influenced by the humidity of the air, when SBC-mediated effects on T_{brain} were excluded. Thus the effects of humidity on $\dot{V}_E$ and RF were transmitted by SBC.
Taken together the coupling between the enthalpy of the inspired air and the activity of the respiratory heat loss mechanisms assigns a function to SBC, which serves the homeothermy not just of the brain, but of the total body core.

The thermoregulatory responses are driven by brain and trunk temperatures, and the brain is providing about 50 % of all internal temperature signals in the goat (Jessen & Feistkorn, 1984). SBC reduces the increase of T_{brain} during hyperthermia and therefore decreases the activation of heat loss mechanisms. This consequence has been shown by comparing experiments with and without the influence of SBC. The effect of SBC was to shift and to smooth the onset of heat loss mechanisms and to reduce the gain of REHL, resulting in widening the interthreshold zone and damping the effector responses. Thermoregulatory mechanisms induced at hypothermic body temperatures like MR were not affected by the SBC mechanism.

SBC is regarded as a mechanism to protect the brain against heat damage, but additionally this mechanism enables the animal to modify the thermoregulatory responses by changing T_{brain}.

ACKNOWLEDGEMENT

This study was supported by DFG Ku 807/1-1.

REFERENCES

Albers, C. (1977): Respiratory control of body temperature: a theoretical model. *Respir.Physiol.* 30, 137-151.
Baker, M.A. (1982): Brain cooling in endotherms in heat and exercise. *Ann.Rev.Physiol.* 44, 85-96.
Jessen, C. & Feistkorn, G. (1984): Some characteristics of core temperature signals in the conscious goat. *Am.J.Physiol.* 274, R456-R464.
Jessen, C., Felde, D., Volk, P.,Kuhnen, G. (1990): Effects of spinal cord temperature on generation and transmission of temperature signals in the goat. *Pflügers Arch.* 416, 428-433.
Kuhnen, G. & Jessen, C. (1991): Threshold and slope of selective brain cooling. *Pflügers Arch.* 418, 176-183.
Kuhnen, G. & Jessen, C. (1992): Effects of selective brain cooling on mechanisms of respiratory heat loss. *Pflügers Arch.* 421, 204-208.
Mitchell, D.,Laburn, H.P., Nijland, M.J.M., Zurowski, Y., Mitchell, G. (1987): Selective brain cooling and survival. *S.Afr.J.Sci.* 83, 598-604.

Integrative and cellular aspects of autonomic functions : temperature and osmoregulation. Eds K. Pleschka, R. Gerstberger. John Libbey Eurotext, Paris © 1994, pp. 347-352.

Selective dilatation of arteriovenous anastomoses by cooling the brood patch of the hen

J.R.S. Hales[(1)], U. Midtgård[(2)] and A.A. Fawcett

C.S.I.R.O., Ian Clunies Ross Animal Research Laboratory, Sydney, Australia. [(1)] Present address : School of Physiology and Pharmacology, University of N.S.W., Sydney, Australia. [(2)] Present address : National Institute of Occupational Health, Copenhagen, Denmark

SUMMARY

To further elucidate thermal influences on the partition of skin blood flow (BF) between different microvascular compartments, cold-induced vasodilatation (CIVD) in thoracic skin of the domestic hen was investigated. Measures of total and capillary BF (difference = arteriovenous anastomotic, AVA) were obtained by the simultaneous injection of radioactive microspheres 50 and 15 μm diameter (respectively). In non-broody hens, there was normally no difference between total and capillary BF in thoracic skin, and local cooling of the skin from 39-40 to 34-35°C had no significant effects on BF. In broody hens in which the thoracic skin was well developed as a brood patch, again there was normally no difference between total and capillary BF; however, local cooling of the brood patches from 39-40 to 36-38°C resulted in total BF increasing approx. 6.5-fold with no change in capillary BF. It is concluded that very mild cooling elicits local increases in skin BF by selectively dilating AVAs without any change in capillary BF in hens with a developed brood patch containing functional AVAs. The mechanism may be comparable to CIVD in the dog's tongue and in tissues of other species including humans, or in view of the milder temperature and shorter time course and transient existence, may represent a special adaptation in birds.

INTRODUCTION

Lewis (1930) first described a "hunting reaction" in which human digits immersed in iced water would increase in temperature by up to around 8°C over about 2 min periods. Cold-induced vasodilation (CIVD) must have been responsible for the temperature increase, and controversy has continued over the role and specific microvascular site of the vasodilatation. That is, is the purpose of CIVD to avoid cold damage of tissues by warming, or is the temperature increase secondary to increased local blood flow necessary to provide or remove metabolites? Dilatation of AVAs would very adequately serve the former purpose, whereas the latter would require increased blood flow through capillaries. To specifically examine the route of blood flow during CIVD, Edwards (1967) combined ^{24}Na uptake and plethysmographic techniques in the anaesthetised rabbit's foot, which indicated restriction of the increased blood flow to AVAs. Krönert et al. (1980) used microsphere and electromagnetic flow measurements to show that dilatation in response to local cooling of the anaesthetised dog's tongue was restricted to AVAs, although this was possibly not 'classical' CIVD. One of us has observed that: (a) The brood patch of the bird is abundant in AVAs (Midtgård, 1985); (b) the AVA nerve supply contains vascoactive intestinal polypeptide (VIP)-immunoreactive fibres which are usually regarded as vasodilatory (Midtgård, 1988); (c) local blood flow in the brood patch, measured by ^{133}Xe washout, increases markedly in response to local cooling (Midtgård et al., 1985). The present study has employed the one technique, radioactive microspheres, to measure both capillary and AVA BF in the brood patch of the conscious hen.

METHODS

Experiments were preformed on two broody and two non-broody domestic hens (*Gallus g. domesticus*) of body weight 1530 ± 75g; the broody hens had been incubating eggs for at least 10 days and had well developed brood patches. Vascular catheterisations were performed about 3 h before experimental observations, with the bird under general anaesthesia induced and maintained using halothane in oxygen. For injection of the microsphere dose, a polyethylene catheter (0.5 mm i.d., 0.8 mm o.d.) was advanced via the right carotid artery into the left ventricle under fluoroscopic guidance; Urografin (Schering, Berlin FRG) was used for visualisation. For withdrawal of a reference blood sample (approx. 7 ml min^{-1}) a vinyl catheter (0.58 mm i.d., 0.96 mm o.d.) in either the right brachial or sciatic artery was used.

Quantitative measurements of skin blood flow rate and cardiac output were obtained using radioactive microspheres as previously described (Hales, 1974, 1981). The dose contained a precisely measured number of microspheres approximating 1×10^6 of nominally 15 μm diameter labelled with ^{46}Sc or ^{113}Sn (NEN-TRAC, DuPont, Sydney) mixed with approximately 2×10^5 of 50 μm diameter labelled with ^{85}Sr or ^{153}Gd (TRACER, 3M Co., St. Paul, Minnesota).

About 2 h after the hen regained consciousness, it was restrained in dorsal recumbency in a room with a dry bulb temperature of 23 - 24°C. Body temperatures were monitored using thermocouples inserted 5 cm into the cloaca (T_c) and glued onto the centres of each of the bilateral thoracic skin areas (T_{sk}) which form the blood patch. One side was cooled by directing a stream of compressed air onto a wet cotton swab applied to the skin; the contralateral area served as a control. This cooling technique was chosen (in preference to, say, a thermode chamber) to avoid pressure effects on the highly sensitive microcirculation. When T_{sk} of the cooled area was at a lowered , stable level (usually within 2 min), the microsphere dose was injected while a reference blood sample was withdrawn. The hen was then killed by an overdose of barbiturate and skin samples and the lungs were removed for gamma assay.

RESULTS AND DISCUSSION

Observations on rabbits (Hales & Cliff, 1977) showed that microspheres of 15μm diameter lodge in the end-arteriolar or pre-capillary sphincter regions of microvasculature; blood flow values obtained using this sized microsphere are therefore usually taken as nutrient or capillary blood flow which will be less than total blood flow if patent AVAs are present. Work on various mammals has shown that significant quantities of even 50 μm microspheres can pass through some AVAs and that if lodgement of microspheres is to be used to measure the total blood flow of certain tissues (e.g. skin) when AVAs are patent, then an intolerably large size needs to be employed (reviews: Hales, 1974, 1981). This appears not to be the case in the domestic fowl, wherein very few 50μm microspheres bypass microvasculature and therefore a measure of total organ blood flow may be obtained (Wolfensen, 1983). Thus, by simultaneously injecting doses of 50 and 15 μm diameter microspheres, measures of total and capillary blood flows (respectively) are obtained, and by difference, flow through AVAs is estimated. Nevertheless, only one injection per bird was made because use of numbers of 50μm microspheres sufficient to reliably estimate total blood flow commonly elicited marked hyper or hypotension, or nystagmus or unconsciousness. This effect is unlikely to have influenced the BF values obtained because, as indicated by renal BF measurements in the baboon (Hoheimer et al., 1983) the microspheres have lodged by the time their untoward effects occur.

Individual data for the four hens are presented in Table 1 and Fig. 1. A markedly higher total than capillary blood flow rate was seen only in the cooled, well-developed brood patch. In the brood patch at normal temperature (control side of hens a and b) or in either cooled or normal temperature thoracic skin of non-broody hens, total and capillary blood flows approximated each other and also approximated capillary blood flow in the cooled brood patch; they were only approximately 13% of the potential total flow as exhibited by the cooled brood patch. Further evidence that the local cooling enhanced perfusion of AVAs is gained from the markedly increased appearance of microspheres (particularly 15μm) in the lungs of broody hens (Table 1). These differences in blood flow rate in thoracic skin of broody compared with non-broody hens when cooled, would also account for the differences in skin temperature (Table 1) despite the same cooling technique being applied to all skin areas. Therefore, a clear role in CIVD is demonstrated for the AVAs which have been observed histologically to be conspicuous in the brood patch but very difficult to identify in skin of non-broody birds (Midtgård, 1985, 1988). Although AVAs might not necessarily be the principal site of heat exchange (Weinbaum et al., 1984), increased blood flow through AVAs is extremely effective in promoting body heat loss through skin (Rübsamen & Hales, 1984). CIVD in the brood patch was associated with skin temperatures in the 30's, a level comparable to similar reactions in the dog tongue (Krönert et al., 1980) and sheep leg (Hales et al., 1985). It remains to be determined whether the same mechanism prevails in the classical CIVD described by Lewis (1930), which occurs at temperatures approaching zero. Blood flow increases immediately in response to mild cooling with the former but increases after a delay of very many seconds in response to severe cooling with the latter. Nevertheless, we conclude that, rather than to meet metabolic demands, the likely purpose of CIVD is to promote local heat delivery (a) specifically in the brood patch, to warm eggs, and (b) in general, to promote body heat loss at 'moderate' temperatures and to prevent cold damage at low temperatures.

This agrees with Haftorn & Reinertsen's (1982) "aim" of increased brood patch BF and with Hales' (1985) heat transport role of AVAs. However, in view of the different temperature and time-course involved with this phenomenon in birds compared with Lewis' (1930) hunting reaction, care must be taken in extrapolating current findings to the latter. A major question remaining to be addressed is the mechanism underlying the transient adaptation of skin microvasculature, i.e., the presence of CIVD in hens only when they are broody.

Table 1. Effects of unilateral cooling of thoracic skin of two broody and two non-broody hens.
T_c (°C): core temperature. T_{sk} (°C): skin surface temperature.
% Lungs: percentage of microsphere dose in lungs.
CO (ml min^{-1}): cardiac output. HR (Beats min^{-1}): heart rate.
MAP (mmHg): mean arterial pressure.

Hen	T_c	T_{sk}		% Lungs		CO	HR	MAP
		cooled	control	50	15			
BROODY								
a	42.1	36.1	39.3	0.24	0.88	304	396	145
b	41.3	38.4	40.1	0.23	0.59	356	426	127
NON-BROODY								
c	41.0	34.8	40.1	0.05	0.10	307	402	117
d	41.0	33.8	39.3	0.07	0.13	379	336	118

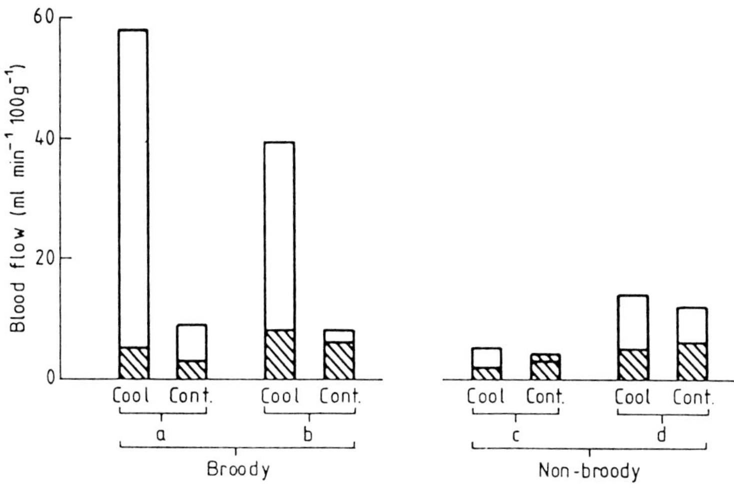

Fig. 1 Total and capillary ▧ blood flow rate in skin of the cooled and non-cooled ("cont" = control) sides of the thorax of two broody (a and b) and two non-broody (c and d) hens. White portion of column is AVA perfusion.

REFERENCES

Edwards, M.A. (1967): The role of arteriovenous anastomoses in cold-induced vasodilation, rewarming and reactive hyperemia as determined by ^{24}Na clearance. *Can.J. Physiol. Pharmacol.,* 45: 39-48.

Haftorn,S. and Reinertsen, R.E. (1982): Regulation of body temperature and heat transfer to eggs during incubation. *Ornis Scand.* 13:1-10.

Hales, J.R.S. (1974): Radioactive microsphere techniques for studies of the circulation. *Clin. Exp. Pharm. Physiol. Suppl.* 1: 31 - 46.

Hales J.R.S. (1981): Use of microspheres to partition the microcirculation between capillaries and arteriovenous anastomoses. In *Progress in Microcirculation Research,* ed. D. Garlick, pp. 394 - 412. Sydney: Unv. N.S.W. Committee in Postgrad. Med. Edn.

Hales, J.R.S. (1985): Skin arteriovenous anastomoses: Control & role in thermoregulation. In *Cardiovascular Shunts*, eds. K. Johansen and W.W. Burggren, pp. 433 - 448. Copenhagen: Munksgaard.

Hales, J.R.S. and Cliff, W. (1977): Direct observations on the behaviour of microspheres in microvasculature. *Biblio. Anat.* 15: 87-91.

Hales, J.R.S., Jessen, C., Fawcett, A.A. and King, R.B. (1985): Skin AVA and capillary dilatation and constriction induced by local skin heating. *Pflügers Arch.* 404: 203 - 207.

Hohimer, A.R., Hales J.R.S., Rowell, L.B. and Smith, O.A. (1983): Regional distribution of blood flow during mild dynamic leg exercise in the baboon. *J. Appl. Physiol.* 55: 1173 - 1177.

Krönert, H., Wurster, R.D., Pierau, Fr. -K, and Pleschka, K. (1980): Vasodilatory response of arteriovenous anastomoses to local cold stimuli in dog's tongue. *Pflügers Arch.* 388: 17 - 19.

Lewis, T. (1930): Observations upon the reactions of the vessels of the human skin to cold. *Heart* 15: 177 - 208.

Midtgård, U. (1985): Arteriovenous anastomoses in the incubation patch of Herring gulls. *Condor* 87: 549 - 551.

Midtgård, U. (1988): Innervation of arteriovenous anastomoses in the brood patch of the domestic fowl. *Cell Tissue Res.* 252: 207 - 210.

Midtgård, U., Sejrsen, P. and Johansen, K. (1985): Blood flow in the brood patch of Bantam hens.: evidence of cold vasodilatation. *J. Comp. Physiol. B.* 155: 703 - 709.

Rübsamen, K. and Hales, J.R.S. (1984): The role of arteriovenous anastomoses and capillaries in determining heat transfer across extremity skin of sheep. In *Thermal Physiology,* ed. J.R.S. Hales, pp. 259 - 262. New York: Raven Press.

Weinbaum, S., Jiji, L.M. and Lemons, D.E. (1984): Theory and experiment for the effect of vascular microstructure on surface tissue heat transfer - Part I: Anatomical foundation and model conceptualisation.*ASME J. Biomech. Engng.* 106: 321 - 330.

Wolfenson, D. (1983): Blood flow through arteriovenous anastomoses and its thermal function in the laying hen. *J. Physiol.* 334: 395 - 407.

Integrative and cellular aspects of autonomic functions : temperature and osmoregulation. Eds K. Pleschka, R. Gerstberger. John Libbey Eurotext, Paris © 1994, pp. 353-360.

Functional evidence for an intrinsic mechanism underlying cold induced vasodilation of arteriovenous anastomoses in canine facial and nasal tissues

Klaus Pleschka and Heinz Fontanji

Max-Planck-Institut für Physiologische and Klinische Forschung, W.G. Kerckhoff-Institut, D-61231 Bad Nauheim, Germany

SUMMARY

This study was designed to elucidate the mechanisms underlying cold-induced vasodilatation in arteriovenous anastomoses (AVA) of the face and nose of the dog. Perfusion pressures and total blood flows of the internal maxillary artery (IMA-FLOW) were recorded bilaterally at moderately warm (+28°C) and severely cold (-32°C) inspired air temperatures in 10 normothermic anesthetized dogs. Control recordings were obtained in the untreated state (C) and after the sequential systemic pretreatment with atropine (A), phenoxybenzamine (Ph), oxprenolol (Ox) and hexamethonium (H). Partitioning of IMA-FLOWs to the capillaries (CAP-FLOW) and arteriovenous anastomoses (AVA-FLOW) was determined by the distribution of tracer microspheres. Cold air exposure led to the following increases in IMA-FLOW: 52.7 per cent in the control group (C), 52.7 per cent after cholinergic blockade (A), 56.8 per cent after α-adrenergic blockade (Ph), 11.0 per cent after ß-adrenergic blockade (Ox), and 20.3 per cent after ganglionic transmission blockade (H). All increases in flow were due to significant decreases in maxillary resistance since perfusion pressures remained nearly constant. For the partitioned flows, AVA-FLOW increased by 82.6 per cent (A), 200.5 per cent (Ph) and 180.9 per cent (Ox), while CAP-FLOW decreased by 6.4 per cent (A), 8.3 per cent (Ph) and 40.8 per cent (Ox) during breathing of cold air. It is concluded that blood flow control through AVA is not only achieved by neurogenic vasomotor efferents, including the co-action of neuropeptides containing fibres, but also by direct temperature effects on mechanisms underlying AVA intrinsic tone.

INTRODUCTION

In the non-shivering dog, blood flow derived from the internal maxillary artery (IMA-FLOW) through facial and nasal arteriovenous anastomoses (AVA-FLOW) has been shown to be inversely related to the temperature of the inspired air, whereas capillary flow (CAP-FLOW) exhibits a direct thermal relationship (Pleschka *et al.*, 1987). This means that with decreasing inspired air temperature, AVA-FLOW increases and CAP-FLOW decreases, provided that no neurogenic shivering is present. In the presence of shivering, interestingly, the AVA-FLOW response reverts to one of cold-induced vasoconstriction, indicative of the complex neuronal control of AVA-FLOW under different situations. For the non-shivering state, the principle cause of the reflex-induced cold vasodilation is usually attributed to a graded decrease of vasoconstrictor activity of the AVA. Alternately, however, increases in nasal

AVA flows during cold exposure may be due to enhanced activities in central cholinergic and non-cholinergic pathways leading to parasympathetic vasodilatation. Considering the growing evidence for the co-action of local vasodilator effector systems in vasomotor control, the question arises as to which predominates in regulating the AVA vasomotor tone during the cold air inhalations, alterations in central autonomic neuronal control or local effects of the cold? To address this issue, perfusion pressure (PP), total blood flow of the IMA, and distribution of flows between the vascular compartments of AVA and CAP were measured at moderately warm (+28°C) and severely cold (-32°C) inspired air temperatures in normothermic, anesthetized and untreated spontaneously breathing dogs (C). Measurements were taken before and after blockade of cholinergic, adrenergic and ganglionic transmission. Plasma catecholamines levels were quantitatively determined to assess any potential interference with regional neurogenic blood flow control under the conditions of these experiments.

MATERIAL AND METHODS

Animals

The experiments were performed on 10 adult mongrel dogs with mean body weights of 27.5 kg ± 2.7 sd. Anesthesia was induced with ketamine hydrochloride (Ketavet, 12 mg/kg i.m.) and acepromacin-maleate (Vetranquil, 1.6 mg/kg i.m.) and maintained with pentobarbital sodium (Nembutal, 30 mg/kg i.v. per hour). Catheters were advanced through the femoral vessels and into the abdominal aorta or inferior vena cava (heart level) for systemic arterial blood pressure monitoring or intravenous injections and blood withdrawal.

Instrumentation of the Internal Maxillary Vascular Bed

The internal maxillary artery was exposed bilaterally immediately cephalic to the site of its terminal branching. Non-occlusive electromagnetic flow probes were then positioned for the continuous recording of maxillary flow in accordance with the procedure of Hashimoto *et al.* (1987). Catheters were advanced retrogradely into the left and right maxillary arteries via their pterygoid branches for the injection of microspheres and measurement of local perfusion pressures.

Recordings and Measurements

Arterial blood pressure, heart rate, total blood flow and perfusion pressure of both maxillary arteries were recorded as described by Hashimoto *et al.* (1987). All data streams were digitalized and stored by means of a data acquisition system. Catecholamine concentrations in plasma samples were determined by means of a high-performance liquid chromatography system with an electrochemical detector. Regional blood flow compartmentalizations were determined under up to six physiologically defined conditions by injecting radioactive microspheres of unique isotope labels unilaterally. At the end of the experiment the test animals were euthanized with an anesthetic overdose and tissue specimens of anatomically defined compartments were recovered for analysis by gamma-spectrometry (Hashimoto *et al.*, 1987).

Experimental Procedure

All physiological variables including bilateral IMA FLOWs were measured during dry-air inspiration at temperatures of +28°C and -32°C in untreated dogs (control group). The breathing air was thermally conditioned by routing compressed air through a heat exchanger cooled by a mixture of mashed dry ice and propylalcohol and then over a heating coil for final temperature adjustment (Pleschka *et al.*, 1987). The same warm-air/cold-air procedure was followed after the sequential systemic pretreatment with 0.8 mg/kg atropine, 1.4 mg/kg phenoxybenzamine, 1.4 mg/kg oxprenolol and 10 mg/kg hexamethonium (experimental groups). To counteract the hypotensive actions of hexamethonium, the ganglionic blocker was delivered in 400 ml of a 10 per cent dextran solution infused within 45 minutes. Blood flow partitioning by injection of different labelled microspheres was determined before and during cold-air breathing only after muscarinic and α- and β-adrenergic receptor blockade due to the limited number of differently labelled microspheres.

Statistics

Mean values and standard deviations were calculated for steady-state periods. Statistical differences between the various groups were tested with Student's *t*-test for paired data and a Bonferroni-Holm correction for multiple comparisons. Probability levels less than the 0.05 level were accepted as statistically significant.

RESULTS

Internal maxillary flows, resistances, and pressures during warm- and cold-air breathing are presented in Fig. 1. The cold-induced, bilateral increases in IMA-FLOWs (C) were not diminished by cholinergic (A), adrenergic (Ph) or ganglionic (H) blockades. Following pretreatment with oxprenolol (Ox) the elevation of IMA-FLOW during warm-air breathing was so marked that the cold-air challenge only resulted in a small, but significant increase in flow. All increases in IMA-FLOW are attributed to parallel falls in IMA resistances since perfusion pressures for any given group remained unchanged. This indicates that during cold-air breathing there is significant release of vasomotor tone in the IMA vascular beds on both sides, resulting in increased perfusions of the regions.

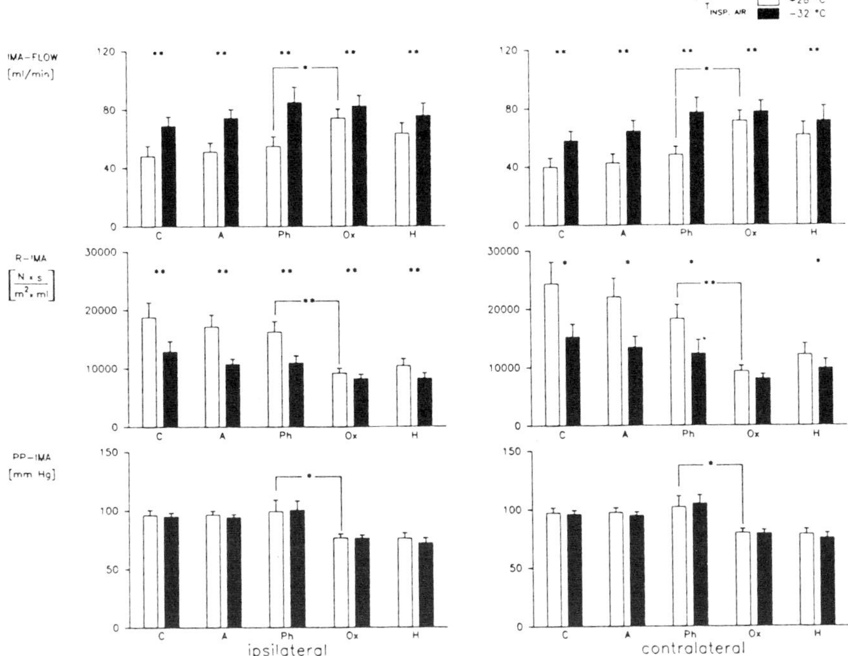

Fig. 1: Effects of inhaled-air temperature and blocker agents on the hemodynamics of the internal maxillary arteries. Blood flows (IMA-FLOW), resistances to flow (R-IMA) and perfusion pressures (PP-IMA) are displayed during warm-air (open bars) or cold-air (black bars) breathing. Five groups are distinguished on sides ipsilateral and contralateral to the route of drug administrations (left IMA): untreated control dogs (C); atropine pretreatment (A); phenoxybenzamine (Ph); oxprenolol (Ox); hexamethonium (H). Values are expressed as means ± sd for 10 dogs. Significant differences within and between groups are designated by asterisks ($P < 0.05$).

The compartmentalizations of internal maxillary arterial flows under different test conditions are shown in Fig. 2. Cold-air breathing again caused an increase in IMA-FLOW after consecutive pretreatment with atropine, phenoxybenzamine and oxprenolol, but in each case this excess flow was carried by the arteriovenous anastomoses (AVA-FLOW), not the capillary beds (CAP-FLOW). The results show that blockade of α-receptors with phenoxybenzamine had no effect on compartmental blood flow rates in the IMA vascular bed. Interestingly, ß-blockade with oxprenolol produced a significant rise in the resting CAP-FLOW rate.

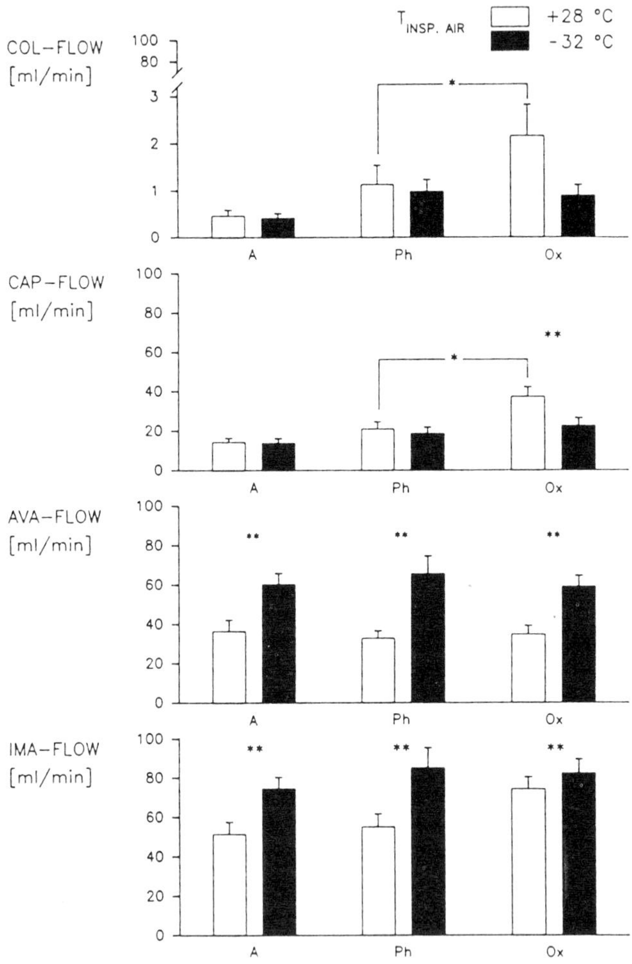

Fig. 2: Effects of inhaled-air temperature and blocker agents on flow distributions of the internal maxillary arteries. Total maxillary flows (IMA-flow) are partitioned into capillary flows (CAP-FLOW), arteriovenous anastomoses flows (AVA-FLOW) and collateral flows (COL-FLOW). Three blocking agents were administered systemically ipsilateral to the flow measurements including atropine (A), phenoxybenzamine (Ph) and oxprenolol (Ox). Values are expressed as means ± sd for 10 dogs. Significant differences between groups are designated by asterisks ($P < 0.05$).

The fact that the enhancement of IMA-FLOW during cold-air breathing resulted exclusively from vasodilatation of the arteriovenous anastomoses is illustrated in Fig. 3. This relationship held irrespective of pretreatments with atropine, phenoxybenzamine and oxprenolol.

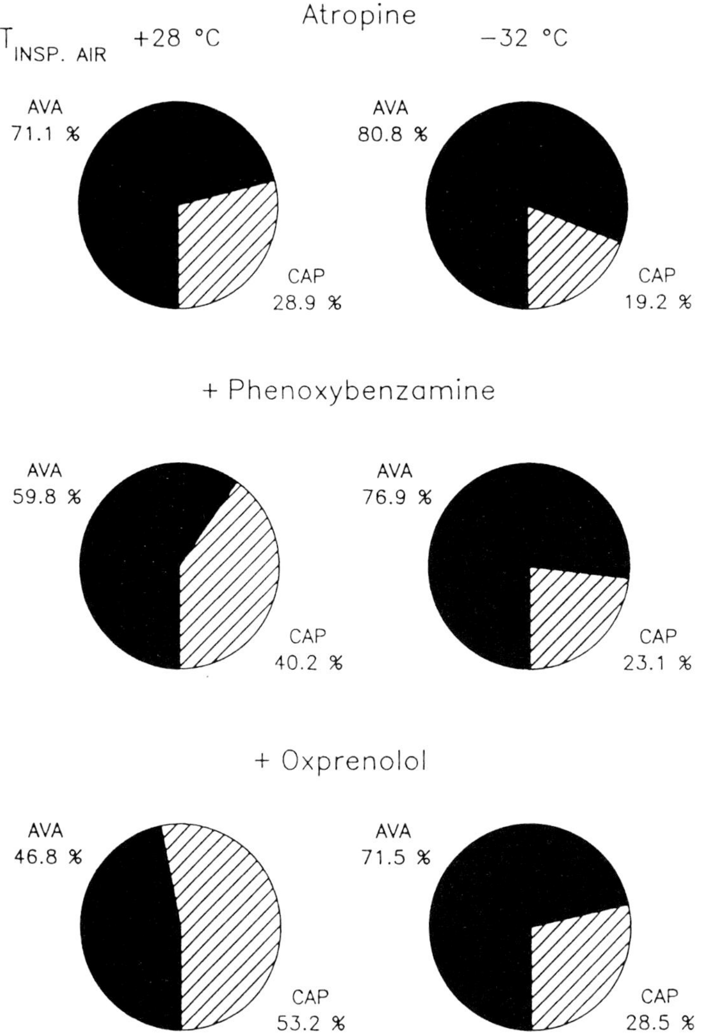

Fig. 3: Effects of inhaled-air temperature and blocker agents on fractional flow distributions through the arteriovenous anastomoses (AVA) and capillary beds (CAP) of the internal maxillary artery. Increases in total AVA-FLOW during cold-air breathing (Fig. 2) is carried principally by the arteriovenous anastomoses, not the capillaries. Successive pretreatments with atropine, phenoxybenzamine and oxyprenolol induce no substantial changes in the responses. Pie charts and numerical percents computed from means of 10 dogs. Drugs injected ipsilateral to the sites of flow measurement.

Of all the general circulatory variables studied, only heart rate was influenced by some of the applied receptor blockers. Atropine caused an increase in heart rate and phenoxybenzamine caused a further increase. After oxprenolol, heart rate tended to fall, but ot significantly. Cold-air breathing had no effect on altering heart rate and systemic arterial pressure was unaffected by the cold challenge. Plasma catecholamine levels in the various groups are depicted in Fig. 4. In no case did cold-air breathing significantly alter the prevailing concentrations of norepinephrine, epinephrine or dopamine, but the reference levels of all three catecholamines did change in response to successive pretreatment with the blocker drugs phenoxybenzamine and oxprenolol in the presence of atropine. As expected, plasma catecholamine concentrations were finally returned to control levels after ganglionic blockade with hexamethonium.

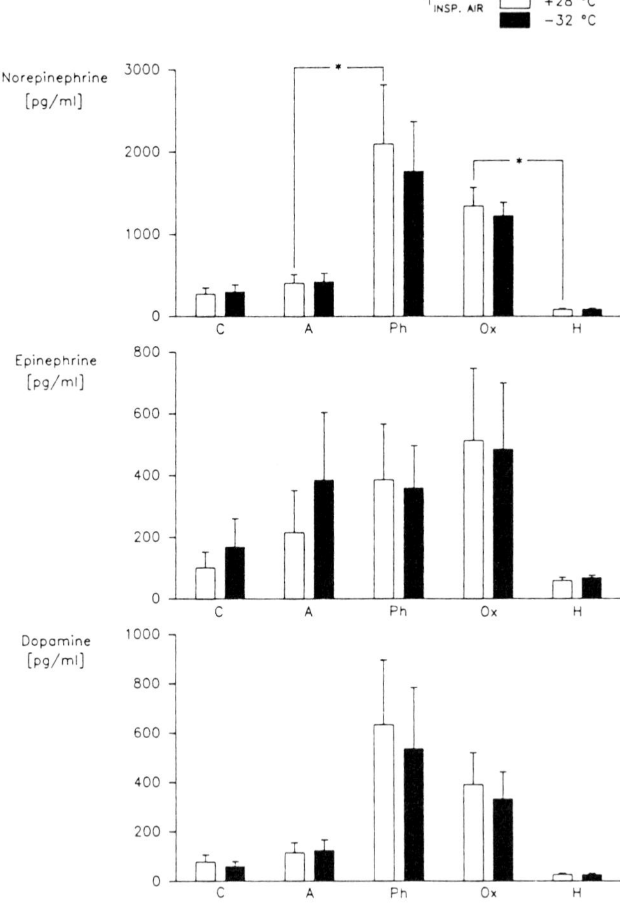

Fig. 4: Effects of inhaled-air temperature and blocker agents on plasma catecholamine concentrations. Norepinephrine, epinephrine and dopamine concentrations are unchanged by cold-air breathing. Compared with untreated control dogs (C), atropine (A) had no effect on the catecholamines, but all three were significantly elevated following pretreatment with phenoxybenzamine (Ph) and oxprenolol (Ox). The levels decreased once again under the influence of hexamethonium (H). Values are expressed as means ± sd for 10 dogs. Significant differences between groups are designated by asterisks ($P < 0.05$).

DISCUSSION

The present data confirm our earlier findings that cold-air breathing increases internal maxillary blood flow in the normothermic, anesthetized dog (Pleschka *et al.*, 1987). It is also confirmed that this IMA-FLOW increase is due exclusively to significant increases in AVA-FLOW with no change or even decreases in CAP-FLOW. Insofar that the local tissue temperature was kept relatively warm, attributing the reflex vasodilatation to cold-induced inhibition of the sympathetic vasoconstrictor system can a priori be excluded. Indeed, the general cold-defense reaction results in vasoconstriction of the AVA (Pleschka *et al.*, 1987), not AVA dilatation as observed in this study.

It was first assumed that cold-induced vasodilatation of the nasal AVA was due to the activation of parasympathetic vasodilator fibers, especially those innervating the AVA (Sugahara & Pleschka, 1992). However, the present study shows that cold-induced AVA vasodilatation occurs in the presence of atropine and, therefore, cannot be mediated by a cholinergic muscarinic mechanism. The present study further excludes any participatory role of adrenergic mechanisms. Neither the loss of the vasoconstrictor input by adrenergic α-receptor blockade nor the loss of the vasodilatory input by ß-receptor blockade significantly affected the vasodilatory response of the AVA during the inspiration of cold air. The only change observed was with resting CAP-FLOW which was significantly increased by the pretreatment with oxprenolol. This response was probably due to an agonistic drug action (Goodman & Gilman, 1985), resulting in a reduced vasodilatory response to the subsequent inspiration of cold air. Since the cold-induced response pattern of IMA-FLOW remained unaffected during the blockade of ganglionic transmission with hexamethonium, the involvement of a nicotinic mechanism in the cold-induced increase in AVA-FLOW appears equally unlikely. The independence of cold-induced AVA vasodilatation from any of the classical autonomic transmitters would suggest that other mediators might be involved. As summarized by Table 1, cold-air breathing is the strongest stimulus for increasing AVA flow of the internal maxillary artery. Other mediators promote only smaller increases in AVA flow while favoring flow distribution patterns to the capillaries.

Table 1: Directional changes in internal maxillary artery blood flow and its partitioned compartments in response to various stimuli. All stimuli enhance IMA flow (++), but only cold-air breathing results in a marked increase in arteriovenous anastomoses (AVA) flow (+++) while decreasing capillary (CAP) flow (-). By contrast, the other stimuli tend to have no (0) or only slight (+) to moderate (++) vasodilatory responses in the AVA with most of the flow being carried by the capillary beds (+++). The agents listed were directly injected into the internal maxillary artery ipsilateral to the site of flow measurements.

stimulus	IMA	AVA	CAP	references
Cold inspired air	(++)	(+++)	(-)	Pleschka *et al.*, 1987
Electrical stimulation of the Vidian nerve	(++)	(++)	(++)	Sugahara & Pleschka, 1992
Vasoactive intestinal polypeptide (VIP)	(++)	(+)	(+++)	Dorrong *et al.*, 1988
Substance P	(++)	(+)	(+++)	Pleschka & Ikeda, 1988
Histamine	(++)	(0)	(+++)	Bari *et al.*, 1993
AMP, ADP, ATP	(++)	(+)	(+++)	Bari *et al.*, 1993

Although the involvement of other neurotransmitters with potent vasodilatory properties cannot be excluded as participatory in the cold-induced vasodilatations of the nasal and facial AVAs (e.g. calcitonin gene related peptide), a direct temperature effect on local mechanisms regulating the intrinsic tone of AVA seems to be a more likely explanation. At present, nitric oxide produced in endothelial cells seems to play a prominent role in the local mediation of smooth muscle relaxation (Iadecola, 1993). Likewise, membrane hyperpolarization due to Ca^{++} activated potassium channels appears to be a viable and effective mechanism for the dilation of arterial vessels (Nelson, 1993).

Acknowledgements: The authors are grateful to Dr. Charles Webber Jr. (Loyola University, Chicago) for reviewing the manuscript and to Mrs. I. Lürkens, Mrs. C. Kammer and Mrs. A. Frank for their excellent technical assistance.

REFERENCES

Bari, F., Ariwodola, J.O. & Pleschka, K. (1993): Histamine responsiveness of the various vascular beds of facial and nasal tissues in the dog. *J. Vasc. Res.* 30, 30-37.

Bari, F., Ariwodola, J.O. & Pleschka, K. (1993): Circulatory effects caused by intraarterial infusion of AMP, ADP and ATP. *J. Vasc. Res.* 30, 125-131.

Dorrong, Y., Fontanji, H. & Pleschka, K. (1988): Vascular tone and reactivity to vasoactive intestinal peptide (VIP) in the internal maxillary vascular beds of the dog. *Pflügers Arch.* 412 [Suppl 1]: 62, R37.

Goodman, L.S. & Gilman, A. (1985): The pharmacological basis of therapeutics, 7th edition. London: Macmillan.

Hashimoto, M., Sommerlad, U. & Pleschka, K. (1987): Sympathetic control of blood flow to AVAs and capillaries in nasal and facial tissues supplied by the internal maxillary artery in dogs. *Pflügers Arch.* 410:589-595.

Iadecola, C. (1993): Regulation of the cerebral microcirculation during neural activity: is nitric oxide the missing link? *TIMS* 16, 206-214.

Nelson, M.T. (1993): Ca^{++}-activated potassium channels and ATP-sensitive potassium channels as modulators of vascular tone. *TCM* 3, 54-60.

Pleschka, K., Sugahara, M., Hashimoto, M., Sommerlad, U., Lürkens, I., & Ernst, C. (1987): Local circulatory control in thermal stress. In *Comparative physiology of environmental adaptations*, ed. Dejours P., Vol. 2, PP 107-122. Basel: Karger.

Sugahara, M. and Pleschka, K. (1992): Nutrient and shunt flow responses to vidian nerve stimulation in nasal and facial tissues of the dog. *Eur. Arch. Otorhinolaryngol.* 249, 79-84.

Integrative and cellular aspects of autonomic functions : temperature and osmoregulation. Eds K. Pleschka, R. Gerstberger. John Libbey Eurotext, Paris © 1994, pp. 361-369.

Effects of benzodiazepines on melatonin secretion in the photosensory pineal organ of the teleost, *Oncorhynchus mykiss*

H. Meissl, P. Ekström*, J. Yáñez and E. Grossmann

*Max-Planck-Institute for Physiology and Clinical Research, W.G. Kerckhoff-Institute, Parkstrasse 1, D-61231 Bad Nauheim, Germany and *Department of Zoology, University of Lund, S-22362 Lund, Sweden*

INTRODUCTION

The pineal organ as a component of photoneuroendocrine systems (cf. Scharrer, 1964) subserve the control of autonomic functions in response to photoperiodic changes in the environment by involving neuronal and neuroendocrine mechanisms. The organ has been shown to be implicated in a number of physiological processes including circadian rhythmicity and seasonal reproduction (Reiter, 1986). In lower vertebrates, the direct photoreception and conversion of photic information into cyclic neuroendocrine and neural signals is an essential feature of the pineal photoreceptor cell (cf. Collin et al., 1986). Direct photoreception and melatonin formation seem to be coupled processes within the photoreceptor cell (Gern and Greenhouse, 1988). Teleosts, for example, display a strong nocturnal cycle in melatonin synthesis and secretion when maintained in a light:dark illumination regimen. In vitro melatonin secretion in superfusion studies yield similar secretion profiles with high melatonin levels in darkness, whereas illumination of the isolated pineal organ results in a suppression of melatonin release that is directly related to the intensity and wavelength of the incident light and, at least in the trout, not under the influence of an endogenous oscillator (Gern and Greenhouse, 1988; Max and Menaker, 1992).

The cyclic, light-dependent melatonin production that conveys information on the light-dark cycle is accompanied by a neuronal sensory mechanism whose properties were extensively investigated by Dodt and coworkers (cf. Dodt and Heerd, 1962; Dodt, 1973; Meissl and Dodt, 1981), but whose functional significance is still enigmatic. The dualism of neuronal and neuroendocrine output of the pineal organ prompted us to investigate the possible interrelationship between neuronal activity and melatonin formation and secretion. Previously we have shown that melatonin may possess a paracrine action in the pineal gland of trouts and suppresses the spike discharges of projecting neurons (Meissl et al., 1990). It was assumed that this paracrine action may be related to an interaction of melatonin with the $GABA_A$/benzodiazepine receptor complex (Meissl et al., 1993). This was shown by electrophysiological recordings from single luminosity neurons of the explanted pineal organ of the trout during perifusion with medium containing diazepam and melatonin and by incubation of the tissue with the fluorescent benzodiazepine receptor ligand Bodipy Ro-1986 simultaneously with melatonin application. Comparable data were previously reported in mammals where melatonin appears to interfere with GABA/benzodiazepine receptor binding (Niles, 1989). The diurnal variations in melatonin binding in the brain appear to be also modified by benzodiazepines (Anis et al., 1992).

In the present study we have investigated the action of benzodiazepines on melatonin formation of pineal photoreceptor cells. These experiments should provide further evidence for a possible interdependence of neuronal and endocrine mechanisms in the photosensory pineal organ.

MATERIAL AND METHODS

Experiments were performed on pineal organs of adult rainbow trouts, *Oncorhynchus mykiss*, obtained from a commercial hatchery. All animals were kept in large aquaria with oxygenated fresh water (ca. 8°C) under a LD cycle of 12:12 for at least three weeks before the experiments, that were conducted throughout the year.

For the experiments, trouts were rapidly decapitated, the pineal organs excised and placed on small pieces of filter paper. The filter papers with the pineal organs were washed in cold Hank's buffer and then transferred to the perifusion apparatus which contained 6 experimental chambers of 100 μl volume. Each chamber held one pineal organ and was continously perifused with modified Hank's buffer (pH 7.4) at a flow rate of 0.5 ml/hr provided by a microprocessor controlled peristaltic pump (Gilson MP3). The temperature was kept constant at 16°C by a LAUDA cryostate throughout the experiment.

Illumination of the isolated pineal organs was provided by a 150 W Xenon arc. Wavelengths were selected by interference band filters (Type AL, Schott, Mainz), light energy was adjusted by neutral density filters. Pineal organs were light adapted, if not otherwise stated, with monochromatic light of 520 nm, which is the maximum photopic spectral sensitivity of the trout pineal photoreceptor cell (cf. Meissl and Ekström, 1988). The energy of the adapting light was 6.48×10^{11} photons/cm^2/s at 520nm.

Pineal organs were light adapted for exactly 2 hours after their removal from the brain before the first collection period started. Samples were collected in intervals of 60 min. Pineal organs were dark adapted overnight for 15 hours in the absence of any light under otherwise identical conditions. Experiments were then repeated on the second day in the dark-adapted preparation. Thirty hours after beginning of the perifusion culture, the experiments were terminated, although the pineal organs still synthesize melatonin at an almost unchanged rate at this time.

Melatonin and the methoxyindoles 5-methoxytryptamine (5-MT), 5-methoxyindoleacetic acid (5-MIAA) and 5-MTOL (5-methoxytryptophol) in the superfusion samples were separated by high performance liquid chromatography (HPLC) and measured by electrochemical detection. Aliquots of the superfusate were injected directly into the chromatograph which consisted of a Gynkotek M480 pump, a Marathon autosampler (Spark, Holland) with a Rheodyne 7010 injector and a 100 μl loop, a Spherisorb ODS II column (125 x 4.6 mm, 5 μm particle size) and a Gynkotek M 20 electrochemical detector. The potential of the glassy carbon working electrode was set at +0.90 V relative to the reference electrode. The solvent system consisted of 18 % acetonitril, 50 mM sodium acetate, 0.05 mM EDTA and 50 mM citric acid (pH 4.3). The indolic compounds were identified on the basis of their retention times and their concentrations by comparison with external standards. The peaks were analyzed by the software program Chromstar (Bruker). Drugs were added to the perifusion medium for one complete sampling period, i.e. for one hour, then the medium was completely exchanged before the organ was perifused with fresh medium. Diazepam and clonazepam were firstly dissolved in dimethyl sulfoxid (DMSO) and then diluted with Hank's buffer (final concentration of DMSO was 0.3% for the 100 μM diazepam solution).

RESULTS

The pineal organ of the trout synthesizes and secretes melatonin under in vitro perifusion culture with elevated levels in darkness and low levels in light. Light-adapted pineal organs exposed to stepwise decreasing irradiances of monochromatic light increase their melatonin release up to a maximum level at about 2.2 x 10^{10} photons/cm^2/s before the melatonin output slightly declines to a plateau when the light intensity was further diminished (Fig. 1).

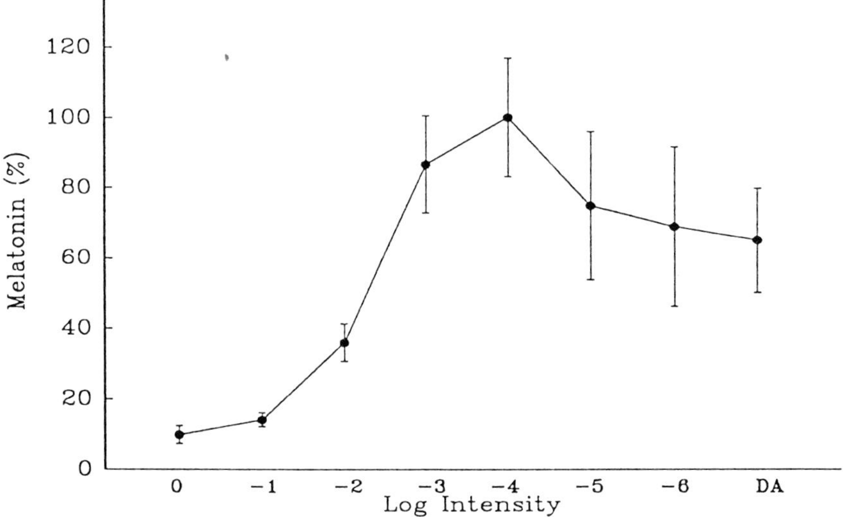

Fig. 1 Melatonin release of trout pineals in vitro in response to decreasing intensities of background illumination of 520 nm. The organs were exposed for one hour to each intensity. Each point plots the average production of melatonin secretion per hour by nine pineal organs. Melatonin production was normalized in respect to the highest melatonin output (-4log). Mean ± SEM, log 0 corresponds to 2 x 10^{14} photons/cm^2/s, DA: dark-adapted.

In the following experiments we used a constant monochromatic illumination (520 nm) of 6.48 x 10^{11} photons/cm^2/s for light adaptation, that is approximately 1.5 log units higher than the light threshold of the melatonin response. Under these conditions, the isolated pineal organ of the trout produces between 40% and 70% melatonin of a dark adapted

preparation. This irradiance was selected for light adaptation because under these conditions the pineal organ produces sufficient melatonin that allows its measurement by HPLC with electrochemical detection. Monochromatic light of 520 nm was used because this wavelength corresponds to the maximum spectral sensitivity of most photoreceptors of the trout pineal organ as determined with intracellular recordings (Meissl and Ekström, 1988).
The release of melatonin into the superfusion medium varied considerably between individual glands, but each gland showed a relatively constant rate of melatonin production. Therefore, to facilitate comparisons between different experimental groups the quantity of melatonin produced by individual glands was normalized in some experiments in respect to the control value. We used as control the average melatonin production in the two hours before the drug was added or the medium changed.

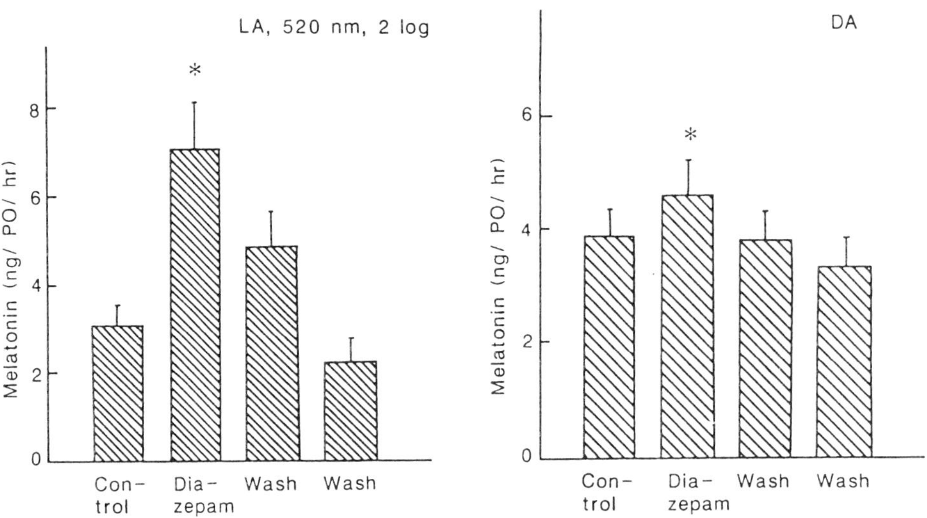

Fig.2 Effect of diazepam (100μM) on melatonin release of pineal organs (n = 15) of trouts *in vitro*. The organs were light adapted (LA; left plot) for two hours (520 nm, 11.81 log photons/cm^2/s) before the first collection period started. Plot on the right side shows melatonin release of dark-adapted (DA) pineal organs before (control) and during application of 100 μM diazepam. Samples of the perifusate were collected in 60 min timed fractions. Diazepam was added for 1 hour, then the organs were washed with Hank's medium. Mean ± SEM, **(LA)** $^*p < 0.001$ vs control, **(DA)** $^*p < 0.05$ vs control.

Figure 2 (left plot) shows the average melatonin release of 15 pineal organs of the trout in vitro under mesopic light conditions. The pineal organs released melatonin in relatively high amounts under these conditions. When diazepam (100 μM) was added to the perifusion medium for one hour, melatonin production and release from the isolated pineal organs increased significantly to values usually observed only under scotopic light conditions. After a subsequent exchange of the perifusion medium with normal Hank's medium, melatonin release declined to the control value. In dark-adapted pineal organs, diazepam caused a comparable increase of melatonin release, but the effect was considerably lower than in light-adapted preparations (Fig. 2, right). The effect was reversible after exchange of the medium. The differences in the average melatonin production in the light- and dark adapted preparations shown in figure 1 are relatively small. This is probably due to the relatively weak illumination used for light adaptation and the long time of dark-adaptation. We have frequently observed that during dark-adaptation melatonin production rises to a peak value in the initial phase of dark-adaptation and then declines to a plateau value.

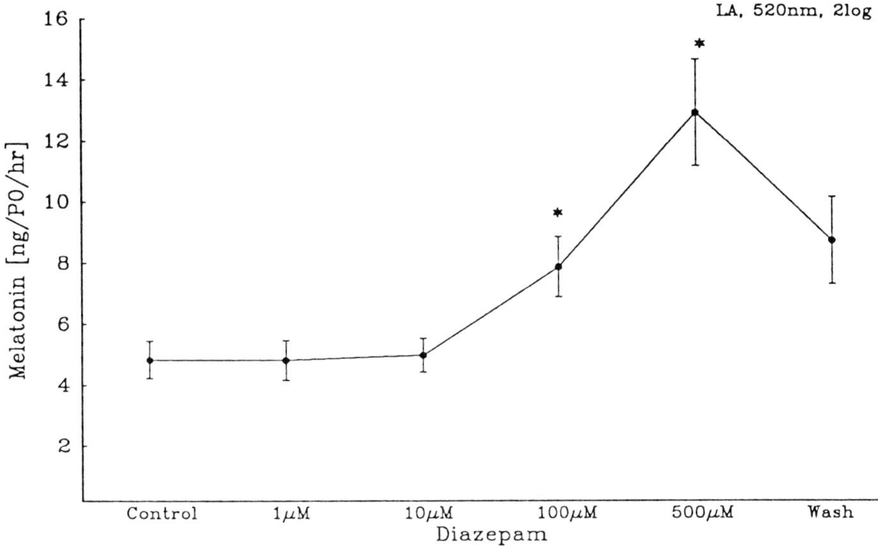

Fig. 3 Effects of increasing concentrations of diazepam on melatonin release of isolated, perifused pineal organs in the light-adapted state (LA). Light conditions: 520 nm, 11.81 log photons/cm^2/s. Mean values ± SEM, $^*p < 0.001$, n=4.

The threshold concentration for diazepam to induce an effect on melatonin production was between 10^{-5} and 10^{-4} M. Figure 3 depicts an experiment where the concentration of diazepam was continuously raised from 1 μM to 500 μM. The pineal organs were exposed for the whole collection time of 1 hour to each concentration. The increase in melatonin release was nearly three-fold for the highest diazepam concentration (500 μM) compared to the control value. The threshold concentration for clonazepam, a short-acting benzodiazepine, seemed to be higher, in the range between 100 and 500 μM.

The possible site of action of benzodiazepines was tested by determining the influence of magnesium, in addition to low calcium levels, on the release of melatonin from perifused glands. We assumed that, as at other chemically-mediated synapses, high levels of extracellular magnesium block neurotransmitter release from the presynaptic terminal (cf. Dowling and Ripps, 1973). Exchange of Hank's medium against a perifusion medium containing low Ca^{2+}- (0.13 mM) and enhanced Mg^{2+}- (6.08 mM) concentrations reduced melatonin production and secretion considerably to 25% of the control value in the mesopic range (Fig. 4A). Subsequent perifusion with normal Hank's medium increased melatonin output to the previous control value. Addition of diazepam to the modified, low Ca^{2+}-, high Mg^{2+}-Hank's medium significantly increased melatonin release (Fig. 4B) irrespectively of the light-adaptation conditions.

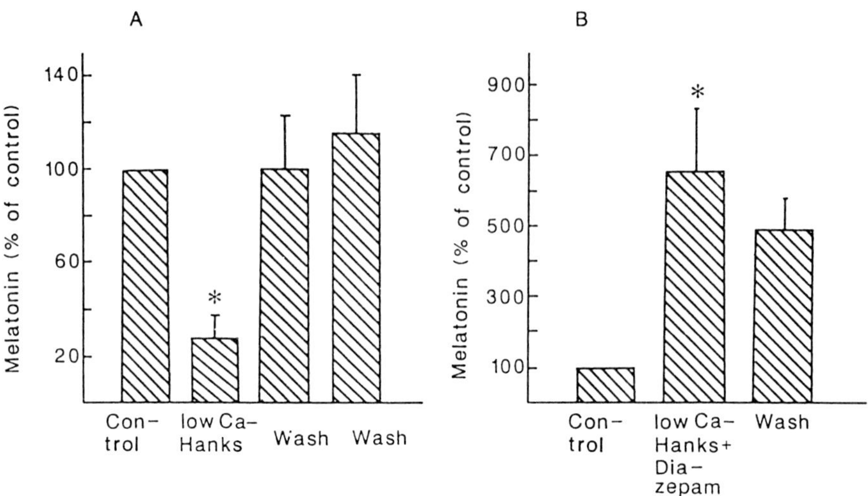

Fig. 4 Effects of blocking synaptic transmission with low extracellular Ca^{2+}- (0.13 mM) and high Mg^{2+}- (6.08 mM) levels on melatonin secretion of isolated, light-adapted pineal organs (A). Addition of diazepam (500 μM) to low Ca^{2+}-medium increased melatonin release by more than 600% (B). n=4, *p < 0.001 vs control.

DISCUSSION

Melatonin production and release of the pineal gland of the rainbow trout *in vitro* depends primarily on intensity and wavelength of the incident light, but also on the temperature (cf. Gern and Greenhouse, 1988; Max and Menaker, 1992), whereas an endogenous, circadian rhythm in melatonin secretion in the isolated trout pineal organ was never detected, contrary to other teleost species (Kezuka et al., 1989; Falcón et al., 1989). However, from the present investigations it appears that melatonin formation and secretion from photoreceptors are also under the possible control of the intrapineal neuronal circuitry, or signals of extrapineal origin. Such a putative influence of the intrapineal neuronal network on photoreceptor function was previously only shown in electrophysiological studies (Meissl and Ekström, 1991). It was suggested that a GABAergic mechanism may participate in the modulation of light sensitivity during light-and dark adaptation processes and that this action is mediated by $GABA_A$ receptors. Our working hypothesis for the present experiments was derived from the assumption that a neurotransmitter induced change in light sensitivity should also cause an influence on the endocrine photoreceptor function.

The present experiments indicate that benzodiazepines that act on the $GABA_A$/benzodiazepine-receptor complex cause an influence on the endocrine functions of pineal photoreceptor cells. This influence is possibly a direct action on photoreceptor cells because blocking synaptic transmission with a medium containing low calcium and high magnesium concentrations that usually block synaptic transmission did not affect the benzodiazepine response. However, in interpreting the results of such experiments some caution should be exercised, since in avians with photoreceptive pineal organs it has been shown that calcium influx through voltage-sensitive channels appear to be involved in regulating melatonin synthesis (Harrison and Zatz, 1989; Zawilska and Nowak, 1990). Also in rat pineal organs α_1-adrenergic stimulation requires external Ca^{2+} (Sugden et al., 1986). However, the effect of diazepam seems to override such a blockade, since the diazepam action is not reduced in low calcium, high magnesium, medium.

In the central nervous system benzodiazepine effects are explained through the potentiation of GABA neurotransmission by acting on central benzodiazepine binding sites (cf. Leonard, 1990) and causing a kind of fine tuning of the GABA receptor (Müller, 1987). The tranquilizing effects of benzodiazepines and barbiturates may be explained by such mechanisms enhancing GABA neurotransmission (cf. DeFeudis, 1983). Several studies have also indicated that melatonin possesses sedative and anticonvulsant effects in mammals including humans (Anton-Tay, 1974) and it was proposed that the pharmacological action of melatonin involve enhancement of central GABAergic neurotransmission (Niles, 1989).

Investigations in the rat indicate a direct action of benzodiazepines on pineal function modulating the nocturnal melatonin concentration, associated with a decrease in melatonin secretion in vivo (Wakabayashi et al., 1991) or an increase as observed in in vitro experiments (Cardinali et al., 1986).

In the present study we observed a significant increase of melatonin release induced by benzodiazepines. This phenomenon was especially pronounced in the light adapted state when melatonin production is low. The nature of the diazepam response in the trout pineal is not clear, because it seems that it is opposite to that of the $GABA_A$-receptor agonist muscimol which seems to decrease melatonin production (Meissl et al., 1993). However, it appears that melatonin synthesis and release in the trout are not exclusively regulated by photic conditions and temperature but also by the intrapineal neuronal circuitry and possibly by blood-borne signals of extrapineal origin, that may have the function of stabilizing and fine tuning of the system.

ACKNOWLEDGEMENT

The authors wish to thank Mrs. Monika Euler for skillful technical assistance in the HPLC-measurements.

REFERENCES

Anis, Y.; Nir, I.; Schmidt, U.; Zisapel, N. (1992): Modification by oxazepam of the diurnal variations in brain ^{125}I-melatonin binding sites in sham-operated and pinealectomized rats. *J. Neural Transm.* 89, 155-166.

Anton-Tay, F. (1974): Melatonin: effects on brain function. *Adv. Biochem. Psychopharmacol.* 11, 315-324.

Cardinali, D.P.; Lowenstein, P.R.; Rosenstein, R.E.; Gonzalez Solveyra, C.; Keller Sarmiento, M.I.; Romeo, H.E.; Acuna Castroviejo, D. (1986): Functional links between benzodiazepine and GABA receptors and pineal activity. In: GABA and Endocrine Function, Racagni, G.; Donoso, A.O. (eds.); pp. 155-164, Raven Press, New York.

Collin, J.-P.; Meissl, H.; Voisin, P.; Brisson, P.; Falcón, J. (1986): Rhythmic signals of pineal transducers: physiological, biochemical and cytochemical evidence. *Adv. Pineal Res.* 1: 41-50.

DeFeudis, F.V. (1983): Psychoactive agents and GABA-receptors. *Pharmacol. Res. Comm.* 15, 29-39.

Dodt, E. (1973): The parietal eye (pineal and parietal organs) of lower vertebrates. In: "Handbook of Sensory Physiology". Vol. VII/3B, Jung, R. (ed.), pp. 113-140, Springer, Berlin, Heidelberg, New York.

Dodt, E. and Heerd, E. (1962): Mode of action of pineal nerve fibers in frogs. *J. Neurophysiol.* 25: 405-429.

Dowling J.E. and Ripps H. (1973): Effect of magnesium on horizontal cell activity in the skate retina. *Nature* 242: 101-103.

Falcón, J.; Brun-Marmillon,J.; Claustrat, B.; Collin, J.-P. (1989): Regulation of melatonin secretion in a photoreceptive pineal organ: an in vitro study in the pike. *J. Neurosci.* 9: 1943-1950.

Gern, W. A. and Greenhouse, S.S. (1988): Examination of in vitro melatonin secretion from superfused trout (*Salmo gairdneri*) pineal organs maintained under diel illumination or continuous darkness. *Gen. Comp. Endocrinol.* 71: 163-174.

Harrison, N.L. and Zatz, M. (1989): Voltage-dependent calcium channels regulate melatonin output from cultured chick pineal cells. *J. Neurosci.* 9, 2462-2467.

Kezuka H.; Aida, K.; Hanyu, I. (1989): Melatonin secretion from goldfish pineal gland in organ culture. *Gen. Comp. Endocrinol.* 75, 217-221.

Leonard, B.E. (1990): Neurochemistry: Modulation of GABAergic function by benzodiazepines. In: Benzodiazepines: Current Concepts, Hindmarch, I.; Beaumont, G.; Brandon, S.; Leonard, B.E. (eds.), pp. 43-59, J. Wiley & Sons, Chichester.

Max,M. and Menaker, M. (1992) Regulation of melatonin production by light, darkness, and temperature in the trout pineal. *J. Comp. Physiol.* 170: 479-489.

Meissl, H.; Anzelius, M.; Östholm, T.; Ekström, P. (1993): Interactions of GABA, benzodiazepines and melatonin in the photosensory pineal organ of salmonid fish. In: Melatonin and the pineal gland. Touitou, Y.; Arendt, J.; Pevet, P. (eds.) pp. 95-98, Elsevier, Amsterdam.

Meissl, H. and Dodt, E. (1981): Comparative physiology of pineal photoreceptor organs. In: "The Pineal Organ: Photobiology - Biochronometry - Endocrinology". A. Oksche, P. Pevet (eds.), pp. 61-80, Elsevier, Amsterdam.

Meissl, H.; Ekström, P. (1988): Photoreceptor responses to light in the isolated pineal organ of the trout, *Salmo gairdneri*. *Neuroscience* 25: 1071-1076.

Meissl, H. and Ekström, P. (1991): Action of γ-aminobutyric acid (GABA) in the isolated photosensory pineal organ. *Brain Res.* 562, 71-78.

Meissl, H.; Martin, C.; Tabata, M. (1990): Melatonin modulates the neural activity in the photosensory pineal organ of the trout: Evidence for endocrine-neuronal interactions. *J. Comp. Physiol.A*, 167: 641-648.

Müller, W.E. (1987): The Benzodiazepine Receptor. Cambridge University Press, Cambridge.

Niles, L.P. (1989): Melatonin and N-acetyl-5-methoxykynurenamine interaction with brain receptors for γ-aminobutyric acid and diazepam. In: Advances in Pineal Res. 3, R. Reiter and S.F. Pang, (eds.), pp. 201-206. J. Libbey, London.

Scharrer, E. (1964): Photo-neuro-endocrine systems: general concepts. *Ann. N.Y. Acad. Sci.* **117**: 13-22.

Sugden, A.L.; Sugden, D.; Klein, D.S. (1986): Essential role of calcium influx in the adrenergic regulation of cAMP and cGMP in rat pinealocytes. *J. Biol. Chem.* **261**, 11608-11612.

Wakabayashi, H.; Shimada, K.; Satoh, T. (1991): Effects of diazepam administration on melatonin synthesis in the rat pineal gland in vivo. *Chem. Pharm. Bull.* 39, 2674-2676.

Zawilska, J.B. and Nowak, J.Z. (1990): Calcium influx through voltage-sensitive calcium channels regulates in vivo serotonin N-acetyltransferase (NAT) activity in hen retina and pineal gland. *Neurosci. Lett.* **118**, 17-20.

II.Osmoregulation

A. Central body fluid homeostasis and circumventricular neuronal structures

Integrative and cellular aspects of autonomic functions : temperature and osmoregulation. Eds K. Pleschka, R. Gerstberger. John Libbey Eurotext, Paris © 1994, pp. 375-382.

Importance of brain sodium concentration for the control of renal sodium excretion

Peter Bie and Claus Emmeluth

Department of Medical Physiology, Division of Pathophysiology, Panum Institute, University of Copenhagen, 3 Blegdamsvej, DK-220 Copenhagen N, Denmark

The regulation of renal sodium excretion is normally dominated by the renin-angiotensin-aldosterone system (Sealey & Laragh 1990). However, there is evidence that other substances and systems can be involved in the final regulation of the amount of sodium excreted. The role of atrial natriuretic peptide in the regulation of renal sodium excretion (DeBold et al. 1981) is still under debate (Goetz et al. 1988, Bie et al. 1990). The role of endogenous inhibitors of the Na^{+}/K^{+}-ATPase is still a theme of much controversy (Kramer et al. 1991), and the situations where renal nerves play an important role in the regulation of renal sodium handling are still not totally clarified.

The tonicity of extracellular fluid is able to influence the rate of renal sodium excretion. It has previously been shown that changes in extracellular tonicity can be perceived by structures within the brain, most likely the circumventricular organs, and that changes in extracellular tonicity were followed by changes in renal excretion of sodium (e.g. Andersson et al. 1967). But, it has until a few years ago not been shown that very small changes in plasma tonicity i.e. changes within the physiological range can influence renal sodium excretion (Chodobski & McKinley 1989, Emmeluth et al. 1990). That structures within the brain could be involved in perception of these small changes in extracellular sodium concentration has just recently been described (Chodobski & McKinley 1989, McKinley et al. 1992, Emmeluth et al. 1992). One possible substance involved in the natriuretic response to an increase in extracellular tonicity has been described as it was shown that a selective 2% increase in the plasma sodium concentration of the blood perfusing the brain caused a substantial increase in renal sodium excretion and a marked increase in renal excretion of urodilatin (Emmeluth et al. 1992) - a renal analogue of atrial natriuretic peptide (Schultz-Knappe et al. 1988).

We have investigated the role the renin-angiotensin-aldosterone system and the role of the renal nerves in the natriuresis caused by a selective and physiologically relevant increase in the sodium concentration of the blood perfusing the brain.

MATERIALS AND METHODS

Animals. Experiments were performed in conscious trained female dogs (Beagle) with a body weight between 10 and 12 kg. They had a daily intake of Na^{+} around 2.6 mmol (kg b.wt)$^{-1}$ and free access to tap water. Prior to the study the common carotid arteries were placed in skin loops by use of general anesthesia and sterile surgical techniques.

Preparation. Sterile catheters were introduced into the saphenous vein, into the v. cava via the external jugular vein and into one or both common carotid arteries. A modified Foley catheter was inserted into the bladder. Arterial blood pressure and heart rate were measured continuously and averaged. (STATHAM P50 transducer, ECG, DANICA MONITOR interrogated by a DEC µVAX computer every 10 seconds). The water excreted was replaced continuously by use of a servo mechanism infusing a glucose-urea solution (40 and 25 mM, respectively). This servo mechanism measures the weight of the dog and keeps it within ± 0.1% (Bie 1976). After 60 or 90 minutes of equilibration the experiments were begun.

The experiments are divided into series infusing NaCl intravenously as either a hypertonic or an isotonic solution and into experiments infusing NaCl as an isotonic solution into both the v. cava and the two carotid arteries or as a 'split-infusion' (se below).

During the experiments infusing the NaCl intravenously to conscious dogs the rate of infusion was 60 µmol kg^{-1} min^{-1}. This was infused as either a hypertonic (770 mM) or an isotonic solution (154 mM). The infusates were placed on the load cell controlling the replacement of water, and changes in Na^+ and water content in these series are illustrated in figure 1a.

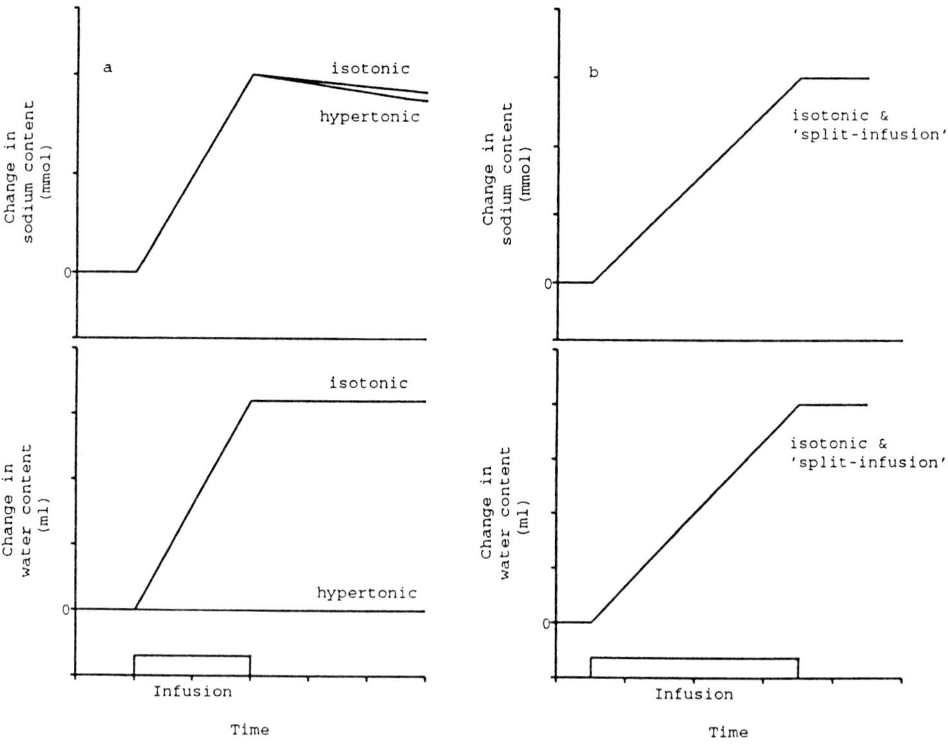

Fig. 1. Diagrammatic presentation of the changes in body water and body sodium content during the intravenous study (a) and the studies with cartotid infusions (b).

In the experiments where infusions where performed into both the carotid arteries and the v. cava two protocols where performed. A 'split-infusion' (SPI) where hypertonic NaCl was infused into the

two carotid arteries increasing the Na^+ concentration of carotid plasma with 3 mM (series 1 and 2, 50 µmol kg^{-1} min^{-1}) or 6 mM (series 3, 90 µmol kg^{-1} min^{-1}). Concomitantly distilled water was administered via the jugular catheter at a rate making the sum of the two infusates isotonic (154 mmol l^{-1} ~ 286 mOsm kg^{-1}). The second protocol was an isotonic volume expansion (IVE) where the same infusion rates were used as in SPI infusing the same total amount of NaCl. However, isotonic NaCl (154 mM) was infused into the carotid arteries as well as into the v. cava. Again infusates were placed on the load cell controlling the replacement of water. In these experiments the Na^+ excreted was replaced by infusing NaCl at the rate of excretion. During these protocols the increases in content of water and NaCl of the body were identical in SPI and IVE series of the experiments despite differences in the excretory patterns (Fig. 1b). The protocols where performed in normal conscious dogs (series 1), in conscious dogs with chronic bilateral renal denervation (series 2), and in conscious dogs with acute blockade of the renin-angiotensin-aldosterone system (series 3).
In none of the series the manipulations were accompanied by hemolysis.

Chemical analyses. Na^+ and K^+ were measured by flame photometry (IL 243, Instrumentation Laboratory, USA). Concentrations of creatinine in plasma and urine were measured by spectrophotometry (SPECTRONIC 1001, Bausch & Lomb, USA) in accordance to Bonsnes and Taussky (1945). The osmolality of plasma and urine was estimated by freezing point depression (GONOTEC OSMOMAT, Germany).

Hormone analysis. The renal excretion of urodilatin was calculated after determination of urinary urodilatin concentration by radioimmunoassay as recently described (Drummer et al. 1990). The antibody, which did not crossreact with ANP, was kindly provided by Prof. Forssmann, Hannover, Germany. Detection limit was 7.1±4.1 pM and the 50% intercept was 67 pM. Intraassay and interassay coefficients of variation were 8 and 12%, respectively. Using ethanol-extraction the recovery of urodilatin from urine was above 90%.
The determination of plasma atrial natriuretic peptide (ANP) was performed in an assay modified after the assay described by Schütten et al. (1987). Briefly, acidified arterial plasma was extracted using C_{18} SEP-PAK® cartridges (Waters, Millipore Corporation, USA). In the radioimmunoassay bound antigen was separated from free using a charcoal-plasma suspension. The detection limit was below 5 pg $tube^{-1}$ and half maximal binding was observed at 27±2 pg $tube^{-1}$. The interassay coefficients of variation was 6% at 20 pg $tube^{-1}$ (n=6). The intra-assay coefficient of variation was 2.6% at 30 pg ml^{-1} plasma (n=5). Recovery of 10 pg synthetic ANP added to one ml of plasma was 68±12% (mean±SD, n=5).
Vasopressin (AVP) in plasma was measured by use of a radioimmunoassay following extraction from acidified plasma using SEP-PAK® C_{18} cartridges. The assay was performed using a specific antibody (Mitsubishi Petrochemical Co., LTD., Japan) (final dilution 1:75,000). Bound antigen was separated from free by a charcoal-plasma suspension. The interassay coefficient of variation was 5% at 2.3 pg $tube^{-1}$ (n=10). The intra-assay coefficient of variation was 11% at 0.25 pg ml^{-1} plasma (n=7). Recovery of 2 pg synthetic AVP added to plasma was 94±8% (mean±SD, n=6). Detection limit was less than 0.15 pg $tube^{-1}$ (B_0-3*SD).

RESULTS

During the series where the NaCl was infused intravenously renal excretion of sodium increased from 2.4±0.6 to 105±27 μmol min^{-1} (Fig. 2) during the hypertonic infusion, while it changed from 3.9±1.3 to 58±17 μmol min^{-1} during the isotonic infusion. The same pattern was observed in the fractional excretion of sodium. This response was

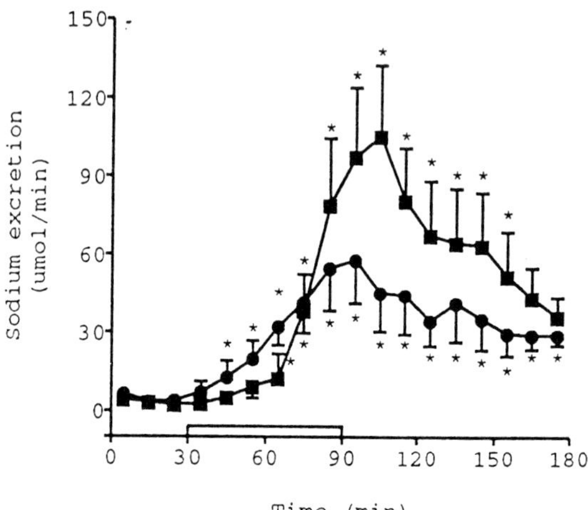

Fig. 2. Changes in renal excretion of sodium during intravenous infusion of hypertonic NaCl (■) and isotonic NaCl (●).

observed without changes in mean arterial blood pressure, plasma ANP, and plasma catecholamines. In the experiments where infusions were per-formed into both carotid arteries and to the v. cava the split-in-fusion in series 1 caused sodium excretion to increase from

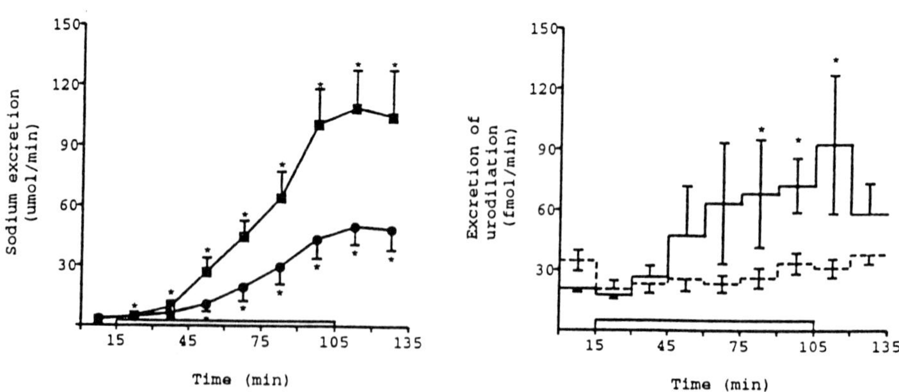

Fig. 3. Changes in renal excretion of sodium (left panel, split-infusion is symbolized by (■) and isotonic volume expansion by (●)) and urodilatin (right panel, split-infusion is symbolized by a full line and isotonic volume expansion by a dashed line) during the series with intracarotid infusions to normal conscious dogs.

3.2±0.4 to 109±19 µmol min^{-1} (Fig. 3) while during the isotonic volume expansion the increase was from 3.7±0.6 to 50±9 µmol min^{-1}. The urinary excretion of urodilatin increased in parallel to the increase in sodium excretion during the split-infusion while there was no increase in the excretion of urodilatin during the isotonic volume expansion.

In series 2 in the dogs with renal denervation the same pattern in renal excretion of sodium was observed (Fig. 4). The data on renal excretion of urodilatin is not available at the moment. In series 2 a small increase in the plasma level of AVP from 0.18±0.04 to 0.53±-0.12 was observed during the splitinfusion, while there was no increase during the isotonic volume expansion. Plasma ANP increased in both series, but the increase was larger during the isotonic volume expansion compared to the split-infusion.

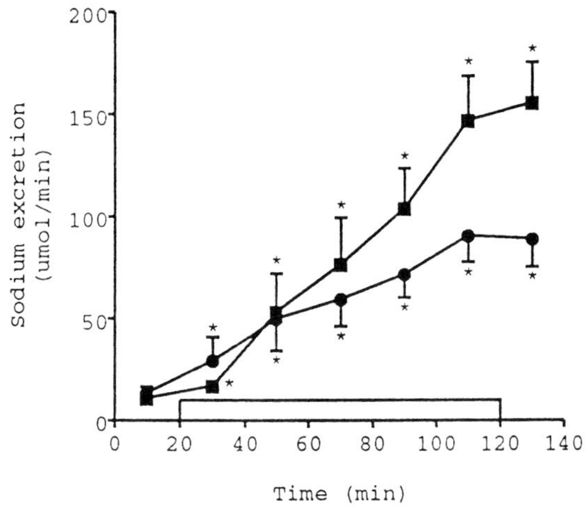

Fig. 4. Changes in renal excretion of sodium (split-infusion is symbolized by (■) and isotonic volume expansion by (●)) during the series with intracarotid infusions to conscious dogs, who prior to the experiments had been subjected to surgical denervation of both kidneys.

In series 3 where the dogs were pretreated with enalaprilat and canrenoate a similar pattern in the change in renal sodium excretion as the one seen in series 1 and 2 was observed (Fig. 5). Due to the blockade of the renin-angiotensin-aldosterone system control level of renal sodium excretion was elevated to around 100 µmol min^{-1}. This increase was not followed by an increase in renal excretion of urodilatin.

However, during the infusions renal excretion of urodilatin increased, and this increase was larger during the split-infusion compared to the isotonic volume expansion. The plasma level of AVP increased in this series during the split-infusion from 0.68±0.11 to 2.4±0.8 pg ml^{-1}. There were no changes during the isotonic volume expansion from 0.51±0.09 pg ml^{-1}. Plasma ANP increased to the same extent during the split-infusion and during the isotonic volume expansion. In the three series with intracarotid infusion there were increases in MABP during the split-infusion which were not observed

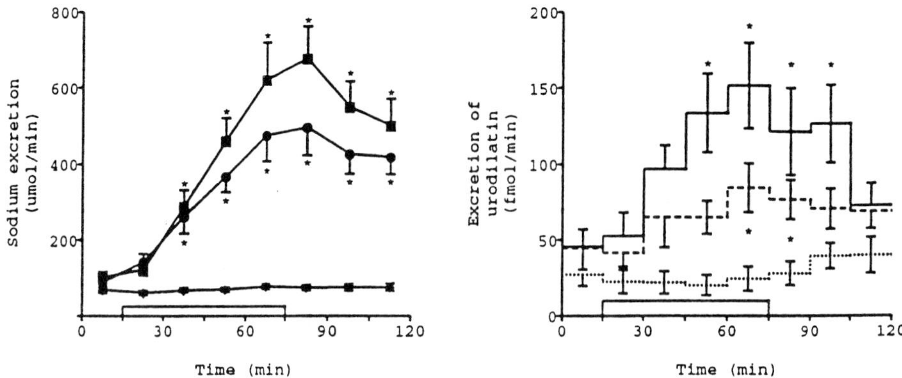

Fig. 5. Changes in renal excretion of sodium (left panel, split-infusion is symbolized by (■) and isotonic volume expansion by (●)) and urodilatin (right panel, split-infusion is symbolized by a full line and isotonic volume expansion by a dashed) during the series with intracarotid infusions to conscious dogs treated with enalaprilat and canrenoate. The third curve is from a time control study.

during the isotonic volume expansion (Table 1).

Table 1. Mean arterial blood pressure (MABP) and heart rate (HR) in the experiments with 'split-infusion (SPI) and isotonic volume expansion (IVE).

	Conscious dogs				Conscious dogs prior subjected to bilateral renal denervation				Conscious dogs treated with enalaprilat and canrenoate			
	MABP (mmHg)		HR (per min)		MABP (mmHg)		HR (per min)		MABP (mmHg)		HR (per min)	
	SPI	IVE	SPI	IVE	SPI	IVE	SPI	IVE	SPI	IVE	SPI	IVE
Before	109 ±1	106 ±2	74 ±3	76 ±3	96 ±5	94 ±4	88 ±4	92 ±10	106 ±2	104 ±2	75 ±4	76 ±4
End-infusion	121 ±2*	113 ±4	98 ±6	95 ±5	109 ±4*	101 ±2	116 ±7*	107 ±3*	113 ±4*	104 ±3	96 ±5	106 ±5*
After	116 ±2*	107 ±3	97 ±6*	93 ±3*	104 ±6	102 ±3	110 ±5*	110 ±5*	111 ±4*	107 ±5	91 ±4*	96 ±4*

I series three plasma ANP and plasma vasopressin levels increased significantly during the split-infusion (Fig. 6). During the isotonic volume expansion plasma ANP increased in a manner not disatinguisable from that seen during the split-infusion. There were no changes in plasma vasopressin during the isotonic volume expansion.

DISCUSSION

Our results demonstrate that a tonicity dependent natriuresis is operating with changes in the plasma concentration of sodium within the physiological relevant range. This natriuresis seems to be independent of changes in arterial blood pressure, glomerular filtration rate, the renin-angiotensin-aldosterone system, the renal

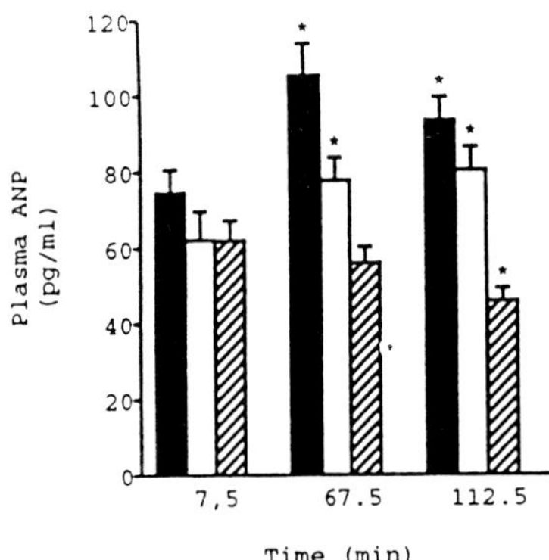

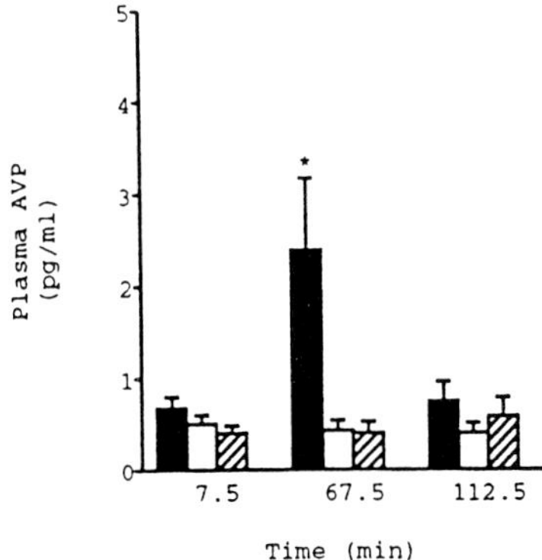

Fig. 6. Plasma level of atrial natriuretic peptide (ANP) and vasopressin (AVP) during split-infusion (■) and isotonic volume expansion (□) in dogs treated with enalaprilat and canrenoate. The third bar denotes the results obtained during time control.

nerves and circulating levels of ANP and catecholamines. As mentioned the concentration stimuli applied were well within the variations which can be seen during normal live. Further, the amount of sodium applied is in the range normally ingested with the daily meal of the dog. Therefore, the risk of inducing artefacts due to the use of large stimuli is smaller during the present setup.

The increases in MABP during the split-infusions indicate that this factor might participate in the excess excretion of sodium. It would be helpfull if it was possible to control renal perfusion pressure without changing the hemodynamics of the rest of the circulation in order to evaluate the role of the increase in MABP in the excess excretion of sodium.

Whether vasopressin is involved in the excess excretion is still to be resolved. The plasma level observed during the split-infusion in the animals with blockade of the renin-angiotensin-aldosterone system is within the range where AVP is able to act as a natriuretic substance in the dog. But there is much evidence in the litterature that demonstrates that tonicity dependent natriuresis is not mediated through AVP (e.g. Park et al. 1985).

The role of urodilatin in the excess excretion of sodium during the split-infusion is still not totally clarified. There is no doubt that the excretion of urodilatin goes up parallel in time to the excess excretion of sodium. However, the evidence that urodilatin is causally involved remains to be determined. The use of a specific blocker could clarify this issue. On the other hand our data show that the increase in the excretion of urodilatin is not only a function of the increase in renal sodium excretion, as the increase in renal sodium excretion induced by the blockade of the renin-angiotensin-aldosterone system was not followed by an increase in the excretion rate of urodilatin.

The results show that the body is able to respond to very minute changes in the plasma concentration of sodium. The sensor involved in this response is situated in the area perfused by the carotid arteries. The result of the change in plasma sodium of carotid blood is an increase in renal sodium excretion, which relates in time to the renal excretion of urodilatin. The respons to the split-infusion is not dependent on a functional renin-angiotensin-aldosterone system or on renal nerves. It thus seems likely that the information perceived by the brain is communicated to the kidneys via a humoral substance or through the increase in arterial blood pressure observed.

REFERENCES

Andersson, B., Dallman, M.F. & Olsson, K. (1969): Evidence of hypothalamic control of renal sodium excretion. *Acta Physiol Scand* 75, 496-510.

Bie, P. (1976): Studieş of cerebral osmoreceptors in anesthetized dogs: The effect of intravenous and intracarotid infusion of hyperosmolar sodium chloride solutions during sustained water diuresis. *Acta Physiol Scand* 96, 306-318.

Bie, P., Wang, B.C., Leadley, R.J., Jr. & Goetz, K.L. (1990): Enhanced atrial peptide natriuresis during angiotensin and aldosterone blockade in dogs. *Am J Physiol* 258, R1101-R1107.

Bonsnes, R.W. & Taussky, H.H. (1945): On the colorimetric determination of creatinine by the Jaffe reaction. *J Biol Chem* 158, 581-591.

Chodobski, A. & M.J. McKinley, M.J. (1989): Cerebral regulation of renal sodium excretion in sheep infused intravenously with hypertonic NaCl. *J Physiol Lond* 418, 273-291.

Drummer, C., Fiedler, F., König, A. & Gerzer, R. (1990): Urodilatin, a kidney-derived atriuretic factor, is excreted with circadian rhythm and stimulated by saline infusion in man. *J Am Soc Nephrol* 1, 1057-1059.

Emmeluth, C., Schütten, H.J., Knigge, U., Warberg, J. & P. Bie. (1990): Increase in plasma sodium enhances natriuresis in response to a sodium load unable to change plasma atrial peptide concentration. *Acta Physiol Scand* 140, 119-127.

Emmeluth, C., Drummer, C., Gerzer, R. & Bie, P. (1992): Roles of cephalic Na^+ concentration and urodilatin in control of renal sodium excretion. *Am J Physiol* 262, F513-F516.

Goetz, K.L., Wang, B.C., Bie, P., Leadley, R.J., Jr. & Geer, P.G. (1988): Natriuresis during atrial distension and a concurrent decline in plasma atriopeptin. *Am J Physiol* 255, R259-R267.

Kramer, H.J., Meyer-Lehnert, H., Michel, H. & Predel H.-G. (1991): Endogenous natriuretic and ouabain-like factors. Their role in body fluid volume and blood pressure regulation. *Am J Hypertens*, 4, 81-89.

McKinley, M.J., Lichardus, B., McDougall, J.G. & Weisinger, R.S. (1992): Periventricular lesions block natriuresis to hypertonic but not isotonic NaCl loads. *Am J Physiol* 262, F98-F107.

Park, R.G., Congiu, M., Denton, D.A. & McKinley, M.J. (1985): Natriuresis induced by arginine vasopressin infusion in sheep. *Am J Physiol* 249, F799-F805.

Schulz-Knappe, P., Forssmann, K., Herbst, F., Hock, D., Pipkorn, R. & Forssmann, W.G. (1988): Isolation and structural analysis of "urodilatin", a new peptide of the cardiodilatin-(ANP)-family, extracted from human urine. *Klin Wochenschr* 66, 752-759.

Schütten, H.J., Johannessen, A.C., Torp-Pedersen, C., Sander-Jensen, K., Bie, P. & Warberg, J. (1987): Central venous pressure - a physiological stimulus for secretion of atrial natriuretic peptide in humans? *Acta Physiol Scand* 131, 265-272.

Sealey, J.E. & Laragh, J.H. (1990): The integrated regulation of electrolyte balance and blood pressure by the renin secretion. In: D.W. Seldin and G. Giebisch (Eds.) *The regulation of sodium and chloride balance*, pp 133-193, Raven Press, New York.

Integrative and cellular aspects of autonomic functions : temperature and osmoregulation. Eds K. Pleschka, R. Gerstberger. John Libbey Eurotext, Paris © 1994, pp. 383-388.

Hypertension and thirst long outlasting a temporary rise in blood angiotensin II concentration

Stefan Eriksson, Bengt Andersson*, Ulf Gunnarsson and Mats Rundgren

*Department of Physiology, Karolinska Institutet, Stockholm and *Department of Physiology, Faculty of Veterinary Medicine, Swedish University of Agricultural Sciences, Uppsala, Sweden*

That a centrally mediated component contributes to the hypertensive effect of blood-borne angiotensin II (Ang II) was originally suggested more than 30 years ago by Bickerton & Buckely (1961) on the basis of results obtained in cross-transfusion experiments. More conclusive evidence for this concept has been provided by subsequent studies in various mammalian species, and it has been shown that the blood pressure rise elicited by a cerebral action of systemic Ang II predominantly is exerted via increased sympathetic tone (*cf* Squire & Reid, 1993). Over the past several years it has gradually become evident that also other effects of systemic Ang II, among them thirst (Epstein *et al.*, 1970), are mediated by the brain.

Circulating Ang II does not to any appreciable extent pass over the blood-brain barrier (BBB), and substantial experimental evidence suggests that the influence of systemic Ang II on cerebral mechanisms is exerted via a binding of the octapeptide to receptors in the circumventricular organs, which are located outside the BBB (*cf* Mc Kinley *et al.*, 1990). The dipsogenic effect seems mainly to be the results of binding of the octapeptide to receptors in the subfornical organ and the organum vasculosum of the lamina terminalis which both are located in the anterior wall of the third cerebral ventricle. A more likely candidate as mediator of the hypertensive effect is another circumventricular organ, the area postrema, located in the bottom of the forth cerebral ventricle (*cf* Squire & Reid, 1993). However, all centrally mediated effects of systemic Ang II can be reproduced from the inside of the BBB, by intracerebroventricular (ICV) administration of small amounts of the octapeptide. As regards ICV infusions of Ang II, it was observed in the goat more than 20 years ago that the dipsogenic and hypertensive effects may outlast the ICV infusion by half an hour or longer (Andersson *et al.*, 1972). Here are summarized the results of two recent studies in conscious sheep (Eriksson *et al.*, 1992, 1993) demonstrating that correspondingly sustained effects can also be obtained in response to a brief elevation of systemic Ang II, and that the centrally mediated vasoconstrictor effect of the octapeptide apparently does not affect the renal blood flow (RBF).

Arterial blood pressure

As illustrated by Fig. 1, our first study (Eriksson *et al*, 1992) revealed that the carotid blood pressure (cBP) remained elevated for more than 30 min after 10 min bilateral intracarotid (i.c.)

infusions of Ang II at 40 pmol kg^{-1} min^{-1}.

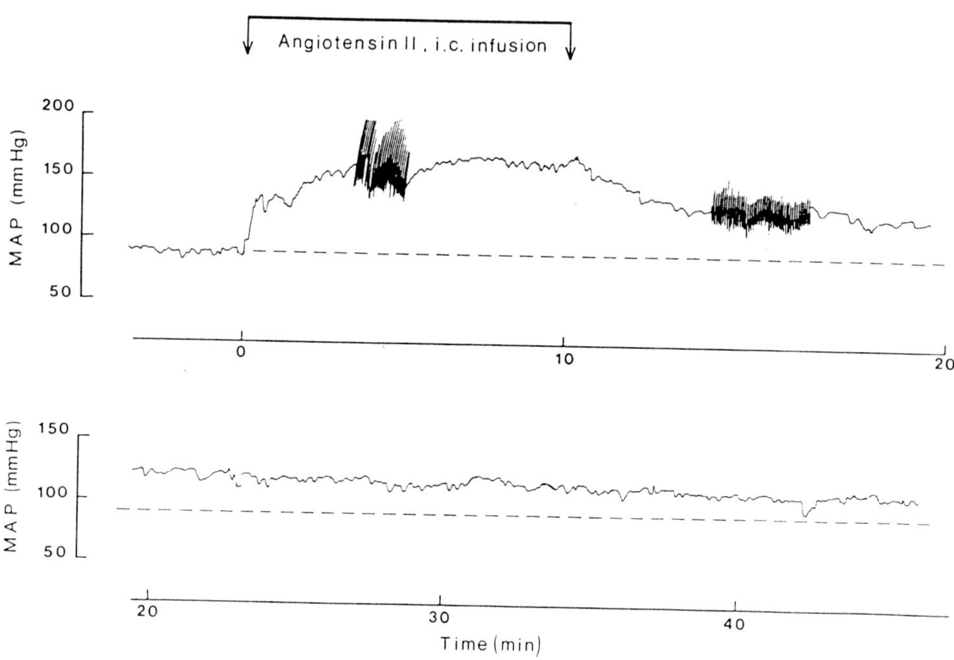

Fig. 1. Continuous recording of the elevation of mean arterial blood pressure (MAP; carotid artery) induced by a 10 min bilateral intracarotid (i.c.) infusion of angiotensin II (40 pmol kg^{-1} min^{-1}) in a sheep. Note that the MAP remains >10mmHg above initial level (- - - - -) as late as 30 min post-infusion. (From : Eriksson *et al* 1992)

In our second study (Eriksson *et al.* 1993) it was found that a similarly sustained rise in cBP was obtained also after corresponding i.c. and intravenous (i.v.) infusions of the octapeptid at 20 pmol kg^{-1}min^{-1}, but that the pressor response induced by the i.v. infusions was less pronounced than that induced by the i.c. infusions (Fig. 2, open symbols). Since the centrally mediated component of the pressor response to systemic Ang II has been shown to be exerted via increased sympathetic vascular tone (*cf* Squire & Reid, 1993) it was of interest to study how pre-treatment with the combined alfa- and beta-adrenoceptor blocker labetalol (Gold *et al.*,1982) would affect the hypertensive effect of the i.c.and i.v. Ang II infusions. It was found that treatment with labetalol somewhat reduced the pronounced rise in cBP occurring during the Ang II infusions and significantly attenuated and shortened the sustained elevation of cBP seen after the infusions (Fig. 2, filled symbols).

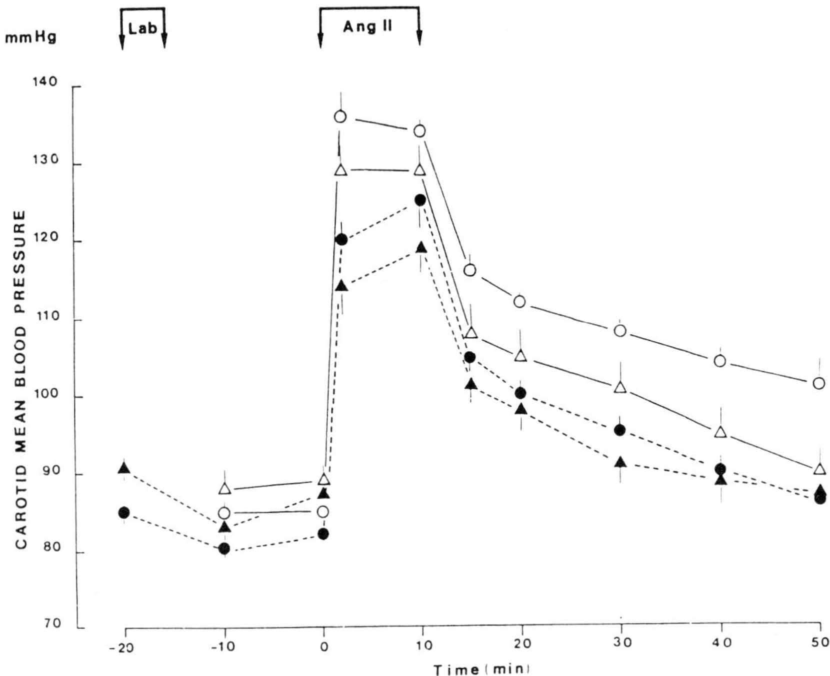

Fig. 2. Elevation of carotid mean blood pressure induced by 10 min bilateral intracarotid (○—○) and intravenous (△—△) infusions of angiotensin II at 20 pmol $kg^{-1}min^{-1}$ compared to the elevation obtained after pre-treatment with the adrenoceptor blocker labetalol (● ---- ●); (▲ ---- ▲). Labetalol was infused intravenously for 4.5 min at 0.5 mg $kg^{-1}min^{-1}$ (**Lab**). Animals and experiments in each of the four series, n=6. Vertical bars show SEM.

A near at hand explanation for the larger rise in cBP obtained during and after the i.c. infusions of Ang II is that, in comparison to the i.v. infusions, more of the octapeptide became available for exerting a centrally mediated pressor effect. The attenuation of the pressor response seen after pre-treatment with the adrenoceptor blocker labetalol provides additional evidence that an action of Ang II at the cerebral level contributed to the pronounced rise in cBP obtained during the intravascular infusions, and was the main cause of the elevation of the cBP seen between 5 and 40 min post-infusion. A further, indirect support for this concept is provided by the simultaneously made observations on RBF (see below).

The blood pressure of unanaesthetized dogs has been reported to remain elevated above normal for up to 24 h after the discontinuation of a 7 day intravertebral infusion of Ang II (Fukiyama *et al.*, 1971). However, with the exception of the studies reviewed here, it has, to our knowledge, not been reported previously that also a brief elevation of systemic Ang II may elicit a sustained, centrally mediated elevation of the arterial blood pressure. This observation appears to be of clinical interest, since it implies that a transient, but intense activation of the renin-angiotensin system may induce a rather sustained hypertension.

As regards the cerebral site of the centrally mediated hypertensive response to blood-borne

Ang II, studies in dogs and rabbits have demonstrated that administration of the octapeptide into the vertebral artery has a significantly greater pressor effect than a corresponding i.v. or i.c. administration, and that this difference can be eliminated by bilateral lesions in the area postrema region (*cf* Squire & Reid, 1993). Whereas in the mentioned, and many other mammalian species the medulla oblongata is supplied with blood from the vertebral arteries, the entire brain (including the medulla oblongata) is normally supplied with carotid blood in goats and sheep (Andersson & Jewell, 1956; Baldwin & Bell, 1963). Hence, the sustained post-infusion elevation of cBP observed here in response to a brief i.c. administration of Ang II may well have been a manifestation of binding of blood-borne Ang II to receptors in the medullary region. Some indirect support for that idea has recently been obtained by the observation (unpublished) that medial forebrain lesions, which eliminate the dispogeneic response to i.c. Ang II, leave the sustained pressor response unaffected in sheep.

Renal blood flow

Ang II is a potent renal vasoconstrictor (*cf* Hall & Brands, 1993). Consequently, in the experiments reviewed here (Eriksson *et al.*, 1993), unilateral ultrasonic recording revealed a conspicuous (60% and 70%, respectively) reduction in RBF during the 10 min i.c. and i.v. Ang II

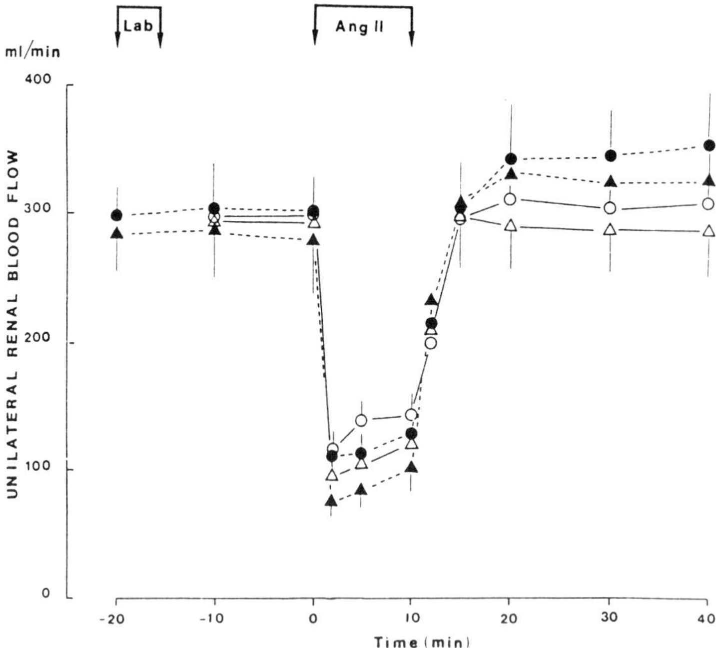

Fig. 3. Reduction of unilateral renal blood flow induced by 10 min bilatral intracarotid or intravenous infusions of angiotensin II at 20 pmol $kg^{-1} min^{-1}$ without (○ — ○ ; △ — △), or with (● ---- ● ; ▲ ---- ▲) pre-treatment with the adrenoceptor blocker labetalol. Labetalol was infused intravenously for 4.5 min at 0.5 mg kg^{-1} min^{-1} (**Lab**). Animals and experimetns in each of the four series, n=5. Vertical bars show SEM.

infusions. This effect of systemic Ang II was not visibly changed by labetalol treatment (Fig.3.). At 5 min after the intravascular Ang II administration, RBF had returned to initial value. Then, the blood concentration of Ang II ought to have become reduced to pre-infusion level, since the biological half life of the octapeptide in the blood of the sheep is less than 1 min (Cain *et al.* 1970). In contrast, 30 min ICV infusions of Ang II at 2 pmol $kg^{-1}min^{-1}$, which previously have been found to elevate the cBP of sheep by about 15 mmHg (Hjelmqvist *et al.*, 1992) did not reduce the RBF. Taken together, these observations imply that no centrally mediated component contributes to the reduction in RBF induced by a rise in systemic Ang II. The pronounced renal vasoconstriction seems to be due exclusively to a local action of Ang II in the kidney, which does not to any appreciable extent involve sensitization of renovascular adrenoceptors.

Thirst

As concerns the dipsogenic effect of systemic or ICV Ang II, sheep are, in comparison with for instance dogs, poor responders (Abraham *et al.*, 1975) and exhibit great interindividual differences (own observations). In our first study (Eriksson *et al.*, 1992), therefore, 20 min i.c. infusion of Ang II at as high a rate as 80 pmol $kg^{-1}min^{-1}$ was employed to demonstrate that the dipsogenic effect persisted for at least 25 min after discontinuation of the Ang II infusion. In our second series of experiments (Eriksson *et al.*, 1993) 12.5% of that dose was administered with the result that only 50% of the animals exhibited an obvious urge to drink during the 10 min i.c. i fusion of Ang II. However, in the responding sheep thirst developed regardless of whether the animals were labetalol-treated or not, and the dipsogenic effect was sustained. Thus, the sheep drank 1,1 to 2,2 l when allowed free access to water at 40 min post-infusion.

The results of previous studies (*cf* McKinley *et al.*, 1990) make likely that the dipsogenic response to the i.c. Ang II infusions was exerted via binding of the octapeptide to receptors in the circumventricular organs which are located outside the BBB in the anterior wall of the third cerebral ventricle, and perhaps also to receptors in nearby forebrain structures.

Concluding remark

The integrated results of the two studies reviewed here suggest that, once bound to circumventricular organs in the forebrain and medualla, initially systemic Ang II may continue to exert centrally mediated effects for a considerable length of time, alternatively, may initiate cerebral processes of long duration.

The studies reviewed were supported by the Swedish Medical Research Council (projects 503 and 6553), and were approved by the Ethical Committee for Animal Experiments.

References

Abraham, S.F., Baker, R.M., Blaine, E.H., Denton, D.A. & McKinley, M.J. (1975): Water drinking induced in sheep by angiotensin ---- a physiological or pharmacological effect? *J.Comp.Physiol.Psychol.* 88,503-518.

Andersson, B., Eriksson, L., Fernández, O., Kolmodin, C.-G. & Oltner, R. (1972): Centrally mediated effects of sodium and angiotensin II on arterial blood pressure and fluid balance. *Acta Physiol.Scand.* 85,398-407.

Andersson, B. & Jewell, P.A. (1956): The distribution of carotid and vertebral blood in the brain and spinal cord of the goat. *Q.J.Exp. Physiol.* 41,462-474.

Baldwin, B.A. & Bell, F.R. (1963): The anatomy of the cerebral circulation of the sheep and ox. The dynamic distribution of the blood supplied by the carotid and vertebral arteries to cranial regions. *J.Anat.Lond.* 97,203-215.

Bickerton, R.K. & Buckley, J.P. (1961): Evidence for a central mechanism in angiotensin induced hypertension. *Proc.Soc.Exp.Biol.Med.* 106, 834-836.

Cain, M.D., Catt, R.J., Coghlan, J.P. & Blair-West, J.R. (1970): Evaluation of angiotensin II metabolism in sheep by radioimmunoassay. *Endocrinology* 86, 955-964.

Epstein, A.N., Fitzsimons, J.T. & Rolls, B.J. (1970): Drinking induced by injection of angiotensin into the brain of the rat. *J.Physiol.* 210, 457-474.

Eriksson, S., Andersson, B. & Rundgren, M. (1992): Thirst and hypertension remaining after intracarotid infusions of angiotension II in the sheep. *Acta Physiol.Scand.* 146, 413-414.

Eriksson, S., Andersson, B., Gunnarsson, U. & Rundgren, M. (1993): Hypertension and thirst outlasting renal vasoconstriction as effects of a brief elevation of systemic angiotension II in sheep. *Acta Physiol.Scand.* Submitted.

Fukiyama, K., McCubbin, J.W. & Page, I.H. (1971): Chronic hypertension elicited by infusion of angiotensin into vertebral arteries of unanaesthetized dogs. *Clin.Sci.* 40, 283-291.

Gold, E.H., Chang, W., Cohen, M., Baum, T., Ehrreich, S., Johnson, G., Prioli, N.& Sybertz E.J. (1982): Synthesis and comparison of some cardiovascular properties of the stereoisomers of labetalol. *J.Med.Chem.* 25, 1363-1370.

Hall, J.E. & Brands, M.W. (1993): Intrarenal and circulating angiotension II and renal function. In *The renin-angiotensin system: Biochemistry, physiology, pathophysiology,therapeutics,* ed. J.I.S. Robertson & M.G. Nichols, Vol. 1, pp. 26.1-26.43. London : Gower Med.Publ.

Hjelmqvist, H., Ullman, J., Hamberger, B. & Rundgren, M. (1992): Cardiovascular and renal effects of intracerebroventricular angiotensin II in conscious sheep. *Acta Physiol.Scand.* 145, 25-32.

McKinley, M.J., McAllen, R.M., Mendelsoh, F.A.O., Allen, A.M., Chai, S.y. & Oldfield, B.J. (1990): Circumventricular organs: neuroendocrine interfaces between the brain and the hemal milieu. *Frontiers in Neuroendocrinology* 11, 91-127.

Squire, I.B. & Reid, J.L. (1993): Interactions between the renin-angiotensin system and the autonomic nervous system. In *The renin-angiotensin system: Biochemistry, Physiology, pathophysiology, therapeutics,* ed. J.I.S. Robertson & M.G. Nichols, Vol. 1. pp. 37.1-37.1 London: Gower Med.Publ.

Integrative and cellular aspects of autonomic functions : temperature and osmoregulation. Eds K. Pleschka, R. Gerstberger. John Libbey Eurotext, Paris © 1994, pp. 389-395.

Effects on tolerance to haemorrhage by changes of cerebrospinal fluid [Na^+] or intracerebroventricular infusion of angiotensin II in conscious sheep

Hans Hjelmqvist, Johan Ullman, Ulf Gunnarsson and Mats Rundgren

Department of Physiology, Karolinska Institutet, S-171 77 Stockholm, Sweden

Hypertonic saline solutions have been used intermittently in resuscitation of hypovolaemic shock for decades and several haemodynamically and metabolically beneficial effects of this treatment have been demonstrated in experimental animals (Baue *et al.*, 1967). Twelve years ago, Velasco and associates reported that intravenous infusion of a relatively small amount (4 ml/kg) of strongly hypertonic NaCl solution (2400 mOsm/kg) effectively reversed severe haemorrhagic shock in dogs (Velasco *et al.*, 1980). With the subsequent report that a similar treatment of patients in terminal hypovolaemic shock had been successful (DeFilippe Jr *et al.*, 1980), the interest in the clinical use of small volume hyperosmotic resuscitation fluid therapy was revived. Since then, numerous studies in animals and man have sought to characterize the haemodynamic and metabolic effects of such a treatment, and tried to identify in what shock situations it may be an advantageous alternative (Haglind & Haljamäe, 1992; Gross *et al.*, 1989; Kreimeier & Messmer, 1992). Still, very little is known about the mechanisms behind the haemodynamic effects of intravenous hypertonic NaCl in hypovolaemic shock. A vagally mediated reflex, elicited by perfusion of hypernatraemic blood through the lungs, has been suggested to be of importance in dogs (Lopes *et al.*, 1981), but this could not be confirmed in studies on sheep (Hands *et al.*, 1986).Furthermore, angiotensin-mediated effects at the cerebral level have recently been suggested to be involved in relaying the effects of intravenous resuscitation with hypertonic NaCl during haemorrhagic shock (Velasco *et al.*, 1990).

The extracellular concentration of both Na and angiotensin II (ANG II) are crucially involved in the cerebral control of fluid balance (Andersson *et al.*, 1984) and possibly also in cardiovascular control. The centrally elicited pressor effect of ANG II has been widely confirmed since its first description by Bickerton & Buckley in 1961. About twenty years ago, Andersson and coworkers showed that intracerebroventricular (ICV) infusion of hypertonic NaCl elevates the blood pressure (Andersson *et al.*, 1971). In analogy with influences on water turnover, the hypertensive effect of elevated CSF [Na] was found to be potentiated by concomitant ICV administration of ANG II (Andersson *et al.*, 1971). Subsequent investigations have demonstrated that the central pressor effects of both Na (Bunag & Miyajima, 1984) and ANG II (Tobey *et al.*, 1983) are mainly mediated via increased sympathetic nervous activity, although in some species increased vasopressin (AVP) release appear to contribute (Rettig *et al.*, 1989; Takata *et al.*, 1989).

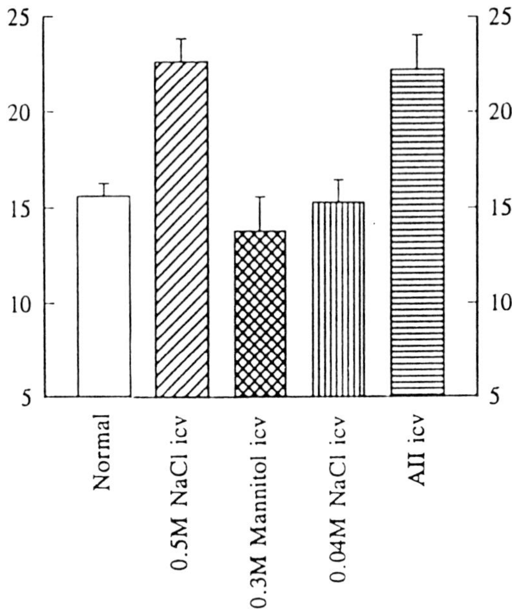

Fig. 1. Summary of the haemorrhage volumes needed to induce acute hypotension during ICV infusion of various solutions. "Normal" represents control experiments where the animals were bled without a concomitant ICV infusion, or received 0.9% NaCl ICV.

With regard to the beneficial effects of intravascular administration of hypertonic NaCl in haemorrhagic shock and the interrelated central effects of Na and ANG II in fluid balance and cardiovascular control, it appeared of interest to study whether changes in the CSF concentration of Na, as well as ANG II, would affect the haemodynamic and humoral responses to haemorrhage. The results of some recent studies in conscious sheep (Hjelmqvist *et al.*, 1992b; Ullman *et al.*, 1993b) on tolerance to blood loss by changes in CSF [Na] and ANG II concentration are presented here.

EFFECTS OF CHANGES IN CSF [Na] / OSMOLALITY

Tolerance to blood loss and haemodynamic effects

In the first study (Hjelmqvist *et al.*, 1992b), elevation of the CSF [Na] by ICV infusion into one lateral ventricle at 0.02 ml/min of 0.5 M NaCl was found to increase the tolerance to haemorrhage (Fig. 1), defined as the volume of blood removal needed to reduce the mean systemic arterial pressure (MSAP) to < 50 mm Hg. In spite of a larger blood loss, animals receiving ICV hypertonic NaCl had an improved recovery of the blood pressure after haemorrhage. This was apparently due to a better maintained cardiac output (CO) concomitant with a delayed and attenuated increase in systemic vascular resistance (SVR) (Fig. 2). Elevation of the CSF [Na] also augmented the tachycardia in response to haemorrhage. In preliminary tests with ICV infusions of a more hypertonic (0.7 M) NaCl solution the animals could be bled 25 ml/kg b.w. without any change in CO and a small reduction (15 mm Hg) in MSAP. In contrast to the infusions with 0.5 M Nacl, these experiments caused some behavioural side effects (restlessness) and therefore were not continued. The haemodynamic effects of elevated CSF [Na] during blood loss led us to study whether iso- or hypo-osmotic lowering of the CSF [Na] would have the opposite effects (Ullman *et al.*, 1993a). This turned out not to be the case. Neither ICV infusion of 0.3 M mannitol (mean reduction CSF [Na] = 18 mM), or 0.04 M NaCl (mean reduction CSF [Na] and osmolality: 13 mM and 25 mOsm/kg, respectively) significantly reduced the volume of haemorrhage required to cause the predefined degree of hypotension (< 50 mm Hg) (Fig. 1.). Furthermore, the cardiovascular responses to during and after haemorrhage did not differ from control experiments.

The haemodynamic responses to increased CSF [Na] during and after haemorrhage are very similar to those observed in dogs (Velasco *et al.*, 1980) and sheep (Nakayama *et al.*, 1984) by intravenous administration of hypertonic NaCl solutions in haemorrhagic shock. The standard procedure for the latter treatment usually increases the plasma [Na] by 10-15 mM. Previous studies in our group (Rundgren *et al.*, 1990) have revealed that ICV infusion of 0.5 M NaCl at the present rate (0.02 ml/min) elevates the CSF [Na] by a similar magnitude (15-20 mM). Therefore, the present results may suggest that the beneficial cardiovascular effects of intravascular hypertonic saline in haemorrhagic shock are mediated via cerebral effects of the hypernatraemia. The observation that cerebral mechanisms involved in fluid balance control are simultaneously acessible to effects of increased extracellular [Na] on both sides of the blood brain barrier (Andersson *et al.*, 1984) lend some support for this idea. Recently, more direct evidence for cerebrally mediated effects of plasma Na in hypertonic saline resuscitation was obtained when lesioning of the anterior wall of the third ventricle was found to abolish the effects of that treatment in rats (Barbosa *et al.*, 1990).

Humoral responses and body fluid composition

The improved cardiovascular responses to haemorrhage during elevated CSF [Na] are not readily explained by the effects on the plasma concentrations of AVP, ANG II and noradrenaline (NA), although the latter was augmented. The normal responses of these parameters to an acute hypotensive haemorrhage in sheep are, largely maintained basal levels during the nonhypotensive stage of blood loss, followed by an exponential rise in plasma AVP concentration, a doubling of the ANG II levels, and still unchanged plasma NA concentration when the blood pressure suddenly falls (Hjelmqvist *et al.*, 1992b). Of possible importance for the increased tolerance to blood loss during ICV infusion of hypertonic NaCl is that in these experiments, the plasma AVP concentration increased significantly already during the nonhypotensive stage of haemorrhage. However, lowering of the CSF [Na] was accompanied by a reduced AVP response to haemorrhage, without a concomitant significant reduction in tolerance to blood loss (Ullman *et al.*, 1993a).

Somewhat surprisingly the ICV infusion of hypertonic NaCl *per se* was found to lower the plasma protein concentration, indicating transfer of fluid to the vascular compartment. Furthermore, the aututransfusion of interstitial fluid normally seen during haemorrhage (Mellander & Johansson, 1968) appeared to be reinforced by the elevated CSF [Na]. Iso- or hypo-osmotic lowering of the CSF [Na] did not have any effect on the plasma protein concentration.

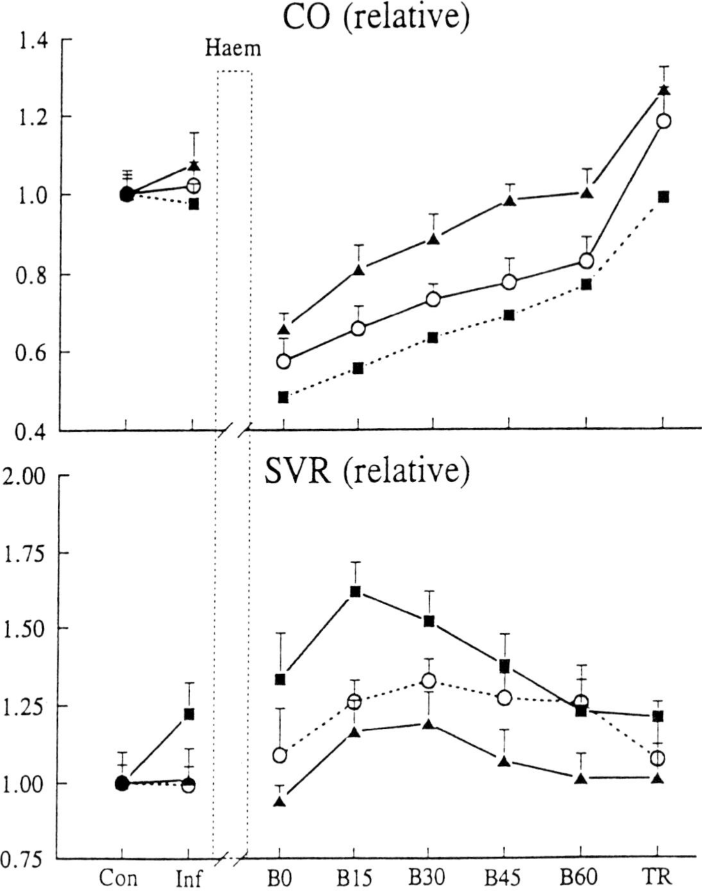

Fig. 2. A summary of data on the effects of ICV infusion of 0.5 M NaCl (filled triangles), respectively, ANG II (filled squares) on cardiac output (CO) and systemic vascular resistance (SVR) before and after a hypotensive haemorrhage (Hem). Open circles show control experiments where the animals were subjected only to haemorrhage or to haemorrhage during an ICV infusion of 0.9% NaCl.

EFFECTS OF ICV ANGIOTENSIN II

Tolerance to blood loss and haemodynamic effects

Similar to the effect of hypertonic NaCl also ICV infusion of ANG II (2pmol/kg/min) increased the tolerance to haemorrhage (Fig. 1) (Ullman *et al.*, 1993b). The haemodynamic responses to blood loss were quite different, however, with a reinforced increase in SVR and a slightly reduced CO (Fig. 2). Further, the recovery of the MSAP after haemorrhage was not affected by the central ANG II administration. The hypokinetic circulatory effect of centrally applied ANG II during haemorrhage agrees with two recent studies in sheep where the hypertensive effect of ICV infusion of the octapeptide was found to be mediated entirely via peripheral vasoconstriction (Breuhaus & Chimoskey, 1990; Hjelmqvist *et al.*, 1992). As mentioned above, the centrally elicited hypertensive effects of both increased [Na] and ANG II appears to be mediated via increased sympathetic nervous activity. The haemodynamic effects of ICV infusions of hypertonic

NaCl and ANG II during haemorrhage suggest that these two rather crude manipulations may initiate differentiated effects on the sympathetic outflow, with functionally dissimilar consequences.

Humoral responses and body fluid composition

The effects of ICV ANG II on the responses of vasoactive humoral agents were similar to those observed by increased CSF [Na]. Thus, the plasma AVP levels increased already during the nonhypotensive stage of haemorrhage, but the peak levels after hypotension did not differ from controls. Also here, the plasma NA response was reinforced.

As observed in a previous study (Hjelmqvist *et al.*, 1992a) the plasma protein concentration decreased in response to ICV ANG II *per se*. Like hypertonic NaCl, ANG II also augmented the haemodilution normally seen during and after blood loss. This possibly reinforced autotransfusion of fluid to the vascular compartment was not reflected by the haemodynamic situation with a relatively low CO. Regarding the mechanisms for this central effect of ANG II it is of interest to note that ICV ANG II has previously been observed to increase intestinal fluid absorption in rats (Brown & Gillespie 1988).

REGIONAL BLOOD FLOW

In further studies with the same haemorrhage and ICV infusion protocol we have measured the renal and femoral arterial blood flow via permanently implanted ultrasonic flow probes. Preliminary results of these investigations show that both the renal and femoral blood flow are rather well maintained during the nonhypotensive stage of blood loss. At the onset of hypotension, the femoral blood flow even transiently increased, whereas renal perfusion dropped by 60-70%, followed by a recovery which was slower than for the MSAP. ICV infusion of ANG II appears to be without effect on femoral and renal blood flow under basal conditions and the described changes in regional blood flow during and after haemorrhage. In contrast, elevation of the CSF [Na] increased pre-haemorrhage femoral blood flow and caused a slower but more persistent increase in association with the blood pressure fall. The renal blood flow was largely unaffected by the ICV infusion of hypertonic NaCl before haemorrhage, but the recovery after end of blood removal was clearly attenuated. These changes in regional blood flow before, during, and after haemorrhage adds to the previous suggestion that the cerebral effects of Na and ANG II on sympathetic nervous activity (or other vasoactive humoral agents) are differentiated.

CONCLUDING REMARKS

The results of the studies reviewed here show that elevated CSF concentration of both Na and ANG II improves the tolerance to blood loss, whereas lowering of the CSF [Na] has no apparent effect in this regard. The centrally elicited effects on haemodynamics and body fluid shifts of Na and ANG II *per se*, and in association with haemorrhage, indicate differentiated effects on the cardiovascular system, probably mediated via the sympathoadrenal system. The haemodynamic effects of increased CSF [Na] appear more favourable than those of ANG II in a hypovolaemic shock situation regarding cardiac load and tissue perfusion, although the latter may not be true for all vascular beds.

The studies reviewed were supported by the Swedish Medical Research Council (project no 6553), the Swedish Society of Medicine, the Laerdal Foundation, the Tornspiran Foundation, and the Medical Faculty at the Karolinska Institute. The studies were approved by the regional Ethical Committee for Animal Experiments.

REFERENCES

Andersson, B., Eriksson, L. & Fernández, O. (1971): Reinforcement by Na^+ of centrally mediated hypertensive response to angiotensin II. *Life Sci.* 10, 633-638.

Andersson, B., Leksell, L.G. & Rundgren, M. (1984): Regulation of body fluids: intake and output. In *Edema*, ed. N.C. Staub & A.E. Tayler, pp. 300-318. New York: Raven Press.

Barbosa, S.P., Saad, W.A., De Arrouda Gamargo, L.A., De Luca Jr, L.A., Fracasso, J.F. & Menani, J.W. (1990): Lesion of the anteroventral third ventricle region abolishes the beneficial effects of hypertonic saline on hemorrhagic shock in rats. *Brain Res.* 520, 342-344.

Baue, A.E., Tragus, E.T. & Parkins, W.M. (1967): A comparison of isotonic and hypertonic solutions and blood on blood flow and oxygen consumption in the initial treatment of hemorrhagic shock. *J. Trauma* 7, 743-756.

Bickerton, R.K. & Buckley, J.P. (1961): Evidence for a central mechanism in angiotensin-induced hypertension. *Proc. Soc. Exp. Biol. Med.* 106, 834-836.

Breuhaus, B.A. & Chimoskey, J.E. 1990. Hemodynamic and behavioral effects of angiotensin II in conscious sheep. *Am. J. Physiol.* 258, R1230-R1237.

Brown, D.R. & Gillespie, M.A. (1988): Actions of centrally administered neuropeptides on rat intestinal transport: enhancement of ileal absorption by angiotensin II. *Eur. J. Pharmachol.* 148, 411-418.

Buñag, R.D. & Miyajima, E. (1984): Sympathetic hyperactivity elevates blood pressure during acute cerebroventricular infusions of hypertonic saline in rats. *J. Cardiovasc. Pharmachol.* 6, 844-851.

DeFilippe Jr, J., Timoner, J., Velasco, I.T., Lopes, O.U. & Rocha E Silva Jr, M. (1980): Treatment of refractory hypovolaemic shock by 7.5 % sodium chloride injections. *Lancet* 2, 1002-1004.

Gross, D., Landau, E.H., Klin, B. & Krausz, M. (1989): Quantitative measurement of bleeding following hypertonic saline therapy in uncontrolled hemorrhagic shock. *J. Trauma* 29, 79-83.

Hands, R.D., Günther, R.A., Perron, P.R., Mertens, S., Holcroft, J.W. & Kramer, G.C. (1986): Periferal injection of hypertonic saline-dextran to resuscitate from hemorrhagic shock. *Circ. Shock* 18, 377-378.

Haglind, E. & Haljamäe, H. (1992): Failure of hypertonic saline to resuscitate intestinal ischemia shock in rat. *Acta Anaesthesiol. Scand.* 36, 410-418.

Hjelmqvist, H., Ullman, J., Hamberger, B. & Rundgren, M. (1992a): Cardiovascular and renal effects of intracerebroventricular angiotensin II in conscious sheep. *Acta. Physiol. Scand.* 145, 25-32.

Hjelmqvist, H., Ullman, J., Gunnarsson, U., Hamberger, B. & Rundgren, M. (1992b): Increased resistance to haemorrhage induced by intracerebroventricular infusion of hypertonic NaCl in conscious sheep. *Acta Physiol. Scand.* 145, 177-186.

Kreimeier, U. & Messmer, K. (1992): Future trends in emergency fluid resuscitation: small volume resuscitation by means of hypertonic saline dextran. *Int. Care World* 9, 16-20.

Lopes, O.U., Pontieri, V., Rocha E Silva Jr, M., Velasco, I.T. (1981): Hyperosmotic NaCl and severe hemorrhagic shock: role of the innervated lung. *Am. J. Physiol.* 241, H883-H890.

Mellander, S. & Johansson, B. (1968): Control of resistance, exchange and capacitance functions in the peripheral circulation. *Pharmacol. Rev.* 20, 117-196.

Nakayama, S., Sibley, L., Günther, R.A., Holcroft, J.W. & Kramer, G.C. (1984): Small-volume resuscitation with impaired cardiovascular reflexes due to idiopathic ortostatic hypotension. *J. Cardiovasc. Pharmacol.* 2, 367-376.

Rettig, R., Gerstberger, R., Meyer, J.-U., Intaglietta, M. & Printz, M. (1989): Central effects of angiotensin II in conscious hamsters: drinking, pressor response, and release of vasopressin. *J. Comp. Physiol. B.* 158, 703-709.

Rundgren, M., Jónasson, H. & Hjelmqvist, H. (1990): Water intake and changes in plasma and CSF composition in response to acute administration of hypertonic NaCl and water deprivation in sheep. *Acta. Physiol. Scand.* 138, 85-92.
Takata, Y., Yamashita, Y., Nakao, Y., Takishita, S. & Fujishima, M. (1989): Pressor response to intracisteral administration of hypertonic NaCl in conscious normotensive and spontaneously hypertensive rats. *Arch. Int. Pharmacodyn.* 299, 110-126.
Tobey, J.C., Fry, H.K., Mizejewski, C.S., Fink, G.D. & Weaver, L.C. (1983): Differential sympathetic responses initiated by angiotensin and sodium chloride. *Am. J. Physiol.* 245, R60-R68.
Ullman, J., Hjelmqvist, H. & Rundgren, M. (1993a): Effects of reduced CSF Na concentration and osmolality on haemodynamic and humoral responses to hypotensive haemorrhage in conscious sheep. *Acta. Physiol. Scand.* 148, 85-91
Ullman, J., Hjelmqvist, H. & Rundgren, M. (1993b): Improved tolerance to hemorrhage by intracerebroventricular angiotensin II in conscious sheep. Submitted to *Acta. Physiol. Scand.*
Velasco, I.T., Pontieri, V., Rocha E Silva Jr, M., Lopes, O.U. (1980): Hyperosmotic NaCl and severe hemorrhagic shock. *Am. J. Physiol.* 239, H664-H673.
Velasco, I.T., Baena, R.C., Rocha E Silva, M. & Loureiro, M.I. (1990): Central angiotensinergic system and hypertonic resuscitation from severe hemorrhage. *Am. J. Physiol.* 259, H1752-H1758.

Integrative and cellular aspects of autonomic functions : temperature and osmoregulation. Eds K. Pleschka, R. Gerstberger. John Libbey Eurotext, Paris © 1994, pp. 397-405.

The role of arterial pressure and arterial baroreceptors in the modulation of the drinking response to centrally-administered angiotensin II

Robert L. Thunhorst and Alan Kim Johnson

Departments of Psychology and Pharmacology and the Cardiovascular Center, University of Iowa, Iowa City, Iowa 52242, USA

Drinking in response to deficits of extracellular fluid is under the control of several physiological mechanisms including the hormone angiotensin II (ANG II) and neural input from cardiopulmonary and arterial baroreceptors. Several lines of evidence suggest that these hormonal and neural mechanisms may act independently to produce thirst (Hosutt *et al*, 1978; Hosutt & Stricker, 1981; Klingbeil *et al*, 1991; Rettig *et al*, 1981; Stricker, 1973). However, under physiological conditions, it is likely that these different sensory mechanisms interact. Such interactions compound the difficulty in establishing their relative roles in controlling water intake (Evered, 1992). For example, intravenously (iv) administered ANG II is only weakly dipsogenic (Van Eekelen & Phillips, 1988), but it stimulates more drinking in rats when its potent pressor effects are buffered by the use of hypotensive agents (Robinson & Evered, 1987) and when arterial pressure is normalized or reduced below resting levels during ANG II infusion (Evered *et al*, 1988). In dogs, the damping effect of arterial pressure on the drinking response to iv infusion of ANG II is accentuated by ganglionic blockade (Kucharczyk, 1988), and diminished by removal of neural input from both cardiopulmonary and arterial baroreceptors (Klingbeil *et al*, 1991). These findings indicate that the strategy of infusing ANG II iv into fluid replete animals to establish its physiological role in drinking is wholly inadequate because the concomitant rise in arterial pressure creates pathophysiological conditions (Evered, 1992; Van Eekelen & Phillips, 1988). Under physiological conditions when circulating levels of ANG II are elevated, blood pressure is in the normo- to hypotensive range.

Most of the work exploring interactions between arterial pressure and ANG II in the control of drinking has focused on the role of circulating ANG II. However, ANG II also produces drinking when it is administered intracerebroventricularly (icv)(Phillips, 1987), and central renin-angiotensin mechanisms have been proposed to be engaged during systemic hypotension and/or hypovolemia (Johnson & Edwards, 1991).

We explored interactions between arterial pressure and centrally-administered ANG II on drinking in rats. This was accomplished by infusing ANG II into the cerebral ventricles under conditions when arterial pressure was at resting levels, and when arterial pressure was decreased or increased from resting levels, respectively, by iv infusions of vasodilator or vasoconstrictor substances.

We first established that reductions in arterial blood pressure increased the drinking response to centrally-administered ANG II (Thunhorst & Johnson, 1993). Groups of rats received infusions of relatively low doses of ANG II into a lateral cerebral ventricle. One group received a dose of 4 ng/hr and the other a dose of 16 ng/hr. The groups received these central infusions twice, once when arterial pressure was at resting levels during the iv infusion of the vehicle, and once when arterial pressure was reduced below resting levels during the iv infusion of the vasodilator substance, minoxidil (25 mg/kg/min). The vehicle included a high dose of the angiotensin-converting enzyme (ACE) inhibitor, captopril (0.33 mg/min), which by itself does not affect arterial blood pressure in sodium replete animals, to prevent the endogenous formation of ANG II during the experiments.

Infusions of the mixture of minoxidil and captopril (MINOXCAP) slowly and progressively produced almost identical reductions in mean arterial pressure (MAP) for the two groups. The icv infusion of ANG II at resting levels of MAP produced dose-dependent water intake (Fig. 1). The dipsogenic response doubled when ANG II was infused icv during hypotensive conditions. Additionally, urine volume and electrolyte excretion were greatly reduced during these hypotensive states, so that the increased ingestion occurred in the face of increased retention of fluid and electrolytes.

These experiments demonstrate that the dipsogenic response to centrally-administered ANG II is increased when MAP is reduced below resting levels. The increased drinking observed in response to icv infusion of ANG II during hypotensive states probably did not result from additional, endogenous formation of either circulating or central ANG II during the hypotensive state because the high dose of captopril used in the vehicle is sufficient to block ACE centrally as well as peripherally (Robinson & Evered, 1987; Thunhorst & Johnson, 1993). Furthermore, the increased drinking in response to icv infusion of ANG II during hypotensive states probably did not reflect dipsogenic properties of the hypotensive treatment itself. Control experiments showed that the small amount of water ingested in response to an iv infusion of minoxidil, by itself, is completely eliminated by the addition of captopril to the infusate (MINOXCAP treatment), even though greater hypotension results. Lastly, the increased drinking to icv infusion of ANG II during hypotensive states probably was not due to buffering of a centrally-mediated pressor response to the icv ANG II. The relatively low doses of ANG II administered icv in these studies failed to produce the large, sustained increases in arterial pressure that have been demonstrated to inhibit drinking to ANG II (Evered *et al*, 1988; Harland *et al*, 1988; Robinson & Evered, 1987).

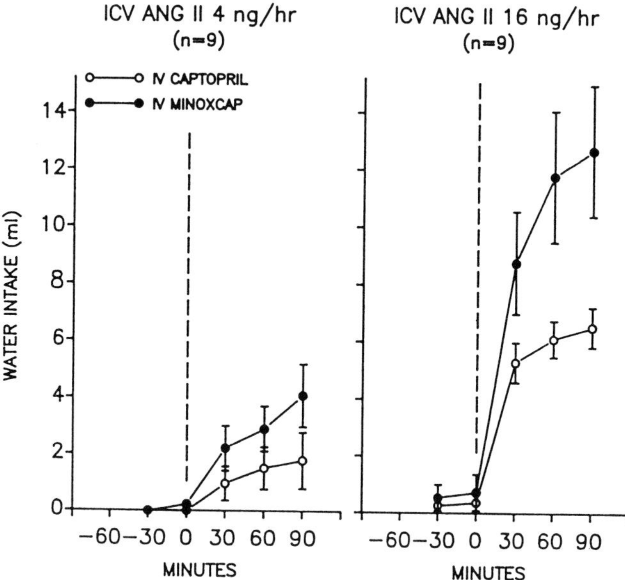

Fig. 1. At each dose of icv ANG II, water intakes were greater when MAP was reduced by iv MINOXCAP, compared to when MAP was normal during iv captopril vehicle. (from Thunhorst and Johnson, 1993.)

The results of the first series of experiments indicates that arterial pressure interacts with ANG II to stimulate drinking. In the absence of icv ANG II, hypotension did not produce drinking. In the presence of icv ANG II, hypotension increased drinking to both doses of ANG II. In other words, hypotension potentiates drinking to centrally-administered ANG II. For example, arterial pressure may influence the gain of the central angiotensin mechanism controlling drinking behavior.

The enhanced dipsogenic response to icv ANG II during hypotension could reflect either the removal of inhibitory influences or the addition of facilitory influences mediated through baroreceptor mechanisms. Both sino-aortic and cardiopulmonary baroreceptors are implicated in the control of drinking (Fitzsimons & Moore-Gillon, 1980; Quillen *et al*, 1988; Zimmerman *et al*, 1981). There is mounting evidence both that increased arterial pressure (Evered *et al*, 1988; Robinson & Evered, 1987; Werber & Fink, 1981) and atrial stretch (Kaufman, 1984; Moore-Gillon & Fitzsimons, 1982) inhibit water intake, and that reduced arterial (Hosutt *et al*, 1978; Hosutt & Stricker, 1981; Pawloski & Fink, 1990; Rettig *et al*, 1981) and atrial (Thrasher *et al*, 1982) pressure reflexly facilitate water intake. Since the present experiments employed manipulations of arterial pressure, and the sino-aortic baroreceptors are the principle monitors of arterial pressure, it seems reasonable to propose that the sino-aortic baroreceptors mediate the enhanced dipsogenic response to centrally-administered ANG II (Thunhorst & Johnson, 1993).

The next series of experiments were designed to determine if disruption of the flow of arterial pressure information to the central nervous system via surgical denervation of the sino-aortic baroreceptors (SAD) affected drinking in response to the icv administration of ANG II at different levels of arterial pressure (Thunhorst *et al*, 1993). Arterial pressure was manipulated by the iv administration of vasoconstrictor or vasodilator substances in the presence of large doses of captopril to ensure that no endogenous formation of ANG II occurred during testing.

Arterial baroreceptor denervation was performed as described by Krieger (1964). Rats were allowed approximately 4-5 wks recovery before receiving icv cannulas. Mean arterial pressure and water intake were measured in the same animals in a series of three tests. Rats were tested with icv ANG II under conditions when MAP was normal, increased, or decreased. The iv test infusates consisted of the vehicle solution of captopril, the combination of minoxidil plus captopril (MINOXCAP) to reduce blood pressure, or a combination of phenylephrine (1 or 10 µg/kg/min) plus captopril (PHENCAP) to increase blood pressure. Central infusions of ANG II at 16 ng/hr began 60 min later and ran concurrently with the iv infusions for another 90 min.

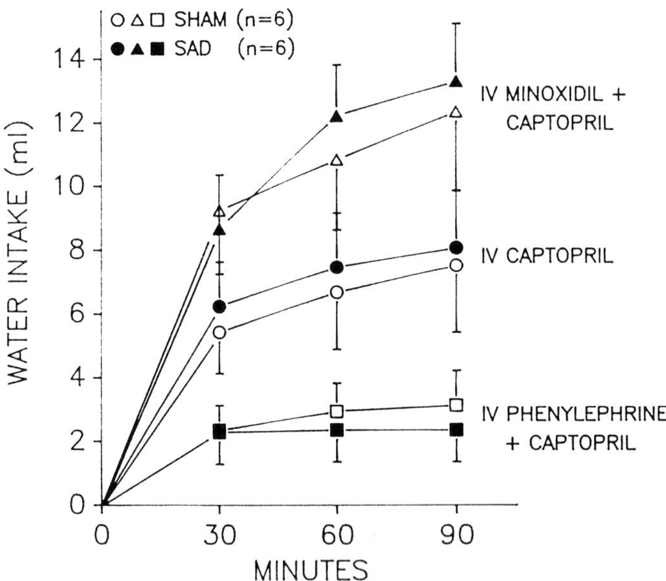

Fig. 2. Effects of iv treatments on water intake in response to icv ANG II (16 ng/hr) in intact (SHAM) and sino-aortic denervated (SAD) rats. (from Thunhorst, Lewis & Johnson, 1993.)

The iv infusions of PHENCAP and MINOXCAP produced exaggerated changes in MAP in SAD rats compared to sham-denervated controls. Rats drank significantly more water when MAP was reduced by iv infusions of MINOXCAP, and significantly less water when MAP was elevated by iv infusions of PHENCAP, in comparison to water intake during normotensive, iv infusions of captopril (Fig. 2). Furthermore, barodenervation, which was functionally demonstrated to abolish the cardiac baroreflex, had no effect on water intake in response to icv ANG II. In other words, water intake was inversely related to MAP in both groups of rats. The results of regression analysis relating water intake to level of MAP showed no differences between controls and SAD rats (Fig. 3). Both lines had slopes that were significantly different from zero, and the slopes were not significantly different from each other (regression equation of the line for intact rats is: $y = -0.086\ x + 5.45$, and for SAD rats is: $y = -0.051\ x + 4.66$).

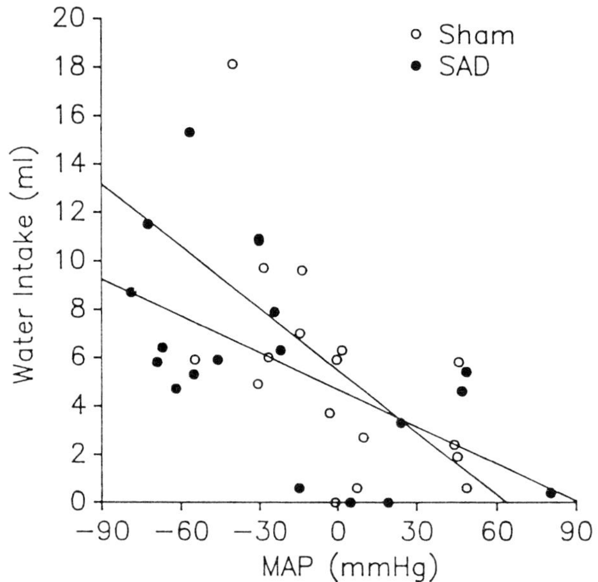

Fig. 3. Water intake in the first 30 min of icv ANG II (16 ng/hr) in relation to changes in MAP from baseline. Each rat contributed three points, one from each of three iv conditions.

The effects of MAP on water balance paralleled those for water intake, and were not different between the groups. Measures of hematocrit and plasma protein concentrations indicated that plasma volume of the groups was increased equivalently (5-6%) during hypotension and decreased equivalently (5-16%) during hypertension compared to normotensive conditions.

The finding that drinking in response to centrally-administered ANG II was still significantly affected by MAP in SAD, as well as sham-denervated rats, indicates either that some mechanism other than afferents from sino-aortic baroreceptors signals the brain of alterations in MAP during icv ANG II, or that the role of arterial baroreceptor responses is somehow masked by compensatory mechanisms in the absence of the baroreceptor input.

As noted above, the iv treatments produced greater effects on MAP in SAD rats compared to sham-denervated rats. Therefore, one could argue that SAD rats were refractory to the modulatory effects of MAP precisely because their water intakes were not similarly affected to a greater degree compared to sham-denervated rats. However, if the carotid sinus and aortic arch baroreceptors were solely responsible for mediating the effects of arterial pressure on water intake in response to icv ANG II, then water intake of SAD rats should not have changed *at all* in response to changes in arterial pressure. Thus, even if SAD "blunts" the modulatory effect of arterial pressure on drinking in response to icv infusions of ANG II, other mechanisms must remain that are operating in the absence of the arterial baroreceptors. Klingbeil *et al* (1991) found it necessary to eliminate both the sino-aortic and cardiopulmonary baroreceptors in dog to prevent the pressor response to iv infusion of ANG II from inhibiting the dipsogenic response. Thus, it may be necessary to remove both sets of baroreceptors to prevent the modulatory effects of MAP on drinking to centrally-administered ANG II.

There are additional beds of baroreceptors within the systemic circulation that potentially could supply the central nervous system with information concerning arterial blood pressure relevant for drinking in the absence of functional sino-aortic baroreceptors. Systemic baroreceptors have been identified, or postulated to reside, within mesenteric (Tuttle & McCleary, 1975), hepatic (Sawchenko & Friedman, 1979), pancreatic (Sarnoff & Yamada, 1959), and renal (Stella & Zanchetti, 1991) vessels. Evidence in favor of renal mechanisms are the findings that renal denervation in rats diminishes both thirst (Sharpe *et al*, 1978) and salt appetite (Thunhorst *et al*, 1992) in response to hypovolemic stimuli. Additionally, the contribution of vagally-mediated reflexes (i.e., from cardiopulmonary baroreceptors) to the control of arterial pressure is increased in the absence of functional arterial baroreflexes, at least in dog (Bishop & Barron, 1980; Raymundo *et al*, 1989). In rat, it has been reported that removal of cardiopulmonary baroreceptor information to the central nervous system via cervical vagotomy affects water intake in experimental procedures that alter blood pressure or volume (Moore-Gillon, 1980; Zimmer *et al*, 1976), although this is controversial (Martin, 1981; Vance, 1970). Further studies will be necessary to determine if other putative vascular baroreceptors provide sufficient input to allow SAD rats to respond to changes in arterial pressure with apparently normal ingestive responses to centrally-administered ANG II.

In summary, water intake in response to icv infusions of ANG II was inversely related to the level of arterial pressure achieved during normotensive, hypotensive and hypertensive treatments. Furthermore, elimination of the sino-aortic baroreceptors, the principle receptors that apprise the central nervous system of arterial pressure, did not markedly alter the modification of water intake by changes in arterial pressure. These studies suggest that other afferent mechanisms mediate the effects of arterial blood pressure on water intake induced by central ANG II in the absence of functional sino-aortic baroreceptors.

ACKNOWLEDGMENTS

We thank Terry Beltz, Brenda Coleman, Jill Michka, and Kim Withrow for their expert technical assistance. We also thank the Squibb Research Institute for the generous donation of captopril, and Ciba-Geigy for the generous donation of angiotensin.

This research was supported by National Heart, Lung, and Blood Institute Grants HL-14388 and HL-44546.

REFERENCES

Bishop, V.S. & Barron, K. (1980): The contribution of vagal afferents in the regulation of the circulation in conscious dogs. In *Arterial Baroreceptors and Hypertension*, ed. P. Sleight, pp. 91-97. Oxford:Oxford University Press.

Evered, M.D. (1992): Investigating the role of angiotensin II in thirst: Interactions between arterial pressure and the control of drinking. *Can. J. Physiol. Pharmacol.* 70, 791-797.

Evered, M.D., Robinson, M.M. & Rose, P.A. (1988): Effect of arterial pressure on drinking and urinary responses to angiotensin II. *Am. J. Physiol.* 254, R67-R74.

Fitzsimons, J.T. & Moore-Gillon, M.J. (1980): Drinking and antidiuresis in response to reductions in venous return in the dog: neural and endocrine mechanisms. *J. Physiol. (Lond.)* 308, 403-416.

Harland, D., Gardiner, S.M. & Bennett, T. (1988): Cardiovascular and dipsogenic effects of angiotensin II administered i.c.v. in Long-Evans and Brattleboro rats. *Brain Res.* 455, 58-64.

Hosutt, J.A., Rowland, N. & Stricker, E.M. (1978): Hypotension and thirst in rats after isoproterenol treatment. *Physiol. Behav.* 21, 593-598.

Hosutt, J.A. & Stricker, E.M. (1981): Hypotension and thirst in rats after phentolamine treatment. *Physiol. Behav.* 27, 463-468.

Johnson, A.K. & Edwards, G.E. (1991): Central projections of osmotic and hypovolaemic signals in homeostatic thirst. In *Thirst*, ed. D.J. Ramsay & D.A. Booth, pp. 149-175. London:Springer-Verlag Press.

Kaufman, S. (1984): Role of right atrial receptors in the control of drinking in the rat. *J. Physiol. (Lond.)* 349, 389-396.

Klingbeil, C.K., Brooks, V.L., Quillen, E.W., Jr., & Reid, I.A. (1991): Effect of baroreceptor denervation on stimulation of drinking by angiotensin II in conscious dogs. *Am. J. Physiol.* 260, E333-E337.

Krieger, E.M. (1964): Neurogenic hypertension in the rat. *Circ. Res.* 15, 511-521.

Kucharczyk, J. (1988): Inhibition of angiotensin-induced water intake following hexamethonium pretreatment in the dog. *Eur. J. Pharmacol.* 148, 213-219.

Martin, J.R. (1981): Effects of partial and complete vagal denervation on spontaneous ingestion and drinking induced with volemic and osmotic regulatory challenges. *J. Neurosci. Res.* 6, 243-250.

Moore-Gillon, M.J. (1980): Effects of vagotomy on drinking in the rat. *J. Physiol. (Lond.)* 308, 417-426.

Moore-Gillon, M.J. & Fitzsimons, J.T. (1982): Pulmonary vein-atrial junction stretch receptors and the inhibition of drinking. *Am. J. Physiol.* 242, R452-R457.
Pawloski, C. M. & Fink, G.D. (1990): Circulating angiotensin II and drinking behavior in rats. *Am. J. Physiol.* 259, R531-R538.
Phillips, M.I. (1987): Functions of angiotensin in the central nervous system. *Ann. Rev. Physiol.* 49, 413-435.
Quillen, E.W., Jr., Reid, I.A. & Keil, L.C. (1988): Cardiac and arterial baroreceptor influences on plasma vasopressin and drinking. In *Vasopressin: Cellular and Integrative Functions*, ed. A.W. Cowley, J.-F. Liard & D.A. Ausiello, pp. 405-411. New York:Raven Press.
Raymundo, H., Scher, A.M., O'Leary, D.S. & Sampson, P.D. (1989): Cardiovascular control by arterial and cardiopulmonary baroreceptors in awake dogs with atrioventricular block. *Am. J. Physiol.* 257, H2048-H2058.
Rettig, R., Ganten, D. & Johnson, A.K. (1981): Isoproterenol-induced thirst: renal and extrarenal mechanisms. *Am. J. Physiol.* 241, R152-R157.
Robinson, M.M. & Evered, M.D. (1987): Pressor action of intravenous angiotensin II reduces drinking response in rats. *Am. J. Physiol.* 252, R754-R759.
Sarnoff, S.J. & Yamada, S.I. (1959): Evidence for reflex control of arterial pressure from abdominal receptors with special reference to the pancreas. *Circ. Res.* VII, 325-335.
Sawchenko, P.E. & Friedman, M.I. (1979): Sensory functions of the liver - a review. *Am. J. Physiol.* 236, R5-R20.
Sharpe, M.D., Mogenson, G.J. & Calaresu, F.R. (1978): The role of renal nerves in the response to dipsogenic stimuli in the rat. *Can. J. Physiol. Pharmacol.* 56, 731-734.
Stella, A. & Zanchetti, A. (1991): Functional role of renal afferents. *Physiol. Rev.* 71, 659-682.
Stricker, E.M. (1973): Thirst, sodium appetite, and complementary physiological contributions to the regulation of intravascular fluid volume. In *The Neuropsychology of Thirst*, ed. A. Epstein, H.R. Kissileff & E. Stellar, pp. 73-98. Washington, D.C.:Winston.
Thrasher, T.N., Keil, L.C. & Ramsay, D.J. (1982): Hemodynamic, hormonal, and drinking responses to reduced venous return in the dog. *Am. J. Physiol.* 243, R354-R362.
Thunhorst, R.L. & Johnson, A.K. (1993): Effects of arterial pressure on drinking and urinary responses to intracerebroventricular angiotensin II. *Am. J. Physiol.* 264, R211-R217.
Thunhorst, R.L., Lewis, S.J. & Johnson, A.K. (1993): Role of arterial baroreceptor input on thirst and urinary responses to intracerebroventricular angiotensin II. *Am. J. Physiol.* in press.
Thunhorst, R.L., Kirby, R.F. & Johnson, A.K. (1992): Effects of renal denervation of salt appetite during hypovolemia. *FASEB J.* 6, A1837.
Tuttle, R.S., & McCleary, M. (1975): Mesenteric baroreceptors. *Am. J. Physiol.* 229, 1514-1519.
Van Eekelen, J.A.M. & Phillips, M.I. (1988): Plasma angiotensin II levels at moment of drinking during angiotensin II intravenous administration. *Am. J. Physiol.* 255, R500-R506.

Vance, W.B. (1970): The effects of vagotomy on the water intake of the white rat. *Psychon. Sci.* 20, 21-22.

Werber, A.H. & G.D. Fink (1981): Cardiovascular and body fluid changes after aortic baroreceptor deafferentation. *Am. J. Physiol.* 240, H685-H690.

Zimmer, L.J., Meliza, L. & Hsiao, S. (1976): Effects of cervical and subdiaphragmatic vagotomy on osmotic and volemic thirst. *Physiol. Behav.* 16, 665-670.

Zimmerman, M.B., Blaine, E.H. & Stricker, E.M. (1981): Water intake in hypovolemic sheep: effects of crushing the left atrial appendage. *Science Wash. DC* 211, 489-491.

Integrative and cellular aspects of autonomic functions : temperature and osmoregulation. Eds K. Pleschka, R. Gerstberger. John Libbey Eurotext, Paris © 1994, pp. 407-418.

Cerebrospinal fluid-contacting neurons : morphological and physiological considerations

H.-W. Korf[(1)], E. Rommel[(2)], K. Hirunagi[(1)] and A. Oksche[(2)]

[(1)] *Center of Morphology, Section on Neurobiology, Johann Wolfgang Goethe-University, Frankfurt/Main, Germany*
[(2)] *Department of Anatomy and Cytobiology, Justus-Liebig-University, Giessen, Germany*

INTRODUCTION

A comprehensive synopsis on intraependymal neurons, which are related to the ventricular system and whose perikarya are located within or close to the ependyma has been provided by E. Agduhr as early as 1922. By means of classical neurohistological techniques, e.g., silver impregnation methods, Agduhr was able to show that 1) these neurons are widespread among all vertebrate classes including mammals and 2) most of them are endowed with an adventricular process. Although functional analysis was not the central aim of his study, the morphological results led Agduhr to suggest that the intraependymal neurons might represent an "ependymal sense organ" ("Ependymsinnesorgan") which, like the cutaneous sense organ, may serve multiple receptive functions.

These early findings (see also Tretjakoff, 1909, 1913; Kolmer, 1921) have been amply confirmed and extended with the use of modern morphological techniques such as electron microscopy and immunocytochemistry (Vigh and Vigh-Teichmann, 1973; Vigh-Teichmann and Vigh, 1974, 1983). Following the nomenclature of Vigh et al. (1969) the intraependymal neurons described by Agduhr and Kolmer are now generally called cerebrospinal fluid-contacting neurons ("Liquorkontaktneurone"). Extensive comparative studies have established that such neurons from aggregates along the walls of the third and fourth ventricles and the central canal of the spinal cord. Accordingly, "hypothalamic CSF-contacting neuronal areas" were distinguished from the "medullospinal CSF-contacting neuronal system" (Vigh and Vigh-Teichmann, 1973). Whereas the medullospinal CSF-contacting neuronal system is well developed in all vertebrate classes including mammals, distinct hypothalamic CSF-contacting areas, e.g., the paraventricular, tuberal and recessus praeopticus organs, are found only in submammalian species, but not in adult mammals. As a consequence, CSF-contacting dendrites are reduced in the mammalian hypothalamus; if present, they originate from perikarya rather remote from the surface of the third ventricle. These findings led to the assumption that, during phylogeny, CSF-contacting neurons are stepwise shifted from their original ependymal or subependymal location toward deeper layers of the brain (Vigh and Vigh-Teichmann, 1973; see also Oksche, 1976).

Most of the cerebrospinal fluid-contacting neurons give rise to a single adventricular process which, according to ultrastructural criteria, can be classified as a dendrite (Vigh and Vigh-

Teichmann, 1973). The CSF-contacting dendrite often displays a bulbous swelling protruding into the ventricular lumen. The ultrastructural features of these terminal swellings differ between the hypothalamic and the medullospinal CSF-contacting neurons. Ventricular terminals of the former are endowed with a cilium of the 9x2+0 type, whereas ventricular terminals of the medullospinal CSF-contacting neurons bear a kinocilium (9x2+2-type) and several stereocilia. The axons of the CSF-contacting neurons originate from the basal pole of the perikaryon or a main basal dendrite and extend toward deeper brain regions.

These structural and ultrastructural features support the view that CSF-contacting neurons serve receptive functions and convey external stimuli or internal signals related to the composition, pressure or flow of the CSF to the neuronal and/or neuroendocrine apparatus of the brain (Vigh-Teichmann and Vigh, 1983). Moreover, several investigators have emphasized that CSF-contacting neurons resemble secretory elements and may be regarded as the source of neuropeptides and other neurotransmitters present in the CSF (Leonhardt, 1980; Rodríguez et al., 1982).

This contribution will focus on two distinct sets of CSF-contacting neurons which occur in (i) the paraventricular nucleus of the hypothalamus (PVN) and (ii) the nucleus accumbens/lateral septum (NAc/LS) of sauropsids and which appear as interesting candidates for correlated functional and structural analyses.

CSF-CONTACTING NEURONS IN THE AVIAN PARAVENTRICULAR NUCLEUS

The avian paraventricular nucleus (PVN) can be regarded as an integrative hypothalamic center involved in the regulation of various autonomic mechanisms (cf. Abel et al., 1975; McNeill et al., 1975; Korf, 1984). One of the best investigated functions controlled by the paraventricular nucleus is the regulation of renal water excretion (Deutsch and Simon, 1980; Simon-Oppermann et al., 1980; Simon, 1982). Inter alia, this depends on the production of arginine-vasotocin (AVT), the antidiuretic principle of birds, and its release into the general circulation. As shown by immunocytochemical studies (Goossens et al., 1977; Bons, 1980; Panzica et al., 1986), AVT immunoreaction is found in a distinct population of magnocellular nerve cells of the supraoptic nucleus and in numerous neurons of the PVN belonging to different size classes (magnocellular and parvocellular elements). AVT-immunoreactive efferent projections of the PVN extend into the posterior lobe of the pituitary, the external zone of the median eminence and several other extrahypothalamic brain regions (cf. Korf, 1984).

Physiological experiments with the Pekin duck (Deutsch and Simon, 1980; Simon-Oppermann and Simon, 1982) have shown that cerebral osmoreceptors participate in controlling the production of AVT and its release from the axon terminals in the neurohypophysis into the general circulation. These studies have suggested that the cerebral osmoreceptors in birds like those in mammals might be periventricular nerve cells monitoring the sodium concentration of the CSF in the third ventricle (Simon, 1982; for mammals, see Andersson, 1977; Andersson and Olsson, 1977). In order to localize these cells and to characterize their structural features, a series of investigations was performed finally resulting in the demonstration of CSF-contacting neurons which are located in the periventricular layer of the avian PVN and project to the magnocellular (AVT-producing) neurons (Korf et al., 1982; 1983; Korf, 1984; Panzica et al., 1986).

The first piece of evidence for the existence of this cell type was obtained from retrograde tracer experiments with the use of horseradish peroxidase (HRP). The location of the PVN was identified prior to the tracer application by observing antidiuretic reactions to electrostimulations of the rostral hypothalamus of conscious, hydrated Pekin ducks. These

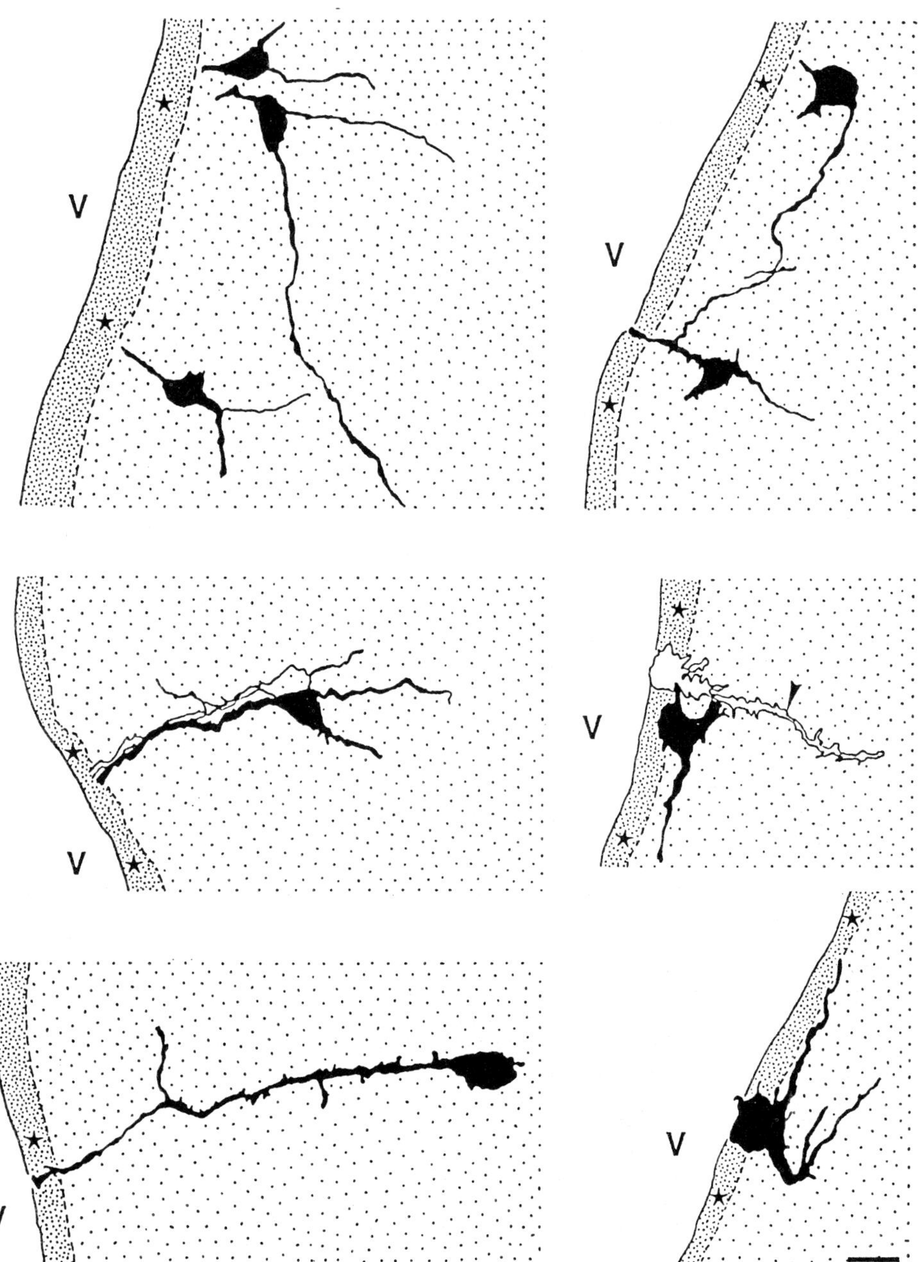

Fig. 1. Camera lucida drawings of Golgi-impregnated intraependymal nerve cells and CSF-contacting neurons in the dorsal portion of the paraventricular nucleus of the Pekin duck. Cross sections. V third ventricle. Stars ependymal layer. Arrowhead impregnated ependymocyte with process. Bar = 20 μm.

responses could be elicited only in circumscribed areas of the rostral hypothalamus. HRP was then injected into such areas whose stimulation had elicited the most pronounced antidiuresis. This procedure resulted in a very delicate injection site restricted to the periventricular portion of the PVN. Among other retrogradely labeled afferents (see Korf et al., 1982; Korf, 1984), the label was observed in small neurons located in the ependymal layer covering the dorsal aspect of the PVN. Unfortunately, the HRP preparations did not allow to analyze the structure of these neurons in greater detail. This, however, seemed to be of importance since the retrogradely labeled neurons would basically satisfy the postulates concerning cerebral osmoreceptors: 1) They establish afferent connections to magnocellular neuroendocrine elements producing AVT. 2) They occupy a position favoring the perception of the composition of the CSF.

In an attempt to further characterize the intraependymal neurons of the avian PVN, the Golgi-impregnation method was applied to the PVN of the duck (Korf et al., 1983). By means of this technique, several intraependymal neurons of the PVN were shown to give rise to dendrites extending into the ventricle (Fig. 1). Moreover, an additional population of parvocellular nerve cells was found in deeper layers of the periventricular PVN, which also gain direct access to the cerebrospinal fluid by means of long dendrites terminating with a bulbous-like swelling in the third ventricle (Fig. 1). The axons of these cells were followed into the magnocellular portion of the PVN. The majority of the magnocellular neurons of the PVN showed an isodendritic arrangement. Their dendrites were endowed with a varying number of spines. Interestingly, several dendrites of the magnocellular neurons reached the ependymal layer where they sometimes branched immediately beneath the ependymal cells or appeared to enter the ependymal layer. Very similar findings were obtained in the PVN of chicken, Japanese quail and house sparrow (Panzica et al., 1986). Immunocytochemical investigations revealed that some of the CSF-contacting neurons are immunoreactive for vasotocin and neurophysin; however, no immunoreactions for corticotropin-releasing factor (CRF), serotonin, vasoactive intestinal peptide (VIP) or angiotensin II were observed in these elements.

The morphological results reviewed suggest a complex arrangement of the pathways facilitating the transmission of signals from the ventricular compartment to the vasotocin-secreting magnocellular neurons. 1) The magnocellular neurons themselves may directly receive signals from the CSF via their dendrites extending into the ependymal regions. Although these dendrites do not penetrate the ependyma, they might be exposed to the CSF-dominated milieu, because the ependymal cells are connected to each other via gap junctions and, thus, do not form a barrier for cations or neuropeptides. These findings raise the possibility that the magnocellular neurons of the PVN represent, in functional terms, CSF-contacting neurons. Since the axons of these cells project to the neural lobe of the hypophysis or the external layer of the median eminence lacking a blood-brain barrier, it would appear that one pole (the subependymal dendrite) of such magnocellular neurons is exposed to the CSF and the other (the axon) to the hemal milieu. 2) The small intra- and subependymal cerebrospinal fluid-contacting neurons may convey signals from the CSF-compartment to the AVT-producing cells via synaptic mechanisms. Recent physiological studies have provided evidence that the CSF-contacting neurons located in the dorsal aspect of the PVN respond to changes in the sodium concentration of the CSF and, thus, are indeed involved in cerebral osmoreception (Kanosue et al., 1990; 1992). Electrophysiological recordings from these neurons in slice-preparations and subsequent intracellular application of tracers such as Lucifer yellow or horseradish peroxidase offer a promising perspective for a more precise correlation of structural and functional parameters of these CSF-contacting neurons.

CSF-CONTACTING NEURONS IN THE LATERAL SEPTUM/NUCLEUS ACCUMBENS OF BIRDS AND REPTILES

Although Agduhr (1922) had already mentioned the existence of CSF-contacting neurons in telencephalic areas, the "telencephalic arrays of CSF-contacting neurons" received little attention until this cell type was demonstrated in the nucleus accumbens/lateral septum (NAc/LS) of the Japanese quail (Yamada et al., 1982) and the Pekin duck (Korf and Fahrenkrug, 1984) by means of immunocytochemical demonstration of vasoactive intestinal peptide (VIP). In the duck, these neurons formed bilateral accumulations; their perikarya were located in the ependymal and subependymal layers of the NAc/LS (Fig. 2). The neurons gave rise to a single CSF-contacting dendrite displaying a bulb-like intraventricular terminal (Fig. 2). Their axons were of beaded appearance and originated from the basal aspect of the perikaryon or a basal dendrite. They penetrated deeply into the nucleus accumbens and the lateral septum. In these areas VIP-immunoreactive processes of beaded appearance were frequently found to encompass VIP-immunonegative perikarya

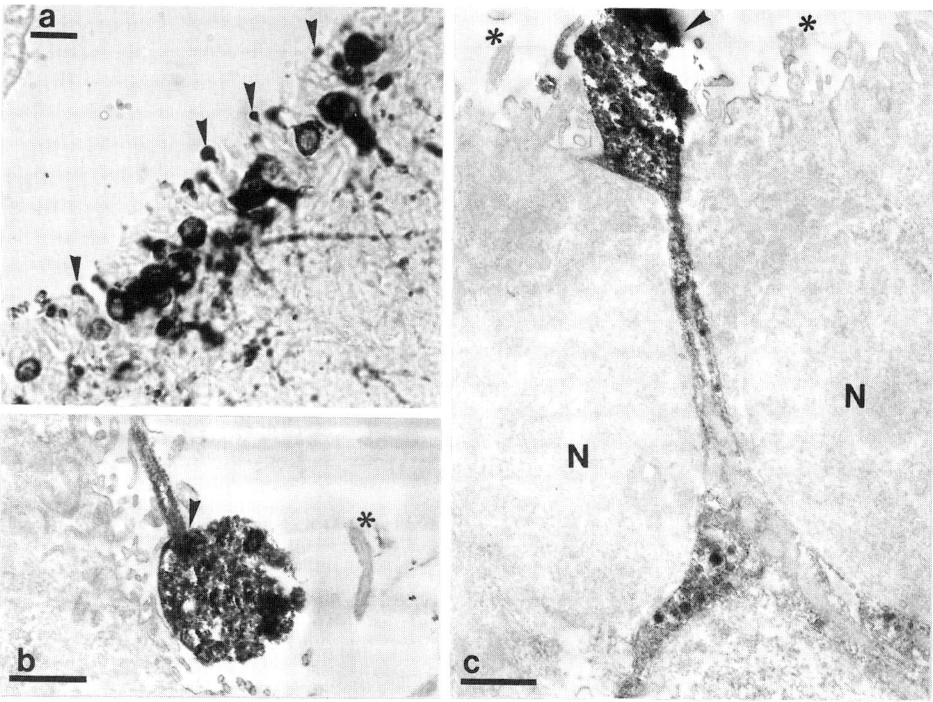

Fig. 2. VIP-immunoreactive CSF-contacting neurons in the nucleus accumbens/lateral septum of the Pekin duck. a Light microscopic demonstration of VIP. Several immunoreactive neurons give rise to ventricular dendrites (arrowheads). Cross section. Bar = 20 μm. b, c Electron microscopic demonstration of VIP. Preembedding method. The ventricular terminal of the CSF-contacting dendrite displays an intense immunoreaction and is apparently endowed with a cilium (arrowhead). Note unlabeled kinocilia (asterisks) and nuclei (N) of ependymocytes. Bar = 1μm

in a basket-like pattern resembling terminal fields. Similar aggregations of VIP-immunoreactive CSF-contacting neurons were found in the NAc/LS of the ring dove (Silver et al., 1988) and the pigeon (Hof et al., 1991).

The region inhabited by the VIP-immunoreactive CSF-contacting neurons also displayed characteristic ependymal and vascular specializations. In the duck, the ependymal layer was extremely folded and occasionally formed deep invaginations resembling ependymal canaliculi. At the tip of the folds, the ependymocytes were rather flat whereas the bottom of the folds was mostly covered by columnar cells. Such folds were lacking in the NAc/LS of the chicken. In this species the ventricular wall of the NAc/LS comprised multiple layers of ependymocytes, most of which were columnar cells (Kuenzel and van Tienhoven, 1982). In both, chicken and duck, the capillaries of the nucleus accumbens showed a wide lumen and prominent pericapillary spaces, thus resembling vessels of circumventricular organs which lack a blood-brain barrier. Detailed analyses of these vessels were performed by means of electron microscopy and injection of horseradish peroxidase into the general circulation (Fig. 3). Electron microscopical investigations in the chicken showed that the capillaries of the nucleus accumbens/lateral septum were endowed with a fenestrated endothelium (Fig. 3) as holds also true for capillaries of a representative number of circumventricular organs, e.g., choroid plexus, subfornical organ or pineal gland (see Leonhardt, 1980, for review and references). The lack of a blood-brain barrier in this region was confirmed by means of horseradish peroxidase injections into the general circulation (Rommel, 1992). HRP was found in the neuropil of the nucleus accumbens of chicken and ducks who had been sacrificed one minute after injection of the tracer into the left ventricle of the heart. At the first glance, these findings might suggest that the VIP-immunoreactive CSF-contacting neurons, like certain magnocellular neurons of the PVN, are exposed to the CSF-dominated compartment via their CSF-contacting dendrites and to the hemal milieu via their axons. However, as shown with the use of light- and electron microscopic immunocytochemistry, VIP-immunoreactive profiles were never located in the vicinity of capillaries, but approached neuronal perikarya of the nucleus accumbens/lateral septum. Thus, the CSF-contacting neurons in the nucleus accumbens/lateral septum obviously transmit their signals via synaptic mechanisms. Interestingly, the ependymal and vascular specializations were not restricted to the area of the CSF-contacting neurons, but extended from the rostral tip of the lateral ventricle to the interventricular foramen.

The neuronal, ependymal and vascular specializations of the NAc/LS point toward the possibility that this region represents a telencephalic circumventricular organ, which may be denominated as the "lateral septal organ" (cf. Kuenzel and van Tienhoven, 1982). Since circumventricular organs are well preserved in the course of phylogeny, it appeared of interest to investigate the region of the NAc/LS also in other vertebrate classes. We therefore analyzed the lateral septal area in eleven species of reptiles belonging to three different orders (Hirunagi et al., 1993). In all individuals investigated, VIP-immunoreactive CSF-contacting neurons were shown to be present in this region. They closely resembled those in the avian NAc/LS. As shown by VIP-immunocytochemistry, their axonal projections penetrated into the NAc and the LS, where they encompassed VIP-immunonegative neurons in a basket-like arrangement. The ependyma adjacent to the CSF-contacting neurons was composed of columnar cells frequently arranged in a stratified manner. Like in birds, these ependymal specializations were not restricted to the area of the CSF-contacting neurons, but extended from the rostral portion of the lateral ventricle to the interventricular foramen. As shown for the lizard, Lacerta sicula, a blood-brain barrier is present in the NAc/LS, since HRP injected into the left ventricle of the heart did not leave the capillaries of this region, whereas the tracer was found in the stroma of the choroid plexus and in the pineal parenchyma (Fig. 3; Rommel, 1992). The comparative analyses showed that the VIP-immunoreactive CSF-contacting neurons can be considered as a basic feature of the NAc/LS

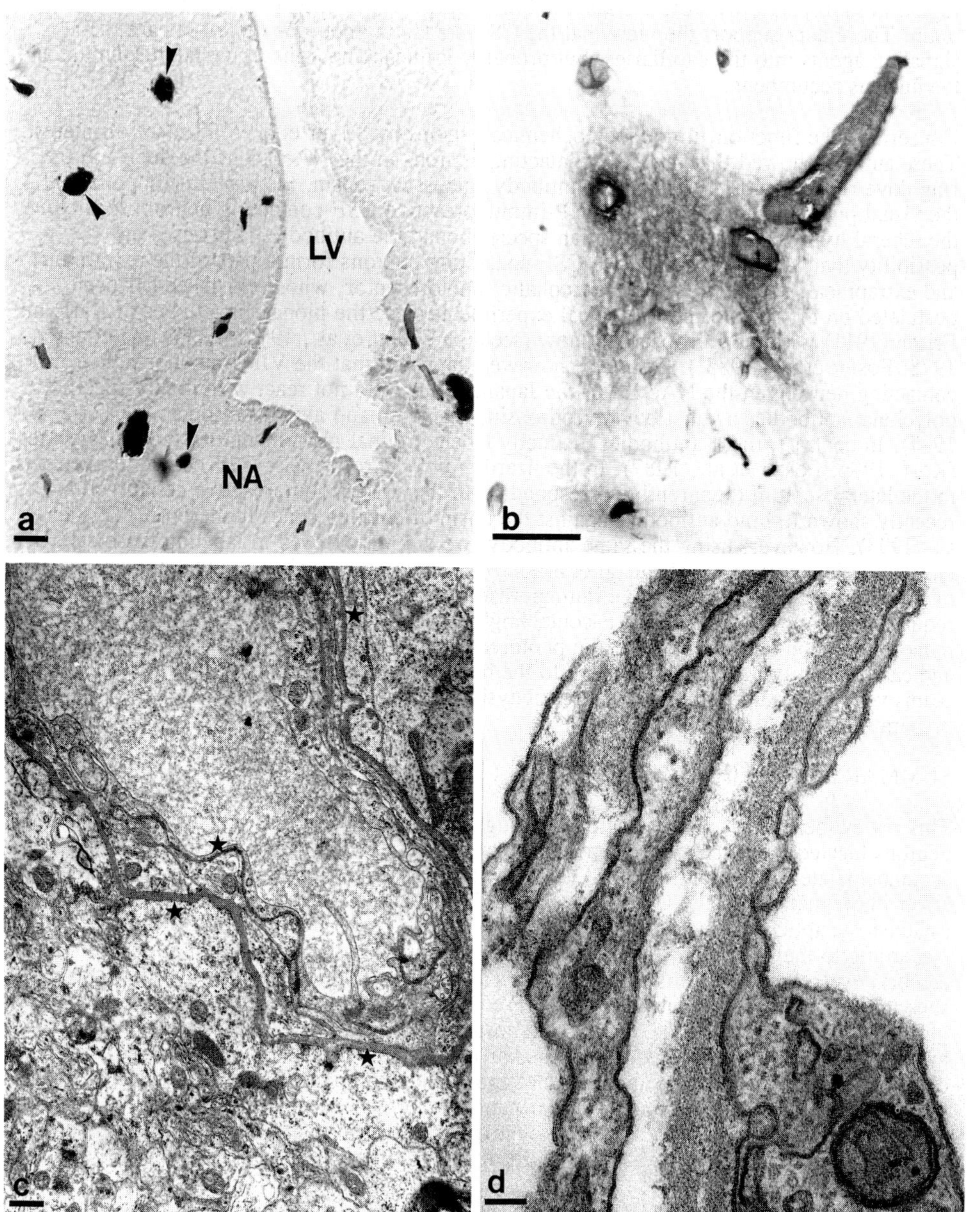

Fig. 3. Properties of the blood-brain barrier and capillary endothelium of the nucleus accumbens of a lizard (Lacerta sicula), Pekin duck and chicken. a No HRP is found in the nucleus accumbens (NA) of the lizard after intracardial injection of the tracer. Arrowheads: erythrocytes containing endogenous peroxidase. LV lateral ventricle. Frozen section. Bar = 50 μm. b Following intracardial injection of HRP, the tracer is found in the neuropil of the nucleus accumbens of the duck. Frozen section. Bar = 20 μm. c Electron microscopic demonstration of HRP (stars) in the neuropil of the nucleus accumbens of the duck after intracardial injection. Bar = 1 μm. d Conventional transmisson electron microscopy reveals fenestrated endothelial cells in the capillaries of the nucleus accumbens of the chicken. Bar = 0,5 μm.

region in sauropsids, whereas the lacking blood-brain barrier is a specialization of the avian brain. These data support the view that the CSF-contacting neurons do not release their signaling agents into the capillaries, but probably influence neurons in the lateral septum and the nucleus accumbens.

Concerning the function, immunocytochemical studies by Silver et al. (1988) are of interest. These authors showed that the CSF-contacting neurons in the NAc/LS of the duck and the ring dove reacted with a monoclonal antibody against rod-opsin, the protein component of the visual pigment rhodopsin. Also VIP-immunoreactive CSF-contacting neurons located in the tuberal hypothalamus of these avian species bound the antibody. This raises the possibility that VIP-immunoreactive CSF-contacting neurons form a part of the extraretinal and extrapineal, so-called "deep" (encephalic) photoreceptor, whose existence has been postulated on the basis of physiological experiments since the pioneering studies of Karl von Frisch (1911) with the European minnow (see also Follett et al., 1975; Yokoyama et al., 1978; Foster et al., 1985;). It should, however, be noted that the VIP-immunoreactive CSF-contacting neurons in the NAc/LS of the Japanese quail did not react with highly specific, polyclonal antibodies against bovine rod-opsin, S-antigen and alpha-transducin (Foster et al., 1987). In contrast, these antibodies distinctly labeled retinal photoreceptors and pinealocytes (Korf, 1986; Foster et al., 1987). In the lizard, *Anolis carolinensis*, CSF-contacting neurons in the lateral septum (apparently corresponding to the VIP-immunoreactive cells) were recently shown to bind antibodies against the opsin of chicken cones (cone-opsin) (Foster et al., 1993). However, using the same antibody, we did not succeed in labeling these cells in eleven species of reptiles (Hirunagi et al., 1993), although retinal and pineal photoreceptors of these species displayed positive immunoreactions. In conclusion, further experiments are required to prove whether the CSF-contacting neurons in the lateral septum of sauropsids are indeed a component of the encephalic photoreceptors. Since these neurons form aggregates and can be readily localized according to the neighbouring ependymal specializations, it seems very promising to perform electrophysiological recordings from these cells in slice preparations.

SUMMARY AND CONCLUSIONS

This review deals with morphological investigations of two different sets of CSF-contacting neurons located (i) in the avian paraventricular nucleus (PVN) and (ii) in the nucleus accumbens/lateral septum (NAc/LS) of birds and reptiles. By means of tracer techniques and silver impregnations of the Golgi-type, the CSF-contacting neurons in the PVN are shown to establish synaptic contacts with the magnocellular, AVT-producing neurons of the PVN. According to their location, these neurons may represent a morphological correlate of cerebral osmoreceptors. This assumption is substantiated by electrophysiological recordings showing that the area inhabited by the CSF-contacting neurons was very sensitive to changes in the sodium composition of the CSF (Kanosue et al., 1990; 1992). Moreover, the morphological studies revealed that particular magnocellular neurons are endowed with dendrites extending into the subependymal and ependymal layers. This arrangement raises the possibility that also the magnocellular neurons themselves may receive signals from the ventricular compartment, since the ependymocytes of the PVN are connected to each other by means of gap junctions and, thus, do not form a barrier for cations and neuropeptides present in the CSF.

CSF-contacting neurons are not restricted to hypothalamic and medullospinal regions, but occur also in the lateral ventricles. A conspicuous accumulation of CSF-contacting neurons is found in the nucleus accumbens/lateral septum of birds and reptiles by means of VIP-immunocytochemistry. These neurons obviously project to nerve cells of the nucleus accumbens and lateral septum. In both, reptiles and birds, the ependyma covering the

NAc/LS region shows specializations, which, however, are not restricted to the area of the CSF-contacting neurons. In birds, but not in reptiles, this region was shown to lack a blood-brain barrier by means of intracardial injections of horseradish peroxidase. These findings suggest that the VIP-immunoreactive CSF-contacting neurons may form the neuronal component of a telencephalic circumventricular organ. In functional terms, it appears interesting that the CSF-contacting neurons in the NAc/LS of the ring dove and the lizard, Anolis carolinensis, reacted with antibodies against photoreceptor-specific proteins (Silver et al., 1988; Foster et al. 1993). This has led to the assumption that these neurons represent a part of the extraretinal, extrapineal photoreceptor. This hypothesis is now in need of thorough electrophysiological investigation, particularly, because the VIP-immunoreactive CSF-contacting neurons in the NAc/LS of the Japanese quail and eleven species of reptiles did not bind any of the antibodies raised against photoreceptor-specific proteins and applied in our studies. The electrophysiological studies would certainly benefit from the fact that (i) the VIP-immunoreactive CSF-contacting neurons are rather concentrated and form an organ-like aggregate and (ii) their location can be readily identified in slice preparations according to the neighbouring specializations of the ependymal wall.

In conclusion, the morphological and immunocytochemical data presented support Agduhr's notion that the system of intraependymal (or CSF-contacting) neurons may serve multiple receptive functions. The CSF-contacting neurons of the PVN and of the NAc/LS appear as ideal candidates for combined studies allowing to correlate structural features with functional characteristics and, thus, to precisely elucidate the function of CSF-contacting neurons of the vertebrate brain.

REFERENCES

Abel, J.H., Takemoto, Hoffman, D., McNeill, T.H., Kozlowsky, G.P., Masken, J.F., & Sheridan, P. (1975): Corticoid uptake by the paraventricular nucleus in the hypothalamus of the duck, Anas platyrhynchos. *Cell Tiss. Res.* 161, 285-291.

Agduhr, E. (1922): Über ein zentrales Sinnesorgan (?) bei den Vertebraten. *Z. Anat. Entw. Gesch.* 66, 223-360.

Andersson, B. (1977): Regulation of body fluids. *Annu. Rev. Physiol.* 39, 185-200.

Andersson, B. & Olsson, K. (1977): Evidence for periventricular sodium-sensitive receptors of importance in the regulation of ADH secretion. In *Neurohypophysis*, ed. A.M. Moses, & L. Share, pp. 118-127. Basel: Karger.

Bons, N. (1980): The topography of mesotocin and vasotocin systems in the brain of the domestic mallard and Japanese quail: Immunocytochemical identification. *Cell Tiss. Res.* 217, 37-51.

Deutsch, H. & Simon, E. (1980): Intracerebroventricular osmosensitivtiy in the Pekin duck. *Pfluegers Arch.* 387, 1-7.

Follett, B.K., Davies, D.T., & Magee, V. (1975): The rate of testicular development in Japanese quail (Coturnix coturnix japonica) following stimulation of the extraretinal photoreceptor. *Experientia* 31, 48-49

Foster, R.G., Follett, B.K., & Lythgoe, J.N. (1985): Rhodopsin-like sensitivity of extra-retinal photoreceptors mediating the photoperiodic response in quail. *Nature* 313, 50-52.

Foster, R.G., Korf, H.-W., & Schalken, J.J. (1987): Immunocytochemical markers revealing retinal and pineal but not hypothalamic photoreceptor systems in quail. *Cell Tiss. Res.* 248, 161-167.

Foster, R.G., Alones, V., Garcia,-Fernandez, J.M., & de Grip, W.J. (1993): Opsin localization and chromophore retinoids identified within the basal brain of the lizard Anolis carolinensis. *J. Comp. Physiol. A.* (in press)

Frisch, K. von (1911): Beiträge zur Physiologie der Pigmentzellen in der Fischhaut. *Pfluegers Arch.* 138, 319-387.

Goossens, N., Blähser, S., Oksche, A., Vandesande, F. & Dierickx, K. (1977): Immunocytochemical investigation of the hypothalamo-neurohypophysial system in birds. *Cell Tiss. Res.* 184, 1-13.

Hirunagi, K., Rommel, E., Oksche, A., & Korf, H.-W. (1993): Vasoactive intestinal peptide-immunoreactive cerebrospinal fluid-contacting neurons in the reptilian lateral septum/nucleus accumbens. *Cell Tiss. Res.* (in press)

Hof, P.R., Dietl, M.M., Charnay, Y., Martin, J.L., Bouras, C., Palacios, J.M., & Magistretti, P.J. (1991): Vasoactive intestinal peptide binding sites and fibers in the brain of the pigeon Columba livia: an autoradiographic and immunohistochemical study. *J. Comp. Neurol.* 305, 393-411.

Kanosue, K., Schmid, H., Simon, E. (1990): Differential osmoresponsiveness of periventricular neurons in duck hypothalamus. *Am. J. Physiol.* R973-R981.

Kanosue, K., Gerstberger, R., Simon-Oppermann, C., Simon, E. (1992): Ionic responsiveness in third ventricular hypertonic stimulation of antidiuresis in ducks. *Brain Res.* 569, 268-274.

Kolmer, W. (1921): Das "Sagittalorgan" der Wirbeltiere. *Z. Anat. Entw. Gesch.* 60, 652-717.

Korf, H.-W. (1984): Neuronal organization of the avian paraventricular nucleus: intrinsic, afferent, and efferent connections. *J. Exp. Zool.* 232, 387-395

Korf, H.-W. (1986): *Zur Frage photoneuroendokriner Zellen und Systeme: Vergleichende Untersuchungen am Pinealkomplex.* Thesis, Fachbereich Humanmedizin, Giessen.

Korf, H.-W. & Fahrenkrug, J. (1984): Ependymal and neuronal specializations in the lateral ventricle of the Pekin duck, Anas platyrhynchos. *Cell Tiss. Res.* 236, 217-227.

Korf, H.-W., Simon-Oppermann, C., & Simon, E. (1982): Afferent connections of physiologically identified neuronal complexes in the paraventricular nucleus of conscious Pekin ducks involved in regulation of salt- and water-balance. *Cell Tiss. Res.* 226, 275-300

Korf, H.-W., Viglietti-Panzica, C., & Panzica, G.C. (1983): A Golgi study on the cerebrospinal fluid (CSF)-contacting neurons in the paraventricular nucleus of the Pekin duck. *Cell Tiss. Res.* 228, 149-163.

Kuenzel, W.J. & Tienhoven, A. van (1982): Nomenclature and location of avian hypothalamic nuclei and associated circumventricular organs. *J. Comp. Neurol.* 206, 293-313.

Leonhardt, H. (1980): Ependym und circumventriculäre Organe. In *Neuroglia I. Handbuch der mikroskopischen Anatomie des Menschen,* ed. A. Oksche & L. Vollrath, pp. 177-666. Berlin: Springer.

McNeill, T.H., Abel, J.H., & Kozlowsky, G.P. (1975): Correlations between brain catecholamine neurosecretion and serum corticoid levels in osmotically stressed mallard ducks (Anas platyrhynchos). *Cell Tiss. Res.* 161, 277-283.

Oksche, A. (1976): The neuroanatomical basis of comparative neuroendocrinology. *Gen. Comp. Endocrinol.* 29, 225-239.

Panzica, G.C., Korf, H.-W., Ramieri, G., & Viglietti-Panzica, C. (1986): Golgi-type and immunocytochemical studies on the intrinsic organization of the periventricular layer of the avian paraventricular nucleus. *Cell Tiss. Res.* 243, 317-344.

Rodríguez E.M., Pena, P., Rodríguez, S., & Aguado, L.I. (1982): Evidence for the participation of the CSF and periventricular structures in certain neuroendocrine mechanisms. *Front. Horm. Res.* 9, 142-158.

Rommel, E. (1992): Neuronale und vasculäre Spezialisierungen im Nucleus accumbens verschiedener Vertebraten. Doctoral Thesis, Fachbereich Humanmedizin, Justus Liebig Universität Giessen.

Silver, R., Witkovsky, P., Horvath, P., Alones, V., Barnstable, C.J., & Lehman, M.N. (1988): Coexpression of opsin- and VIP-like immunoreactivity in CSF-contacting neurons in the avian brain. *Cell Tiss. Res.* 253, 189-198.

Simon, E. (1982): The osmoregulatory system of birds with salt glands. *Comp. Biochem. Physiol. A* 71, 547-556.

Simon-Oppermann, C. & Simon, E. (1982): Osmotic and volume control of diuresis in conscious ducks. *J. Comp. Physiol.* 146, 17-25.

Simon-Oppermann, C., Simon, E., Deutsch, H., Möhring, J., & Schoun, J. (1980): Serum arginine-vasotocin (AVT) and afferent and central control of osmoregutlation in conscious Pekin ducks. *Pfluegers Arch.* 387, 99-106.

Tretjakoff, D.(1909): Das Nervensystem von Ammocoetes. II. Das Gehirn. *Arch. Mikrosk. Anat.* 73, 607-680.

Tretjakoff, D. (1913): Die zentralen Sinnesorgane bei Petromyzon. *Arch. Mikrosk. Anat.* 83, 68-117.

Vigh, B. & Vigh-Teichmann, I. (1973): Comparative ultrastructure of CSF-contacting neurons. Int. Rev. Cytol. 35, 189-251.

Vigh, B., Vigh-Teichmann, I., & Aros, B. (1969): Das Paraventrikularorgan und das Liquorkontakt-Neuronensystem. *Anat. Anz.* 125, 683-688.

Vigh-Teichmann, I. & Vigh, B. (1974): The infundibular cerebrospinal fluid-contacting neurons. *Adv. Anat. Embryol. Cell Biol.* 50, 1-90.

Vigh-Teichmann, I. & Vigh, B. (1983): The system of cerebrospinal fluid-contacting neurons. *Arch. Histol. Jpn.* 46, 427-468.

Yamada, S., Mikami, S., & Yanaihara, N. (1982): Immunohistochemical localization of vasoactive intestinal polypeptide (VIP)- containing neurons in the hypothalamus of the Japanese quail,Coturnix coturnix. *Cell Tiss. Res.* 226, 13-26.

Yokoyama, K., Oksche, A., Darden, T.R., & Farner, D.S. (1978): The sites of encephalic photoreception in photoperiodic induction of the growth of the testes in the White-crowned sparrow, Zonotrichia leucophrys gambelii. Cell Tiss. Res. 189, 441-467.

ACKNOWLEDGEMENTS

K.H. was supported by a fellowship from the Deutscher Akademischer Austauschdienst. The authors are grateful to Mrs. E. Laedtke and Mrs. I. Habazettl for technical assistance, to I. Szász for graphical work and to Mrs. G. Müller for secretarial help.

Integrative and cellular aspects of autonomic functions : temperature and osmoregulation. Eds K. Pleschka, R. Gerstberger. John Libbey Eurotext, Paris © 1994, pp. 419-428.

Effects of osmotic stimulation of the supraoptic nucleus on central and peripheral release of vasopressin and on baro- and thermoregulation

Rainer Landgraf, Thomas Horn[(1)], Inga Neumann[(2)], Quentin J. Pittman[(2)] and Mike Ludwig[(1)]

Max Planck Institute of Psychiatry, Clinical Institute, Kraepelinstrasse 2, D-80804 Munich, Germany. [(1)] Neurology Group, Section of Biosciences, University of Leipzig, Leipzig, Germany and [(2)] Neuroscience Research Group, University of Calgary, Calgary, Canada

1. Introduction

There is a growing body of evidence which indicates that neuropeptides are critically involved in the central regulation of autonomic functions. Of all the neuropeptides, vasopressin (AVP) has been one of the most widely studied. In most cases, synthetic AVP or its antagonists have been administered peripherally and centrally, respectively, with subsequent determination of effects on autonomic functions.
It has proven very difficult to define brain actions of peptide-containing pathways that can be conclusively shown to be mediated by the endogenous neuropeptides released centrally from these pathways, including axons, dendrites and cell bodies. The availability of sophisticated microperfusion and radioimmunoassay techniques, however, has recently enabled the characterization of the dynamic pattern of basal and stimulated AVP release in numerous areas of the rat brain known to be involved in autonomic regulation. These include the supraoptic (SON) and paraventricular nuclei of the hypothalamus as well as the limbic and rhombencephalic regions (for review see Landgraf, 1992). With regard to behavioral changes following direct osmotic stimulation of the SON-area, which are causally related to central release of AVP (Landgraf et al., 1993), it is of particular interest to examine parallel changes in autonomic parameters. Therefore, in the present study, (1) the intra-SON and peripheral release of AVP was monitored by microdialysis under basal conditions and in response to direct osmotic stimulation of the SON-area using NaCl, mannitol or urea, and (2) an attempt was made to correlate this stimulation and the subsequent release of AVP within the nucleus and into blood with changes in baro- and thermoregulation.

2. Methodological considerations

Microdialysis has rapidly become popular in recent years as an *in vivo* approach to monitor the changes of endogenous mediator levels in the extracellular fluid; it also allows the simultaneous administration of exogenous substances. In addition to brain microdialysis, we established microdialysis in blood to monitor peripheral AVP release.

Microdialysis offers a number of advantages:

- It allows collection of biologically active mediators from the extracellular space of a local circumscribed brain area.
- It causes minimal damage to the tissue, including the blood-brain barrier.
- The perfusion is simple (no problems in balancing inflow and outflow of perfusate, leading to erroneous values; Pittman et al., 1985).
- The dialysis membrane prevents large molecules (e.g., enzymes or peptides which could interfere with assays) from diffusing into the dialysate.
- Microdialysis in blood (recovery of AVP in vivo: 65 per cent) avoids frequent blood sampling which might compromise homeostasis, especially when studying the vasopressinergic system which is sensitive to cardiovascular alterations.

The major disadvantages of microdialysis are:

- The recovery of AVP is very low (in vitro: 1.6 per cent). We use a highly sensitive radioimmunoassay (detection limit 0.1 pg/dialysate, half-maximal displacement of ^{125}I-AVP at 3 pg) and are usually able to measure both AVP and oxytocin in the same dialysate without extraction prior to assay.
- Microdialysis of neuropeptides has poor time resolution. While this technique is capable of "integrating" all fluctuations in AVP concentration occurring during the 30-min sampling period, it provides only an average over that time; thus, any sharp changes in central and peripheral release over short intervals are blunted. This, however, should be less disadvantageous for non-synaptically released mediators, than for synaptically released putative neurotransmitters.
- A major concern about microdialyis derives from the changes in mediator levels observed on different days following the implantation of the probe, i.e. recovery declines with time. These changes are thought to be related to the development of tissue reactions around the probe. Fig. 1 provides a typical example of how AVP release within the SON is attentuated during the post-implantation period and in dependence on direct osmotic stimulation of the SON. Consequently, we decided to study AVP release patterns and autonomic responses to SON stimulation in more acute experiments.

Ketamine- or urethane-anesthetized male rats were used throughout the studies. Home-made U-shaped probes (Landgraf and Ludwig, 1991) were stereotaxically implanted with their tips in, or adjacent to, the SON. Additionally, concentric blood microdialysis probes (Neumann et al., 1993) were placed in the external jugular vein. About 2 h after the implantation, consecutive 30-min dialysates were collected at a flow rate of 3 μl artificial CSF (aCSF)/min and subsequently analyzed for AVP by radioimmunoassay. Either NaCl, mannitol, or urea were added to the dialysis medium as indicated in the legends to the Figs. In some animals, blood pressure or body temperature was recorded. At the end of each experiment, brains were removed and cut to locate the placement of the probes histologically. Data were analysed statistically using parametric and non-parametric tests, as appropriate.

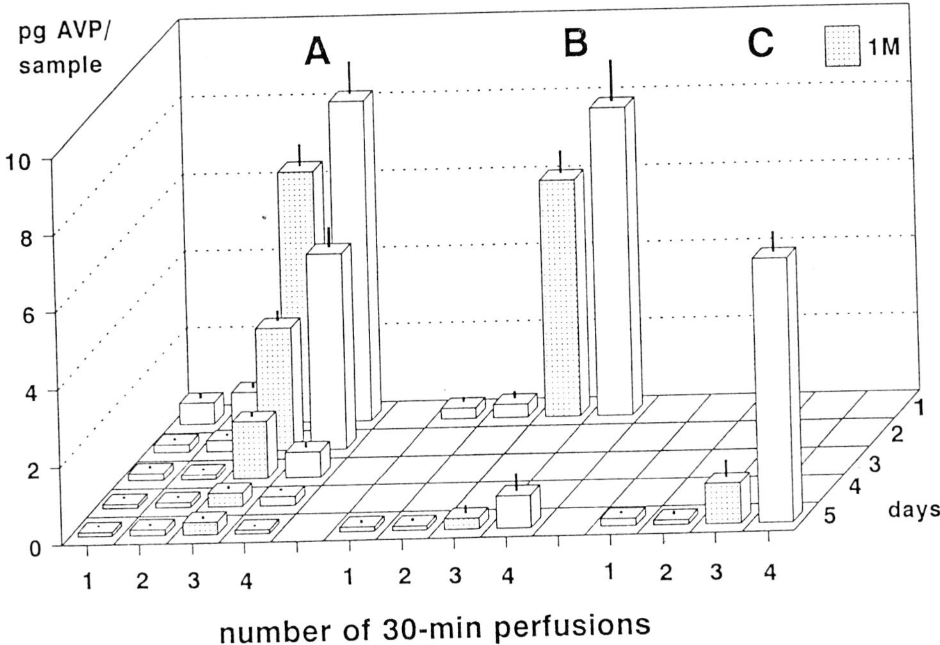

Fig. 1. AVP contents in 30-min dialysates collected consecutively within the SON before, during and after hypertonic stimulation (1M NaCl-aCSF, dotted columns). Animals were dialysed (A) on five successive days, (B) on the 1st and 5th, and (C) on the 5th day post implantationem (n=6 each).

3. Stimulus-dependent central and peripheral release of AVP

Previous studies have shown Ca^{2+}-dependency and K^{+}-stimulation of central AVP release suggesting exocytosis from intact neuronal structures in the SON rather than diffusion from plasma or cerebrospinal fluid (Ludwig and Landgraf, 1992). SON neurons are responsive to changes in osmolality of the dialysis medium, being apparently more sensitive to decreases in osmolality. As shown in Figs. 1-4, AVP levels increased significantly during the hypertonic pulse, but peaked only in the 30-min period following the stimulation, i. e. when osmolality was decreasing. A second hypertonic pulse given 120 min later produced similar initial and rebound release of AVP from the SON (Fig. 2). Thus, release does not appear to have been due to destruction of SON neurons by the initial hypertonic dialysis. The response was dose-dependent in that 0.5M NaCl-aCSF produced a lower response than 1 M or 2M NaCl-aCSF. A significant increase in intra-SON release was also seen when isotonic (0.15 M) or hypertonic (1M) aCSF was replaced with hypotonic (0.01M) aCSF (Fig. 2).
AVP contents in dialysates did not change in response to osmotic stimulation, if the microdialysis probe was placed outside the SON-area; this holds true for all herein mentioned experiments.

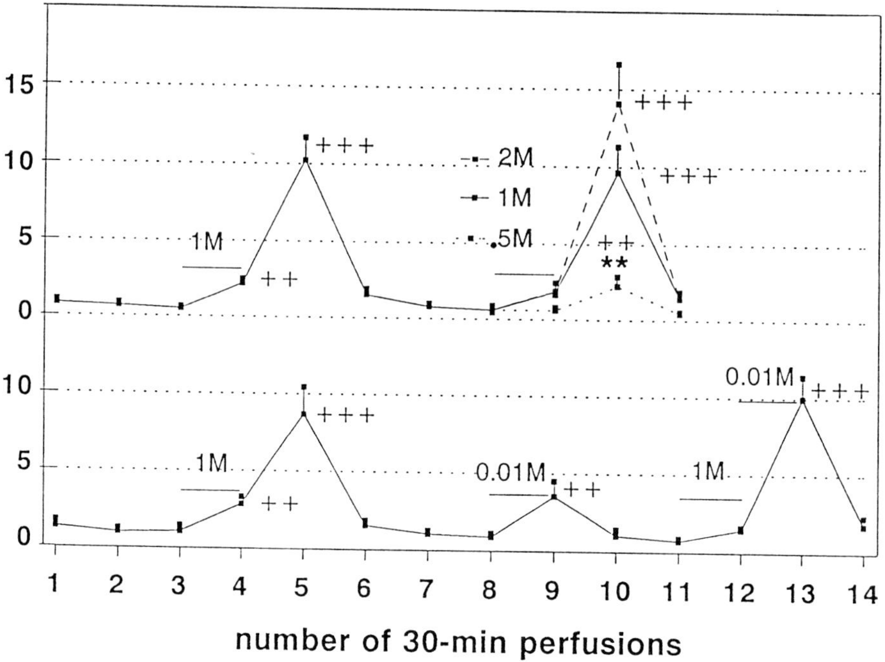

Fig. 2. AVP contents in 30-min dialysates from the SON collected consecutively before, during (horizontal bars) and after administration of hyper- or hypotonic aCSF. AVP contents in dialysates in response to hypertonic (0.5M, n=12; 1M, n=6; 2M NaCl-aCSF, n=12, upper panel) and hypotonic pulses (0.01M aCSF, dialysates 9 and 13, n=8, lower panel) were compared to levels before, during and after the first (1M NaCl-aCSF) stimulation. Means + S.E.M., ++ $p<0.01$, +++ $p<0.001$ compared to basal (dialysates 1 to 3), ** $p<0.01$ compared to AVP release following the first 1M pulse (from Ludwig and Landgraf, 1992).

As shown in Fig. 3A, simultaneous microdialysis in brain and blood indicates that during osmotic stimulation of the SON-area with 1M NaCl-aCSF, central AVP release is accompanied by a peripheral release from the neurohypophysis into blood. After replacement of hypertonic aCSF with isotonic aCSF, a further rise in intra-SON AVP release occurred, whereas neuropeptide content in dialysates taken from blood returned to basal levels, indicating that the rebound effect is a phenomenon characteristic of the SON.

Intraperitoneal application of hypertonic saline resulted in a drastically increased AVP content in dialysates taken from blood which peaked in the 30-min period after the stimulus (Fig. 3 B). In contrast, the delayed intra-SON release increased step-wise over a 2.5-h period. Similar results were obtained after i.v. osmotic stimulation (data not shown), further confirming that

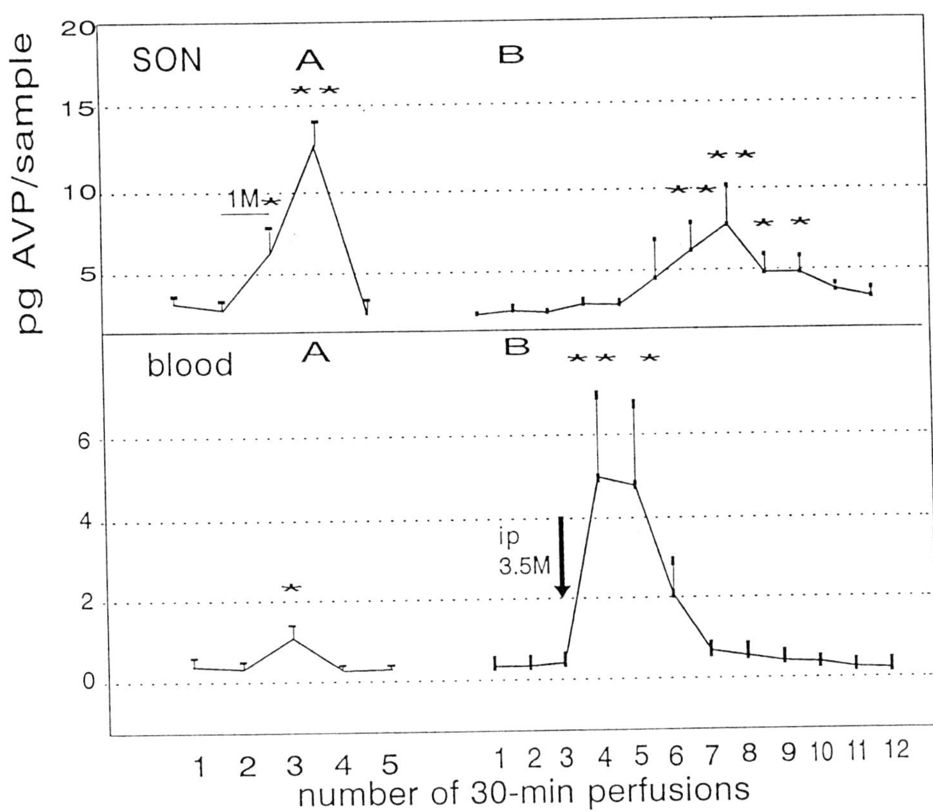

Fig 3. AVP contents in 30-min dialysates collected consecutively within the SON and in blood. (A) Microdialysis administration of 1M NaCl-aCSF (horizontal bar, n=6) into the SON during the 3rd collection interval. (B) Peripheral administration of 3.5M saline (1.5ml i.p., arrow, n=8). Means + S.E.M., * $p < 0.05$, ** $p < 0.01$ compared to pre-stimulation values.

central and peripheral release of AVP may be independently regulated.

The SON consists almost entirely of magnocellular neurons which project to the neurohypophysis (Hatton, 1990). The data of this study are consistent with the finding that virtually all parts of their membrane seem to be competent to release AVP (Pow and Morris, 1989). It is tempting to speculate that, under certain experimental conditions, the same supraoptic neuron is capable of regulating the release from different areas of its plasmalemma in a different manner.

As shown in Fig. 4, microdialysis was used to clarify whether the response of AVP neurons in the SON-area to direct stimulation is dependent on NaCl or hypertonicity _per se_. AVP release within the SON was similarly increased during the stimulation period regardless of whether NaCl, mannitol or urea was used. In the post-stimulation period, mannitol treatment resulted in a further increase in AVP release, being nearly identical to that after NaCl. In contrast, urea failed to evoke a rebound increase;

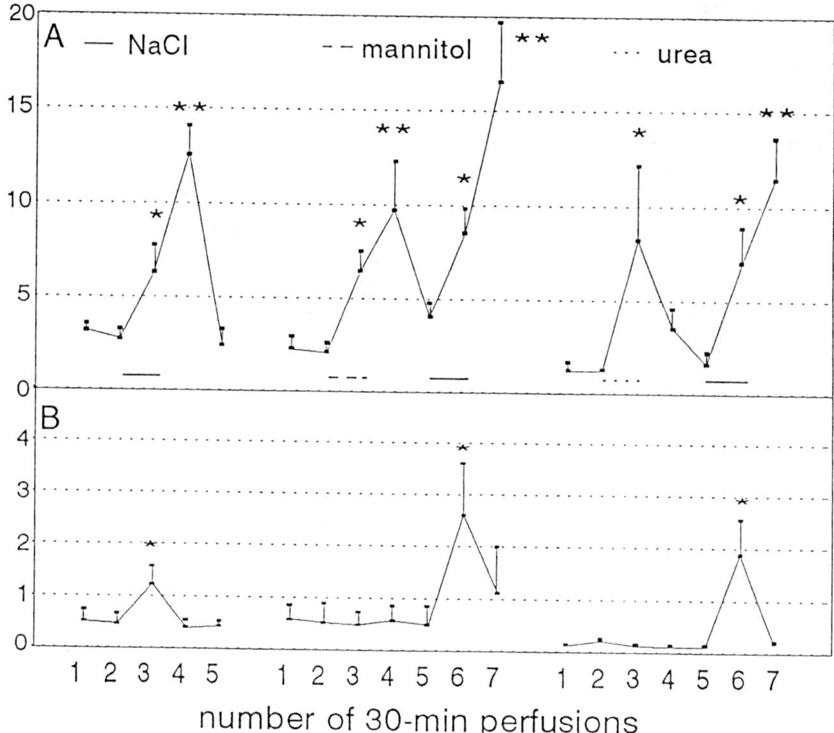

Fig. 4. AVP contents in 30-min dialysates collected simultaneously from the SON **(A)** and blood **(B)** before, during and after bilateral dialysis of the SON with hypertonic aCSF containing 1M NaCl, 1M mannitol or 1M urea (n=6 each). 1h after treatment with mannitol or urea, a second hypertonic pulse (1M NaCl-aCSF) was given via the microdialysis probe. Means + S.E.M., * $p<0.05$, ** $p<0.01$ compared to pre-stimulation values.

extracellular AVP levels returned to basal values during the post-stimulation period.

When mannitol or urea was dialysed bilaterally into the SON, no measurable increases in plasma AVP were observable. The neuropeptide contents in the dialysates taken from the blood were increased significantly only during a second pulse using 1M NaCl-aCSF, indicating that the failure to respond to mannitol and urea was due to the characteristics of the stimulus rather than to a non-responsive AVP system (Fig. 4). Taken together, the observed central AVP release within the SON is likely to be due to specific dynamics of local osmotic changes. Peripheral release, however, seems to be primarily dependent on Na^+ rather than on osmotic pressure _per se_ (Ludwig et al., 1993). While further studies are in progress to elucidate the mechanisms of the local release process, the osmotic activation of SON neurons nonetheless remains a useful tool for examining possible functional consequences of central peptide release.

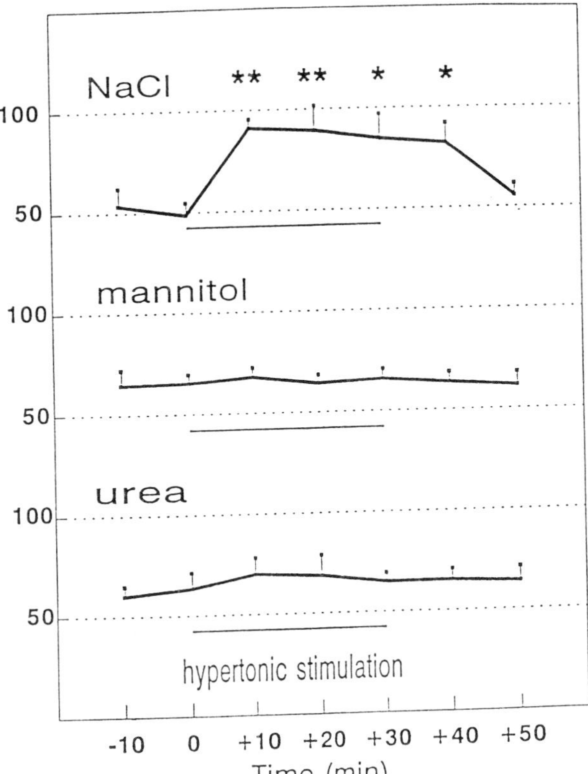

Fig. 5. Mean arterial pressure (MAP) responses to bilateral hyperosmotic stimulation of the SON (microdialysis with 1M NaCl-aCSF, 1M mannitol-aCSF or 1M urea-aCSF, horizontal bars; see also Fig. 4, n=6 each). Means + S.E.M., * $p<0.05$, ** $p<0.01$ compared to pre-stimulation with isotonic aCSF

4. Autonomic effects of direct osmotic stimulation of the SON

Baroregulation

In the 1M NaCl-aCSF group (Fig. 4), blood pressure increased significantly during the initial 10 min after the start of the hypertonic pulse. It then remained at this level before returning to basal levels within the next 20 min after the replacement of hypertonic with isotonic aCSF. In contrast, during dialysis of the SON-area with either hypertonic mannitol- or urea-aCSF, no significant changes in blood pressure could be observed (Fig. 5). However, blood pressure was increased significantly in these animals during a second stimulation with 1M NaCl-aCSF confirming the responsiveness of the system (data not shown).
Despite this close correlation between changes in plasma AVP and blood pressure, plasma AVP appears not to play a major role in the associated cardiovascular response. The blood pressure was not

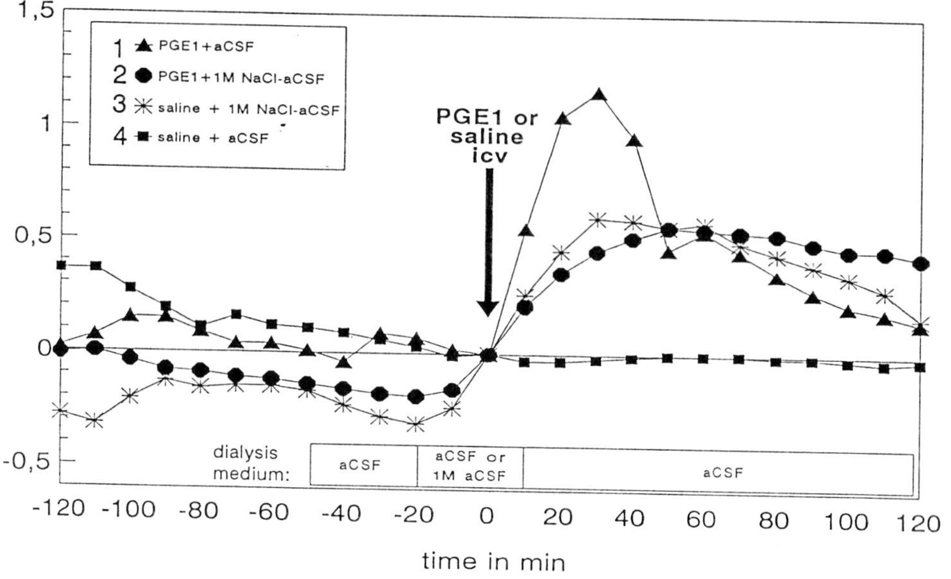

Fig. 6. Changes in body (colonic) temperature in response to (1) 120 ng PGE1/5 ul icv during simultaneous dialysis of the SON with isotonic aCSF (PGE1+aCSF, n=5), (2) PGE1 icv during hypertonic stimulation of the SON (PGE1+1M NaCl-aCSF, n=7), (3) saline icv during hypertonic stimulation of the SON (saline+1M NaCl-aCSF, n=5) or (4) saline icv during dialysis of the SON with isotonic aCSF (saline+aCSF, controls, n=5). The arrow and the bar at the bottom indicate the time course of the treatments. Significant differences ($p<0.05$): 1 vs 2,3 (20-40 min), 2 vs 4 (20-120 min), 1,3 vs 4 (20-70 min)

influenced by blockade of vascular V1 AVP receptors which, nevertheless, resulted in a total block of blood pressure responses to synthetic AVP (Ludwig et al., 1993). Our preliminary data (Callaghan and Ludwig, unpublished), however, indicate that the V1 receptor antagonist $d(CH_2)_5Tyr(Me)AVP$, given into the 4th ventricle, can reduce the rise in blood pressure observed after microdialysis of the SON with 1M NaCl-aCSF. This finding further confirms the hypothesis that endogenous AVP released within the brain is critically involved in neural control of blood pressure (Landgraf et al., 1990a; Pittman and Landgraf, 1991).

Thermoregulation

There is considerable evidence that the SON contributes to endocrine and cardiovascular responses via the sympathetic nervous system, but little is known about the involvement of these pathways in the homeostasis of body temperature and in febrile thermogenesis. In order to test the described model utilizing hypertonic NaCl stimulation of the SON via the microdialysis probe for its effect on body temperature, we examined the effects of the NaCl stimulus under normal conditions and prostaglandin E1 (PGE1)-induced fever, i. e. when the temperature setpoint is shifted to a higher level.

In 4 groups of animals, the SON was dialyzed with isotonic or hypertonic aCSF and PGE1 or saline was infused icv according to Fig. 6. Controls showed a stable body temperature during the recording period indicating that neither the icv injection nor microdialysis with isotonic aCSF affected normal body temperature. The delivery of PGE1 into the ventricle resulted in the known febrile response. When such febrile animals were dialysed in the SON with 1M NaCl-aCSF, the peak of the fever response to PGE1 was significantly reduced to approx. 50 per cent. Interestingly, this antipyretic effect was not blocked if the V1 receptor antagonist was bilaterally infused into the ventral septal area, which is well known to be involved in antipyresis, 10 to 30 min before PGE1 (data not shown). The delivery of 1M NaCl-aCSF *per se* resulted in a significant increase in body temperature of 0.6±0.15 ^{o}C when compared to controls (Fig. 6).

These results indicate that the stimulation of the SON-area with hypertonic NaCl-aCSF results in an activation of thermogenesis in the normal animal, and in a reduced febrile response to central application of PGE1. While there is now good evidence that endogenous AVP may act as an antipyretic agent in the brain (Pittman et al., 1988; Landgraf et al., 1990b), its involvement, if any, in the described effects requires further detailed investigations.

5. Conclusions

The data presented show that

- microdialysis is an appropriate *in vivo* technique to monitor the dynamic pattern of basal and evoked release of AVP within the SON-area and into blood;
- changes in the osmolality of the dialysis medium result in increased intra-SON release of AVP; a very robust increase is associated with a switch to a decreased osmolality;
- release of AVP within the SON and into blood may be independently regulated;
- central release of AVP seems to depend primarily on local osmotic changes within the SON, whereas the peripheral release pattern into blood depends on local Na^+ levels;
- direct stimulation of the SON area with hypertonic NaCl, but not mannitol and urea, results in a transient rise in blood pressure;
- direct stimulation of the SON area with hypertonic NaCl causes a transient increase in body temperature in normal rats and a reduced fever in febrile animals.

The mechanisms underlying the effects of hypertonic stimulation of the SON on baro- and thermoregulation, as well as behavioral consequences of altered autonomic functions, remain to be determined.

Acknowledgement: We thank T.J. Malkinson and Mrs. Brigitte Wolff for expert technical assistance. These studies were supported by VW, MRC and NATO.

REFERENCES

Hatton, G.I. (1990): Emerging concepts of structure-function dynamics in adult brain: the hypothalamo-neurohypophysial system. *Prog. Neurobiol.* 34, 437-504.

Landgraf, R., (1992): Central release of vasopressin: stimuli, dynamics, consequences. *Prog. Brain Res.* 91, 29-39.

Landgraf, R. & Ludwig, M. (1991): Vasopressin release within the supraoptic and paraventricular nuclei of the rat brain: osmotic stimulation via microdialysis. *Brain Res.* 558, 191-196.

Landgraf, R., Ludwig, M. & Engelmann, M. (1993): Central release of vasopressin and its role in behavioral performance. In *Vasopressin IV*, Berlin, May 23-27, Abstr. No. 53.

Landgraf, R., Malkinson, T.J., Horn, T., Veale, W.L., Lederis, K. & Pittman, Q.J. (1990a): Release of vasopressin and oxytocin by paraventricular stimulation in rats. *Am. J. Physiol.* 258, R155-R159.

Landgraf, R., Malkinson, T.J., Veale, W.L., Lederis, K., & Pittman, Q. (1990b): Vasopressin and oxytocin in rat brain in response to prostaglandin fever. *Am. J. Physiol.* 259, R1056-R1062.

Ludwig, M., Horn, T., Callahan, M.F., Grosche, A., Morris, M. & Landgraf, R. (1993): Osmotic stimulation of the supraoptic nucleus of the rat brain: comparison of central and peripheral vasopressin release and effects on blood pressure. *Am. J. Physiol.* (submitted)

Ludwig, M. & Landgraf, R. (1992): Does the release of vasopressin within the supraoptic nucleus of the rat brain depend upon changes in osmolality and Ca^{2+}/K^{+}? *Brain Res.* 576: 231-234.

Neumann, I., Ludwig, M., Engelmann, M., Pittman, Q.J. & Landgraf, R. (1993): Simultaneous microdialysis in blood and brain: Oxytocin and vasopressin release in response to central and peripheral osmotic stimulation and suckling in the rat. *Neuroendocrinology* (submitted).

Pittman, Q.J., Disturnal, J., Riphagen, C., Veale, W.L. & Bauce, L. (1985): Perfusion techniques for neural tissue. In *Neuromethods, Vol.1. General neurochemical techniques,* ed. A.A. Boulton & G.B. Baker, pp.279-303. Clifton, New Jersey: Humana Press.

Pittman, Q.J. & Landgraf, R. (1991): Vasopressin in thermoregulation and blood pressure control. In *Vasopressin*, ed. S. Jard & R. Jamison, pp. 177-184. London: John Libbey & Company.

Pittman, Q.J., Naylor, A., Poulin, P., Disturnal, J., Veale, W.L., Martin, S.M., Malkinson, T.J. & Mathieson, B. (1988): The role of vasopressin as an antipyretic in the ventral septal area and its possible involvement in convulsive disorders. *Brain Res. Bull.* 20, 887-892.

Pow, D.V. & Morris, J.F. (1989): Dendrites of hypothalamic magnocellular neurons release neurohypophysial peptides by exocytosis. *Neuroscience* 32: 435-440.

Integrative and cellular aspects of autonomic functions : temperature and osmoregulation. Eds K. Pleschka, R. Gerstberger. John Libbey Eurotext, Paris © 1994, pp. 429-438.

Development and function of the chicken arginine-vasotocin system

Roland Grossmann, Bin Xu, Eckhard Mühlbauer and Franz Ellendorff

Institute for Small Animal Research, Federal Research Centre of Agriculture, Dörnbergstrasse 25-27, 29223 Celle, Germany

SUMMARY
The neurohypophysial hormone Arginine-Vasotocin (AVT) is involved in osmoregulation in avian species. AVT is synthesized in magnocellular neurons of the hypothalamic nuclei supraopticus and paraventricularis (PVN) and axonally transported to the neurohypophysis. We have investigated the embryonic development of AVT gene transcription, hormone synthesis and hormone secretion. Furthermore, we have designed a method to record the electrical activity of single neurones, identified as magnocellular cells within the PVN of chicken embryos at different developmental stages and newly hatched chicks (D1), since electrical activity of neurosecretory neurones is closely related to hormone secretion.

Our results show that the onset of gene expression and peptide synthesis is detectable as early as day 6 (E6) of embryonal development. A continuous increase in transcript and AVT concentration in the hypothalamus was observed up to E18. Despite the early onset of the peptide synthesizing machinery, osmoregulatory function of AVT may not occur before the end of the second week of incubation. Rapid maturational processes during the last week of embryonic development lead to a 'close-to-mature' response of the AVT system to physiological stimuli in the D1 chick.

Changes in electrophysiological characteristics of the hypothalamo-neurohypophysial system (HNS) are obvious and include a tremendous increase in spontaneous electrical activity as well as in conduction velocity of the axons. During osmotic challenge not only plasma osmolality, plasma sodium and AVT concentration is elevated but also AVT gene expression is upregulated. In addition, neuronal firing activity increases up to tenfold and more within 40 min after osmotic stimulation in the D1 chick.

The hypothalamo-neurohypophysial system (HNS) consists of magnocellular neurones, the vast majority of them are located within the hypothalamic nuclei supraopticus (SON) and paraventricularis (PVN). Among neuroendocrine circuitries in the brain the HNS is, as far as its "classical" functions are concerned, one of the best understood neuroendocrine structures. Morphologically well defined and biologically discretely localized, the HNS has been the subject of a variety of intensive investigations for many years. In birds Arginine-Vasotocin (AVT) is by far the more important of the two nonapeptide hormones produced in the HNS, since for the second, Mesotocin, up to date no clear physiological function has been described.

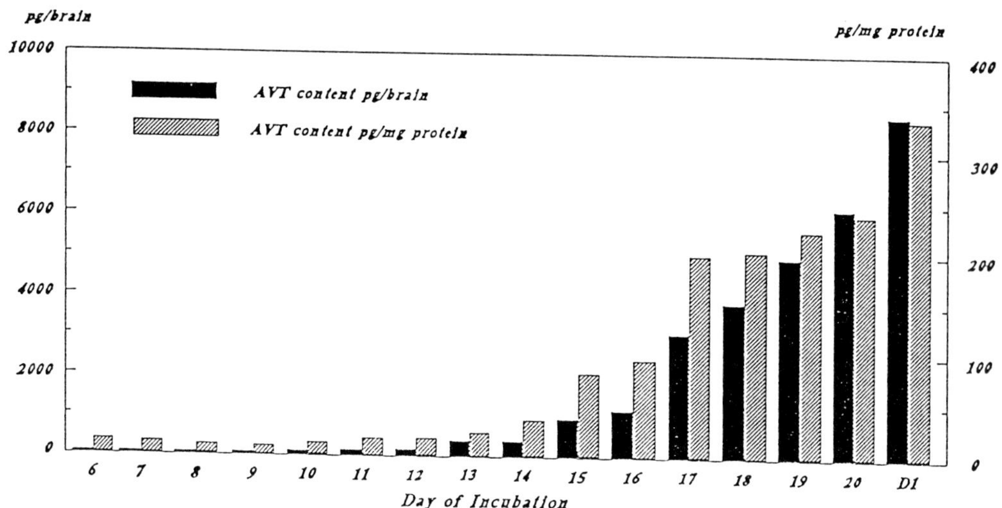

Fig. 1. Brain AVT concentration during embryonal development in the chicken

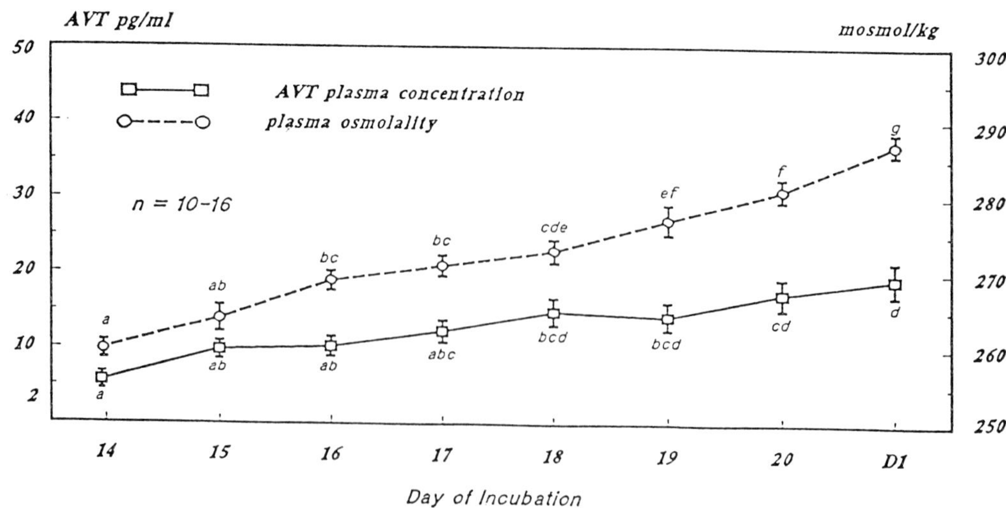

Fig. 2. Plasma osmolality and plasma AVT concentration in the chick embryo and newly hatched chicken (D1)

Both nonapeptides are synthesized as large precursor molecules and excised by specific cleavage enzymes before and during axonal transport along the hypothalamo-neurohypophysial tract (Acher & Chauvet, 1988). The mature hormone is stored in axon terminals in the neural lobe and released by adequate stimuli. The release itself is, in turn, under the control of the hypothalamic perikarya. Apart from the general agreement that electrial activity of neurones is closely correlated to their secretory dynamics it has been shown in some mammalian species, that oxytocinergic and vasopressiniergic cells are distinguishable by their specific firing pattern which accompanies the secretion of larger amounts of hormones (for review see Poulain & Wakerley, 1982; Dyball et al., 1988).

In birds, AVT has not only osmoregulatory function, but is also involved in reproductive events such as oviposition, which is comparable to the action of oxytocin on the mammalian myometrium. The physiology of AVT is also reflected by its structure being somehow between Arg-Vasopressin and Oxytocin (Acher et al., 1970).

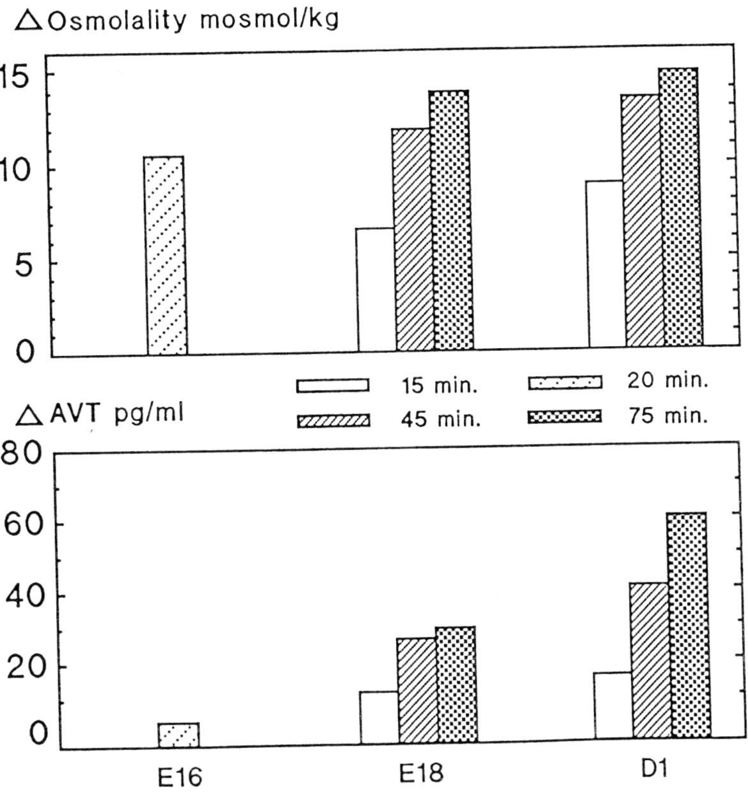

Fig. 3. Functional maturation of the AVT system during the perinatal period: increase in plasma osmolality and AVT concentration after i.p. injection of hypertonic saline. At E16 only two blood samples were withdrawn via indwelling catheter in the jugular vein, while four samples were taken at E18 and D1.

It is obvious that during embryonal and fetal life, the prenatal neuroendocrine brain must attain a sufficient degree of differentiation and development to cope with the change from a well protected prenatal life to the environmental conditions of postnatal life. Some systems are less essential for initial survival, such as reproduction, others are essential, such as thermoregulation or body fluid hormeostasis. For the chicken AVT synthesizing system we have shown that synthesis and storage is established early during development (Ellendorff & Grossmann, 1988; Xu et al. 1992; Grossmann et al. 1993). AVT mRNA is demonstrable at day 6 of embryonal life (E6) by in situ hybridization techniques (Milewski et al., 1989). The gene transcripts are further processed even at this early embryonal stage, since measurable amounts of AVT are present in the chicken brain at E6 (Fig. 1; Mühlbauer et al., 1993). Immunostaining showed the related peptide first around E8 (Tennyson et al., 1986). It is however unlikely that the nonapeptide serve in systemic osmoregulation at this early stage since neither axon terminals nor the neurohypophysis have been developed (Wingstrand, 1954; Lawzewitsch, 1969). Immunoreactive AVT has been observed not before E10 in neurohypophyseal tissue (Tennyson et al., 1986). In the plasma AVT has been first detected at E14 by RIA (Fig. 2; Mühlbauer et al., 1993), although due to technical limitations in collection of sufficient amounts of plasma in the embryo, first appearance might be even earlier.

Age Days	Spontaneously active cells % (n)	Constant latency msec ± SEM (n)	Conduction velocity m/sec
E17	20[a] (25)	20.0 ± 0.9[a] (25)	0.30
E17.5	33 (15)	17.1 ± 1.1[c] (13)	0.36
E18	46[b] (104)	16.3 ± 0.6[b] (103)	0.40
D1	69[c] (43)	15.5 ± 0.5[b] (43)	0.45

Different letters in each row indicate significancy of P at least < 0.05.

Fig. 4A. Maturation of some neurophysiological characteristics of magnocellular neurones in the PVN of the chick embryo and newborn chicken.

At least from E16 onwards the chick embryo is able to respond to osmotic challenge by increased AVT secretion into the blood (Fig. 3; Klempt at al., 1992; Xu et al., 1992). Compared to the adult chicken the D1 chick showed a 'close-to-mature' response (Mühlbauer et al., 1992). Remarkable changes in the electrophysiological characteristics of the hypothalamo-neurohypophysial system of the chick embryo and the newborn chicken were observed during the perinatal period (Fig. 4A; Grossmann & Ellendorff, 1986a,b).

Direct electrical stimulation of the neurohypophysis elicited higher AVT plasma concentrations in the D1 chick compared to the E18 embryo (Fig 5).

	AD-neurones	non-AD neurones
basic frequency (spikes/sec.)	0.68 ± 0.19 n = 12	0.85 ± 0.32 n = 7
10 to 14 min. p.i. (spikes/sec.)	1.24 ± 0.31 n = 12	1.28 ± 0.43 n = 7
25 to 30 min. p.i. (spikes/sec.)	2.03 ± 0.58 n = 8	1.98 ± 0.62 n = 5

Fig. 4B. Changes in firing rate (X ± SEM) of osmoresponsive hypothalamic neurones (AD = antidromically identified) recorded in D1 chicks before and after (p.i.) intraperitoneal injection of hypertonic saline (n = number of recorded cells).

Identified magnocellular neurones of the hypothalamic nucleus paraventricularis in newly hatched chickens are clearly osmosensitive. Osmotic challenge increases neuronal firing activity up to tenfold and more within 40 min after i.p. osmotic stimulation (Fig. 6; Grossmann & Ellendorff, 1990). However, as demonstrated in Fig. 4B not only identified magnocellular neurones are osmosensitive but also other hypothalamic neurones within the paraventricular nucleus respond to osmotic challenge with similar increase in firing frequencies and similar latencies to i.p. injection of hypertonic saline. Thus, the question raised some time ago whether or not osmoreceptors are localized directly on magnocellular neurones remains still to be answered (Leng et al., 1982; Russell et al., 1988) and may apply also for the osmoregulatory system in avian species. The modulatory input to the AVT secretory system also seems to be functional already during the perinatal period. The endogenous opioid peptide system act very sensitive to inhibit the osmotically stimulated AVT release (Xu et al., 1991). Similarly central administration of Angiotensin II is able to release AVT and modulates the osmotically stimulated AVT secretion (Gerulat & Grossmann, manuscript in preparation). During osmotic challenge not only plasma osmolality, plasma sodium and AVT concentration is elevated but also AVT gene expression is upregulated (Fig. 7). Although our neurophysiological and endocrine data point to an advanced developmental stage of the osmoregulatory function of AVT in the one-day-old chicken, maturational processes particularly of the modulatory input to AVT synthesizing neurones may continue well into the postnatal life.

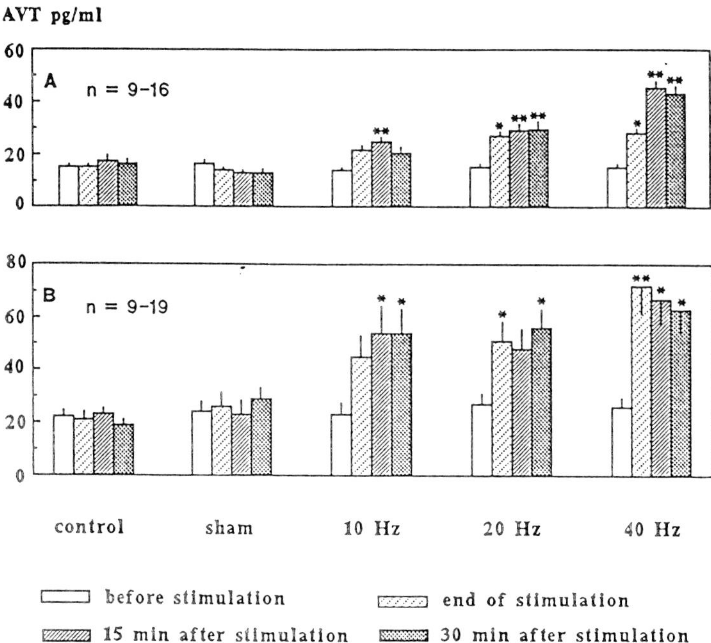

Fig. 5. Comparison of AVT plasma concentrations before and after electrical stimulation (0,8mA peak-to-peak, 1ms, 20sec on/off for 10 min) of the neurohypophysis in the E18 embryo (A) and D1 chick (B).

In conclusion, the chicken embryo provides an useful model for investigating the functional ontogeny of neuroendocrine systems in the absence of some of the problems encountered in mammalian species. The magnocellular neurosecretory system in the chicken is comparable to the mammalian system in many characteristics and thus provides an additional model to investigate a variety of concepts in developmental maturation of the neuroendorcrine brain. The relative inaccessability of the fetal mammalian brain in utero can be overcome by using non-placental animal species like the chick. HNS hormones are synthesized in well-characterized and anatomically discrete cell groups, yet associated with different functions of independent circuitries. The considerable time difference between appearance of first gene expression in hypothalamic neurones and detectable amounts of hormone in the plasma implies that despite the known functions the nonapeptides may also be involved in embryonal processes like cell migration and, thus, in brain development. Finally, basic principles of hormone secretion seem to be established early during development.

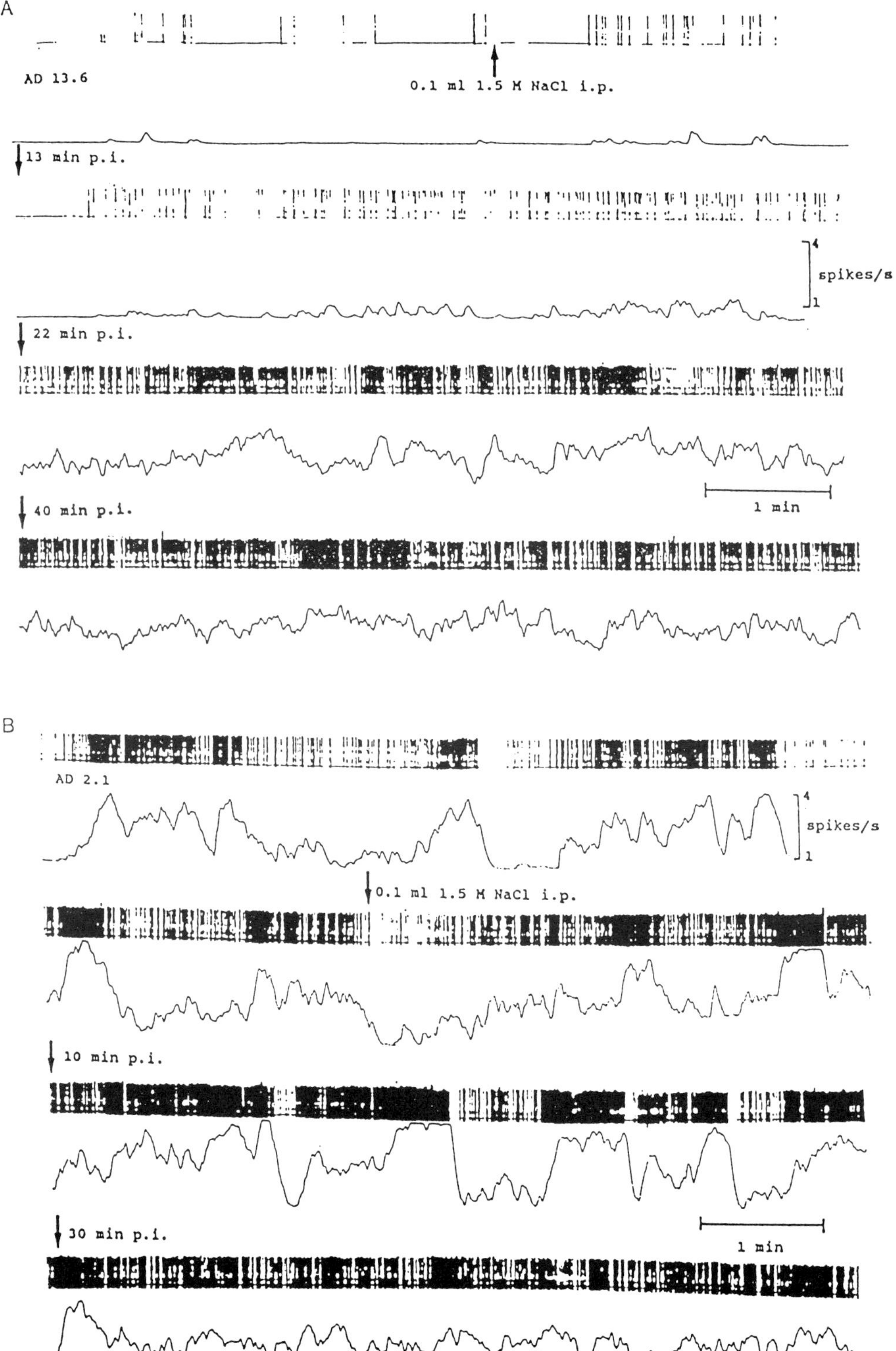
A
AD 13.6
0.1 ml 1.5 M NaCl i.p.
13 min p.i.
spikes/s
4
1
22 min p.i.
1 min
40 min p.i.
B
AD 2.1
spikes/s
4
1
0.1 ml 1.5 M NaCl i.p.
10 min p.i.
1 min
30 min p.i.

Fig. 6. Chart records taken from two magnocellular neurones in the PVN of the D1 chick before and after i.p. injection of hypertonic saline.

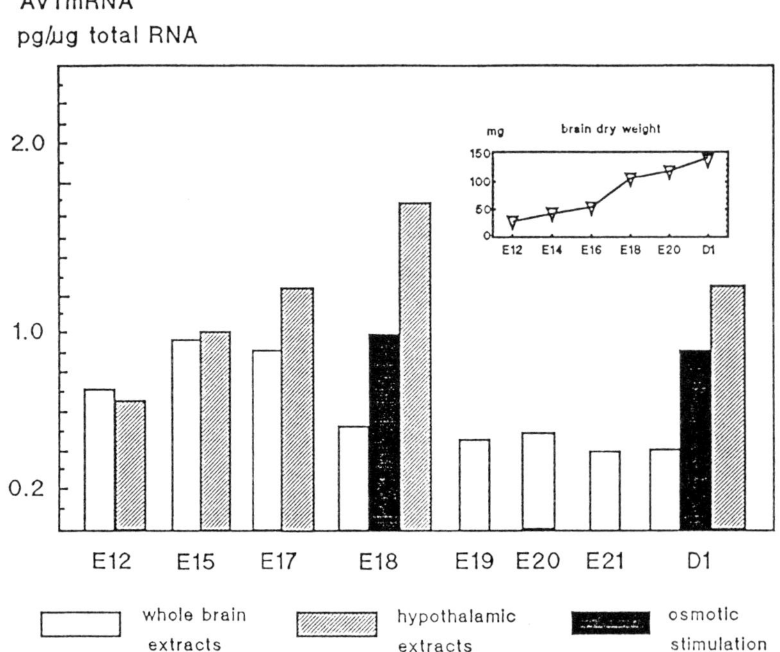

Fig. 7. AVT -mRNA content during embryonal development in the chicken measured in whole brain and hypothalamic extracts, respectively. Responses in the genetic level (whole brain extracts) to osmotic challenge is less pronounced at E18, compared to D1, indicating maturational processes in the gene expressing machinery during this period.

REFERENCES

Acher, R., Chauvet, J., and Chauvet, M.T. (1970): Physiology of neurohypophyseal hormones: the avian active peptides. Eur. J. Biochem. 17: 509-513.

Acher, R., and Chauvet, J. (1988): Structure, processing and evolution of the hypophysial hormone-neurophysin-vasopressin. Biochemie 70: 1197-1207.

Dyball, R.E.J., Grossmann, R., Leng, G., and Shibuki, K. (1988): Spike propagation and conduction failure in the rat neural lobe. J. Physiol. 401: 241-256.

Ellendorff, F., and Grossmann, R. (1988): Development of neuroendocrine pathways of the hypothalamo-neurophypophysial system (HNS) in the prenatal chick and pig. In Fetal and Neonatal Development, ed C. T. Jones, Perinatology Press, Ithaca, N.Y.: 215-221.

Ellendorff, F., Grossmann, R., Milewski, N., Klempt, M., and Ma. E. (1990): Development of the fetal neuroendocrine brain for adaption. In Fetal autonomy and adaptation, eds G.S. Dawes, A. Zacutti, F. Borruto and A. Zacutti, jr., pp. 137-143. John Wiley & Sons Ltd., London.
Grossmann, R., and Ellendorff, F. (1986a): Functional development of the prenatal brain.I. Recording of extracellular action potentials from the magnocellular systems of the 18-day-old chicken embryo. Exp. Brain Res. 62: 635-641.
Grossmann, R., and Ellendorff, F. (1986b): Functional development of prenatal brain. II. Ontogeny of the hypothalamo-neurohypophysial axis in the pre- and perinatal chicken brain. Exp. Brain Res. 62: 642-647.
Grossmann, R., and Ellendorff, F. (1990): Neural control of posterior pituitary. In Research Trends in Fetal Physiology, eds K.G. Rosen and H. Lagerkrantz. Chalmers University of Technology, Göteborg, Sweden, pp. 59-62.
Grossmann R., Xu, B., Dublecz, K., Mühlbauer, E., and Ellendorff, F. (1993): Hypothalamic regulation of fluid balance. In Proceedings of the 4th European Symposium on Poultry Welfare, eds C.J. Savory and B.O. Hughes, pp. 67-78.
Klempt, M., Ellendorff, F., and Grossmann R. (1992): Functional maturation of arginine vasotocin secretory responses to osmotic stimulation in the chick embryo and the newborn chicken. J. Endocrinol. 133: 269-274.
Leng, G., Mason, W.T., and Dyer, R.G. (1982): The supraoptic nucleus as an osmoreceptor. Neuroendocrinol. 34: 75-82.
Lawzewitsch, I. von (1969): Development of the neurosecretory hypothalamic-hypophyseal system of the chick embryo as evidenced in the total preparation. Acta anat. 72: 83-93.
Milewski, N., Ivell, R., Grossmann, R., and Ellendorff, F. (1989): Ontogeny of Vasotocin/Mesotocin gene expression in the hypothalamo-neurohypophyseal system of the chicken embryo. J. Neuroendocrinol. 1: 473-484.
Mühlbauer, E., Hamann, D., Xu, B., Ivell, R., Ellendorff, F., and Grossmann R. (1992): Arginine vasotocin gene expression during osmotic challenge in the chicken. J. Neuroendocrinol. 4: 347-351.
Mühlbauer, E., Hamann, D., Xu, B., Ivell, R., Udovic, B., Ellendorff, F., and Grossmann, R. (1993): Arg-Vasotocin gene expression and hormone synthesis during ontogeny of the chicken embryo and the newborn chicken. J. Neuroendocrinol. 5: 281-288.
Nouwen, E.J., Decuypere, E., Kühn, E.R., Michels, H., Hall, T.R., and Chadwick, A. (1984): Effects of dehydration, haemorrhage and oviposition on serum concentrations of Vasotocin, Mesotocin and prolactin in the chicken. J. Endocrinol. 102: 345-351.
Poulain, D.A., and Wakerley, J.B. (1982): Electrophysiology of hypothalamic magnocellular neurones secreting oxytocin and vasopressin. Neuroscience 7: 773-808.
Russell, J.A., Blackburn, R.E., and Leng, G. (1988): The role of the AV3V region in the control of magnocellular oxytocin neurones. Brain Res. Bull. 20: 803-810.
Tennyson, V.M., Nilaver, G., Hou-Yu, A., Valiquette, G., and Zimmerman, E.A. (1986): Immunocytochemical study of the development of vasotocin/mesotocin in the hypothalamo-hypophysial system of the chick embryo. Cell Tiss. Res. 243: 15-31.
Wingstrand, K.G. (1954): Neurosecretion and antidiuretric activity in chick embryos with remarks on the subcommissural organ. Arkiv Zool. 6: 41-67.

Xu, B., Ellendorff, F., and Grossmann, R. (1991): Endogenous opioid peptides are involved in regulation of Arg-vasotocin secretion in the newly hatched chicken. Acta Endocrinol. 124: 1: 4.

Xu, B., Ellendorff, F., and Grossmann, R. (1992): Neurohypohyseal hormones are synthesized very early during embryonic life in the chicken. Acta Endocrinol. 126: 4: 128.

Integrative and cellular aspects of autonomic functions : temperature and osmoregulation. Eds K. Pleschka, R. Gerstberger. John Libbey Eurotext, Paris © 1994, pp. 439-449.

Osmosensitive circumventricular structures connected to the paraventricular nucleus in the duck brain

Andreas R. Müller, Frank Schäfer, Herbert A. Schmid and Rüdiger Gerstberger

Max-Planck-Institut für Physiologische und Klinische Forschung, W.G. Kerckhoff-Institut, Parkstrasse 1, D-61231 Bad Nauheim, Germany

CENTRAL OSMOSENSITIVITY INSIDE AND OUTSIDE THE BLOOD-BRAIN BARRIER

Representing the main integrative and an important sensory center for homeostatic systems including body fluid homeostasis, the hypothalamus is composed of nuclear and regional neuronal entities within the blood-brain barrier (BBB) and circumventricular structures (CVOs) lacking a pronounced BBB and therefore being accessible to blood-born agents. With regard to the central control of drinking and antidiuretic hormone (ADH) release from the paraventricular (PVN) and supraoptic (SON) nuclei, the pioneering studies of Verney (Jewell & Verney, 1957) suggested the anterior hypothalamus as a major perceptive site. Intrahypothalamic or intracerebroventricular (icv) microinjections of hypertonic artificial cerebrospinal fluid (aCSF) in the rat (Buggy et al., 1979), goat (Andersson, 1978), sheep (Rundgren et al., 1986, McKinley et al., 1990), dog (Eriksson et al., 1987), monkey (Swaminathan, 1980) and Pekin duck (Gerstberger et al., 1984b) suggested the existence of osmo-, and possibly sodium sensitive structures in the vicinity of the third cerebral ventricle (V-III). The magnocellular ADH-synthesizing SON itself was long believed to serve as the osmosensor within the brain responsible for the control of ADH release (Leng et al., 1982). In addition, the anterior ventral wall of the third cerebral ventricle (AV3V region) proved to be a promising candidate for the central osmoreceptor (Johnson et al., 1992). Just recently the orthodromic excitation of SON neurones non-responsive to osmotic stimuli *per se* could be verified after either electrical stimulation of the AV3V region or pressure ejection of hypertonic saline directly to AV3V neurones in the anaesthetized rat. Lidocain injection into the AV3V region or the interconnected median preoptic nucleus (PMN) resulted in the suppression of SON activation following a systemic hypertonic saline stimulus, indicative of a functional SON-AV3V-PMN-SON osmoreceptor complex (Honda et al., 1990). The well characterized cytoarchitecture of the avian hypothalamus with its topographical separation of PVN subnuclei (Korf, 1984) enabled the identification of periventricular osmosensitive elements in the anterodorsal part of the V-III, employing *in vivo* animal physiology and *in vitro* electrophysiology (Gerstberger et al., 1984b; Kanosue et al., 1990,1992). Icv perfusion experiments indicated the presence of a brain-intrinsic cation (sodium)-selective mechanism monitoring changes in extracellular fluid (ECF) tonicity, similar to results obtained in mammals.

Lesioning of the lamina terminalis containing not only structures within the BBB but also two CVOs ouside the BBB, the subfornical organ (SFO) and the organum vasculosum laminae terminalis (OVLT), led to the pioneering report by Andersson et al. (1975) describing the important sensory role of these structures within the lamina terminalis in mechanisms controlling water and sodium balance. Localized ablations of the OVLT, AV3V region and PMN in the dog, sheep or rat resulted in at least temporarily impaired osmoregulation (Thrasher, 1989; Johnson et al., 1992; McKinley et al., 1992). Besides the presence of intracerebral osmo- or sodium sensitive elements, the putative osmosensory function of these CVOs outside the BBB was supported by experiments carried out in the sheep, dog and duck. Intracarotid infusions of various hyperosmolar electrolyte and non-electrolyte solutions resulted in stimulated release of ADH into the blood, water intake or duck salt gland secretion (McKinley et al., 1978; Deutsch et al., 1979; Thrasher et al., 1980). Despite the fact that all compounds were equally effective in elevating intracerebral sodium concentration (McKinley et al., 1978), the magnitude of the effector responses proved to be different for the various hyperosmotic stimuli. Neural tracing and electrophysiological studies clearly revealed complex reciprocal connections between the SFO, OVLT or AV3V region and the PVN, SON and PMN in the rat and sheep (McKinley et al., 1990,1992; Ferguson, 1992). With regard to the presence of putative osmosensitive neurones within the OVLT and SFO, studies have been performed almost exclusively in the rat, indicating cells in the SFO responsive to hypertonic saline and mannitol, but not urea (Sibbald et al., 1988). Whereas these data obtained for the SFO remain rather questionable, extracellular recordings from rat OVLT neurones *in vitro* indicate osmo- or sodium sensitivity of most of the neurones tested (Sayer et al., 1984; Vivas et al., 1990). In the duck, retrograde tracing studies and Golgi impregnations demonstrated monosynaptic axonal connections of osmosensitive, periventricular cells to magnocellular PVN neurones (Korf et al., 1982,1983). With limited information available for the SFO (Korf, 1984), neuroanatomical connections of the OVLT to the PVN or SON have not been demonstrated for this species, and neurones of both structures have not been thoroughly analyzed with regard to putative osmo- or sodium sensitivity. The present study was therefore carried out to elucidate both aspects, employing techniques of neuronal tract-tracing and *in vitro* electrophysiology.

NEURAL CONNECTIONS BETWEEN THE MAGNOCELLULAR PVN AND SENSORY CVOs OF THE LAMINA TERMINALIS: A DiI TRACING STUDY.

The carbocyanine fluorescent dye DiI was employed to identify neural connections between the magnocellular component of the PVN (mPVN) and putative osmosensory structures inside and outside the BBB of the duck hypothalamus, namely the periventricular nucleus (PeN), cerebrospinal fluid contacting neurones, or the SFO and OVLT. This neural tract-tracing technique has proven useful to mark both antero- and retrograde neuronal connections within the central nervous system of higher vertebrates (Honig & Hume, 1989; Holmqvist et al., 1992).

Methodology

For tracing of afferent and efferent neural connections to and from the avian PVN, adult Pekin ducks (n=15) heparinized by intravenous (iv) injection of 4,000 IE of Vetren R (Promonta, FRG) were transcardially perfused under pentobarbital anaesthesia (60 mg/kg bw iv) with 800 ml 0.2 M PBS, pH 7.4, followed by 800 ml of freshly prepared 4% paraformaldehyde in 0.1 M phosphate buffer, pH 7.4 at 4°C. The quickly dissected brains were postfixed for 24-48 hrs using the same fixative, and embedded in 3% agarose (Sigma, FRG) at 40°C. The brains were cut in half by transversal sectioning at the level of the mPVN (Kuenzel & Masson, 1988). For histological identification of the mPVN, serial sections of one half of the brain were counterstained with methylene blue. The second half of the brain was used for application of a minute crystal of the fluorescent carbocyanine dye DiI (Molecular

Probes, USA) into the mPVN using the counterstained sections as reference (Fig. 1). Applications were performed either unilateral into the mPVN or adjacent tissue for control of labelling specificity. To allow transport of DiI, the half-brain was incubated in 4% paraformaldehyde solution at 40°C in the dark for 1-7 days. The tissue block was then immersed in 30% sucrose for 3 days at 4°C, frozen in powdered dry-ice and mounted in a cryostat. Serial, quickly air-dried sections were analyzed by fluorescence microscopy using a rhodamine filter set (Zeiss, FRG).

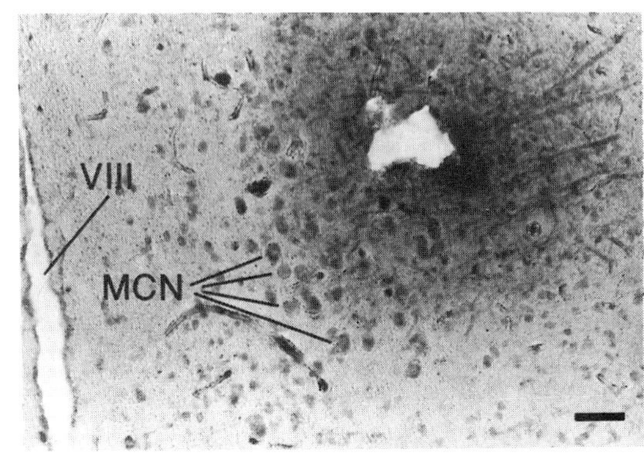

Fig. 1: Micro-application of a minute DiI cristal into the magnocellular portion of the right paraventricular nucleus (mPVN) in the duck brain prefixed with 4% paraformaldehyde. Counterstainings of the brain tissue with methylene blue emphasize the discrete localization of the tracer application including its halo of diffusion after 3 days of incubation. MCN = magnocellular neurone, VIII = third cerebral ventricle. Bar represents 80 μm.

Results and Discussion

Application of the DiI crystal into the mPVN *in vitro* resulted in retro- and anterograde labelling of neurones and fibers, respectively, along the entire lamina terminalis of the duck brain, extending from the optic recess of the V-III to the main body of the SFO located dorso-caudally to the anterior commissure (Fig. 2). Embedded within the rostoventral component of the lamina, the OVLT of the duck exhibits specialized tanycytes, small neurones and capillaries of both the fenestrated and non-fenestrated type (Bosler, 1977). Throughout the ventral lamina, densely stained fiber networks could be observed running either in parallel to the ventricular wall or perpendicular to it towards the adjacent brain parenchyma. Both within the OVLT and in adjacent sections located caudally of the lamina, small bipolar neurones and periventricular cells proved to be DiI positive. Fibers could be traced dorsally to the OVLT, with a distinct and dense network containing a few cells aggregated within a section of the upper lamina of yet unknown neuroanatomical description (Kuenzel, personal communication), but possibly representing a dorsoventral component of the avian SFO. Small, losely distributed cells containing DiI were also observed within the main body of the SFO surrounding its central blood vessel. Fibers could be detected in the SFO as well as running between the mPVN and the SFO along the ventricle. Thus, both CVOs lining the lamina terminalis appear to send and receive information to/from the mPVN, whereas the dorsally oriented paraventricular organ appeared not to be connected with the mPVN (data not shown). Retrograde tracing using horseradish peroxidase (HRP) injected into the duck PVN did not reveal labelling of the OVLT-AV3V region, but showed staining of SFO elements when using large amounts of the tracer (Korf et al., 1982). Dense DiI positive fiber networks could additionally be traced running in rostro-caudal direction within the N. anterior hypothalami medialis. Transsection of these fibers markedly impaired osmoregulatory function in the pigeon (Takei, 1977). Interestingly, laminar structures in the duck hypothalamus labelled with DiI and thus connected to the mPVN, as well as the PVN itself, are endowed with angiotensin II (ANG-II) receptors, with the clear-cut exception of the OVLT (Gerstberger et al., 1992). Mimicking brain-intrinsic transmitter action, *in vivo* application of ANG-II into the rostral component of the V-III exclusively caused dose-dependent rises in blood pressure, release of ADH, antidiuresis and inhibition of salt gland function. Intracarotid infusions of ANG-II did not change blood pressure and plasma ADH, but inhibited salt gland secretion *via* a clearly SFO-mediated mechanism (Gerstberger et al., 1992; Simon et al., 1992). In mammals, centrally acting ANG-II plays an important role in thirst mechanism, ADH release and the control of blood pressure (McKinley et al., 1992).

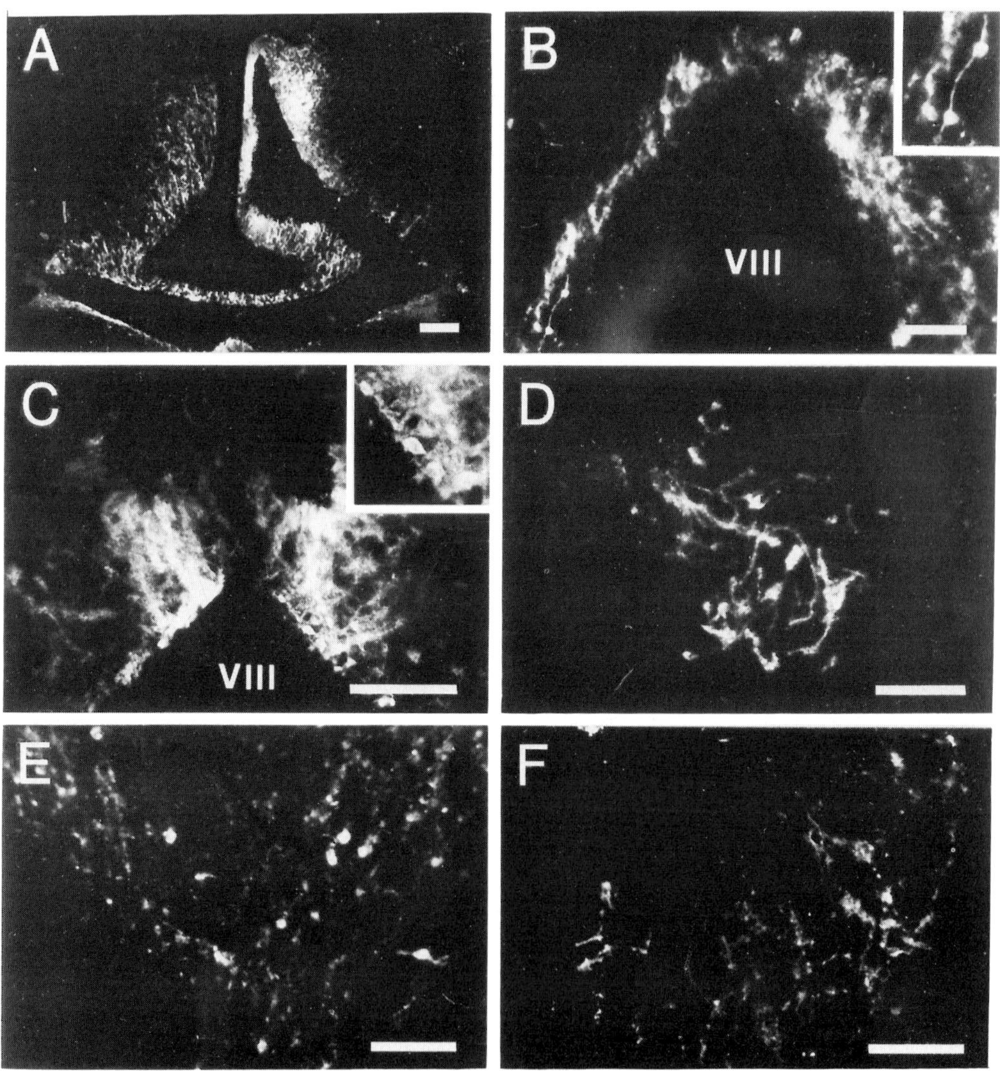

Fig. 2 : Tracing of cellular and fiber components along the lamina terminalis of the duck hypothalamus after DiI application into the mPVN. Cells and fibers could be located within the entire lamina (A), the OVLT (B), dorsocaudal to the OVLT (C) and a yet unidentified component of the dorsal lamina possibly representing a rostroventral component of the SFO (D). Single cells (E) and fibers (F) were stained within the main body of the SFO. Insets in B and C show small bipolar (B) and periventricular (C) cells stained with DiI within or close to the OVLT. Bar represents 200 μm in A, 100 μm in C,D and 50 μm in B,E,F.

ANG-II might therefore be a prime candidate for central mediation of afferent but also efferent signals in the control of body fluid homeostasis, relaying information from the SFO to the PVN, as also indicated for the rat (Bains et al., 1992; Tanaka et al., 1989).

Tracing studies performed both in the rat and sheep brain have put forward the hypothesis that osmoreceptors in the SFO and OVLT might directly, e.g. monosynaptically, transmit signals to the magnocellular neurones of the PVN and SON (McKinley et al., 1992). Application of cholera toxin labelled HRP and colloidal gold into the SON resulted in marked labelling of neurones within the OVLT, SFO and PMN, with axons originating in both CVOs being relayed to the SON *via* the PNM (Oldfield et al., 1991; McKinley et al., 1992). The simultaneous injection of fluoro-gold or rhodamine labelled microspheres into the SON and ipsilateral PVN resulted in double-labelling of neurones in the rat SFO, with only a few cells present in the PMN and OVLT. Single cells within structures of the lamina terminalis might therefore provide input to more than one magnocellular neuroendocrine nucleus via branching axons (Weiss & Hatton, 1990). The morphological features of rat OVLT neurones injected with Lucifer Yellow and characterized electrophysiologically revealed axonal projections to the PMN and the ipsilateral PVN and SON. Connections to the SON could additionally be verified by antidromic activation

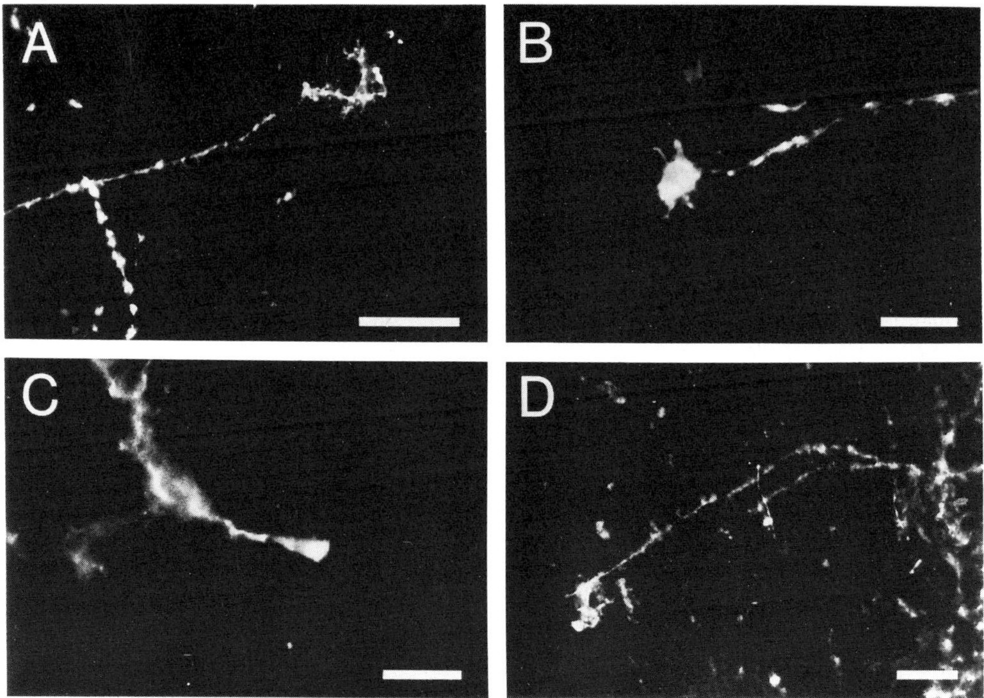

Fig. 3 : Tracing of periventricluar cells in the third ventricular region after DiI application into the mPVN. Labelled neurones could be localized at rather high density in the anteroventral portion of the VIII and dorsocaudally to the mPVN. These cells were either located subependymally with long, sometimes branching axonal projections running perpendicularly towards the mPVN and with dendritic processes sent into the cerebrospinal fluid (A,B), or intraependymally (D). Tanycytes proved also to be labelled with lipophil DiI (C). Bar represents 40 μm in A,D and 20 μm in B,C.

(Nissen et al., 1993). Efferent projections from periventricular parvocellular components of the rat PVN could be followed up to the OVLT, and to a lesser extent also to the SFO, employing anterograde tracing with *Phaseolus lectin* (Larsen et al., 1991).

In addition to CVO components connected with the mPVN of the duck brain, labelled CSF-contacting neurones and fibers in both the AV3V region and the PeN with cells directly responsive to osmotic stimuli (Kanosue et al., 1990,1992) indicate afferent connections to the mPVN (Fig. 3). Periventricular cells containing DiI were found subependymally or positioned within the ependymal layer, with dendritic processes sent into the CSF also stained for some of these neurones. In most cases their axonal projections were running towards the PVN, with axonal branching observed in a few occasions. Bell-shaped intraependymal cells labelled with DiI were often found in closely attached doublets, triplets or quadruplets, and tanycytes appeared to be marked. Microinjection of retrogradely transported HRP into the PVN revealed discrete labelling of cells within the dorsoventral ependymal layer of the V-III representing both CSF-contacting neurones and tanycytes. Direct contact of these cells with magnocellular neurones of the PVN could be revealed employing the Golgi impregnation technique (Korf et al., 1982,1983).

OSMO-RESPONSIVENESS OF NEURONES IN THE OVLT AND SFO

The putative sensory function of OVLT and SFO neurones, located outside the BBB but neurally connected to the ADH-synthesizing mPVN, in the osmoregulatory circuit remains to be elucidated for the Pekin duck as a model system to study body fluid homeostasis. The responsiveness of single neurones within the OVLT and SFO to alterations in the osmolality or sodium concentration of their extracellular environment could be investigated *in vitro*, using the brain slice technique.

Methodology

For the extracellular recording of neuronal activities within the OVLT and SFO of the duck hypothalamus, coronal brain slices of adult duck brains (400 µm thick) containing the optic chiasm (OC) (for OVLT recordings) or the pallial commissure (for SFO recordings) were used (Schmid & Simon, 1992). The brain slices were trimmed to a final size of 3 x 3 mm, preincubated at 34°C for 2 hrs and finally transferred to the recording chamber. The chamber was perfused at a rate of 1.6 ml/min with artificial cerebrospinal fluid (aCSF) containing 124 mM NaCl, 5 mM KCl, 1.2 mM KH_2PO_4, 1.3 mM $MgSO_4$, 1.2 mM $CaCl_2$, 26 mM $NaHCO_3$ and 10 mM glucose, yielding an osmotic pressure of 290 mOsm/kg. The aCSF solution was equilibrated with 95% O_2 and 5% CO_2, pH 7.40, and the temperature was kept constant at 39.0°C by means of a Peltier element (NPI, FRG). Extracellular recordings of OVLT as well as SFO neurones were performed using glass-coated platinum-iridium electrodes. With the body of the SFO being easily accessible (Schmid & Simon, 1992), the cellular component of the duck OVLT could be identfied directly above the OC using the prominent vascularization of the OVLT as a marker (Kuenzel & Masson, 1988). Care was taken to avoid recordings in the adjacent wall of the third ventricle and the median preoptic nucleus. The recorded action potentials were amplified, displayed on a storage oscilloscope (Gould, FRG) and computer-analyzed (Spike2; CED, UK) after passing a window discriminator (WPI, USA). For osmotic stimulation, NaCl was added to a small reservoir of aCSF (10 ml) resulting in a final increase in aCSF osmolality of 20-40 mOsm/kg as proven by vapour pressure osmometry (Wescor 5100C; Wescor, USA). At unaltered perfusion kinetics, the slice was superfused with this hypertonic aCSF for 4.5 to 6.0 min. In addition, the sensitivity of the recorded neurones was determined by superfusing ANG-II (10^{-7} M) for the same length of time. A neurone was considered as being sensitive or responsive to the applied stimulus, when the average change in discharge rate during the response time proved to be both larger than 0.5 Hz and 20% as compared to base-line values.

Results and Discussion

Single unit recordings obtained from the highly vascularized duck OVLT (Bossler, 1977) *in vitro* proved to reveal a lower number of spontaneously active neurones when compared to e.g. the subfornical organ. Out of 29 spontaneously active neurones recorded from duck OVLT, 31% changed their firing rate (FR) in response to an increase in ECF tonicity, with five neurones revealing a significantly enhanced, and four neurones a diminshed neuronal activity. The mean basal FR of all OVLT neurones tested proved to be 3.5±0.7 Hz, which was comparable to recordings obtained from duck SFO neurones (FR = 3.8±0.5 Hz). Similar frequencies were reported for recordings of rat OVLT neurones *in vitro* with a basal discharge rate of 2.7±0.4 Hz (Vivas et al., 1990). The spontaneous activity was not significantly different in osmosensitive vs. non-osmosensitive duck OVLT neurones (2.4±0.7 vs. 3.9±0.9 Hz). Fig. 4 demonstrates examples of three different OVLT neurones responding to hyperosmotic stimuli with an increase, a decrease, or no change in activity. Dose-dependency of the increase in activity could be demonstrated for the neurone in the upper part of Fig.4. With a threshold of about +18 mOsm/kg for its osmoresponsiveness, the hyperosmotically induced stimulation proved to be quickly reversible, and showed no signs of tachyphylaxis but variations in the onset of the response. Extracellular recordings from rat OVLT neurones *in vitro* indicate osmo- or sodium sensitivity in the majority of neurones tested (Knowles & Phillips, 1980; Sayer et al., 1984; Vivas et al., 1990). Intracellular recordings of rat OVLT neurones projecting to the ipsilateral SON revealed membrane depolarization during superfusion of the hypothalamic explant with aCSF of elevated (+10 to +40 mOsm/kg) osmolality due to NaCl or mannitol (Nissen et al., 1993). ANG-II has been discussed as an important hormone with afferent signal tranduction characteristics or as a hypothalamic neuromodulator within the osmoregulatory circuit of mammals and birds (McKinley et al., 1990; Saavedra, 1992; Simon et al., 1992). The rat OVLT is densely endowed with ANG-II specific binding sites (Saavedra, 1992), and neurones of the OVLT proved to be responsive to this neuropeptide (Knowles & Phillips, 1980; Sayer et al., 1984). In the Pekin duck, the OVLT may not be important for ANG-II mediated modulation of the osmoregulatory circuit due to the fact, that neither the presence of ANG-II binding sites (Gerstberger et al., 1992), nor responsiveness of OVLT neurones to ANG-II application could be demonstrated (Fig. 4; Table 1).

	Osmo-responsive	ANG-II responsive
OVLT	9/29 (31%)	0/10 (0%)
SFO	11/35 (31%)	47/64 (73%) [c]
PeN	29/48 (60%) [a]	1/37 (3%) [b]
mPVN	3/30 (10%) [a]	3/81 (4%) [b]

Table 1 : Significant alterations in the firing rate of neurones recorded *in vitro* in the organum vasculosum laminae terminalis (OVLT), subfornical organ (SFO), periventricular subnucleus of the PVN (PeN) and magnocellular PVN (mPVN) of the duck hypothalamus during stimulation with hypertonic artificial CSF (Osmo-responsive) or angiotensin II (ANG-II responsive). Percentages represent the portion of responsive cells (No 1) of all neurones tested (No 2) within one region. Data derived from literature: (a) Kanosue et al., 1990 (b) Matsumura et al., 1990 (c) Schmid & Simon, 1992.

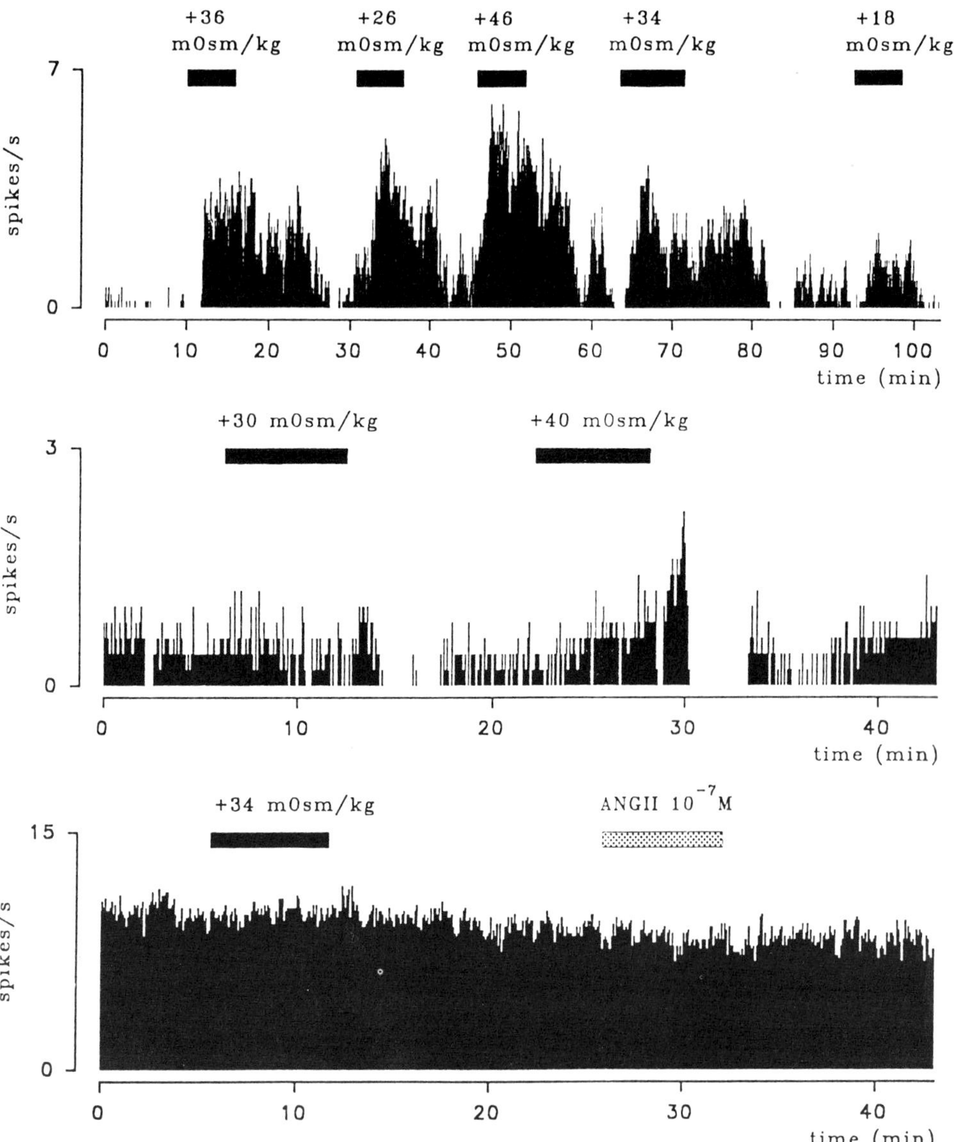

Fig. 4: Continuous rate meter recordings of 3 different OVLT neurones from duck brain slices. The neurone in the upper trace reveals a low basal firing rate and displays a dose-dependent increase in its FR to hypertonic stimuli (at indicated concentrations) by superfusing the slice with aCSF of different NaCl content. The middle trace shows a neurone which slightly decreases its FR in response to a +30 mOsm/kg hyperosmotic stimulus, and reveals a biphasic response, i.e. a short increase followed by a delayed decrease due to a hypertonic stimulus of +40mOsm/kg. The bottom trace demonstrates the ineffectiveness of ANG-II to stimulate an OVLT neurone, which was also not responsive to hypertonic aCSF.

With regard to the SFO, the sensitivity of its neurones to changes in ECF tonicity remains still questionable. Employing the brain slice technique, Buranarugsa & Hubbard (1979) did not report any evidence for the hypothesis that neurones in the rat SFO are excited or inhibited by alterations in extracellular osmotic pressure with sodium concentrations varying from 74 to 174 mM. A few years later, the same research group published, that hypertonic stimulation of rat SFO neurones *in vitro* demonstrated that 30-40% of the cells tested were responsive to small alterations in ECF osmolality with sodium or mannitol as osmotically active substances, whereas none of the cells changed the firing rate after hypertonic urea stimulation (Sibbald et al., 1988). This finding would support the proposal of a central osmosensitivity outside the BBB put forward by McKinely et al. (1978, 1992). Antidromically identified neurones of the rat SFO *in vivo* increased their firing frequency after intracarotid infusion of strongly hypertonic saline (Gutman et al., 1988). Single unit recordings obtained from either magnocellular PVN neurones or SFO neurones in the urethane-anaesthetized rat showed, that the responsiveness of phasic PVN neurones to elevated carotid blood tonicity was dependent on an intact SFO (Tanaka et al., 1989). For the Pekin duck, 30% of the SFO neurones investigated *in vitro* responded to the application of hypertonic aCSF (Table 1). Half of these cells answered with transient excitation, whereas the remaining cells showed a delayed inhibitory response. Preliminary experiments also indicate responsiveness of SFO cells to hyperosmotic stimulation with the rise in tonicity due to mannitol, thus speaking in favour of a general osmo- vs. sodium sensitivity of some of these neurones. ANG-II specific binding sites have been described for the duck as well as rat SFO with putative physiological implications for water intake, blood pressure control, ADH release and avian salt gland function (Saavedra, 1992; Simon et al., 1992). For both classes of vertebrates, electrophysiological recordings revealed a high percentage of ANG-II sensitive neurones within the SFO (Okuya et al., 1987; Matsumura & Simon 1990; Schmid & Simon 1992) (Table 1), with a few neurones of the duck SFO responding to both an osmotic stimulus and ANG-II application.

CONCLUSION

Neurones in the duck SFO and OVLT, two circumventricular structures outside the blood-brain barrier, are accessible to alterations in systemic extracellular fluid as well as brain-intrinsic ion and hormone/transmitter concentrations. Serving as a sensory "window" to the brain, both structures play an important role in afferent signal transduction within the osmoregulatiry circuit. The combined application of the DiI neuronal tract-tracing technique and the *in vitro* brain slice technique for extracellular recording of neuronal activity allowed the identification of osmo-responsive cells in the OVLT and SFO, and neural connections between SFO/OVLT neurones and the magnocellular component of the PVN.

REFERENCES

Andersson, B., Leksell, L.G. & Lishajko F. (1975): Perturbations in fluid balance induced by medially placed forebrain lesions. *Brain Res.* 99, 261-275.

Andersson, B. (1978): Regulation of water intake. *Physiol. Rev.* 58, 582-603.

Bains J.S., Potyok A & Ferguson A.V. (1992): Angiotensin II actions in paraventricular nucleus: functional evidence for neurotransmitter role in efferents originating in subfornical organ. *Brain Res.* 599, 223-229.

Bosler O. (1977): The organum vasculosum laminae terminalis. A cytophysiological study in the duck, *Anas platyrhynchos*. *Cell Tissue Res.* 182, 383-399.

Buggy, J., Hoffman, W.E., Philipps, M.I., Fisher, A.E. & Johnson, A.K. (1979): Osmosensitivity of rat third ventricle and interactions with angiotensin. *Am. J. Physiol.* 236, R75-R82.

Buranarugsa, P. & Hubbard, J.I. (1979): The neural organization of the rat subfornical organ in vitro and a test of the osmo- and morphine-receptor hypothesis. *J. Physiol. (Lond)* 291, 101-116.
Deutsch, H., Hammel, H.T., Simon, E. & Simon-Oppermann, C. (1979): Osmolality and volume factors in salt gland control of Pekin ducks after adaptation to chronic salt loading. *J. Comp. Physiol.* 129, 301-308.
Eriksson, S., Simon-Oppermann, C., Simon, E. & Gray, D.A. (1987): Interaction of changes in 3rd ventricular CSF tonicity, central and systemic AVP concentrations and water intake. *Acta Physiol. Scand.* 130, 575-583.
Ferguson A.V. (1992): Neurophysiological analysis of mechanisms for subfornical organ and area postrema involvement in autonomic control. *Prog. Brain Res.* 91, 413-421.
Gerstberger R., Gray D.A. & Simon, E. (1984a): Circulatory and osmoregulatory effects of angiotensin II perfusion of the third ventricle in a bird with salt glands. *J. Physiol. (Lond)* 349, 167-182.
Gerstberger R., Simon E. & Gray D.A. (1984b): Salt gland and kidney responses to intracerebral osmotic stimulation in salt- and water-loaded ducks. *Am. J. Physiol.* 247, R1022-R1028.
Gerstberger R., Müller A.R. & Simon-Oppermann C. (1992): Functional hypothalamic angiotensin II and catecholamine receptor systems inside and outside the blood-brain barrier. *Prog. Brain Res.* 92, 423-433.
Gutman, M.B., Ciriello, J. & Mogenson, G.J. (1988): Effects of plasma angiotensin II and hypernatremia on subfornical organ neurons. *Am. J. Physiol.* 254, R746-R754.
Holmqvist B.I., Östholm T. & Ekström P. (1992): DiI tracing in combination with immunocytochemistry for analysis of connectivities and chemoarchitectonics of specific neural systems in a teleost, the Atlantic salmon. *J. Neurosci. Meth.* 42, 45-63.
Honda K., Negoro H., Dyball R.E.J., Higuchi T & Takano S. (1990): The osmoreceptor complex: evidence for interactions between the supraoptic and other diencephalic nuclei. *J. Physiol. (Lond)* 431, 225-242.
Honig M.G. & Hume R.I. (1989): DiI and DiO: versatile dyes for neuronal labeling and pathway tracing. *Trends Neurosci.* 12, 333-341.
Jewell P.A. & Verney E.B. (1957): An experimental attempt to determine the site of the neurohypophyseal osmoreceptors in the dog. *Phil. Trans. Roy. Soc. Lond.* 240, 197-324.
Johnson A.K., Zardetto-Smith A.M. & Edwards G.L. (1992): Integrative mechanisms and the maintenance of cardiovascular and body fluid homeostasis: the central processing of sensory input derived from the circumventricular organs of the lamina terminalis. *Prog. Brain Res.* 91, 381-393.
Kanosue K., Schmid H. & Simon E. (1990): Differential osmoresponsiveness of periventricular neurons in duck hypothalamus. *Am. J. Physiol.* 258, R973-R981.
Kanosue K., Gerstberger R., Simon-Oppermann C. & Simon E. (1992): Ionic responsiveness in third ventricular hypertonic stimulation of antidiuresis in ducks. *Brain Res.* 569, 268-274.
Knowles, W.D. & Phillips, M.I. (1980): Angiotensin II responsive cells in the organum vasculosum lamina terminalis (OVLT) recorded in hypothalamic brain slices. *Brain Res.* 197, 256-259.
Korf H-W., Simon-Oppermann C. & Simon E. (1982): Afferent connections of physiologically identified neuronal complexes in the paraventricular nucleus of conscious Pekin ducks involved in regulation of salt and water balance. *Cell Tissue Res.* 226, 275-300.
Korf H-W., Viglietta-Panzica C. & Panzica G.C. (1983): A Golgi study on the cerebrospinal fluid (CSF)-contacting neurons in the paraventricular nucleus of the Pekin duck. *Cell Tissue Res.* 226, 149-163.
Korf H-W. (1984): Neuronal organization of the avian paraventricular nucleus: intrinsic, afferent, and efferent connections, *J. Exp. Zool.* 232, 387-395.
Kuenzel W.J. & Masson M. (1988): A Stereotaxic Atlas of the Brain of the Chick (*Gallus domesticus*). The Johns Hopkins University Press, Baltimore London.
Larsen P.J., Moller M. & Mikkelsen J.D. (1991): Efferent projections from the periventricular and medial parvicellular subnulei of the hypothalamic paraventricular nucleus to circumventricular organs of the rat: a phaseolus vulgaris-leucoagglutinin (PHA-L) tracing study. *J. Comp. Neurol.* 306, 462-479.
Leng G, Mason W.T. & Dyer R.G. (1982): The supraoptic nucleus as an osmoreceptor. *Neuroendocrinol.* 34, 75-82.
Matsumura K. & Simon E. (1990): Locations and properties of angiotensin II-responsive neurones in the circumventricular region of the duck brain. *J. Physiol. (Lond)* 429, 281-296.
McKinley M.J., Denton, D.A. & Weisinger, R.S. (1978): Sensors for antidiuresis and thirst - osmoreceptors or CSF sodium detectors. *Brain Res.* 141, 89-103.
McKinley M.J., McAllen R.M., Mendelsohn F.A.O., Allen A.M., Chai S.Y. & Oldfield B.J. (1990): Circumventricular organs: neuroendocrine interfaces between the brain and the hemal milieu. *Front. Neuroendocrinol.* 11, 91-127.

McKinley M.J., Bicknell R.J., Hards, D., McAllen R.M., Vivas L., Weisinger R.S. & Oldfield B.J. (1992): Efferent neural pathways of the lamina terminalis subserving osmoregulation. *Prog. Brain Res.* 91, 395-402.

Nissen R., Bourque C.W. & Renaud L.P. (1993): Membrane properties of organum vasculosum lamina terminalis. *Am. J. Physiol.* 264, R811-R815.

Okuya S., Inenaga K., Kaneko T. & Yamashita H. (1987): Angiotensin II-sensitive neurons in the supraoptic nucleus, subfornical organ and anterolateral third ventricle of rats. *Brain Res.* 402, 58-67.

Oldfield B.J., Hards D.K. & McKinley M.J. (1991): Projections from the subfornical organ to the supraoptic nucleus in the rat: ultrastructural identification of an interposed synapse in the median preoptic nucleus using a combination of neuronal tracers. *Brain Res.* 558, 13-19.

Rundgren, M., Denton, D.A., McKinley, M.J. & Weisinger, R.S. (1986): The dipsogenic effect of intracerebroventricular infusion of hypertonic NaCl in the sheep is mediated mainly by the Na ion. *Acta Physiol. Scand.* 127, 433-436.

Saavedra J.M. (1992): Brain and pituitary angiotensin. *Endocr. Rev.* 13, 329-380.

Sayer R.J., Hubbard J.I. & Sirett N.E. (1984): Rat organum vasculosum laminae terminalis in vitro: responses to transmitters. *Am. J. Physiol.* 247, R374-R379.

Schmid H.A. & Simon E. (1992): Effect of angiotensin II and atrial natriuretic factor on neurons in the subfornical organ of ducks and rats in vitro. *Brain Res.* 588, 324-328.

Sibbald, J.R., Hubbard, J.I. & Sirett, N.E. (1988): Responses from osmosensitive neurons of the rat subfornical organ in vitro. *Brain Res.* 461, 205-214.

Simon E., Gerstberger R. & Gray D.A. (1992): Central nervous angiotensin II responsiveness in birds. *Prog. Neurobiol.* 39, 179-207.

Swaminathan, S. (1980): Osmoreceptors or sodium receptors: an investigation into ADH release in the Rhesus monkey. *J. Physiol. (Lond)* 307, 71-83.

Takei Y. (1977): The role of the subfornical organ in drinking induced by angiotensin in the Japanese quail *(Coturnix coturnix japonica)*. *Cell Tissue Res.* 185, 175-181.

Tanaka, J., Saito, H. & Yagyu, K. (1989): Impaired responsiveness of paraventricular neurosecretory neurons to osmotic stimulation in rats after local anesthesia of the subfornical organ. *Neurosci. Lett.* 98, 51-56.

Thrasher T.N., Brown C.J., Keil L.C. & Ramsay D.J. (1980): Thirst and vasopressin release in the dog: an osmoreceptor or sodium receptor mechanism? *Am. J. Physiol.* 238, R333-R339.

Thrasher T.N. (1989): Role of forebrain circumventricular organs in body fluid balance. *Acta Physiol. Scand.* 136, *(Suppl.* 583), 141-150.

Vivas L., Chiaraviglio E. & Carrer H.F. (1990): Rat organum vasculosum laminae terminalis in vitro: responses to changes in sodium concentration. *Brain Res.* 519, 294-300.

Weiss M.L. & Hatton G.I. (1990): Collateral input to the paraventricular and supraoptic nuclei in rat. I. Afferents from the subfornical organ and the anteroventral third ventricle region. *Brain Res. Bull.* 24, 231-238.

Integrative and cellular aspects of autonomic functions : temperature and osmoregulation. Eds K. Pleschka, R. Gerstberger. John Libbey Eurotext, Paris © 1994, pp. 451-459.

NADPH-diaphorase staining and NO-synthase immunoreactivity in circumventricular organs of the rat brain

Mirek Jurzak, Herbert A. Schmid and Rüdiger Gerstberger

Max-Planck-Institut für Physiologische und Klinische Forschung, W.G. Kerckhoff-Institut, Parkstrasse 1, D-61231 Bad Nauheim, Germany

SUMMARY

The neuroanatomical localization of the nitric oxide (NO) generating enzyme (NO-synthase = NOS) in neurones of the circumventricular organs (CVO) in the rat brain was the aim of the present study. Histochemical and immunohistochemical techniques using the NADPH-diaphorase reaction and NOS specific antibodies were performed in parallel. The presence of NOS could be confirmed within the subfornical organ (SFO), the organum vasculosum of the lamina terminalis (OVLT), the median eminence (ME) and the neurohypophysis using both methods. Other neuronal structures involved in body fluid homeostasis like the paraventricular and supraoptic nuclei (PVN and SON), the median nucleus of the solitary tract (mNTS), as well as single periventricular neurones were also stained. Cellular elements of other CVOs, like the pineal gland, the subcommissural organ (SCO), the area postrema (AP) and the choroid plexus (CP) remained unstained. The labelling of the SFO and OVLT and of the entire hypothalamo-neurohypophyseal axis implies a central role of NO in homeostatic control.

INTRODUCTION

Besides its host defence properties in macrophages and its function as "endothelial-derived relaxing factor", the free radical nitric oxide (NO) is an important neuronal messenger (Vincent & Hope, 1992; Nathan, 1992; Moncada & Higgs, 1991). According to these different functions distinct proteins account for NO synthesis as revealed by biochemical characterization (Dawson et al., 1991; Hope et al., 1991) and molecular cloning (Lamas et al., 1992; Lowenstein et al., 1992; Bredt et al., 1991). Neuronal NO is synthesized from arginine in a Ca^{++}-dependent fashion by the constitutively expressed isoform of the nitric oxide synthase (NOS-type I) and activates the soluble form of guanylyl cyclase (Nathan, 1992). Ca^{++}-dependent NO production is believed to be an intermediate step in N-methyl-D-aspartate-receptor activation and cyclic GMP production (Bredt et al., 1990; Gally et al., 1990; Hoyt et al., 1992).

The recent discovery of the identity of NOS with the neuronal "nicotinamide adenine dinucleotide phosphate" (NADPH)-diaphorase (Hope et al., 1991; Bredt et al., 1990) has powerfully stimulated the anatomical localization of sites of NO synthesis in the nervous system (Vincent & Kimura, 1992). The term *diaphorase* is used in histochemistry for enzymes that mediate the reduction of tetrazolium salts, but cannot be applied biochemically to a specific enzymatic activity (Arévalo et al., 1992).
Within the hypothalamus, intense NADPH-diaphorase staining has been reported for the supraoptic (SON) and paraventricular (PVN) nuclei and the posterior pituitary (Dawson et al. 1991; Arévalo et al., 1992; Sagar & Ferriero, 1987). The intensity of the NADPH-staining of the SON (Pow, 1992) and the posterior pituitary (Sagar & Ferriero, 1987) has been proposed to be dependent on the osmotic status of the animals, suggesting an involvement of the NO messenger system in vasopressin/oxytocin release and osmoregulation.
This study was undertaken to investigate the presence and distribution of NO generating cells within the circumventricular organs (CVO). The CVOs are characterized by the lack of a blood-brain barrier (BBB) due to their fenestrated capillaries, which enables CVO cells to respond to blood-born stimuli (McKinley et al., 1990; Leonhardt, 1980). The CVOs involved in body fluid homeostasis, such as the subfornical organ (SFO) and the organum vasculosum of the lamina terminalis (OVLT) are both regarded as sites of osmoreception which are connected with the PVN via afferent and efferent fibres (Johnson et al., 1992).
Since other enzyme systems are also capable to reduce tetrazolium salts (Summy-Long et al., 1984; Dawson et al., 1991; Pow, 1992) the labelling observed by the histochemical NADPH-diaphorase reaction was compared with immunohistochemistry using NOS-type I specific antibodies. Both techniques performed in parallel underline the specificity of the labelling and might serve as an additional characteristic marker for CVO neurones.

MATERIAL AND METHODS

Male and female Wistar rats (n=8) of 170-220 g body weight were used in this study. The animals were anaesthetized with 60 mg/kg BW pentobarbitone injected intraperitoneally (Nembutal^R; CEVA, Düsseldorf, FRG). Transcardial perfusion with 200 ml ice-cold phosphate-buffered saline (PBS, pH 7.4) containing 8,000 IU/l heparin (Vetren^R; Promonta, Hamburg, FRG) was followed by perfusion with 200 ml 4 % paraformaldehyde and 5 % picric acid in 0.1 M phosphate buffer, pH 7.4, at constant pressure. The brains were dissected immediately after the perfusion, postfixed in the same fixative for 2-3 h at 4°C, and infiltrated with 20 % sucrose at 4°C for 48 h. After mounting onto cryostat chucks (Tissue-Tek; Cambridge Instr., Nußloch, FRG), serial coronal brain sections (20 µm) were cut at -20°C, mounted onto poly-L-lysine coated slides and air-dried. For the immunohistochemical detection of NOS, specific rabbit antiserum (Dr. Bernd Mayer, Institut für Pharmakologie und Toxikologie, Graz, Austria) was used in a final dilution of 1:1,000. The antibody was raised to the NOS from pig cerebellum and was found to recognize the type-I NOS (from neurones) of all species investigated so far. It does not cross-react with inducible NOS of macrophages (type II). Immunocytochemistry was performed as follows: Air-dried tissue sections were transferred to ice-cold isosmotic PBS, pH 7.4, containing 0.1 % Triton X-100 (Sigma, Deisenhofen, FRG) for 4 h to prepare the rehydrated sections for antibody permeation.

Sections were subsequently rinsed three times in PBS-Triton, followed by a 60 min incubation in 5 % fetal calf serum (Biochrom, Berlin, FRG) diluted in PBS containing 0.3 % Triton X-100 (blocking buffer). Incubation with the specific antibodies diluted in blocking buffer was carried out for 48 h at 4°C in a humidified chamber. Unbound antibodies were removed by three 10-min washes in PBS-Triton. The tissue sections were incubated for 1-2 h at RT with DTAF-coupled 2nd antisera (Dakopatts, Hamburg, FRG) at 1:100 to 1:200 dilutions in blocking-buffer. Unbound antibodies were rinsed off by three 10-min· washes in PBS-Tritonand embedded for light microscopy using a glycerol/PBS solution (Citi-Flour, Citifluor LTD, London, UK). Sections were examined under a Nikon TMD-epifluorescent microscope equipped with the appropriate filter set and photographed using Ilford FP4+ film. (Subsequent) NADPH-diaphorase staining was carried out after three 10 min washes with 0.1 M phosphate buffer, pH 8.0. For enzymatic reaction the sections were incubated in 0.1 M phosphate buffer, pH 8.0 containing 0.3 % Triton, 1 mM reduced β-NADPH, and 0.12 mM nitroblue tetrazolium (both Sigma) for 2 h at 37°C. After incubation, the sections were rinsed in 0.1 M phosphate buffer, pH 8.0, dehydrated and covered with Entellan (Merck, Darmstadt, FRG) for examination by light microscopy.

RESULTS

Single cells or clustered populations of neuronal cell bodies exhibiting NADPH-diaphorase staining and NOS immunoreactivity were found throughout the brain. Concerning the CVOs, labelling with both techniques was observed within the OVLT, the SFO, the median eminence and the neurohypophysis. The pineal gland and the area postrema remained unstained (Fig.2). Cells within the subcommissural organ (SCO) and the choroid plexus, which are CVOs without neuronal cell bodies, were not labelled but dye deposits were sometimes observed above the columnar layer of ependymocytes in the SCO, which might reflect fibre staining.

Figure 1 shows examples of experiments, where NOS immunostaining was followed by a NADPH-diaphorase reaction performend on the same section. For distinct periventricular neurones of the hypothalamic third ventricle (V III), intense labelling with both methods was achieved on cell bodies and dendritic or axonal processes, although a few single cells were stained only by the NADPH-diaphorase histochemistry (Fig. 1A and B). Nevertheless, the colocalization in the staining patterns was found to be very high in the periventricular region as well as in the cortex. Within the SON, both methods exhibited cell bodies stained with different intensities (Fig. 1C and D). Most of the labelled cells exhibited unstained nuclei. Compared to prior NOS immunohistochemistry, the portion of heavily labelled cells was found to be lower in the NADPH-diaphorase reaction in the same tissue section (Fig. 1 C and D). The same holds true for the stainings obtained in the CVOs like the SFO and OVLT (Fig. 1E, F, G and H). The loss of staining intensity in the subsequently performed NADPH-diaphorase reaction was especially pronounced in the OVLT (Fig. 1F compared with 1G).

Higher intensities of the NADPH-diaphorase reaction were obtained on tissue sections, which were exclusively stained with this technique, without a preceeding immunolabelling (Fig. 2). Moderately and densly stained neurones with dendrites and axons in the OVLT-area are shown in Fig. 2A. It is also evident that NADPH-diaphorase staining extends to the adjacent anteroventral wall of the third ventricle (AV3V-region). The same effect of enhanced NADPH-diaphorase reaction

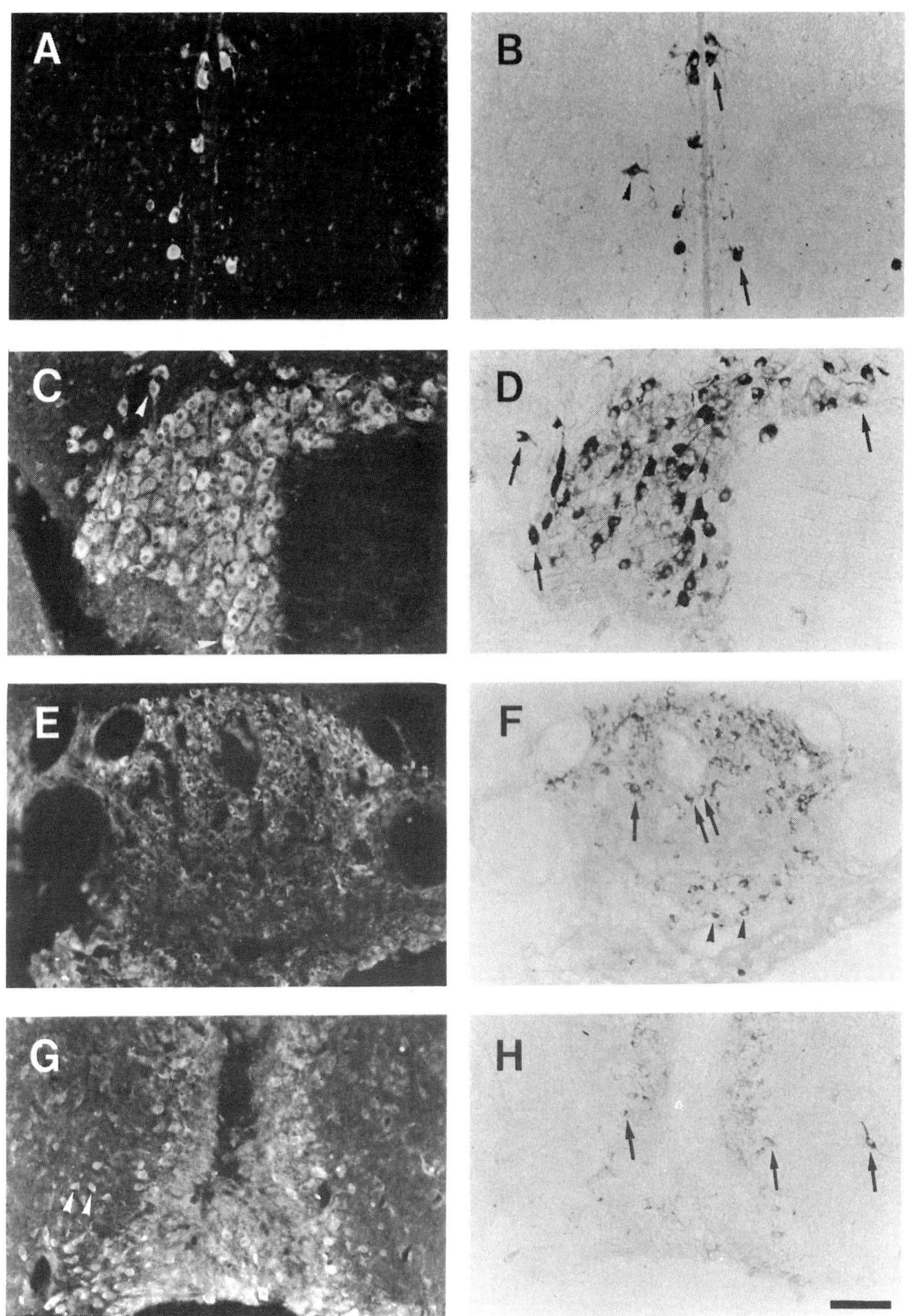

Fig. 1.: Colocalization of neuronal NOS immunoreactivity with NADPH-diaphorase staining in identical sections of the rat hypothalamus. Cells labelled by both techniques are marked by arrows, whereas cells labelled exclusively with one of the methods are marked by

arrowheads. NOS immunoreactive periventricular (V III) neurones (A) colocalize with darkly NADPH-diaphorase stained cells (B), with the exception of two cells. Colabelled cells within the SON are shown in C and D, while arrowheads in (C) mark cells which were heavily stained by the NOS antibody exclusively. NOS immunohistochemistry and NADPH-diaphorase are shown for the SFO in E and F, respectively. Application of both methods in the OVLT (G and H) revealed clearly colocalized labelling, except a population of cells in the ventral part exclusively stained by the NOS antibody. Bar represents 75 µm.

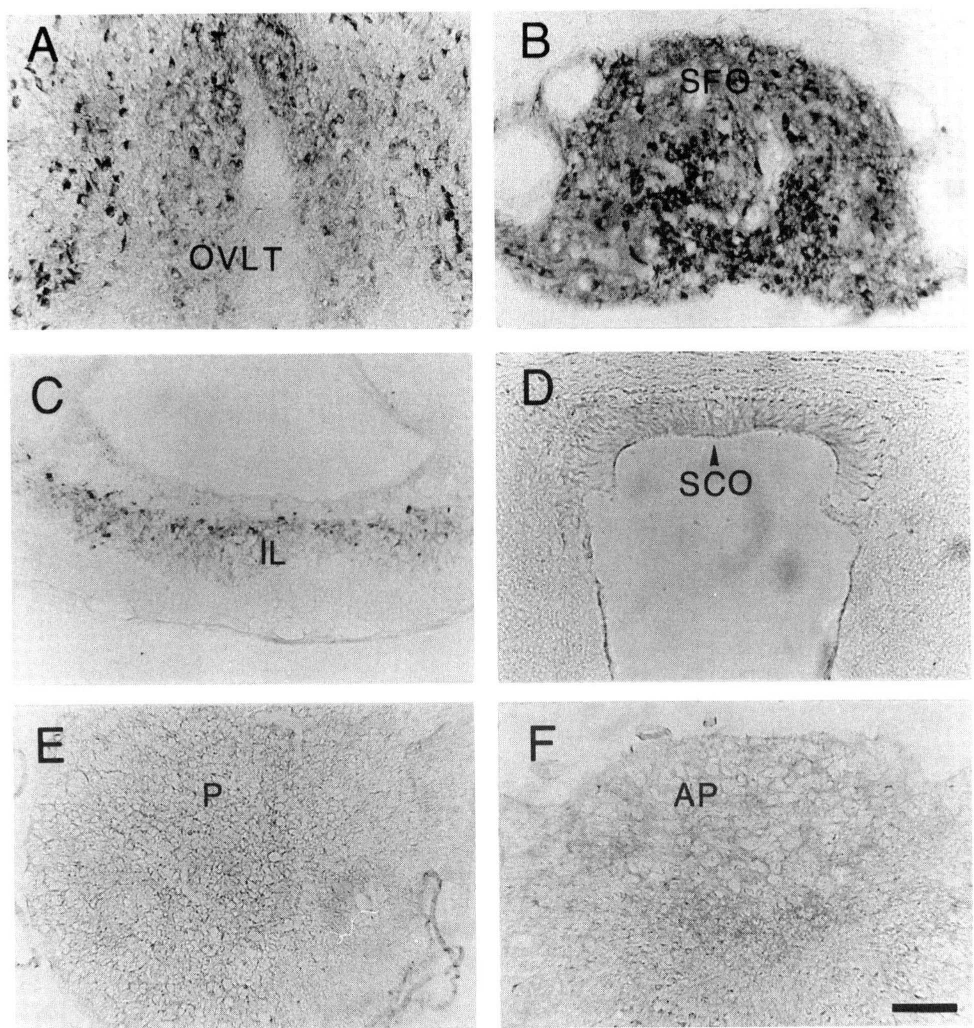

Fig. 2.: NADPH-diaphorase reaction in tissue sections, which were not previously immunolabelled. Higher intensities of staining within the OVLT (A) and SFO (B) in comparison to Fig. 1 were observed. Staining of fibre varicosities in the internal layer of the ME is shown in (C), whereas cell bodies in the SCO (D), the pineal (E) and the area postrema (F) remained unlabelled. Bar represents 75 µm.

in sections not previously immunolabelled with NOS antibodies was observed in the SFO, where dense staining of small neurones was obtained throughout the whole organ (Fig. 2B). Small cell bodies with unstained nuclei were heavily labelled in the core of the SFO, whereas the periphery contained also moderately stained cells. Figure 2D shows thick fibres stained in the internal layer of the median eminence (ME). The lack of NADPH-diaphorase staining within cellular structures of the SCO, the pineal gland and the area postrema is demonstrated in Fig. 2D, E and F, respectively. Labelling of thin fibres and synaptic patches, however, cannot be ruled out in these organs.

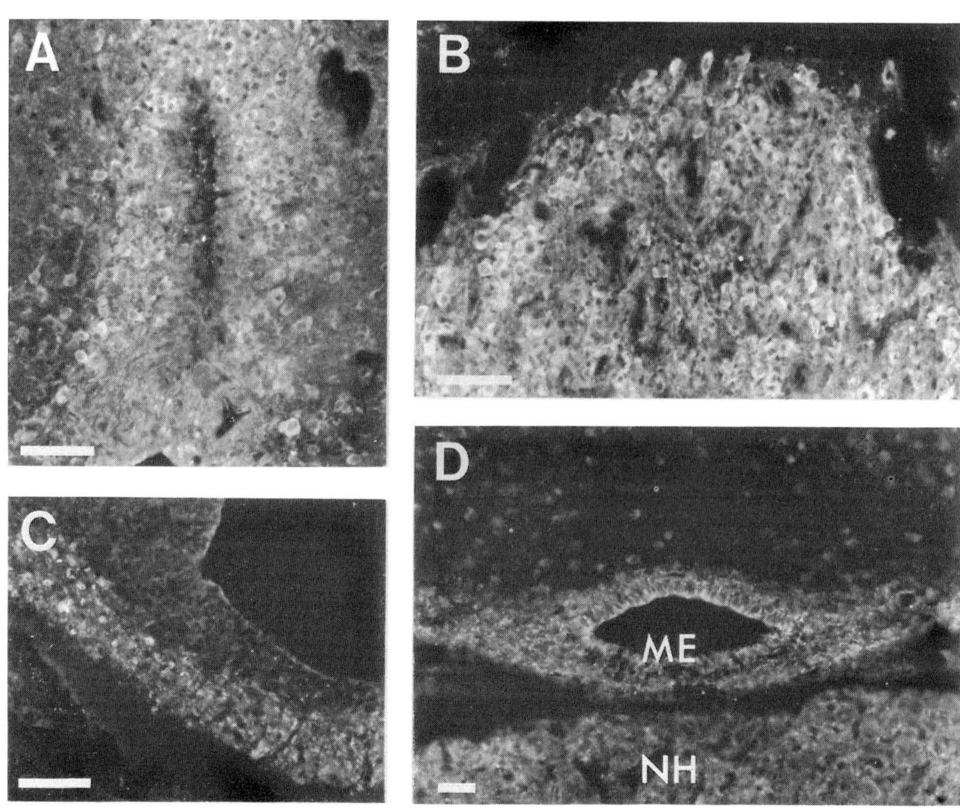

Fig 3. The CVOs of the rat labelled using the indirect immunofluorescence procedure with the antibody specific for the neuronal NOS. Neuronal cell bodies were densely stained within the OVLT (A) and SFO (B). Fibre varicosities in the internal layer (IL) of the ME (C) are stained, as well as cellular elements in the neurohypophysis (NH in D). Bars represent 75 µm.

Strong fluorescence due to NOS immunoreactivity within the CVOs is shown in Fig. 3. Neurones of the OVLT (Fig. 3A) and SFO (Fig. 3B) displayed the typical strong to moderate NOS labelling in the soma with unstained nuclei. The positive staining of fibres within the internal layer of the ME shown for the NADPH-diaphorase reaction (shown in Fig 2C) correlates with the positive NOS immunoreaction in Fig. 3C. Positive immunolabelling was also found in the neural lobe of the pituitary shown together with the ME in Fig. 3D.

DISCUSSION

Circumventricular organs are characterized by the lack of a functional blood-brain barrier and are involved in homeostatic functions. Despite of the numerous reports demonstrating the distribution of NADPH-diaphorase and therefore neuronal NOS in the brain, its localization in the CVOs generally escaped notice.

This is the first combined study of NADPH-diaphorase staining and NOS immunohistochemistry in the CVOs of the rat brain. Using both techniques, labelling of cellular elements was obtained within the OVLT, the SFO, the median eminence and the neural lobe of the pituitary. Both methods revealed different staining intensities of CVO neurones, in contrast to other brain areas like the striatum, olfactory bulb and cortex, where an uniformly dark and Golgi-like labelling was observed. This heterogeneity of NADPH-diaphorase staining has already been reported for the SON (Pow, 1992) and for the posterior pituitary (Sagar & Ferriero, 1987) and was proposed to be activity dependent. However, since the NADPH-diaphorase reaction is sensitive to heavy or prolonged fixation (Pow, 1992; Dawson et al., 1991), and the highly vascularized CVOs are located on ventricular surfaces, they might be more exposed to fixation conditions than structures within the brain. Consistent with the NADPH-diaphorase reaction, the heterogeneity of the staining was also observed with NOS immunohistochemistry. Both techniques might therefore rather reflect different levels of the NOS expression (or preservation of the antigenic sites), than a missleading contribution of other enzymes to the enhanced NADPH-diaphorase staining, as discussed by Pow (1992).

Comparison of stainings at the cellular level obtained with both methods in the same section, revealed extensive but not exclusive colocalization. The degree of colabelling varied for the individual structures, being very high in the cortex (not shown) and for periventricular neurones, and lowest in the SFO and OVLT. Direct NADPH-diaphorase staining (without previous NOS immunohistochemistry) resulted in enhanced overall staining intensity and indicates therefore, that the combination of both methods can reduce the number of positive cells. Obviously, antibody-occupation of the enzyme, or simple washout during the prolonged procedure can lead to a reduced NADPH-diaphorase reaction intensity. Furthermore, the optimal degree of cross-linking during the fixation can differ for both methods, which might be more important for structures accessible to the circulating fixative such as the CVOs, than for cell clusters within the BBB.

In the hypothalamus, the presence of NADPH-diaphorase activity has been reported for the PVN, SON and accessory magnocellular nuclei (Arévalo et al. 1992; Paw, 1992; Dawson et al., 1991; Vincent & Kimura, 1992; Alonso et al., 1992). In a previous study, which used tetrazolium-histochemistry for the determination of total hydrogen generated by glucose-6-phosphate-dehydrogenase, positive NADPH-diaphorase reaction (performed as control) was found in all CVOs including the pineal and the SCO (Summy-Long et al., 1984). Enhanced glucose-6-phosphate-dehydrogenase activity in the SFO after a thirst challenge was also reported already for the rat (Sarrat, 1968).

The NADPH-diaphorase staining pattern reported in this paper is in good agreement with that found by Vincent and Kimura (1992). These authors described the labelling of the hypothalamo-neurohypophyseal axis, the median nucleus of the solitary tract and the SFO, and a lack or only scarce staining in the pineal, suprachiasmatic nucleus and AP, without mentioning the OVLT.

The colocalization of NADPH-diaphorase histochemistry and the NOS immunoreactivity within these structures is shown here for the first time. Both techniques performed in parallel underline the specificity of the labelling and might serve as additional characteristic markers for CVO neurones of the SFO and OVLT. Furthermore, the neuroanatomical distribution of NOS along the hypothalamo-neurohypophyseal axis as well as within the interconnected SFO and OVLT underlines the putative role of NO in the central control of autonomic functions.

ACKNOWLEDGEMENT

The authors are very grateful to Dr. B. Mayer, Graz, Austria, for providing the NOS antibodies and to Dr. M. Schemann for discussion.

REFERENCES

Alonso, J.R., Sanchez, F., Arevalo, R., Carretero, J., Aijon, J. & Vazquez, R. (1992): CaBP D-28k and NADPH-diaphorase coexistence in the magnocellular neurosecretory nuclei. *Neuroreport*. 3, 249-252.

Arévalo, R., Sánchez, F., Alonso, J.R., Carretero, J., Vázquez, R. & Aijón, J. (1992): NADPH-diaphorase activity in the hypothalamic magnocellular neurosecretory nuclei of the rat. *Brain Res. Bull*. 28, 599-603.

Bredt, D.S., Hwang, P.M., Glatt, C.E., Lowenstein, C., Reed, R. & Snyder, S. (1991): Cloned and expressed nitric oxide synthase structurally resembles cytochrome P-450 reductase. Nature. 351, 714-718.

Bredt, D.S., Hwang, P.M. & Snyder, S.H. (1990): Localization of nitric oxide synthase indicating a neural role for nitric oxide. *Nature*. 347,768-770.

Dawson, T.M., Bredt, D.S., Fotuhi, M., Hwang, P.M. & Snyder, S.H. (1991): Nitric oxide synthase and neuronal NADPH diaphorase are identical in brain and peripheral tissues. *Proc. Natl. Acad. Sci. USA*. 88, 7797-7801.

Gally, J.A., Montague, P.R., Reeke Jr., G.N. & Edelman G.M. (1990): The NO hypothesis: Possible effects of a short-lived, rapidly diffusible signal in the development and function of the nervous system. *Proc. Natl. Acad. Sci. USA*. 87, 3547-3551.

Hope, B.T., Michael, G.J., Knigge, K.M. & Vincent, S.R. (1991): Neuronal NADPH diaphorase is a nitric oxide synthase. *Proc. Natl. Acad. Sci. USA*. 88, 2811-2814.

Hoyt, K.R., Tang, L.-H., Aizenman, E. & Reynolds, I.J. (1992): Nitric oxide modulates NMDA-induced increases in intracellular Ca^{++} in cultured rat forebrain neurons. *Brain Res*. 592, 310-316.

Johnson, A.K., Zardetto-Smith, A.M. & Edwards, G.L. (1992): Integrative machanisms and the maintenance of cardiovascular and body fluid homeostasis: the central processing of sensory input derived from the circumventricular organs of the lamina terminalis. In *circumventricular organs and brain fluid environment*, ed. A. Ermisch, R. Landgraf & H.-J. Rühle, pp. 381-392. *Progress in brain research*, 91. Amsterdam-London-New York-Tokyo: Elsevier.

Lamas, S., Marsden, P.A., Li, G.K., Tempst, P. & Michel T. (1992): Endothelial nitric oxide synthase: Molecular cloning and characterization of a distinct constitutive enzyme isoform. *Proc. Natl. Acad. Sci. USA*. 89, 6348-6352.

Leonhardt, H. (1980): Ependym und Circumventriculäre Organe. In *Handbuch der Mikroskopischen Anatomie des Menschen*, ed. A. Oksche & L. Vollrath, pp. 177-666. Vol. 10/4, Berlin: Springer.

Lowenstein, C.J., Glatt, C.S., Bredt, D.S. & Snyder, S. (1992): Cloned and expressed macrophage nitric oxide synthase contrasts with the brain enzyme. *Proc. Natl. Acad. Sci. USA*. 89, 6711-6715.

Moncada, S. & Higgs, E.A. (1991): Endogenous nitric oxide: physiology, pathology and clinical relevance. *Eur. J. Clin. Invest*. 21, 361-374.

McKinley, M.J., McAllen, R.M., Mendelsohn, F.A.O., Allen, A.M., Chai, S.Y. & Oldfield, B.J. (1990): Circumventricular organs: Neuroendocrine interfaces between the brain and the hemal milieu. *Front. Neuroendocrinol*. 11, 91-127.

Nathan, C. (1992): Nitric oxide as a secretory product of mammalian cells. *FASEB J*. 6, 3051-3064.

Pow, D.A. (1992): NADPH-diaphorase (nitric oxide synthase) staining in the rat supraoptic nucleus is activity-dependent: possible functional implications. *J. Neuroendocrinol*. 4, 377-380.

Sagar, S.M. & Ferriero, D.M. (1987): NADPH diaphorase activity in the posterior pituitary: relation to neuronal function. *Brain Res*. 400, 348-352.

Sarrat, R. (1968): Enzymhistochemische Untersuchungen am Subfornikalorgan der Ratte. *Experientia*. 24. 1239-1240.

Summy-Long, J.Y., Salisbury, R., Marietta, M. P., Hartman, R.D. & Weisz, J. (1984): Pathways of hydrogen utilization from NADPH generated by glucose-6-phosphate dehydrogenase in circumventricular organs and the hypothalamo-neurohypophyseal system: a cytochemical study. *Brain Res*. 294, 23-35.

Vincent, S.R. & Hope, B.T. (1992): Neurons that say NO. *TINS*. 15, 108-113.

Vincent, S.R. & Kimura, H. (1992): Histochemical mapping of nitric oxide synthase in the rat brain. *Neuroscience*. 46, 755-784.

Integrative and cellular aspects of autonomic functions : temperature and osmoregulation. Eds K. Pleschka, R. Gerstberger. John Libbey Eurotext, Paris © 1994, pp. 461-470.

Hypothalamic integration of afferent osmo- and thermoregulatory signals and the release of osmoregulatory hormones

Ralf Keil, Rüdiger Gerstberger and Eckhart Simon

Max-Planck-Institut für Physiologische und Klinische Forschung, William G. Kerckhoff-Institut, Parkstrasse 1, D-6350 Bad Nauheim, Germany

INTRODUCTION

In homeothermic vertebrates, the hypothalamus serves as the main integrative control center for multiple autonomic functions such as the homeostasis of body temperature and extracellular fluid (ECF) balance, providing both perceptive and signal processing functions (Simon et al., 1986; Hatton, 1990). Receiving afferent signals from thermo-, osmo-, baro-, sodium and volume receptors located in the periphery and/or the central nervous system itself, hypothalamic units appear to be involved in the "central meshing" of both regulatory systems. With only limited information available concerning putative diencephalic interrelations of these two autonomic control circuits, physiological studies reported alterations in thermoregulatory effector responses such as reduced respiratory water loss at augmented body temperature due to reduction of ECF volume (Schmidt-Nielsen et al., 1957; Taylor, 1970; Baker & Turlejska, 1989). The possible elevation of the thermoregulatory set-point has been put forward as a working hypothesis (Baker & Doris, 1982; Baker et al., 1983). With regard to temperature influences on osmoregulatory function, studies in goats (Andersson & Larsson, 1961), monkeys (Hayward & Baker, 1968) and ducks (Simon-Oppermann et al., 1980) described the phenomenon of cold-induced diuresis, using hypothalamic cooling as an experimental approach. Integration of both regulatory control circuits at the hypothalamic level is additionally indicated by studies in rabbits (Sadowski et al., 1977) and dogs (Szczepanska-Sadowska, 1974) reporting antidiuresis during hypothalamic warming. Although alterations in the plasma concentration of the neurohypophyseal antidiuretic hormone (ADH) during hypothalamic cooling or warming have been suggested, direct evidence of thermally induced changes in plasma ADH for mammals is lacking while indicated for the duck (Simon-Oppermann et al., 1980; Simon & Nolte, 1990). The aim of present study was therefore to characterize possible influences of altered hypothalamic or deep body temperature on neurohypophyseal ADH release during both isosmotic and hyperosmotic status of the extracellular fluid in the conscious rabbit.

THERMAL AND OSMOTIC STIMULATIONS

Whole body warming as well as hypothalamic warming and cooling were chosen as appropriate thermal stimuli in combination with systemic osmotic stimulation to challenge both the thermo- and osmoregulatory system in the conscious rabbit. To characterize the modulatory role of locally altered hypothalamic temperature on thermoregulatory effector parameters and plasma concentrations of the osmoregulatory hormones vasopressin (AVP), angiotensin II (ANGII), aldosterone (ALD) and corticosterone (COR), six experimental series (S1-S6) were performed in conscious rabbits (n=14). Animals were provided with a chronically implanted hypothalamic perfusion thermode straddling the preoptic anterior hypothalamic region as verified histologically. During the experimental series S3 to S6 (see Fig. 1) the thermode array consisting of four blind ending stainless-steel tubes (1.0 mm OD, 5 mm apart) was perfused for 3 hrs with water at a rate of 20 ml/min per tube by means of indwelling cannulas. Perfusion with water at 34°C (hypothalamic cooling = HC) or 43.5°C (hypothalamic warming = HW) resulted in corresponding hypothalamic tissue temperatures (T_{hyp}) of 36.9°C and 41.0°C, respectively. To elucidate the influence of whole body warming (HE) on thermo- and osmoregulatory effector parameters, conscious rabbits (n=8) were exposed for 3 hrs to an ambient temperature of 31.5°C at 30-40 % relative humidity in a climatic chamber during both isosmotic and hyperosmotic conditions (series S7 and S8). The hyperosmotic stimulation performed in series S1,S5,S6, and S7 consisted of a continuous systemic infusion of hyperosmotic saline (8.8% NaCl) at 90 µEq NaCl/min/kg b.w. into an ear vein (i.v.) for 3 hrs, and led to a significant and steady rise in plasma osmolality by 26 mOsm/kg, finally reaching elevated steady-state values of about 310 mOsm/kg in the second half of the experiment. The increase in plasma osmolality was due to elevated plasma sodium and chloride concentrations at slightly reduced plasma potassium levels. Studies carried out under isosmotic control conditions in series S2,S3,S4 amd S8 with animals receiving isotonic saline i.v. at identical rates showed no changes in plasma electrolytes and osmolality. The hyperosmotic infusion caused a mild ECF volume expansion as indicated by the net fall in hematocrit of 1.5% and the constantly low ANGII plasma concentrations (7 pg/ml) (Fig. 7b).

THERMOREGULATORY EFFECTOR RESPONSES DURING HYPOTHALAMIC AND WHOLE BODY THERMAL STIMULATION

With regard to the rabbit`s thermoregulatory status, the measured parameters shown in Fig. 1-4 clearly demonstrate that the magnitude of physiological thermal stimulation prooved to be sufficient to produce thermoregulatory effector responses during both local stimulation of the hypothalamus (±2°C) (Fig. 1-3) and exposure of the animal to an ambient temperature of 31.5°C (Fig. 4). At least with regard to the respiratory frequency, local warming of the hypothalamus (Fig. 1) appeared to be a more potent stimulus than high ambient temperature (Fig. 4). Despite ECF volume expansion, the hyperosmotic infusion induced a significant decrease in respiratory frequency not only at thermoneutrality but also under conditions of hypothalamic or whole body thermal stimulation. Known to be closely related to evaporative heat loss (REHL) in rabbits (Gonzales et al., 1971; Stitt, 1976), this reduction in respiratory frequency may be regarded as an osmoregulatory induced water-saving mechanism.

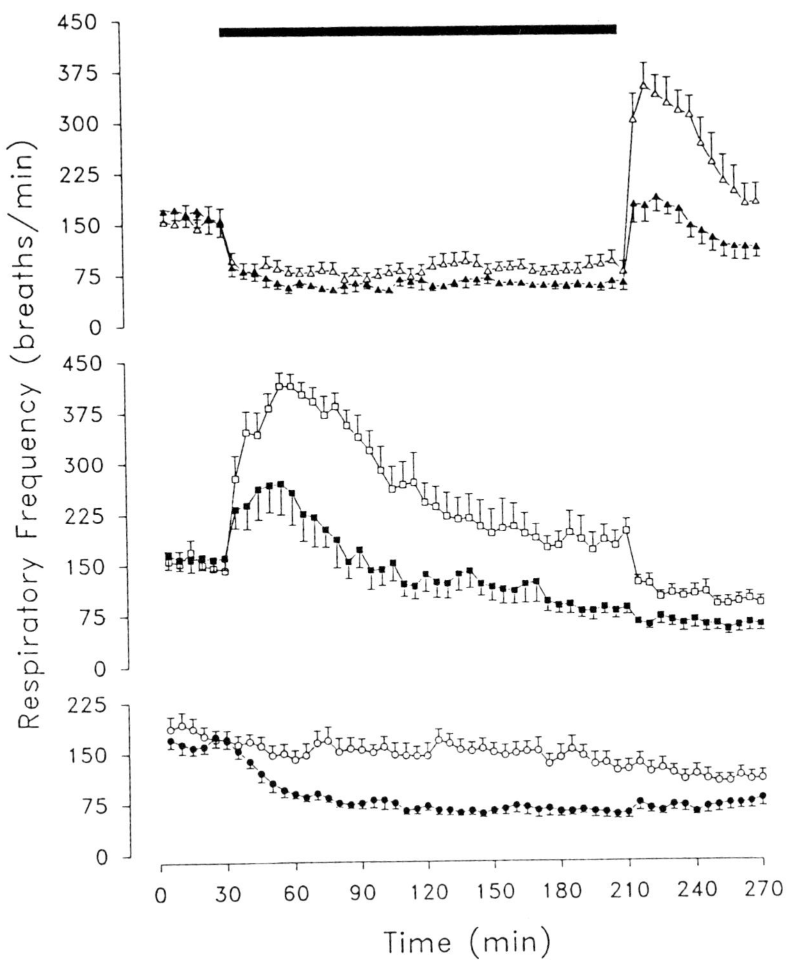

Figure 1: Effects of hypothalamic thermal stimulation during isosmotic and hyperosmotic saline loading on respiratory frequency. Bar (▬) represents the experimental period (180 min). Values are presented as means ± SEM. Iso- or hyperosmotic saline was infused into a peripheral ear vein punctured percutaneously (Braunula G20) under local anaesthesia (1.0% xylocaine) and catheterized with PP-20 tubing directed towards the heart (15 cm). Hyperosmotic (8.8% NaCl at 90 µEq NaCl/min/kg b.w.) and isosmotic (0.9% NaCl) stimulations were carried out for 3 hrs.

In *series 1* (S1) (●) only the osmoregulatory system of the animals was stimulated by continuous i.v. infusion of hyperosmotic saline for 3 hrs. In the subsequent post-experimental period (60 min), the hyperosmotic infusion was replaced by isosmotic saline at unchanged rate. *Series 2* (S2) (○) served as isosmotic control experiment with i.v. infusion of isosmotic saline for 3 hrs. *Series 3* (S3) (△) and *series 4* (S4) (□) were based on the infusion schedule of S2 (=isosmotic), and *series 5* (S5) (▲) and *series 6* (S6) (■) were based on the infusion schedule of S1 (=hyperosmotic). During the experimental period (▬) of S3 to S6, the animals were subjected to either hypothalamic cooling (S3,S5; T_{hyp}=36.9°C) or warming (S4,S6; T_{hyp}=41.0°C).

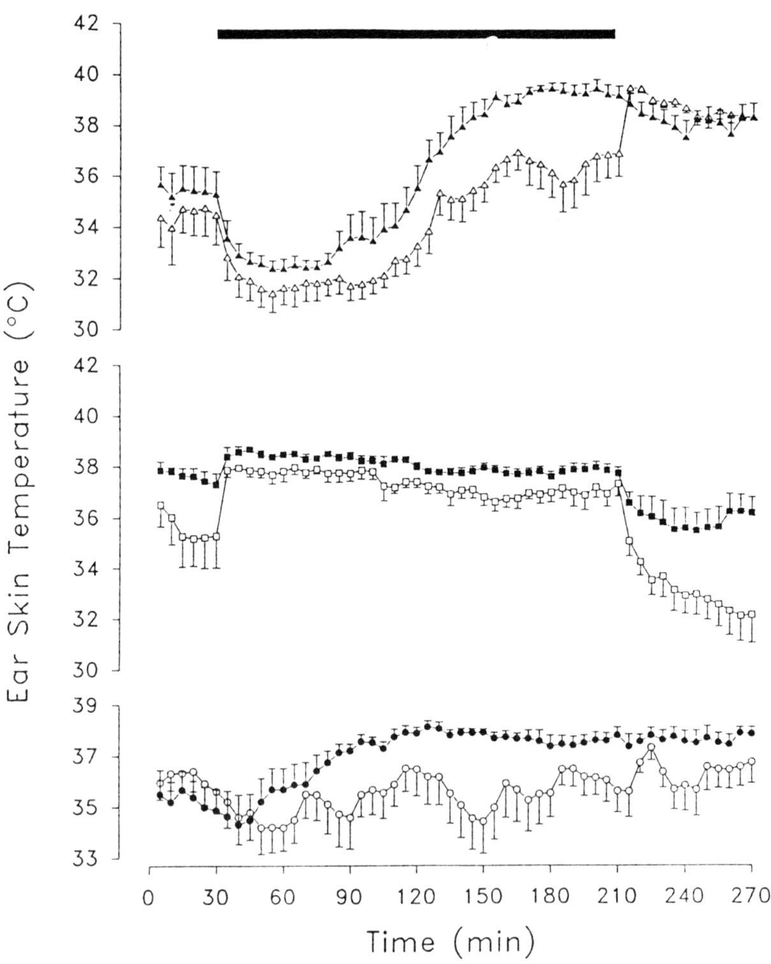

Figure 2: Effects of hypothalamic thermal stimulation during isosmotic and hyperosmotic saline loading on ear skin temperature. Bar (■■■) represents the experimental period (180 min) consisting of: S1 (●) hyperosmotic stimulation; S2 (○) isosmotic infusion; S3 (△) isosmotic infusion at T_{hyp}=36.9°C; S4 (□) isosmotic infusion at T_{hyp}=41.0°C; S5 (▲) hyperosmotic stimulation at T_{hyp}=36.9°C and S6 (■) hyperosmotic stimulation at T_{hyp}=41.0°C. Values are presented as means ± SEM.

This findings are in good agreement with previous studies pointing to a pronounced reduction in REHL due to alterations of ECF osmolality or sodium concentration (Baker & Turlejska, 1989) despite hyperthermal stimulation in rabbits (Turlejska-Stelmasiak, 1974) and dogs (Baker & Dawson, 1985). Using the ear skin temperature (T_{ear}) (Fig. 2,4) as an indicator of peripheral vasomotor activity involved in convective heat dissipation, both modes of thermal stimulations produced typical patterns of cold or heat defence, respectively. The hyperosmotic infusion *per se* significantly induced vasodilation

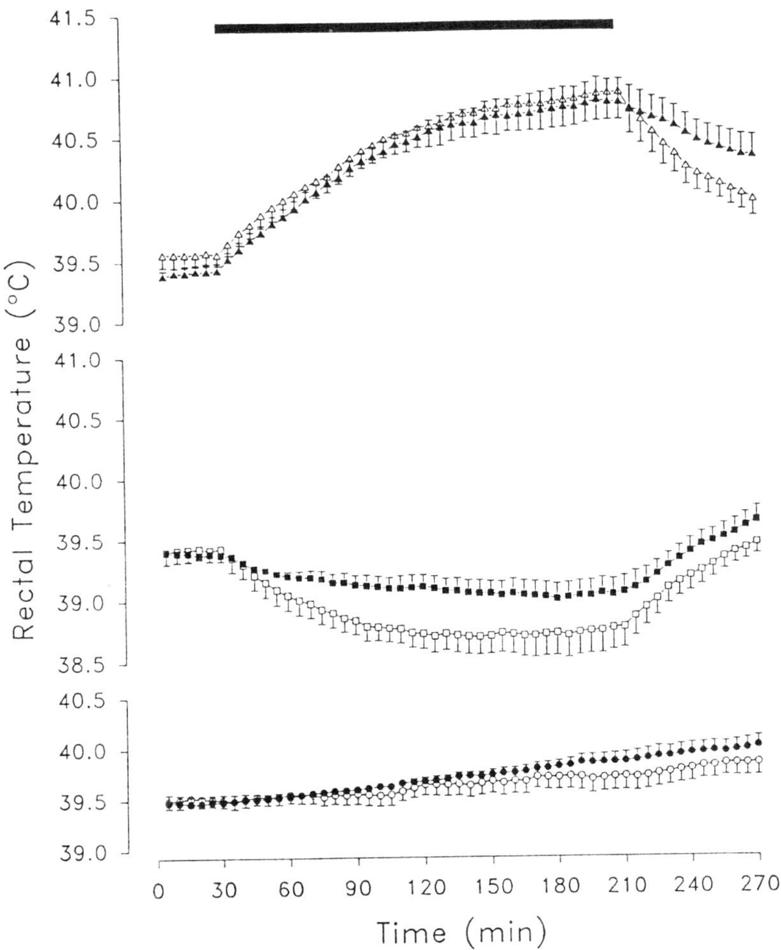

Figure 3: Effects of hypothalamic thermal stimulation during isosmotic and hyperosmotic saline loading on rectal temperature. Bar (▬) represents the experimental period (180 min) consisting of: S1 (●) hyperosmotic stimulation; S2 (○) isosmotic infusion; S3 (△) isosmotic infusion/T_{hyp}=36.9°C; S4 (□) isosmotic infusion/T_{hyp}=41.0°C; S5 (▲) hyperosmotic stimulation/T_{hyp}=36.9°C and S6 (■) hyperosmotic stimulation/T_{hyp}=41.0°C. Values are presented as means ± SEM.

despite reduced respiratory frequency and REHL. This implies marked differentiation of two major heat loss mechanisms, REHL and convective heat loss, already under thermoneutral conditions. Interestingly, this result was underlined and strengthened by the data obtained in experiments in which the animal's thermoregulatory system was stimulated either by HW or whole body heat exposure (Fig. 1,4). Consideration of a simple upward shift in the thermoregulatory threshold as postulated from experiments with partially dehydrated and therefore hypovolemic cats (Baker & Doris, 1982) and dogs (Baker et al., 1983) does not seem to be an adequate explanation. The results rather support recent data obtained in pigeons (Brummermann & Rautenberg, 1989), in which a comparable and significant differentiation of two heat defense mechanisms was observed during hyperosmotic saline infusion. The time course of the rectal temperature (T_{rec}) (Fig. 3,4) mirrors the sum of the thermoregulatory adjustments caused by thermal and/or osmotic stimulations. Due to the osmotically reduced respiratory frequency during the first half of the experiment, the hyperosmotically stimulated animals additionally subjected to HW (S6) (Fig. 3) remained at a significantly higher T_{rec}, when compared to isosmotically infused and hypothalamically warmed rabbits. The greater increase in T_{rec} in experiments in which the animals were hyperosmotically stimulated whilst exposed to high ambient temperature (S7) can also be traced back to a diminished respiratory frequency (Fig. 4).

PLASMA CONCENTRATIONS OF OSMOREGULATORY IMPORTANT HORMONES

With regard to osmoregulatory hormonal effector systems, the hyperosmotic saline infusion produced the expected rise in plasma AVP (Fig. 5a). The osmotic sensitivity of the AVP system with values of about 0.3 pg/ml per mOsm/kg proved to be comparable to data reported for other species (Wade et al., 1983; Gray & Simon, 1983). Modification of neurohypophyseal antidiuretic hormone release due to thermal stimulation was implicated in previous studies describing the phenomenon of cold-induced diuresis (Hayward & Baker, 1968; Simon-Oppermann et al., 1980) and warm-induced antidiuresis (Sadowski et al., 1977; Szczepanska-Sadowska, 1974). Assuming the osmotic status of the animal to be the main regulatory component of systemic AVP release, hypothalamic warming/cooling or whole body warming in addition to the long-term continuous infusion of hypertonic saline did not significantly influence plasma vasopressin concentrations despite the tendency of plasma AVP to decrease during HC and to increase during HW (Fig. 5a). Whereas these studies with slow, dynamic changes in the ECF compartment associated with parallel rises in plasma AVP did not reveal thermal influences on hormone release, studies in rabbits subjected to short-term (20 min) HW and HC under hyperosmotic steady-state conditions clearly demonstrated the existence of thermal modulatory effects on neurohypophysal AVP release in the rabbit (Fig. 6a) (Keil et al., 1993). Control experiments with thermal stimulation of the hypothalamus or the whole body performed under isosmotic conditions with no need for the animal to conserve body water appear to rule out non-specific temperature effects on AVP release (Fig. 6a).

Long-term osmotic stimulation of the conscious rabbits not only led to a rise in plasma AVP, but also corticosterone and aldosterone (Fig. 5b,7a). The parallel changes of plasma AVP and adrenal steroid release during hyperosmotic infusion may be explained by co-activation of the AVP system and of the hypothalamo-pituitary-adrenal axis. AVP and corticotrophin-releasing factor (CRF) have been reported to be co-localized in parvocellular neurones of the paraventricular nucleus (Piekut & Joseph, 1986). In addition to possible stimulation of the hormonal cascade with CRF, adrenocorticotrophin (ACTH) and corticosterone as sequential components, AVP might have acted as a secretagogue for ACTH (Rivier & Veale, 1983). Independent of the mode of COR release, elevation of plasma COR in the present experiments reflects a hormonal stress response due to

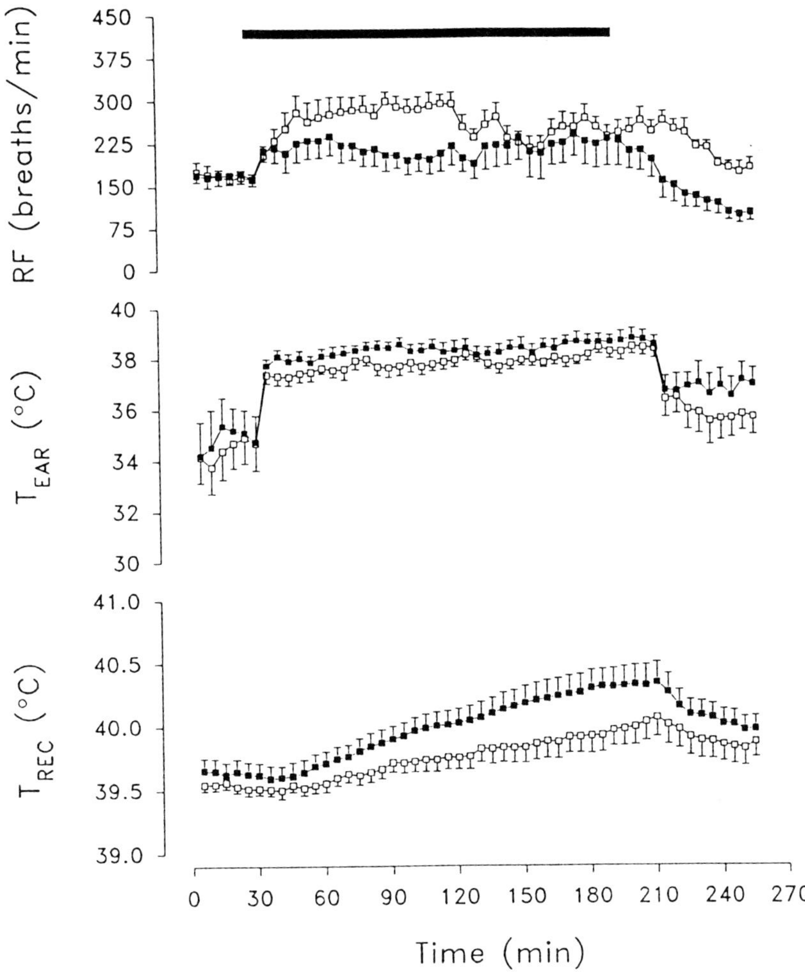

Figure 4: Effects of whole body warming during isosmotic and hyperosmotic saline loading on respiratory frequency (RF), ear skin temperature (T_{EAR}) and rectal temperature (T_{REC}). Bar (▬) represents the experimental period (180 min), during which the animals were exposed to 31.5°C ambient temperature in a climatic chamber, and were also infused with hyperosmotic (8.8% NaCl) (■) and isosmotic (0.9% NaCl) (□) saline at rates of 90 and 9 µEq NaCl/min/kg b.w., respectively. Values are presented as means ± SEM.

the hyperosmotic loading (Besch & Brigmon, 1991). Neither hypothalamic thermal stimulation nor whole body warming *per se* under isosmotic control conditions induced any changes in the plasma concentration of both corticosterone and aldosterone, thus ruling out stress-induced or non-specific effects of the applied thermal stimuli on steroid secretion (Fig. 5b,6b,7a). Superimposing thermal and hyperosmotic stimulations, however, induced a more pronounced COR release than either one of the applied stimuli alone, especially when exposing the rabbit to high ambient temperatures (Fig. 5b).

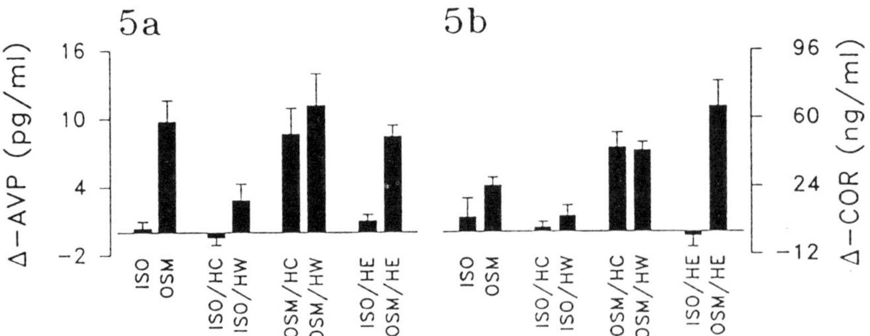

Figure 5: Alterations in the plasma hormone concentrations of vasopressin (AVP) (5a) and corticosterone (COR) (5b) during the experimental periods in experiments of series S1 to S8. Values are given as deviations from basal hormone concentrations at the beginning of the experiment and presented as means ± SEM. S1 with hyperosmotic infusion (OSM), S2 with isosmotic infusion (ISO), S3 with hypothalamic cooling during isosmotic infusion (ISO/HC), S4 with hypothalamic warming during isosmotic infusion (ISO/HW), S5 with hypothalamic cooling during hyperosmotic infusion (OSM/HC), S6 with hypothalamic warming during hyperosmotic infusion (OSM/HW), S7 with whole body warming during hyperosmotic infusion (OSM/HE) and S8 with whole body warming during isosmotic infusion (ISO/HE).
Plasma hormone concentrations were determined in arterial blood samples collected by cannulation of a central ear artery under local anaesthesia (1.0% xylocain), using radioimmunoassay methodology. Plasma AVP was detected according to Gray & Simon (1983) with ^{125}I-AVP as radioligand, plasma COR was measured according to Gerstberger et al. (1992) using commercially available antisera and ^{3}H-labelled tracer.

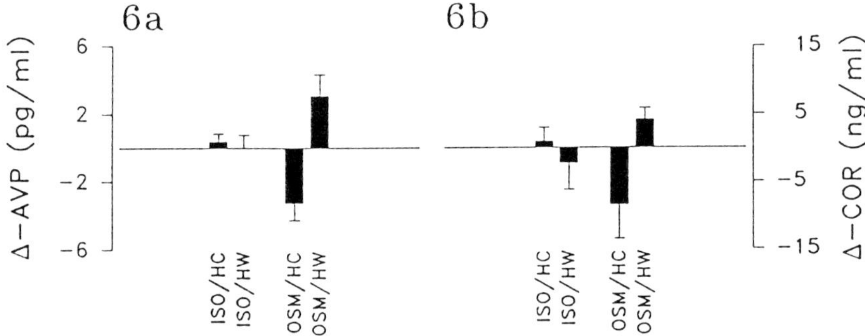

Figure 6: Alterations in the plasma hormone concentrations of vasopressin (AVP) (6a) and corticosterone (COR) (6b) during experiments of short-term (20 min) hypothalamic cooling/warming (HC/HW) under isosmotic (ISO) or hyperosmotic (OSM) steady-state conditions. These were established after 180 min of i.v. infusion of hyperosmotic (8.8% NaCl) and isosmotic (0.9% NaCl) saline at 90 and 9 µEq NaCl/min/kg b.w., respectively. Values are given as deviations from basal hormone concentrations at the beginning of the experiment and presented as means ± SEM. Data adopted from Keil et al.(1993).

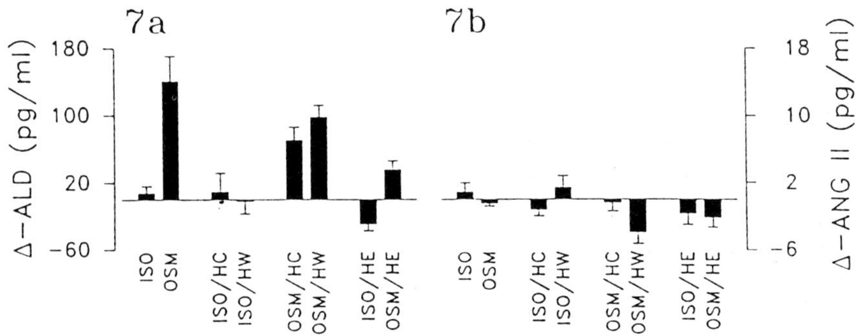

Figure 7: Alterations in the plasma hormone concentrations of aldosterone (ALD) (7a) and angiotensin II (ANGII) (7b) during the experimental periods in experiments of series S1 to S8. Values are given as deviations from basal hormone concentrations at the beginning of the experiment and presented as means ± SEM. Abbreviations are identical to those used in the legends of Fig. 6.
Plasma ANGII was detected according to Gray and Simon (1985) with ^{125}I-ANGII as radioligand, plasma ALD was measured according to Gerstberger et al. (1992) using commercially available antisera and ^{3}H-labelled tracer.

Short-term hypothalamic thermal stimulation during hyperosmotic steady-state conditions caused pronounced alterations in COR secretion (Fig. 6b). The proposed co-localization and -activation of AVP and CRF during thermal stimulation, and their putative combined action on ACTH secretion might be a good explaination. Activation of the CRF-ACTH system could also be responsible for the rise in plasma ALD during hyperosmotic stimulation despite augmented plasma sodium and reduced potassium concentrations (Fig. 7a). This is supported by the presence of AVP receptors in zona glomerulosa cells (Balla et al.,1985) and recently demonstrated AVP-induced ALD-release *in vitro* (Woodcock et al., 1986).

REFERENCES

Andersson B. & Larsson B. (1961): Influence of local temperature changes in the preoptic area and rostral hypothalamus on the regulation of food and water intake. *Acta Physiol. Scand.* 52, 75-89.

Baker M.A. & Doris P.A. (1982): Effect of dehydration on hypothalamic control of evaporation in the cat. *J. Physiol.* 322, 457-468.

Baker M.A., Doris P.A. & Hawkins M.J. (1983): Effect of dehydration and hyperosmolality on thermoregulatory water losses in exercising dogs. *Am. J. Physiol.* 244, R516-R521.

Baker M.A. & Dawson D.D. (1985): Inhibition of thermal panting by intracarotid infusion of hypertonic saline in dogs. *Am. J. Physiol.* 249, R787-R791.

Baker M.A. & Turlejska E. (1989): Thermal panting in dehydrated dogs: effects of plasma volume expansion and drinking. *Pflügers Arch.* 413, 511-515.

Balla T., Enyedi P., Spat A. & Antoni F.A. (1985): Pressor-type vasopressin receptors in the adrenal cortex: Properties of binding, effects on phosphoinositide metabolism and aldosterone secretion. *Endocrinology* 117, 421-423.

Besch E.L. & Brigmon R.L. (1991): Adrenal and body temperature changes in rabbits exposed to varying effective temperatures. *Lab. Animal Sci.* 41(1), 31-34.
Brummermann M. & Rautenberg W. (1989): Interaction of autonomic and behavioral thermoregulation in osmotically stressed pigeons (*Columba livia*). *Physiol. Zool.* 62(5), 1102-1116.
Gerstberger, R., Schütz, H., Luther-Dyroff, Keil, R. & Simon, E. (1992): Inhibition of vasopressin and aldosterone release by atrial natriuretic peptide in conscious rabbits. *Exp. Physiol.* 77, 587-600.
Gonzales, R.R., Kluger, M.J. & Hardy, J.D. (1971): Partitional calorimetry of the New Zealand white rabbit at temperatures 5-35°C. *J. Appl. Physiol.* 31, 728-734.
Gray D.A. & Simon E. (1983): Mammalian and avian antidiuretic hormon: studies related to possible species variation in osmoregulatory systems. *J. Comp. Physiol.* 151, 241-246.
Gray D.A. & Simon E. (1985): Control of plasma angiotensin II in a bird with salt glands (*Anas platyrhynchos*). *Gen. Com. Endocrin.* 60, 1-13.
Hayward J.N. & Baker M.A. (1968): Diuretic and thermoregulatory responses to preoptic cooling in the monkey. *Am. J. Physiol.* 214(4), 843-850.
Hatton G.I. (1990): Emerging concepts of structure-function dynamics in adult brain: The hypothalamo-neurohypophysial system. *Prog. Neurobiol.* 34, 437-504.
Keil, R., Gerstberger, R & Simon, E. (1993): Hypothalamic thermal stimulation modulates vasopressin release in hyperosmotically stimulated rabbits. *J. Physiol. (Lond)*, (submitted).
Piekut D.T. & Joseph S.A. (1986): Coexistence of CRF and vasopressin immunoreactivity in parvocellular paraventricular neurons of rat hypothalamus. *Peptides* 7, 891-898.
Rivier C. & Vale W. (1983): Interaction of corticotropin-releasing factor and arginine vasopressin on adrenocorticotropin secretion in vivo. *Endocrinology* 113, 939-941.
Sadowski J., Kruk B. & Chwalbinska-Moneta J. (1977): Renal function changes during preoptic-anterior hypothalamic heating in the rabbit. *Pflügers Arch.* 370, 51-57.
Schmidt-Nielsen K., Schmidt-Nielsen B., Jarnum S.A. & Houpt T.R. (1957): Body temperature of the camel and its relation to water economy. *Am. J. Physiol.* 188, 103-112.
Simon E., Pierau F.K. & Taylor D.C.M. (1986): Central and peripheral thermal control of effectors in homeothermic temperature regulation. *Physiol. Rev.* 66, 235-289.
Simon E. & Nolte P. (1990): Temperature dependence of thermal and nonthermal regulation: Hypothalamic thermo- and osmoregulation in the duck. In: *Thermoreception and Temperature Regulation*, ed. J. Bligh & K. Voigt, pp. 191-199. Heidelberg, Springer Press.
Simon-Oppermann C., Simon E., Deutsch H., Möhring J. & Schoun J. (1980): Serum arginine-vasotocin (AVT) and afferent and central control of osmoregulation in conscious Pekin ducks. *Pflügers Arch.* 387, 99-106.
Stitt, J.T. (1976): The regulation of respiratory evaporative heat loss in the rabbit. *J. Physiol. (Lond)* 258, 157-171.
Szczepanska-Sadowska E. (1974): Plasma ADH increase and thirst suppression elicited by preoptic heating in the dog. *Am. J. Physiol.* 226, 155-161.
Taylor C.R. (1970): Dehydration and heat: effects on temperature regulation of East African ungulates. *Am. J. Physiol.* 219, 1136-1139.
Turlejska-Stelmasiak E. (1974): The influence of dehydration on heat dissipation mechanisms in the rabbit. *J. Physiol. (Paris)* 68, 5-15.
Turlejska E. & Baker M.A. (1986). Elevated CSF osmolality inhibits thermoregulatory heat loss responses. *Am. J. Physiol.* 251, R749-R754.
Wade, C.E., Keil, L.C. & Ramsey, D.J. (1983): Role of volume and osmolality in the control of plasma vasopressin in dehydrated dogs. *Neuroendocrinology* 37, 349-353.
Woodcock E.A., McLeod J.K. & Johnston C.I. (1986): Vasopressin stimulates phosphatidylinositol turnover and aldosterone synthesis in rat adrenal glomerulosa cells: Comparison with angiotensin II. *Endocrinology* 118, 2432-2436.

Integrative and cellular aspects of autonomic functions : temperature and osmoregulation. Eds K. Pleschka, R. Gerstberger. John Libbey Eurotext, Paris © 1994, pp. 471-475.

Effects of temperature on supraoptic osmosensitive neurons in hypothalamic slices *in vitro*

Toshihiro Nakashima, Kazumi Ofuji, Seiji Miyata and Toshikazu Kiyohara

Department of Applied Biology, Kyoto Institute of Technology Matsugasaki, Sakyo-ku, Kyoto 606, Japan

There are some evidences that considerable interference between the thermoregulatory system and osmoregulatory system (Turlejska and Baker, 1986; Simon-Oppermann, Hammel and Simon, 1979). For example, the majority of warm-sensitive neurons in the medial preoptic area which is a most important control center of thermoregulatory system decreased their activity in response to local and peripheral hyperosmotic stimulation in tissue slices in vitro and anesthetized rat (Nakashima, Hori, Kiyohara and Shibata, 1985; Koga, Hori, Inoue, Kiyohara and Nakashima, 1987). The reduced activity of medial preoptic warm-sensitive neurons in a hyperosmotic environment may explain the phenomenon of reduced evaporative heat loss in dehydrated mammals.

On the other hand, neurosecretory cells in the supraoptic (s.o.) nucleus and paraventricular (p.v.) nucleus are known to regulate body water balance releasing vasopressin from the posterior pituitary into the systemic circulation. It is reported that vasopressin secretion has been related to changes in hypothalamic temperature as well as plasma osmotic pressure (Forsling, Ingram and Stanier, 1976). The thermosensitivity of the s.o. and p.v. cells were studied under normal osmotic pressure condition (Matsumura and Nakayama, 1987; Inenaga, Osaka and Yamashita, 1987) but not under hyperosmotic conditions.

The present study concerns the effect of temperature on the s.o. neurosecretory cells at hyperosmotic environment. We have investigated thermosensitivity of the s.o. cells during perfusion of normal and hyperosmotic pressure solution in hypothalamic slices in vitro.

METHODS

Male rats of Wistar strain, weighing 150-180 g, were stunned and their brains were quickly removed. Thin (400 μm thick), coronal hypothalamic slices containing the s.o. nucleus and its vicinity were cut from each brain with a tissue slicer. After pre-incubation in oxygenated (95% O_2 + 5% CO_2) Krebs-Ringer solution (300 mOsm/kg, pH 7.4) at 37℃ for about 1 h, a selected slice was submerged in a recording chamber which was perfused with the same solution at a flow rate of 2.0 ml/min. The Krebs-Ringer solution contained (in mM) NaCl 124, KCl 5, KH_2PO_4 1.24, $MgSO_4$ 1.3, $CaCl_2$ 2.6, $NaHCO_3$ 26 and glucose 10. Extracellular single neural activities were recorded with a glass micropipette filled with 0.5 M sodium acetate containing 2% Pontamine sky blue (resistance, 15-40 MΩ) from s.o. nucleus. After passing a high input impedance pre-amplifier the potentials were amplified and displayed on an oscilloscope. A ratemeter processed this impulse activity and the output was recorded on a chart recorder. The slice temperature was maintained at, or changed to, any temperature within the range of 34-40℃ by controlling the perfusate temperature, and was continuously measured with a fine copper-constantan thermocouple placed on the slice.

The firing activity was observed at 38℃ for at least 5 min and only the neurons showing stable firing rate were subjected for further studies. Neural responses to temperature changes were determined in the standard medium (300 mOsm/kg), then the thermal responses in hyperosmotic pressure solutions (maximum change, +20 mOsm/kg) were investigated. The osmotic pressure of the perfusate was adjusted by changing the NaCl concentration. After a change from one perfusion medium to the other, at least 15 min were allowed to elapse for equilibration between medium and tissue. The neurons were classified according to their thermal coefficients. Cells exhibiting thermal coefficients of > 0.6 impulses/s・℃ were classified as warm-sensitive. Cells with a negative thermal coefficient of < -0.6 impulses/s・℃ were cold-sensitive (Nakashima, Pierau, Simon and Hori, 1987). At the end of each recording, the recording site was marked with Pontamine sky blue by passing a current through the electrode and verified in the formalin fixed brain slice.

RESULTS AND DISCUSSION

A total of 20 cells in the s.o. nucleus were studied for their responses to changes in slice temperature. Of these, 9 (45%) cells showed warm sensitivity which increased the firing rate with a rise in slice temperature and 2 (10%) cells were cold-sensitive neurons showing the opposite type of response to temperature changes. The remaining 9 (45%) cells were thermally insensitive. The population ratio of thermosensitive neurons to total number of cells is almost same with previous work (Matsumura and Nakayama, 1987).

Table 1. Number of neurons in the s.o. nucleus classified according to firing pattern and thermosensitivity. Number in parenthesis indicates neurons changed their firing patterns during change in temperature from those observed at 38℃

	Warm-sensitive	Cold-sensitive	Thermally insensitive
Phasic	3	0	3
Non-phasic	6 (2)	2 (1)	6
Total	9 (2)	2 (1)	9

Regarding as the firing patterns, 6 cells showed phasically firing pattern and the remaining 14 cells fired non-phasically i.e. fast continuous or slow irregular at 38℃ during perfusion with standard medium (300 mOsm/kg). The distribution of thermosensitive and thermally insensitive neurons of each firing patterns is shown in Table 1. Two warm-sensitive and one cold-sensitive neurons firing non-phasically at 38℃ changed their firing patterns to phasic and decreased their firing rates during cooling and warming, respectively. No cell firing phasically at 38℃ changed its firing pattern during changes in temperature between 34 and 40℃. It is proposed by <u>in vivo</u> studies that the phasic activity is a feature of vasopressin neurons and cells firing non-phasically are regarded as putative oxytocin neurons (Poulain and Wakerley, 1982). The suppression of a half of phasic neurons by cooling may explain cold induced plasma vasopressin reduction.

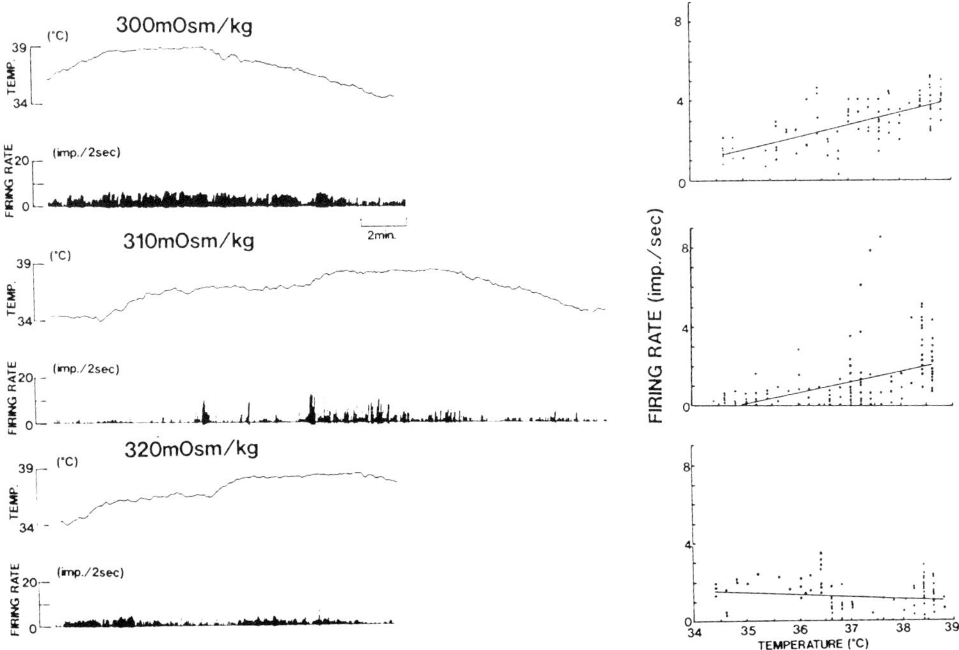

Fig.1. A typical warm-sensitive neuron in the s.o. nucleus. Left panels show firing rate responses to changes in temperature. Right panels show relationship between firing rate and temperature of slice. Upper, middle and lower panels indicate the responses to temperature changes during perfusion with solution of different osmotic pressure of 300, 310 and 320 mOsm/kg, respectively.

Further, we studied thermosensitivity of s.o. neurons in the hyperosmotic conditions (310, 320 mOsm/kg or both). An example of a non-phasically warm-sensitive neuron is shown in Fig.1. The firing rate of this neuron at 38℃ was 3.3 impulses/s in standard medium and which decreased to 1.8 and 1.1 impulses/s when the osmolarity was increased to 310 and 320 mOsm/kg, respectively. The thermosensitivity of this neuron at 310 mOsm/kg (mean ± SE, 0.54 ± 0.09 impulses/s・℃) did not differ from that in the standard medium (0.62 ± 0.06 impulses/s・℃)($P > 0.1$) but lost in more hyperosmotic medium of the 320 mOsm/kg (-0.11 ± 0.06 impulses/s・℃)($P < 0.05$). Same tendency was shown in other one warm- and one cold-sensitive neurons. This finding tempts us to a supposition that thermosensitive neurons retain thermosensitivity at mild hyperosmotic environment but lose it by an excessive hyperosmotic condition.

Twelve s.o. cells were challenged to hyperosmotic stimulations. Of these, 4 cells showd phasically firing pattern and 8 cells were non-phasic neurons. Three of 4 phasic neurons increased their firing rate during hyperosmotic stimulations and remaining one did not change. This result well explains the elevation of plasma vasopressin level during dehydration. On the other hand, 4 of 8 non-phasic neurons decreased their activities when the osmotic pressure was increased and remaining 4 cells did not respond. According to the criteria of Poulain and Wakerley (1982), non-phasically firing cells were thought to be oxytocin neurons. Many previous studies revealed that cells in the s.o. and p.v. nucleus respond to increase in the osmotic pressure with a increase in firing rate and dehydration causes elevation of plasma levels of oxytocin. In the present study, activities of half putative oxytocin neurons were decreased by hyperosmotic pressure. It may be explainable this discrepancy by inhibitory synaptic inputs on the non-phasic neurons from osmoreceptor cells insie or outside of s.o. nucleus. Since we did not block synaptic transmission within the slice, inherent property of s.o. cells to osmolarity changes is still unknown. It is necessary to clarify the effect of hyperosmotic pressure on the activity of non-phasic s.o. cells under synaptic blockage condition.

REFERENCES

Forsling, M.L., Ingram, D.L. and Stanier, M.W. (1976): Effects of various ambient temperatures and heating and cooling the hypothalamus and cervical spinal cord on antidiuretic hormone secretion and urinary osmotic pressure in pigs. J. Physiol. (London), 257, 673-686.

Inenaga, K., Osaka, T. and Yamashita, H. (1987): Thermosensitivity of neurons in the paraventricular nucleus of the rat slice preparation. Brain Res. 424, 126-132.

Koga, H., Hori, T., Inoue, T., Kiyohara, T. and Nakashima, T. (1987): Convergence of hepatoportal osmotic and cardiovascular signals on preoptic thermosensitive neurons. Brain Res. Bull. 19, 109-113.

Matsumura, K. and Nakayama, T. (1987): Local cooling suppresses supraoptic neurosecretory cells in vivo and in vitro. J. therm. Biol. 12, 109-111.

Nakashima, T., Hori, T., Kiyohara, T. and Shibata, M. (1985): Osmosensitivity of preoptic thermosensitive neurons in hypothalamic slices in vitro. Pflugers Arch. 405, 112-117.

Nakashima, T., Pierau, Fr.-K, Simon, E. and Hori, T. (1987): Comparison between hypothalamic thermoresponsive neurons from duck and rat slices. Pflugers Arch. 409, 236-243.

Poulain, D.A. and Wakerley, J.B. (1982): Electrophysiology of hypothalamic magnocellular neurons secreting oxytocin and vasopressin. Neuroscience, 7, 773-808.
Simon-Oppermann, C., Hammel, H.T. and Simon, E. (1979): Hypothalamic temperature and osmoregulation in the pekinduck. Pflugers Arch. 378, 213-221.
Turlejska, E. and Baker, M.A. (1986): Elevated CSF osmolality inhibits thermoregulatory heat loss responses. Am. J. Physiol. 251, R749-754.

B. Endocrine control mechanisms in body fluid homeostasis

Integrative and cellular aspects of autonomic functions : temperature and osmoregulation. Eds K. Pleschka, R. Gerstberger. John Libbey Eurotext, Paris © 1994, pp. 479-485.

The role of uric acid in fluid and ion balance of birds

Eldon J. Braun, Stephani L.B. Boykin and Myra M. Pacelli

Department of Physiology, College of Medicine, University of Arizona, Tucson, 85724 Arizona, USA

INTRODUCTION

The regulation of fluid and electrolyte balance in birds is somewhat more elaborate than it is in mammals mainly because more organs can be involved in the process. In mammals, this balance is regulated solely by the kidney, whereas in birds the kidney must be regulated in concert with the lower gastrointestinal tract and in some cases also with nasal salt glands. This regulation in birds is further complicated because they excrete principally uric acid as the end product of metabolism. On average, birds excrete 70-75% of their excess nitrogen as uric acid. This compound is generally considered a very efficient form in which to excrete nitrogen because of its very low aqueous solubility (Table 1) and the number of nitrogen atoms per molecule (4).

Table 1. Aqueous solubility of uric acid and some of its salts

Compound	Solubility (mM/l)
Uric Acid	0.384
Sodium Urate	6.76
Potassium Urate	12.06

The low aqueous solubility of uric acid can be both an asset and a liability. The low solubility facilitates excretion with a minimal amount of water, but this same property can cause uric acid to precipitate from solution and form crystalline masses that could block flow within the renal tubules.

There are additional intriguing aspects of the biology of uric acid excretion that are of interest to examine more closely. However, most of these are related to its very low aqueous solubility. These questions are centered around how the avian kidney processes uric acid and consequently how the presence of uric acid in the urine affects the regulation of fluid and electrolyte balance in birds.

PROCESSING OF URIC ACID BY THE AVIAN KIDNEY

Uric acid is avidly secreted by the avian kidney. This secretion is thought to be primarily localized to the proximal tubule of the nephrons and this process is probably facilitated by the organic anion secretory system of these nephrons. Because of its low aqueous solubility, questions can be posed about the mechanism of secretion by the proximal renal tubule. A model for the transcellular secretion of uric acid in birds can be constructed based primarily on data obtained from whole animal studies on starlings (Laverty and Dantzler, 1983 and Fig. 1)..

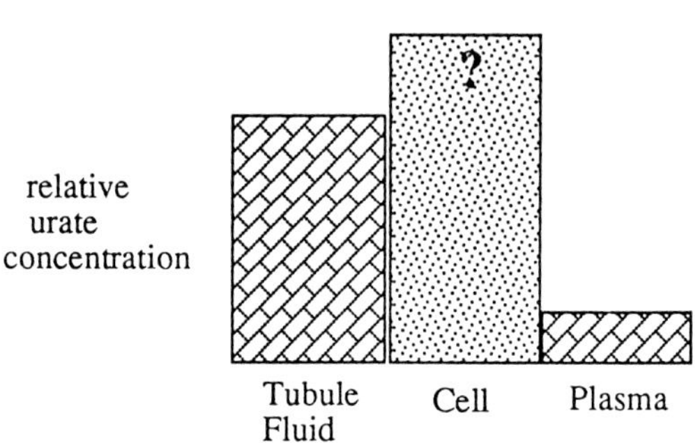

Fig. 1. Model of uric acid secretion by the avian kidney. Data are from Laverty and Dantzler, 1983. The plasma concentration of uric acid was 0.72 mM/l and the tubule fluid-to-plasma ratio for uric acid was 4.98. Therefore the tubule fluid concentration of uric acid was 3.6 mM/l. The cell concentration of uric acid was not measured.

Data from a very recent study of uric acid transport by nephron segments isolated from the chicken kidney and perfused in vitro support the whole animal studies (Brokl et al., submitted). In the model depicting the secretion of uric acid (urate), the energy requiring step appears to be at the basolateral membrane where urate must be taken into the cell against an electrical and chemical gradient. The model suggests that urate builds to a relatively high concentration within the cell, then moves across the apical membrane down a concentration gradient into the tubule lumen by a process that appears to be passive. The uptake at the basolateral membrane (BLM) is probably one of secondary or tertiary active transport and is an anion exchange process which maintains electroneutrality. The results of the avian perfused tubule studies are qualitatively similar to data for perfused segments of snake renal tubules, but with some quantitative differences (Dantzler, 1973). In snakes the anions exchanged for urate may be chloride and to a lesser extent the dicarboxcylic acid, α-ketoglutaric acid (Benyajati and Dantzler, 1988). The process appears to be quite similar in birds in that chloride may be exchanged for urate, but it may not be as tightly coupled to an exchange with α-ketoglutaric acid (Brokl et al., submitted). For birds and snakes, the model further suggests that there is a significant accumulation of urate within the proximal tubule cells as a result of the active uptake across the BLM. The high intracellular concentration permits uric acid to move passively across the brushborder membrane (BBM) into the tubule lumen.

The movement of uric acid across the cell and its exit across the BBM are worth examining more closely. For starlings, the plasma concentration of uric acid (0.72 mM/l) is almost two times greater than its aqueous solubility (Laverty and Dantzler, 1983). Furthermore, if as the models of the transcellular movement of uric acid suggest, the intracellular concentration exceeds both the plasma and lumen concentrations by several fold (Laverty and Dantzler reported a tubule fluid-to-plasma ratio of 4.98), the concentration within the cells must be much in excess of the solubility of uric acid. In addition, in vitro studies have been reported where the intracellular concentration of uric acid can reach levels 22 times greater than the medium (100 μm) in which the cells are bathed, therefore exceeding the solubility limit of uric acid by 5.7 fold (Tanner et al., 1983).

How are cells protected from concentrations of uric acid (or the salts of uric acid) that should precipitate from solution ? Recent studies of the transport of organic aniōns (and organic cations) may help to explain how this process can occur. It has been suggested that once the transported species crosses the BLM, it is incorporated into membrane bound vesicles (Miller et al., 1993). As described, this is not a process of endocytosis, as the molecules appear first to be transported by the BLM to the cell interior where they are vesicularized. It is the vesicles containing potentially toxic material (in this case uric acid) that move across the cell interior. This leaves open the question of how the substances are vesicularized and is there an area of the cell interior that is exposed to the transported species ? However, at a first approximation such a process could explain how potentially toxic molecules move transcellularly without damaging cells.

The studies directed toward describing the mechanism by which uric acid crosses the BBM have yielded somewhat puzzling results. First, it is suggested that uric acid moves passively across this membrane. However, when the apparent permeabilities of the BBM and the BLM are compared, the BBM has a permeability (0.75×10^{-5} cm sec^{-1}) that is four times lower than that of the BLM (3.10×10^{-5} cm sec^{-1}; Dantzler, 1993). Based on the model describing the transcellular movement of uric acid, the opposite might be expected. Furthermore, studies on BBM vesicles prepared from snakes kidneys do little to clarify how uric acid crosses the BBM. These vesicles failed to show a concentrative uptake of uric acid as they do for the prototypical organic acid para-aminohippuric acid (Benyajati and Dantzler, 1988). What modest uptake of uric acid that was seen could not be inhibited by the classical anion exchange inhibitor SITS (4-acetamido-4'-iosthiocyanostilbene-2-2'-disulfonic acid) or other organic acids and was not sensitive to a sodium gradient. (Benyajati and Dantzler, 1988).

From these observations, it is clear that it is very unclear how uric acid crosses the BBM and enters the proximal tubule to complete the process of tubular secretion. An interesting question is whether free urate or uric acid can be liberated into the lumen without precipitating from solution. Three processes contribute to increasing the concentration of uric acid in the fluid of the proximal tubule. First, it enters the tubule by filtration. Whereas there may be some slight plasma binding of uric acid in birds, it is for the most part freely filterable through the glomerular filtration barrier (Laverty and Dantzler, 1983). Second, there appears to be no reabsorption of uric acid in the proximal tubule (Laverty and Dantzler, 1983). Thus the normal fluid reabsorption that goes on in the proximal tubule should contribute to increasing the uric acid concentration in the tubule. Third, avid secretion occurs in the proximal tubule. These processes raise the tubule fluid concentration to levels much beyond the solubility of uric acid. Thus, it can be suggested that it would not be physiologically advantageous for free uric acid to enter the proximal tubule lumen. It indeed may be the case that free uric acid does not enter the lumen. It may be complexed to some other molecule before or as it crosses the BBM. Such complexing may be one explanation for the somewhat anomalous results obtained from the permeability and vesicle studies cited above. In these studies, the uric acid used may not have been in the same chemical form as it occurs in vivo, which may have prevented the membrane transport proteins from recognizing the species to be transported.

URIC ACID IN THE URINE

Avian ureteral urine, when freshly collected, has a very cloudy, turbid appearance and when samples of this urine are centrifuged, very pronounced pellets are produced. Such pellets, when dried to a constant mass and analyzed, are found to consist of about 65% uric acid (Braun and Pacelli, 1991). Prior to drying, when samples of a pellet are examined with the scanning electron microscope, they are found to consist of small spheres that average about 3 µm in diameter with some of the spheres being as large as 25 µm (Braun and Pacelli, 1991; Fig. 2).

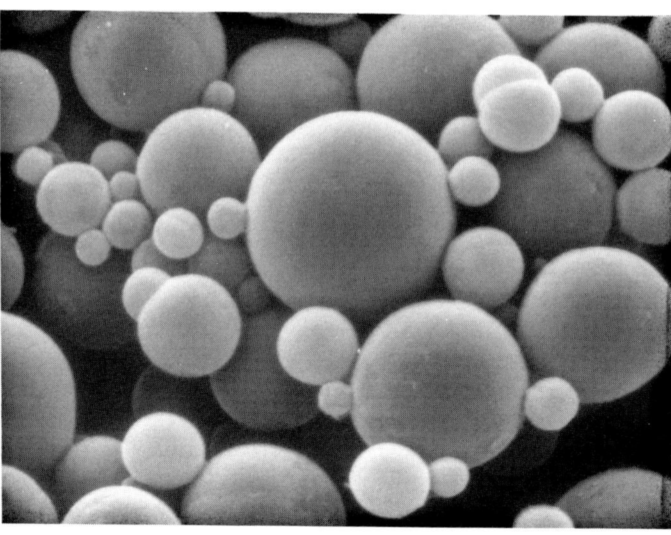

Fig. 2. Scanning electron micrograph (SEM) of avian urine. For this picture, a drop of fresh bird urine was allowed to air dry on an SEM stub, gold coated and viewed with an SEM. Note absence of crystals and the 1 micrometer scale bar at the lower left.

Thus, it is apparent that the uric acid is "packaged" in a way that prevents crystal formation. At this time, the chemical form of uric acid within the spheres is uncertain. Exactly where this process begins in the nephron (or cells) is unknown at this time, but it may be part of the process or mechanism by which uric acid crosses the BBM of the cells. Because it is unlikely that free urate or uric acid can enter the lumen of the proximal tubule where conditions are not suitable for its existence in solution, urate may be complexed to a stabilizing molecule before or as it crosses the bounds of the apical cell membrane.

In addition to the high concentration of uric acid there is also a very high concentration of protein in the tubule fluid (Boykin and Braun, 1993). The urine of white leghorn domestic fowl typically contains 5 - 15 mg/ml of protein whereas mammalian (human) urine contains 0.1 mg/ml protein by comparison (Boykin and Braun, 1993). Preliminary analysis (SDS-PAGE) suggests that the major component of this protein has a molecular weight of about 45 Kd (Boykin and Braun, 1993). How this protein gets into the urine and how it functions is uncertain at this time. However, some suggestions can be made. First, how does the protein get into the urine ? While the porosity of the avian glomerular basement membrane (GBM) has not been determined as for mammals (Bohrer et al., 1978), the apparent size (MW) of the major protein in avian urine (45 Kd) is possibly at the cut-off for molecules to pass through the filtration barrier (based on mammalian standards). A second source of the protein could be the glycocalyx of the brush border. However, whereas there is a high turnover of cells in renal tubules, the quantity of protein in the urine makes this an unlikely source. Beyond the proximal tubule (and loops of Henle for those nephrons that have loops) the avian nephron has large numbers of goblet cells that could be the source of the protein (glycoprotein; McNabb et al., 1973b). However, if the protein does not appear in the tubule fluid until the distal nephron, it leaves open the question of how the fluid in the early portion of the nephron is protected from uric acid precipitation. As has been eluded to, the primary functions of the protein may be to prevent crystal formation and to serve as a lubricant to keep the small spheres moving along the course of the nephron.

It has been suggested that uric acid (urate) functions to buffer hydrogen ions that are secreted by the proximal tubule (Long & Skadhauge, 1983a). This supposed function is based on the assumption that uric acid exists as uric acid dihydrate in avian tubule fluid (Lonsdale and Sutor, 1971). This suggestion may not be tenable. Uric acid has two pKa's, one at 5.75 and the other at 10.3. The pH of the fluid in the proximal tubule has been measured in one species, the starling and found to be about 7.5 (Laverty and Alberici, 1987). Thus, one H^+ could be buffered, and if this were to happen, uric acid could be formed and because of its very low aqueous solubility, it would likely precipitate from solution. This appears not to happen as avian ureteral urine does not contain crystals of uric acid (Boykin and Braun, 1993).

Because the buffering capacity of the urine has been correlated with its uric acid concentration, this has been interpreted as support for the argûment that uric acid functions as a buffer. We have observed that the protein and uric acid concentrations of the urine are positively correlated (Boykin and Braun 1993). Thus, because of the net negative charge that exists on most proteins it is quite possible that the protein in the urine, in addition to preventing uric acid precipitation, may also serve to buffer hydrogen ions. Therefore, the correlation between urine uric acid concentration and buffering capacity of avian urine may be fortuitous.

The protein that we have measured in the urine of birds has been for urine collected directly from the ureters. As birds do not have a urinary bladder (the ostrich has been reported to have a bladder but this may actually be a diverticulum of the cloaca), the ureteral urine enters the cloaca, the terminal portion of the gastrointestinal tract. In most birds, the urine does not remain in the cloaca until excreted, but is moved forward by peristaltic activity into the colon (rectum) and digestive ceca (in those birds that possess ceca) (Akester, et al, 1967; Ohmart, et al, 1970; Brummermann and Braun, submitted). In these regions of the GI tract, the urine and its contents are brought into contact with a large and varied population of anaerobic bacteria (Barnes and Impey, 1974) and epithelial tissues capable of solute transport (Thomas, et al, 1979). While we have not yet quantified the protein content of avian excreta, we believe that the majority of the protein found in avian urine is broken down by bacteria and absorbed as smaller elements by the epithelia of the lower GI tract. We do know that 68% of the uric acid of avian urine is broken down and several of the products (short chain volatile fatty acids and glutamine) are absorbed by the lower GI tract (Campbell and Braun 1986; Karasawa and Nakata, 1989). Thus, if the protein is binding uric acid, it is also likely to be broken down when uric acid is degraded. While we have not carried out energy balance studies, the excretion (loss) of the large amount of protein found in urine would represent a significant energy loss to a bird.

Thus, while in a classical sense uric acid is a very efficient form in which to excrete excess nitrogen, this form of nitrogen excretion is not without significant cost, for it is intricately involved in the processes by which fluid and electrolyte balance is regulated. Fortunately, some of these costs are recouped by the kidney and lower GI tract functioning in concert to regulate the fluid and electrolyte balance. Finally, much remains to be learned with respect to the biology of uric acid as a form in which to excrete nitrogen.

Supported by NSF IBN-9220241

REFERENCES

Akester, A.R., Anderson, R.S., Hill, K.J. & Osbaldiston, G.W. (1967): A radiographic study of urine flow in the domestic fowl. *Br. Poultry Sci.* 8, 209-212.

Barnes, E.M. & Impey, C.S. (1974): The occurrence and properties of uric acid decomposing anaerobic bacteria in the avian caecum. *J. Appl. Bact.* 37, 393-409.

Benyajati, S. & Dantzler, W.H. (1988): Enzymatic and transport characteristics of isolated snake renal brush-border membranes. *Am. J. Physiol.* 255, R52-R60.

Bohrer, M.P., Baylis, C., Humes, H.D., Glassock, R.J., Robertson, C.R. & Brenner, B.M. (1978): Permselectivity of the glomerular capillary wall. Facilitated filtration of circulating polycations. *J. Clin. Invest.* 61, 72-78.

Boykin, S.L.B. & Braun, E.J. (1993): Relationship between protein and uric acid in avian urine. *FASEB J.* 7(3), A5.

Braun, E.J. & Pacelli, M.M. (1991): The packaging of uric acid in avian urine. *FASEB J.* 5, A1408.

Brokl, O.H., Braun, E.J. & Dantzler, W.H. (submitted for publication): Organic anion, organic cation, and fluid transport by isolated perfused and nonperfused avian renal proximal tubules.

Brummermann, M. & Braun, E.J. (submitted for publication): Effect of water deprivation on colonic motility of White Leghorn roosters.

Campbell, C.E. & Braun, E.J. (1986): Cecal degradation of uric acid in Gambel quail. *Am. J. Physiol.* 251, R59-R62.

Dantzler, W.H. (1993): Mechanisms of transport of glucose, amino acids, organic acids (or anions) and organic bases (or cations) in reptilian nephrons. In *New insights in vertebrate kidney function*, ed. J.A. Brown, R.J. Balment & J.C. Rankin, pp. 145-165. Cambridge: University Press.

Dantzler, W.H. (1973): Characteristics of urate transport by isolated perfused snake proximal renal tubules. *Am. J. Physiol.* 224(2), 445-453.

Karasawa, Y. & Nakata, C. (1989): Incorporation of intraportally infused [^{15}N] ammonia into urinary uric acid in cockerels pretreated with methionine sulfoximine. *Br. Poultry Sci.* 30, 947-952.

Laverty, G. & Alberici, M. (1987): Micropuncture study of proximal tubule pH in avian kidney. *Am. J. Physiol.* 253, R587-R591.

Laverty, G. & Dantzler, W.H. (1983): Micropuncture study of urate transport by superficial nephrons in avian (*Sturnus vulgaris*) kidney. *Pflügers Arch* 397, 232-236.

Long, S. & Skadhauge, E. (1983a): Renal acid excretion in domestic fowl. *J. Exp. Biol.* 104, 51-58.

Lonsdale, K. & Sutor, D.J. (1971): Uric acid dihydrate in bird urine. *Science* 172, 958-959.

McNabb, F.M.A., McNabb, R.A. & Steeves, H.R. III. (1973b): Renal mucoid materials in pigeons fed high and low protein diets. *The Auk* 90, 14-18.

Miller, D.S., Stewart, D.E. & Pritchard, J.B. (1993): Intracellular compartmentation of organic anions within renal cells. *Am. J. Physiol.* 264, R882-R890.

Ohmart, R.D., McFarland, L.Z. & Morgan, J.P. (1970): Urographic evidence that urine enters the rectum and ceca of the roadrunner (*Geococcyx californianus*) Aves. *Comp. Biochem. Physiol.* 35, 487-489.

Tanner, R.J., Chonko, A.M., Edwards, R.M. & Grantham, J.J. (1983): Evidence for an inhibitor of renal urate and PAH secretion in rabbit blood. *Am. J. Physiol.* 244, F590-F598.

Thomas, H.D., Skadhauge, E. & Read, M.W. (1979): Acute effects of aldosterone on water and electrolyte transport in the colon and coprodeum of the domestic fowl (*Gallus domesticus*) *in vivo. J. Endocr.* 83, 229-237.

Integrative and cellular aspects of autonomic functions : temperature and osmoregulation. Eds K. Pleschka, R. Gerstberger. John Libbey Eurotext, Paris © 1994, pp. 487-495.

Blood pressure, urinary protein excretion and kidney function in rats after renal transplantation

Bernhard Schmitt, Christiane Graf and Rainer Rettig

Department of Pharmacology, University of Heidelberg, and German Institute for High Blood Pressure Research, Im Neuenheimer Feld 366, D-69120 Heidelberg, Germany

SUMMARY

In renal transplantation studies bilaterally nephrectomized recipients of a kidney from stroke-prone spontaneously hypertensive rats (SHRSP) developed posttransplantation hypertension, but not recipients of a kidney from normotensive Wistar-Kyoto rats (WKY). The underlying mechanisms are not well understood. To investigate whether differential alterations of the transplanted kidneys from the respective donor strains may play a role we subjected rats after successful renal transplantation to a dietary regimen consisting of a low-salt diet for ten days followed by a high-salt diet for 14 days. Arterial blood pressure was lowest in recipients of a WKY kidney, significantly higher in recipients of an F1 hybrid (F1H) kidney, and highest in recipients of an SHRSP kidney. Body weight, glomerular filtration rate (GFR), and renal plasma flow (RPF) were similar in all three groups. Daily urinary protein excretion and cumulative renal sodium retention were higher in recipients of an SHRSP kidney than in the two other groups, although the former consumed significantly less food than the latter. These data indicate that renal function is impaired in transplanted SHRSP kidneys compared to transplanted kidneys from normotensive donor strains. These alterations are likely genetically determined and may contribute to posttransplantation hypertension.

INTRODUCTION

Spontaneously hypertensive rats (SHR) and their stroke-prone substrain (SHRSP) are the most widely used animal models of genetic hypertension (Folkow 1982). The particular genetic abnormalities determining the development of hypertension in these strains are not well understood.

When grafted with a kidney from SHR (Kawabe et al., 1979) or SHRSP donors (Rettig et al., 1989), bilaterally nephrectomized recipients with normal or only slightly elevated blood pressures develop posttransplantation hypertension. This effect is not observed with a renal graft from normotensive Wistar-Kyoto rats (WKY) from which SHR and SHRSP were originally bred (Louis et al., 1990). Posttransplantation hypertension also develops in recipients of a renal graft from SHRSP donors without hypertension and without hypertension-induced organ damages. This could be shown in two studies in which SHRSP kidney donors were used that either had been treated with antihypertensive drugs to permanently normalize blood pressure before removal of the kidney (Rettig et al., 1990a), or that were young enough not to have developed elevated blood pressures (Kopf et al., 1993).

These findings have been interpreted to indicate that a genetic defect residing in SHR and SHRSP kidneys is responsible for posttransplantation hypertension in recipients of kidneys from these strains (Cowley, 1992; DeWardener, 1990; Rettig et al., 1990b). The mechanisms by which a transplanted SHR or SHRSP kidney increases blood pressure in the recipients are currently poorly understood. Glomerular filtration rate (GFR) and renal blood flow (RBF) have been reported to be similar in transplanted WKY and SHRSP kidneys (Rettig et al., 1990a,c; Graf et al., 1993; Kopf et al., 1993). However, more sensitive parameters of renal function such as renal protein excretion and sodium handling have not yet been investigated in transplanted kidneys from these strains.

Our study was designed to further elucidate renal function in transplanted WKY and SHRSP kidneys. Specifically, GFR, RBF, renal protein excretion, and sodium balance were measured in transplanted rats kept in metabolic cages and fed a low-salt diet for ten days followed by a high-salt diet for another 14 days.

METHODS

Animals

Experiments were conducted in adult male stroke-prone spontaneously hypertensive rats (SHRSP), normotensive Wistar-Kyoto rats (WKY), and F1 hybrids (F1H) bred from SHRSP and WKY parents as kidney donors, as well as in male F1 hybrids as renal graft recipients. Rats were obtained at birth from the rat breeding facilities of the University of Heidelberg, FRG, where inbred WKY and SHRSP strains have been maintained since 1975. Animals were housed in plastic cages in a temperature and humidity controlled environment with lights on at 0600 h and off at 1800 h. If not otherwise indicated, standard rat food (Altromin pellets) containing 0.6% NaCl and tap water were available to the rats ad libitum. All experiments were preapproved by a governmental committee on animal welfare.

Surgery

Renal Transplantation

The microsurgical technique used in this study has been described in detail elsewhere (Rettig et al., 1990a). It is a modification of the technique first described by Fisher and Lee (1965) and Lee (1967). Briefly, a kidney donor and a recipient were anesthetized simultaneously with pentobarbital (60 mg/kg, i.p.). The abdomen of the donor was opened by a long midline incision. Blood vessels and ureter of the left kidney were exposed, renal blood circulation was interrupted, and the kidney was immediately perfused with 1 ml of an ice-cold isotonic electrolyte solution (Euro-Collins, Fresenius, Bad Homburg, FRG). The kidney was removed and transferred to the recipient rat which had been prepared by left unilateral nephrectomy and exposing the abdominal aorta and vena cava just caudal to the renal vessels through a long abdominal midline incision. During surgery, the graft was repeatedly rinsed with ice-cold saline. Blood flow through the anastomotic area was transiently interrupted and the grafted blood vessels were anastomosed end-to-side to the abdominal aorta and vena cava of the recipient, using 9-0 polyamid suture material (Ethilon, Ethicon Comp., Norderstedt, FRG). The grafted ureter was directly inserted into the recipient's urinary bladder through a small hole in the bladder wall. Total graft ischemia lasted 40-50 min. Following surgery, rats were treated with 50 mg ampicilline (Binotal, Bayer AG, Leverkusen, FRG) i.p. per rat per day for ten days.

Vascular and Ureter Catheters

For clearance measurements, rats were instrumented under ether anesthesia with chronic catheters in the right femoral artery and vein as well as in the graft ureter. The femoral artery catheter consisted of two pieces of PE10 and PE50 tubing (Portex Corp., Hythe, UK) sealed together under hot air. The catheter for the femoral vein was a single piece of PE25 tubing. Both catheters were inserted about 3 cm into the respective blood vessels and tunnelled under the skin to exit through the scruff of the neck. When not in use, catheters were filled with heparinized saline (20 U/ml) and closed with a stainless steel pin.

The ureter catheter was a single piece of silicone tubing (LHD, Heidelberg, FRG) with internal and external diameters of 0.3 mm and 0.5 mm, respectively. The procedure for catheter implantation has been described in detail elsewhere (Horst et al., 1988). Briefly, the graft ureter was exposed through an abdominal midline incision and the catheter was inserted about 5 mm into the ureter. The free end of the catheter was carefully guided through the lateral abdominal wall and tunnelled under the skin to exit at the nape of the neck. The catheter was anchored to the skin of the neck in a way that urine dripping from its orifice would fall to the cage floor without incommodating the animal. After all catheters had been implanted, animals were allowed to recover for 48 h.

Glomerular Filtration Rate and Renal Plasma Flow

Glomerular filtration rate (GFR) and renal plasma flow (RPF) were determined in conscious rats as inulin and para-aminohippuric acid (PAH) clearances. Animals were instrumented with indwelling vascular and ureteral catheters as described above. Catheters were connected to extension lines to be handled from outside the cage without disturbing the animal. Rats received an 1 ml i.v. bolus injection containing 40 mg inulin and 5 mg PAH, immediately followed by a continuous i.v. infusion of 2 mg/kg/min inulin and 0.25 mg/kg/min PAH. Infusion volume was 17 µl/min for a total infusion time of 135 min, delivered by an automatic infusion pump (Braun-Melsungen, FRG). Urine was quantitatively collected into preweighed tubes during five consecutive sampling intervals of 15 min each, starting after 1 h of infusion. Urine flow (µl/min) was determined gravimetrically. At the end of each urine collection period, about 300 µl of blood were obtained through the arterial line. Plasma and urine samples were stored at -20 °C until assayed. Inulin and PAH concentrations in plasma and urine were determined photometrically. All measurements were done in duplicate. Inulin and PAH clearances were calculated according to standard procedures.

Blood Pressure

Blood pressure was directly measured via an arterial line at the end of the protocol. Rats were conscious and unrestrained in their home cages. The arterial line was connected to a Statham P23Db pressure transducer, the latter to a Gould Brush pressure computer and a Gould Brush 2400 recorder (all Gould Inc., Oxnard, California, USA).

Urinary Protein and Sodium Excretion

Urinary protein concentrations were determined using the bicinchoninic acid (BCA) method (Pierce, Oud Beijerland, The Netherlands). Proteins were separated from interfering substances by TCA precipitation according to a method adapted from Bensadoun and Weinstein (1976). Sodium concentrations were determined by flame photometry. Cumulative sodium retention was calculated as the sum of the daily differences between dietary sodium intake and renal sodium excretion over 14 days on high-salt diet.

Experimental Protocol

The left kidneys from 12 SHRSP, 14 WKY, and 13 F1H donors were removed and transplanted to 39 unilaterally nephrectomized F1 hybrids. Renal graft recipients were returned to their home cages and placed on a standard rat diet containing 0.60% NaCl. Seven days after renal transplantation, the remaining native kidney was also removed from renal graft recipients and the animals were transferred to metabolic cages. They were immediately put on a low-salt diet (0.18% NaCl) for ten days, followed by a high-salt diet (1.80% NaCl) for another 14 days. During special salt diets, rats were offered distilled water for drinking fluid instead of tap water. The diets consisted of sodium-free granular rat chow supplemented with NaCl according to the protocol. The sodium content of the diet was verified by flame photometry. Every day rats were offered a known amount of chow in a special trough equipped with a food trap to minimize spillage. After each 24 h period, food remaining in the trough or the food trap was weighed and 24 h food consumption was calculated. Daily sodium intake was calculated as the product of sodium concentration in the chow times the mass of chow consumed. Twentyfour-hour urine samples were quantitatively collected from each rat into volumetric cylinders containing mineral oil. Urine volume was recorded and samples were stored at -20 ^{o}C until further analysis. Daily urinary protein and sodium excretions, respectively, were calculated as the products of their concentrations times 24 h urinary volume.

At the end of the high-salt period, rats were instrumented with indwelling catheters in the right femoral artery and vein as well as in the grafted ureter. After two days of recovery from surgery, arterial blood pressure, glomerular filtration rate, and renal plasma flow were measured in conscious animals.

Statistics

Data are expressed as means $\pm$ SEM. Statistical comparisons were made as indicated either by Student's t-test for unpaired samples, by one-way analysis of variance, or by two-way analysis of variance with repeated-measures on one factor where "time" was the within-subjects factor and "donor strain" the between-subjects factor. When appropriate, analysis of variance was followed by post-hoc Newman-Keuls tests. Results were regarded to be statistically significant when $p<0.05$.

RESULTS

Mean arterial pressure (Table I) in recipients of an F1H kidney 31 days after renal transplantation was 127±7 mmHg. Compared to this group, recipients of a WKY kidney had significantly lower and recipients of an SHRSP kidney significantly higher mean arterial pressures. Body weight (Table I), GFR (Fig. 1) and RPF (Fig. 1) were similar in all three groups of transplanted rats. Posttransplantation hypertension in recipients of an SHRSP kidney on a high-salt diet was associated with enhanced renal sodium retention (Table I), decreased daily food intake (Table I), and increased proteinuria (Fig. 2). Daily urine volume on high-salt diet was only slightly and transiently elevated in F1H and WKY kidney recipients, but permanently in recipients of an SHRSP kidney. In the latter group, urinary protein concentration increased progressively following transplantation, whereas recipients of F1H and WKY kidneys exhibited only a small increment which reached a stable maximum on a relatively low level soon after the onset of the high-salt diet.

Table I: Body weight, mean arterial pressure, cumulative sodium retention, and daily food intake at the end of a 14 days regimen of high-salt diet in bilaterally nephrectomized F1H recipients of a WKY, F1H, or SHRSP kidney, respectively.

Kidney Donor	**WKY** n=14	**F1H** n=13	**SHRSP** n=12
Body Weight (g)	332±7	330±9	328±10
Mean Arterial Pressure (mmHg)	108±8	127±7*	173±9**§
Cumulative Sodium Retention (mmol)	5.0±0.8	6.6±1.4	10.4±1.4*§
Daily Food Intake (g)	29.9±1.4	27.8±1.9	22.0±1.6**§§

* $p < 0.05$; ** $p < 0.01$ compared to WKY
§ $p < 0.05$; §§ $p < 0.01$ compared to F1H

DISCUSSION

The present study shows that the genetic predisposition of the kidney donor to high blood pressure is a strong determinant of blood pressure in the recipients. Similar findings have been obtained before by this group (Rettig et al., 1989, 1990a; Graf et al. 1993; Kopf et al., 1993) and others (Bianchi et al., 1974; Dahl et al., 1974, 1975; Fox and Bianchi, 1976; Kawabe et al., 1979). Though these studies give good evidence for the hypothesis that genetically determined alterations of the kidney are largely responsible for the development of posttransplantation hypertension in this model, they all failed to identify the particular mechanisms.

We have previously reported that GFR and renal blood flow (RBF) do not differ between transplanted WKY and SHRSP kidneys (Rettig et al., 1990a,c; Graf et al., 1993; Kopf et al., 1993). The present study confirms these reports and extends our observations to transplanted F1H kidneys which exhibited GFR and RPF values similar to those of transplanted WKY and SHRSP kidneys. While GFR and RPF, generally used as the only measures of renal function, suggested unimpaired renal function in all groups, the assessment of urinary protein concentration detected great differences between them. Since urinary protein concentration is affected by urinary volume, while protein loss chiefly depends on the extent of glomerular damage and renal plasma flow, the latter being similar in all groups and presumably constant during the protocol, total protein excretion per day is a more valid indicator of glomerular damage. The fact that differences in urinary protein excretion among the groups were only apparent under high-salt diet, whereas similar values were measured under low-salt diet, indicates that glomerular structure was not grossly affected in transplanted SHRSP kidneys under normal conditions. An increased dietary salt load, however, somehow led to disturbance of the structural integrity of the glomeruli and to subsequent proteinuria in recipients of an SHRSP kidney. These findings suggest that transplanted kidneys from genetically hypertensive donors may be more sensitive to adverse environmental stimuli than renal grafts from normotensive donors. Some exogenous stimuli including high dietary salt are well known to trigger an elevation of arterial pressure in

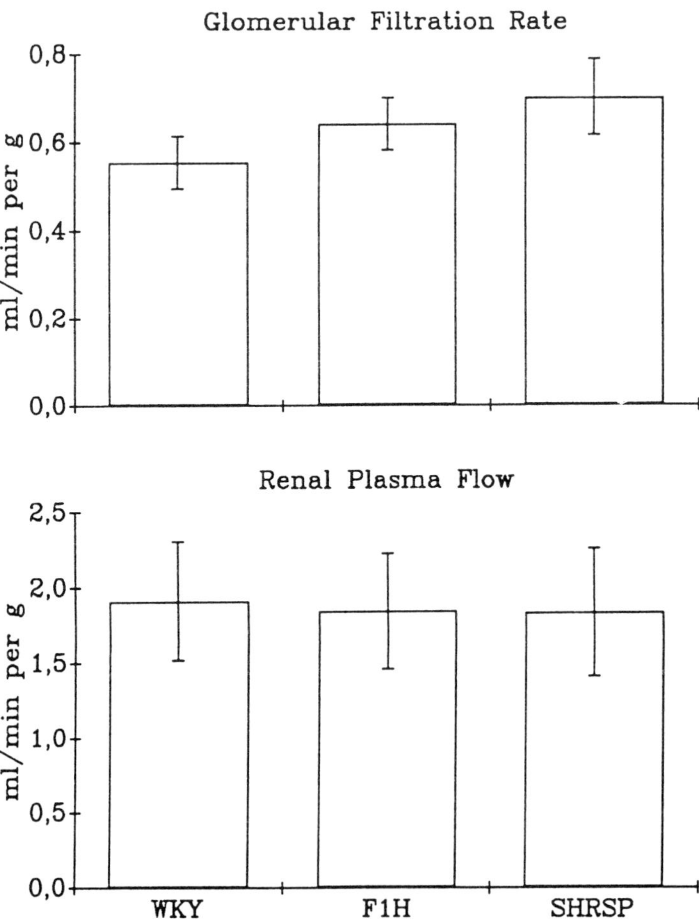

Fig. 1: Glomerular filtration rate and renal plasma flow in renal transplanted rats (F1H) with a kidney from WKY (n=14), F1H (n=13), or SHRSP donors (n=12), respectively.

susceptible individuals (Folkow, 1992). These observations are in keeping with the hypothesis that a genetic defect in the transplanted SHRSP kidneys causes posttransplantation hypertension.

A mechanism by which transplanted SHRSP kidneys might bring about hypertension in the recipients is renal sodium retention (Beierwaltes et al., 1982; Roman and Cowley, 1985). In our study, cumulative renal sodium retention through 14 days on a high-salt diet was significantly higher in hypertensive recipients of an SHRSP kidney than in recipients of an F1H or WKY kidney with slightly elevated or normal blood pressures, respectively. It should

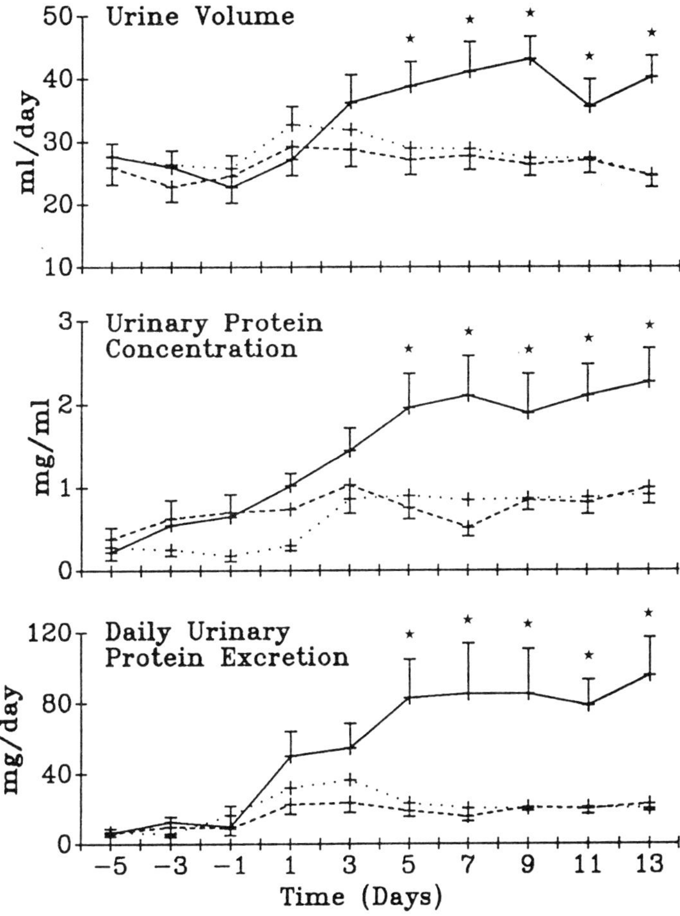

Fig. 2: Urine volume, urinary protein concentration and daily urinary protein excretion in renal transplanted rats (F1H) with a kidney from WKY (dashed lines, n=14), F1H (dotted lines, n=13), or SHRSP donors (solid lines, n=12), respectively. Day 0 corresponds to transition from low-salt to high-salt diet. Renal transplantations were performed 17 days before Day 0. ★ $p<0.05$ vs. WKY and F1H.

be noted that recipients of an SHRSP kidney exhibited excess sodium retention despite an increased systemic blood pressure which can be assumed to be associated with an elevated renal perfusion pressure and subsequent pressure natriuresis (Roman and Cowley, 1985). Even with the inappropriately high losses of isotonic volume with urine substituted by distilled water and, consequently, contributing to the excretion of a dietary salt load, they did not succeed in maintaining sodium balance when they were put on a high-salt diet.

Excess renal sodium retention in recipients of an SHRSP kidney amounted to approximately 4-5 mmol over 14 days, corresponding to a calculated weight gain of about 30-35 g, if isotonic retention is assumed. Measured body weights at the end of the high-salt regimen were, however, almost identical in the three groups of transplanted rats. The failure of recipients of an SHRSP kidney to gain excess body weight may be explained by several factors. First, a substantial amount of sodium may have been excreted via the feces. In adult WKY and SHR on a standard diet containing 0.5% salt, the fraction of sodium excreted via the feces has been reported to be approximately one third of the total amount excreted, with no significant differences between the two strains (Beierwaltes et al., 1982). Furthermore, in SHR, the absolute amount of sodium excreted via the feces appears to be relatively stable over a wide range of dietary salt intake (Lundin et al., 1982). Thus, this factor is unlikely to have contributed to the failure of sodium-retaining recipients of an SHRSP kidney to gain extra body weight.

A more likely explanation for this phenomenon is the observation that daily food intake was significantly lower in recipients of an SHRSP kidney than in recipients of a WKY or F1H kidney. Finally, in contrast to recipients of a WKY or an F1H kidney, recipients of an SHRSP kidney exhibited marked proteinuria while they were on a high-salt diet. In a recent study (Hirano et al., 1991) in rats, puromycin aminonucleoside-induced proteinuria of 400-500 mg/day for ten days accounted for a deficit in body mass of approximately 50 g. Although in our study proteinuria was far less severe, it may still have been largely responsible for the failure of sodium-retaining recipients of an SHRSP kidney to gain more body weight during a high-salt diet than recipients of an F1H or WKY kidney.

Increased proteinuria and sodium retention in recipients of a kidney from SHRSP donors reported in this study are the first particular functional and structural defects detected in these grafts. Further studies should strive to elucidate the genetic basis of these defects and to answer the question whether they are, at least in this model, substantially related to the etiology of posttransplantation hypertension.

REFERENCES

Beierwaltes WH, Arendshorst WJ, Klemmer PJ: Electrolyte and water balance in young spontaneously hypertensive rats. Hypertension 1982; 4:908-915

Bensadoun A, Weinstein D: Assay of proteins in the presence of interfering materials. Anal Biochem 1976; 70:241-252

Bianchi G, Fox U, DiFrancesco GF, Giovanetti AM, Pagetti D: Blood pressure changes produced by kidney cross-transplantation between spontaneously hypertensive and normotensive rats. Clin Sci Mol Med 1974; 47:435-448

Cowley AW: Long-term control of arterial blood pressure. Physiol Rev 1992; 72:231-300

DeWardener HE: The primary role of the kidney and salt intake in the aetiology of essential hypertension: parts I and II. Clin Sci 1990; 79:193-200 and 289-297

Dahl LK, Heine M: Primary role of renal homografts in setting chronic blood pressure levels in rats. Circ Res 1975; 36:692-696

Dahl LK, Heine M, Thompson K: Genetic influence of the kidneys on blood pressure. Evidence from chronic renal homografts in rats with opposite predisposition to hypertension. Circ Res 1974; 34:94-101

Fisher B, Lee S: Microvascular surgical techniques in research, with special reference to renal transplantation in the rat. Surgery 1965; 58:904-914

Folkow B: Physiological aspects of primary hypertension. Physiol Rev 1982; 62:347-504

Folkow B: Critical review of studies on salt and hypertension. Clin Exper Hypertension 1992; A14:1-14

Fox U, Bianchi G: The primary role of the kidney in causing the blood pressure diference between the Milan hypertensive strain (MHS) and normotensive rats. Clin Exp Pharmacol Physiol 1976; Suppl. 3:71-74

Graf C, Maser-Gluth C, Muinck Keizer W, Rettig R: Sodium retention and hypertension after kidney transplantation in rats. Hypertension 1993; 21:724-730

Hirano T, Mamo JCL, Nagano S, Sugisaki T: The lowering effect of probucol on plasma lipoprotein and proteinuria in puromycin aminonucleoside-induced nephrotic rats. Nephron 1991; 58:95-100

Horst PR, Veelken R, Unger T: A new method for collecting urine directly from the ureter in conscious unrestrained rats. Renal Physiol Biochem 1988; 11:902-908

Kawabe K, Watanabe TX, Shiono K, Sokabe H: Influence on blood pressure of renal isografts between spontaneously hypertensive and normotensive rats, utilizing the F1 hybrids. Jap Heart J 1979; 20:886-894

Kopf D, Waldherr R, Rettig R: Posttransplantation hypertension in recipients of kidneys from young SHR. Am J Physiol 1993; in press

Lee S: An improved technique of renal transplantation in the rat. Surgery 1967; 61:771-773

Louis WJ, Howes LG: Genealogy of the spontaneously hypertensive rat and Wistar-Kyoto rat strains: Implications for studies of inherited hypertension. J Cardiovasc Pharmacol 1990; 16(Suppl. 7):S1-S5

Lundin S, Herlitz H, Hallbäck-Nordlander M, Ricksten SE, Göthberg G, Berglund G: Sodium balance during development of hypertension in the spontaneously hypertensive rat (SHR). Acta Physiol Scand 1982; 115:317-323

Rettig R, Folberth C, Stauss H, Kopf D, Waldherr R, Unger T: Role of the kidney in primary hypertension: A renal transplantation study in rats. Am J Physiol 1990a; 258:F606-F611

Rettig R, Folberth CG, Kopf D, Stauss H, Unger T: Role of the kidney in the pathogenesis of primary hypertension. Clin Exp Hypertens 1990b; A12:967-1002

Rettig, R, Folberth CG, Stauss H, Kopf D, Waldherr R, Baldauf G, Unger T: Hypertension in rats induced by renal grafts from renovascular hypertensive donors. Hypertension 1990c; 15:429-435

Rettig R, Stauss H, Folberth C, Ganten D, Waldherr R, Unger T: Hypertension transmitted by kidneys from stroke-prone SHR. Am J Physiol 1989; 257:F197-F203

Roman RJ, Cowley AW: Abnormal pressure-diuresis-natriuresis response in spontaneously hypertensive rats. Am J Physiol 1985; 248:F199-F205

ACKNOWLEDGEMENTS

This work was supported by a grant from the Deutsche Forschungsgemeinschaft (Re 522/7-1). We thank Manuela Ritzal for expert technical assistance.

Integrative and cellular aspects of autonomic functions : temperature and osmoregulation. Eds K. Pleschka, R. Gerstberger. John Libbey Eurotext, Paris © 1994, pp. 497-508.

Salt gland excretion enhanced during cross circulation of blood between two Pekin ducks : evidence for positive feedback

Harold T. Hammel and Eckhart Simon

Max-Planck-Institut für Physiologische und Klinische Forschung, W.G. Kerckhoff-Institut, D-6350 Bad Nauheim, Germany

ABSTRACT

Salt gland excretion by salt adapted Pekin ducks can be induced by an increase in tonicity of plasma and interstitial fluid (ISF) and by an increase in volume of ISF space. However, it has not been possible to relate the rate of excretion to any combination of changes in these properties of body fluids. Increasing tonicity of exctracellular fluid (ECF) and/or volume of ISF space above their threshold values initiates excretion. But stimuli which initiate excretion do not sustain it as these stimuli soon decrease to or below initial threshold values. The rate of salt excretion may exceed the rate of salt infusion and continue at this rate even after more salt was excreted than was infused. It is as if excretion is initiated by negative feedback loops and then enhanced and sustained by some kind of positive feedback loop. We investigated whether a blood-borne substance enhanced and sustained salt secretion. Blood was exchanged between two ducks at equal rates (7 $ml \cdot min^{-1}$). The experiments began after both ducks had received a priming salt infusion but were no longer excreting salt. In one series of experiments, one duck received an isotonic saline infusion (286 mosm $NaCl \cdot kg^{-1}$ H_2O at 1.4 $ml \cdot min^{-1}$). Only the volume of its ECF increased and stimulated excretion at 1.6 times the excretion rate for the same infusion into a single duck. Moreover, the other duck receiving no direct infusion excreted salt at half the rate of a single duck receiving an isotonic saline infusion. Since the second duck continuously received blood at the rate its blood was removed, its tonicity and ECF volume changed little. Nevertheless, it was stimulated to excrete salt, suggesting that the first duck released some substance into its blood that stimulated salt secretion in the second duck. At the same time, the second duck released something into its blood which enhanced excretion in the first duck. Other experiments with hypertonic infusates and with infusion of angiotensin II confirmed these results.

INTRODUCTION

Salt gland excretion by salt adapted Pekin ducks relates to increases in certain intensive and extensive properties of the body fluids when a saline solution is intravenously (i.v.) infused. Depending on the concentration of the saline infusate, 1) the tonicity of the plasma and interstitial fluid (ISF) is increased, unchanged or diminished, 2) the osmolality of all body fluid is increased, unchanged or decreased and 3) the volume of the plasma and ISF are always increased. Since volume is an extensive property and only intensive properties can be transduced, tension receptors, which monitor the magnitude of one or both of the ISF and plasma volumes in some organ tissues are stimulated by the infusate.

However, it has not been possible to relate the rate of salt gland excretion to any combination of changes in the intensive properties of the appropriate receptors caused by changes in the body fluids (Hammel, 1989a,b, 1990). Increasing the tonicity of the extracellular fluid (ECF) and/or the volume of the ECF (and ISF) above the threshold values of these parameters initiates salt gland excretion. The excreted solution is between 900 and 1100 mosm·kg^{-1} H_2O and is about 97% NaCl and 3% KCl (Peaker & Linzell, 1975). But the signals which can initiate excretion can not sustain it as these signals decrease to or below their intitial threshold values. The rate of salt excretion soon exceeds the rate of salt infusion and continues at this rate even after more salt has been excreted than was infused. Members of this laboratory have been searching for an explanation of this unusual homeostatic response for several years (Hammel et al., 1980). It appears as if the excretion rate is initiated by a control system with negative feedback and is enhanced and sustained by some kind of positive feedback. In this context, the question arises, is there a blood-borne facilitation of the system controlling salt gland function? A preliminary study provided a negative answer to this question (Hammel, 1986). Removing 15 ml of blood from a salt stressed donor duck did not effect a duck receiving this blood. Of course this result could not be viewed as certain that nothing in the blood of the donor could influence the response of the recipient under every circumstance. To obtain a definitive answer to the question, especially if the answer were positive, we undertook to exchange blood between two ducks at equal rates and continuously (Hammel et al., 1990). This required the development of a machine to cross circulate blood between the two ducks so that both became donors and both became recipients of blood at equal rates. The machine could not employ peristaltic pumping of blood because the nucleated erythrocytes of avian blood are damaged as the blood passes between the rollers of the peristaltic pump. Therefore a machine was developed in collaboration with Walter Klein and Manfred Wasserhess, Masters in the Electronic Shop and Instrument Shop, respectively, of this Institute.

ANIMALS AND METHODS

Treatment of animals:

Young Pekin ducks were slowly adapated to thrive on an aqueous solution of sodium chloride as the only drinking solution (400 mosm NaCl·kg^{-1} H_2O). After eight weeks the ducks were able to excrete a 1000 mosm·kg^{-1} H_2O solution at a rate up to 0.6 ml·min^{-1} from their nasal salt glands. When ducks were to be used in an experiment they were isolated from the flock and deprived of food for 24 hours. Also, on the day before an experiment, a PE 90 polyethylene catheter was inserted 7 cm into the brachial artery on the left or right side as described previously (Simon-Oppermann et al., 1984). Immediately before the experiment, one brachial vein/and or foot vein were cannulated by percutaneous venipuncture. During an experiment, salt gland excretions were aspirated from the external nares and collected in vials for periods of 15 minutes, as previously described (Hammel et al., 1980). When urine was formed during an experiment, it was caught in a container secured to the rear end of the duck and was aspirated into a graduate clyinder for periods of up to 30 minutes. The nasal secretion and the urine formed during the collection periods were weighed and their osmolalities were measured. In addition, the concentrations of Na^+, K^+ and Cl^- were measured in the urine. In some experiments, blood samples were obtained one minute prior to and one minute before the end of an interval of salt loading. These blood samples were analyzed for Na^+, K^+ and Cl^- concentrations, colloid osmotic pressure and hematocrit.

During an experiment, the ducks were supported in a sling with their legs protruding through openings in the sling. When a pair of ducks were used in experiments involving cross circulation of blood, the ducks were held end to end at the same level. Blood from both ducks was obtained from the cannulae in the brachial arteries and was returned to the brachial veins in the opposite wing or to a vein in the foot. Blood was exchanged at the rate of 7 ml·min^{-1}.

Cross Circulator:

To insure exchange of blood between the two ducks at equal rates, two pair of all plastic 20 ml syringes were mounted back to back in the cross circulating machine, as illustrated in Fig. 1. The plungers of one pair of syringes, mounted side by side, were driven in one direction (filling) as the plungers of the other pair of syringes were driven in the opposite direction (emptying) at the same time. When

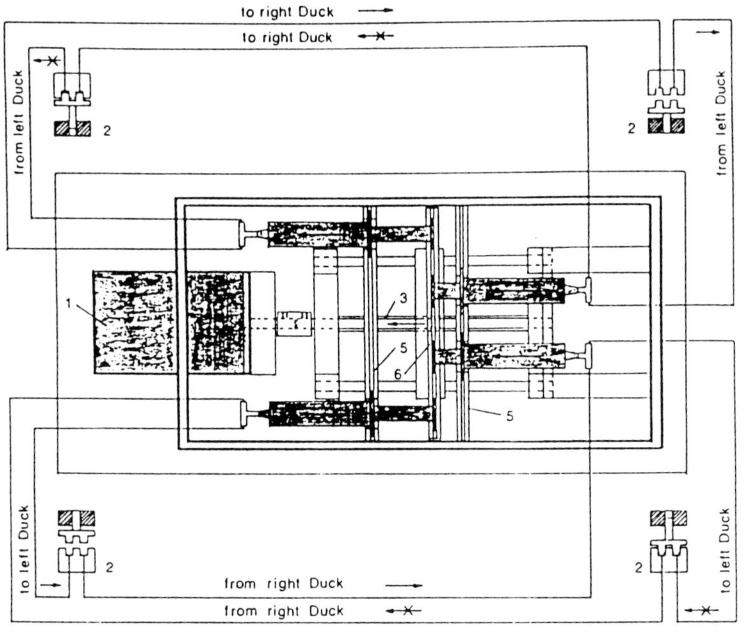

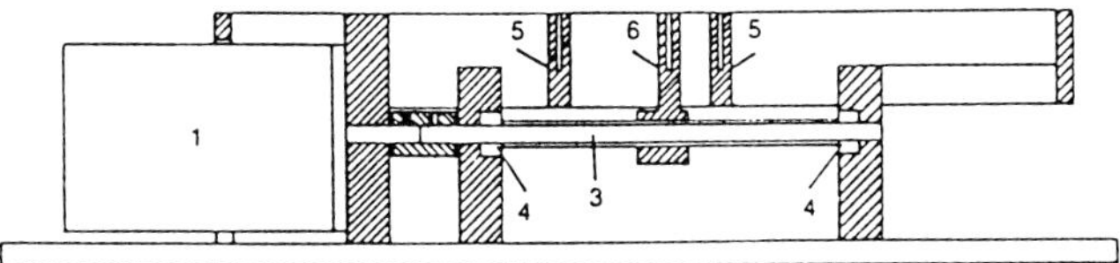

1 5-PHASE-STEPPER-MOTOR

2 SOLENOID CLAMPS + TIMER

3 DRIVE SCREW

4 ROLLER BEARINGS

5 RIGID SUPPORT OF BARREL OF PLASTIC SYRINGES

6 DRIVER OF SYRINGE PLUNGERS

Figure 1: Schema for cross circulation of blood. Upper part: Top view of two pairs of 20 ml syringes mounted back to back. Flow of blood between solenoid activated clamps and syringes is depicted by arrows without x's when plungers of all syringes are moving from right to left. Flow of blood between solenoid clamps and ducks is not depicted. Lower part: Side view of cross circulator driving mechanism and of solenoid clamps.

the plungers arrived at the end of their excursions, the driving motor reversed direction and drove all plungers in directions opposite to that they had been moving. When the plungers arrived at the end of their excursions in this direction, the driving motor again reversed direction. Each syringe was joined to a 16 gauge plastic cannula which in turn was joined to the stem of a plastic T by a short length of Dow Corning silastic tubing (0.125 in OD, 0.062 in ID). The branches of the T were each joined to a long length of thin walled silastic tubing (0.095 in OD, 0.062 in ID). Tubing from one branch of each of the four Ts passed through one of four channels and beneath an open clamp that could be closed by activating a solenoid. Tubing from the other branch of each of the four Ts passed through one of four channels and beneath a closed clamp which was closed by another active solenoid. The four thin walled tubings from the open clamp were joined to one branch of another set of four plastic Ts and the four thin walled tubings from the closed clamp were joined to the other branch of the same four plastic Ts. One of the four stems of those Ts was then joined by a long length of thick walled silastic tubing (0.125 in OD, 0.062 in ID) to the arterial cannula of one duck. Another stem was joined by a long silastic tubing to the arterial cannula of the other duck. The third T stem was joined to the venous cannula of one duck and the fourth T stem was joined to the venous cannula of the other duck by long silastic tubings. As the blood flowed through the cross circulator, it was always enclosed in silastic or plastic material. There was also a plastic T in the long silastic tubing carrying arterial blood from each duck to the machine. Heparin could be injected into these two streams of blood from a syringe connected to each of the stems of the two Ts.

Prior to connecting the ducks to the cross circulation machine, all components containing fluid were cleaned and sterilized with a detergicide, rinsed with distilled water and then filled with 28.5 ml of 0.9% saline solution containing 4000 I.U. of heparin per liter of solution. At the time the ducks were connected to the cross circulator, 0.2 ml of heparin solution (4000 I.U. per 2 ml) was injected into each of the streams of arterial blood from the ducks to the circulator. Also, 0.15 ml of the same heparin solution was injected every hour into these blood flows. This was necessary to prevent coagulation of blood external to the ducks. By experience we found that this amount of heparin was just sufficient to prevent clotting and also to avoid internal or external bleeding from any lesions that might be in the vascular systems of the ducks. The volume of the isotonic saline in the machine at the time of connection with the ducks was 28.5 ml. Therefore, the hematocrits of both ducks were diminished about 2% as this saline solution was mixed with the blood volume and diluted the hematocrit and colloid osmotic pressure of their blood. At the rate of flow used in this study, there was no evidence of hypoxia, since 7 ml·min^{-1} is less than 2% of the cardiac output (Brummermann & Simon, 1990).

Analytical methods:

Samples of plasma were obtained from the blood samples by spinning at 4000 g for 10 min. The osmolality of the excretion by the salt glands, of the urine and of plasma samples were determined with a Wescor vapor pressure osmometer model 5100C. Blood plasma samples were measured four times so that the average is reliable within 2 mOsm·kg^{-1} H_2O. The colloid osmotic pressure of plasma samples was measured with a Wescor colloid osmometer, Model 4100, and duplicate measurements were averaged so that the average was reliable within 0.2 Torr. The concentrations of sodium and potassium ions in urine and plasma samples were measured in duplicate with an Instrumentation Laboratory flame photometer, Model 943, and the results were reliable within 1 meq $Na^+ \cdot l^{-1}$ and 0.1 meq $K^+ \cdot l^{-1}$. The chloride concentrations of urine and plasma samples were measured in duplicate with a Corning-EEL chloride meter, Model 920, and were reliable within 1 meq $Cl^- \cdot l^{-1}$. Osmolalities of salt gland excretions and urine were converted by a variable factor into milliosmoles per kg solution, then multiplied by the weights of the samples and divided by the collection time in order to obtain the excretion rate in milliosmols per min. The milliequivalents per minute of ions in the urine is the sum of the sodium, potassium and chloride concentrations in the urine.

When differences between results were compared, the difference was tested by paired t statistic or by t statistic for two means and the significance was expressed for each t value as the value of 2 p less than the value stated.

RESULTS

Preliminary experiments:
Figure 2 shows the ability of the salt glands of individually infused salt adapted Pekin ducks to respond to salt loading with a hypertonic aqueous solution of 1025 mosm·kg^{-1} H_2O at rates of 0.2, 0.4 and 0.6 ml·min^{-1}. All ducks could easily excrete at rates greater than 0.4 ml·min^{-1} but the higher rate of 0.6 ml·min^{-1} was, in some ducks, a greater rate of salt stress than they could excrete. The rate of excretion of the 0.4 ml·min^{-1} in Fig. 2 will subsequently be referred to as the control salt response of a singly infused duck to a hypertonic infusion.

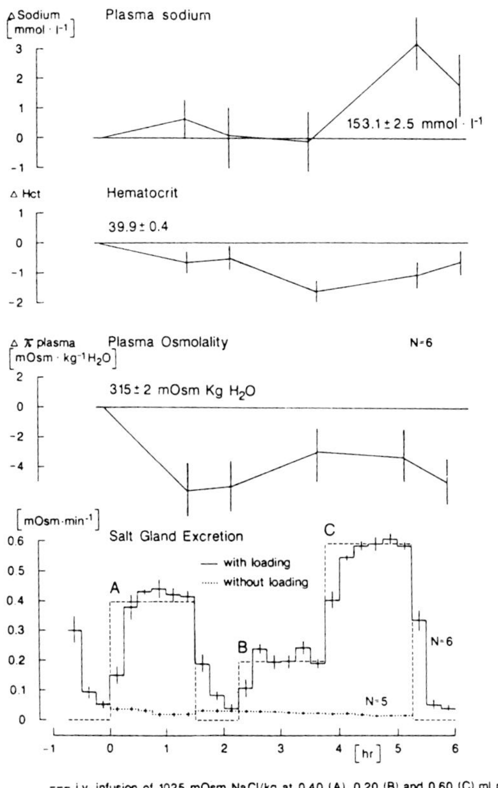

Figure 2: Mean rate of salt gland excretion (+ or - SEM) of six salt adapted Pekin ducks in response to an i.v. infusion of a hypertonic saline aqueous solution (1025 mosm NaCl·kg^{-1} H_2O) at 0.4, 0.2 and 0.6 ml·min^{-1}. The hematocrit, plasma sodium concentration and osmolality were measured at the times indicated and plotted as a change from their respective values after priming and before the start of the experimental infusion. Prior to the experimental infusion, each duck was primed with a solution of 1025 mosm NaCl·kg^{-1} H_2O at 0.4 ml·min^{-1} and until it was excreting. The priming was stopped and excretion was allowed to diminish to threshold for 45 min. The mean rate of salt excretion is also illustrated when no infusion followed the priming infusion (dotted line in the lower diagram).

Figure 2 also illustrates the rate salt is excreted by the salt glands when nothing is infused following a priming infusion of 1000 mosm NaCl·kg^{-1} H_2O at 0.4 ml·min^{-1}. The priming infusion was made until the salt glands were fully active and the osmolality of the salt gland excretion was between 1000 and 1160 mosm·kg^{-1} H_2O. This infusion was then terminated and excretion decreased for the next 45 min until time 0. The subsequent response of the salt glands will be referred to as the control response to zero infusion during which the osmolality of the salt gland excretion was between 400 and 700 mosm·kg^{-1} H_2O.

In another preliminary experiment involving two ducks simultaneously, Fig. 3, 15 ml of blood was obtained from a donor duck which was stressed with an i.v. infusion of an aqueous solution of 4000 mosm NaCl·kg^{-1} H_2O; and this blood was then administered to a recipient duck receiving a continuous i.v. of a slightly hypotonic solution of 250 mosm NaCl·kg^{-1} H_2O at 1.6 ml·min^{-1}. At a later time, after the donor duck had been continuously infused with an aqueous solution of glucose (200 mosm·kg^{-1} H_2O) for 3 hours to stop salt gland secretion completely, another 15 ml blood sample was obtained from the donor and infused into the recipient duck receiving the same hypotonic NaCl solution as before. The recipient duck did not respond perceptibly to these two injections of 15 ml blood. Under the circumstances of this experiment, there was no evidence that the blood from the donor duck, when it was excreting, contained a substance stimulating salt secretion that was not in the blood when it was not excreting. For later reference, it should be noted that the recipient duck was excreting salt continuously at the rate of about 0.2 mosm·min^{-1} in response to the infusion of salt in a hypotonic solution at a rate of 0.4 mosm·min^{-1}. That is, about half of the infused salt was excreted by the salt glands and the other half was eliminated by the kidneys. Water was obtained and available for urine formation because the osmolality of the salt gland excretion was between 900-1000 mosm·kg^{-1} H_2O, about three times the osmolality of the infusate. This rate of excretion by the salt glands of a nearly isotonic infusate at 1.6 ml·min^{-1} will be subsequently referred to as the control response of a singly infused duck to an isotonic infusion.

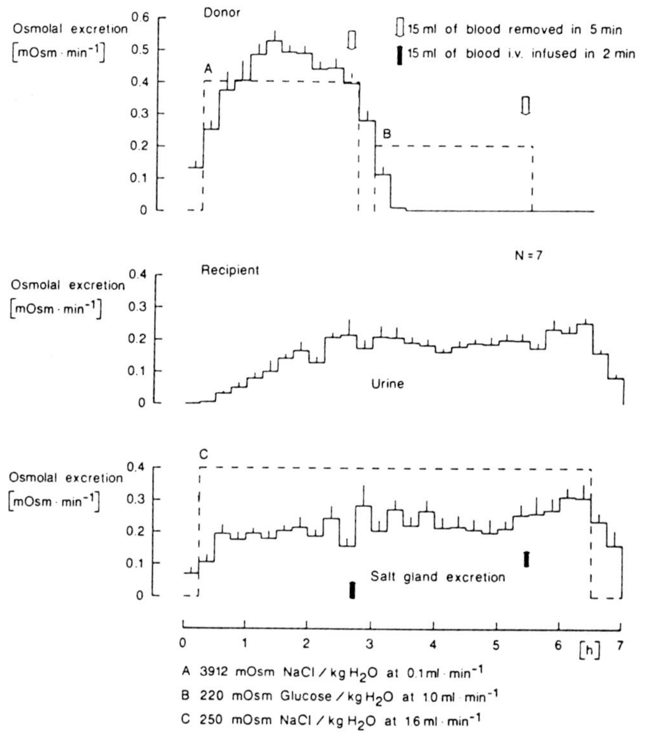

Figure 3: Upper ordinate: Mean rates of salt excretion of a duck (donor) receiving first a hypertonic salt load applied by infusion A followed by a water load applied by infusion B. Middle and lower ordinate: Osmolal excretions in the urine and salt gland fluid of a simultaneously investigated duck (recipient) in response to a slightly hypotonic salt load applied by infusion C. Blood samples were removed from the donor duck at the times indicated by the open arrows and immediately infused into the recipient at the times indicated by the filled arrows.

Cross circulated ducks receiving staggered infusions of isotonic saline:

Both ducks in this series of six experiments were primed by receiving a 1000 mosm NaCl·kg^{-1} H_2O solution at 0.4 ml·min^{-1}. When the salt glands of both ducks were excreting the salt load at the rate they received it, the priming infusions were terminated and the excretions of both ducks were allowed one hour to return to a low rate, i.e. threshold for excretion. All the while, the ducks were continuously cross circulated at 7 ml.min^{-1}. As cross circulation continued, one duck, the left duck in Fig. 4, received an infusion of isotonic saline at 1.4 ml·min^{-1} starting at 0.5 hours on the abscissa. During the first hour (0.5 to 1.5 hours), the right duck received no infusion other than isovolumetrically exchanging blood with the left duck. The salt glands of the left duck were responding to the isotonic salt load by excreting a solution between 900 and 1000 mosm.kg^{-1} H_2O. Since it was thereby gaining water from the infusate and retained part of it, as indicated by the rate of urinary fluid excretion, right part of Fig. 4, this lessened the

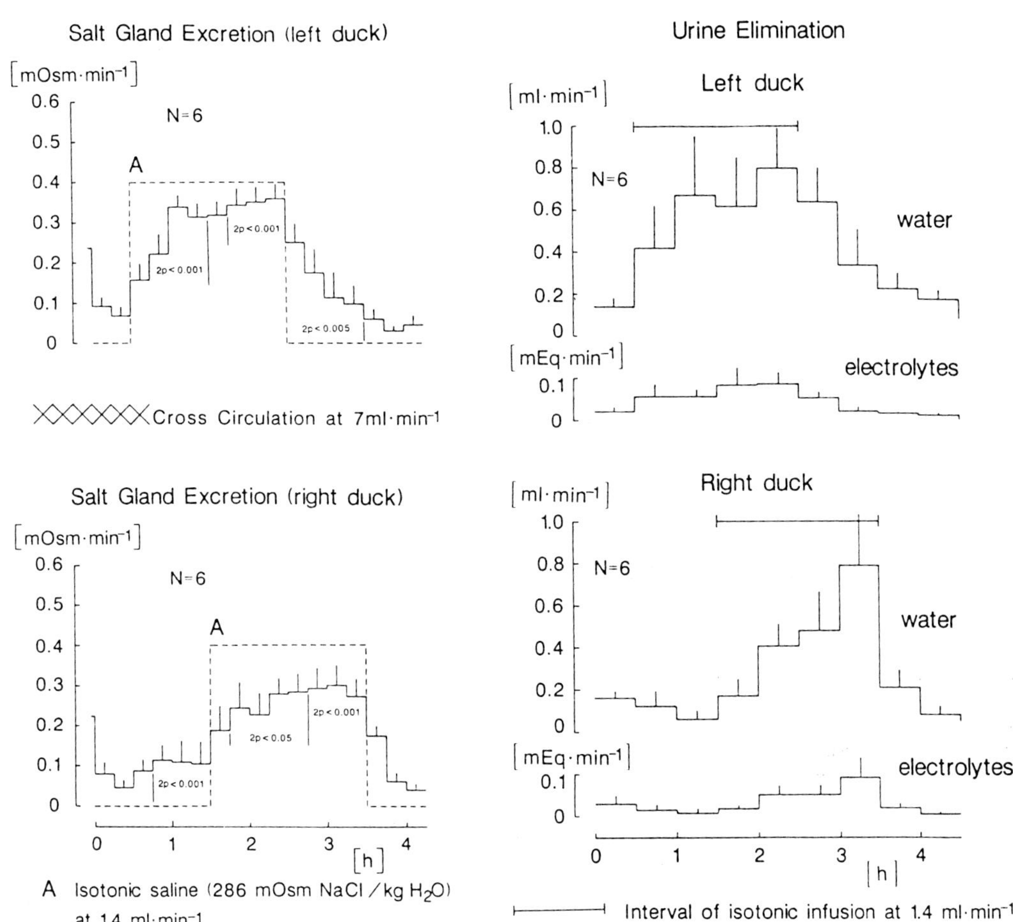

Figure 4: Left hand diagrams: Mean rates of salt gland excretion from two ducks in response to staggered infusions of isotonic saline at 1.4 ml·min^{-1}. Blood was exchanged between the two ducks at rates of 7 ml·min^{-1} for more than 4 hours. Right hand diagrams: Mean rates of elimination of water and electrolytes in urine from the same two ducks. The values of 2p denote the signficance of the difference between the responses of the cross circulated ducks and the control responses of the singly infused ducks (Fig. 3) and ducks receiving no infusion (Fig. 2).

osmolality and salt concentration of its ECF as well as the plasma delivered to the right duck. Thus, if no other factor was involved, the right duck should have been inhibited by the blood it received. In fact, as the left part of Fig. 4 illustrates, the right duck was stimulated to excrete at the rate of about 0.1 $mosm \cdot min^{-1}$. The salt gland responses of the ducks during the second, third and fourth 15 min sampling periods in the left part of Fig. 4 were compared with the control response to zero infusion of a singly infused duck during the same periods illustrated in Fig. 2. For these three periods the responses of the right cross circulated duck was significantly greater ($2p<0.001$) than a single duck receiving no infusion. Some factor in the blood of the left duck must have facilitated salt excretion in the right duck which also excreted a solution between 900 and 1000 $mosm \cdot kg^{-1}$ H_2O. Both ducks were hydrating themselves during this hour by excreting a solution that was three times more concentrated than the infusate received by the left duck.

Moreover, during the second to fourth 15 min period of the first hour of salt loading the left duck was excreting at an average rate of 0.33 $mosm \cdot min^{-1}$ which is 1.6 times the rate a single duck would excrete receiving the same infusate. Comparing the salt gland response of the left duck in Fig. 4 with the control response of a singly infused duck to an isotonic infusion illustrated in Fig. 3, for the same periods, the difference is highly significant ($2p<0.001$). Also, in more than five series of published and unpublished experiments in which single ducks received isotonic saline, their salt glands excreted at a rate of 0.2 $mosm \cdot min^{-1}$. Thus, there must have been something added to the blood by the right duck which was facilitating excretion in the left duck as well. The release of a blood-borne facilitator by both ducks enhanced salt excretion in both ducks which further hydrated both ducks, reduced the tonicity of the ECF of both ducks and further increased the ECF (and ISF) volumes of both ducks.

During the next hour (1.5 to 2.5 hours), both ducks received an i.v. infusion of isotonic saline each at the rate of 1.4 $ml \cdot min^{-1}$ so that NaCl was infused into each duck at a rate of 0.4 $mosm \cdot min^{-1}$. Both ducks excreted this salt at rates significantly greater than either would have excreted it had they received it as single ducks ($2p<0.001$, left duck; $2p<0.05$, right duck).

During the last hour (2.5 to 3.5 hours) only the right duck received an infusion of isotonic saline. Nevertheless, the left duck continued to excrete at a significantly higher rate ($2p<0.005$) even though both ducks were hydrating their body fluids. Again it appears that each duck received a blood-borne facilitator of salt gland secretion from the other. Water was recovered from the infusate because the salt received in the isotonic infusate was excreted as a salt solution between 900-100 $mosm \cdot kg^{-1}$ H_2O. This water was used in part to form urine and was thereby eliminated, right part of Fig. 4. Some sodium, chloride and potassium ions were also eliminated in the urine but their sum was excreted at a low rate.

<u>Cross circulation between two ducks as one duck received isotonic saline and the other received angiotensin II (A II):</u>

Another series of experiments was a continuation of the four hour series described above. Between hours 4.5 and 6 the left duck received an isotonic solution at the rate of 1.4 $ml \cdot min^{-1}$ as the right duck received an i.v. infusion of avian angiotensin II (val^5 angiotensin II, 80% purity) in an isotonic solution infused at 0.05 $ml \cdot min^{-1}$ so that it received 200 ng A II per kg body weight per minute, left part of Fig. 5. Excretion from the right duck was completely inhibited by A II until after the A II infusion was terminated. Excretion from the left duck was somewhat inhibited when compared with the excretion rate of the same left duck when the same right duck received no A II infusion (compare Fig. 4). Salt accumulated due to this moderate inhibition was then excreted by both ducks after the infusion of A II to the right duck was stopped. The total salt excreted by both ducks was 1.9 times the amount of salt that would have been excreted by both ducks when singly infused with the same infusions. The salt excreted by the salt glands of both ducks was in a solution of 900-1000 $mosm \cdot kg^{-1}$ H_2O so that water was retained, the ducks were hydrated and their ECF (and ISF) volumes were increased. The hematocrit and colloid osmotic pressure of the plasma were decreased significantly. The amount of water eliminated in the urine was increased, right part of Fig. 5.

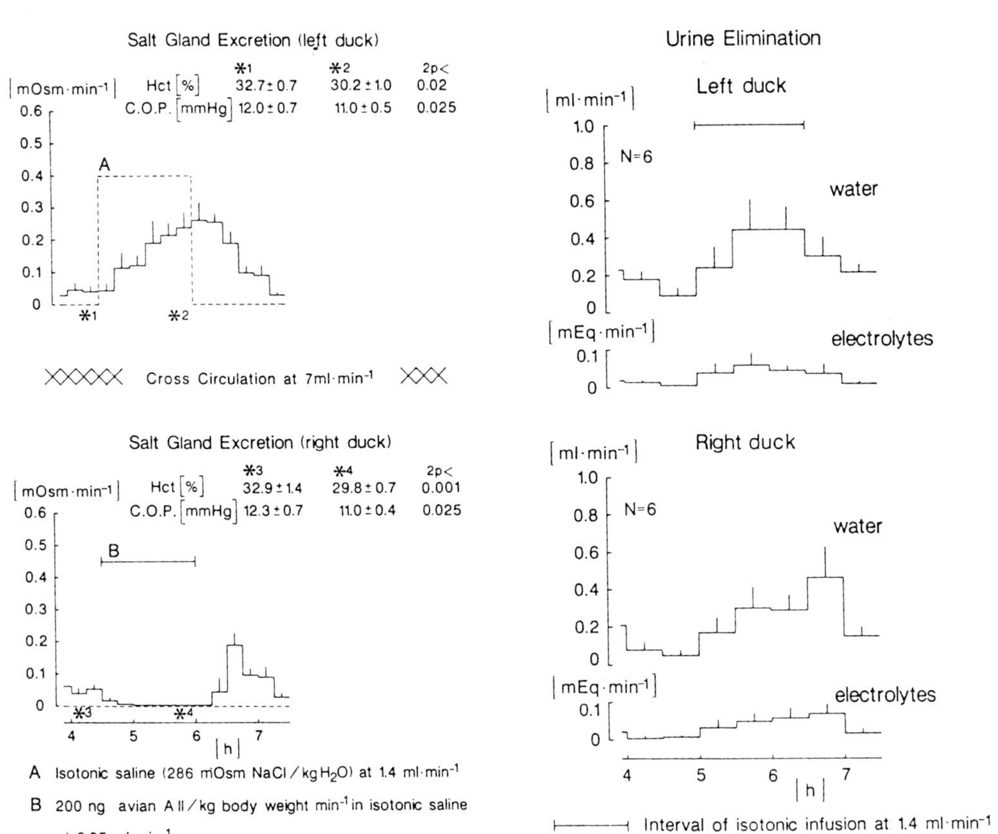

Figure 5: Left hand diagrams: mean rates of salt gland excretion from two ducks. The left duck (upper diagram) received isotonic saline at 1.4 $ml \cdot min^{-1}$ (A) as the right duck (lower diagram) received Avian angiotensin II at 200 ng A II per kg body weight per min (B, purity of the A II was 80%). These experiments followed the experiments in Fig. 4, i.e. from the 4th to 8th hour of cross circulation. Right hand diagrams: Mean rates of water and electrolyte elimination in urine from the same two ducks

Cross circulation between two ducks as one duck received a hypertonic saline solution and the other received angiotensin II (A II):

Both ducks were primed to excrete salt and then allowed one hour to stop excreting. All the while, they were cross circulated with blood from the other at equal rates. As cross circulation continued at 7 $ml \cdot min^{-1}$, the left duck received an infusion of a solution containing 1037 mosm $NaCl \cdot kg^{-1}$ H_2O (from 0.75 to 2.25 hours). At the same time the right duck received A II at 200 $ng \cdot min^{-1} \cdot kg^{-1}$, Fig. 6. The right duck excreted almost nothing during the 90 minutes it received A II and for another 15 min after the A II infusion was ended. From 15 to 90 minutes after A II infusion, excretion commenced in the right duck and increased up to 0.16 $mosm \cdot min^{-1}$ and then decreased to a low rate. Note in Figs. 5 and 6 that a duck singly infused with A II alone would never excrete salt from its salt glands. The left duck, receiving the hypertonic saline, excreted it at a rate exceeding by 20% the rate a single duck receiving the same load excreted it. Compare the rate of excretion of the left duck in Fig. 6 with the rate of excretion of a hypertonic saline solution by a single duck in Fig. 2 infused at 0.4 mosm $NaCl \cdot kg^{-1}$ $H_2O \cdot min^{-1}$. It might appear that the excretion rate of the left duck was not lessened by the A II it received in the blood from the right duck. However, the large excretion post infusion by the left duck indicates that it was inhibited by the low concentration of A II it received from the right duck. Moreover, it appears that as the right duck was inhibited by A II, it also released a facilitating substance which enhanced excretion in the left duck, rendering its excretion rate excessive, even as it was somewhat inhibited by A II. The total amount of salt excreted by both ducks was 1.8 times the amount of salt infused. The total amount excreted by the left duck alone was 1.54 times the amount of salt it received. In Fig. 2, the singly infused ducks excreted only 1.09 times the amount of salt infused.

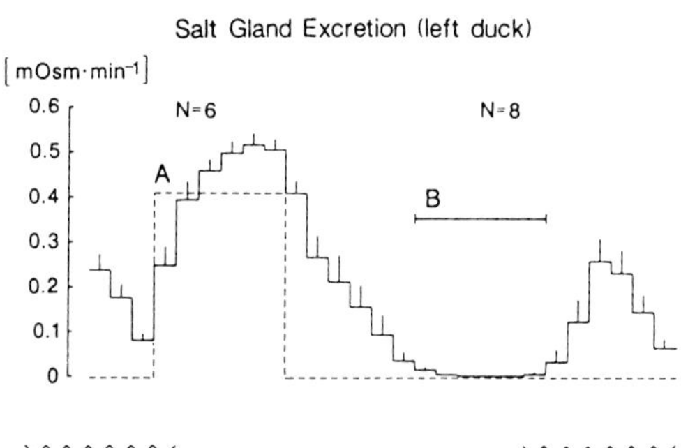

Cross Circulation at 7ml·min⁻¹

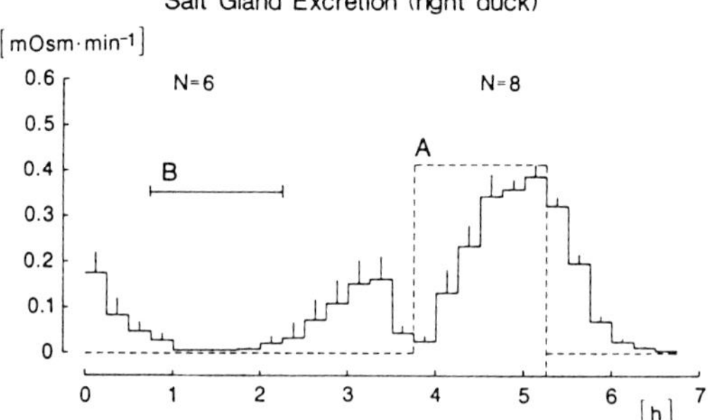

A 1037 mOsm NaCl/kgH_2O at 0.4 ml·min^{-1}

B 200ng avian AII kg^{-1}·min^{-1} in isotonic saline at 0.05 ml·min^{-1}

Figure 6: Mean rates of salt gland excretion from two ducks between which blood was cross circulated at 7 $ml \cdot min^{-1}$ for more than 7 hours. During the first experiment, the left duck (upper diagram) received an i.v. infusion of hypertonic saline (A) as the right duck (lower diagram) received 200 ng A II per kg body weight per min (B). During the second experiment, the right duck received the hypertonic infusion as the left duck received the A II infusion.

When both ducks returned to a low level of excretion at 3.75 hours, Fig. 6, they returned to threshold. However, this threshold was certainly higher than the threshold at the beginning of the experiment, since the total salt excreted by both ducks was 80% more than the salt infused between 0.75 and 2.25 hours. As cross circulation continued, the right duck received the infusion of hypertonic saline for 90 minutes as the left duck received an A II infusion (from 3.75 to 5.25 hours). Again the duck receiving the hypertonic saline responded by excreting much of the salt it received during the infusion; and, post-infusion, it continued to excrete at an excessive rate compared with excretion of a single duck following an infusion of the same solution (see Fig. 2). Salt gland secretion by the left duck was inhibited during the infusion of A II and the duck did not excrete again until after the A II was cleared. In this case, the total salt excreted by both ducks was 1.23 times the amount infused. These two series of experiments back to back, Fig. 6, differed in that excretion in the latter series was less than in the former, and the former excretions were excessive compared with excretion from a single duck receiving the same hypertonic infusate. Although the ducks had returned to threshold conditions at the onset of the experiments in both series, the threshold was higher in the latter series, which accounts for the lesser responses.

DISCUSSION

Blood borne substances which facilitate or inhibit the response of an organ are essential for control of certain organs involved in homeostasis. Cross circulation of blood between two animals of the same species has been a successful method for demonstrating the presence or absence of these blood borne substances. By this method, Loewi (1921) demonstrated the presence of acetyl choline and discovered its role in inhibition of the frog heart. The possibility that a blood-borne substance might influence salt excretion was suggested by the fact that until now we have not been able to account for the sustained rate of salt gland secretion with any combination of stimuli known to initiate excretion. The orbital salt glands of salt adapted ducks serve almost exclusively to rid body fluids of excess NaCl ingested with salty food and water. For this reason, the salt gland is an attractive organ for investigating the stimuli and factors involved in its activation as well as sustaining its activity when NaCl is ingested and absorbed or is infused into the ECF. For example, a single duck infused with isotonic saline at 0.4 $mosm \cdot min^{-1}$ excretes the salt at 0.2 $mosm \cdot min^{-1}$. This relationship between the rate of infusion and excretion may last for hours, Fig. 3. Since the excreted solution was about 1000 $mosm \cdot kg^{-1}$ H_2O, the duck gained water as it excreted the salt, hydrated itself and, thereby, increased its ECF (and ISF) volumes. Some of the water retained in this process was later eliminated by the kidneys along with residual salts and osmolytes. The isotonic infusate could only increase the ECF (and ISF) volumes so that associated tension receptors with afferent nerves (possibly the vagi, Hanwell et al., 1972; Simon-Oppermann et al., 1980) going to the osmoregulatory center in the brainstem were stimulated. In any case, as water was retained, it lessened the NaCl concentration and tonicity of the ECF as well as the osmolality of all body fluids. This effect alone, acting on tonicity and osmo- receptors in the brainstem and elsewhere, diminishes the rate of excretion. Nevertheless, the rate of excretion was sustained. This suggests that some factor was involved in sustaining excretion other than neural receptors stimulated by increases in ECF tonicity and volume. Our expectation was confirmed by the results depicted in Figs. 4, 5 and 6.

For example, in Fig. 4, blood was exchanged at 7 $ml \cdot min^{-1}$ between two ducks. The left duck received an infusion of isotonic saline for one hour at 1.4 $ml \cdot min^{-1}$. During this hour the right duck was receiving blood from the left duck. Nevertheless, the right duck excreted salt at 0.1 $mosm \cdot min^{-1}$. Moreover, the left duck excreted at 1.6 times the rate expected from a single duck receiving this same infusion. Both ducks must have released something into their blood which enhanced excretion in the other. The excessive excretion in both ducks further hydrated both animals, decreased tonicities and osmolalities in both ducks and further increased ECF (and ISF) volumes in both ducks. The same inferences apply to the second and third hours of this experiment. These results seem to demonstrate the presence of a messenger in the blood which enhances and sustains salt gland excretion once it is initiated by a primary stimulus such as increasing the ECF (and ISF) volumes. Perhaps, stretching the interstitial space of some unidentified tissue releases the blood-borne facilitator. Release of the facilitator continues even as tonicity and osmolality of the blood decline, Fig. 4, and even as excretion is inhibited by angiotensin II, Figs. 5 and 6. We note that an infusion of water or a solution of glucose, Fig. 3, did not elicit salt gland excretion, even though the ECF (and ISF) volumes were expanded. Increasing these volumes elicits excretion only if this stimulus is not inhibited, initially, by a decrease in ECF tonicity and osmolality.

An alternate account of the results depicted in Figs. 4 and 5 would be that a blood-borne substance inhibited salt excretion from the kidneys of left and right cross circulated ducks, thereby, requiring the salt glands of both ducks to eliminate the salt load that would otherwise be eliminated by the kidneys of ducks singly infused with isotonic saline (see Fig. 3). However, this account can not explain the excessive excretion in Fig. 6 wherein the infusate was hypertonic saline and little or none of its salt was eliminated by the kidneys, even in singly infused ducks.

Our results do not indicate the nature of the putative facilitator molecules nor the tissue from which it was released. It does not seem to be any of the hormonal substances presently identified as influencing salt excretion by the kidneys.

Acknowledgements: M. Wasserheß and K. Burk designed and constructed the cross circulation machine and drafted Fig. 1. W. Klein and N. Bucher designed and constructed the control and valve system for the circulator. Daniela Luther-Dyroff, Roswitha Bender, Elke Ratajczak and Anette Grieb assisted in the laboratory and in drafting Figs. 2 to 6. H.T. Hammel wishes to express special gratitude to the Max-Planck-Gesellschaft for enabling him to serve as foreign scientific member of the MPI for Physiological and Clinical Research in Bad Nauheim, Germany.

REFERENCES

Brummermann, M. & Simon, E. (1990) Arterial hypotension in ducks adapted to high salt intake. *J. Comp. Physiol. B* 160, 127-136.

Hammel, H.T. (1986) Enhanced salt excretion without blood-borne facilitation. *Proc. Int. Union Physiol. Sci.* 16, 379.

Hammel, H.T. (1989a) Homeostasis: embracing negative feedback enhanced and sustained by positive feedback. In *Physiological Functions in Special Environments*, ed. C.V. Paganelli & L.E. Farhi, pp. 191-202. New York: Springer.

Hammel, H.T. (1989b) Neural control of salt gland secretion: enhanced and sustained by autofacilitation. In *Progress in Avian Osmoregulation*, ed. M.R. Hughes & A.C. Chadwick, pp. 163-181. Leeds: Leeds Philosophical and Literary Society.

Hammel, H.T. (1990) Negative plus positive feedback. In *Thermoreception and Temperature Regulation*, ed. J. Bligh & K. Voigt, pp. 174-182. Berlin-Heidelberg: Springer.

Hammel, H.T., Gray, D.A. & Simon E. (1990) Blood from salt loaded donor ducks induces salt gland excretion in recipient ducks. *FASEB J.* 4, 413.

Hammel, H.T., Simon-Oppermann, Ch. & Simon, E. (1980) Properties of body fluids influencing salt gland secretion in Pekin ducks. *Am. J. Physiol.* 239, R489-R496.

Hanwell, A., Linzell, J.L. & Peaker, M. (1972) Nature and location of the receptors for secretion by the salt glands of the goose. *J. Physiol. (Lond.)* 226, 453-472.

Loewi, O. (1921) Über die humorale Übertragbarkeit der Herznervenwirkung. I. Mitteilung. *Pflügers Arch.* 189, 239-242.

Peaker, M. & Linzell JL (1975) *Salt Glands in Birds and Reptiles*. Cambridge: Cambridge University Press.

Simon-Oppermann, Ch., Simon, E., Deutsch, H., Möhring, J. & Schoun J. (1980) Serum arginine vasotocin (AVT) and afferent and central control of osmoregulation in conscious Pekin ducks. *Pflügers Arch.* 387, 99-106

Simon-Oppermann, Ch., Gray, D., Szczepanska-Sadowska, E. & Simon, E. (1984) Blood volume changes and arginine vasotocin (AVT) blood concentration in conscious, fresh water and salt water adapted ducks. *Pflügers Arch.* 400, 151-159.

Integrative and cellular aspects of autonomic functions : temperature and osmoregulation. Eds K. Pleschka, R. Gerstberger. John Libbey Eurotext, Paris © 1994, pp. 509-516.

Atrial natriuretic peptide in lactating, conscious goats

Kerstin Olsson, Helmuth Schütz*, Katarina Cvek and Rüdiger Gerstberger*

*Department of Animal Physiology, Swedish University of Agricultural Sciences, P.O.B. 7045, S-750 07 Uppsala, Sweden and *Max-Planck-Institute for Physiological and Clinical Research, W.G. Kerckhoff-Institute, D-6350 Bad Nauheim, Germany*

SUMMARY

The aim of this study was to investigate the role of atrial natriuretic peptide (ANP) in the regulation of body fluid homeostasis, arterial blood pressure and plasma volume during lactation and nonlactation in goats, and to relate physiological reactions to morphological distribution of ANP receptor sites. ANP was infused intravenously at 5, 20, and 45 ng/kg/min in two different lactation periods, and infusions at the dose of 33 ng/kg/min during a third lactation has been included for comparison (Olsson ***et al.***, 1989). The results show that ANP did not induce significantly increased renal Na excretion in lactating goats, and that a small natriuresis was observed only for the two highest doses in nonlactating goats. In contrast, plasma protein concentration and hematocrit increased at all doses, and arterial blood pressure dropped during infusions at the highest dose both in lactating and nonlactating goats. Receptor sites for ANP were found in vessel walls and glandular tissue of the mammary glands, and in glomeruli and medullary region of the kidneys. Displacement experiments implied existence of "biologically active" receptors, also known as ANP(R1) in the mammary glands, whereas both ANP(R1) and "clearance" receptors or ANP(R2) appeared to exist in the kidney. It is suggested that effects of ANP on the vascular bed in the milk-producing mammary gland could participate in regulating blood supply to this organ probably at the expense of the blood flow to other tissues; e.g. kidneys. Such effects could explain the inconsistent natriuretic response to ANP infusions in lactating goats.

INTRODUCTION

Atrial natriuretic peptide (ANP) is a term applied to a group of peptides produced by the mammalian atrial myocytes (Bovy 1990). ANP is released from the atria in response to local wall stretch (e.g. increased intravascular volume), and exogenously administered ANP induces natriuresis and diuresis, transudation of fluid from the intravascular compartment to the interstitium, and hypotension (de Bold ***et al.***, 1981; Atlas ***et al.***, 1984; Brenner ***et al.***, 1990; Awazu & Ichikawa, 1993). These observations point to a potentially important physiological role of ANP. However, the physiological and/or pathophysiological settings in which ANP plays a significant role in body fluid homeostasis are yet to be identified.

Pregnancy and lactation are conditions characterized by increased demands on cardiovascular and body fluid regulatory mechanisms. The dairy goat can produce up to 10 kg of milk per day with 90 % water and a Na content of about 15 mmol/kg, which requires enlarged plasma volume, increased cardiac output, and abundant blood supply to the udder (Linzell, 1974). This made it of interest to study the role of ANP in the regulation of body fluid homeostasis during lactation.

In the first series of experiments, we found that an intravenous infusion of ANP at the dose of 1 μg/min (mean= 33 ng/kg/min) did not induce natriuresis when goats were lactating, but it did so when the same goats no longer produced milk. The ANP infusions caused decreased plasma volume both during lactation and the anestrus period (Olsson *et al.*, 1989). The arterial blood pressure was not measured in those experiments. These observations initiated studies to find out the threshold dose for natriuresis, hypovolemia, and hypotension, respectively, in nonlactating and lactating goats. Moreover, receptor binding studies with $[^{125}I]$-hANP as radioligand were done in the mammary glands and kidneys in order to correlate physiological effects with the distribution of ANP receptor sites.

MATERIAL AND METHODS

Fourteen goats of the Swedish landrace breed were used in two series of experiments. The goats were fed at 07.00 h and 15.30 h according to their requirements. In order to keep them in a sodium replete condition, they were given 3 g of NaCl mixed with concentrates in the morning and afternoon. The milk yield was 1- 3.5 kg/day.

In one series of experiments ANP was infused for 90 min at the doses of 5 ng/kg/min and 20 ng/kg/min. Four lactating and four nonlactating goats were used at random. In the second series infusions of ANP were given at a higher dose, (2 μg/min; mean=45 ng/kg/min) for 60 min. These infusions were not prolonged due to the fall in arterial blood pressure that occurred. The same six goats were used both during lactation and nonlactation conditions.

The ANP used was α-human $ANP_{(1-28)}$ purchased from Bachem, Bubendorf, Switzerland.

Experimental procedure

The goats had one or both carotid arteries exteriorized subcutaneously on the neck in principle as described by Dueck *et al.*, (1982). On the day of the experiment a permanent catheter was introduced into the artery and connected to a saline-containing syringe. To prevent clotting 0.2 ml/min of physiological saline was infused into the artery. The catheter was connected to a pressure transducer and the arterial blood pressure registered on a Grass polygraph. The heart rate was calculated from the systolic/diastolic tracings. Catheters were placed in each jugular vein, one for blood sampling and one for infusions. Blood was obtained in heparinized tubes for measurements of hematocrit and total plasma protein concentration. For determination of hematocrit the blood was centrifuged in capillary tubes for 5 min in triplicate. Total plasma protein concentration was estimated by refractometry.

Urine was collected via a bladder catheter. The urine and milk Na concentrations were analyzed using ion selective electrodes (E2A Electrolyte Analyzer, Beckman Instruments, Stockholm,

Sweden). Glomerular filtration rate (GFR) was determined by constant intravenous infusion of ^{3}H-inulin, and effective renal plasma flow by infusion of para-aminohippurate-sodium (PAH) as described previously (Olsson ***et al.***, 1992).

In vitro receptor autoradiography

For the autoradiographical demonstration of ANP binding sites, tissue blocks from kidneys and mammary gland were taken out. From three lactating and three nonlactating goats, biopsies of mammary gland tissue was obtained during general anesthesia using a combination of ketamin (Ketalar®, Parke-Davis, Morris Plains, NJ, USA) and detomidin (Domosedan®, Orion-Farmos, Åbo, Finland). In addition, mammary gland tissue was taken from three nonlactating goats at slaughter. Three of the nonlactating goats had never been pregnant. Kidney tissue was also obtained from the three nonlactating goats at slaughter. The tissue blocks were immediately frozen in -50°C hexane and stored at - 20°C. The tissue was prepared for receptor autoradiography as described earlier (Schütz & Gerstberger, 1990).

Receptor binding studies with [^{125}I]-hANP

To characterize membrane-intrinsic binding sites specific for $ANP_{(1-28)}$ in both the mammary gland and the kidney, serial tissue sections of the organs were incubated with 200 nM monoradioiodinated [^{125}I]-hANP in the absence or presence of unlabeled hANP as well as the $hANP_{(4-23)}$ fragment in concentrations from 10^{-12} to 10^{-6} M. The relative potencies of both unlabeled ANP analogues (IC $_{50}$ values) to competitively displace the radioligand from its specific binding sites in both organs were analyzed by means of computerized optical density measurements of the autoradiograms obtained.

Data evaluation

The experimental results are presented as means with standard error of the means (s.e.m.). Statistical analysis was done using the Statistical Analysis System (SAS), procedure GLM (1988).

RESULTS

Physiological responses to ANP

Prolonged infusions of ANP at the low dose of 5 ng/kg/min caused no change in renal Na excretion neither in lactating nor in nonlactating goats. When the dose was increased to 20 ng/kg/min, renal Na excretion rose higher during infusions in the nonlactating goats, but the rise during lactation was not significant. A further increase of the ANP dose to 45 ng/kg/min increased renal Na excretion significantly only in nonlactating goats (Fig.1).

Urine flow rose during infusions of the highest dose of ANP. It increased from 2.0±0.2 to 3.2±0.7 ml/min ($P<0.05$) in lactating goats, and from 0.9±0.1 to 2.0±0.5 ml/min ($P<0.01$) in nonlactating goats. Renal free water clearance did not turn positive in any of the experimental series. The GFR and effective renal plasma flow were measured during infusions of the highest dose of ANP. GFR increased from 77±7 to 96±7 ml/min ($P<0.05$) in nonlactating goats. In lactating goats, GFR was 78±8 ml/min before, and 75±10 ml/min at the end of the infusion (NS). Effective renal plasma flow was 466±25 ml/min before, and 535±66 ml/min at the end of the infusion (NS) in nonlactating goats, and 616±49 ml/min and 590±73 ml/min (NS), respectively in lactating goats.

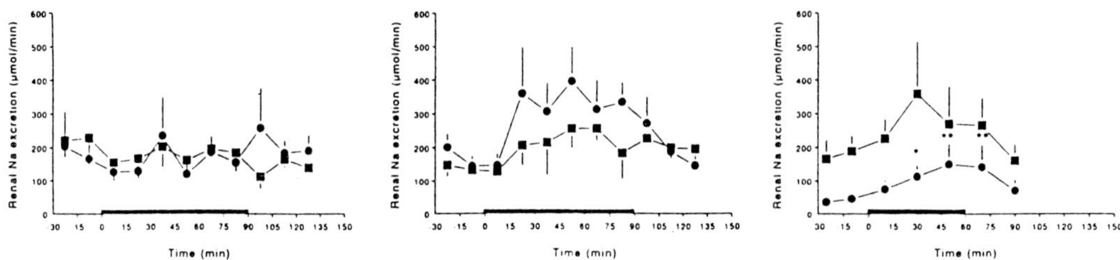

Fig. 1. Renal Na excretion during intravenous infusions of ANP (horizontal bar) in nonlactating ( ) and lactating ( ) goats. The response to the dose of 5 ng/kg/min is illustrated in the left panel, to 20 ng/kg/min in the middle panel (N=4, nonlactating goats; N=4 lactating goats), and to 45 ng/kg/min in the right panel (N=6; the same goats were used in both periods). *P<0.05; **P<0.01.

Taken together the present results and those obtained earlier in goats (Olsson *et al.,* 1989; Olsson **et al.**, 1991) indicate that the threshold dose of ANP to elicit increased renal Na excretion is about 30 ng/kg/min. As shown in Fig. 1, neither prolongation of the infusion from 60 min to 90 min at the low doses, nor increase of the dose to 45 ng/kg/min caused any further rise of the natriuretic response. Furthermore, the change in renal Na excretion during ANP infusions was small compared to the variation of basal levels that normally occur in sodium replete goats (Fig. 2).

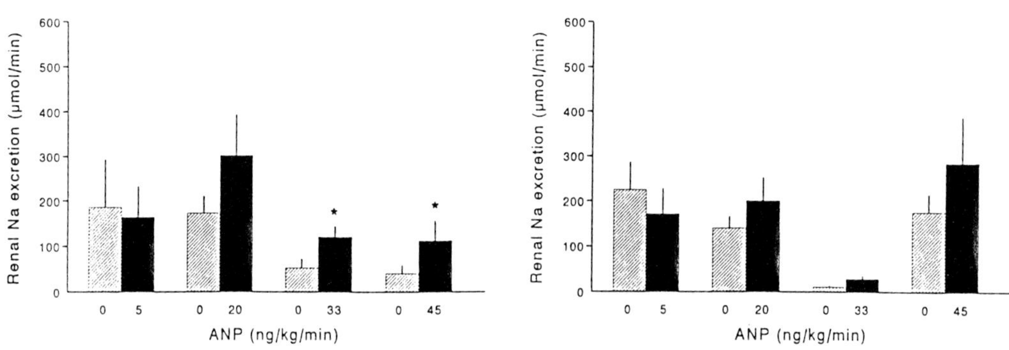

Fig. 2. Summary of the effects on renal Na excretion during intravenous infusions of ANP at four different doses in nonlactating (left) and lactating goats (right). Means and s.e.m. Dashed bars denote values during 30 min before infusion and filled bars values obtained during 60 min of infusion. *P<0.05 (paired t-test). (Values at 33 ng/kg/min from Olsson **et al.**, 1989, with permission).

The infusions of ANP at the two lower doses caused no change in mean arterial blood pressure. The heart rate was 70±8 beats/min before and at the end of the infusions. When the dose of ANP was increased to 45 ng/kg/min the mean arterial blood pressure fell from 105±2 mmHg to 98±2 mmHg (nonlactating goats; $P<0.5$), and from 98±2 mmHg to 88±3 mmHg (lactating goats; $P<0.01$). Heart rate rose from 85±5 beats/min before, to 105±10 beats/min at the end of the infusions ($P<0.05$) in nonlactating goats. In lactating goats the heart rate was 119±4 beats/min before, and 123±6 beats/min at the end of the infusion (NS).

In contrast to the effects on renal Na excretion, intravenous infusions of ANP increased total plasma protein concentration both in lactating and nonlactating goats even at the lowest dose examined here. The response was augmented when the dose was elevated to 20 ng/kg/min, but infusions of the highest dose caused no further increase (Fig.3). The hematocrit changed in parallel to the plasma protein concentration (data not shown), indicating that the effects were due to a reduction of plasma volume.

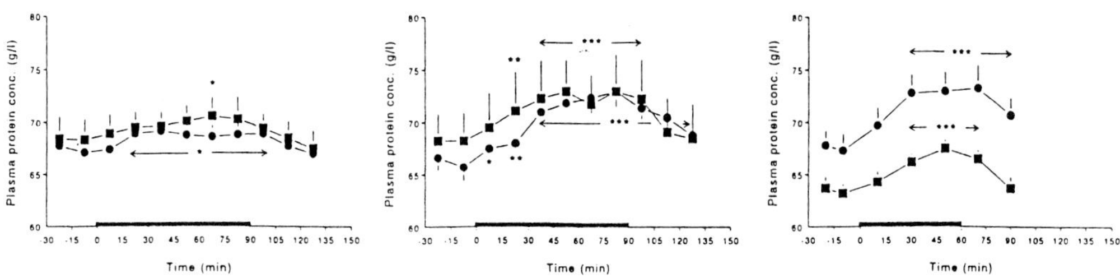

Fig. 3. Increase in total plasma protein concentration during intravenous infusions of ANP. Left panel: 5 ng/kg/min; middle panel: 20 ng/kg/min, and right panel: 45 ng/kg/min. Symbols as in Fig. 1. *$P<0.05$; **$P<0.01$; ***$P<0.001$.

ANP-specific binding sites in kidney and mammary gland

Receptor autoradiography using $[^{125}I]$-hANP as radioligand was employed to study the distribution of putative ANP binding sites in the kidneys and mammary glands of the goats.

In the kidneys the autoradiograms revealed specifically labeled hANP binding sites both in the glomeruli and the medullary region in nonlactating goats (lactating goats have not yet been examined). In those regions the radiolabeled $hANP_{(1-28)}$ was readily displacable by increasing concentrations of unlabeled $hANP_{(1-28)}$ and also by the $hANP_{(4-23)}$fragment. This indicates presence of both ANP(R1) and ANP(R2) in the kidneys of nonlactating goats.

In the mammary glands binding sites were found both in lactating and nonlactating goats. The radiolabeled $hANP_{(1-28)}$ was displacable by increasing concentrations of unlabeled $hANP_{(1-28)}$, but not by the $hANP_{(4-23)}$ fragment. This indicates that only ANP(R1) are present in the mammary gland of goats.

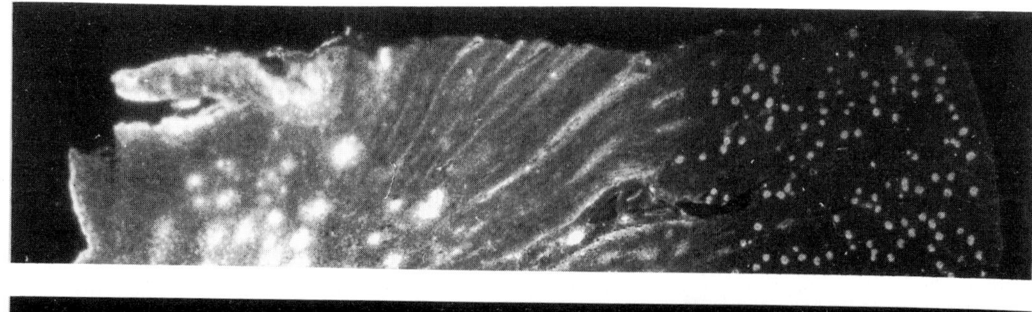

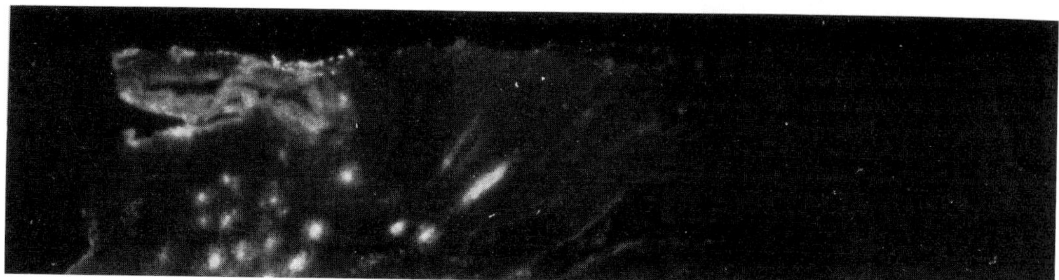

Fig. 4. Histological distribution of ANP-binding sites in the kidney. Receptor autoradiogram of $[^{125}I]$-hANP in the kidney cortex and medullary region of a nonlactating goat. Tissue sections were incubated with 200 nM $[^{125}I]$-hANP$_{(1-28)}$ in the absence (above) or presence (below) of 10^{-6}M unlabeled ANP. Magnification X 8.

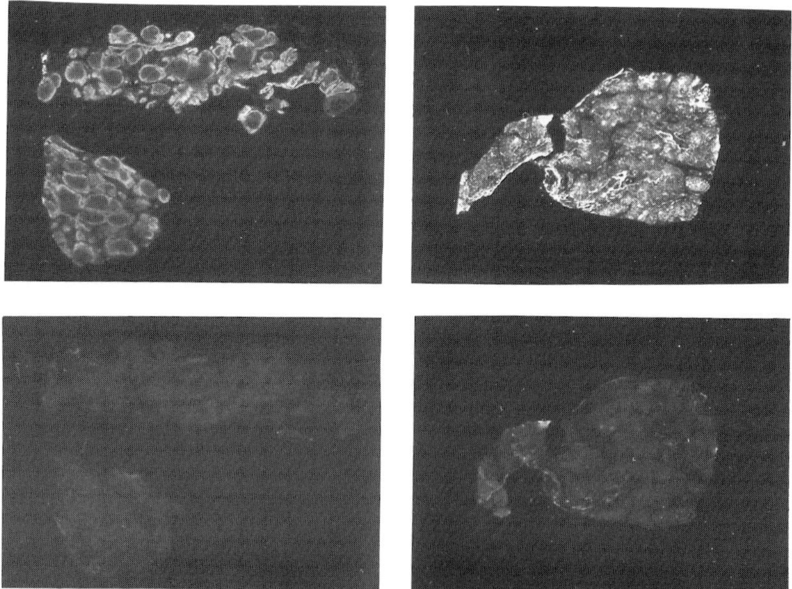

Fig. 5. Histological distribution of ANP-binding sites in mammary gland tissue. Receptor autoradiogram of $[^{125}I]$-hANP in the mammary gland of a nonlactating (left) and a lactating goat (right). Tissue sections were incubated with 200nM $[^{125}I]$-hANP in the absence (above) or presence (below) of 10^{-6}M unlabeled ANP. Magnification X 3.5.

DISCUSSION

The results of this study show that the dose interval at which ANP increased renal Na and water excretion in nonlactating goats is narrow and that the maximal natriuresis and diuresis were small. Likewise, the natriuretic response was modest also in sheep given 60 min infusions of ANP at a dose corresponding to the highest dose used here (Parkes et al. 1987). The finding of binding sites both to ANP(R1) and ANP(R2) in the kidneys support observations in other species (Maack **et al.,** 1987, Maack,1992). The localization of receptors in the glomeruli is consistent with the increased GFR, which was found during ANP-induced natriuresis in the nonlactating goats. Number and affinity of the ANP receptors in the kidneys have not yet been studied in goats during lactation, but will be of interest because no rise in GFR was found during ANP infused in that period. The fact that the renal Na excretion did not rise more in response to higher doses of ANP was by all probability related to simultaneous reduction in plasma volume and drop in blood pressure. If, however, the blood pressure was kept above normal by means of acute angiotensin II infusions, combination of that infusion with ANP resulted in an exaggerrated natriuretic response, with the Na excretion increasing to 3000 μmol/min in nonlactating goats (Olsson ***et al.,*** 1992). Furthermore, such combined infusions resulted in the same tremendous renal Na excretion also during lactation (Olsson & Hossaini-Hilali, unpublished observations). Large pressor doses of angiotensin II have been shown to promote endogenous ANP release in several species (Ruskoaho, 1993). Therefore, it is possible that the regulation of body fluid losses is affected during pathological conditions in which increased levels of angiotensin II and ANP coexist.

Lactation is a period characterized by markedly elevated water and electrolyte turnover in the dairy goat (Olsson ***et al.,*** 1982). In the lactating goats infusions of ANP reduced plasma volume at all doses similar to the response observed during nonlactating conditions. This support the notion that ANP is an important physiological regulator of intravascular volume in mammals (Redfield & Burnett 1989; Brenner ***et al.,*** 1990), and that this effect prevails during lactation. In the present study the goats excreted between 20 to 70 μmol/min of Na via the milk. Consequently, this route of Na loss may well exceed that of the kidneys. In this context the localization of specific binding sites of ANP in vessel walls and tissue of the mammary glands was of great interest. In the mammary glands we found binding sites for ANP(R1), but not ANP(R2). This was a surprising observation, because the ANP (R2) appear to outnumber the ANP(R1) in all other organs (Maack, 1992). Our findings indicate that ANP can be involved in local blood flow regulation and/or capillary permeability, which potentially may influence milk secretion. Since mammary blood flow accounts for about 1/5 of the total cardiac output, small changes in this vascular bed may have a great influence on the blood distribution to other organs among which the kidneys plays a central role. A redistribution of blood flow between mammary glands and the kidneys under the influence of ANP might explain the inconsistent natriuretic response to ANP during lactation in our goats. It may be argued that the specific labeling of the receptors was found also during nonlactation. However, the tissue mass and number of capillaries are much smaller during nonlactating conditions (Cvek ***et al.,*** this volume). We suggest that ANP could be of importance in the regulation of blood supply to the udder. If ANP is directly involved in the milk secretory process of goats remains to be shown.

ACKNOWLEDGEMENTS

This work was supported by the Swedish Medical Research Council (project nr 3392).

REFERENCES

Atlas, S. A., Kleinert, H. D., Camargo, M.J., Januszewicz, J. E., Sealey, J. E., Laragh, J. H., Schilling, J. W., Lewicki, J. A., Johnson, L. K. & Maack, T. (1984): Purification, sequencing and synthesis of natriuretic and vasoactive rat atrial peptide. *Nature Lond.* 309, 717-719.

Awazu, M. & Ichikawa, I. (1993): Biological significance of atrial natriuretic peptide in the kidney. *Nephron.* 63, 1-14.

Bovy, P. R. (1990): Structure activity in the atrial natriuretic peptide (ANP) family. *Med. Res. Rev.* 10, 115-142.

Brenner, B.M., Ballermann, B.J., Gunning, M. E. & Zeidel, M.L. (1990): Diverse biological actions of atrial natriuretic peptide. *Physiol. Rev.* 70, 665-698.

de Bold, A. J., Borenstein, H. b:, Veress, A.T. & Sonnenberg, H. (1981): A rapid and potent natriuretic response to intravenous injection of atrial myocardial extract in rats. *Life Sciences.* 28, 89-94.

Dueck, R., Schroeder, J. P., Parker, H. R., Rathbun, M. & Smolen, K. (1982): Carotid artery exteriorization for percutaneous catherizatiiion in sheep and dogs. *Am. J. Vet. Res.* 43, 898-901.

Linzell, J.L. (1974): Mammary blood flow and methods of identifying and measuring precursors of milk. In *Lactation. A comprehensive treatise.* Larson, B.L. & Smith, V.R. Academic Press, New York and London.

Maack, T., Suzuki, M., Almeida, F.A., Nussenzveig, D., Scarborough, R.M., McEnroe, G.A., & Lewicki,, J.A. (1987): Physiological role of silent receptors of atrial natriuretic factor. *Science* 238, 675-678.

Maack, T. (1992): Receptors of atrial natriuretic factor. *Annu. Rev. Physiol.* 54, 11-27.

Olsson, K., Benlamlih, S., Dahlborn, K. & Örberg, J. (1982): A serial study of fluid balance during pregnancy, lactation and anestrus in goats. *Acta Physiol. Scand.* 115, 39-45.

Olsson, K., Karlberg, B.E. & Eriksson, L. (1989): Atrial natriuretic peptide in pregnant and lactating goats. *Acta endocrin. (Copenh).* 120, 519-525.

Olsson, K, Dahlborn, K., Karlberg, B.E. & Eriksson, L. (1991): Atrial natriuretic peptide (ANP) in the goat. In *Physiological aspects of digestion and metabolism in ruminants,* ed. T.Tsuda, Y.Sasaki & R. Kawashima. Proc. 7th Int. Symp. on Ruminant Physiology, pp 257-275. Academic Press, San Diego.

Olsson, K., Hossaini-Hilali, J. & Eriksson, L. (1992): Atrial natriuretic peptide attenuates pressor but enhances natriuretic responses to angiotensin II in pregnant conscious goats. *Acta Physiol. Scand.* 145, 385-394.

Parkes, D.G., Coghlan, J.P., McDougall, J.G. & Scoggins, B.A. (1987): Hemodynamic effects of atrial natriuretic peptide in conscious sheep. *Clin. and Exper. Hyper.-Theory and Practise.* A9, 2143-2155.

Ruskoaho, H. (1992): Atrial natriuretic peptide. Synthesis, release and metabolism. *Pharmacol. Rev.* 44, 479-602.

SAS Institute Inc. (1988). SAS User's guide: Statistics, 1988. SAS Institute Inc., Cary, NC, USA.

Schütz, H. & Gerstberger, R. (1990). Atrial natriuretic factor controls salt gland secretion in the Pekin Duck (*Anas platyrhynchos*) through interaction with high affinity receptors. 127, 1718-1726.

Integrative and cellular aspects of autonomic functions : temperature and osmoregulation. Eds K. Pleschka, R. Gerstberger. John Libbey Eurotext, Paris © 1994, pp. 517-525.

The role of the lower intestinal tract in avian osmoregulation

Margarethe Brummermann and Eldon J. Braun

Department of Physiology, College of Medicine, University of Arizona, Tucson, Arizona 85712, USA

SUMMARY

In birds, after reaching the cloaca, ureteral urine can be refluxed into the colon and the ceca by retrograde peristalsis. This brings the urine into contact with epithelia of high transport capacities for sodium and water. Because of this, the lower gastro-intestinal (GI) tract should have a decisive impact on the avian salt and water balance, inevitably resulting in an integration of the excretory function of GI tract and kidneys. One way to control the modification of the final urine by the lower GI tract would be to control the retrograde peristalsis and, thus, the amount of urine refluxed.

In this study we observed lower GI tract motility by radiographic methods and quantified the mechanic activity of the colon *in vivo* using strain gauges which were surgically attached to the serosa. Two types of retrograde contraction waves were observed, small fast peristaltic contractions (15 min $^{-1}$) and large slow contractions, probably of the longitudinal musculature (3 min $^{-1}$).

The effects of water deprivation and of intravenous volume and salt loads on colonic motility were analysed. All manipulations that increased urine flow rates resulted in increased propagation speed of the large contraction waves. All manipulations that increased urine osmolality resulted in decreases in activity time for the large waves. These results suggest a hydration state related control of the slow retrograde colonic motility in birds. Direct stimulation of central osmoreceptors had no such effects, indicating that local volume and osmolality receptors of the cloaca are most important for this adjustment of colonic motility.

INTRODUCTION

In the sixties, several authors (Koike & McFarland, 1966; Akester et al., 1967; Ohmart et al., 1969) demonstrated the physical basis for a possible postrenal modification of urine in birds: Ureteral urine, after reaching the bird's cloaca, can be refluxed into colon and ceca by retrograde peristalsis of the lower intestine. The surfaces of cloaca, colon and ceca (lower GI tract) show considerable capacities for transport of sodium and potassium and water reabsorption (Bindslev & Skadhauge, 1971). Urine coming into contact with these epithelia results in an inevitable integration of the excretory functions of kidney and lower GI tract. Because of this, the lower GI tract should have a decisive impact on electrolyte and water balance of the birds. The aspects of the interorgan communication that must occur to allow for the efficient regulation of fluid and electrolyte balance are not well understood.

In vivo and *in vitro* studies have shown, that the tissues of cloaca and lower intestine are capable of net sodium absorption and sodium-linked water absorption against small osmotic gradients only (Bindslev & Skadhauge, 1971, Goldstein & Braun, 1986). This means that solute linked water reabsorption will occur only as long as urine osmolality is close to plasma osmolality. This could cause a problem during severe dehydration or high sodium loads, when the ureteral urine can be quite concentrated relative to the osmolality of the extracellular body fluid. Under these conditions water could be osmotically drawn into the intestinal lumen to be voided and lost, thus, defeating all attempts to conserve water. The transport processes in the tissues of the cloaca and lower intestine appear to be controlled with regard to dietary sodium intake (Bindslev & Skadhauge, 1971, Arnason & Skadhauge, 1991), the control being mediated by changes in plasma levels of aldosterone (Thomas et al., 1979) and arginine vasotocin, the antidiuretic hormone of birds (Skadhauge et al., 1985). But the described change in the number of sodium transport sites was due to long-term dietary adaptation rather than short-term water deprivation. It appears that a rapid and direct way of controlling GI salt and water reabsorption *vs.* excretion would be to control the amount of ureteral urine that is brought into contact with absorbing surfaces - that means to control the retrograde peristalsis which propels the urine from the cloaca towards the ceca.

This paper will describe the contraction waves of the colon and the cloaca that are responsible for the retrograde transport of urine into the lower intestine. Ten subadult White Leghorn roosters (BM 1600-1900g) were use as experimental animals. The retrograde fluid movement was documented using radiographic techniques. Strain gauges chronically attached to the colon (method of Lai & Duke, 1978) were used to monitor the mechanical activity of the colon. The study reported here will show how the colonic motility changes when birds are deprived of water and discuss the questions:

A: Is the refluxing of urine indeed regulated with regard to the bird's hydration state.

B: Which physiological parameters that are affected by water deprivation are the signals critical for the adjustment of colonic motility .

C: What are the receptors and pathways are involved.

RESULTS AND DISCUSSION

Radiographic study: In five roosters, aqueous radioopaque material (Pent-o-pake) was infused through the chronically catheterized ductus deferens which enters the cloaca in direct proximity of à ureter. After accumulation of the fluid in the cloaca, regular peristaltic contractions, originating in the coprodeum and moving through the colon towards the ceca, started to force bolus amounts of fluid in retrograde direction. The cycle length of these peristaltic waves was about 4 sec. Although the radiographic method showed the process of urine refluxing very clearly, it was not suited to study the impact of water deprivation on the colonic motility. Because of the necessity to infuse a certain amount of radioopaque fluid, it is not possible to monitor the true conditions of dehydration by this method.

Measurements of mechanical activity: Small strain gauges were sutured to the outside of the colon about 1 and 3 cm proximal of the cloaca. After this surgery, the birds were allowed to recover for five days. Recordings from the implanted strain gauges showed three different types of contractions (Fig. 1):

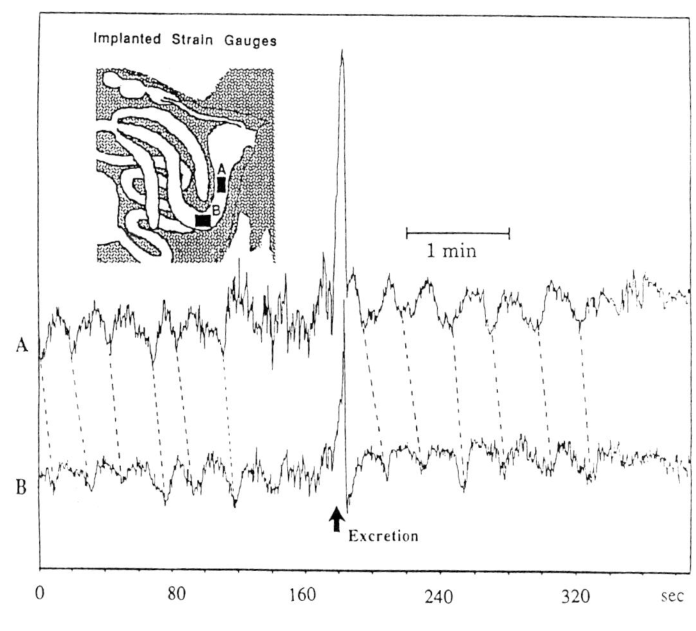

Fig.1: Strain gauge recordings from normally hydrated rooster. Inset: Schematic drawing of the lower GI tract to indicate the position of the gauges.

1.) Single giant contractions, which were recorded by both gauges simultaneously as sharp peaks, were correlated with the excretion of cloacal contents.

2.) Small irregular peaks could be recorded continuously. They seemed to correlate with the 4 sec peristaltic waves responsible for refluxing the urine as seen in the radiographic study. It was not possible, though, to filter out the impact of breathing expansions which coincide with 80% of the recorded peaks. Furthermore the gauges, due to their seize, were too far apart to show a phase shift between recorded small waves which would allow to characterize them by the direction of their proceeding.

3.) Large regular contraction waves occurred during distinct activity phases of about 80% of the recording time. Periods of activity were clearly separated by periods of quiescence. The cycle length of these large contractions was interindividually different, but in normally hydrated roosters always between 15 and 25 seconds. As shown by the phase shift in Fig. 2, these waves are moving from gauge A to gauge B, which means from the cloaca towards the ileo-caeco-colic junction. These waves are probably manifestations of rhythmical contractions of the longitudinal muscles in the colon. These were the only spontaneous contractions which we could observe in *in vitro* studies of isolated colon preparations. As these large waves proceed in the proximal direction, they may support urine refluxing and modulate the amount refluxed by the smaller peristaltic waves.

Water deprivation: 65 and 83 hours of water deprivation decreased the body weight of five roosters by 7.1±0.8 and 9.7±0.9%, respectively. After maximal water deprivation all individuals were feeding and excreting, though food intake was reduced.

Table 1 shows changes in plasma and urine composition evoked by water deprivation: As expected, plasma osmolality increased significantly after 65 and even more so after 83 h, mainly due to an increase in plasma sodium (and probably chloride which was not measured), whereas plasma potassium did not change significantly. Urine osmolality was more than doubled after 65h and increased just slightly more after 83h of water deprivation. This increase in urine osmolality was based on increases in both potassium and sodium concentration. The flow rate of ureteral urine was half the normal rate after 65h and did not further decrease after 83h.

Of the three described types of colonic activity, the large regular retrograde contraction wave was affected most clearly by water deprivation (Fig. 3). After 65h of water deprivation, with 56±8.4% of the total recording time, this wave pattern was not as prominent any more as during normal hydration when it was present for 82±5%. This reduction, though, was not yet statistically significant. Sustained water deprivation for 83h reduced its activity

Table 1. Effect of dehydration on physiological parameters in 5 roosters

	Control	Water deprivation 65 h	Water deprivation 83 h
Body mass (g)	1731.0 ±47.7	1610.0 ±52.2	1555.8* ±51.0
Plasma osmolality ($mOsm \cdot kg^{-1}$ H_2O)	314.0 ±3.05	337.8* ±1.11	355.9* ±3.65
Plasma Na ($m\ mol \cdot l^{-1}$)	153.2 ±2.7	165.5* ±1.7	171.4* ±1.1
Plasma K ($m\ mol \cdot l^{-1}$)	3.7 ±0.6	3.8 ±0.26	3.8 ±0.24
Urine flow ($ml \cdot min^{-1}$)	0.04 ±0.003	0.022* ±0.007	0.019* ±0.005
Urine osmolality ($mOsm \cdot kg^{-1}$ H_2O)	331.8 ±67.4	555.8* ±20.5	571.4* ±20.9
Urine Na ($m\ mol \cdot l^{-1}$)	54.0 ±11.2	88.7 ±7.7	83.6 ±17.7
Urine K ($m\ mol \cdot l^{-1}$)	68.3 ±17.98	147.9* ±10.46	176.9* ±24.1

Values are mean ± SE; * indicates mean values significantly different from controls

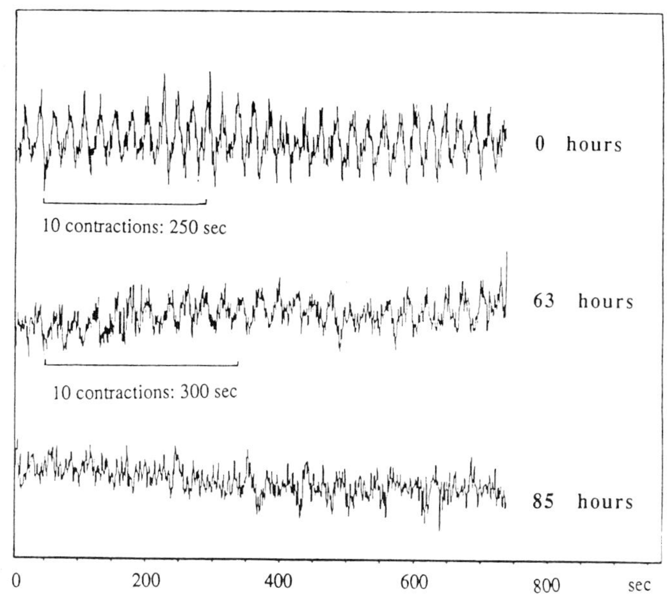

Fig.2: Recordings from one strain gauge during normal hydration and after 63 and 85 hours of water deprivation

time significantly to 40.3±11%, less than half of the activity time of the large contractions during normal hydration. In one of the five roosters, the large wave pattern all but disappeared. As shown by the significantly increased cycle length from 20.6±0.4 to 26.4±1.3 sec, the big waves were also significantly and increasingly slowed by progressive dehydration of the roosters.

The following experimental series was designed to single out the parameters that are affected by water deprivation to determine the ones responsible for the observed change in the large wave pattern of colonic motility.

Systemic hyperosmotic load: Fig. 3 compares the effects of hyperosmotic intravenous infusions (1000 mOsm·kg^{-1} H_2O NaCl at 0.5 ml· min $^{-1}$·kg^{-1}, N=5) to those of isotonic saline infusions of the same volume and control conditions without infusion. To mimic plasma osmolalities measured during water deprivation the hyperosmotic infusion had to last for at least 45 min. By this time not only plasma but also urine osmolality and flow rate were significantly increased, and the birds had experienced a considerable volume load. As a result, the cycle length of the large wave pattern decreased, but - as during water deprivation - the percentage of active

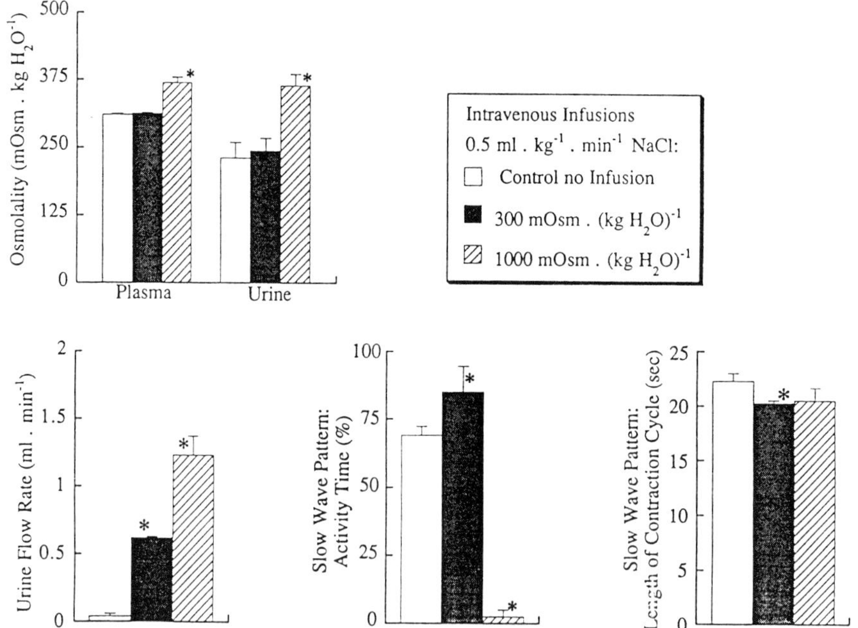

Fig. 3: Effects of intravenous saline infusions in five normally hydrated roosters.

wave pattern was extremely reduced. Isotonic NaCl infusion of the same rate and duration did not change blood or urine osmolality, but increased the urine flow rate and decreased the cycle length of the large wave, whereas the activity time increased significantly from about 70 to 85 % of recording time. Thus, increased urine flow due to isoosmotic volume loads seems to stimulate the large wave motility by increasing its speed and activity time. During hyperosmotic infusions increased urine flow also stimulates the speed, but the high osmolality of plasma and/or urine decreases the activity time.

Hyprosmotic central infusions: In this series, infusions were applied through the carotid artery to directly stimulate central osmoreceptors which are known to be important for the control of renal and extrarenal osmoregulation in birds. Again a hyperosmotic load (1000 mOsm·kg H_2O NaCl at 0.1ml· min $^{-1}$·kg^{-1}, N=5) was compared to control conditions without any infusion and to an isoosmotic saline load to detect volume and pressor effects. By infusing directly towards the targeted receptors, extensive volume loads could be avoided, as the osmolality of the plasma of the jugular vein which drains the cranial area was increased within the first ten min of hypertonic infusion. At this time the osmolality of the systemic plasma - collected from the brachial vein - and of the urine were not increased significantly, and urine flow was unaltered. Inspite of the

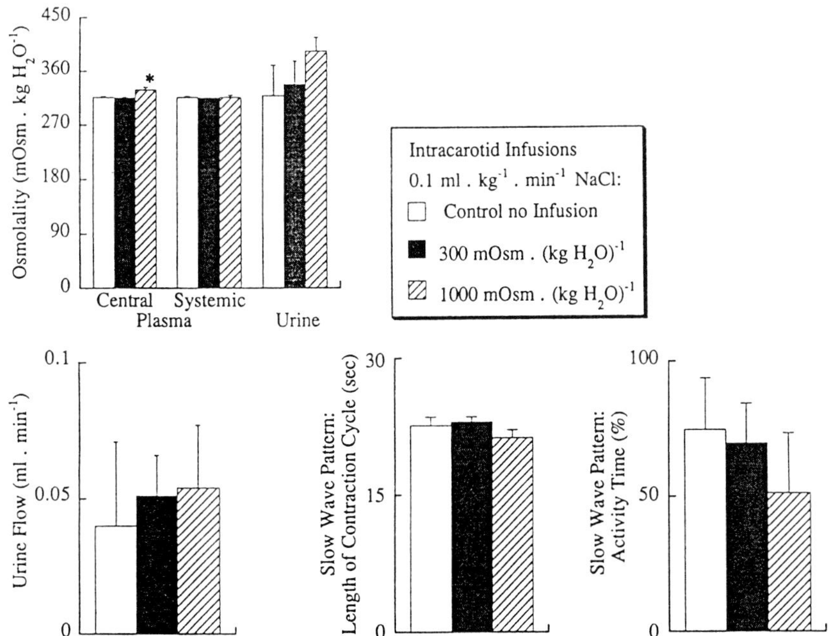

Fig.4: Effects of intracarotid saline infusions in five normally hydrated roosters.

increased central osmolality neither the cycle length nor the percentage of activity time of the large slow wave were significantly affected (Fig. 4). Thus, changes in central osmolality do not seem to be critical for the control of the large wave pattern of colon motility.

CONCLUSIONS

Our methods did not produce clear data on the small peristaltic wave of 4 sec cycle length which appear to be the ones directly responsible for urine refluxing. The recordings of the large retrograde wave pattern, though, clearly show the influence of the bird's hydration state on its colonic motility. The effects of central and peripheral volume loads indicate that the decisive signal is perceived by local volume and osmoreceptors of the cloaca rather than by central osmoreceptors. At this point we can only speculate how the large wave activity influences the amount of urine refluxed. But, if it is supportive or permissive to the retrograde transport of urine, its reduced speed or total disappearance during water deprivation and high urine osmolality due to high salt loads could indicate reduced urine refluxing under these conditions.

Considering the limited ability of the transporting surfaces of the colon to overcome high osmotic gradients, this reduction in contact between urine and colonic tissue could serve to avoid water loss from the tissue drawn into the lumen by the high osmolality of its contents. The finding of Arnason and Skadhauge (1991), that the transport capacity of cloacal, but not of colonic tissue can be greatly changed by osmoregulatory hormone concentrations, supports this idea: contact of the urine to cloacal tissue is inevitable. Contact of the urine to colonic tissue can be regulated by the refluxing activity.

REFERENCES

Akester, A.R., R.S. Anderson, K.J. Hill & G.W.Osbaldiston (1967): A radiographic study of urine flow in the domestic fowl. Br. Poult . Sci. 8:209-212.

Arnason, S.S. & E. Skadhauge (1991): Steady-state sodium absorption and chloride secretion of colon and coprodeum, and plasma levels of osmoregulatory hormones in hens in relation to sodium intake. J. Comp. Physiol. B 161: 1-14.

Arnason, S.S., G.Rice, & E. Skadhauge (1986): Plasma levels of arginine vasotocin, prolactin, aldosterone and corticosterone during prolonged dehydration in the domestic fowl: effect of dietary NaCl. J. Comp. Physiol. B 156: 386-397.

Bindslev, N., & E. Skadhauge (1971): Sodium chloride absorption and solute -linked water flow across the epithelium of the coprodeum and large intestine in the normal and dehydrated fowl (*Gallus domesticus*). *In vivo* perfusion studies. J. Physiol. 216:753-768.

Goldstein, D., L (1989): Absorbtion by the cecum of wild birds: Is there intraspecific variation? J. Exp. Zool. Suppl. 3: 103-110.

Koike, T.L. and L.Z. McFarland. Urography in the unanesthetized hydropenic chicken. Am. J. Vet. Res. 27:1130-1133, 1966

Lai, H. C. and G. E. Duke. Colonic motility in domestic turkeys. Am . J. Digest. Diseases 23(8):673-681, 1978

Ohmart R.D., L.Z. McFarland and J.P. Morgan Urographic evidence that urine enters the rectum and ceca of the roadrunner (*Geococcyx californianus*). Aves. Comp. Biochem. Physiol. 35:487-489. 1970

Thomas, H. D., M. Jallageas, B. G. Munck, and E. Skadhauge. Aldosterone effects on electrolyte transport of the lower intestine (coprodeum and colon) of the fowl (*Gallus domesticus*) *in vitro*.. Gen. Comp. Endocrinol. 40: 44-51, 1979

Thomas, H. D, E. Skadhauge.and M.W. Read. Acute effects of aldosterone on water and electrolyte transport in the colon and croprodeum of the domestic fowl (*Gallus domesticus*) *in vivo*. J. Endocr. 83:229-237, 1979

Integrative and cellular aspects of autonomic functions : temperature and osmoregulation. Eds K. Pleschka, R. Gerstberger. John Libbey Eurotext, Paris © 1994, pp. 527-532.

Carbonic anhydrase activity in the goat mammary tissue during lactogenesis

Katarina Cvek, Yvonne Ridderstråle, Kristina Dahlborn and Kerstin Olsson

Department of Animal Physiology, Swedish University of Agricultural Sciences, P.O.B. 7045, S-750 07 Uppsala, Sweden

SUMMARY

The aim of this study was to histochemically localize carbonic anhydrase (CA) in the mammary gland of goats and to study possible changes during lactogenesis. Biopsies were taken from the mammary glands of 10 lactating goats at days 2, 5, 9, 14, 15, 18, 23, 25, 30, and at three months of lactation, respectively. In addition, biopsies were taken from three late pregnant and five nonlactating goats. A cobalt precipitation method was used to visualize CA. The results show that neither the tissue from the nonlactating nor the late pregnant goats had any CA activity except for a faint reaction in the nuclei of the alveolar epithelium and the apical membrane of a few cells in some larger ducts. Two days after onset of lactation staining of the capillaries could still not be seen, but on day 5 and 9 of lactation some of the capillaries were weakly stained. From day 14 and onwards all the capillaries were stained with the most intense staining from day 18 until 3 months. It is suggested that CA activity in the capillaries of the goat mammary tissue increases in parallel to the rise in milk production.

INTRODUCTION

The enzyme carbonic anhydrase (CA) catalyzes the first step of the reversible reaction: $CO_2 + H_2O \leftrightarrow H_2CO_3 \leftrightarrow H^+ + HCO_3^-$. When this reaction takes place in the red blood cells, the reaction between CO_2 and H_2O is accelerated and the red blood cells can carry CO_2 away from metabolically active organs most efficiently. CA has been found in a variety of tissues, but few studies concerning CA activity in the mammary gland have been published. By biochemical methods Brown & Baily (1963) found CA activity in the rat mammary tissue. They showed that the enzyme activity in the rat mammary tissue started to increase during late pregnancy, was high immediately postpartum, but then gradually declined as nursing proceeded. CA was later histochemically localized in the lactating rat mammary tissue, mainly in the cytoplasm and nuclei of the alveolar cells and in the capillaries (Ridderstråle & Knutsson, 1980). Since lactogenesis is a process that involves a great increase in secretory function in the mammary gland cells, it was of interest to study the localization of CA during the different stages of lactogenesis in another species. The species chosen was the Swedish landrace goat since the composition of its milk differs from that of the rat and it has been selected for high milk yield.

MATERIALS AND METHODS

Animals.

Biopsy specimens were taken from the mammary glands of 10 healthy goats in different stages of early lactation (days 2, 5, 9, 14, 15, 18, 23, 25, 30, and after 3 months of lactation). Biopsies were also taken from three pregnant goats and from five dry goats, whereof one had never been lactating. One animal occurs in all three groups and another one in both the lactating and the dry group.
Before the biopsy was taken, the animal was sedated by an intravenous injection of acepromazin (Plegicil®, Pherrovet , Malmö, Sweden). The area where the biopsy was to be taken was locally anaesthetized by a subcutaneous injection of lidocain (Xylocain®, Astra läkemedel, Södertälje, Sweden). A biopsy needle (Biopty™, Radiplast AB, Uppsala, Sweden) was used to obtain small biopsy specimens. Larger tissue samples were taken from two animals during general anaesthesia using ketamin (Ketalar®, Parke-Davis, Morris Plains, USA) in combination with detomidin (Domosedan®, Orion-Farmos, Åbo, Finland). A small incision through the skin of the udder was made and a piece of the mammary tissue was cut out (0.5 X 0.5 cm). The wound was closed with two to three stitches.

Preparation of tissue for microscopy.

The tissue samples were fixed in phosphate buffered 2.5% glutaraldehyde, pH 7.2 for six hours and then rinsed in phosphate buffer, pH 7.2. After the rinsing in buffer, the samples were dehydrated through graded ethanols, and embedded in the water-soluble resin Historesin for histochemistry, as described by Ridderstråle (1976, 1991).

Histochemical demonstration of CA activity.

The cobalt precipitation method of resin embedded tissue described by Ridderstråle (1976, 1991) was used. From each biposy, 2 µm thick sections were cut. The sections were incubated for 3 or 6 min floating on the surface of an incubating medium containing 157 mM $NaHCO_3$, 3.5 mM $CoSO_4$, 11.7 mM KH_2PO_4 and 52.6 mM H_2SO_4. After incubation the sections were rinsed on 0.67 mM phosphate buffer, pH 5.9, for 1 min, treated with 0.5% $(NH_4)_2S$ for 3 min, and finally rinsed in two successive baths of distilled water, 1 min in each. Controls were run with the specific CA inhibitor acetazolamide (10^{-5}M) in the incubation medium. Before mounting, some of the sections were weakly counterstained with azure blue.

RESULTS

Nonlactating goats

The tissues from all the nonlactating animals were of similar appearance. The capillaries surrounding the alveoli were unstained. A faint reaction could be seen in the nuclei of the alveolar epithelium, but apart from this, the alveolar cells did not show any staining. Some staining of the epithelial apical membranes in a few ducts was noted. Stained erythrocytes were seen in all specimens (Fig. 1).

Late pregnant goats

In two of the three biopsies no staining of the capillaries could be seen (Fig. 2), but the mammary biopsy from the third late pregnant goat showed membrane bound staining of the capillaries.

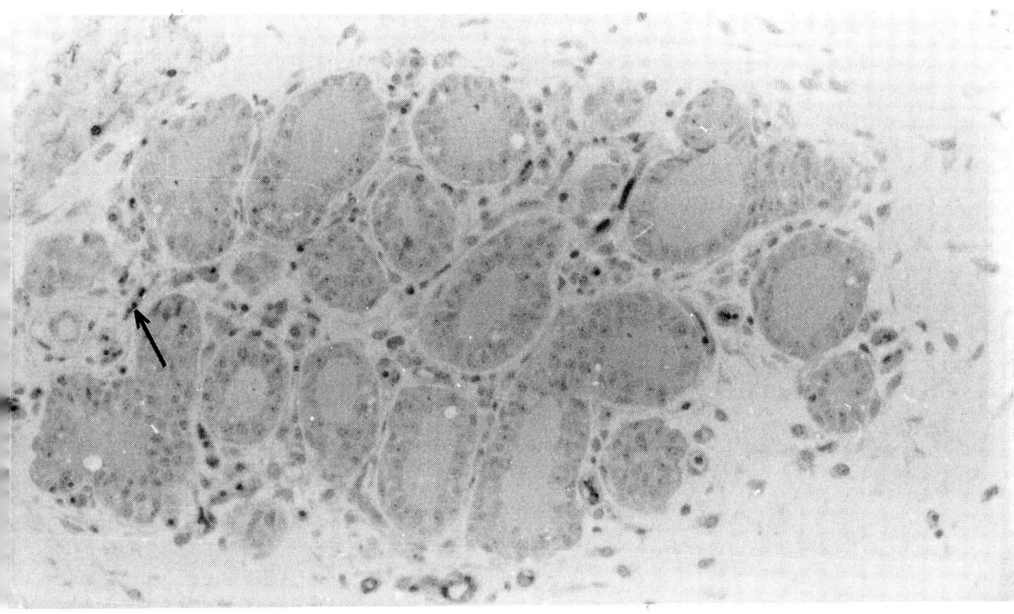

Fig. 1. Nonlactating goat mammary gland incubated 6 min for demonstration of CA activity. No staining showing CA is present except for the stained erythrocytes (arrow). Magnification X 300.

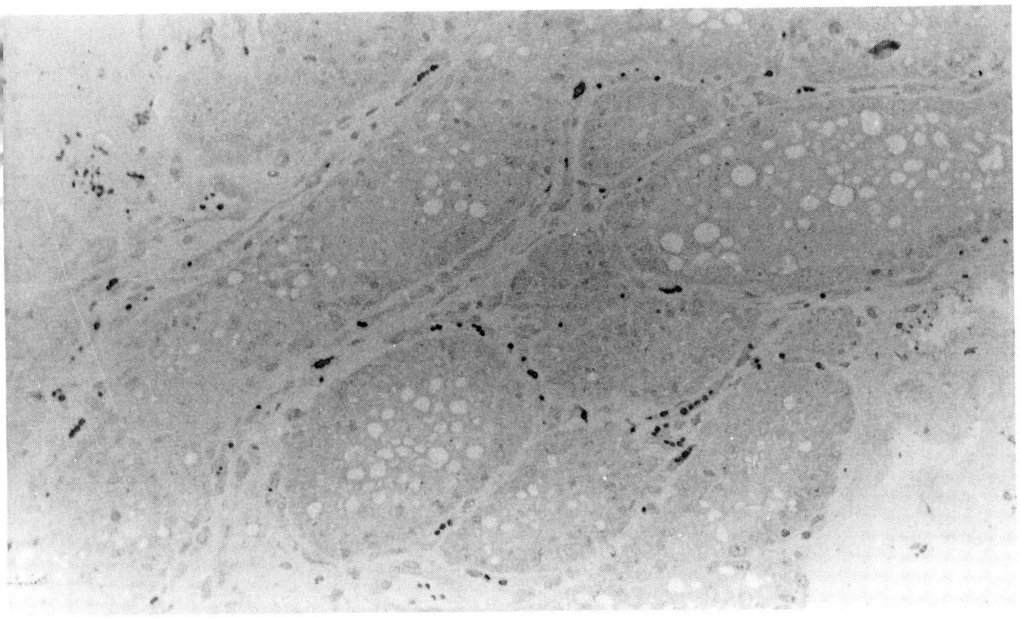

Fig. 2. Late pregnant goat mammary gland incubated 6 min for demonstration of CA activity. Neither the alveoli nor the capillaries show staining. Erythrocytes are stained. Magnification X 300.

Lactating goats

Tissue from the one goat that produced colostrum (day two of lactation) showed no staining of the capillaries. Biopsies taken at day 5 and 9 of lactation showed weak staining of some capillaries (fig. 3a). At day 14 and 15 of lactation all capillaries were stained with the largest moiety of the capillary network showing weak staining and about one third intense staining. From day 18 of lactation and onwards (day 23, 25, 30 and 3 months - different goats) all the capillaries were heavily stained (fig. 3b). The number of capillaries in the mammary tissues from the lactating goats were much greater than in the nonpregnant, nonlactating goats.

The erythrocytes that could be seen in the lumen of the capillaries showed a similar staining in all specimens. No staining was found in the alveolar cells, nor in the epithelial cells of the larger ducts in any of the samples examined.

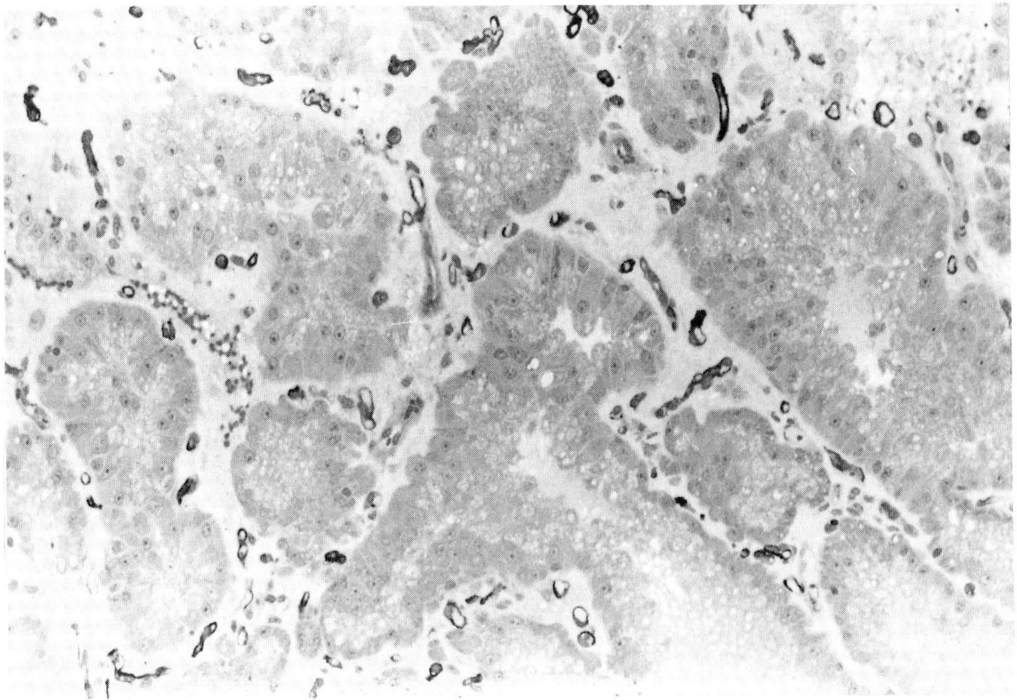

Fig. 3a. Goat mammary gland day 9 of lactation incubated 6 min for demonstration of CA activity. The capillaries show staining, but some of them are only partly or weakly stained and some are unstained. The alveoli show no staining. Magnification X 300.

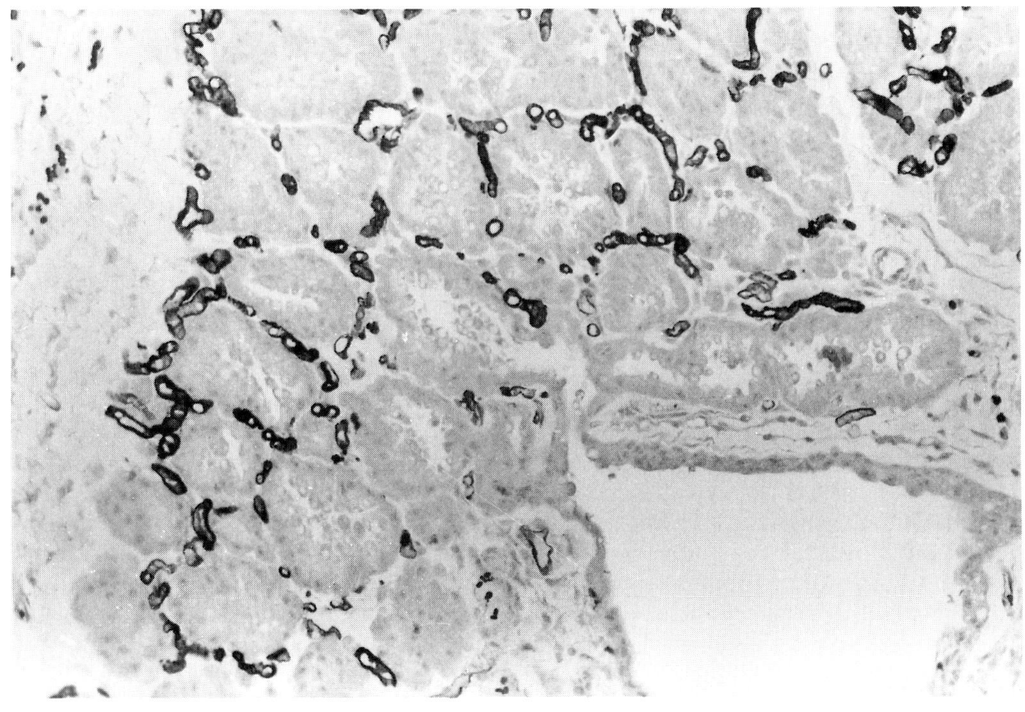

Fig. 3b. Goat mammary gland day 18 of lactation incubated 6 min for demonstration of CA activity. The capillaries show intense staining. The alveoli are unstained. Magnification X 300.

DISCUSSION

In the present study the most intense staining was associated with capillaries of the mammary tissue taken from goats at peak lactation (18 days - 3 months of lactation), whereas no staining of the capillaries could be seen during the colostrum phase and only weak staining days 5 and 9 of lactation. Since the goat reaches peak lactation about three weeks after partus, it appears that the activity of CA in the mammary capillaries is enhanced in parallel to the increase in milk yield. During the increasing milk production the alveolar cells enhance their metabolic activity and more CO_2 is produced. CA in the capillaries of the lactating mammary tissue of the goat could play an active part in facilitated CO_2 diffusion to maintain a stable intracellular pH. CA in the mammary gland probably also has other functions, because in the rat, but not in the goats, CA was found in the alveolar cells and the activity of the enzyme did not increase in parallel with milk production (see Introduction). It is known that CA is involved in intermediary metabolism. In kidney and liver cells CA supplies substrate for gluconeogenesis (Dodgson & Contino, 1988; Dodgson & Forster, 1986), and it has a role in fatty acid synthesis in the reptilian liver cells (Herbert & Coulson, 1983). Therefore, it is not unlikely that CA could be involved in milk fat synthesis. The fatty acid synthesis of milk fat and the precursors of the fatty acids differ between ruminant and non ruminant species. If CA is involved in milk fat synthesis, this could be a reason for the differing CA activity and localization in the mammary gland between rats and goats which were found. The mechanism that triggers the onset of CA activity in the mammary gland is unknown.

ACKNOWLEDGEMENTS

This study was supported by the Medical Research Council (project nr 3392) and the Swedish Council for Forestry and Agricultural Reseach (project nr 984.0809/92).

REFERENCES

Brown, D.W.C. & Baily, G. (1963): Carbonic anhydrase activity of mammary tissues. *Endocrinology* 72:662-663.

Dodgson, S.J. & Contino, L.C. (1988): Rat kidney mitochondrial carbonic anhydrase., *Arch. Biochem. Biophys.* 260: 334-341.

Dodgson, S.J. & Forster II, R.E. (1986): Inhibition of CA V decreases glucose synthesis from pyruvate. *Arch. Biochem. Biophys.* 251: 198-204.

Herbert, J.D. & Coulson, R.A. (1983): A role for carbonic anhydrase in *de novo* fatty acid synthesis in liver. *Ann. N.Y. Acad. Sci.* 429: 525-527.

Ridderstråle, Y. (1976): Intracellular localization of carbonic anhydrase in the frog nephron. *Acta Physiol. Scand.* 98:465-469.

Ridderstråle, Y. (1991): Localization of carbonic anhydrase by chemical reactions. In *The carbonic anhydrases: cellular physiology and molecular genetics,* eds. S.J. Dodgson, R.E. Tashian, G. Gros and Carter, N., pp.133. New York: Plenum Press.

Ridderstråle, Y & Knutsson, P.G. (1980): Distribution of carbonic anhydrase in the mammary gland. In *Proc. 28th Int. Congr. Physiol. Sci.* p. 663. Budapest: Hungarian Physiological Society.

Integrative and cellular aspects of autonomic functions : temperature and osmoregulation. Eds K. Pleschka, R. Gerstberger. John Libbey Eurotext, Paris © 1994, pp. 533-542.

Modulation of avian renal salt and water elimination by arginine vasotocin, mesotocin and ADH receptor subtype-specific analogues

Rüdiger Gerstberger, Roswitha Bender, Mirek Jurzak, Ralf Keil, Irene Küchenmeister and Helmuth Schütz*

*Max-Planck-Institute für Physiologische und Klinische Forschung, W.G. Kerckhoff-Institut, Parkstrasse 1, D-61231 Bad Nauheim, and *Bayer AG, Institut für Herz-Kreislauf und Artheriosklerose Forschung, P.O. Box 101709, D-5600 Wuppertal, Germany*

SUMMARY

The characteristics of renal ADH receptors in the avian kidney have been elucidated employing physiological studies in conscious, unrestrained Pekin ducks *in vivo*, and receptor binding techniques using an enriched plasma membrane preparation of the avian kidney *in vitro*. Significant down-regulation of the renal ADH receptor could be shown by the decrease in receptor density at unaltered affinity during acute or chronic extracellular fluid volume depletion. AVT, AVP, mesotocin and various V_1- and V_2-receptor subtype-specific AVP analogues were either infused systemically under conditions of steady-state diuresis *in vivo*, or tested in membrane binding experiments for their potency to displace [H]AVP from putative AVT binding sites *in vitro*. AVT and AVP dose-dependently led to a decrease in renal water elimination and osmolal excretion despite an elevated urine osmolality with 3-10 fold higher potency of AVT as compared to AVP. *In vitro*, both peptides competed for the tritiated radioligand with comparable affinity (K_D = 2-3 nM). Mesotocin revealed a 400-fold lower affinity in receptor binding studies, while inducing a dose-dependent natriuresis and delayed diuresis in the conscious animal, indicative of an AVT-independent mesotocin-specific receptor system being present in the duck kidney. High concentrations of the V_2-receptor subtype-specific agonist dDAVP reduced urine flow to half-maximal values at markedly enhanced urinary sodium concentration, yielding an overall unchanged level of osmolal excretion, although only weak competition with [^{3}H]AVP binding to kidney membranes was observed. V_2-receptor subtype-specific antagonists, however, did neither inhibit the AVT-induced antidiuresis and antinatriuresis even when applied at high concentrations, nor exhibit competitive binding potency to displace [^{3}H]]AVP. Similar results were obtained for V_1- and oxytocin-specific receptor antagonists. Both physiological and receptor binding approaches therefore suggest the existence of yet unidentified ADH receptor-subtypes in the avian kidney, which differ markedly from mammalian ADH-receptive systems.

INTRODUCTION

Regulatory functions of the hypothalamo-neurohypophyseal antidiuretic hormone (ADH) with regard to renal salt and water elimination have been described for all classes of vertebrates (Braun & Dantzler, 1974; Dantzler, 1980; Pang et al., 1982; Cree, 1988; Dantzler, 1989), with the most detailed knowledge available concerning actions of mammalian-specific Arg^8-vasopressin (AVP). In additon to its well known modulation of water permeability in the collecting duct of the mammalian kidney, AVP has been proposed to influence glomerular filtration, renal medullary blood flow or sodium transport in the ascending loop of Henle (Bankir et al., 1989; Kirk & Schafer, 1992). The characterization of binding sites for radiolabelled AVP and AVP analogues revealed the presence of two major AVP receptor subtypes in the rat kidney (Tribollet et al., 1988; Gerstberger & Fahrenholz, 1989). Antidiuretic and (anti-)natriuretic AVP effects in the kidney appear therefore mediated by V_{1a}- (mesangium, cortical thick ascending limb) and V_2- (collecting duct) specific AVP receptors (Jard et al., 1987; Leite & Suki, 1990; Rose et al., 1991; Ammar et al., 1992). The classification of the $V_{1a/b}$- and V_2-receptor subtypes in mammals is based on different ligand specificity and second messenger formation. Whereas $V_{1a/b}$-subtype mediated AVP actions are due to phosphoinositol pathway activation with a subsequent rise in intracellular Ca^{2+}-concentration (Jard et al., 1987; Nitschke et al., 1991), G-protein mediated stimulation of adenylate cyclase occurs *via* V_2-subtype specific interaction (Caltabiano & Kinter, 1991; Rose et al., 1991). Recently the molecu-lar cloning of the renal V_2-subtype of mammalian AVP receptors has been reported (Birnbaumer et al., 1992; Lolait et al., 1992).

ADH-induced renal actions in non-mammalian vertebrates have been tested in amphibian and avian species using the endogenous hormone Arg^8-vasotocin (AVT) (Dantzler, 1989). In both anuran and urodele amphibians, AVT induced antidiuresis via action on the glomerular vasculature (Pang et al., 1982; Cree, 1988; Boyd & Moore, 1990). Autoradiographical studies revealed [^{3}H]AVP-specific binding sites exclusively within the glomerular compartment (Boyd & Moore, 1990; Kloas & Hanke, 1992), and V_1- and V_2-receptor subtype-specific ADH analogues (according to their classification in mammals) revealed only low potencies to displace [^{3}H]AVP binding (Kloas & Hanke, 1992). This suggested the existence of a neurohypophyseal nonapeptide receptor in the frog renal glomeruli different from mammalian ADH receptor subtypes. As far as avian species are concerned, systemically administered AVT reduced renal water elimination in the Pekin duck, Kelp gull and the domestic fowl (Gerstberger et al., 1985; Stallone & Braun, 1985; Gray & Erasmus, 1988). Tracer dilution techniques indicated both glomerular and tubular mechanisms mediating the antidiuretic AVT activity. Taking the complex bird-specific renal cytoarchitecture of both looped and loopless nephrons into account (Morild et al., 1985; Wideman, 1988), single-nephron glomerular filtration rates as determined in the starling and quail support physiological targeting of renal glomeruli belonging to loopless nephrons by AVT (Braun & Dantzler, 1974; Braun, 1991). ADH receptor subtypes in birds responsible for the AVT-induced antidiuresis at reduced glomerular filtration and the mild antinatriuresis, however, have not yet been characterized. The kidney of the Pekin duck was therefore chosen to elucidate the avian ADH receptive system.

LOCALIZATION AND CHARACTERIZATION OF THE AVIAN RENAL ADH RECEPTOR

Antidiuresis, antinatriuresis despite augmented urinary sodium concentration, reductions in glomerular filtration rate and renal plasma flow could be observed in the domestic fowl, gull and duck after systemic infusion of AVT in physiologically relevant concentrations, thus indicating both glomerular and tubular actions of the neurohormone (Gerstberger et al., 1985; Stallone & Braun, 1985; Gray & Erasmus, 1988). To localize putative ADH-specific binding sites in the duck kidney, receptor autoradiography was performed using [^{3}H]AVP as radioligand (Fig. 1). Labelling was confined to the glomeruli of loopless nephrons, areas containing distal tubules surrounding the central veins and distinct patches of medullary tissue, but not proximal tubules or ureteral branches of the duck kidney. For all nephronal entities shown, specificity of the labelling was indicated by full displacement of [^{3}H]AVP using high concentrations of either unlabelled AVP or AVT. With the glomeruli exclusively stained in the amphibian kidney (Boyd & Moore, 1990; Kloas & Hanke, 1992), and both inner and outer medullary structures, possibly proximal tubules and mesangial cells labelled in the rat kidney (Jard et al., 1987; Jung & Endou, 1991; Ammar et al., 1992), the ADH-receptive system of birds might represent an evolutionary state of transition from pure glomerular to medullary tubular function.

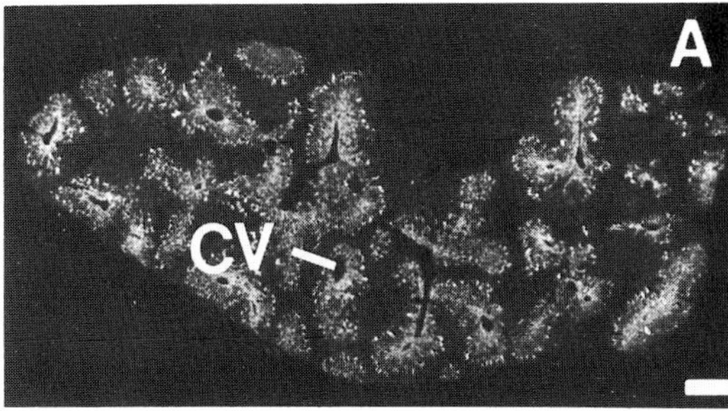

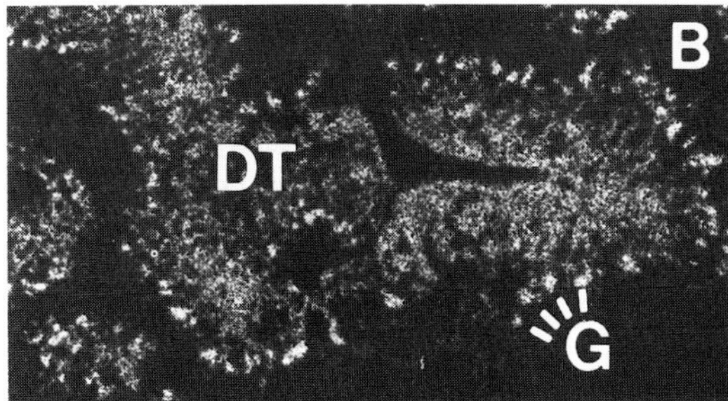

Fig. 1 : Receptor autoradiographical localization of [^{3}H]AVP specific binding sites in a cross-section of the duck kidney. Unfixed renal cryostat sections of the avian kidney were incubated in the presence of 10 nM radiolabelled AVP in 30 mM Tris/HCl buffer, pH 8.0, containing 5 mM NaCl, 200 mM sucrose, 10 mM $MgCl_2$ and 0.1 mM PMSF, for 90 min at 22°C. Control experiments for the determination of non-specific binding were performed in the presence of 1 µM unlabelled bird-specific AVT. Dry radiolabelled tissue sections were exposed to radiation-sensitive Ultrofilm. Counter-stained sections were used for histological identification. CV: central vein, DT: distal tubules, G: glomeruli of loopless nephrons.

Regulation of peptide hormone receptors during various physiological conditions with regard to either number of binding sites per cell or binding affinity for the respective hormone supports functionality of a hormone-specific "binding site". With regard to the role of ADH in body fluid homeostasis, both the plasma concentration of AVT and the ADH receptor density and affinity in the kidney were determined under conditions of euhydration, dehydration and chronic salt acclimation in the duck. Scatchard and Hill analysis of these binding data clearly indicated the existence of a single class of high affinity ADH-specific binding sites. Short- and long-term induced elevations in extracellular fluid (ECF) tonicity induced an enhanced release of neurohypophyseal AVT and a concomitant decrease in the number of [^{3}H]AVP-specific binding sites in the kidney at unaltered binding affinity, with more pronounced effects after 48 h of water withdrawal (Schütz et al., 1993). A comparable down-regulation of ADH binding sites was reported for the baso-lateral membranes of the rat kidney collecting tubules after 72 h of dehydration (Steiner & Phillips, 1988).

For detailed pharmacological characterization of the avian renal ADH receptive system, neurohypophyseal hormones such as AVT, AVP, mesotocin, oxytocin or the hypothalamic fragment [pGlu,Cyt^6]AVP$_{(4-9)}$ were tested for their potency to competitively displace [^{3}H]AVP from its renal binding sites. With AVT and mesotocin representing avian-specific, and AVP and oxytocin mammalian-specific peptides, the fragment AVP$_{(4-9)}$ could be detected in both classes of vertebrates (Burbach, 1986). AVP and AVT proved to be equipotent to compete with the radioligand at high affinity, whereas oxytocin and mesotocin revealed potencies 60 and 400 times lower, respectively (Table 1). Binding experiments performed for the glomeruli of the anuran *Xenopus laevis* exhibited comparable data for all four neuropeptides (Kloas & Hanke, 1992). Employing the tritiated V_2-specific agonist dDAVP as a suitable radioligand to characterize renal V_2-receptors in medullopapillary membranes of the rat kidney, oxytocin proved to have very low affinity for the mammalian V_2-receptor subtype (Marchingo et al., 1988). V_{1a}-vasopressin receptors in the rat cortical collecting duct and cultured mesangial cells, however, showed only 50-100 times lower affinity for oxytocin than AVP to displace non-selective ADH receptor ligands (Jard et al., 1987; Ammar et al., 1982). The AVT/AVP fragment peptide did not bind to putative ADH receptors in the duck nor rat kidney (Marchingo et al., 1988) (Table 1). Employing numerous AVP receptor subtype-specific antagonists as well as the V_2-specific agonist dDAVP in duck kidney binding assays, none of the ligands was able to displace [^{3}H]AVP from its binding sites, with IC_{50} values at least 2,000 times higher than for AVP or AVT. These data were again consistent with those obtained for the *Xenopus* kidney (Kloas & Hanke, 1992), but different from the situation in urodele amphibians (Boyd & Moore, 1990) and the mammalian kidneys (Fahrenholz et al., 1984; Rose et al., 1991). The avian kidney ADH receptor therefore appears to exhibit ligand binding characteristics quite different from those reported for various mammalian species.

(ANTI-)DIURESIS INDUCED BY AVT, AVP AND MESOTOCIN

Physiological experiments were performed in conscious ducks (n=24) to evaluate the (anti-)diuretic and (anti-)natriuretic actions of AVT, AVP and mesotocin (Gerstberger et al., 1985). Ducks suspended in a cotton sling allowing free movement of the neck and the legs received a continuous intravenous infusion of 0.7 ml/min of 200 mOsm/kg D-glucose and 0.3 ml/min of 200 mOsm/kg NaCl solutions. During the subsequently established steady-state diuresis, renal clearances of water, sodium and chloride amounted to 95, 65 and 58 percent, respectively, at stable plasma concentrations of AVT, angiotensin II and atrial natriuretic factor as important osmoregulatory hormones.

Table 1 : Competitive displacement of 2 nM [^{3}H]AVP from ADH-specific binding sites in an enriched membrane fraction of the duck kidney using various endogenous AVP analogues, AVP/AVT fragment peptides and V_1- or V_2-AVP receptor subtype-specific ligands. The potencies of the various peptides to displace the radioligand are presented as the IC_{50} values (mean of 3-10 experiments) in nanomolar concentration.

	IC_{50} (nM)	(n)
ADH-like hormones		
[Arg^8]vasotocin	2.9	(10)
[Arg^8]vasopressin	2.7	(9)
[Ile^8]oxytocin (=Mesotocin)	1100	(7)
[Leu^8]oxytocin	156	(3)
AVP/AVT fragment		
[pGlu,Cyt^6]AVP(4-9)	> 10000	(3)
V_1-specific antagonists		
[d(CH$_2$)$_5$,Sar^7,Arg^8]VP	> 10000	(3)
[d(CH$_2$)$_5$,Tyr^2(Me),Arg^8]VP	5270	(7)
V_2-specific agonist		
[Deamino,D-Arg^8]AVP (dDAVP)	5920	(9)
V_2-specific antagonists		
[d(CH$_2$)$_5$,D-Phe^2,Ile^4,Ala^9-NH$_2$]AVP	> 10000	(5)
[Adamantaneacetyl^1,D-Tyr^2(Et),Val^4, Abu^6, Arg^{8,9}]VP	9900	(7)
Oxytocin/AVP antagonist		
[d(CH$_2$)$_5$,Tyr^2(Me),Thr^4,Orn^8, Tyr^9-NH$_2$]AVT	8500	(3)

Bird-specific AVT infused systemically for 10 min during steady-state diuresis in concentrations of 0.1, 0.5, 1.0 and 5.0 ng/min/kg b.w. resulted in plasma AVT concentrations of 5.5±0.6, 11.8±2.1, 18.5±3.1 and 58.0±8.8 pg/ml, respectively, as compared to basal values of 4.3±0.5 pg/ml (determined by specific RIA according to Gray & Simon, 1983). Urine flow rate was dose-dependently reduced at concomitantly elevated urine osmolality due to enhanced concentrations of sodium and chloride mainly (Fig. 2). Both absolute (Fig. 2) and time-integrated (Fig. 3) values of antidiuresis revealed saturation kinetics for hormone action, with 1.0 ng/min/kg b.w. representing the AVT concentration almost causing anuria. With regard to osmolal excretion, antinatriuresis proved to be maximal already at 0.5 ng/min/kg b.w. concerning both absolute and time-integrated values (Fig. 3). Mammalian-specific AVP infused at 0.5, 1.0 and 5.0 ng/min/kg b.w. induced comparable effects with regard to renal water and sodium elimination at, however, reduced activity (Fig. 2). AVP at 5 ng/min/kg b.w. exhibited renal antidiuresis equipotent to 0.5 ng/min/kg b.w. AVT at even weeker antinatriuretic activity (Fig. 3). The high affinity of both ligands to bind to duck renal ADH receptors (Table 1), and their similar activity pattern to induce antidiuresis and -natriuresis strongly suggest that AVT and AVP bind to identical receptive structures in the avian kidney, with AVP being a weaker agonist in the system (Schütz et al., 1993).

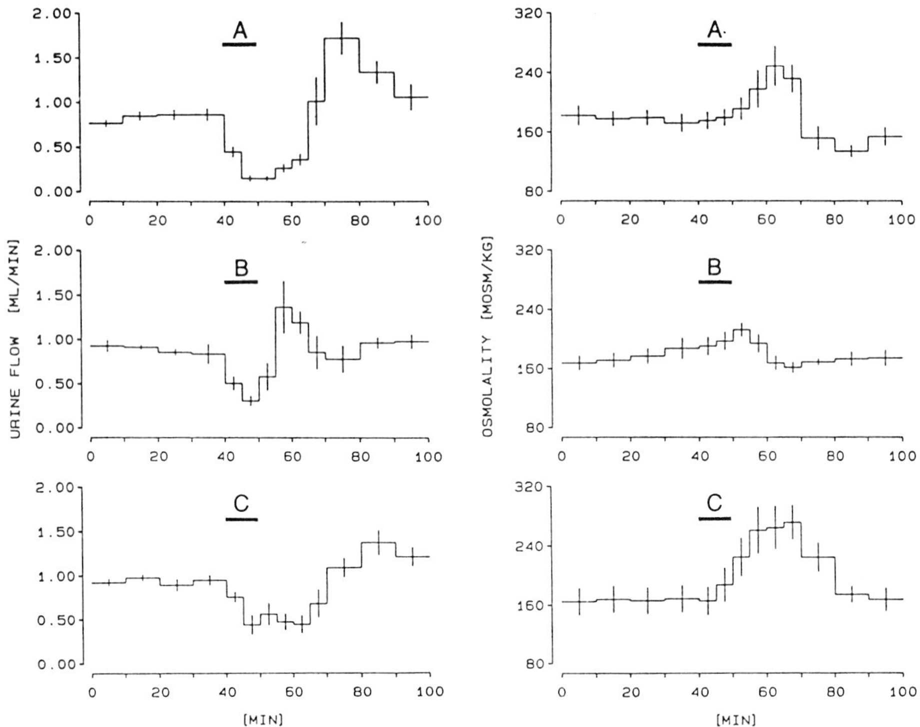

Fig. 2 : Antidiuretic response patterns of freshwater-adapted ducks during steady-state diuresis, driven by continuous systemic infusion of 0.7 ml/min of 200 mOsm/kg D-glucose solution and 0.3 ml/min of 200 mOsm/kg NaCl solution, to the intravenous application (10 min) of **(A)** 0.5 ng/min/kg b.w. AVT (n=10), **(B)** 0.5 ng/min/kg bw AVP (n=6) and **(C)** 10 ng/min/kg b.w. dDAVP (n=7). Urine drained from a perforated bulb inserted into the cloaca was continuously aspirated into graded cylinders and collected at 5-10 min intervals. Urine osmolality was determined by vapour pressure osmometry.

Labelling of other neurohypohyseal peptide receptors in the duck kidney using [^{3}H]AVP as radioligand appear to be ruled out by the very low affinity of mesotocin in the competitive displacement studies. Administered systemically to conscious diuretic ducks in concentrations of 1, 10 and 100 ng/min/kg b.w., mesotocin led to a dose-dependent increase in urine flow rate with remarkably delayed onset and long duration of action, quite in opposite to AVT-induced antidiuresis (Fig. 4). In addition, osmolality, sodium and chloride concentrations of the urine were augmented immediately after beginning of peptide application and decreased to basal values at the time of diuresis onset, indicating direct tubular natriuretic actions of mesotocin in the duck. The physiological pattern of renal mesotocin actions therefore appeared to be different from AVT/AVP effects with marked diuretic and natriuretic as compared to antidiuretic and antinatriuretic responses (Fig. 4). Applying graded doses of oxytocin, the mammalian mesotocin analogue, systemically to conscious rats, renal sodium excretion proved to be linearily correlated to plasma oxytocin concentrations within physiological limits. In the presence of an oxytocin-, but not a V_1-specific receptor antagonist, natriuresis was inhibited (Verbalis et al., 1991).

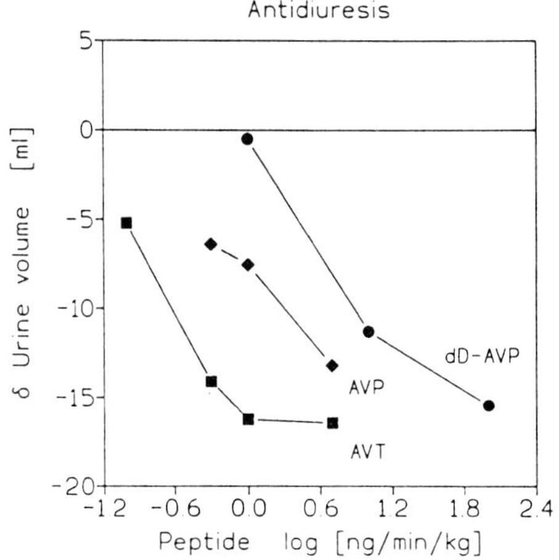

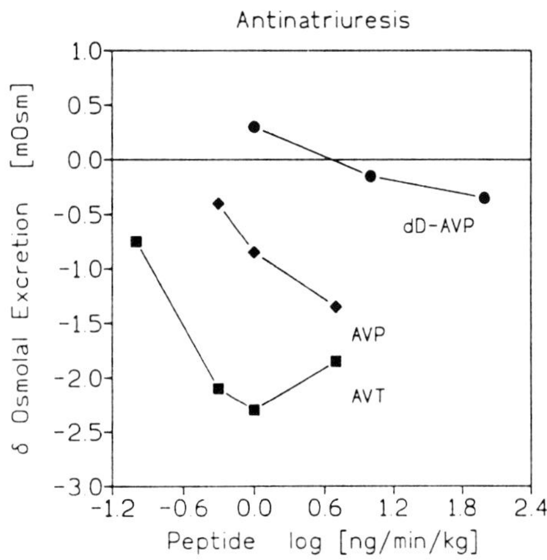

Fig. 3 : Time-integrated, dose-dependent antidiuretic and antinatriuretic effects of AVT, AVP and dDAVP applied systemically (10 min) to freshwater ducks during steady-state diuresis (see Fig. 2). AVT was administered at 0.1/0.5/1/5 ng/min/kg b.w., AVP at 0.5/1/5 ng/min/kg b.w. and dD-AVP at 1/10/100 ng/min/kg b.w. (n=4-10; mean values presented). Both responses were calculated as the amount of water and sodium retained over time due to hormone action.

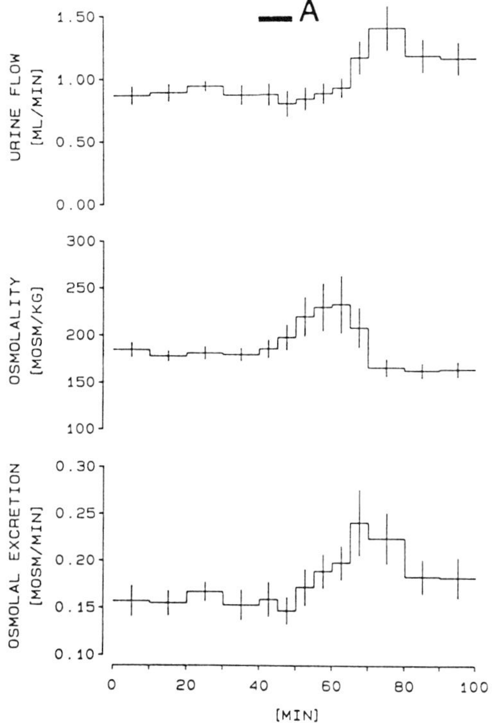

Fig. 4 : Effect of systemically applied mesotocin on duck renal function.
Renal diuretic and natriuretic response pattern of freshwater-adpated ducks during steady-state diuresis, driven by continuous systemic infusion of 0.7 ml/min of 200 mOsm/kg D-glucose solution and 0.3 ml/min of 200 mOsm/kg NaCl solution, to the intravenous application (10 min) of 10 ng/min/kg b.w. mesotocin (**A**). Urine flow, urine osmolality and kidney osmolal excretion are recorded. Data are presented as mean values ± SEM (n=10).

EVALUATION OF PUTATIVE CLASSICAL ADH RECEPTOR SUBTYPES IN THE AVIAN KIDNEY

The physiological role of ADH receptors of both the V_{1a}- and V_2-subtype with regard to renal hemodynamics and water elimination in the mammalian kidney became apparent only recently (Rose et al., 1991). Vasopressin V_2-receptors are located along the entire cortical collecting duct, whereas V_1-receptors could be described for cortical collecting duct cells and glomerular mesangial cells (Kirk & Schafer, 1992), inducing the release of calcium from intracellular stores upon ADH/receptor interaction (Burnatowska-Hledin & Spielman, 1989). As demonstrated, receptor binding studies in the duck kidney have not revealed the presence of either of these subtypes employing various V_1- and V_2-specific receptor antagonists and the V_2-specific agonist dDAVP. Binding of [^{3}H]AVP to the renal membranes was reduced in the presence of non-hydrolyzable GTP analogues suggestive of a G-protein coupled receptor. AVT did, however, not induce the formation of cyclic AMP as second messenger in preparations otherwise responsive to forskolin, GTP and parathyroid hormone (Schütz et al., 1993), as indicated already 20 years ago (Dousa, 1974). To further elucidate the participation of classical ADH receptors of the V_2-subtype in ADH-induced antidiuresis, the V_2-specific agonist dDAVP was administered systemically to conscious diuretic ducks in concentrations of 1, 10 and 100 ng/min/kg bw. Only at high concentrations of 10 and 100 ng/min/kg b.w., dDAVP induced a marked increase in urinary osmolality and sodium concentration (Fig. 2). Urine flow rate was reduced to half-maximal values of 0.4-0.5 ml/min as compared to 0.9 ml/min during control steady-state

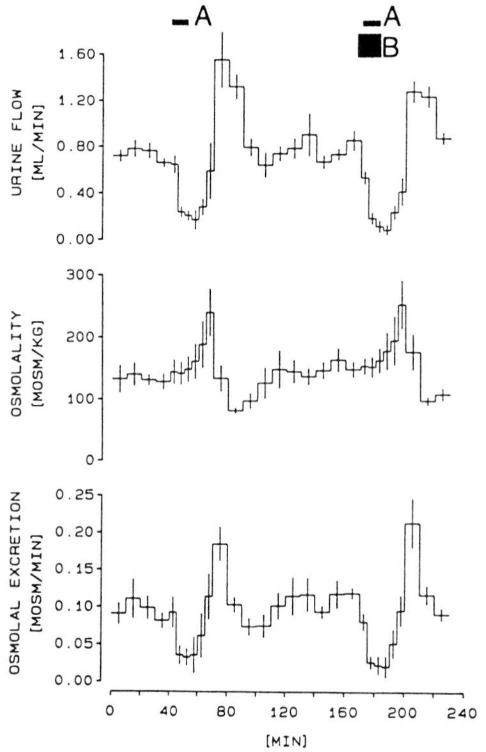

Fig. 5 : Antinatriuretic and antidiuretic response pattern to the systemic application (10 min) of (A) 1 ng/min/kg b.w. AVT to freshwater-adpated ducks during steady-state diuresis driven by continuous systemic infusion of 0.7 ml/min of 200 mOsm/kg D-glucose solution and 0.3 ml/min of 200 mOsm/kg NaCl solution. The antidiuretic hormone was infused alone (left) or in the presence of the V_1-specific antagonist [d(CH$_2$)$_5$, Sar^7,Arg^8]VP or the V_2-specific antagonists [d(CH$_2$)$_5$,D-Phe^2,Ile^4, Ala^9-NH$_2$]AVP or [Adamantaneacetyl^1,D-Tyr^2(Et),Val^4,Abu^6,Arg^{8,9}]VP (B). None of the antagonists administered for 15 min at 100 ng/min/kg b.w. could block the AVT-induced antidiuresis and antinatriuresis. Urine flow, urine osmolality and kidney osmolal excretion are recorded. Data are presented as mean values ± SEM (n=6).

diuresis. With regard to time-integrated antidiuresis and antinatriuresis, dDAVP did not alter the elimination of Na^+ and Cl^-, but led to a partial retention of water (Fig. 3). To elucidate the involvement of a V_2-like ADH receptor subtype in AVT-induced antidiuresis, the V_1-, V_2- and oxytocin-specific antagonists employed in the membrane binding studies were infused at high concentration (100 ng/min/kg b.w.) systemically prior and during the application of AVT (1 ng/min/kg b.w.). Having no agonistic potency themselves as tested beforehand, neither V_2- nor V_1-antagonists inhibited the marked antidiuresis and -natriuresis induced by the endogenous hormone AVT (Fig. 5). Despite the partial antidiuresis but not antinatriuresis caused by dDAVP when administered at high concentrations, (1) the lack of ADH receptor interaction with various V_1- and V_2-specific ligands, (2) the ineffectiveness of V_1- and V_2-specific antagonists to block renal AVT actions and (3) the failure of AVT to stimulate cyclic AMP production suggest the existence of an avian-specific ADH receptor which differs markedly from known mammalian receptor subtypes (Kirk & Schafer, 1992). Further pharmcological and molecular biological studies might give better insight into the cellular mechanisms of AVT-induced antidiuresis in birds.

REFERENCES

Ammar, A., Roseau, S. & Butlen, D. (1992): Pharmacological characterization of V_{1a} vasopressin receptors in the rat cortical collecting duct. *Am. J. Physiol.* 262: F546-F553.

Bankir, L., Bouby, N. & Trinh-Trang-Tan, M.M. (1989): The role of the kidney in the maintenance of water balance. *Baillière's Clin. Endoc. Metab.* 3, 249-311.

Birnbaumer, M., Seibold, A., Gilbert, S., Ishido, M., Barberis, C. Antaramian, A., Brabet, P. & Rosenthal, W. (1992): Molecular cloning of the receptor for human antidiuretic hormone. *Nature* 357, 333-335.

Boyd, S.K. & Moore,F.L. (1990): Autoradiographic localization of putative arginine vasotocin receptors in the kidney of a urodele amphibian. *Gen. Comp. Endocrin.* 78, 344-350.

Braun, E.J. & Dantzler W.H. (1974): Effects of ADH on single-nephron glomerular filtration rates in the avian kidney. *Am. J. Physiol.* 226, 1-8.

Braun, E.J. (1991): The role of the avian kidney in fluid and electrolyte balance. *Acta XXth Congr. Internat. Ornith.*, Symp. 38, 2130-2137.

Burbach, J.P.H. (1986): Proteolytic conversion of oxytocin, vasopressin and related peptides in the brain. In: *Neurobiology of Oxytocin*, ed. D. Ganten & D. Pfaff, pp. 55-90. Heidelberg, Springer Press.

Burnatowska-Hledin, M.A. & Spielman, W.S. (1989): Vasopressin V_1 receptors on the principal cells of the rabbit cortical collecting tubule. *J. Clin. Invest.* 83, 84-89.

Caltabiano, S. & Kinter, L.B. (1991): Up-regulation of adenylate cyclase-coupled vasopressin receptors after chronic administration of vasopressin antagonists to rats. *J. Pharm. Exp. Ther.* 258, 1046-1054.

Cree, A. (1988): Effects of arginine vasotocin on water balance of three leiopelmatid frogs. *Gen. Comp. Endocrin.* 72, 340-350.

Dantzler, W.H. (1980): Comparative nephron function in reptiles, birds, and mammals. *Am. J. Physiol.* 239, R197-R213.

Dantzler, W.H. (1989): *Comparative Physiology of the Vertebrate Kidney.* Springer, New York.

Dousa, T.P. (1974): Effects of hormones on cyclic AMP formation in kidneys of nonmammalian vertebrates. *Am. J. Physiol.* 226, 1193-1197.

Fahrenholz, F., Boer, R., Crause, P., Fritzsch, G. & Grzonka, Z. (1984): Interactions of vasopressin agonists and antagonists with membrane receptors. *Eur. J. Pharmacol.* 100, 47-58.

Gerstberger, R., Kaul, R., Gray, D.A. & Simon, E. (1985): Arginine vasotocin and glomerular filtration rate in saltwater-acclimated ducks. *Am. J. Physiol.* 248, F663-F667.

Gerstberger, R. & Fahrenholz, F. (1989): Autoradiographic localization of V_1 vasopressin binding sites in rat brain and kidney. *Eur. J. Pharmacol.* 167, 105-116.

Gray, D.A. & Simon, E. (1983): Mammalian and avian antidiuretic hormone: studies related to possible species variation in osmoregulatory systems. *J. Comp. Physiol.* 151 B, 241-246.

Gray, D.A. & Erasmus, T. (1988). Glomerular filtration changes during vasotocin-induced antidiuresis in kelp gulls. *Am. J. Physiol.* 255, R936-R939.

Jard, S., Lombard, C., Marie, J. & Devilliers, G. (1987): Vasopressin receptors from cultured mesangial cells resemble V_{1a} type. *Am. J. Physiol.* 253: F41-F49.

Jung, K.Y. and H. Endou. A novel vasopressin receptor in rat proximal tubule. *Biochem. Biophys. Res. Comm.* 180: 131-137, 1991.

Kirk, K.L. & Schafer, J.A. (1992): Water transport and osmoregulation by antidiuretic hormone in terminal nephron segments. In: *The Kidney: Physiology and Pathophysiology*, Sec. Eds., ed D.W. Seldin & G. Giebisch, pp. 1693-1725. New York, Raven Press.

Kloas, W. & Hanke, W. (1992): Localization and quantification of nonapeptide binding sites in the kidney of *Xenopus laevis*: Evidence for the existence of two different nonapeptide receptors. *Gen. Comp. Endocrin.* 85, 71-78.

Lolait, S.J., O'Carroll, A-M., McBride, O.W., Konig, M., Morel, A. & Brownstein, M. (1992): Cloning and characterization of a vasopressin V_2 receptor and possible link to nephrogenic diabetes insipidus. *Nature* 257, 336-339.

Leite, M. & Suki, W.N. (1990): AVP and dDAVP in rabbit cortical collecting tubule: a comparative time-course study. *Am. J. Physiol.* 258, R99-R103.

Marchingo, A.J., Abrahams, J.M., Woodcock, E.A., Smith, A.I., Mendelsohn, F.A.O. & Johnston, C.I. (1988): Properties of (^{3}H)-1-Desamino-8-D-arginine vasopressin as a radioligand for vasopressin V_2-receptors in rat kidney. *Endocrinology* 122, 1328-1336.

Morild, I., Bohle, A. & Christensen, J.A. (1985): Structure of the avian kidney. *The Anatom. Rec.* 212, 33-40.

Nitschke, R., Fröbe, U. & Greger, R. (1991): Antidiuretic hormone acts via V_1 receptors on intracellular calcium in the isolated perfused rabbit cortical thick ascending limb. *Eur. J. Physiol.* 417, 622-632.

Pang, P.K.T., Uchiyama, M. & Sawyer, W.H. (1982): Endocrine and neural control of amphibian renal functions. *Fed. Proc.* 41, 2365-2373..

Rose, C.E., Rose, K.Y. & Kinter, L.B. (1991): Effect of V_1/V_2-receptor antagonism on renal function and response to vasopressin in conscious dogs. *Am. J. Physiol.* 260, F273-F282.

Schütz, H., Fuchs, D., Jungbluth, D., Keil, R., Pena Salazar, P., Jurzak, M. & Gerstberger, R. (1993): Characterization and regulation of renal [^{3}H]AVP binding in the duck. *Am. J. Physiol.* (submitted).

Stallone, J. & Braun, E.J. (1985): Contributions of glomerular and tubular mechanisms to antidiuresis in conscious domestic fowl. *Am. J. Physiol.* 249, F842-F850.

Steiner, M. & Phillips, M.I. (1988): Renal tubular vasopressin receptors downregulated by dehydration. *Am. J. Physiol.* 254, C404-C410.

Tribollet, E., Barberis, C., Dreifuss, J.J. & Jard, S. (1988): Autoradiographic localization of vasopressin and oxytocin binding sites in the rat kidney. *Kidney Int.* 33,959-965.

Verbalis, J.G., Mangione, M.P. & Stricker, E.M. (1991): Oxytocin produces natriuresis in rats at physiological plasma concentrations. *Endocrinology* 128, 1317-1322.

Wideman, R.F. (1988): Avian kidney anatomy and physiology. *CRC Crit. Rev. Poultry Biol.* 1, 133-176.

Integrative and cellular aspects of autonomic functions : temperature and osmoregulation. Eds K. Pleschka, R. Gerstberger. John Libbey Eurotext, Paris © 1994, pp. 543-551.

Enhanced AVP release during recovery from arterial hypertension induced by IV infusion of NE in conscious dogs

Maria Szczypaczewska*, Eckhart Simon, Christa Simon-Oppermann, David A. Gray and Daniela Jungbluth

Max-Planck-Institute für Physiologische und Klinische Forschung, W.G. Kerckhoff-Institut, D-6350 Bad Nauheim, Germany
**On leave from the Department of Clinical and Applied Physiology, Medical Academy of Warsaw, 00-790 Warsaw, Poland*

ABSTRACT
The observation of constant levels of plasma arginine vasopressin (AVP) during hypertensive infusions of norepinephrine (NE), however, with a strong post-infusion rise of AVP when the NE infusion was stopped, was analyzed in conscious chronically instrumented dogs by applying NE infusions of different durations. The degree of post-infusion rise in plasma AVP increased with increasing duration of the NE infusions. Measurements of arterial mean pressure (MAP) and heart rate (HR) showed that the early post-infusion decreases in MAP and reflex increases of HR became more pronounced with increasing duration of the NE infusion. On the other hand, the post-infusion release of AVP seen after 30 min of NE infusion disappeared completely when the rate of NE application was gradually reduced and, as a result, the post-infusion readjustments of MAP and HR occurred more slowly. The results of this study are compatible with the hypothesis that systemically infused NE exerts a strong stimulatory effect on AVP release which is counteracted by baroreceptor reflex activation caused by the rise in MAP. The activating NE effect on AVP release persists long enough after the end of NE infusion to become disinhibited by the rapid post-infusion decrease in MAP and/or rise in HR. On the other hand, this effect disappears, if the cardiovascular readjustments are retarded by gradual reduction of the NE application.

INTRODUCTION

In mammals the hypothalamo-neurohypophysial peptide hormone arginine vasopressin (AVP) is an essential regulator of renal water excretion. In addition, it is a hormone relevant in cardiovascular control because of its peripheral vasoconstrictor action and its modulatory influence on the baroreflex system (Cowley & Barber, 1982). Arterial blood pressure and the state of cardiovascular filling, in turn, influence AVP release. Arterial hypotension is a strong stimulator, probably because of the diminishing inhibitory influence of the baroreceptor input (Robertson, 1977). Less obvious are inhibitory actions of the baroreceptor input on AVP release in the normal to hypertensive blood pressure range. The low normal AVP levels make it difficult to demonstrate further inhibition by elevating blood pressure. However, such an inhibition might be responsible for the weak or conflicting effects of putative stimulators of AVP release, like angiotensin II (ANGII) and noradrenaline (NE), because of their hypertensive action. Indeed, these hormones were shown to stimulate AVP release strongly, when the accompanying arterial hypertension was approximately balanced by simultaneous sodium nitroprusside infusion (Szczypaczewska et al., 1993).

In these studies an unexpected large rise in plasma AVP was seen early after the end of NE infusions when arterial hypertension and reflex inhibition of heart rate subsided rapidly. The observation suggested that this transient effect might be due to the interference of the decaying stimulatory effect of NE with the rapid release of baroreceptor inhibition in the control of AVP release. In the present study the dynamics of this post-infusion response of AVP to NE were investigated in more detail.

METHODS

Experimental animals. The experiments were carried out on 6 dogs of the beagle strain which had been accustomed to the laboratory environment to tolerate repeated experimental sessions of up to 5 h duration without any signs of stress. Animals had a subcutaneously dislocated common carotid artery, as described previously (Szczepanska-Sadowska et al., 1992). At minimum intervals of 1 week catheters (1.1 mm o.d.) filled with heparinized saline (200 IU/ml), were placed into both the carotid artery and external jugular vein under sterile precautions and during short-lasting anesthesia induced in the afore-mentioned way. Experiments were carried out in the conscious animals one and/or two days after catheter insertion and the catheters were subsequently removed.

Course of the experiment. On the morning of the experimental day the animal was transferred to the climatic chamber in which the experiments were carried out at constant, thermoneutral ambient conditions (25-27°C). The dog was placed in a canvas sling which permitted comfortable standing or resting. The arterial catheter was connected to a Statham pressure transducer. A flexible, medical-grade cannula (Viggo G18, LIC, Sweden) which caused no sustained discomfort was inserted into a leg vein (iv) after local superficial anesthesia (xylocaine spray, Lidocain, Hoechst, FRG). The iv cannula was connected to an infusion line to apply 1.0 ml/min of either sterile saline alone as the control infusion or, as test infusions, saline containing NE to apply a dose of 1.6 µg/min/kg. Arterial and jugular vein catheters were kept patent by slow counterinfusions of 0.1 ml/min heparinized (200 IU/ml) saline. For control purposes, plasma osmolality was determined with a vapor pressure analyzer (Wescor 5100-C, USA) and core temperature was measured in the rectum by means of a thermocouple.

Cardiovascular recordings. The blood pressure signal was continuously monitored and evaluated on line with a transducer-microcomputer system (Brush-Gould, USA) to obtain mean arterial pressure (MAP), arterial pulse pressure (APP) and heart rate (HR). Cardiovascular parameters were displayed on the monitor and stored on the hard-disk of the micro-computer.

Hormone measurements. Freely flowing blood samples were taken from the arterial catheter to ensure rapid collection of about 3 ml blood in polypropylen tubes kept on ice, the blood volume loss being replaced by injecting the same volume of saline. In part of the experiments parallel collections were made from the jugular vein. Plasma was separated from erythrocytes by centrifugation (2000 g; +2°C, 10 min), and duplicates of plasma aliquots of 1 ml were stored at -18°C until assayed. AVP and ANG II were measured by radioimmunoassays (RIA) with previously described methods (Gray & Simon, 1983; Simon-Oppermann et al., 1986). For control measurements of circulating NE by means of HPLC with an electrochemical detector (Waters, USA), plasma samples were repeatedly collected during standard NE infusions at 1.6 µg/min/kg.

Statistics. Average values are presented as means with standard errors of means (SE). Changes in AVP plasma concentrations and of MAP and HR determined during and after periods of NE infusions were tested for time-dependent changes and for differences between experimental series by one- or multi-factor ANOVA and post-hoc multiple range testing (Scheffé; $P<0.05$).

RESULTS

In an exploratory series of 7 experiments on 4 dogs in which the rates of NE infusion had been increased in 4 successive 30 min periods (0.4, 0.8, 1.6 and 3.2 µg/min/kg), MAP rose in a parallel fashion from 103 ± 2 mm Hg prior to infusion to a final level of 145 ± 4 mm Hg. Plasma ANG II concentration also rose successively from its pre-infusion level of 12.2 ± 1.3 pg/ml to 163.4 ± 40.8 pg/ml at the end of the last infusion period with 3.2 µg/min/kg NE. This effect was clearly dose dependent ($F_{5,22}$ P<0.01). The dose of 1.6 µg/min/kg NE was chosen as the standard stimulus and produced degrees of arterial hypertension similar to those induced previously in dogs with 2 µg/min/kg NE (Casals-Stenzel et al. 1982).

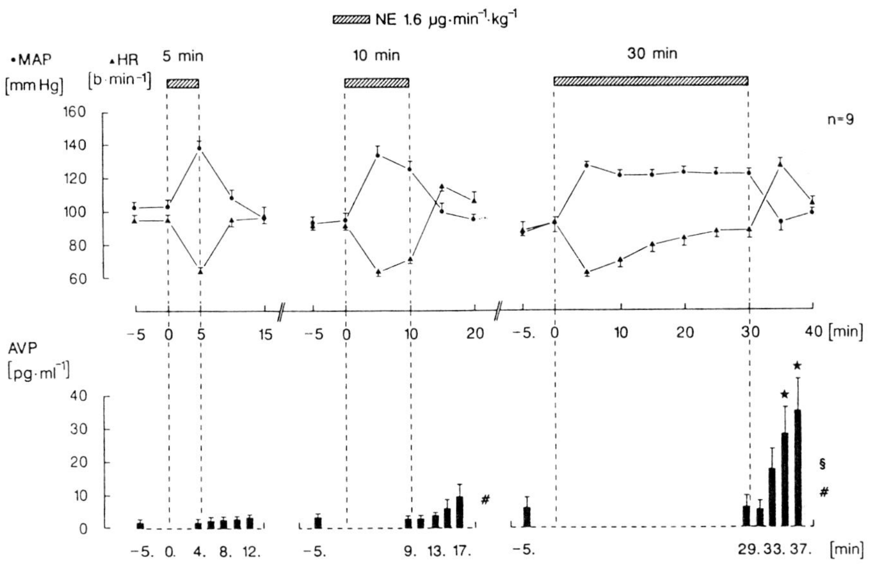

Fig. 1: Cardiovascular parameters and AVP plasma concentrations during and after NE infusions of different durations; 5 min (left), 10 min (middle) and 30 min (right). Upper diagram: courses of mean arterial pressure (MAP) and heart rate (HR) determined at 5 min intervals. Lower diagram: AVP plasma concentrations measured before, at the end and 1, 3, 5, and 7 min after the end of NE infusions. Means with SE, 9 experiments. #: significant post-infusion rise in plasma AVP, §: post-infusion response significantly larger after 30 as compared to 10 min of NE infusion, ANOVA (details see text); * significantly larger than value before end of infusion (Scheffé test).

The lower diagrams of Fig. 1 show the average infusion and early post-infusion effects of NE on plasma AVP concentrations, as determined after periods of 5, 10 and 30 min of NE infusion, with AVP measurements made before, 1 min before the end and 1, 3, 5, and 7 min after the end of infusions. At each duration of NE infusions, there were no significant increases in plasma AVP 1 min prior to the end of infusion. The time courses of the AVP concentrations from the 1th min before to the 7th min after the end of infusion were tested for significant changes by analysis of variance for each NE infusion series. For the 5 min series no post-infusion change in plasma AVP was found ($F_{4,42}$=2.059; P=0.11). After 10 min of NE infusion the tendency for plasma AVP to increase during the post-infusion period was significant ($F_{4,32}$=2.812; P<0.05). This was found also after 30 min of NE infusion ($F_{4,30}$=10.166; P<0.0001); specifically, the AVP levels determined 5 and 7 min after the end of NE infusion had risen significantly (Scheffé test; P<0.05). Comparison of the different durations of infusion periods by analysis of variance supported the conclusion that the post-infusion AVP response increased with the duration of the NE infusion ($F_{4,118}$=7.121; P>0.0001), the average rise after 30 min being significantly larger than after 10 min (Scheffé-test; P<0.05).

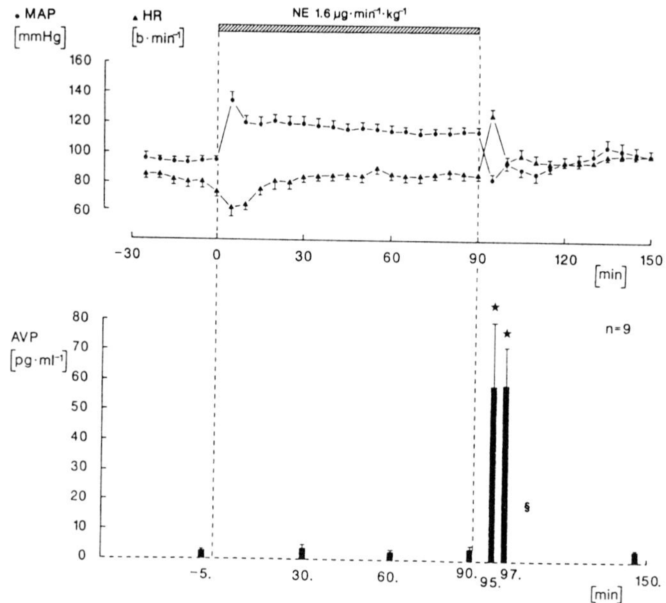

Fig. 2: Cardiovascular parameters and AVP plasma concentrations during and after the end of NE infusions of 90 min duration. Upper diagram: mean arterial pressure (MAP) and heart rate (HR) determined at 5 min intervals. Lower diagram: AVP plasma concentrations measured before, during, at the end, and 5 and 7 min after the end of NE infusions.
Means with SE, 9 experiments. §: post-infusion response at the 5th and 7th min significantly larger than at the corresponding times after 30 min of NE infusion (Fig. 1), ANOVA (details see text); * significantly larger than value before end of infusion.

In order to ascertain the observed relationship between the duration of the NE infusion periods and post-infusion response of AVP, a second series of 9 experiments with 90 min of NE infusion were carried out and plasma AVP was measured before the end and 5 and 7 min after the end of infusion. As shown by Fig. 2, there was no significant change in plasma AVP concentration during the 30th, 60th and 90th min of NE infusion. However, AVP concentrations determined 5 and 7 min after the end of 90 min of NE infusion rose to higher values than those found 5 and 7 min after the end of the 30 min infusions, as confirmed by analysis of variance ($F_{1,31}$=4.719; P<0.05).

The upper diagrams of Fig. 1 present the general course of MAP and HR determined in 9 experiments for the NE infusions of 5, 10 and 30 min duration. The curves show that the initial rises of MAP in response to NE tended to subside with increasing duration of the NE infusions. Similarly, the reflex inhibition of HR was most pronounced early after the start of infusion and progressively subsided with increasing duration of the NE infusion. After the end of NE infusions, the level to which MAP decreased became lower and the post-infusion rise of HR became higher with increasing duration of NE infusion.

The upper diagram of Fig. 2 shows for the NE infusion of 90 min duration the same tendency for MAP to decline and of HR to rise gradually in the course of the NE infusion. Further, pronounced post-infusion decreases of MAP and increases in HR occurred after the end of NE infusion.

Table 1, A - C, presents the time courses of MAP at each minute from 1 min before the end to 5 min after the end of NE infusions of 5, 10 and 30 min duration. MAP levels before the end of NE infusions became lower with increasing duration of the infusion and decreased to lower levels after the end of NE infusion. This tendency was found to be significant, when the data of table 1 obtained from the series of 5, 10 and 30 min of NE infusion were compared by analysis of variance ($F_{2,122}$=25.731; P<0.0001). Part D of table 1 presents the course of MAP from 1 min before to 5 min after the end of NE infusion of 90 min duration. Analysis of variance confirmed ($F_{1,93}$=6.309; P<0.02) that the MAP levels during this post-infusion phase were significantly lower than during the post-infusion phase after 30 min of NE infusion (part C of table 1).

Table 1: Time course of mean arterial pressure from 1 min before to 5 min after the end of NE infusions of different durations.

		1 min before end of NE infusion	after the end of NE infusion				
			1 min	2 min	3 min	4 min	5 min
Part A	Mean	135.7	113.0	107.8	99.9	112.2	97.9
5 min of NE	SE	± 5.0	±10.1	± 6.1	± 4.4	± 4.1	± 4.8
infusion	n	9	5	8	7	9	8
Part B	Mean	120.3	104.1	101.7	104.0	97.3	91.4
10 min of NE	SE	± 5.7	± 5.5	± 3.9	± 6.5	± 6.3	± 7.0
infusion	n	9	7	9	6	7	7
Part C	Mean	119.4	95.5	91.3	90.1	89.6	88.8
30 min of NE	SE	± 3.4	± 6.4	± 6.0	± 4.9	± 7.4	± 5.7
infusion	n	9	8	8	9	8	6
Part D	Mean	116.1	93.8	85.8	80.6	76.9	80.0
90 min of NE	SE	± 3.8	± 5.9	± 3.7	± 4.2	± 4.0	± 6.1
infusion	n	9	8	9	9	9	8
Part E	Mean	128.4	122.6	120.2	116.0	114.0	108.3
30 min of NE	SE	± 3.6	± 4.9	± 4.3	± 6.8	± 6.7	± 4.5
infusion with gradual ending	n	5	5	5	4	4	4

[mm Hg]; Means ± SE. Statistics see text (ANOVA). Pre-infusion control values see Figs. 1 - 3.

Table 2, A - C, presents the time courses of HR at each minute from 1 min before the end to 5 min after the end of NE infusions of 5, 10 and 30 min duration. HR levels before the end of NE infusions tended to be higher with increasing duration of the infusion and rose to higher levels after the end of NE infusion. This tendency was found to be significant, when the data of table 2 obtained from the series of 5, 10 and 30 min of NE infusion were compared by analysis of variance ($F_{2,113}$=67.311; P<0.0001). Part D of table 2, presents the course of HR from 1 min before the end to 5 min after the end of NE infusion of 90 min duration. It did not differ significantly ($F_{1,91}$=0.106; P=0.75) from the course of HR during the post-infusion phase after NE infusion for 30 min. (part C of table 2).

Table 2: Time course of heart rate from 1 min before to 5 min after the end of NE infusions of different durations.

		1 min before end of NE infusion	after the end of NE infusion 1 min	2 min	3 min	4 min	5 min
Part A	Mean	64.0	71.6	94.5	98.9	98.0	102.4
5 min of NE	SE	± 3.1	± 4.0	± 5.1	± 3.4	± 4.9	± 7.0
infusion	n	6	8	6	8	6	8
Part B	Mean	75.5	83.7	110.5	113.6	112.7	106.3
10 min of NE	SE	± 2.2	± 3.8	± 6.8	± 3.5	± 4.0	± 4.3
infusion	n	8	7	6	7	7	7
Part C	Mean	87.4	96.0	123.3	125.4	127.5	116.4
30 min of NE	SE	± 2.3	± 1.2	± 5.2	± 3.2	± 4.5	± 3.4
infusion	n	7	8	8	8	6	8
Part D	Mean	88.0	119.8	123.5	118.5	111.8	105.5
90 min of NE	SE	± 3.9	± 5.0	± 5.7	± 5.7	± 6.1	± 5.8
infusion	n	9	9	9	9	9	8
Part E	Mean	96.0	96.9	94.5	98.8	106.2	106.4
30 min of NE	SE	± 5.6	± 5.3	± 6.1	± 6.4	± 4.8	± 5.0
infusion with gradual ending	n	5	5	5	5	5	5

[min^{-1}]; Means ± SE. Statistics see text (ANOVA). Pre-infusion control values see Figs. 1 - 3.

The data obtained for the degrees of post-infusion increases in AVP plasma concentrations on the one hand, and for differences in the post-infusion time courses of MAP, and partially HR, on the other, indicated that the hormonal and cardiovascular responses both depended on the duration of the preceding NE infusion. These observations suggested that the hormonal responses might be causally related to the cardiovascular responses. To test this assumption, 5 experiments were carried out in which NE was infused for 30 min and its rate of application was subsequently reduced in 5 steps of 2 min duration from 1.6 to 1.2, 0.8, 0.4 and 0.0 µg/min/kg. As shown by the upper diagram in Fig. 3, the adjustments of MAP and HR after 30 min were reduced and delayed. As shown by the lower diagram in Fig. 3, the post-infusion rise in plasma AVP was completely abolished.

Parts E of tables 1 and 2 present, in more detail, the time courses of MAP and HR at each minute from 1 min before the end to 5 min after the end of 30 min of NE infusions with gradual withdrawal of NE. It is obvious that the decrease in MAP and the rise in HR during the post-infusion phase were greatly retarded in comparison to the courses of the two parameters when the NE application had been suddenly stopped after 30 min of NE infusion.

In the experimental series with 90 min of NE infusion additional parameters were measured in 6 experiments in blood samples taken from the jugular vein. NE plasma concentration was 120 ± 17 pg/ml prior to NE infusion. During NE infusion its plasma concentration was determined as 19.0 ± 2.3 ng/ml at the 30th min and 16.0 ± 1.3 ng/ml at the 90th min, but it had decreased to 341 ± 31 pg/ml at the 7th min after the end of NE infusion. ANG II plasma concentration was 19.0 ± 0.8 pg/ml prior to NE infusion. With 41.4 ± 12.4 pg/ml it had approximately doubled at the 30th min of infusion, but tended to decrease during the further course of infusion, with 36.3 ± 12.4 pg/ml at the 60th and 31.6 ± 6.8 pg/ml at the 90th min. Plasma osmolality was found unchanged at the 30th min (287 ± 2 mosmol/kg) and 90th min (285 ± 2 mosmol/kg) as compared to the control value prior to NE infusion (287 ± 2 mosmol/kg). Core temperature was measured in the rectum and was not affected by NE infusion. Its level was 38.0 ± 0.1 °C prior to infusion and, during infusion, 38.0 ± 0.1 °C at the 30th min and 38.1 ± 0.1 at the 90th min.

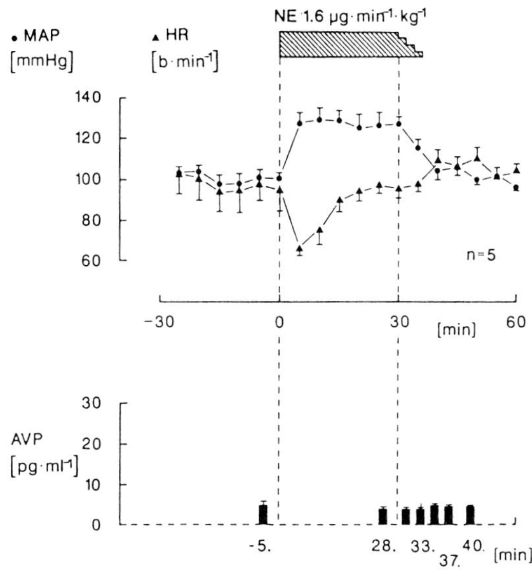

Fig. 3: Cardiovascular parameters and AVP plasma concentrations during and after NE infusions when, after 30 min, the infusion rate of NE was reduced in 2 min steps from 1.6 µg/kg to 1.2 µg/kg (min 30 and 31), 0.8 µg/kg (min 32 and 33), 0.4 µg/kg (min 34 and 35) and subsequently 0 µg/kg. Upper diagram: mean arterial pressure (MAP) and heart rate (HR) determined at 5 min intervals. Lower diagram: AVP plasma concentrations determined before, at the 28. min of infusion and at the 31th, 33th, 35th, 37th and 40th min.
Means with SE, 5 experiments.

DISCUSSION

The degree of rise in plasma AVP immediately after the end of NE infusions depended on the duration of the preceding infusion. However, this effect disappeared completely when NE infusion was ended gradually by stepwise reduction of the NE infusion and MAP gradually returned from the elevated to its normal level. The observation that AVP release was enhanced when MAP fell more rapidly from an elevated to a normal or slightly subnormal level would demonstrate, for the hypertensive to normotensive range the dominant role of baroreflex control of AVP release which has been documented frequently for the normotensive to hypotensive blood pressure range (Reid & Schwartz, 1984).

In agreement with the earlier observation that NE infusion in dogs induced clear elevations of MAP only when NE plasma levels exceeded 10 ng/ml (Casals-Stenzel et al., 1982), the dose of 1.6 μg/kg/min NE was chosen as the standard stimulus in this study and caused a moderate initial rise of MAP by some 25 mm Hg. According to the results of a preceding investigation (Szczypaczewska et al., 1993), it is assumed that this NE dose exerts a stimulatory action on AVP release which is counteracted by the enhanced arterial baroreceptor input. Moderate NE-induced rises in plasma ANG II, as confirmed also in the present study, are not assumed to mediate NE-induced stimulation of AVP release, considering that measurable increases in plasma AVP induced by iv. ANG II infusions required increases in plasma ANG II concentration which exceeded the physiological range (Szczypaczewska et al., 1993). For these reasons, it is assumed that the courses of plasma NE concentration and changes in cardiovascular control by the hypertensive action of NE were the main determinants for the reactions of the antidiuretic system.

While the post-infusion rises of plasma AVP increased with increasing durations of the preceding NE infusions (Figs. 1 and 2), it is also clear that this effect was eliminated, when gradual cessation of the NE infusion reduced the speed of the post-infusion readjustments of MAP and HR (Fig. 3). This suggests, as an explanation, that the post-infusion AVP response was determined by the dynamics of the changes in baroreceptor input. Considering the levels of MAP and HR attained at the end of NE infusions of different durations, the level of baroreflex activity at this time, as well as the level to which it decreased subsequently, seemed to be important for the degree to which plasma AVP rose during the early post-infusion phase. Generally, the rise in MAP and, consequently, the degree of baroreceptor activation tended to subside progressively with the duration of the NE infusion and, as a result, the level decreased to which baroreceptor input was presumably reduced during the post-infusion phase. The detailed analysis of the time courses of MAP from 1 min before to 5 min after the end of the NE infusions in table 1 shows that, after durations of 5, 10, 30 and 90 min, the postinfusion decrease of MAP started from progressively lower levels and reached progressively lower post-infusion values during the time of observation, although distinctly hypotensive levels were not attained. The courses of HR during the same time periods may be considered as indicators for the changes in baroreceptor reflex control of the central sympathetic and/or vagal innervation of the heart. At least for the 5, 10 and 30 min infusion periods, the data of table 2 confirm a progressive reduction of inhibition and a progressively larger degree of post-infusion disinhibition of the pacemaker of the heart with increasing duration of NE infusion. In addition, the dynamic component of the post-infusion readjustment seems to have been critical. This follows from the absence of the post-infusion AVP response in the experiments with stepwise reduction of the NE infusion which was paralleled by the gradual subsidence of baroreflex inhibition according to the courses of MAP and HR in this experimental series. Similarly retarded cardiovascular adjustments have previously been observed after the end of ANG II infusions and were also not followed by post-infusion increases in plasma AVP (Szczypaczewska et al., 1993).

The above considerations do not exclude the possibility that circulating NE exerted a direct, centrally mediated stimulatory action on AVP release which was, however, offset by inhibitory effects resulting from cardiovascular reflex activation, due to the rise of MAP. Assuming that this effect was also involved in the observed post-infusion response of AVP, then the increasing response with increasing duration of the preceding NE infusion would suggest that a time dependent accumulation of a stimulating agent, NE or a metabolite, may have occurred. Its putative action should have persisted long enough, after stopping the NE infusion, to become effective during the post-infusion phase, when the baroreceptor-induced reflex inhibition of AVP subsided rapidly with falling MAP. On the other hand, the presumed NE-induced central stimulation of enhanced AVP release should have diminished rapidly enough, with falling plasma NE levels, to become ineffective when the baroreceptor-mediated inhibition of AVP release was gradually reduced by stepwise reduction of the NE infusion.

From the results of the present investigation, it is difficult to draw suggestions about the central sites of action of high levels of circulating NE with respect to the blood brain barrier (BBB). Mechanical opening of the BBB does not seem likely, considering the moderate rises of MAP, however, sustained high levels of plasma NE may have inactivated the enzymatic component of the BBB by saturating the monoamine degrading enzymes (Rapoport, 1976). On the other hand, NE actions on structures near to sites where the BBB is leaky, i.e., in the vicinity of the circumventricular organs, cannot be excluded. Among these structures the median eminence might be a likely physiological target for NE, considering the demonstration of key enzymes for adrenergic transmitter generation in this structure (Saavedra et al. 1974) and the histochemical evidence for adrenergic receptors (Unnerstall et al., 1985).

Acknowledgement: The autors are indebted to Mrs. Irene Küchenmeister and Mrs. Roswitha Bender for their excellent technical assistance.

REFERENCES

Casals-Stenzel, J., Tree, M., Brown, J.J., Fraser, R., Lever, A.F., Millar, J.A., Morton, J.J., Robertson, J.I.S., Reid, J.L., and Hamilton, C. (1982) Prolonged infusion of norepinephrine in the conscious dog: effects on blood pressure, heart rate, renin, angiotensin II, and aldosterone. J. Cardiovasc. Pharmacol. 4 (Suppl 1), S114-S121.

Cowley jr, A.W., and Barber, B.J. (1983) Vasopressin vascular and reflex effects - a theoretical analysis. In The Hypophysis: Structure, Function and Control (Progress in Brain Research series, vol. 60), ed B.A. Cross & G. Leng, pp. 415-424 . Amsterdam, New York,: Elsevier.

Gray, D., and Simon, E. (1983) Mammalian and avian antidiuretic hormone: studies related to possible species variation in osmoregulatory systems. J. Comp. Physiol. 151, 241-246.

Rapoport, S.I. (1976) Blood Brain Barrier in Physiology and Medicine, 316 p. New York: Raven Press.

Reid, I.A., and Schwartz, J. (1984) Role of vasopressin in the control of blood pressure. In Frontiers in Neuroendocrinology (vol. 8), ed L.Martini & F.W. Ganong, pp. 177-197. New York: Raven Press.

Robertson, G.L. (1977) Vasopressin function in health and disease. Rec. Progr. Horm. Res. 33, 333-374.

Saavedra, J.M., Palkovits, M., Brownstein, M.J., and Axelrod, J. (1974) Localisation of phenylethanolamine N-methyltransferase in the rat brain nuclei. Nature 248, 695-696.

Simon-Oppermann, C., Gray, D.A., and Simon, E. (1986) Independent osmoregulatory control of central and systemic angiotensin II concentration in dogs. Amer. J. Physiol. 250, R918-R925.

Szczepanska-Sadowska, E., Simon-Oppermann, C., Simon, E., Gray, D.A., Pleschka, K., and Szczypaczewska, M. (1992) Central ANP administration in conscious dogs responding to dehydration and hypovolemia. Am. J. Physiol. 262, R746-R753.

Szczypaczewska, M., Simon. E., Simon-Oppermann, C., and Gray, D.A. (1993) Disinhibition of AVP release during NE and A II infusions in dogs by maintaining normotension with sodium nitroprusside. Pflügers. Arch. 423, 238-244.

Unnerstall, J.R., Kopajtic, T.A., and Kuhar, M.J. (1984) Distribution of α_2 agonist binding sites in the rat and human central nervous system: analysis of some functional, anatomic correlates of the pharmacologic effects of clonidine and related adrenergic agents. Brain. Res. Rev. 7, 69-101.

Author index

LOUIS-JEAN
avenue d'Embrun, 05003 GAP cedex
Tél. : 92.53.17.00
Dépôt légal : 40 — Janvier 1994
Imprimé en France